Materials Selection in Mechanical Design

Materials Selection in Mechanical Design

Fourth Edition

Michael F. Ashby

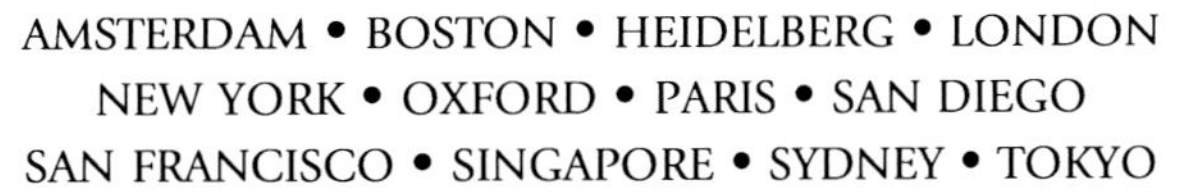

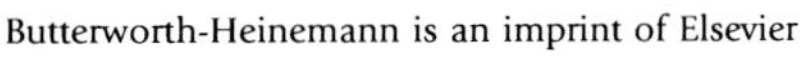

Butterworth-Heinemann is an imprint of Elsevier
30 Corporate Drive, Suite 400
Burlington, MA 01803, USA

The Boulevard, Langford Lane
Kidlington, Oxford, OX5 1GB, UK

Library of Congress Cataloging-in-Publication Data
Ashby, M. F.
Materials selection in mechanical design / Michael F. Ashby. — 4th ed.
p. cm.
Includes index and readings.
ISBN 978-1-85617-663-7
1. Materials. 2. Engineering design. I. Title.
TA403.6.A74 2011
620.1'1–dc22 201002069

British Library Cataloguing-in-Publication Data
A catalogue record for this book is available from the British Library.

For information on all Butterworth–Heinemann publications
visit our website at *www.elsevierdirect.com*

Typeset by: diacriTech, India

Contents

Preface

Materials, of themselves, affect us little; it is the way we use them which influences our lives.

Epictetus, AD 50–100, *Discourses*, Book 2, Chapter 5

Materials influenced lives in Epictetus' time and continue to do so today. In his day, the number of materials was small; today it is vast. The opportunities for innovation that materials offer now are equally immense. But advance is possible only if a procedure exists for making a rational choice from the materials on this great menu, and—if they are to be used—a way of identifying ways to shape, join, and finish them. This book develops a systematic procedure for selecting materials and processes, leading to the subset that best matches the requirements of a design. It is unique in the way that the information it contains has been structured. The structure gives rapid access to data and allows the user great freedom in exploring potential choices. The method is implemented in software* to provide greater flexibility.

The approach here emphasizes design with materials rather than materials "science," although the underlying science is used whenever possible to help with the structuring of selection criteria. The first six chapters require little prior knowledge: A first-year grasp of materials and mechanics is enough. The chapters dealing with shape and multiobjective selection are a little more advanced but can be omitted on a first reading. As far as possible, the book integrates materials selection with other aspects of design; the relationships with the stages of design and optimization and with the mechanics of material, are developed throughout. At the teaching level, the book is intended as a text for third- and fourth-year engineering courses on Materials for Design: A 6- to 10-lecture unit can be based on Chapters 1 through 6, 13, and 14; a full 20-lecture course, with project work using the associated software, will require use of the entire book.

---

* The *CES Edu* materials and process selection platform is a product of Granta Design (*www.grantadesign.com*).

Beyond this, the book is intended as a reference of lasting value. The method, the charts, and the tables of performance indices have application in real problems of materials and process selection; and the table of data and the catalog of "useful solutions" (Appendices A and B) are particularly helpful in modeling—an essential ingredient in optimal design. The reader can use the content (and the software) at increasing levels of sophistication as his or her experience grows, starting with the material indices developed in the book's case studies and graduating to the modeling of new design problems, leading to new material indices and penalty functions, as well as new—and perhaps novel—choices of material. This continuing education aspect is helped by the "Further readings" at the end of each chapter and Appendix E—a set of exercises covering all aspects of the text. Useful reference material is assembled in Appendices A, B, C, and D.

As in any other book, the contents in this one are protected by copyright. Generally, it is an infringement to copy and distribute materials from a copyrighted source. However, the best way to use the charts that are a central feature of the book, for readers to have a clean copy on which they can draw, try out alternative selection criteria, write comments, and so forth; presenting the conclusion for a selected exercise is often most easily done in the same way. Although the book itself is copyrighted, instructors or readers are authorized to make unlimited copies of the charts and to reproduce these for teaching purposes, provided a full reference to their source is given.

ACKNOWLEDGMENTS

Many colleagues have been generous with discussion, criticism, and constructive suggestions. I particularly wish to thank Professor Yves Bréchet of the University of Grenoble in France, Professor Anthony Evans of the University of California at Santa Barbara, Professor John Hutchinson of Harvard University, Professor David Cebon, Professor Norman Fleck, Professor Ken Wallace, Professor John Clarkson, Dr. Hugh Shercliff of the Engineering Department of Cambridge University, Professor Amal Esawi of the American University in Cairo, Professor Ulrike Wegst of Drexel University, Dr. Paul Weaver of the Department of Aeronautical Engineering at the University of Bristol, and Professor Michael Brown of the Cavendish Laboratory in Cambridge, UK.

Mike Ashby

Features of the Fourth Edition

Since publication of the third edition of this book, changes have occurred in the field of materials and their role in engineering, as well as in the way these subjects are taught in university- and college-level courses. There is increasing emphasis on *materials efficiency*—design that uses materials effectively and with as little damage to the environment as possible. All this takes place in a computer-based environment; teaching, too, draws increasingly on computer-based tools. This new edition has been comprehensively revised and reorganized to address these. The presentation has been enhanced and simplified; the figures, many of them new, have been redrawn in full color; worked in-text examples illustrate methods and results in chapters that are not themselves collections of case studies; and additional features and supplements have been added. The key changes are outlined next.

Key changes

- Chapter 1, Introduction, has been completely rewritten and illustrated to develop the history of materials and the evolution of materials in engineering.
- Engineering Design, introduced in Chapter 2, has been edited, with a full revision of all figures.
- Material Properties and Property Charts—a unique feature of the book, which appear in Chapters 3 and 4, have been redrawn in full color.
- Chapter 5 and 6—the central chapters that describe and illustrate selection methods—have been extensively revised with new explanations of the essential selection strategy.
- Chapters 7 and 8 (Multiple Constraints) have been revised, with in-text examples and more illuminating case studies.
- Chapters 9 and 10 (Materials and Shape) have been rewritten for greater clarity, with numerous in-text examples in Chapter 9.

- Chapters 11 and 12, Hybrid Materials, represent a further development of what was in the earlier edition, with a new development of the treatment of sandwich structures and with enhanced case studies.
- Chapters 13 and 14, Processing, contain sections and figures that emphasize the influence of processing on properties.
- Chapter 15, Materials and the Environment, is revised, with improved examples and links to the new information.[1]
- Chapter 16, Industrial Design, is updated and linked to the second edition of the related text[2] on this subject.
- Chapter 17, Forces for Change, has been updated.
- Appendices with Tables of Materials Properties, Useful Solutions, Indices, and Data Sources are updated, enlarged and reillustrated.
- The final appendix contains Exercises that are listed by chapter number.

Material selection charts

Full color versions of a number of the Material Selection Charts presented in this book are available. Samples can be found at *www.grantadesign.com/ashbycharts.htm*. This web page also provides a link to a page where users of CES EduPack (details follow) can download further charts and other teaching resources, including PowerPoint lectures. Although the author retains the copyright for the charts, users of this book are authorized to download, print, and make unlimited copies of those available on the site; in addition, they can be reproduced for teaching purposes (but not for publication), with proper reference to their source.

Instructor's manual and Image Bank

The book ends with a comprehensive set of exercises in Appendix E. Worked-out solutions to the exercises are available, free of charge, to teachers, lecturers, and professors who adopt the book.

The Image Bank provides tutors and lecturers who have adopted this book with PDF versions of the figures contained in it; they can be used for lecture slides and class presentations.

To access the instructor's manual and Image Bank, please visit *www.textbooks.elsevier.com* and follow the onscreen instructions.

---

[1] *Materials and the Environment—Eco-informed materials choice* (2009) by M.F. Ashby, Butterworth-Heinemann, ISBN 978-1-85617-608-8.

[2] *Materials and Design—The art and science of materials selection in Product Design*, 2nd edition (2009), by M.F. Ashby and K. Johnson, Butterworth-Heinemann, ISBN 978-1-85617-497-8.

The CES EduPack

The CES EduPack is a widely used software package that implements the methods developed here. The book does not rely on the software, but the learning experience is enhanced by using the two together to create an exciting teaching environment that stimulates exploration, self-teaching, and design innovation. For further information, see the last page of this book or visit *http://www.grantadesign.com/education/*.

CHAPTER 1

Introduction

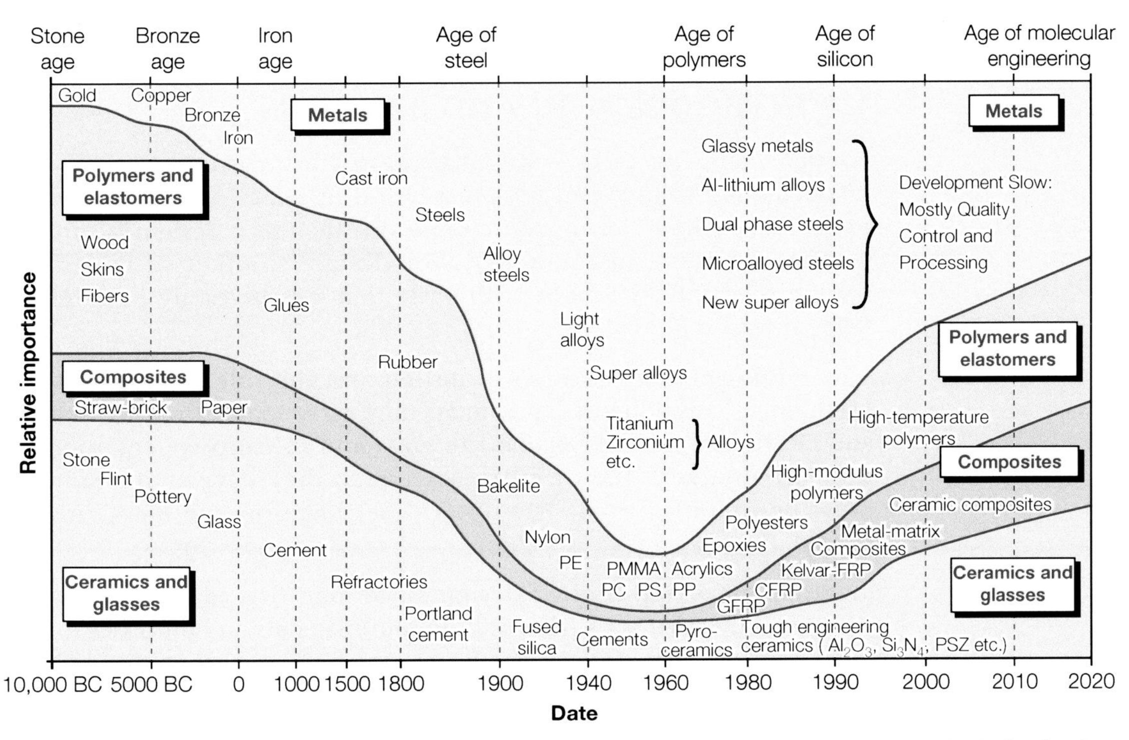

The evolution of engineering materials with time. "Relative importance" is based on information contained in the books listed under "Further reading"; plus, from 1960 onward, data for the teaching hours allocated to each material family at U.K. and U.S. universities. The projections to 2020 rely on estimates of material usage in automobiles and aircraft by manufacturers. The time scale is nonlinear. The rate of change is far faster today than at any previous time in history.

Materials Selection in Mechanical Design. DOI: 10.1016/B978-1-85617-663-7.00001-1

CONTENTS

1.1 INTRODUCTION AND SYNOPSIS

"Design" is one of those words that mean all things to all people. Every manufactured thing, from the most lyrical of ladies' hats to the greasiest of gearboxes, qualifies, in some sense or other, as a design. It can mean yet more. Nature, to some, is divine design; to others it is design by natural selection. The reader will agree that it is necessary to narrow the field, at least a little.

This book is about mechanical design and the role of materials in it. Mechanical components have mass; they carry loads; they conduct heat and electricity; they are exposed to wear and to corrosive environments; they are made of one or more materials; they have shape; and they must be manufactured. The book describes how these activities are related.

Materials have had limited design since man first made clothes, built shelters, and waged wars. They still do. But materials and processes to shape them are developing faster now than at any time in history; the challenges and opportunities they present are greater than ever before. This book develops a strategy for confronting such challenges and seizing those opportunities.

1.2 MATERIALS IN DESIGN

Design is the process of translating a new idea or a market need into the detailed information from which a product can be manufactured. Each of its stages requires decisions about the materials of which the product is to be made and the process for making it. Normally, the choice of material is dictated by the design. But sometimes it is the other way around: The new product, or the evolution of the existing one, was suggested or made possible by a new material.

The number of materials available to engineers is vast: 160,000 or more. Although standardization strives to reduce the number, the continuing appearance of new materials with novel and exploitable properties expands the options further. How, then, do engineers choose, from this vast menu, the material best suited to their purpose? Do they rely on their experience? In the past that was how it was done, passing on this precious commodity to apprentices who, much later in their lives, might themselves assume the role of in-house materials guru.

There is no question of the value of experience. But many things have changed in the world of engineering, and all of them work against the success of this model. There is the drawn-out time scale of apprentice-based learning. There is job mobility, meaning that the guru who is here today is usually gone tomorrow. And there is the rapid evolution of materials information, as already mentioned. A strategy that relies on experience is not in tune with today's computer-based environment. We need a *systematic* procedure—one with steps that can be taught quickly, that is robust in the decisions it reaches, that allows computer implementation, and that is compatible with the other established tools of engineering design.

The choice of material cannot be made independently of the choice of process by which the material is to be shaped, joined, and finished. Cost enters the equation, both in the choice of material and in the way the material is processed. So, too, does the influence of material usage on the environment in which we live. And it must be recognized that good engineering design alone is not enough to sell products. In almost everything from home appliances to automobiles and aircraft, the form, texture, feel, color, beauty, and meaning of the product—the satisfaction it gives the person who owns or uses it—are important. This aspect, known confusingly as *industrial design*, is one that, if neglected, can lose markets. Good design works; excellent design also gives pleasure.

Design problems are almost always open-ended. They do not have a unique or "correct" solution, though some solutions will clearly be better than others. They differ from the analytical problems used in teaching mechanics, or structures, or thermodynamics, which generally do have single, correct answers. So the first tool a designer needs is an open mind: a willingness to consider all possibilities. But a net cast widely draws in many different fish. A procedure is necessary for selecting the excellent from the merely good.

This book deals with the materials aspects of the design process. It develops a methodology that, properly applied, gives guidance through the forest of complex choices the designer faces. The ideas of *material* and *process attributes* are introduced. They are mapped on material and process *selection charts* that show the lay of the land, so to speak, and that simplify the initial survey for potential candidate materials. Real life always involves *conflicting objectives*—minimizing mass while at the same time minimizing cost is an example—requiring the use of *trade-off methods*. The interaction between *material* and *shape* can be built into the method. Taken together, these suggest schemes for expanding the boundaries of material performance by creating *hybrids*—combinations of two or more materials, shapes, and configurations with unique property profiles. None of this can be implemented without *data* for material properties and process attributes: Ways to find them are described. The role of *aesthetics* in engineering design is discussed. The *forces driving change* in the materials world are surveyed, the most obvious of which is the force dealing with environmental concerns. The Appendices contain additional useful information.

The methods lend themselves readily to implementation as computer-based tools; one, the *CES Edu* materials selection platform,[1] is used for many of the case studies and figures in this book. They offer, too, potential for interfacing with tools for computer-aided design, finite element analysis, optimization routines, and product data management software.

All this is found in the following chapters, with case studies illustrating applications. But first, a little history.

1.3 THE EVOLUTION OF ENGINEERING MATERIALS

Throughout history, materials have had limited design. The ages of man are named for the materials he used: stone, bronze, iron. And when a man died, the materials he treasured were buried with him: Tutankhamun in his enameled sarcophagus, Agamemnon with his bronze sword and mask of gold, Viking chieftains in their burial ships—each treasure representing the high technology of their day.

If these men had lived and died today, what would they have taken with them? Their titanium watch, perhaps; their carbon-fiber–reinforced tennis racquet; their metal-matrix composite mountain bike; their shape-memory alloy eyeglass frame with lenses coated with diamond-like carbon; their polyether-ethyl-ketone crash helmet; their carbon nanotube reinforced iPod? This is not the age of one material; it is the age of an immense range of materials. There has never been an era in which their evolution was faster and the range of their properties more varied. The menu of materials has expanded so rapidly that designers who left college 20 years ago can be forgiven for not knowing that many of them exist. But not to know is, for the

[1] Granta Design Ltd., Cambridge, U.K.—*www.grantadesign.com*.

designer, risking disaster. Innovative design often means the imaginative exploitation of the properties offered by new or improved materials. And for the man on the street, the schoolboy even, not to know is to miss one of the great developments of our age: the age of advanced materials.

This evolution and its increasing pace are illustrated on the cover page and, in more detail, in Figure 1.1. The materials of prehistory (before 10,000 BC, the Stone Age) were ceramics and glasses, natural polymers, and composites. Weapons—always the peak of technology—were made of wood and flint; buildings and bridges of stone and wood. Naturally occurring gold and silver were available locally and, through their rarity, assumed great influence as currency, but their role in technology was small. The development of rudimentary thermo-chemistry allowed the extraction of, first, copper and bronze, then iron (the Bronze Age, 4000–1000 BC and the Iron Age, 1000 BC–1620 AD), stimulating enormous advances in technology.[2] Cast iron technology (1620s) established the dominance of metals in engineering; since then the evolution of steels (1850 onward), light alloys (1940s), and special alloys has consolidated their position. By the 1950s, "engineering materials" meant "metals." Engineers were given courses in metallurgy; other materials were barely mentioned.

There had, of course, been developments in the other classes of material. Improved cements, refractories, and glasses; and rubber, Bakelite, and polyethylene among polymers; but their share of the total materials market was small. Since 1950 all that has changed. The rate of development of new metallic alloys is now slow; demand for steel and cast iron has in some countries actually fallen.[3] The polymer and composite industries, on the other hand, are growing rapidly, and projections of the growth of production of new high-performance ceramics suggests continued expansion here also.

The material developments documented in the timeline of Figure 1.1 were driven by the desire for ever greater performance. One way of displaying this progression is by following the way in which properties have evolved on *material-property charts*. Figure 1.2 shows one of them—a *strength-density chart*. The oval bubbles plot the range of strength and density of materials; the larger colored envelopes enclose families. The chart is plotted for six successive points in historical time, ending with the present day. The materials of pre-history, shown in (a), cover only a tiny fraction of this strength-density

---

[2] There is a cartoon on my office door, put there by a student, showing an aggrieved Celt confronting a swordsmith with the words, "You sold me this bronze sword last week and now I'm supposed to upgrade to iron!"

[3] Do not, however, imagine that the days of steel are over. Steel production accounts for 90% of all world metal output, and its unique combination of strength, ductility, toughness, and low price makes it irreplaceable.

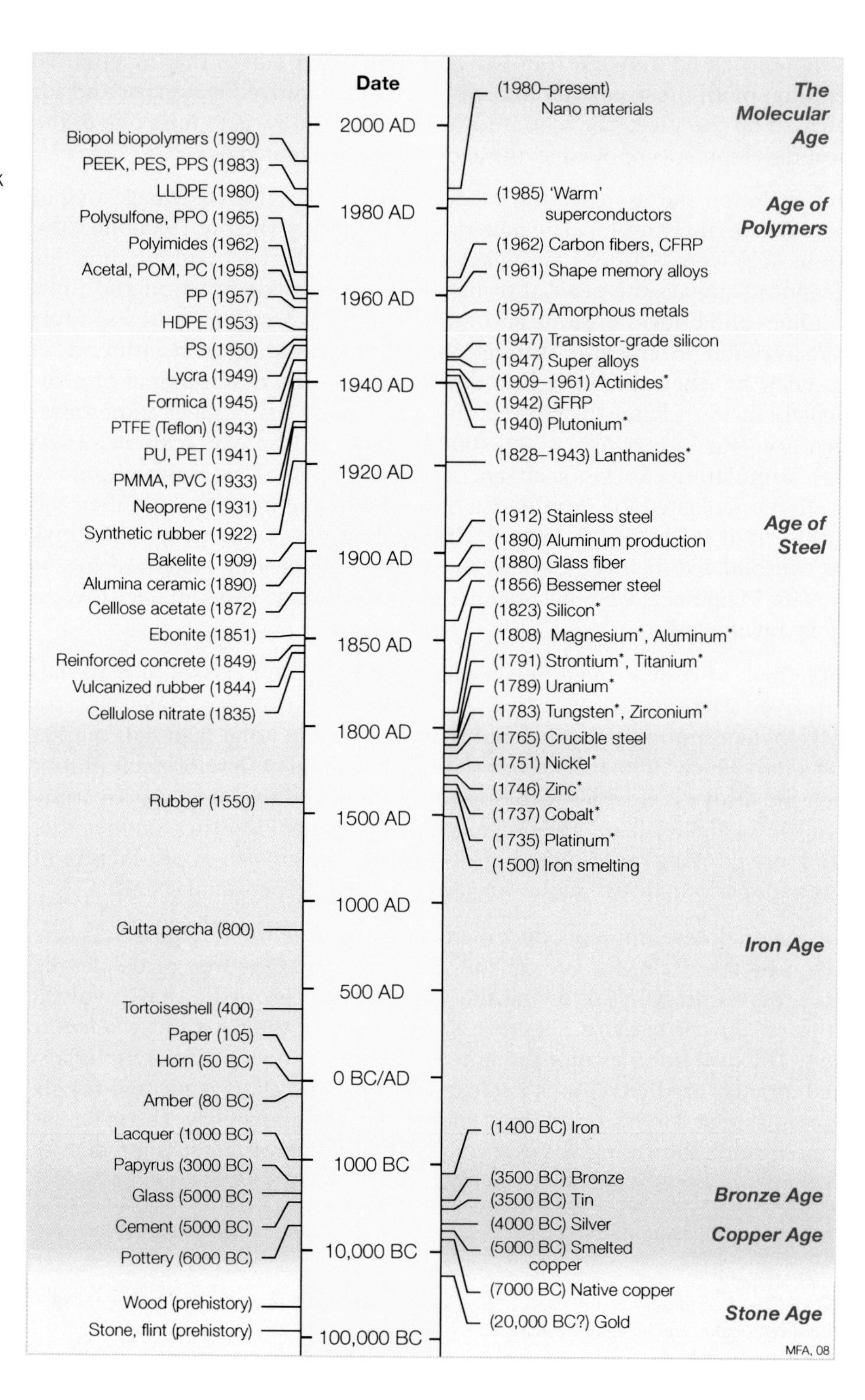

FIGURE 1.1
A materials timeline. The scale is nonlinear, with big steps at the bottom, small ones at the top. An asterisk (*) indicates the date at which an element was first identified. Labels without asterisks note the time at which the material became of practical importance.

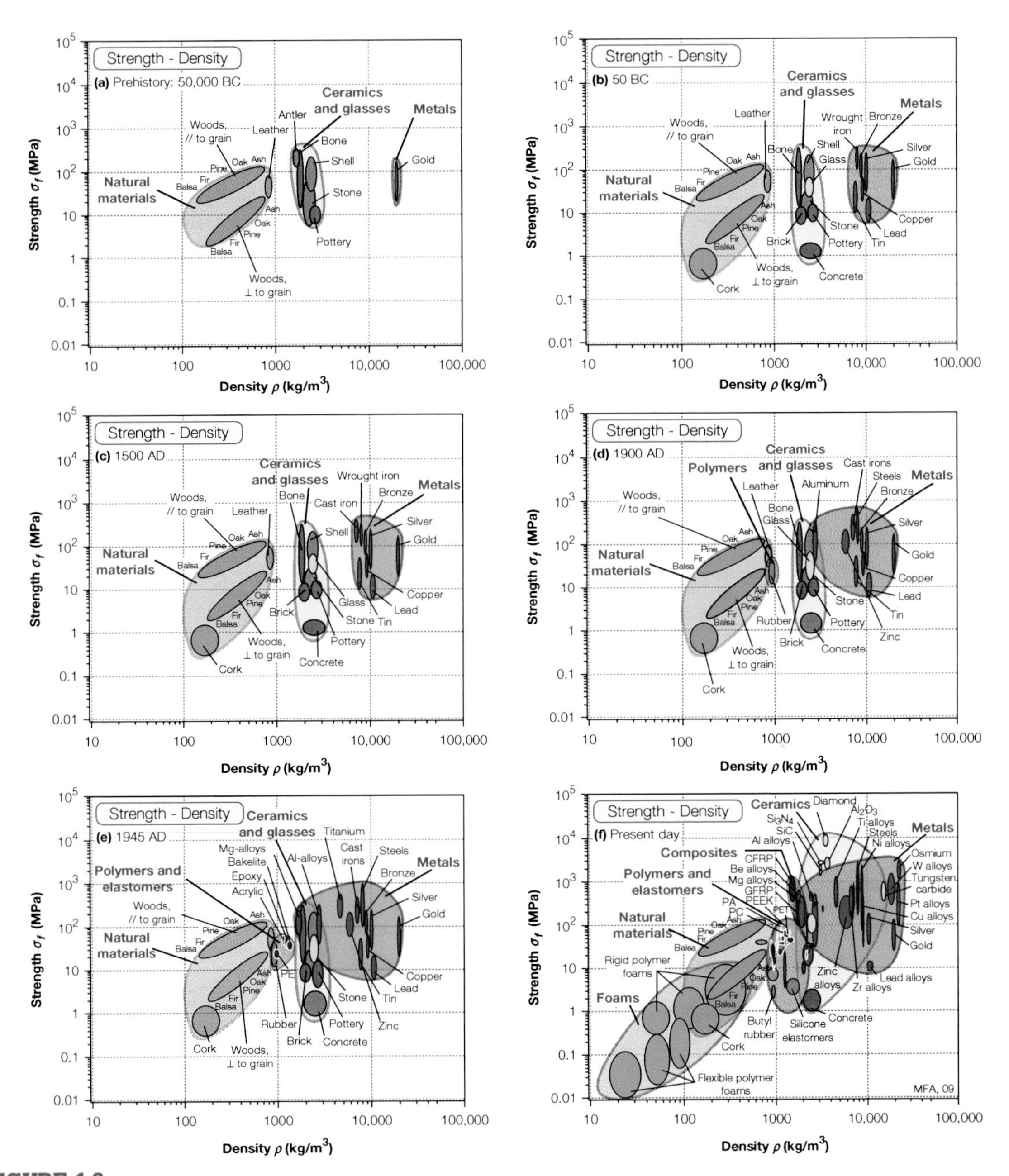

FIGURE 1.2

The progressive filling of material-property space over time (the charts list the date at the top left) showing the way the materials have been developed over time to meet demands on strength and density. Similar time plots show the progressive filling for all material properties.

space. By the time of the peak of the Roman Empire, around 50 BC (b), the area occupied by metals had expanded considerably, giving Rome critical advantages in weaponry and defense. The progress thereafter was slow: 1500 years later (c) not much has changed, although, significantly, cast iron started to appear. Even 500 years after that (d), expansion of the occupied area of the chart is small; aluminum only just starts to creep in. Then things accelerate. By 1945 the metals envelope has expanded considerably and a new envelope—that of synthetic polymers—occupies a significant position. Between then and the present day the expansion has been dramatic. The filled area now starts to approach some fundamental limits (not shown here) beyond which it is difficult to go.

Any slice through material property space (we will encounter many) shows development like this. How can we expand the filled area further? And what would we gain if we did so? These are fascinating questions that will be answered in Chapters 13 and 14. We end this chapter, instead, with a look at the way material developments have been absorbed into products.

1.4 THE EVOLUTION OF MATERIALS IN PRODUCTS

In this section we consider four examples of the changes in the way materials are used, each spanning about 100 years—not much more than a single life time. Bear in mind that in preceding generations, change was far slower. The horse-drawn carriage has a history of 2000 years; the automobile only a little more than 100.

The kettle is the oldest of household appliances and the one found in more homes than any other; there is evidence (not entirely convincing) of a 4000-year-old kettle. Early kettles, heated directly over a fire, were of

FIGURE 1.3
Kettles: cast iron, bronze, polypropylene.

necessity made of materials that could conduct heat well and withstand exposure to an open flame: iron, copper, or bronze (Figure 1.3). Electric kettles, developed in the 1890s, had external heating elements to replace the flame, but were otherwise much like their predecessors. All that changed with the introduction, by the Swan company (1922), of a heating element sealed in a metal tube placed within the water chamber. The kettle body no longer had to conduct heat—indeed for safety and ease of use it was much better made of a thermal and electrical insulator. Today almost all kettles are made of plastic, allowing economic manufacture with great freedom of form and color.

> "Sweeping and dusting are homicidal practices: they consist of taking dust from the floor, mixing it in the atmosphere, and causing it to be inhaled by the inhabitants of the house. In reality it would be preferable to leave the dust alone where it was."

That was a doctor, writing about 100 years ago. More than any previous generation, the Victorians and their contemporaries in other countries worried about dust. They were convinced that it carried disease and that dusting merely dispersed it, when, as the doctor said, it became yet more infectious. Their response was to invent the vacuum cleaner (Figure 1.4).

The vacuum cleaners of that time were human-powered. The materials were, by today's standards, primitive. The cleaner was made almost entirely from natural materials: wood, canvas, leather, and rubber. The only metal parts

FIGURE 1.4
Vacuum cleaners: a hand-powered cleaner from 1880, the Electrolux cylinder cleaner of 1960, and the Dyson centrifugal cleaner of 2010. *Source: Early cleaner (left) courtesy of Worcester News.*

FIGURE 1.5
Cameras: a plate camera from 1900, a Leica from 1960, and a plastic camera from 2006.

were the straps that link the bellows (soft iron) and the can containing the filter (mild steel sheet, rolled to make a cylinder). It reflects the use of materials in that era. Even a car, in 1900, was mostly made of wood, leather, and rubber; only the engine and drive train had to be metal.

The electric vacuum cleaner first appeared around 1908.[4] By 1950 the design had evolved into the cylinder cleaner. Air flow was axial, drawn through the filter by an electric fan. One advance in design was, of course, the electrically driven air pump. But there were others: This cleaner is almost entirely made of metal—the case, the end-caps, the runners, and even the tube to suck up the dust are mild steel. Metals entirely replaced natural materials.

Developments since then have been rapid, driven by the innovative use of new materials. Power and air flow rate are much increased, and dust separation is centrifugal rather than by filtration. This is made possible by higher power density in the motor reflecting improved magnetic materials. The casing is entirely polymeric, and makes extensive use of snap fasteners for rapid assembly. No metal is visible anywhere; even the straight part of the suction tube, which was metal in all earlier models, is now polypropylene.

The optics of a camera are much older than the camera itself (Figure 1.5). Lenses capable of resolving the heavens (Galileo, 1600) or the microscopic (Hooke, 1665) predate the camera by centuries. The key ingredient, of course, was the ability to record the image (Joseph Nicéphore Niépce, 1814). Early cameras were made of wood and constructed with the care and finish of a cabinetmaker. They had well-ground glass lenses and, later, a metal aperture and shutter manufactured by techniques already well developed for watch- and clockmaking.

---

[4] Inventors: Murray Spengler and William B. Hoover. The second name has become part of the English language, along with the names of such luminaries as John B. Stetson (the hat), S.F.B. Morse (the code), Leo Henrik Baikeland (Bakelite), and Thomas Crapper (the flush toilet).

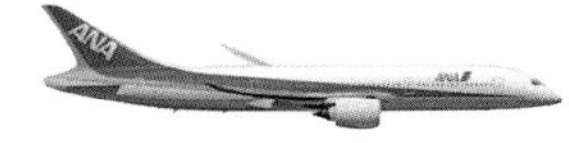

FIGURE 1.6
Aircraft: the Wright biplane of 1903, the Douglas DC3 of 1935, and the Boeing 787 Dreamliner of 2010.

For the next 90 years photography was practiced by a specialized few. Invention of celluloid-backed film and the cheap box camera around 1900 moved it from a specialized to a mass market, with competition between camera makers to capture a share. Wood, canvas, and leather (with brass and steel only where essential) were quickly replaced by precision-engineered steel bodies and control mechanisms; then from the 1960s on, by aluminum, magnesium, or titanium for low weight, durability, and prestige. Digital technology fractionated the market further. High-end cameras now have optical systems with compound lenses combining glasses with tailored refractive indices, manufactured with the precision and electronic sophistication of scientific instruments. At the other end of the range, point-and-shoot cameras with molded polypropylene or ABS bodies, and acrylic or polycarbonate lenses fill a need.

Perhaps the most dramatic example of the way material usage has changed is found in airframes. Early planes were made of low-density woods (spruce, balsa, and ply), steel wire,[5] and silk. Wood remained the principal structural material of airframes well into the twentieth century, but as planes got larger it became less and less practical. The aluminum airframe, exemplified by the Douglas DC3, was the answer. It provided the high bending stiffness and strength at low weight necessary for scale-up and extended range. Aluminum remained the dominant structural material of civil airliners for the remainder of the twentieth century. By the end of the century, the pressure for greater fuel economy and lower carbon emissions had reached a level that made composites an increasingly attractive choice, despite their higher cost and greater technical challenge. The future of airframes is exemplified by Boeing's 787 Dreamliner (80% carbon-fiber–reinforced plastic by volume), claimed to be 30% lighter per seat than competing aircraft. (See Figure 1.6.)

All this has happened within one lifetime. Competitive design requires the innovative use of new materials and the clever exploitation of their special properties, both engineering and aesthetic. Many manufacturers of kettles, cleaners, and cameras failed to innovate and exploit; now they are extinct. That somber thought prepares us for the chapters that follow, in which we consider what they forgot: the optimum use of materials in design.

[5] "Piano wire"—drawn high-carbon-steel wire—was developed for harpsichords around 1350.

1.5 SUMMARY AND CONCLUSIONS

What do we learn? There is an acceleration in material development and of the ways in which materials are used in products. One of the drivers for change, certainly, is performance: The displacement of bronze by iron in weapons of the Iron Age, and that of wood by aluminum in airframes of the twentieth century, had their origins in the superior performance of the new materials. But performance is not the only factor.

Economics exerts powerful pressures for change: the use of polymers in the jug kettle, the vacuum cleaner, and the cheap camera derives in part from the ease with which polymers can be molded to complex shapes. Technical change in other fields—digital imaging technology, for example—can force change in the way materials are chosen. And there are many more drivers for change that we will encounter in later chapters, among them a concern for the environment, restrictive legislation and directives, aesthetics, and taste.

Engineering materials are evolving faster, and the choice is wider, than ever before. Examples of products in which a new material has captured a market are as common as, well, plastic bottles. Or aluminum cans. Or polycarbonate eyeglass lenses. Or carbon-fiber golf club shafts. It is important in the early stage of design, or of redesign, to examine the full materials menu, not rejecting options merely because they are unfamiliar. That is what this book is about.

1.6 FURTHER READING

The history and evolution of materials

Singer, C. et al. (1954–2001). *A history of technology* (21 volumes). Oxford University Press, ISSN 0307-5451.

A compilation of essays on aspects of technology, including materials.

Delmonte, J. (1985). *Origins of materials and processes*. Technomic Publishing Company, ISBN 87762-420-8.

A compendium of information on when materials were first used and by whom.

Dowson, D. (1998). *History of tribology*. Professional Engineering Publishing Ltd., ISBN 1-86058-070-X.

A monumental work detailing the history of devices limited by friction and wear, and the development of an understanding of these phenomena.

Emsley, J. (1998). *Molecules at an exhibition*. Oxford University Press, ISBN 0-19-286206-5.

Popular science writing at its best: intelligible, accurate, simple, and clear. The book is exceptional for its range. The message is that molecules, often meaning materials, influence our health, our lives, the things we make, and the things we use.

Michaelis, R.R. (Ed.). (1992). Gold: Art, science and technology, and focus on gold. *Interdisciplinary Science Reviews* 17(3, 4), ISSN 0308-0188.

A comprehensive survey of the history, mystique, associations, and uses of gold.

The Encyclopaedia Britannica, 11th Edition (1910). The Encyclopaedia Britannica Company.
Connoisseurs will tell you that in its 11th edition the Encyclopaedia Britannica reached a peak of excellence which has not since been equaled, though subsequent editions are still usable.

Tylecoate, R.F. (1992). *A history of metallurgy* (2nd ed.). The Institute of Materials, ISBN 0-904357-066.
A total-immersion course in the history of the extraction and use of metals from 6000 BC to 1976, told by an author with forensic talent and a love of detail.

Vacuum cleaners

Forty, A. (1986). *Objects of desire—Design in society since 1750*. Thames and Hudson, ISBN 0-500-27412-6. p. 174 et seq.
A refreshing survey of the design history of printed fabrics, domestic products, office equipment and transport systems. The book is mercifully free of eulogies about designers, and focuses on what industrial design does, rather than who did it. The black and white illustrations are disappointing, mostly drawn from the late nineteenth or early twentieth centuries, with few examples of contemporary design.

Cameras

Rosenblum, N. (1989). *A world history of photography* (2nd ed.). Abbeville Press, The University of Michigan, ISBN 1-9-781-558-59054-0.
A study that spans the history of the recorded image, from the camera Lucida to the latest computer technology, including discussion of the aesthetic, documentary, commercial, and technical aspects of its use.

Aircraft

Grant, R. G. (2007). *Flight, the complete history*. Dorling Kindersley Ltd., ISBN 978-1-4053-1768-9.
A lavishly illustrated history, particularly good on the materials of early aircraft.

CHAPTER 2

The Design Process

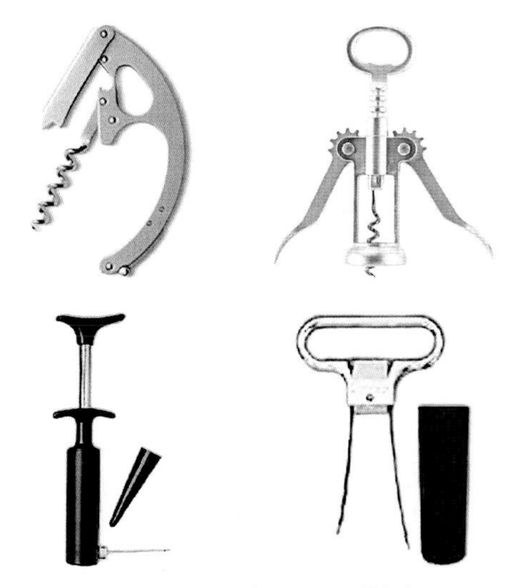

Cork removers. (Image courtesy of A-Best Fixture Co., Akron, Ohio.)

CONTENTS

Materials Selection in Mechanical Design. DOI: 10.1016/B978-1-85617-663-7.00002-3

2.1 INTRODUCTION AND SYNOPSIS

We are primarily concerned here with *mechanical design*: the physical principles, the proper functioning, and the production of mechanical systems. This does not mean that we ignore *industrial design*—pattern, color, texture, and (above all) consumer appeal—but that comes later. The optimum starting point in product development is good mechanical design, and the ways in which the selection of materials and processes contribute to it.

Our aim is to develop a methodology for selecting materials and processes that is *design-led*; that is, it uses, as inputs, the functional requirements of the design. To do so we must first look briefly at the design process itself. Like most technical fields, mechanical design is encrusted with its own special jargon, some of it bordering on the incomprehensible. We need very little, but it cannot all be avoided. This chapter introduces some of the words and phrases—the vocabulary—of design, the stages in its implementation, and the ways in which materials selection links with these.

2.2 THE DESIGN PROCESS

The starting point of a design is a *market need* or a *new idea*; the end point is the full *specification* of a product that fills the need or embodies the idea. A need must be identified before it can be met. It is essential to define the need precisely—that is, to formulate a *need statement*, often in this form: "A device is required to perform task X," expressed as a set of *design requirements*. Writers on design emphasize that the need statement should be solution-neutral (that is, it should not imply how the task will be performed) to avoid narrow thinking constrained by preconceptions. Between the need statement and the product specification lie the stages shown in Figure 2.1: *concept, embodiment*, and *detailed design*, explained in a moment.

The product itself is called a *technical system*. A technical system consists of *subassemblies* and *components*, put together in a way that performs the required task, as in the breakdown of Figure 2.2. It is like describing a cat (the system) as made up of one head, one body, one tail, four legs, and so on (the subassemblies), each composed of components: femurs, quadriceps, claws, fur. This decomposition is a useful way to analyze an existing design, but it is not of much help in the design process itself, that is, in devising new designs. Better, for this purpose, is one based on the ideas of systems analysis, which considers the inputs, flows, and outputs of information, energy, and materials, as in Figure 2.3.

This design converts the inputs into the outputs. An electric motor, for example, converts electrical into mechanical energy; a forging press takes

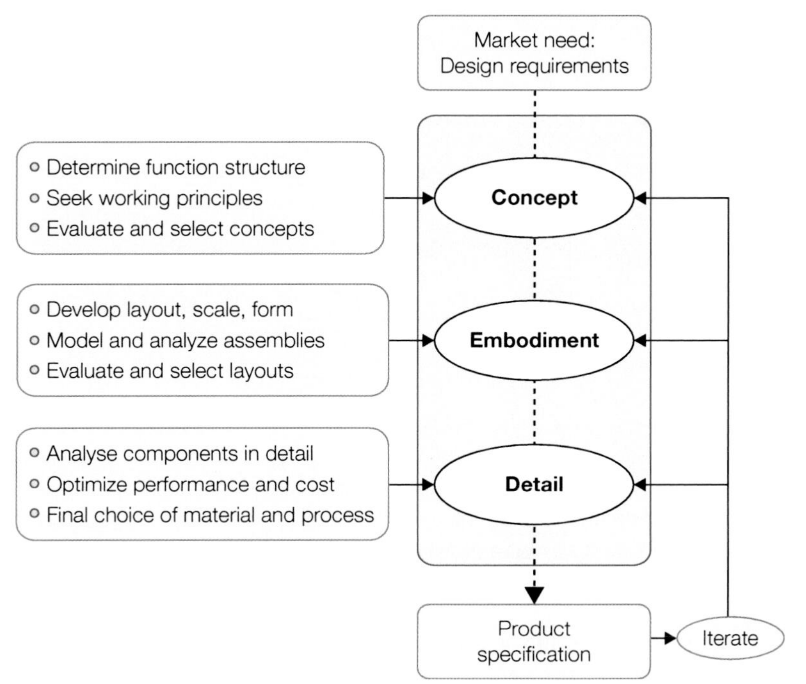

FIGURE 2.1
The design flow chart. The design proceeds from the identification of a *market need*, clarified as a set of *design requirements*, through *concept*, *embodiment*, and *detailed analysis* to a *product specification*.

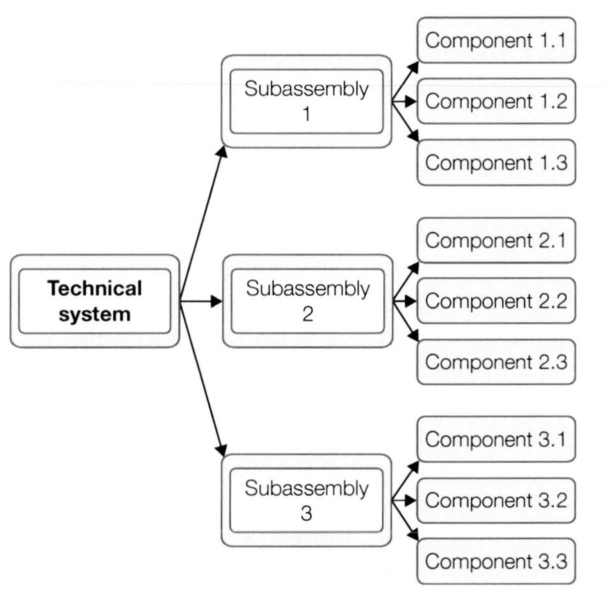

FIGURE 2.2
The analysis of a technical system as a breakdown into assemblies and components. Material and process selection is at the component level.

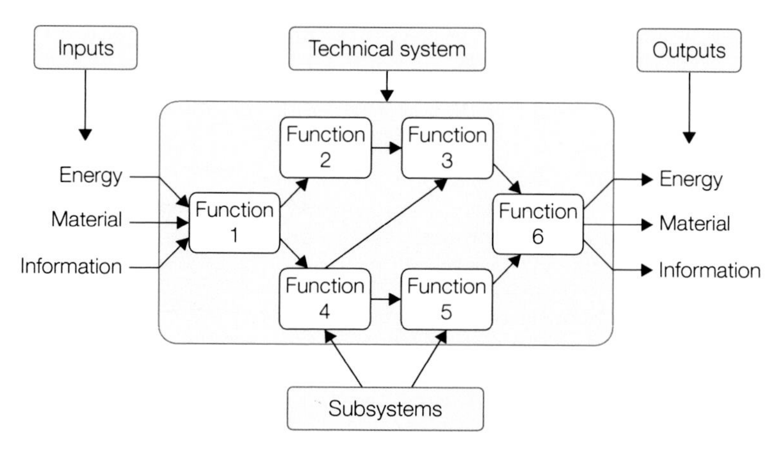

FIGURE 2.3
The function structure is a systems approach to the analysis of a technical system, seen as transformation of energy, materials, and information (signals). This approach, when elaborated, helps structure thinking about alternative designs.

and reshapes material; a burglar alarm collects information and converts it to noise. In this approach, the system is broken down into connected subsystems, each of which performs a specific function, as shown in Figure 2.3. The resulting arrangement is called the *function structure* or *function decomposition* of the system. It is like describing a cat as an appropriate linkage of a respiratory system, a cardio-vascular system, a nervous system, a digestive system, and so on. Alternative designs link the unit functions in alternative ways, combine functions, or split them. The function structure gives a systematic way of assessing design options.

The design proceeds by developing concepts to perform the functions in the function structure, each based on a *working principle*. At this, the conceptual design stage, all options are open: The designer considers alternative concepts and the ways in which these might be separated or combined. The next stage, embodiment, takes the promising concepts and seeks to analyze their operation at an approximate level. This involves sizing the components and selecting materials that will perform properly in the ranges of stress, temperature, and environment suggested by the design requirements, examining the implications for performance and cost. The embodiment stage ends with a feasible layout, which is then passed to the detailed design stage. Here specifications for each component are drawn up. Critical components may be subjected to precise mechanical or thermal analysis. Optimization methods are applied to components and groups of components to maximize performance. A final choice of geometry and material is made and the methods of production are analyzed and costed. The stage ends with a detailed production specification.

All that sounds well and good. If only it were so simple. The linear process suggested by Figure 2.1 obscures the strong coupling between the three stages. The consequences of choices made at the concept or the embodiment stages may not become apparent until the detail is examined. Iteration, looping back to explore alternatives, is an essential part of the design process.

Think of each of the many possible choices that *could* be made as an array of blobs in design space, as shown in Figure 2.4. Here C1, C2... are possible concepts, and E1, E2... and D1, D2... are possible embodiments and detailed elaborations of them. The design process becomes one of creating paths and linking compatible blobs until a connection is made from the top (market need) to the bottom (product specification). Some trial paths have dead ends, some loop back. It is like finding a track across difficult terrain—it may be necessary to go back many times to go forward in the end. Once a path is found, it is always possible to make it look linear and logical (and many books do this), but the reality is more like Figure 2.4 than Figure 2.1.

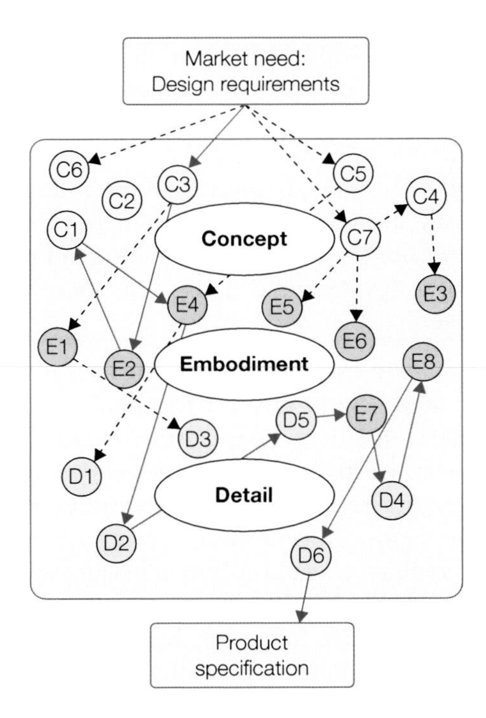

FIGURE 2.4
The convoluted path of design. Here the C-blobs represent concepts; the E-blobs, embodiments of the Cs; and the D-blobs, detailed realizations of the Es. The process is complete when a compatible path from "need" to "specification" can be identified. It is a devious path (the full red line) with back loops and dead ends (the broken lines). This creates the need for tools that allow fluid access to materials information at differing levels of breadth and detail.

Thus a key part of design, and of selecting materials for it, is *flexibility*, the ability to explore alternatives quickly, keeping the big picture as well as the details in mind. Our focus in later chapters is on the selection of materials and processes, where exactly the same need arises. This requires some kind of mapping of the "universes" of materials and processes to allow quick surveys of alternatives while still providing detail when it is needed. The selection charts of Chapter 4 and the methods of Chapter 5 help do this.

Described in the abstract, these ideas are not easy to grasp. An example will help—it comes in Section 2.6. First, a look at types of design.

2.3 TYPES OF DESIGN

It is not always necessary to start, as it were, from scratch. *Original design* does: it involves a new idea or working principle (the ballpoint pen, the compact disc). New materials can offer new, unique combinations of properties that enable original design. Thus high-purity silicon enabled the transistor; high-purity glass, the optical fiber; high coercive-force magnets, the miniature earphone; solid-state lasers the compact disc. Sometimes the new material suggests the new product. Sometimes, instead, the new product demands the development of a new material: Nuclear technology drove the development of a series of new zirconium alloys and low-carbon stainless steels; space technology stimulated the development of light weight composites; gas turbine technology today drives development of high-temperature alloys and ceramic coatings. Original design sounds exciting, and it is. But most design is not like that.

Almost all design is *adaptive* or *developmental*. The starting point is an existing product or product range. The motive for redesigning it may be to enhance performance, to reduce cost, or to adapt it to changing market conditions. Adaptive design takes an existing concept and seeks an incremental advance in performance through a refinement of the working principle. It, too, is often made possible by developments in materials: polymers replacing metals in household appliances; carbon fiber replacing wood in sports equipment. The appliance and the sports equipment markets are fast-moving and competitive. These markets have frequently been won (and lost) by the way in which the manufacturer has adapted the product by exploiting new materials.

Finally, *variant design* involves a change of scale or dimension or detailing without a change of function or the method of achieving it: the scaling up of boilers, or of pressure vessels, or of turbines, for instance. Change of scale or circumstances of use may require change of material: Small boats are made of fiberglass, large ships are made of steel; small boilers are made of copper, large ones of steel; subsonic planes are made of one alloy, supersonic of another—all for good reasons, as detailed in later chapters.

2.4 DESIGN TOOLS AND MATERIALS DATA

To implement the steps of Figure 2.1, use is made of *design tools*. They are shown as inputs, attached to the left of the main backbone of the design methodology in Figure 2.5. The tools enable the modeling and optimization of a design, easing the routine aspects of each phase. Function modelers suggest viable function structures. Configuration optimizers suggest or refine shapes. Geometric and 3D solid modeling packages allow visualization and create files that can be downloaded to numerically controlled prototyping and manufacturing systems. Optimization, DFM, DFA,[1] and cost estimation software allows manufacturing aspects to be refined. Finite element (FE) and computational fluid dynamics (CFD) packages allow precise mechanical and thermal analysis even when the geometry is complex, deformations are

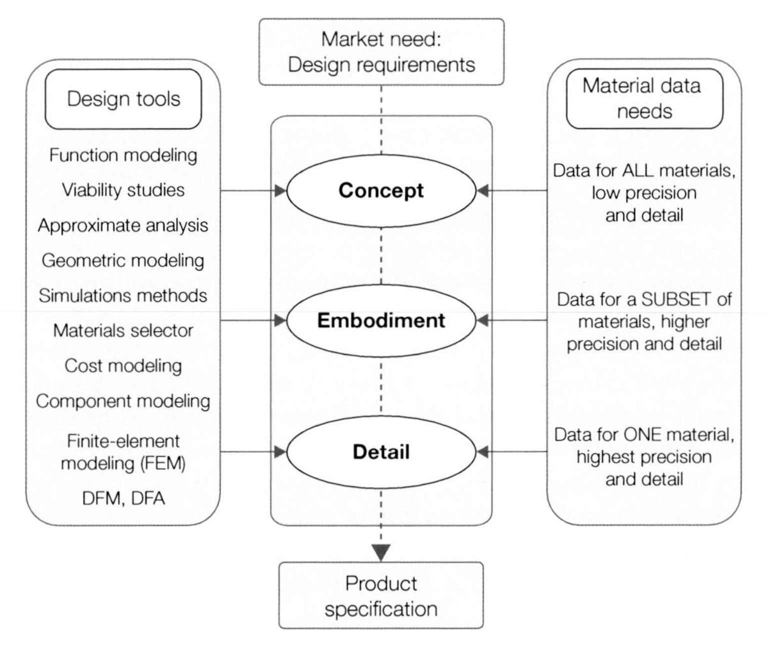

FIGURE 2.5
The design flow chart, showing how design tools and materials selection enter the procedure. Information about materials is needed at each stage, but at very different levels of breadth and precision. Iteration is part of the process.

[1] Design for Manufacture and Design for Assembly.

large, and temperatures fluctuate. There is a natural progression in the use of the tools as the design evolves: approximate analysis and modeling at the conceptual stage; more sophisticated modeling and optimization at the embodiment stage; and precise ("exact"—but nothing is ever that) analysis at the detailed design stage.

Tools for material selection play a major part in each stage of the design. The nature of the data needed in the early stages differs greatly in its level of precision and breadth from that needed later on (Figure 2.5, *right*). At the concept stage, the designer requires approximate property values, but for the widest possible range of materials. All options are open: A polymer may be the best choice for one concept, a metal for another, even though the function is the same. The problem, at this stage, is not precision and detail; it is breadth and speed of access: How can the vast range of data be presented to give the designer the greatest freedom in considering alternatives?

At the embodiment stage the landscape has narrowed. Here we need data for a subset of materials, but at a higher level of precision and detail. These are found in more specialized handbooks and software that deal with a single class or subclass of materials—*metals* or just *aluminum alloys*, for instance. The risk now is that of losing sight of the bigger spread of materials to which we must return if the details don't work out; it is easy to get trapped in a single line of thinking—a single set of "connections" in the sense of Figure 2.4—when other connections may offer a better solution.

The final stage, that of detailed design, requires a still higher level of precision and detail, but for only one or a very few materials. Such information is best found in the datasheets issued by the material producers themselves and in detailed databases for restricted material classes. A given material (polyethylene, for instance) has a range of properties that derive from differences in the ways different producers make it. At the detailed design stage, a supplier must be identified, and the properties of their product used in the design calculations; a product from another supplier may differ. And sometimes even this is not good enough. If the component is a critical one (meaning that its failure could, in some sense or another, be disastrous), then it is prudent to conduct in-house tests to measure the critical properties, using a sample of the material that will be used to make the product itself.

The materials input does not end with the establishment of production. Products fail in service, and failures contain information. It is an imprudent manufacturer who does not collect and analyze data on failures. Often this points to the misuse of a material, one that redesign or reselection can eliminate.

So material choice depends on *function*. But that is not the only constraint.

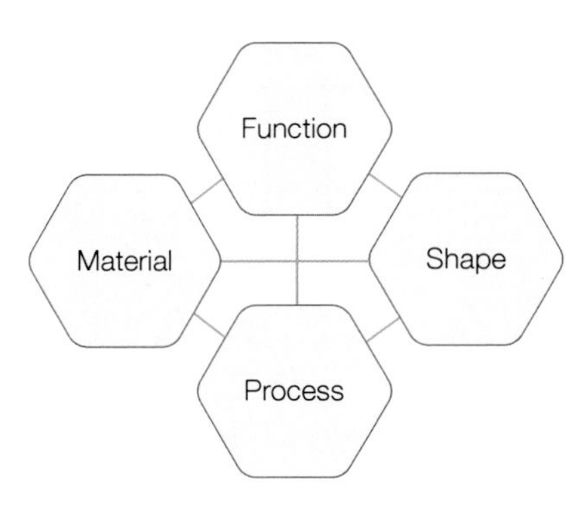

FIGURE 2.6
The central problem of materials selection in mechanical design: the interaction between function, material, process, and shape.

2.5 FUNCTION, MATERIAL, SHAPE, AND PROCESS

The selection of material is tied in with process and shape. To make a shape, the material is subjected to processes that, collectively, are called *manufacture*: these include primary forming processes (e.g., casting and forging), material removal processes (machining, drilling), joining processes (e.g., welding) and finishing processes (e.g., painting or electroplating). Function, material, shape, and process interact (Figure 2.6). Function, as already described, influences material choice. Material choice influences processes through the material's ability to be cast or molded or welded or heat-treated. Process determines shape; size; precision; and, of course, cost. These interactions are two-way: specification of shape restricts the choice of material and process; but equally the specification of process limits the material choice and the accessible shapes. The more sophisticated the design, the tighter the specifications and the greater the interactions.

The interaction between function, material, shape, and process lies at the heart of the material selection process. It is a theme we will return to throughout this book, visiting each of the hexagons of Figure 2.6 in turn. But first we look at a case study to illustrate the design process.

2.6 CASE STUDY: DEVICES TO OPEN CORKED BOTTLES

Wine, like cheese, is one of man's improvements on nature. And as long as humans have cared about wine, people have cared about corks to keep it safely sealed in flasks and bottles. "Corticum... demovebit amphorae...,"

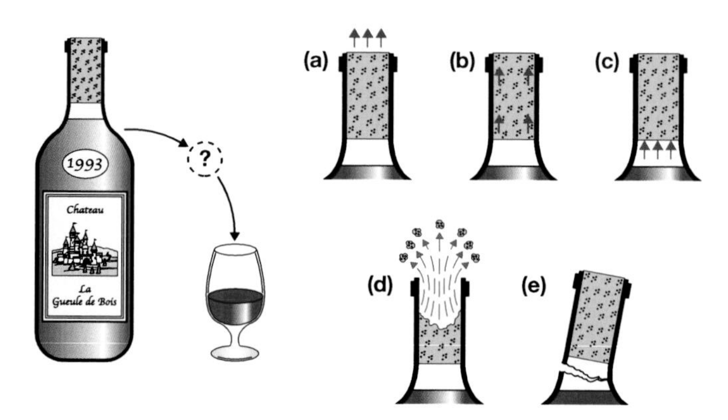

FIGURE 2.7

Left: the market need; a device is sought to allow access to wine contained in a corked bottle. *Right:* five possible concepts, illustrating physical principles, to fill the need.

("Uncork the amphora...)" sang Horace[2] to celebrate the anniversary of his miraculous escape from death by a falling tree. But how did he do it?

A corked bottle creates a market need: the need to gain access to the wine inside. We might state it thus: "A device is required to pull corks from wine bottles." But hold on. The need must be expressed in solution-neutral form, and this is not. The aim is to gain access to the wine; our statement implies that this will be done by removing the cork, and that the cork will be removed by pulling. There could be other ways. So we will try again: "A device is required to allow access to wine in a corked bottle" (Figure 2.7), and one might add, "with convenience, at modest cost, and without contaminating the wine."

Five concepts for doing this are shown on the right of Figure 2.7. In order, the devices act to remove the cork by axial traction (pulling); to remove it by shear tractions; to push it out from below; to pulverize it; and to bypass it altogether by knocking the neck off the bottle.[3]

Numerous devices exist that use the first three of these concepts. The others are used too, though generally only in moments of desperation. We shall eliminate these on the grounds that they might contaminate the wine, and examine the others more closely, exploring working principles. Figure 2.8 shows one example for each of the first three concepts: In the first, a screw

[2] Horace, Q. 27 BC, *Odes*, Book III, Ode 8, line 10.

[3] A Victorian invention for opening old port, the cork of which may have become brittle with age and alcohol absorption, involved ring-shaped tongs. The tongs were heated red on an open fire, then clamped onto the cold neck of the bottle. The thermal shock removed the neck cleanly and neatly.

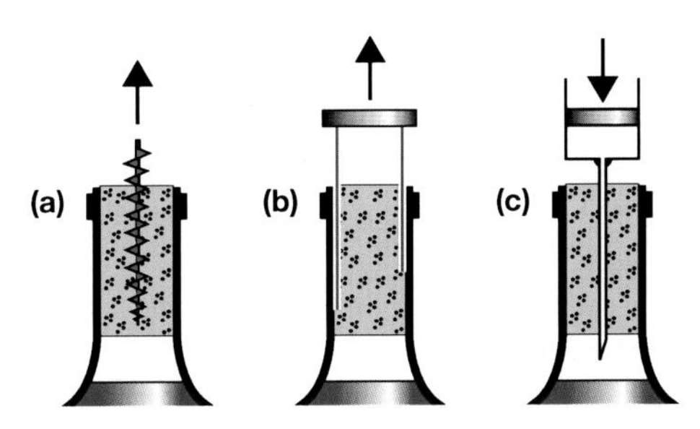

FIGURE 2.8
Working principles for implementing the first three concepts of Figure 2.7. Examples of all of these appear in the cover picture of this chapter.

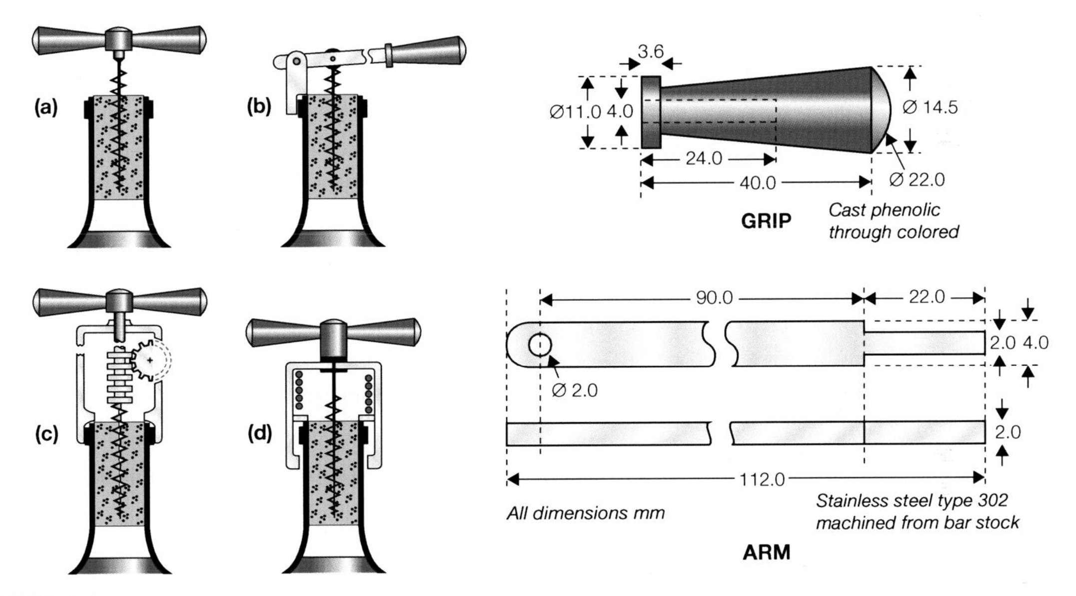

FIGURE 2.9
Left: embodiments: (a) direct pull; (b) lever-assisted pull; (c) gear-assisted pull; (d) spring-assisted pull (a spring in the body is compressed as the screw is driven into the cork). *Right:* detailed design of the lever of embodiment with material choice.

is threaded into the cork to which an axial pull is applied; in the second, slender elastic blades inserted down the sides of the cork apply shear tractions when twisted and pulled; and in the third, the cork is pierced by a hollow needle through which a gas is pumped to push the cork out. Examples of all three appear in the cover picture of this chapter.

Figure 2.9 shows embodiment sketches for devices based on just one concept—that of axial traction. The first is a direct pull; the other three use some sort of mechanical advantage—levered pull, geared pull, and spring-assisted pull. The embodiments identify the *functional requirements*

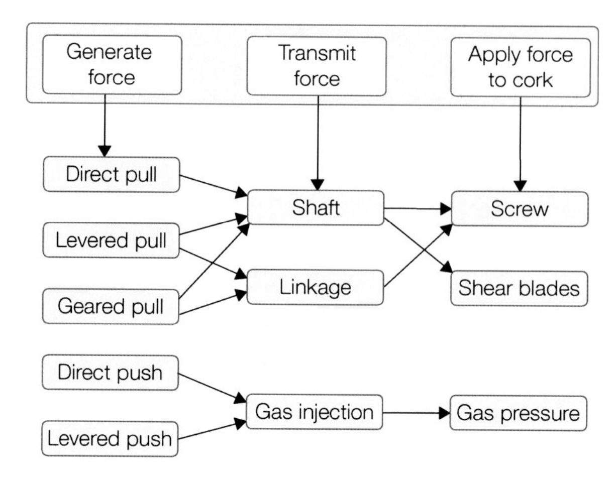

FIGURE 2.10
The function structure and working principles of cork removers.

of each component of the device, which might be expressed in statements such as

- A cheap screw to transmit a prescribed load to the cork
- A light lever (that is, a beam) to carry a prescribed bending moment
- A slender elastic blade that will not buckle when driven between the cork and the bottle neck
- A thin, hollow needle, stiff and strong enough to penetrate a cork

The functional requirements of each component are the inputs to the materials selection process. They lead directly to the *property limits* and *material indices*, as described in Chapter 5, in which we examine procedures with requirements such as "light, strong beam" or "slender, elastic, blade" and use them to identify a subset of materials that perform this function particularly well. The final choice of material and process forms part of the detailed stage of design (Figure 2.9), leading to full specifications to enable manufacture.

We conclude by returning to the idea of function structure. That for the cork remover is sketched in the upper part of Figure 2.10: Generate a force, transmit a force, apply the force to the cork. The alternative designs differ in the working principle by which these functions are achieved, as indicated in the lower part of the figure. Others could be devised by making other links.

2.7 SUMMARY AND CONCLUSIONS

Design is an iterative process. The starting point is a *market need* captured in a set of *design requirements. Concepts* for a product that meet the need are devised. If initial estimates and exploration of alternatives suggest that the concept is viable, the design proceeds to the *embodiment* stage: Working principles are selected, size and layout are decided, and initial estimates of performance and cost are made. If the outcome is successful, the designer proceeds to the *detailed design* stage: optimization of performance, full analysis of critical components, preparation of detailed production drawings (usually as CAD files), specifications of tolerance, precision, assembly, and finishing methods.

Materials selection enters at each stage, but at different levels of breadth and precision. At the conceptual stage all materials and processes are potential candidates, requiring a procedure that allows rapid access to data for a wide range of each, although without the need for great precision. The preliminary selection passes to the embodiment stage, the calculations and optimizations of which require information at a higher level of precision and detail. They eliminate all but a small shortlist of candidate materials and processes for the final, detailed stage of the design. For these few candidates, highest-quality data are required.

Data exist that meet the needs of all these levels. Each level requires its own data management scheme, described in the following chapters. The management system must be design-led, yet must recognize the richness of choice and embrace the complex interaction between the material, its shape, the process by which it is given that shape, and the function it is required to perform. And it must allow rapid iteration—back-looping when a particular path proves to be unprofitable. Tools now exist to help with all of this. We will meet one—the CES materials and process selection platform—later in this book.

But given this complexity, why not opt for the safe bet: Stick to what you used before? Many have chosen that option. Few are still in business.

2.8 FURTHER READING

A chasm exists between books on design methodology and those on materials selection: each largely ignores the other. The book by French is remarkable for its insights, but the word "material" does not appear in its index. Pahl and Beitz have near-biblical standing in the design camp, but their text is heavy going. Ullman and Cross take a more relaxed approach and are easier to digest. The books by Budinski and Budinski, by Charles, Crane, and Furness, and by Farag present the materials case well, but are less good on design. Lewis illustrates material selection through case studies, but does not develop a systematic procedure. The best compromise, perhaps, is Dieter.

General texts on design methodology

Cross, N. (2000). *Engineering design methods* (3rd ed.). Wiley, ISBN 0-471-87250-4.

A durable text describing the design process, with emphasis on developing and evaluating alternative solutions.

Dieter, G.E., & Schmidt, L.C. (2009). *Engineering design* (4th ed.). McGraw-Hill, ISBN 978-0-07-283703-2.

A clear introduction from authors with a strong materials background.

French, M.J. (1985). *Conceptual design for engineers.* The Design Council, London, and Springer, ISBN 0-85072-155-5 and 3-540-15175-3.

The origin of the "Concept–Embodiment–Detail" block diagram of the design process. The book focuses on the concept stage, demonstrating how simple physical principles guide the development of solutions to design problems.

Pahl, G., & Beitz, W. (1997). *Engineering design* (2nd ed.). Translated by K. Wallace & L. Blessing, The Design Council, London, and Springer Verlag, ISBN 0-85072-124-5 and 3-540-13601-0.

The Bible—or perhaps more exactly the Old Testament—of the technical design field, developing formal methods in the rigorous German tradition.

Ullman, D.G. (1992). *The mechanical design process.* McGraw-Hill, ISBN 0-07-065739-4.

An American view of design, developing ways in which an initially ill-defined problem is tackled in a series of steps, much in the way suggested by Figure 2.1 of the present text.

Ulrich, K.T., & Eppinger, S.D. (1995). *Product design and development.* McGraw-Hill, ISBN 0-07-065811-0.

A readable, comprehensible text on product design, as taught at MIT. Many helpful examples but almost no mention of materials.

General texts on materials selection in design

Ashby, M., Shercliff, H., & Cebon, D. (2010). *Materials: Engineering, science, processing and design* (2nd ed.). Butterworth-Heinemann, ISBN 978-1-85617-895-2. North American edition: ISBN 13:978-1-85617-743-6.

An introductory text introducing ideas that are developed more fully in the present text.

Askeland, D.R., & Phulé, P.P. (2006). *The science and engineering of materials* (5th ed.). Thomson, ISBN 0-534-55396-6.

A well-established materials text that deals well with the science of engineering materials.

Budinski, K.G., & Budinski, M.K. (2010). *Engineering materials, properties and selection* (9th ed.). Prentice Hall, ISBN 978-0-13-712842-6.

Like Askeland, this is a well established materials text that deals well with both material properties and processes.

Callister, W.D. (2010). *Materials science and engineering: An introduction* (8th ed.). John Wiley & Sons, ISBN 978-0-470-41997-7.

A well-established text taking a science-led approach to the presentation of materials teaching.

Charles, J.A., Crane, F.A.A., & Furness, J.A.G. (1997). *Selection and use of engineering materials* (3rd ed.). Butterworth-Heinemann, ISBN 0-7506-3277-1.

A materials science rather than a design-led approach to the selection of materials.

Dieter, G.E. (1999). *Engineering design, a materials and processing approach* (3rd ed.). McGraw-Hill, ISBN 9-780-073-66136-0.

A well-balanced and respected text focusing on the place of materials and processing in technical design.

Farag, M.M. (2008). *Materials and process selection for engineering design* (2nd ed.). CRC Press, Taylor and Francis, ISBN 9-781-420-06308-0.

A materials science approach to the selection of materials.

Lewis, G. (1990). *Selection of engineering materials.* Prentice-Hall, ISBN 0-13-802190-2.

A text on material selection for technical design, based largely on case studies.

Shackelford, J.F. (2009). *Introduction to materials science for engineers* (7th ed.). Prentice Hall. ISBN 978-0-13-601260-4.
A well-established materials text with a design slant.

And on corks and corkscrews

The Design Council. (1994). Teaching aids program EDTAP DE9. The Design Council, London.

McKearin, H. (1973). On "stopping," bottling and binning. *International Bottler and Packer*, (April), 47–54.

Perry, E. (1980). *Corkscrews and bottle openers*. Shire Publications.

Watney, B.M., & Babbige, H.D. (1981). *Corkscrews*. Sotheby's Publications, ISBN 0-85667-113-4.

CHAPTER 3

Engineering Materials and Their Properties

Acrylonitrile butadiene styrene (ABS)

General properties				
Density	1100	-	1200	kg/m^3
Price	2.1	-	2.5	USD/kg
Mechanical properties				
Young's modulus	1.1	-	2.9	GPa
Yield strength	19	-	51	MPa
Tensile strength	28	-	55	MPa
Elongation	15	-	40	%
Hardness—Vickers	5.6	-	15	HV
Fatigue strength at 10^7 cycles	11	-	22	MPa
Fracture toughness	1.2	-	4.3	$MPa.m^{1/2}$
Thermal properties				
Glass temperature	360	-	400	K
Maximum service temperature	340	-	350	K
Minimum service temperature	150	-	200	K
Thermal conductivity	0.19	-	0.34	W/m.K
Specific heat capacity	1400	-	1900	J/kg.K
Thermal expansion coefficient	85	-	230	$10^{-6}/°C$
Electrical properties				
Electrical resistivity	3.3×10^{21}	-	3×10^{22}	μohm.cm
Dielectric constant	2.8	-	3.2	
Dielectric loss tangent	0.003	-	0.007	
Dielectric strength	14	-	22	MV/m
Eco-properties				
Embodied energy	91	-	1e2	MJ/kg
CO_2 footprint	3.3	-	3.6	kg/kg

The material

ABS (Acrylonitrile-butadiene-styrene) is tough, resilient, and easily molded. It is usually opaque, although some grades can now be transparent, and it can be given vivid colors. ABS-PVC alloys are tougher than standard ABS and, in self-extinguishing grades, are used for the casings of power tools. The picture shows that ABS allows detailed moldings, accepts color well, and is nontoxic and tough enough to survive the worst that children can do.

Typical uses

Safety helmets; camper tops; automotive instrument panels and other interior components; pipe fittings; home-security devices and housings for small appliances; communications equipment; business machines; plumbing hardware; automobile grilles; wheel covers; mirror housings; refrigerator liners; luggage shells; tote trays; mower shrouds; boat hulls; large components for recreational vehicles; weather seals; glass beading; refrigerator breaker strips; conduit; pipe for drain-waste-vent (DWV) systems.

Tradenames

Claradex, Comalloy, Cycogel, Cycolac, Hanalac, Lastilac, Lupos, Lustran ABS, Magnum, Multibase, Novodur, Polyfabs, Polylac, Porene, Ronfalin, Sinkral, Terluran, Toyolac, Tufrex, Ultrastyr.

Part of a record of the properties of an engineering material, ABS.

Materials Selection in Mechanical Design. DOI: 10.1016/B978-1-85617-663-7.00003-5

CONTENTS

3.1 INTRODUCTION AND SYNOPSIS

Materials, one might say, are the food of design. This chapter presents the menu: the materials shopping list. A successful product—one that performs well, is good value for money, and gives pleasure to the user—uses the best materials for the job, and fully exploits their potential and characteristics. It brings out their flavor, so to speak.

The families of materials—metals, polymers, ceramics, and so forth—are introduced in Section 3.2. What do we need to know about them if we are to design with them? That is the subject of Section 3.3, in which distinctions are drawn between various types of materials information. But it is not, in the end, a *material* that we seek; it is a certain *profile of properties*—the one that best meets the needs of the design. Properties are the currency of the materials world. They are the bargaining chips—the way you trade off one material against another. The properties important in thermo-mechanical design are defined briefly in Section 3.4. It makes boring reading. The reader who is confident in the definitions and units of moduli, strengths, damping capacities, thermal and electrical conductivities, and the like, may wish to skip this, using it for reference, when needed, for the precise meaning and units of the data in the property charts that come later. Don't, however, skip Section 3.2. It sets up the classification structure that is used throughout the book. The chapter ends, in the usual way, with a summary.

3.2 THE FAMILIES OF ENGINEERING MATERIALS

It is conventional to classify the materials of engineering into the six broad families shown in Figure 3.1: metals, polymers, elastomers, ceramics, glasses, and hybrids. The members of a family have certain features in

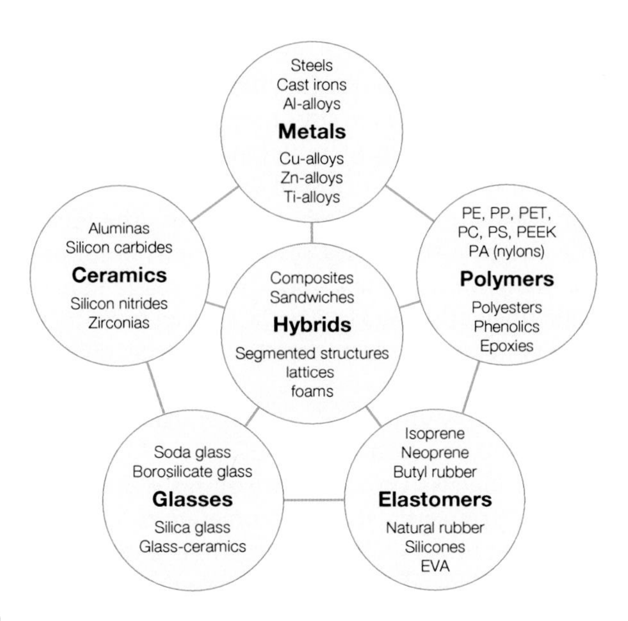

FIGURE 3.1
The menu of engineering materials. The basic families of metals, ceramics, glasses, polymers, and elastomers can be combined in various geometries to create hybrids.

common: similar properties, similar processing routes, and, often, similar applications.

Metals are stiff. They have relatively high elastic moduli. Most, when pure, are soft and easily deformed. They can be made strong by alloying and by mechanical and heat treatment, but they remain ductile, allowing them to be formed by deformation processes. Certain high-strength alloys (spring steel, for instance) have ductilities as low as 1%, but even this is enough to ensure that the material yields before it fractures and that fracture, when it occurs, is of a tough, ductile type. Partly because of their ductility, metals are prey to fatigue and of all the classes of material, they are the least resistant to corrosion.

Ceramics, too, have high moduli, but unlike metal, they are brittle. Their "strength" in tension means the brittle fracture strength; in compression it is the brittle crushing strength, which is about 15 times greater. And because ceramics have no ductility, they have a low tolerance for stress concentrations (like holes or cracks) or for high-contact stresses (at clamping points, for instance). Ductile materials accommodate stress concentrations by deforming in a way that redistributes the load more evenly, and because of this, they can be used under static loads within a small margin of their yield strength. Ceramics cannot. Brittle materials always have a wide scatter in strength, and the strength itself depends on the volume of material under load and the time over which it is applied. So ceramics are not as easy to design with as metals. Despite this, they have attractive features. They are stiff, hard, and abrasion-resistant (hence their use for bearings and cutting tools); they retain their strength to high temperatures; and they resist corrosion well.

Glasses are noncrystalline ("amorphous") solids. The most common are the soda-lime and borosilicate glasses familiar as bottles and ovenware, but there are many more. Metals, too, can be made noncrystalline by cooling them sufficiently quickly. The lack of crystal structure suppresses plasticity, so, like ceramics, glasses are hard, brittle, and vulnerable to stress concentrations.

Polymers are at the other end of the spectrum. They have moduli that are low, roughly 50 times lower than those of metals, but they can be strong—nearly as strong as metals. A consequence of this is that elastic deflections can be large. They creep, even at room temperature, meaning that a polymer component under load may, with time, acquire a permanent set. And their properties depend on temperature so that a polymer that is tough and flexible at 20°C may be brittle at the 4°C of a household refrigerator, yet may creep rapidly at the 100°C of boiling water. Few have useful strength above 200°C. Some polymers are mainly crystalline, some are amorphous (noncrystalline), some are a mix of crystalline and amorphous—transparency goes with the amorphous structure. If these aspects are allowed for in the design, the advantages of polymers can be exploited. And there are many. When combinations of properties, such as strength per unit weight, are important, polymers can compete with metals. They are easy to shape. Complicated parts performing several functions can be molded from a polymer in a single operation. The large elastic deflections allow the design of polymer components that snap together, making assembly fast and cheap. And by accurately sizing the mold and precoloring the polymer, no finishing operations are needed. Polymers resist corrosion (paints, for instance, are polymers) and have low coefficients of friction. Good design exploits these properties.

Elastomers are long-chain polymers above their glass-transition temperature, T_g. The covalent bonds that link the units of the polymer chain remain intact, but the weaker Van der Waals and hydrogen bonds that, below T_g, bind the chains to each other, have melted. This gives elastomers unique properties: Young's moduli as low as 10^{-3} GPa (10^5 times less than that typical of metals) increase with temperature (all other solids show a decrease), and have enormous elastic extension. Their properties differ so much from those of other solids that special tests have evolved to characterize them. This creates a problem: If we wish to select materials by prescribing a desired attribute profile, as we do later in this book, then a prerequisite is a set of attributes common to all materials.

To overcome this, we use a common set of properties in the early stages of design, estimating approximate values for anomalies like elastomers. Specialized attributes, representative of one family only, are stored separately; they are for use in the later stages.

Hybrids are combinations of two or more materials in a predetermined configuration and scale. They combine the attractive properties of the other families of materials while avoiding some of their drawbacks. Their design is the subject of Chapters 11 and 12. The family of hybrids includes fiber and particulate composites, sandwich structures, lattice structures, foams, cables, and laminates; almost all the materials of nature—wood, bone, skin, and leaf—are hybrids. Fiber-reinforced composites are, of course, the most familiar. Most of those at present available to the engineer have a polymer matrix reinforced by fibers of glass, carbon, or Kevlar (an aramid). They are light, stiff, and strong, and they can be tough. They, and other hybrids using a polymer as one component, cannot be used above 250°C because the polymer softens, but at room temperature their performance can be outstanding. Hybrid components are expensive, and they are relatively difficult to form and join. So, despite their attractive properties, the designer will use them only when the added performance justifies the added cost. Today's growing emphasis on high performance and fuel efficiency provides increasing drivers for their use.

These, then, are the material families. What do we need to know about them?

3.3 MATERIALS INFORMATION FOR DESIGN

The engineer, in selecting materials for a developing design, needs data for the materials' properties. Engineers are often conservative in their choice, reluctant to consider materials with which they are unfamiliar, and with good reason. Data for the old, well-tried materials are established, reliable, and easily found. Data for newer, emerging, materials may be incomplete or untrustworthy. Yet innovation is often made possible by new materials. So it is important to know how to judge data quality.

If you're going to design something, what sort of materials information do you need? Figure 3.2 draws relevant distinctions. On the left a material is tested and the *data* are captured. But these raw data—unqualified numbers—are, for our purposes, useless. To make data useful requires statistical analysis. What is the mean value of the property when measured on a large batch of samples? What is the standard deviation? Given these, it is possible to calculate *allowables*: values of properties that, with a given certainty (say, one part in 10^6) can be guaranteed. Material texts generally present test data; by contrast, data in most engineering handbooks are allowables.

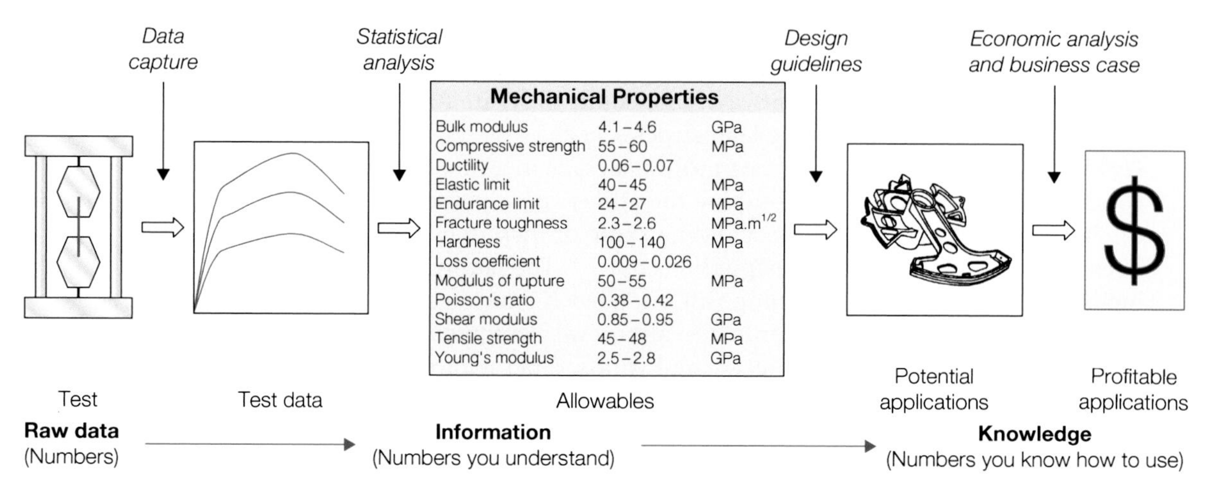

FIGURE 3.2

Types of material information. We are interested here in the types found in the center of this schematic: structured data for design "allowables" and the characteristics of a material that relate to its ability to be formed, joined, and finished; records of experience with its use; and design guidelines for its use.

One can think of data with known precision and provenance as *information*. Information can generally be reported as tables of numbers, as yes/no statements or as rankings: that is, it can be *structured*. Many attributes that can be structured are common to all materials; all have a density, an elastic modulus, a strength, a thermal conductivity. Structured information can be stored in a database and—since all materials have values—it is the starting point for selecting between them. The cover picture of this chapter shows part of a record for the polymer ABS with structured data on the left, reported as ranges that derive from differences in the way different producers make it.

This is a step forward, but it is not enough. To design with a material, you need to know its real character, its strengths, and its weaknesses. How do you shape it? How do you join it? Who has used it before and for what? Did it fail? Why? This information exists in handbooks, is documented as design guidelines, and is reported in failure analyses and case studies. It consists largely of text, graphs and images, and while certain bits of it may be available for one material, for another they may not. It is messier, but it is essential in reaching a final selection. We refer to this supporting information as *documentation*. The image and text on the right of the ABS cover are examples of documentation.

There is more. Material uses are subject to standards and codes. These rarely refer to a single material but to classes or subclasses. For a material to be used in contact with food or drugs, it must carry FDA approval or the equivalent. Metals and composites for use in U.S. military aircraft must have military specification approval. To qualify for best-practice design for the environment, material usage must confirm to ISO 14040 guidelines. And so forth. This, too, is a form of documentation (Table 3.1). The ensemble of information about a material, structured and unstructured, constitutes *knowledge*.

There is yet more (Figure 3.2, *right*). To succeed in the marketplace, a product must be economically viable and compete successfully, in terms of performance, consumer appeal, and cost, with the competition. All of these

Table 3.1 Basic Design-Limiting Material Properties and Their Usual SI Units*

Class	Property	Symbol and Units
General	Density	ρ (kg/m^3 or Mg/m^3)
	Price	C_m ($/kg)
Mechanical	Elastic moduli (Young's, shear, bulk)	E, G, K (GPa)
	Yield strength	σ_y (MPa)
	Tensile (ultimate) strength	σ_{ts} (MPa)
	Compressive strength	σ_c (MPa)
	Failure strength	σ_f (MPa)
	Hardness	H *(Vickers)*
	Elongation	ε (–)
	Fatigue endurance limit	σ_e (MPa)
	Fracture toughness	K_{1c} (MPa.m$^{1/2}$)
	Toughness	G_{1c} (kJ/m^2)
	Loss coefficient (damping capacity)	η (–)
	Wear rate (Archard) constant	K_AMPa^{-1}
Thermal	Melting point	T_m (°C or K)
	Glass temperature	T_g (°C or K)
	Maximum service temperature	T_{max} (°C or K)
	Minimum service temperature	T_{min} (°C or K)
	Thermal conductivity	λ (W/m.K)
	Specific heat	C_p (J/kg.K)
	Thermal expansion coefficient	α (K^{-1})
	Thermal shock resistance	ΔT_s (°C or K)
Electrical	Electrical resistivity	ρ_e (Ω.m or $\mu\Omega$.cm)
	Dielectric constant	ε_r (–)
	Breakdown potential	V_b (10^6 V/m)
	Power factor	P (–)
Optical	Refractive index	n (–)
Eco-properties	Embodied energy	H_m (MJ/kg)
	Carbon footprint	CO_2 (kg/kg)

* *Conversion factors from metric to imperial and cgs units appear inside the back and front covers of this book.*

depend on material choice and the way the material is processed. Much can be said about this, but not here; for now the focus is on structured data and documentation.

That's the essential background. Now for the properties themselves.

3.4 MATERIAL PROPERTIES AND THEIR UNITS

Each material can be thought of as having a set of attributes or properties. The combination that characterizes a given material is its *property profile*. Property profiles are assembled by systematic testing. In this section we scan the nature of the tests and the definition and units of the properties (see Table 3.1). Property values are listed in Appendix A. Units are given here in the SI system. Conversion factors to other systems are printed on the inside front and back cover of the book.

General properties

The *density*, ρ (units: kg/m^3), is the mass per unit volume. We measure it today as Archimedes did: by weighing in air and in a fluid of known density.

The *price*, C_m (units: \$/kg), spans a wide range. Some cost as little as \$0.2/kg, others as much as \$1,000/kg. Prices, of course, fluctuate, and they depend on the quantity you want and on your status as a "preferred customer" with your chosen vendor. Despite this uncertainty, it is useful to have an approximate price in the early stages of material selection.

Mechanical properties

The *elastic modulus*, E (units: GPa or GN/m^2), is the slope of the initial, linear-elastic, part of the stress-strain curve (Figure 3.3). Young's modulus, E, describes response to tensile or compressive loading; the shear modulus, G, describes response to shear loading; and the bulk modulus, K, describes the response to hydrostatic pressure. Poisson's ratio, ν, is the negative of the ratio of the lateral strain, ε_2, to the axial strain, ε_1, in axial loading:

$$\nu = -\frac{\varepsilon_2}{\varepsilon_1}$$

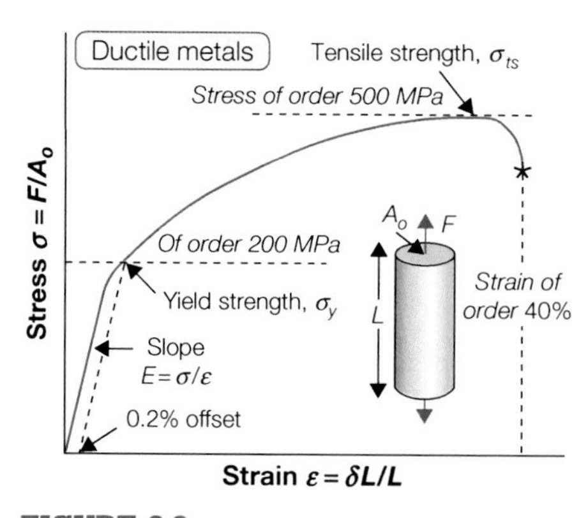

FIGURE 3.3
The stress-strain curve for a metal, showing the modulus, E, the 0.2% yield strength, σ_y, and the ultimate strength, σ_{ts}.

In reality, moduli measured as slopes of stress-strain curves are inaccurate, often low by a factor of 2 or more, because of contributions to the strain from inelasticity, creep, and other factors. Accurate moduli are measured dynamically: by exciting the natural vibrations of a beam or a wire or by measuring the velocity of sound waves in the material.

In an isotropic material, the moduli are related in the following ways:

$$E = \frac{3G}{1 + G/3K} \quad G = \frac{E}{2(1+\nu)} \quad K = \frac{E}{3(1-2\nu)} \tag{3.1}$$

Commonly,

$$\nu \approx 1/3$$

when

$$G \approx \frac{3}{8}E$$

and

$$K \approx E \tag{3.2a}$$

Elastomers are exceptional. For these,

$$\nu \approx 1/2$$

when

$$G \approx \frac{1}{3}E$$

and

$$K >> E \tag{3.2b}$$

Data sources like those described in Appendix D list values for all four moduli. In this book we examine data for E; approximate values for the others can be derived from the (3.2) equations when needed.

Estimating moduli

Young's modulus E for copper is 124 GPa; its Poisson's ratio ν is 0.345. What is its shear modulus, G?

Answer

Inserting the values for E and ν in the central equation (3.1) gives G = 46.1 GPa. The measured value is 45.6 GPa, a difference of only 1%.

The *strength*, σ_f (units: MPa or MN/m^2), of a solid requires careful definition. For metals, we identify σ_f with the 0.2% offset yield strength σ_y (see Figure 3.3), that is, the stress at which the stress-strain curve for axial loading deviates by a strain of 0.2% from the linear-elastic line. It is the same in tension and compression. For polymers, σ_f is identified as the stress at which the stress-strain curve becomes markedly nonlinear, at a strain typically of 1% (Figure 3.4). This may be caused by shear yielding: the irreversible slipping of molecular chains; or it may be caused by crazing: the formation of low-density, crack-like volumes that scatter light, making the polymer look white. Polymers are a little stronger ($\approx$ 20%) in compression than in tension.

Strength, for ceramics and glasses, depends strongly on the mode of loading (Figure 3.5). In tension, "strength" means the fracture strength, σ_t. In compression it means the crushing strength σ_c, which is much greater; typically

$$\sigma_c = 10 \text{ to } 15\, \sigma_t \tag{3.3}$$

When a material is difficult to grip, as is a ceramic, its strength can be measured in bending. The *flexural strength* or *modulus of rupture*, σ_{flex} (units: MPa) is the maximum surface stress in a bent beam at the instant of failure (Figure 3.6). One might expect this to be the same as the strength measured in tension, but for ceramics it is greater by a factor of about 1.3 because the volume subjected to this maximum stress is small and the probability of a large flaw lying in it is small also; in simple tension all flaws see the same stress.

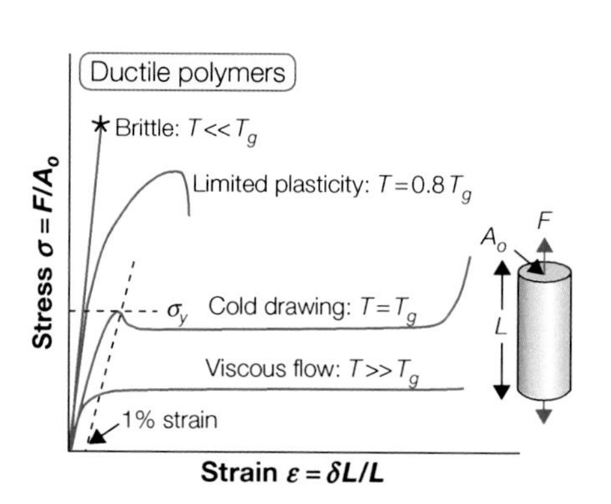

FIGURE 3.4

Stress-strain curves for a polymer below, at, and above its glass transition temperature, T_g.

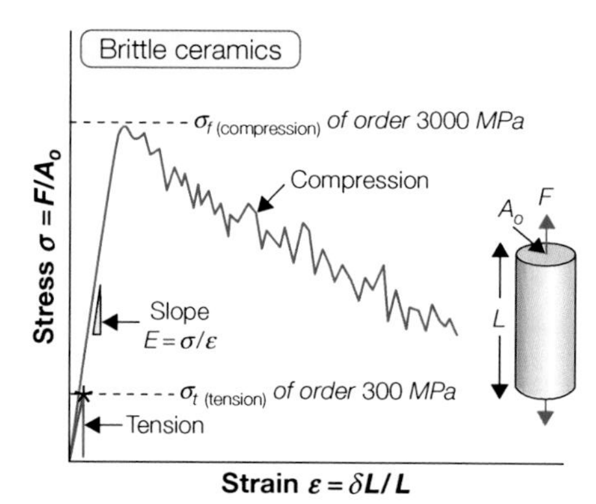

FIGURE 3.5

Stress-strain curves for a ceramic in tension and in compression. The compressive strength, σ_c, is 10 to 15 times greater than the tensile strength, σ_t.

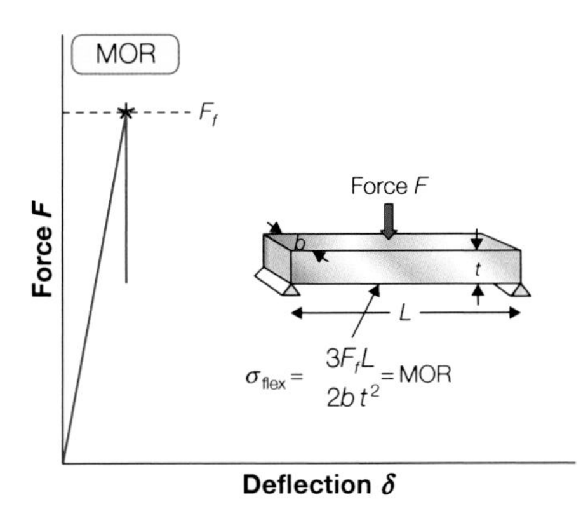

FIGURE 3.6
The modulus of rupture (MOR) is the surface stress at failure in bending. It is equal to, or slightly larger than, the failure stress in tension.

The strength of a composite is best defined by a set deviation from linear-elastic behavior; often an offset of 0.5% is taken. Composites that contain fibers, including natural composites such as wood, are a little weaker (up to 30%) in compression than tension because fibers buckle. In subsequent chapters, σ_f for composites means the tensile strength.

Strength, then, depends on material class and on mode of loading. Other modes of loading are possible: shear, for instance. Yield under multiaxial loads is related to that in simple tension by a *yield function*. For metals, the Von Mises yield function is a good description:

$$(\sigma_1 - \sigma_2)^2 + (\sigma_2 - \sigma_3)^2 + (\sigma_3 - \sigma_1)^2 = 2\,\sigma_f^2 \tag{3.4}$$

where σ_1, σ_2, and σ_3 are the principal stresses, positive when tensile; σ_1, by convention, is the greatest or most positive; σ_3, the smallest or least positive. For polymers the yield function is modified to include the effect of pressure:

$$(\sigma_1 - \sigma_2)^2 + (\sigma_2 - \sigma_3)^2 + (\sigma_3 - \sigma_1)^2 = 2\,\sigma_f^2 \left(\frac{1 + \beta p}{K} \right)^2 \tag{3.5}$$

where K is the bulk modulus of the polymer, $\beta \approx 2$ is a numerical coefficient that characterizes the pressure dependence of the flow strength, and the pressure p is defined by

$$p = -\frac{1}{3}(\sigma_1 + \sigma_2 + \sigma_3)$$

For ceramics, a Coulomb flow law is used:

$$\sigma_1 - B\sigma_2 = C \tag{3.6}$$

where B and C are constants.

The *tensile* (or *ultimate*) *strength* σ_{ts} (units: MPa) is the nominal stress at which a round bar of the material, loaded in tension, separates (Figure 3.3). For brittle solids—ceramics, glasses, and brittle polymers—it is the same as the failure strength in tension. For metals, ductile polymers, and most composites, it is greater than the yield strength, σ_y, by a factor of between

Using yield functions

A metal pipe of radius r and wall thickness t carry an internal pressure p. The pressure generates a circumferential wall stress of $\sigma_1 = pr/t$, an axial wall stress $\sigma_2 = pr/2t$. At what pressure will the pipe first yield?

Answer

Setting $\sigma_2 = \sigma_1/2$, $\sigma_3 = 0$ and $\sigma_f = \sigma_y$ in Equation (3.4) gives the yield condition $\sigma_1 = (2/\sqrt{3})\sigma_y$. Thus the pressure p^* that just causes first yield is $p^* = \frac{2}{\sqrt{3}}\frac{t}{r}\sigma_y$.

1.1 and 3 because of work hardening or, in the case of composites, load transfer to the reinforcement.

Cyclic loading can cause a crack to nucleate and grow in a material, culminating in fatigue failure. For many materials there exists a fatigue or *endurance limit*, σ_e (units: MPa), illustrated by the $\Delta\sigma - N_f$ curve of Figure 3.7. It is the stress amplitude $\Delta\sigma$ below which fracture does not occur, or occurs only after a very large number ($N_f > 10^7$) of cycles.

Tensile and compression tests are not always convenient: A large sample is needed and the test destroys it. The hardness test gives an approximate, nondestructive, measure of the strength. The *hardness, H* (SI units: MPa) of a material is measured by pressing a pointed diamond or hardened steel ball into the material's surface (Figure 3.8). The hardness is defined as the indenter force divided by the projected area of the indent. It is related to the quantity we have defined as σ_f by

$$H \approx 3\sigma_f \tag{3.7}$$

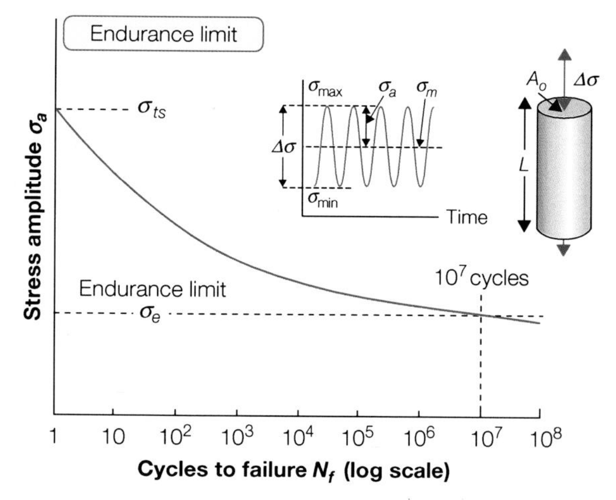

FIGURE 3.7
The endurance limit, σ_e, is the cyclic stress that causes failure in $N_f = 10^7$ cycles.

This, in the SI system, has units of MPa. Hardness is commonly reported in a bewildering array of other units, the most common of which is the Vickers H_v scale with units of kg/mm^2. It is related to H in the units used here by

$$H_v = \frac{H}{10}$$

A conversion chart for five hardness scales, relating them to yield strength, appears in Figure 3.9.

The *toughness*, G_{1c} (units: kJ/m^2), and the *fracture toughness*, K_{1c} (units: MPa/m$^{1/2}$ or MN/m$^{1/2}$), measure the resistance of a material to the propagation of a crack. The fracture toughness is measured by loading a sample containing a deliberately introduced crack of length $2c$ (Figure 3.10), recording the tensile stress σ^* at which the crack propagates. The quantity K_{1c} is then calculated from

$$K_{1c} = Y\sigma^* \sqrt{\pi c} \quad (3.8)$$

and the toughness from

$$G_{1c} = \frac{K_{1c}^2}{E(1+v)} \quad (3.9)$$

where Y is a geometric factor, near unity, that depends on details of the

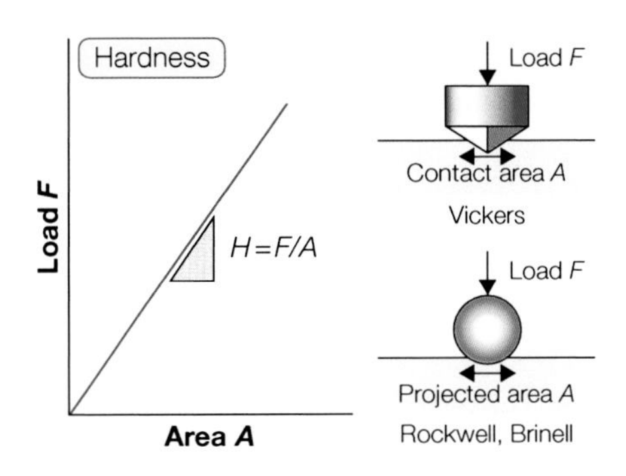

FIGURE 3.8
Hardness is measured as the load, F, divided by the projected area of contact, A, when a diamond-shaped indenter is forced into the surface.

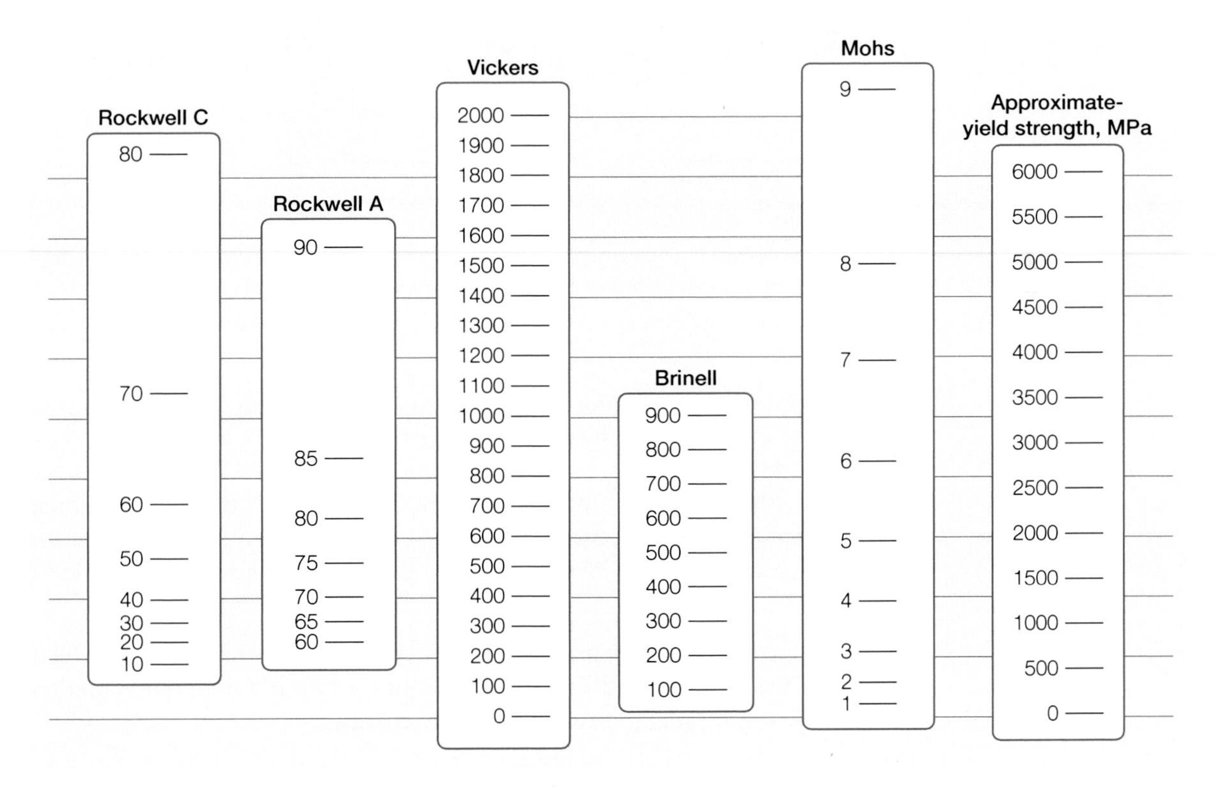

FIGURE 3.9
Commonly used scales of hardness related to each other and to the yield strength.

Strength from hardness

A steel has a hardness of 50 on the Rockwell C scale. Approximately what is its Vickers hardness and yield strength?

Answer

The chart of Figure 3.9 shows that the Vickers hardness corresponding to a Rockwell C value of 50 is approximately $H_v = 500$ and the yield strength is approximately 1,700 MPa.

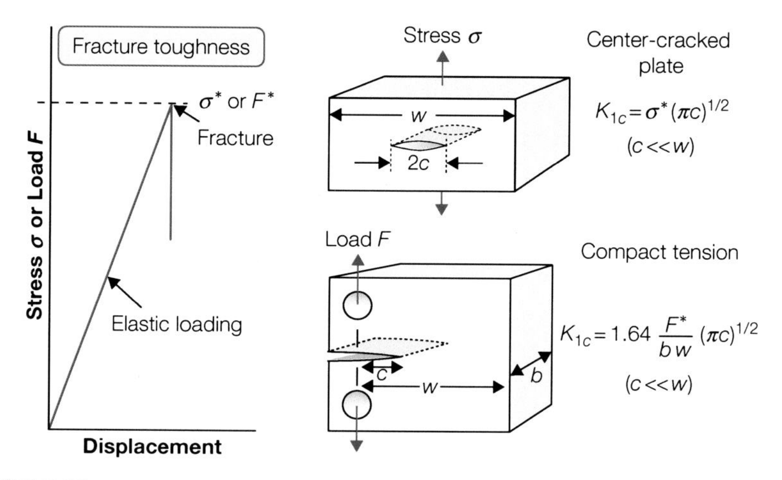

FIGURE 3.10

The fracture toughness, K_{1c}, measures the resistance to the propagation of a crack. The test specimen containing a crack of length $2c$ fails at stress σ^*. The fracture toughness is then $K_{1c} = Y\sigma^*\sqrt{\pi c}$ where Y is a constant near unity.

sample geometry, E is Young's modulus, and ν is Poisson's ratio. Measured in this way K_{1c} and G_{1c} have well-defined values for brittle materials (ceramics, glasses, and many polymers). In ductile materials a plastic zone develops at the crack tip, introducing new features into the way in which cracks propagate that necessitate more involved characterization. Values for K_{1c} and G_{1c} are, nonetheless, cited, and are useful as a way of ranking materials.

The *loss coefficient*, η (a dimensionless quantity), measures the degree to which a material dissipates vibrational energy (Figure 3.11). If a material is loaded elastically to a stress, σ_{max}, it stores an elastic energy

$$U = \oint_0^{\sigma_{max}} \sigma d\varepsilon \approx \frac{1}{2}\frac{\sigma_{max}^2}{E}$$

Using fracture toughness

A glass floor panel contains micro-cracks up to 2 microns in length. Glass has a fracture toughness of $K_{1c} = 0.6$ MPa.m$^{1/2}$. When the panel is walked upon, stresses as high as 30 MPa appear in it. Is it safe?

Answer

The stress required to make a 2-micron crack (so $c = 10^{-6}$ m) propagate in glass with a fracture toughness of $K_{1c} = 0.6$ MPa.m$^{1/2}$, using Equation 3.8 with $Y = 1$, is

$$\sigma_c = K_{1c}/\sqrt{\pi c} = 339 \text{ MPa}$$

The panel is safe.

per unit volume. If it is loaded and then unloaded, it dissipates an energy:

$$\Delta U = \oint \sigma d\varepsilon$$

The loss coefficient is

$$\eta = \frac{\Delta U}{2\pi U_{max}} \qquad (3.10)$$

where U_{max} is the stored elastic energy at peak stress. The value of η usually depends on the time scale or frequency of cycling.

Other measures of damping include the *specific damping capacity*, $D = \Delta U/U$, the *log decrement*, Δ (the log of the ratio of successive amplitudes of natural vibrations), the *phase lag*, δ, between stress and strain, and the "Q" *factor* or *resonance factor*, Q. When damping is small ($\eta < 0.01$) these measures are related by

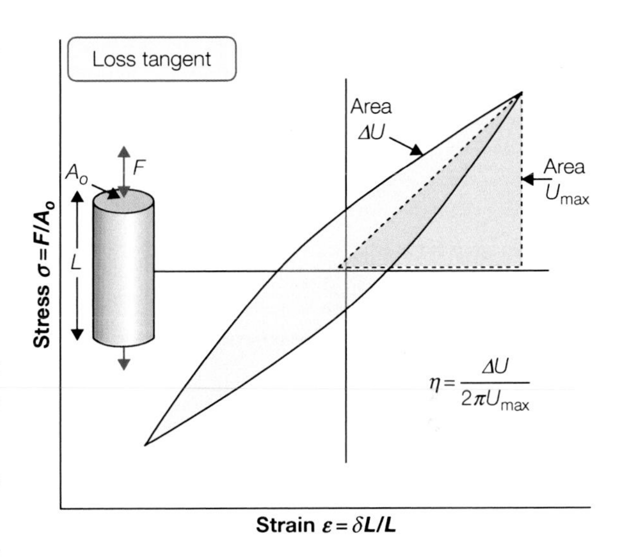

FIGURE 3.11
The loss tangent η measures the fractional energy dissipated in a stress-strain cycle.

$$\eta = \frac{D}{2\pi} = \frac{\Delta}{\pi} = \tan\delta = \frac{1}{Q} \qquad (3.11)$$

but when damping is large, they are no longer equivalent.

Wear, the loss of material when surfaces slide against each other, is a multibody problem. Nevertheless, it can, to a degree, be quantified. When solids slide (Figure 3.12), the volume of material lost from one surface, per unit distance slid, is called the wear rate, W (units: m^2). The wear resistance of the

Using loss coefficients

A bell with a natural frequency of $f = 1{,}000$ Hz is made of a material with a loss coefficient of $\eta = 0.01$. For how long will it ring after being struck? If the material is replaced by one of low damping with $\eta = 10^{-4}$ how long will it ring? (Assume the ring has ended when the amplitude of oscillation A has fallen to one-hundredth of its initial value.)

Answer

Let A and $A + dA$ be the amplitudes of the successive cycles (dA is negative). Then $\text{Log} = \left(\frac{A}{A + dA}\right) = \Delta = \pi\eta$ from which

$$\frac{dA}{Ad\eta} = \frac{1}{10^{\pi\eta}} - 1$$

Integrating over n cycles gives $\ln\frac{A}{A_0} = \left(\frac{1}{10^{\pi\eta}} - 1\right)n$ where A_0 is the initial amplitude. When A has fallen to 0.01 A_0, the term $\ln(A/A_0) = -4.6$, giving $n = 4.6\left(\frac{10^{\pi\eta}}{10^{\pi\eta} - 1}\right)$. Thus a bell with $\eta = 0.01$ will ring for $n = 66$ cycles, giving a time $n/f = 66$ milliseconds. A bell with $\eta = 10^{-4}$ will ring for $n = 6{,}400$ cycles and a time of n/f 6.4 seconds.

surface is characterized by the *Archard wear constant*, K_A (units: 1/MPa) defined by the equation

$$\frac{W}{A} = K_A P \tag{3.12}$$

where A is the area of the slider surface and P is the normal force pressing it onto the other surface. Approximate data for K_A appear in Chapter 4, but must be interpreted as the property of the sliding couple, not of just one member of it.

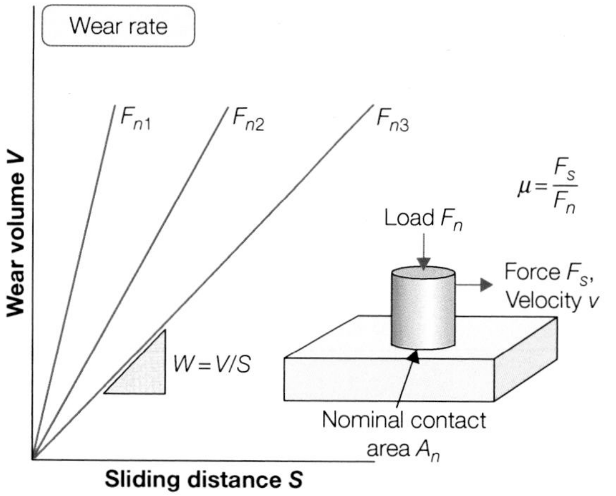

FIGURE 3.12
Wear is the loss of material from surfaces when they slide. The wear resistance is measured by the Archard wear constant, K_A, defined in the text.

Thermal properties

Two temperatures, the *melting temperature*, T_m, and the *glass temperature*, T_g (units for both: K or C), are fundamental because they relate directly to the strength of the bonds in the solid. Crystalline solids have a sharp melting point, T_m. Noncrystalline solids do not; the temperature T_g characterizes the transition from true solid to very viscous liquid. It is helpful, in engineering design, to define two further temperatures: the *maximum* and *minimum service temperature*, T_{max} and T_{min} (both: K or C). The first tells us the highest temperature at which the material can reasonably be used without oxidation, chemical change,

Calculating wear

A steel slider oscillates on a dry steel substrate at frequency $f = 0.2$ Hz and an amplitude $a = 2$ mm under a normal pressure $P = 2$ MPa. The Archard wear constant for steel on steel is $K_A = 3 \times 10^{-8}$ (MPa)$^{-1}$. By how much will the surface of the slider have been reduced in thickness after a time $t = 100$ hours?

Answer

The distance slid in 100 hours is $d = 4\ a\ f\ t$ m. The thickness x removed from the slider over the time $t = 3.6 \times 10^5$ is

$$x = \frac{Volume\ removed}{Area\ A} = 4\,a\,f\,t\,K_A\,P = 3.5 \times 10^{-5}\text{m} = 36\,\mu\text{m}$$

or excessive creep becoming a problem. The second is the temperature below which the material becomes brittle or otherwise unsafe to use.

It costs energy to heat a material. The *heat capacity* or *specific heat* (units J/kg.K) is the energy to heat 1 kg of a material by 1 K. The measurement is usually made at constant pressure (atmospheric pressure) so it is given the symbol C_p. When dealing with gases, it is more usual to measure the heat capacity at constant volume (symbol C_v), and for gases this differs from C_p. For solids the difference is so slight that it can be ignored, and we shall do so here. The heat capacity is measured by calorimetry (Figure 3.13), which is also the standard way of measuring the glass temperature, T_g. A measured quantity of energy (here, electrical energy) is pumped into a sample of material of known mass. The temperature rise is measured, allowing the energy/kg.K to be calculated. Real calorimeters are more elaborate than this, but the principle is the same.

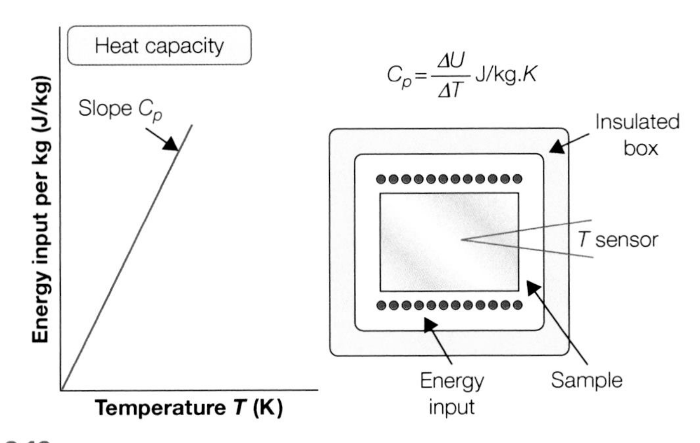

FIGURE 3.13
The heat capacity—the energy to raise the temperature of 1 kg of material by 1°C.

Using specific heat

How much energy is required to heat a 100 mm cube of copper from room temperature (20°C) to its melting point?

Answer

Data for melting point, T_m, specific heat, C_p, and density, ρ, are listed in Appendix A. The values for copper are $T_m = 1{,}082°C$, $C_p = 380$ J/kg.K and $\rho = 8{,}930$ kg/m^3. The mass of copper in the cube is $\rho V = 8.93$ kg. The energy to heat it to $\Delta T = 1{,}062°C$ is

$$\rho V\, C_p\, \Delta T = 3.6 \text{ MJ}$$

(The energy in a liter of gasoline is 35 MJ.)

The rate at which heat is conducted through a solid at steady state (meaning that the temperature profile does not change with time) is measured by the *thermal conductivity*, λ (units: W/m.K). Figure 3.14 shows how it is measured: by recording the heat flux q (W/m^2) flowing through the material from a surface at higher temperature T_1 to a lower one T_2 separated by a distance X. The conductivity is calculated from Fourier's law:

$$q = -\lambda \frac{dT}{dX} = \lambda \frac{(T_1 - T_2)}{X} \tag{3.13}$$

The measurement is not, in practice, easy (particularly for materials with low conductivities), but reliable data are now generally available.

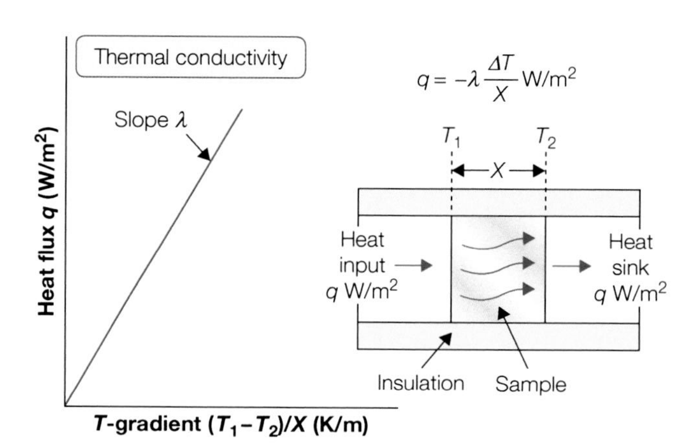

FIGURE 3.14
The thermal conductivity λ measures the flux of heat driven by a temperature gradient dT/dX.

Steady state heat flow

A heat exchanger has an exchange area of $A = 0.5\ m^2$. It passes heat from a fluid at temperature $T_1 = 100°C$ to a second fluid at $T_2 = 20°C$. The exchange wall is made of copper sheet (thermal conductivity $\lambda = 350$ W/m.K) with a thickness $X = 2$ mm. How much energy flows from one fluid to the other in one hour?

Answer

The temperature gradient $dT/dX = 80/0.002 = 40{,}000°C$ /m. The total energy Q that passes across the area A over a time $t = 3{,}600$ seconds is

$$Q = A\,t\,q = A\,t\,\lambda\,\frac{dT}{dX} = 2.5 \times 10^{10}\,\mathrm{J} = 25\,\mathrm{GJ}$$

When heat flow is transient, the flux depends instead on the *thermal diffusivity*, a (units: m^2/s), defined by

$$a = \frac{\lambda}{\rho\,C_p} \tag{3.14}$$

where ρ is the density and C_p is the heat capacity. The thermal diffusivity can be measured directly by measuring the decay of a temperature pulse when a heat source, applied to the material, is switched off; or it can be calculated from λ, via the last equation. The distance x heat diffuses in a time t is approximately

$$x \approx \sqrt{2\,\alpha\,t} \tag{3.15}$$

Most materials expand when they are heated (Figure 3.15). The thermal strain per degree of temperature change is measured by the *linear thermal-expansion coefficient*, α (units: K^{-1} or, more conveniently, "microstrain/°C" or $10^{-6}°C^{-1}$). If the material is thermally isotropic, the volume expansion,

Transient heat flow

You pour boiling water into a tea-glass with a wall thickness $x = 3$ mm. How many seconds have you got to carry it to the table before it becomes to hot to hold? (The thermal conductivity of glass is $\lambda = 1.1$ W/m.K, its density is $\rho = 2{,}450\ kg/m^3$ and its heat capacity $C_p = 800$ J/kg.K.)

Answer

Inserting the data into Equation (3.14) gives a thermal diffusivity for glass of $a = 5.6 \times 10^{-7}\ m^2/s$. Inserting this into Equation (3.13) gives the approximate time

$$t \approx \frac{x^2}{2a} = 8\ \mathrm{seconds}$$

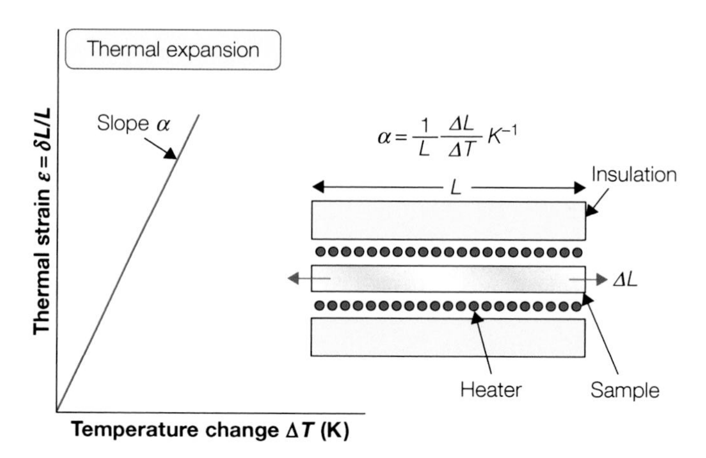

FIGURE 3.15
The linear-thermal expansion coefficient α measures the change in length, per unit length, when the sample is heated.

Thermal stress

An aluminum pipe is rigidly clamped to the face of a concrete building. On a hot day the face of the building in direct sun rises to 80°C, and because the expansion of the aluminum is greater than that of concrete, stress appears in it. What is the value of the stress if the original clamping was done on a day when the temperature was 20°C?

Answer

The expansion coefficient of aluminum is $\alpha = 22.5 \times 10^{-6}$/°C, that of concrete is $\alpha = 9 \times 10^{-6}$/°C, using means of the ranges in Appendix A. The aluminum pipe is rigidly clamped, so the difference in thermal strain $\Delta \alpha \Delta T = 13.5 \times 10^{-6} \times 60 = 8.1 \times 10^{-4}$. This has to be accommodated by elastic compression of the aluminum (modulus $E = 75$ GPa from Appendix A), giving a stress $\Delta \alpha \Delta T E = 61$ MPa. This is enough to cause a soft aluminum to yield.

per degree, is 3α. If it is anisotropic, two or more coefficients are required, and the volume expansion becomes the sum of the principal thermal strains.

The *thermal shock resistance* ΔT_s (units: K or C) is the maximum temperature difference through which a material can be quenched suddenly without damage. I, and the *creep resistance* are important in high-temperature design. Creep is the slow, time-dependent deformation that occurs when materials are loaded above about $\frac{1}{3}T_m$ or $\frac{2}{3}T_g$. Design against creep is a specialized subject. Here we rely instead on avoiding the use of a material above its maximum service temperature, T_{max}, or , for polymers, the "heat deflection temperature."

Electrical properties

The *electrical resistivity*, ρ_e (SI units Ω.m or, commonly, μΩ.cm), is the resistance of a unit cube with unit potential difference between a pair of its faces (Figure 3.16). It has an immense range, from a little more than 10^{-8} in units of Ω.m for good conductors (equivalent to 1 μΩ.cm) to more than 10^{16} Ω.m (10^{24} μΩ.cm) for the best insulators. The electrical conductivity, κ_e (units Siemens per meter, S/m or $(\Omega.\text{m})^{-1}$), is simply the reciprocal of the resistivity.

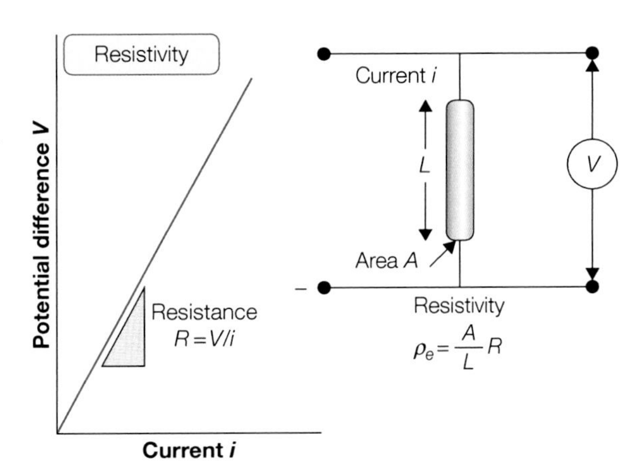

FIGURE 3.16
Electrical resistivity, ρ_e, is measured as the potential gradient, V/L, divided by the current density, i/A. It is related to resistance, R, by $\rho_e = AR/L$.

When an insulator (or *dielectric*) is placed in an electric field, it becomes polarized and charges appear on its surfaces that tend to screen the interior from the electric field. The tendency to polarize is measured by the *dielectric constant*, ε_r, a dimensionless quantity (Figure 3.17). Its value for free space and, for practical purposes, for most gasses, is 1. Most insulators have values between 2 and 30, though low-density foams approach the value 1 because they are largely air.

What does ε_r measure? Two conducting plates separated by a dielectric make a capacitor. Capacitors store charge. The charge, Q (units: coulombs), is directly proportional to the voltage difference between the plates, V (volts):

$$Q = CV \tag{3.16}$$

where C (farads) is the capacitance. The capacitance of a parallel plate capacitor of area A, separated by empty space (or by air), is

$$C = \varepsilon_o \frac{A}{t} \tag{3.17}$$

Resistivity and resistance

Tungsten has a conductivity of $\kappa_e = 8.3 \times 10^6$ Siemens. What is the resistance of a tungsten wire of radius $r = 100$ microns in diameter and length $L = 1$ m?

Answer

The resistivity of tungsten $\rho_e = 1/\kappa_e = 1.2 \times 10^{-7}$ Ω.m. The resistance R of the wire is

$$R = \rho_e \frac{L}{\pi r^2} = 3.8\,\Omega$$

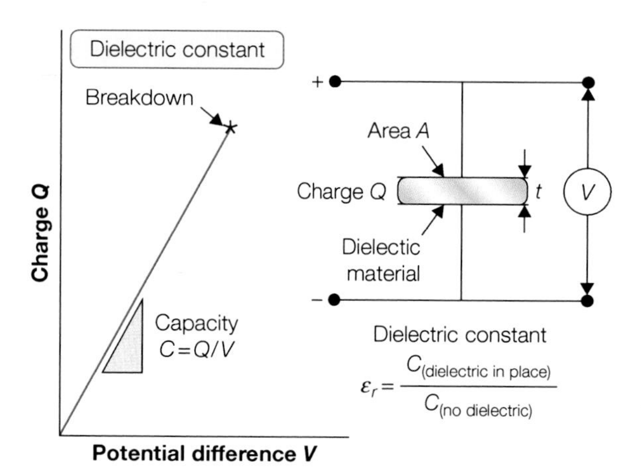

FIGURE 3.17
Dielectric constant: a measure of the ability of an insulator to polarize.

where ε_0 is the *permittivity of free* space (8.85×10^{-12} F/m, where F is farads). If the empty space is replaced by a dielectric, capacitance increases because of its polarization. The field created by the polarization opposes the field E, reducing the voltage difference V needed to support the charge. Thus the capacity of the condenser is increased to the new value:

$$C = \varepsilon \frac{A}{t} \tag{3.18}$$

where ε is the *permittivity of the dielectric* with the same units as ε_0. It is usual to cite not this but the *relative permittivity* or *dielectric constant*, ε_r:

$$\varepsilon_r = \frac{C_{with\ dielectric}}{C_{no\ dielectric}} = \frac{\varepsilon}{\varepsilon_o} \tag{3.19}$$

making the capacitance

$$C = \varepsilon_r \varepsilon_o \frac{A}{t} \tag{3.20}$$

When charged, the energy stored in a capacitor is

$$\frac{1}{2} Q V = \frac{1}{2} C V^2 \tag{3.21}$$

and this can be large: "Super-capacitors" with capacitances measured in farads store enough energy to power a hybrid car.

Polarization involves the small displacement of charge (of either electrons or ions) or of molecules that carry a dipole moment when an electric field is applied to the material. An oscillating field drives the charge between

Stray capacitance

The time constant τ for charging or discharging a capacitor is

$$\tau = RC$$

where R is the resistance of the circuit. That means that stray capacitance in an electronic circuit (capacitance between neighboring conducting lines or components) slows its response. What material choices minimize this?

Answer

Choosing materials with low resistivity ρ_e for the conductors (to minimize R) and choosing insulators with low dielectric constant ε_r to separate them (to minimize C), minimizes τ.

two alternative configurations. This charge motion is like an electric current that—if there were no losses—would be 90° out of phase with the voltage. In real dielectrics this current dissipates energy, just as a current in a resistor does, giving it a small phase shift, δ (Figure 3.18). The *loss tangent, tan δ,* also called the *dissipation factor, D,* is the tangent of the loss angle. The *power factor, P_f,* is the sine of the loss angle. When δ is small, as it is for the materials of interest here, all three are essentially equivalent:

$$P_f \approx D \approx \tan\delta \approx \sin\delta \qquad (3.22)$$

More useful, for our purposes, is the *loss factor L*, which is the loss tangent times the dielectric constant:

$$L = \varepsilon_r \tan\delta \qquad (3.23)$$

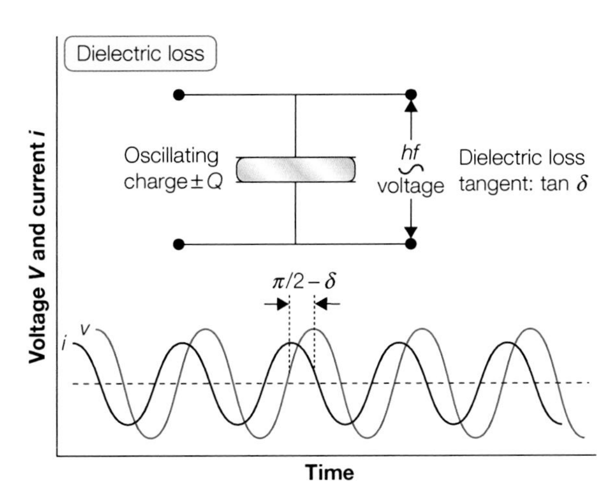

FIGURE 3.18
Dielectric loss, important in dielectric heating, as explained in the text.

It measures the energy dissipated by a dielectric when in an oscillating field. If you want to select materials to minimize or maximize dielectric loss, then the measurement you want is *L*.

When a dielectric material is placed in a cyclic electric field of amplitude *E* and frequency *f*, power *P* is dissipated and the field is correspondingly attenuated. The power dissipated per unit volume (W/m^3) is

$$P \approx f\,E^2 \varepsilon \tan\delta = f\,E^2 \varepsilon_o \varepsilon_r \tan\delta = f\,E^2 \varepsilon_o L \qquad (3.24)$$

Dielectric heating

A nylon component is placed in a microwave cavity with a field strength $E = 10^4$ V/m and a frequency $f = 10^{10}$ Hz for a time $t = 100$ s. The dielectric loss factor for nylon is $L = 0.1$, its density is $\rho = 1130$ kg/m^3 and its heat capacity is $C_p = 1650$ J/kg.K. Assuming no heat loss, how hot will the component become?

Answer

The heat generated by the field is $Q = Pt = f\,E^2 \varepsilon_0\, L\, t = 8.85 \times 10^7$ J/m^3. The heat capacity of nylon per unit volume is $C_p \rho = 1.86 \times 10^6$ J/m^3.K. The temperature rise, ΔT is

$$\Delta T = \frac{f\,E^2 \varepsilon_0\, L\, t}{C_p \rho} = 47.6\,°\mathrm{C}$$

where, as before, ε_r is the dielectric constant of the material and *tan* δ is its loss tangent. This power appears as heat; the higher the frequency or the field strength and the greater the loss factor $L = \varepsilon_r \tan \delta$, the greater the heating and energy loss. Sometimes this dielectric loss is exploited in processing—for example, in radio frequency welding of polymers.

The *breakdown potential* (units: MV/m) is the electrical potential gradient at which an insulator breaks down and a damaging surge of current flows through it. It is measured by increasing, at a uniform rate, a 60-Hz alternating potential applied across the faces of a plate of the material until breakdown occurs, typically at a potential gradient between 1 and 100 million volts per meter (units: MV/m).

Optical properties

All materials allow for some passage of light, although for metals it is exceedingly small. The speed of light when in the material, v, is always less than that in vacuum, c. A consequence is that a beam of light striking the surface of such a material at an angle of incidence, α, enters the material at an angle β, the angle of refraction. The *refractive index*, n (dimensionless), is

$$n = \frac{c}{v} = \frac{\sin \alpha}{\sin \beta} \tag{3.25}$$

It is related to the dielectric constant, ε_r, at the same frequency by

$$n \approx \sqrt{\varepsilon_r}$$

The refractive index depends on wavelength and thus on the color of the light. The denser the material, and the higher its dielectric constant, the greater the refractive index. When $n = 1$, the entire incident intensity enters the material, but when $n > 1$, some is reflected. If the surface is smooth and polished, it is reflected as a beam; if rough, it is scattered. The percentage reflected, R, is related to the refractive index by

$$R = \left(\frac{n-1}{n+1}\right)^2 \times 100 \tag{3.26}$$

As n increases, the value of R approaches 100%.

Eco-properties

The *embodied energy* (units MJ/kg) is the energy required to extract 1 kg of a material from its ores and feedstock. The associated CO_2 *footprint* (units: kg/kg) is the mass of carbon dioxide released into the atmosphere during the production of 1 kg of material. These and other eco-attributes are the subject of Chapter 15.

3.5 SUMMARY AND CONCLUSIONS

There are six important families of materials for mechanical design: metals, ceramics, glasses, polymers, elastomers, and hybrids (which combine the properties of two or more of the others). Within a family there is a certain commonality. Ceramics and glasses as a family are hard, brittle, and corrosion-resistant. Metals are ductile, tough and good thermal and electrical conductors. Polymers are light, easily shaped, and electrical insulators. Elastomers have the ability to deform elastically to large strains. That is what makes the classification useful. But in design we wish to escape from the constraints of family and think, instead, of the material name as an identifier for a certain property profile—one that will in later chapters be compared with an "ideal" profile suggested by the design, guiding our choice. To that end, the properties important in thermo-mechanical design were defined in this chapter. In the next chapter we develop a way of displaying these properties so as to maximize the freedom of choice.

3.6 FURTHER READING

Definitions of material properties can be found in numerous general texts on engineering materials, among them those listed here.

Ashby, M.F., & Jones, D.R.H. (1996). *Engineering materials 1, an introduction to properties, applications and design* (3rd ed.). Elsevier–Butterworth-Heinemann, ISBN 0-7506-6380-4.
An introduction to materials, taking a design-led approach.

Ashby, M.F., Shercliff, H.R., & Cebon, D. (2010). *Materials: engineering, science, processing and design* (2nd ed.). Butterworth-Heinemann, ISBN 978-1-85617-895-2.
An elementary text introducing materials through material property charts and developing selection methods through case studies.

Askeland, D.R., & Phulé, P.P. (2006). *The science and engineering of materials* (5th ed.). Thomson, ISBN 0-534-55396-6.
A widely used introductory text.

ASM Handbooks, Vol. 8 (2004). *Mechanical testing and evaluation.* ASM International.
An online, subscription-based resource, detailing testing procedures for metals and ceramics.

ASM Engineered Materials Handbook (2004). *Testing and characterization of polymeric materials.* ASM International.
An online, subscription-based resource detailing testing procedures for polymers.

ASTM Standards, Vol. 08.01 and *08.02* (1988). *Plastics, Vol. 04.02* (1989). *Concrete, Vols. 01.01* to *01.05* (1990). *Steels, Vol. 02.01. Copper alloys, Vol. 02.03. Aluminum alloys, Vol. 02.04. Nonferrous alloys, Vol. 02.05. Coatings, Vol. 03.01. Metals at high and low temperatures, Vol. 04.09. Wood, Vol. 09.01* and *09.02*. American Society for Testing Materials, ISBN 0-8031-1581-4.
The ASTM set standards for materials testing.

Budinski, K.G., & Budinski, M.K. (2010). Engineering materials, properties and selection (9th ed.). Prentice Hall, ISBN 978-0-13-712842-6.
A well-established materials text that deals well with both material properties and processes.

Callister, W.D. (2010). *Materials science and engineering, an introduction* (8th ed.). John Wiley, ISBN 978-0-470-41997-7.

A well-respected materials text, now in its 7th edition, widely used for materials teaching in North America.

Charles, J.A., Crane, F.A.A., & Furness, J.A.G. (1997). *Selection and use of engineering materials* (3rd ed.). Butterworth-Heinemann, ISBN 0-7506-3277-1.

A materials science approach to the selection of materials.

Dieter, G.E. (1999). *Engineering design, a materials and processing approach* (3rd ed.). McGraw-Hill, ISBN 9-780-073-66136-0.

A well-balanced and respected text focusing on the place of materials and processing in technical design.

Farag, M.M. (2008). *Materials and process selection for engineering design* (2nd ed.). CRC Press, Taylor and Francis: ISBN 9-781-420-06308-0.

A materials science approach to the selection of materials.

Shackelford, J.F. (2009). *Introduction to materials science for engineers* (7th ed.). Prentice Hall, ISBN 978-0-13-601260-3.

A well-established materials text with a design slant.

Material Property Charts

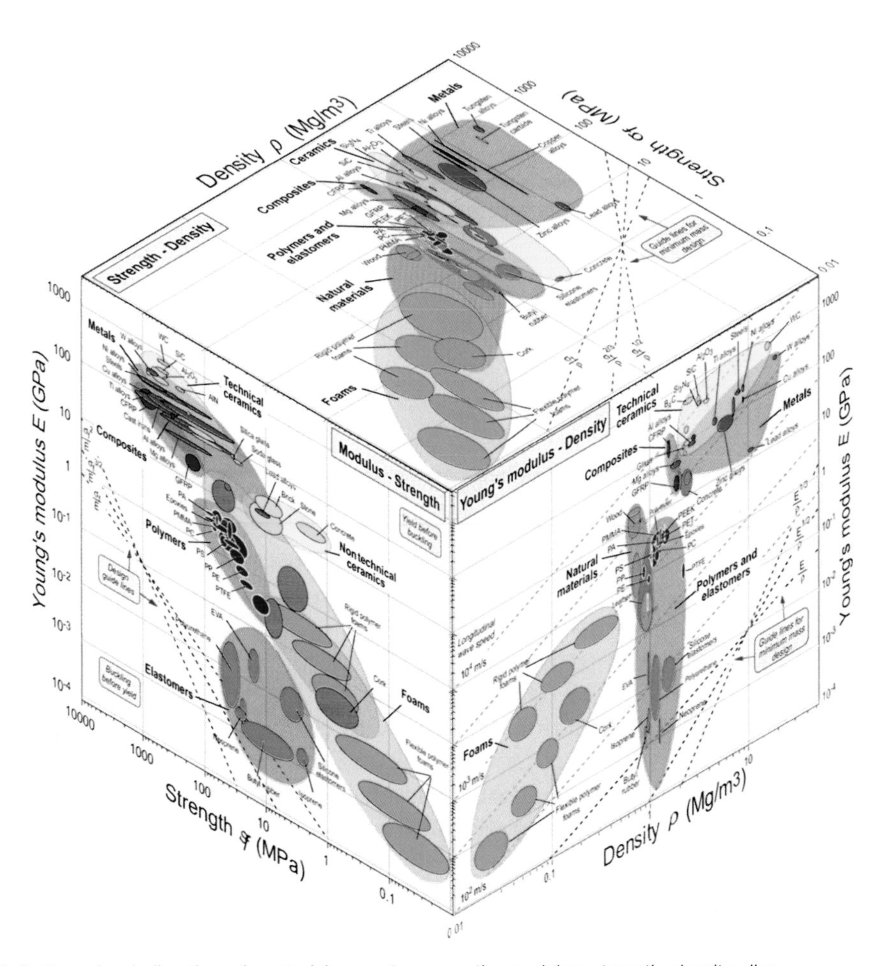

A 3-dimensional slice through material property space: the modulus–strength–density slice.

Materials Selection in Mechanical Design. DOI: 10.1016/B978-1-85617-663-7.00004-7

CONTENTS

4.1 INTRODUCTION AND SYNOPSIS

Material properties limit performance. We need a way of surveying them to get a feel for the values that design-limiting properties can have. One property can be displayed as a ranked list or bar chart, but it is seldom that the performance of a component depends on just one property. More often it is a combination of properties that matter: the need for stiffness at low weight, for thermal conduction coupled with corrosion resistance, or for strength combined with toughness, for example. This suggests the idea of plotting one property against another, mapping out the fields in property space occupied by each material class and the subfields occupied by individual materials.

The resulting charts are helpful in many ways. They condense a large body of information into a compact but accessible form; they reveal correlations between material properties that aid in checking and estimating data; and, as examined in later chapters, they become tools for materials selection, for exploring the effect of processing on properties, for demonstrating how shape can enhance structural efficiency, and for suggesting directions for further material development.

The ideas behind materials selection charts are described briefly in Section 4.2. Section 4.3 introduces the charts themselves. It's not necessary to read it all, but it is helpful to persist far enough to be able to read and interpret the charts fluently, and to understand the meaning of the design guide lines that appear in them. If, later, you use a particular chart, you should read the background to it, given here, to be sure of interpreting it correctly.

As explained in the Preface, the charts can be copied and distributed for teaching purposes without infringing the copyright.[1]

4.2 EXPLORING MATERIAL PROPERTIES

Each property of an engineering material has a characteristic range of values. The span can be large: many properties have values that range over five or more decades. One way of displaying this is as a bar chart like that of Figure 4.1 for Young's modulus. Each bar describes one material; its length shows the range of modulus exhibited by that material in its various forms. The materials are segregated by class. Each class shows a characteristic range: Metals and ceramics have high moduli; polymers have low; hybrids have a wide range, from low to high. The total range is large—it spans a factor of about 10^6—so logarithmic scales are used to display it.

More information is displayed by an alternative plot, illustrated in the schematic of Figure 4.2. Here, one property (the modulus, E, in this case) is plotted against another (the density, ρ). The range of the axes is chosen to include all materials, from the lightest, flimsiest foams to the stiffest, heaviest metals, and it is large, again requiring log scales. It is found that data for a given family of materials (polymers, for example) cluster together; the *subrange* associated with one material family is, in all cases, much smaller than the *full* range of that property. Data for one family can be enclosed in a property envelope—envelopes are shown on this schematic. A real $E - \rho$ chart is shown in Figure 4.3. The family envelopes appear as in the schematic. Within each envelope lie white bubbles enclosing classes and subclasses.

All of this is simple enough—just a helpful way of plotting data. But by choosing the axes and scales appropriately, more can be added. The speed

[1] A set of charts can be downloaded from *www.grantadesign.com*. All charts shown in this chapter were created using Granta Design's CES Edu Materials Selection software. With it you can make charts with any pair (or combination) of properties as axes.

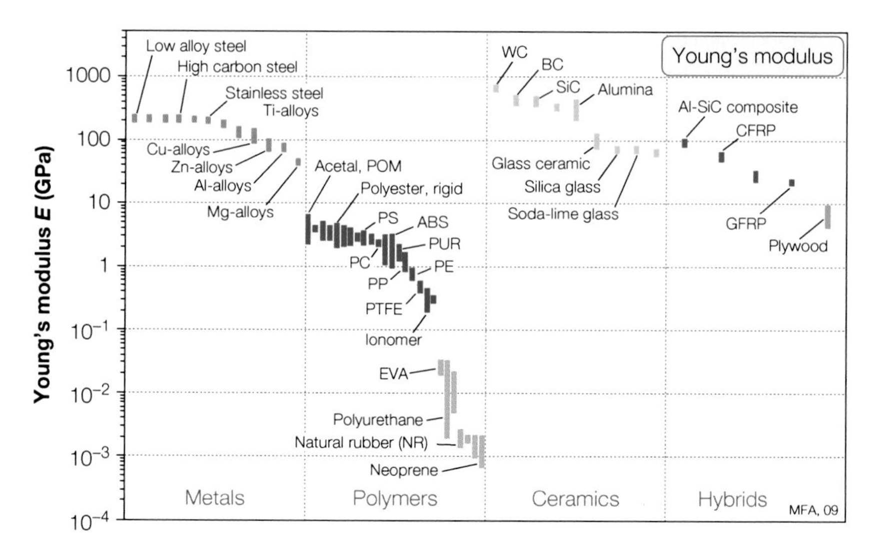

FIGURE 4.1
A bar chart showing modulus for families of solids. Each bar shows the range of modulus offered by a material, some of which are labeled.

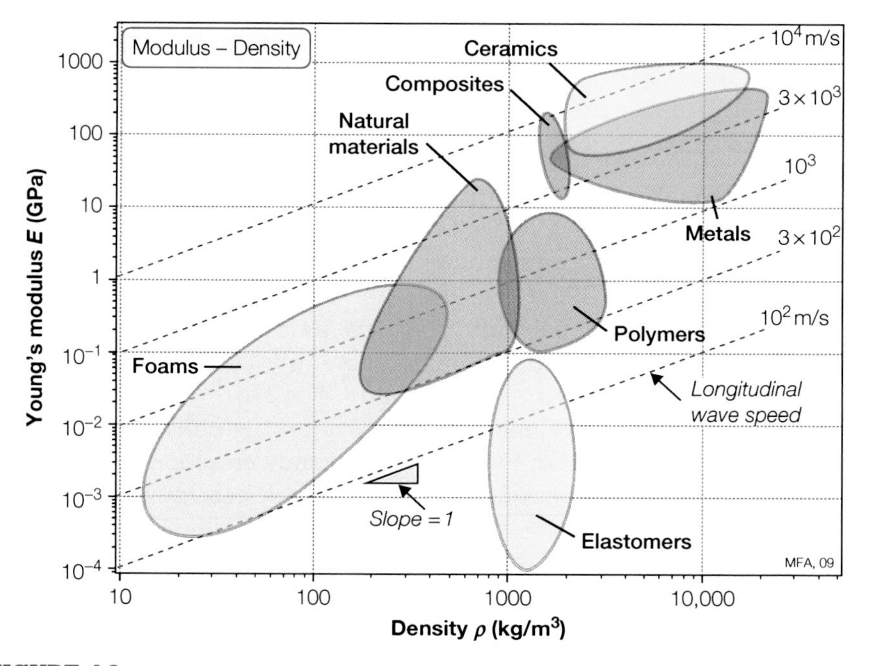

FIGURE 4.2
The idea of a material property chart: Young's modulus E is plotted against the density ρ on log scales. Each material class occupies a characteristic field. The contours show the longitudinal elastic wave speed $v = (E/\rho)^{1/2}$.

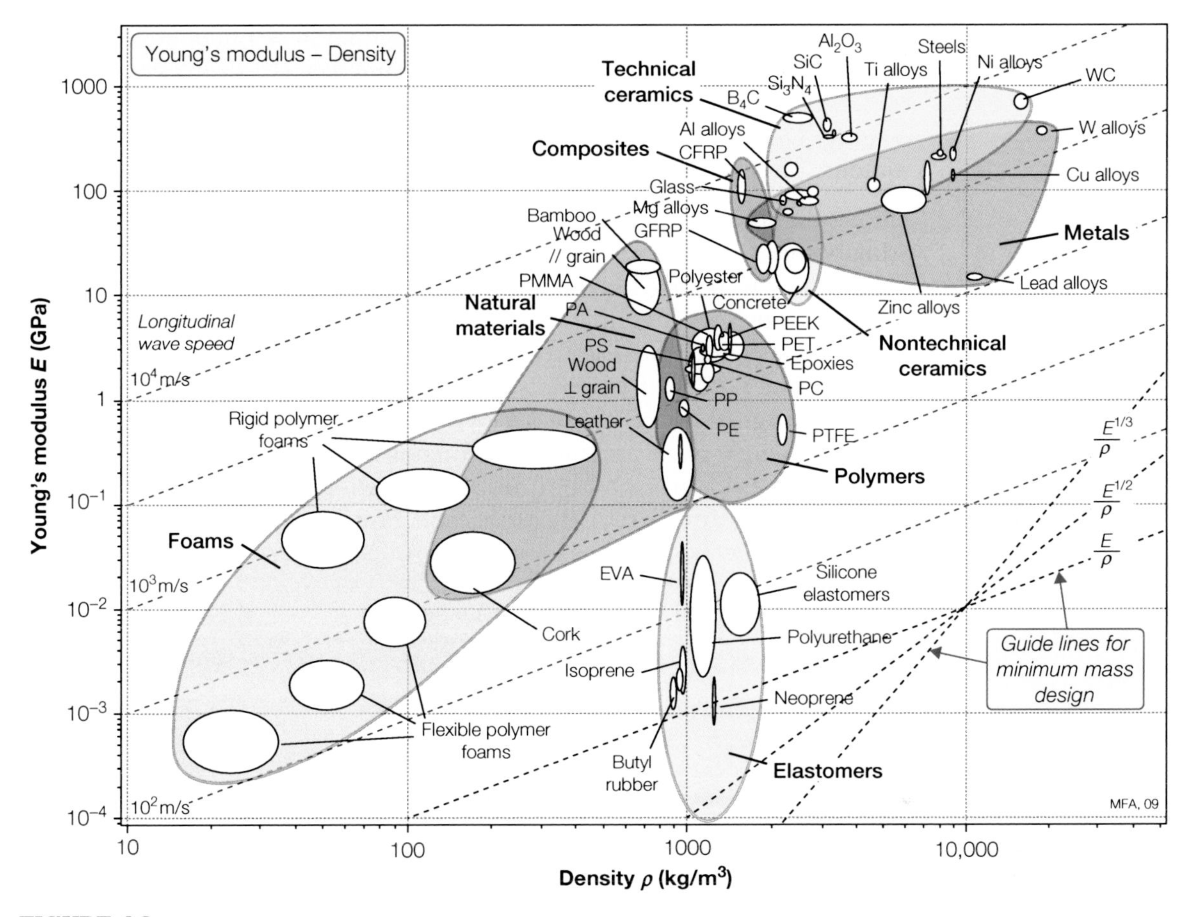

FIGURE 4.3

Young's modulus E plotted against density ρ. The heavy envelopes enclose data for a given class of material. The diagonal contours show the longitudinal wave velocity. The guide lines of constant E/ρ, $E^{1/2}/\rho$, and $E^{1/3}/\rho$ allow selection of materials for minimum weight, deflection-limited, design.

of sound in a solid depends on E and ρ; the longitudinal wave speed v, for instance, is

$$v = \left(\frac{E}{\rho}\right)^{1/2}$$

or (taking logs)

$$\log E = \log \rho + 2 \log v$$

For a fixed value of v, this equation plots as a straight line of slope 1 in Figures 4.2 and 4.3. This allows us to add *contours of constant wave velocity* to the chart: They are the family of parallel diagonal lines, linking materials in which longitudinal waves travel with the same speed. All the charts allow additional fundamental relationships of this sort to be displayed. And there is more: Design-optimizing parameters called *material indices* also plot as contours on the charts (see Chapter 5).

Among the mechanical and thermal properties, there are 30 or so that are of primary importance, both in characterizing the material and in engineering design. They were listed in Table 3.1 and include density, moduli, strength, hardness, toughness, thermal and electrical conductivities, expansion coefficient, and specific heat. The charts display data for these properties for the families and classes of materials listed in Table 4.1. The list is expanded from the original six of Figure 3.1 by distinguishing *composites* from *foams* and from *natural materials,* and by distinguishing the high-strength *technical ceramics* (e.g., silicon carbide) from the low-strength, *nontechnical ceramics* (e.g., concrete and brick). Within each family, data are plotted for a

Table 4.1 Material Families and Classes

Family	Classes	Short Name
Metals (the metals and alloys of engineering)	Aluminum alloys	Al alloys
	Copper alloys	Cu alloys
	Lead alloys	Lead alloys
	Magnesium alloys	Mg alloys
	Nickel alloys	Ni alloys
	Carbon steels	Steels
	Stainless steels	Stainless steels
	Tin alloys	Tin alloys
	Titanium alloys	Ti alloys
	Tungsten alloys	W alloys
	Lead alloys	Pb alloys
	Zinc alloys	Zn alloys
Ceramics, *technical ceramics* (fine ceramics capable of load-bearing application)	Alumina	Al_2O_3
	Aluminum nitride	AlN
	Boron carbide	B_4C
	Silicon carbide	SiC
	Silicon nitride	Si_3N_4
	Tungsten carbide	WC
Ceramics, *nontechnical ceramics* (porous ceramics of construction)	Brick	Brick
	Concrete	Concrete
Glasses	Soda-lime glass	Soda-lime glass
	Borosilicate glass	Borosilicate glass
	Silica glass	Silica glass
	Glass ceramic	Glass ceramic

Table 4.1 *continued*

Family	Classes	Short Name
Polymers (the thermoplastics and thermosets of engineering)	Acrylonitrile butadiene styrene	ABS
	Cellulose polymers	CA
	Ionomers	Ionomers
	Epoxies	Epoxy
	Phenolics	Phenolics
	Polyamides (nylons)	PA
	Polycarbonate	PC
	Polyesters	Polyester
	Polyetheretherkeytone	PEEK
	Polyethylene	PE
	Polyethylene terephthalate	PET or PETE
	Polymethylmethacrylate	PMMA
	Polyoxymethylene (Acetal)	POM
	Polypropylene	PP
	Polystyrene	PS
	Polytetrafluorethylene	PTFE
	Polyvinylchloride	PVC
Elastomers (engineering rubbers, natural and synthetic)	Butyl rubber	Butyl rubber
	EVA	EVA
	Isoprene	Isoprene
	Natural rubber	Natural rubber
	Polychloroprene (Neoprene)	Neoprene
	Polyurethane	PU
	Silicone elastomers	Silicones
Hybrids: composites	Carbon-fiber–reinforced polymers	CFRP
	Glass-fiber–reinforced polymers	GFRP
	SiC-reinforced aluminum	Al-SiC
Hybrids: foams	Flexible polymer foams	Flexible foams
	Rigid polymer foams	Rigid foams
Hybrids: natural materials	Cork	Cork
	Bamboo	Bamboo
	Wood	Wood

representative set of materials, chosen both to span the full range of behavior for the class and to include the most common and most widely used members of it. In this way the envelope for a family encloses data not only for the materials listed in Table 4.1 but for virtually all other members of the family as well.

The charts that follow show a *range* of values for each property of each material. Sometimes the range is narrow: The modulus of copper, for instance, varies by only a few percent about its mean value, influenced by purity, texture, and the like. Sometimes, however, it is wide: The strength of metals can vary by a factor of 100 or more, influenced by composition and the state of work hardening or heat treatment. Crystallinity and degree of cross-linking greatly influence the modulus of polymers. Porosity influences the strength of ceramics. These *structure-sensitive* properties appear as elongated bubbles within the envelopes in the charts.

4.3 THE MATERIAL PROPERTY CHARTS

The modulus–density chart

Modulus and density are familiar properties. Steel is stiff; rubber is compliant: These are effects of modulus. Lead is heavy; cork is buoyant: these are effects of density. Figure 4.3 shows the range of Young's modulus, E, and density, ρ, for engineering materials. Data for members of a particular family of materials cluster together and can be enclosed by a colored envelope. The same family envelopes appear on all the diagrams: They correspond to the main headings in Table 4.1.

The *density* of a solid depends on the atomic weight of its atoms or ions, their size, and the way they are packed. The size of atoms does not vary much: Most have a volume within a factor of two of 2×10^{-29} m^3. Packing fractions do not vary much either—a factor of two more or less. Close packing gives a packing fraction of 0.74; open networks such as that of the diamond-cubic structure give about 0.34. The spread of density comes mainly from the spread of atomic weight, ranging from 1 for hydrogen to 238 for uranium. Metals are dense because they are made of heavy atoms, packed closely together; polymers have low densities because they are largely made of carbon (atomic weight: 12) and hydrogen (atomic weight: 1) in more open amorphous or crystalline packings. Ceramics, for the most part, have lower densities than metals because they contain light O, N or C atoms. Even the lightest atoms, packed in the most open way, give solids with a density of around 1000 kg/m^3, the same as that of water. Materials with lower densities than this are *foams*—materials made up of cells containing a large fraction of pore space.

The *moduli* of most materials depend on two factors: bond stiffness and the number of bonds per unit volume. A bond is like a spring, and, like a spring, it has a spring constant, S (units: N/m). Young's modulus, E, is roughly

$$E = \frac{S}{r_o} \tag{4.1}$$

where r_o is the "atom size" (r_o^3 is the mean atomic or ionic volume). The wide range of moduli is largely caused by the range of values of S. The covalent bond is stiff ($S = 20\text{–}200$ N/m); the metallic and the ionic a little less so ($S = 15\text{–}100$ N/m). Diamond has a very high modulus because the carbon atom is small, giving a high bond density, and its atoms are linked by strong covalent springs ($S = 200$ N/m). Metals have high moduli because close packing gives a high bond density and the bonds are strong, though not as strong as those of diamond. Polymers contain both strong diamond-like covalent bonds and weak hydrogen or Van der Waals bonds ($S = 0.5\text{–}2$ N/m). It is the weak bonds that stretch when the polymer is deformed, giving low moduli.

But even large atoms ($r_o = 3 \times 10^{-10}$ m) bonded with the weakest bonds ($S = 0.5$ N/m) have a modulus of roughly

$$E = \frac{0.5}{3 \times 10^{-10}} \approx 1\ \text{GPa} \tag{4.2}$$

This is the *lower limit* for true solids. The chart shows that many materials have moduli that are lower than this: They are either elastomers or foams. Elastomers have a low E because their weak secondary bonds have melted as their glass temperature, T_g, is below room temperature, leaving only the very weak "entropic" restoring force associated with tangled, long-chain molecules. Foams have low moduli because the cell walls bend easily when the material is loaded. More on this in Chapter 11.

The chart shows that the modulus of engineering materials spans seven decades,[2] from 0.0001 GPa (low-density foams) to 1,000 GPa (diamond). The density spans a factor of 2,000, from less than 0.01 to 20 Mg/m^3. Ceramics as a family are very stiff, metals a little less so—but none have a modulus less than 10 GPa. Polymers, by contrast, all cluster between 0.8 and 8 GPa.

The log scales allow more information to be displayed. As explained in the last section, the velocity of elastic waves in a material and the natural vibration frequencies of a component made of it are proportional to $(E/\rho)^{1/2}$. Contours of this quantity are plotted on the chart, labeled with the longitudinal wave speed. The speed varies from less than 50 m/s (soft elastomers) to a little more than 10^4 m/s (stiff ceramics). We note that aluminum and glass, because of their low densities, transmit waves quickly despite their low moduli. One might have expected the wave velocity in foams to be low because of the low modulus, but the low density almost compensates. That in wood, across the grain, is low; but along the grain, it is high—roughly the same as steel—a fact made use of in the design of musical instruments.

[2] Very-low-density foams and gels (which can be thought of as molecular-scale, fluid-filled foams) can have lower moduli than this. As an example, gelatin (as in Jell-O) has a modulus of about 10^{-5} GPa. Their strengths and fracture toughness, too, can be below the lower limit of the charts.

Surveying material properties

Which class of metallic alloy is the lightest? Which the heaviest? Which the stiffest? Which the least stiff?

Answer
A glance at Figure 4.3 reveals that the lightest class is magnesium alloys and the heaviest is tungsten alloys; tungsten alloys are also the stiffest, and lead alloys are the least stiff.

All the charts that appear in this and subsequent chapters can be used for this fast access to comparisons.

Comparing sound velocities

A metal is needed in which longitudinal waves travel at 300 m/s. Use Figure 4.3 to identify candidates.

Answer
The chart shows that zinc alloys, copper alloys, and tungsten alloys all have a longitudinal wave speed close to 300 m/s.

The chart helps in the common problem of material selection for applications in which mass must be minimized. Guide lines corresponding to three common geometries of loading were shown in Figure 4.3. Their use in selecting materials for stiffness-limited design at minimum weight is described in Chapters 5 and 6.

The strength–density chart

The modulus of a solid is a well-defined quantity with a sharp value. The strength is not. It is shown, plotted against density, ρ, in Figure 4.4.

The word "strength" needs definition (see also Section 3.3). For metals and polymers, it is the *yield strength*, but since the range of materials includes those that have been worked or hardened in some other way as well as those that have been softened by annealing, the range is large. For brittle ceramics, the strength plotted here is the *modulus of rupture*: the flexural strength. It is slightly greater than the tensile strength, but much less than the compression strength, which for ceramics is 10 to 15 times greater than the strength in tension. For elastomers, strength means the tensile *tear strength*. For composites, it is the *tensile failure strength* (the compressive strength can be less by up to 30% because of fiber buckling). We will use

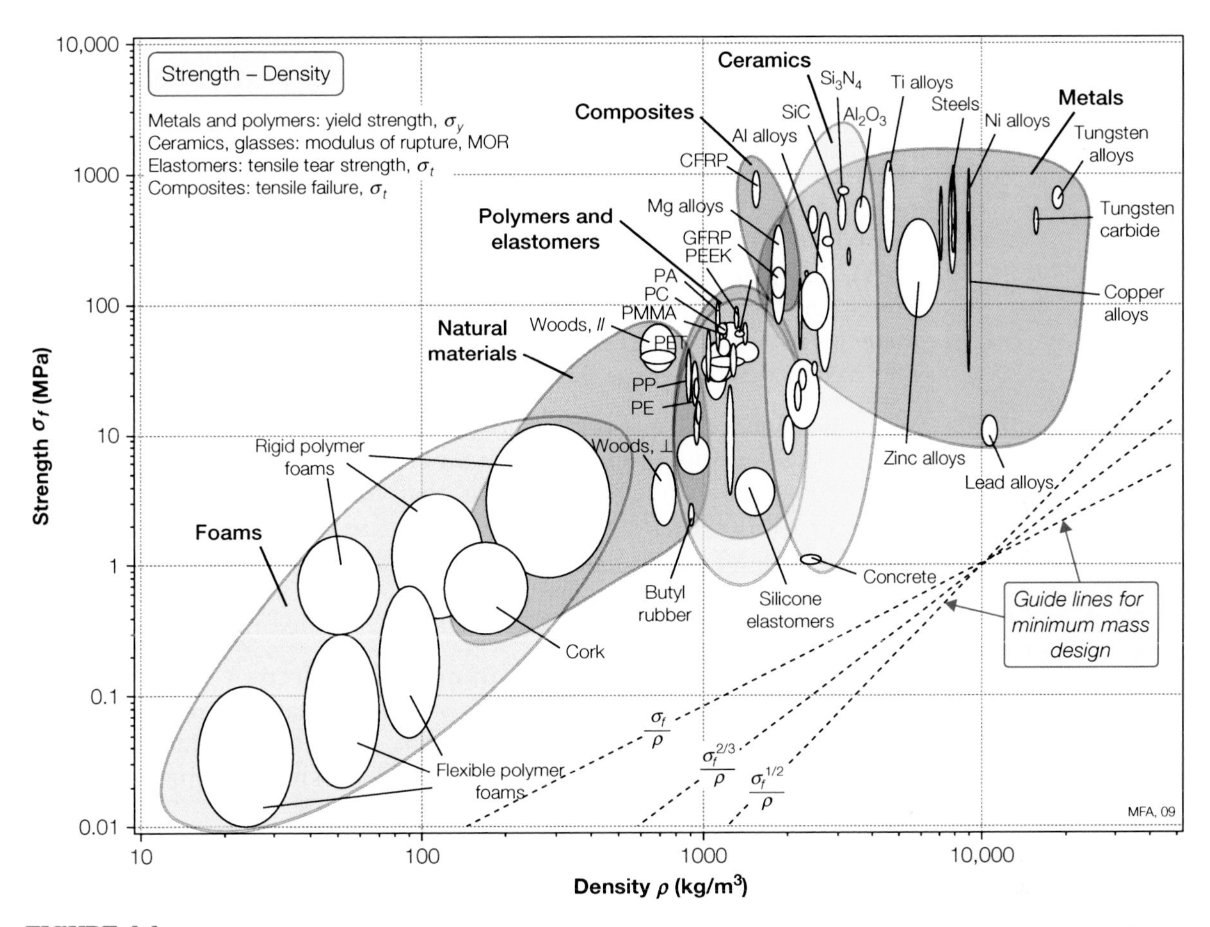

FIGURE 4.4

Strength σ_f plotted against density ρ (yield strength for metals and polymers, compressive strength for ceramics, tear strength for elastomers, and tensile strength for composites). The guide lines of constants σ_f/ρ, $\sigma_f^{2/3}/\rho$, and $\sigma_f^{1/2}/\rho$ are used in minimum weight, yield-limited, design.

the symbol σ_f for all of these, despite the different failure mechanisms involved, to allow a first-order comparison.

The range of strength for engineering materials, like the range for the modulus, spans many decades: from less than 0.01 MPa (foams, used in packaging and energy-absorbing systems) to 10^4 MPa (the strength of diamond, exploited in the diamond-anvil press). The single most important concept in understanding this wide range is the *lattice resistance* or *Peierls stress*. It is the intrinsic resistance of the structure to plastic shear. Plastic shear in a crystal involves the motion of dislocations. Pure metals are soft because the nonlocalized metallic bond does little to hinder dislocation motion, whereas ceramics are hard because their more localized covalent and ionic bonds (which must be broken and reformed when the structure is sheared) lock the dislocations in place. In noncrystalline solids we think instead of the energy associated with the unit step of the flow process as the relative slippage of two segments of a polymer chain, or the shear of a small molecular cluster in a glass network. The strength of noncrystalline solids has the same origin as that underlying the lattice resistance. Thus if the unit step involves breaking strong bonds (as in an inorganic glass), the materials will be strong. If it only involves the rupture of weak bonds (the Van der Waals bonds in polymers for example), it will be weak. Materials that fail by fracture do so because the lattice resistance or its amorphous equivalent is so large that atomic separation (fracture) happens first.

When the lattice resistance is low, the material can be strengthened by introducing obstacles to slip. In metals this is achieved by adding alloying elements, particles, grain boundaries, and other dislocations ("work hardening"); in polymers, by cross-linking or by orienting the chains so that strong covalent bonds, as well as weak Van der Waals bonds, must be broken when the material deforms. When, on the other hand, the lattice resistance is high, further hardening is superfluous—the problem becomes that of suppressing fracture.

An important use of the chart is in materials selection for lightweight strength-limited design. Guide lines are shown for materials selection in the minimum-weight design of ties, columns, beams, and plates, and for yield-limited design of moving components in which inertial forces are important. Their use is described in Chapters 5 and 6.

The modulus–strength chart

High tensile steel makes good springs. But so does rubber. How is it that two such different materials are both suited to the same task? This and other questions are answered by Figure 4.5, one of the most useful of all the charts. It shows Young's modulus, E, plotted against strength, σ_f. The qualifications for "strength" are the same as before. It means yield strength for metals and

High strength at low weight

Which material has the highest ratio of strength σ_f to density ρ? Use Figure 4.4 to find out.

Answer

The materials with the largest values of σ_f/ρ are those toward the top left of the figure. The ratio plots as a line of slope 1 on the chart. There is a guide line with this slope among the three at the lower right. The materials with the highest ratio are the ones furthest above such a line. Carbon-fiber–reinforced polymers (CFRPs) stands out as meeting this criterion.

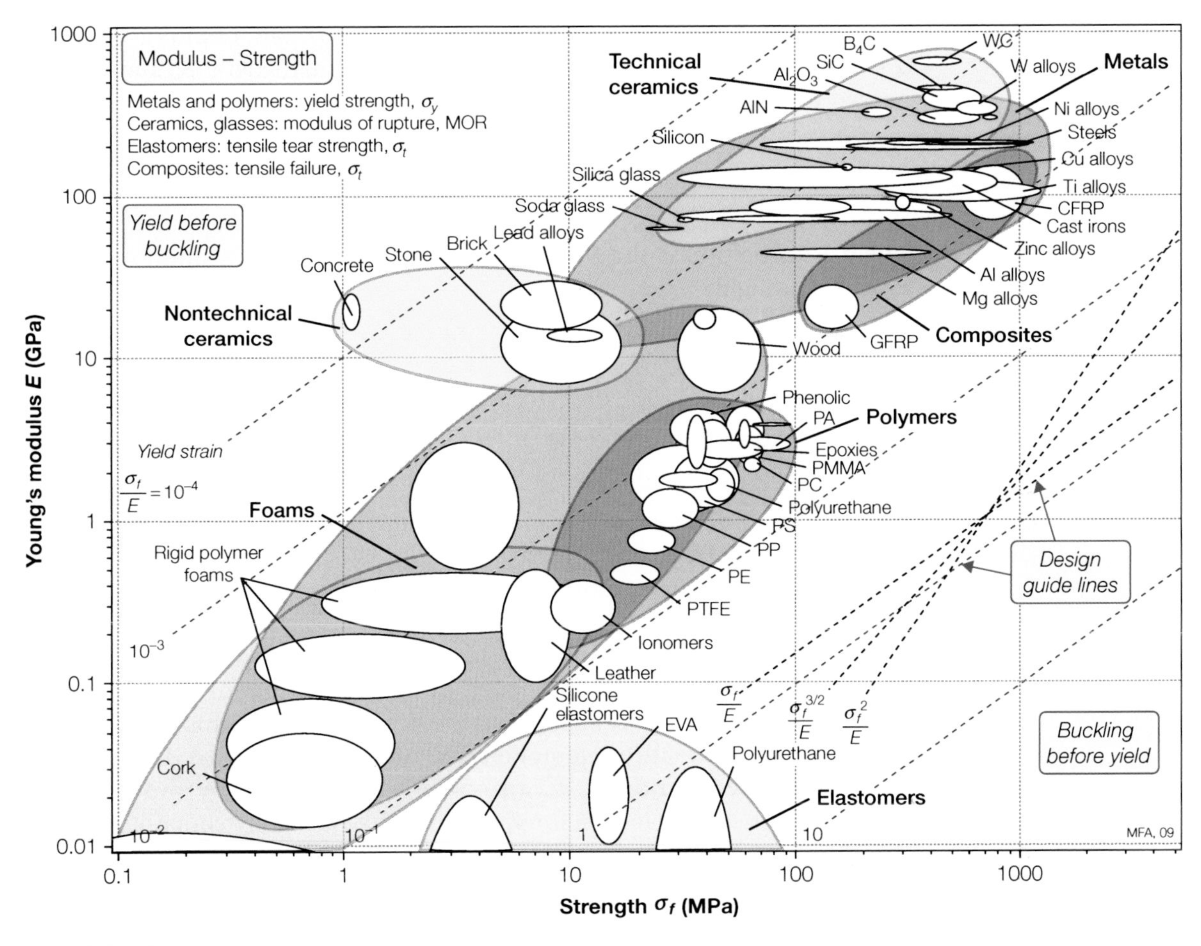

FIGURE 4.5

Young's modulus E plotted against strength σ_f. The design guide lines help with the selection of materials for springs, pivots, knife-edges, diaphragms, and hinges; their use is described in Chapters 5 and 6.

polymers, flexural strength (modulus of rupture) for ceramics, tear strength for elastomers, and tensile strength for composites and woods; the symbol σ_f is used for them all. Contours of *yield strain* or *fracture strain*, σ_f/E (meaning the strain at which the material ceases to be linearly elastic), appear as a family of straight parallel lines.

Examine these first. Engineering polymers have large yield strains of between 0.01 and 0.1; the values for metals are at least a factor of 10 smaller. Composites and woods lie on the 0.01 contour, as good as the best metals. Elastomers, because of their exceptionally low moduli, have values of σ_f/E larger than any other class of material: typically 1 to 10.

The distance over which inter atomic forces act is small; a bond is broken if it is stretched to more than about 10% of its original length. So the force F^* needed to break a bond is roughly

$$F^* \approx \frac{S r_o}{10} \tag{4.3}$$

where S, as before, is the bond stiffness. Then the failure strain of a solid should be roughly

$$\frac{\sigma_f}{E} \approx \frac{F^*}{r_o^2} / \left(\frac{S}{r_o}\right) = \frac{1}{10} \tag{4.4}$$

The chart shows that, for some polymers, the failure strain approaches this value. For most solids it is less, for two reasons.

First, nonlocalized bonds (those in which the cohesive energy derives from the interaction of one atom with large number of others, not just with its nearest neighbors) are not broken when the structure is sheared. The metallic bond, and the ionic bond for certain directions of shear, act in this way. Very pure metals, for example, yield at stresses as low as $E/10{,}000$, and strengthening mechanisms are needed to make them useful in engineering. The covalent bond *is* localized, and covalent solids, for this reason, have yield strengths that, at low temperatures, are as high as $E/10$. It is difficult to measure them (though it can sometimes be done by indentation) because of the second reason for weakness: They generally contain defects—concentrators of stress—from which fractures can propagate at stresses well below the "ideal" $E/10$. Elastomers are anomalous (they have strengths of about E) because the modulus does not derive from bond stretching, but from the change in entropy of the tangled molecular chains when the material is deformed.

It has not yet explained how to choose good materials to make springs. This involves the design guide lines shown on the chart, and it will be examined closer in Section 6.7.

Strong solids

Use the strength-density chart of Figure 4.4 to identify three material classes with members having strengths exceeding 1,000 MPa.

Answer

The material classes of steels, titanium alloys, and carbon-fiber composites (CFRPs) all have members with strengths exceeding 1,000 MPa.

The specific stiffness–specific strength chart

Many designs, particularly those for things that move, call for stiffness and strength at minimum weight. To help with this, the data of the previous chart are replotted in Figure 4.6 after dividing, for each material, by the density; it shows E/ρ plotted against σ_f/ρ. These are measures of "mechanical efficiency," meaning the use of the least mass of material to do the most structural work.

Composites, particularly CFRP, lie at the upper right. They emerge as the material class with the most attractive specific properties, one of the reasons for their increasing use in aerospace. Ceramics have exceptionally high stiffness per unit weight, and their strength per unit weight is as good as

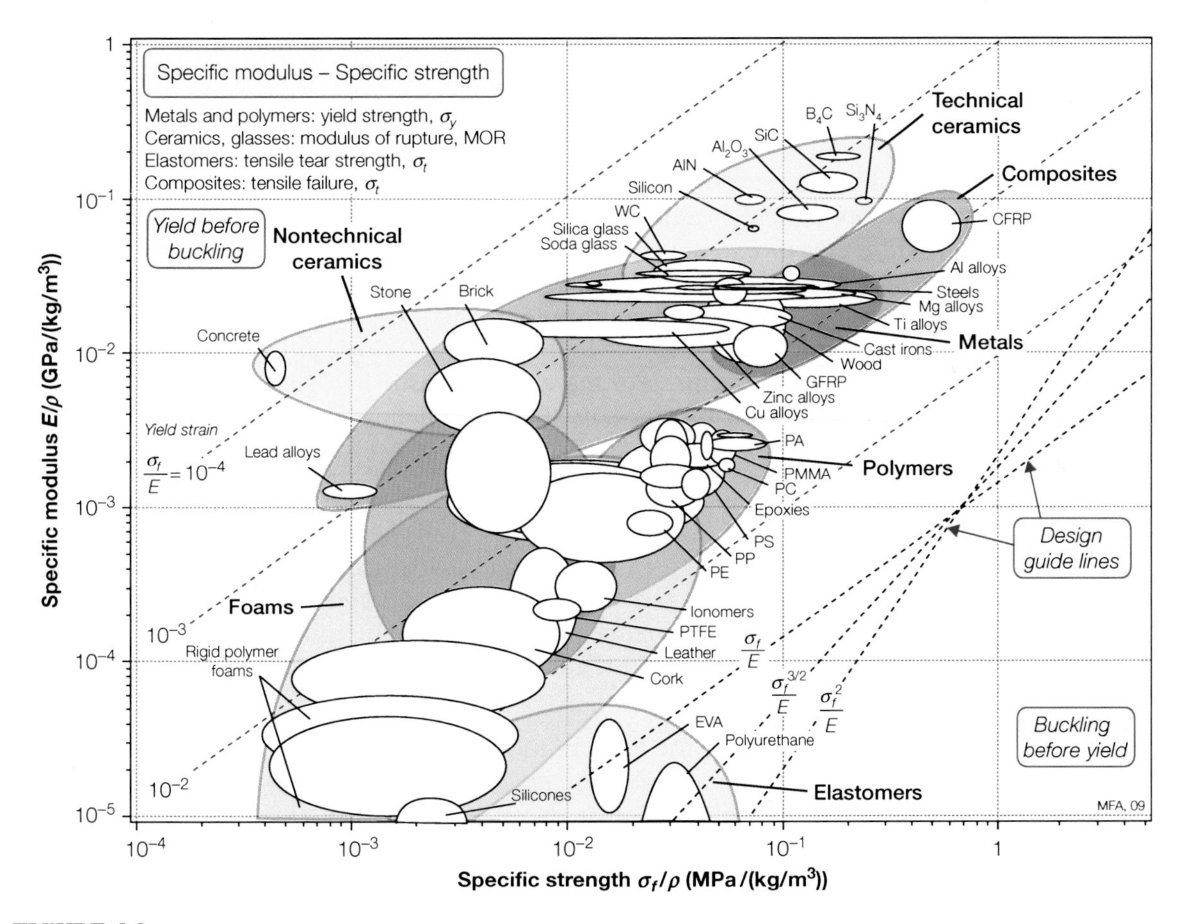

FIGURE 4.6

Specific modulus E/ρ plotted against specific strength σ_f/ρ. The design guide lines help with the selection of materials for lightweight springs and energy-storage systems.

High strength at low weight

High-quality mountain bikes are made of materials with particularly high values of the ratio σ_f/ρ, making them strong and light. Which metal class has the highest value of this ratio?

Answer

Figure 4.6 shows that titanium alloys have the highest value.

that of metals, but their brittleness excludes them from much structural use. Metals are penalized because of their relatively high densities. Polymers, because their densities are low, do better on this chart than on the last one.

The chart shown earlier in Figure 4.6 has application in selecting materials for light springs and energy storage devices. We will examine this in Section 6.7.

The fracture toughness–modulus chart

Increasing the strength of a material is useful only as long as the material remains plastic and does not become brittle; if it does, it is vulnerable to failure by fast fracture initiated from any tiny crack or defect it may contain. The resistance to the propagation of a crack is measured by the *fracture toughness*, K_{1c}, the units of which are $MPa.m^{1/2}$. It is plotted against modulus E in Figure 4.7. Values range from less than 0.01 to over 100 $MPa.m^{1/2}$. At the lower end of this range are brittle materials, which, when loaded, remain elastic until they fracture. For these, linear-elastic fracture mechanics works well, and the fracture toughness itself is a well-defined property.

At the upper end lie the super-tough materials, all of which show substantial plasticity before they break. For these the values of K_{1c} are approximate, derived from critical J-integral (J_c) and critical crack-opening displacement (δ_c) measurements, by writing $K_{1c} = (EJ_c)^{1/2}$, for instance. They are helpful in providing a ranking of materials. The figure shows one reason for the dominance of metals in engineering; they almost all have values of K_{1c} above 18 $MPa.m^{1/2}$, a value often quoted as a minimum for conventional design.

As a general rule, the fracture toughness of polymers is about the same as that of ceramics and glasses. Yet polymers are widely used in engineering structures; ceramics, because they are "brittle," are treated with much more caution. Figure 4.7 helps resolve this apparent contradiction. Consider first the question of the *necessary condition for fracture*. It is that sufficient external work be done, or elastic energy released, to supply the surface energy, γ per unit area, of the two new surfaces that are created. We write this as

$$G \geq 2\gamma \tag{4.5}$$

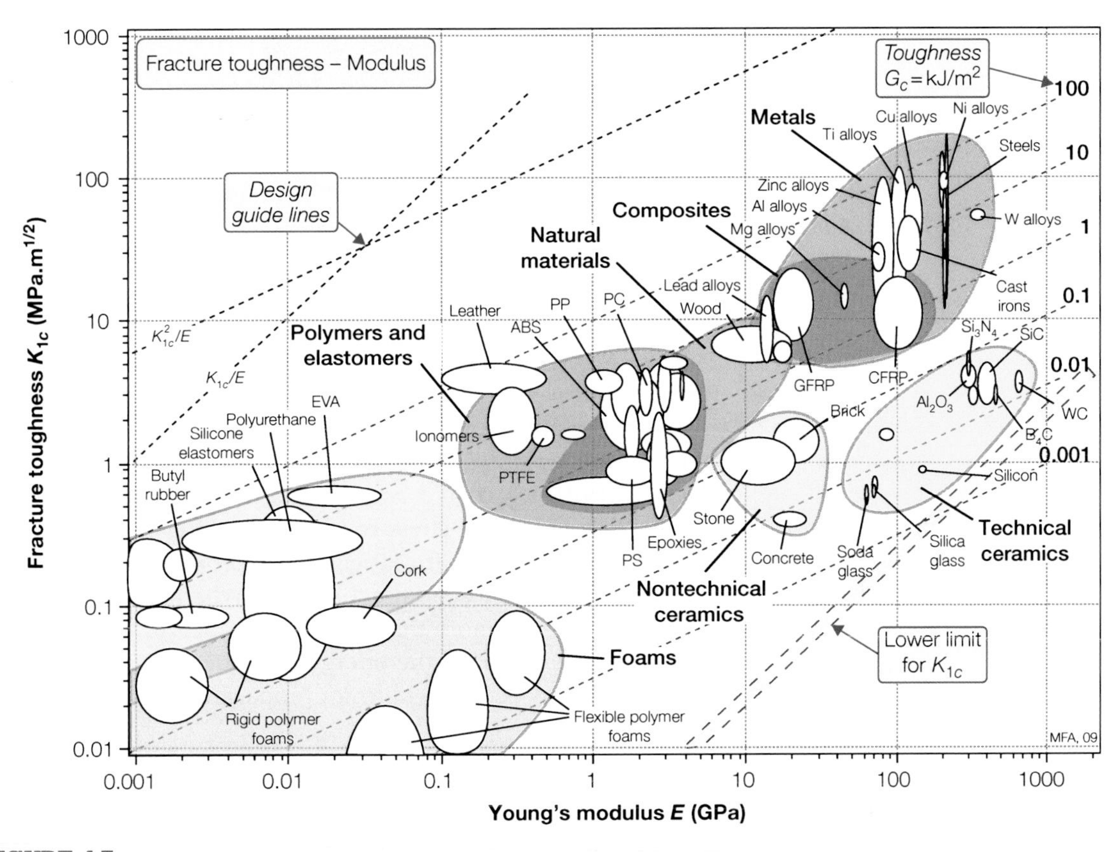

FIGURE 4.7

Fracture toughness K_{1c} plotted against Young's modulus E. The family of lines is of constant K_{1c}^2/E (approximately G_{1c}, the fracture energy or toughness). These, and the guide line of constant K_{1c}/E, help in design against fracture. The shaded band shows the lower limit for K_{1c}.

where G is the energy release rate. Using the standard relation $K = (EG)^{1/2}$ between G and stress intensity K, we find

$$K \geq (2E\gamma)^{1/2} \tag{4.6}$$

Now the surface energies, γ, of solid materials scale as their moduli; to an adequate approximation $\gamma \approx Er_o/20$, where r_o is the atom size, giving

$$K \geq E\left(\frac{r_o}{20}\right)^{1/2} \tag{4.7}$$

We identify the right-hand side of this equation with a lower-limiting value of K_{1c}, when, taking r_o as 2×10^{-10} m,

$$\frac{(K_{1c})_{\min}}{E} = \left(\frac{r_o}{20}\right)^{1/2} \approx 3\times10^{-6}\ \mathrm{m}^{1/2} \tag{4.8}$$

Comparing materials by toughness

The fracture toughness K_{1c} of polypropylene (PP) is about 4 MPa.m$^{1/2}$. That of aluminum alloys is about 10 times greater. But in deflection-limited design it is toughness G_c that is the more important property. Use Figure 4.7 to compare the two materials by toughness.

Answer

Aluminum and PP have almost exactly the same values of G_c: approximately 10 kJ/m^2.

This criterion is plotted on the chart as a shaded, diagonal band near the lower right corner. It defines a *lower limit* for K_{1c}. The fracture toughness cannot be less than this unless some other source of energy such as a chemical reaction, or the release of elastic energy stored in the special dislocation structures caused by fatigue loading, is available, when it is given a new symbol such as $(K_1)_{scc}$ meaning "the critical value of K_1 for stress-corrosion cracking" or $(\Delta K_1)_{threshold}$ meaning "the minimum range of K_1 for fatigue-crack propagation." We note that the most brittle ceramics lie close to the threshold. When they fracture, the energy absorbed is only slightly more than the surface energy. When metals, polymers, and composites fracture, the energy absorbed is vastly greater, usually because of plasticity associated with crack propagation.

Plotted on Figure 4.7 are contours of *toughness*, G_c, a measure of the apparent fracture surface energy ($G_c \approx K_{1c}^2/E$). The true surface energies, γ, of solids lie in the range 10^{-4} to 10^{-3} kJ/m^2. The diagram shows that the values of the toughness start at 10^{-3} kJ/m^2 and range through almost five decades to over 100 kJ/m^2. On this scale, ceramics (10^{-3}–10^{-1} kJ/m^2) are much lower than polymers (10^{-1}–10 kJ/m^2); this is part of the reason polymers are more widely used in engineering than ceramics. This point is developed further in Section 6.10.

The fracture toughness–strength chart

The stress concentration at the tip of a crack generates a *process zone*: a plastic zone in ductile solids, a zone of micro-cracking in ceramics, and a zone of delamination, debonding, and fiber pull-out in composites. Within the process zone, work is done against plastic and frictional forces; it is this that accounts for the difference between the measured fracture energy, G_c, and the true surface energy 2γ. The amount of energy dissipated must scale roughly with the strength of the material within the process zone and with its size, d_y. This size is found by equating the stress field of the crack ($\sigma = K/\sqrt{2\pi r}$) at $r = d_y/2$ to the strength of the material, σ_f, giving

$$d_y = \frac{K_{1c}^2}{\pi \sigma_f^2} \qquad (4.9)$$

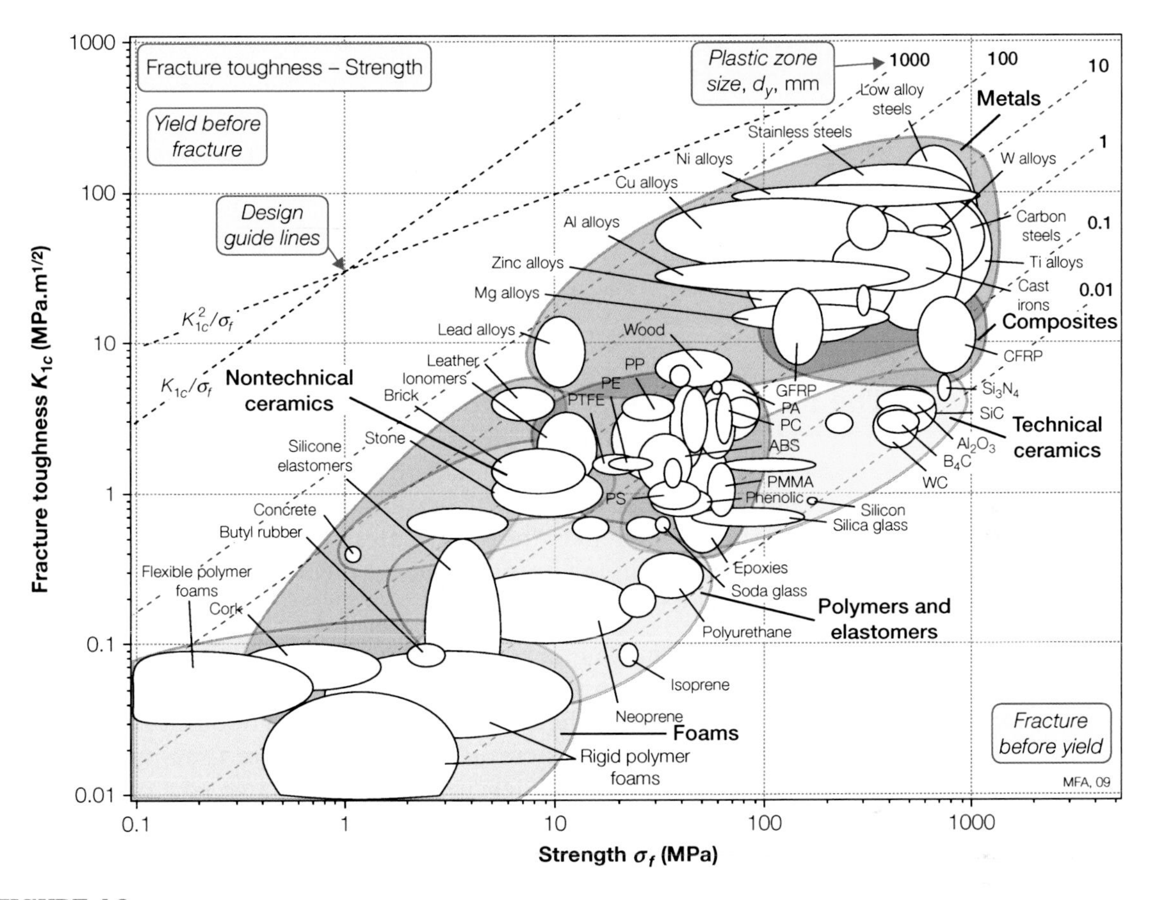

FIGURE 4.8

Fracture toughness K_{1c} plotted against strength σ_f. The contours show the value of $K_{1c}^2/\pi\sigma_f^2$—roughly the diameter d_y of the process zone at a crack tip. The design guide lines are used in selecting materials for damage-tolerant design.

Figure 4.8, fracture toughness plotted against strength, shows that the size of the zone, d_y (broken lines), varies from atomic dimensions for very brittle ceramics and glasses to almost 1 meter for the most ductile of metals. At a constant zone size, fracture toughness tends to increase with strength, as expected. It is this that causes the data plotted in Figure 4.8 to be clustered around the diagonal of the chart.

Materials toward the bottom right have high strength and low toughness; they *fracture before they yield*. Those toward the top left do the opposite: they *yield before they fracture*.

The diagram has application in selecting materials for the safe design of load-bearing structures. Examples are given in Sections 6.10 and 6.11.

Valid toughness testing

A valid fracture toughness test requires a sample with dimensions at least 10 times larger than the diameter of the process zone that forms at the crack tip. Use Figure 4.8 to estimate the sample size needed for a valid test on ABS.

Answer

The process zone size for ABS is approximately 1 mm. A valid test requires a sample with dimensions exceeding 10 mm.

The loss coefficient–modulus chart

Bells, traditionally, are made of bronze. They can be made of glass, and they could (if you could afford it) be made of silicon carbide. Metals, glasses, and ceramics all, under the right circumstances, have low intrinsic damping or "internal friction," an important material property when structures vibrate. Intrinsic damping is measured by the *loss coefficient*, η, which is plotted in Figure 4.9.

There are many mechanisms of intrinsic damping and hysteresis. Some (the "damping" mechanisms) are associated with a process that has a specific time constant; the energy loss is centered about a characteristic frequency. Others, the "hysteresis" mechanisms, are time-independent; they absorb energy at all frequencies. In metals a large part of the loss is hysteretic, caused by dislocation movement: It is high in soft metals like lead and pure aluminum. Heavily alloyed metals like bronze and high-carbon steels have low loss because the solute pins the dislocations; these are the materials for bells. Exceptionally high loss is found in the Mn-Cu alloys, because of a strain-induced martensite transformation, and in magnesium, perhaps because of reversible twinning. The elongated bubbles for metals span the large range made accessible by alloying and work hardening. Engineering ceramics have low damping because of the enormous lattice-resistance pins dislocations that are in place at room temperature.

Porous ceramics, on the other hand, are filled with cracks, the surfaces of which rub, dissipating energy when the material is loaded. The high damping of flake-graphite cast irons has a similar origin. In polymers, chain segments slide against each other when loaded; the relative motion dissipates energy. The ease with which they slide depends on the ratio of the ambient temperature, T, in this case room temperature, to the glass temperature, T_g, of the polymer. When $T/T_g < 1$, the secondary bonds are "frozen," the modulus is high, and the damping is relatively low. When $T/T_g > 1$, the secondary bonds have melted, allowing easy chain slippage; the modulus is

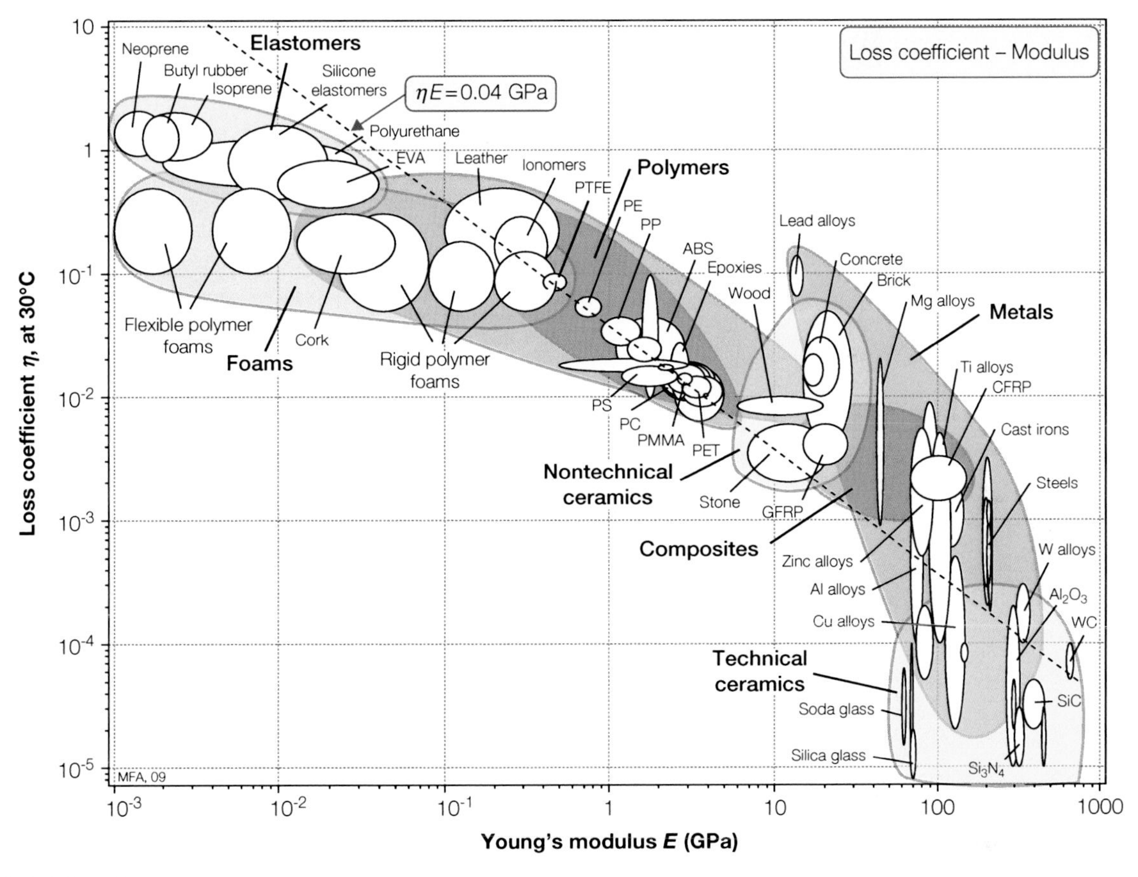

FIGURE 4.9

The loss coefficient η plotted against Young's modulus E. The guide line corresponds to the condition $\eta = CE$.

low, and the damping is high. This accounts for the obvious inverse dependence of η on E for polymers in Figure 4.9; indeed, to a first approximation,

$$\eta = \frac{4 \times 10^{-2}}{E} \tag{4.10}$$

(with E in GPa) for polymers, woods, and polymer-matrix composites.

Vibration damping

A metal is sought for mountings to damp vibration of a small machine tool. Use Figure 4.9 to find the metal with the greatest value of the damping coefficient η to use for the mountings.

Answer

Lead or lead alloys are the best choice.

Vibrational heating

A centrifuge rotating at f = 5,000 rpm is attached to a PTFE (Teflon) mounting. Poor balancing causes the centrifuge to vibrate, loading the PTFE up to a peak stress of σ_{max} = 8 MPa on each cycle. If a centrifuge run lasts 10 minutes and no heat is lost from the PTFE, by how much will its temperature rise? Take the volume-specific heat for PTFE (which can be read from Figure 4.11) to be $\rho C_p = 2 \times 10^6$ J/m^3.K, and retrieve the other material properties you need from Figure 4.9.

Answer

Figure 4.9 shows that the modulus of PTFE is E = 0.4 GPa and its loss coefficient is η = 0.08. The peak elastic energy stored in the PTFE in any one cycle is

$$U_{max} = \frac{\sigma_{max}^2}{2E} = 10^4 \text{ J/m}^3$$

of which $\Delta U = 2\pi\eta U_{max}$ (Equation (3.10)) is lost on each cycle. Thus the energy dumped into the PTFE over 5 minutes is

$$U_{10mins} = 2\pi\eta U_{max}(5f) = 2.52 \times 10^8 \text{ J/m}^3$$

Dividing this by the volume-specific heat of PTFE reveals a temperature rise of 126 °C. It will be necessary to ensure that heat can be conducted away from the mount to prevent it overheating.

The thermal conductivity–electrical resistivity chart

The material property governing the flow of heat through a material at steady state is the *thermal conductivity*, λ (units: W/m.K). (See Figure 4.10.) The valence electrons in metals are "free," moving like a gas within the lattice of the metal. Each electron carries a kinetic energy, $\frac{3}{2}kT$, where k is Boltzmann's constant. It is the transmission of this energy, via collisions, that conducts heat in metals. The thermal conductivity is described by

$$\lambda = \frac{1}{3} C_e \bar{c} \ell \tag{4.11}$$

where C_e is the electron specific heat per unit volume, $\bar{c}$ is the electron velocity (2×10^5 m/s), and ℓ is the electron mean-free path, typically 10^{-7} m in pure metals. In heavily alloyed solid solutions (stainless steels, nickel-based superalloys, and titanium alloys) the foreign atoms scatter electrons, reducing the mean free path to atomic dimensions ($\approx 10^{-10}$ m), much reducing λ.

These same electrons, when in a potential gradient, drift through the lattice, giving electrical conduction. The electrical conductivity, κ, is measured here

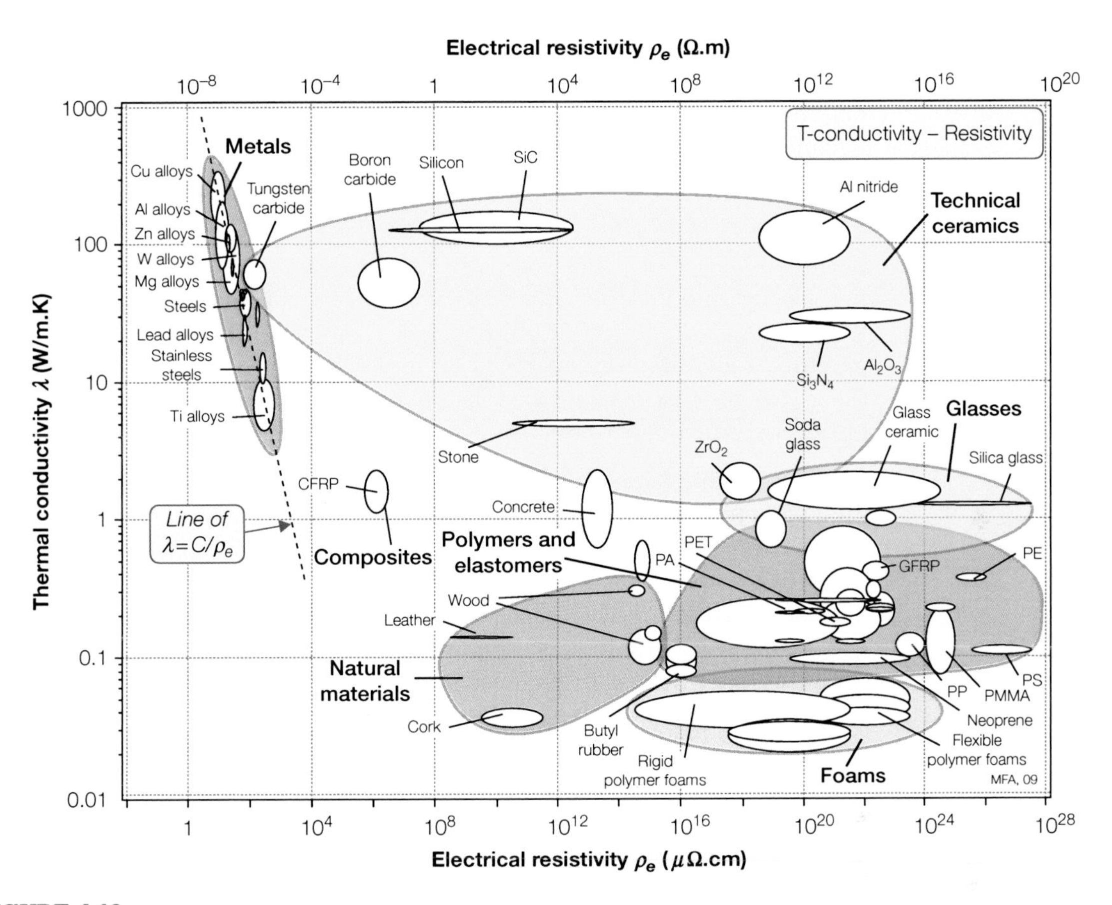

FIGURE 4.10

Thermal conductivity λ plotted against electrical resistivity ρ_e. For metals the two are related.

by its reciprocal, the *resistivity* ρ_e (SI units: Ω.m, units of convenience μΩ.cm). The range is enormous: a factor of 10^{28}, far larger than that of any other property. As with heat, the conduction of electricity in metals is proportional to the density of carriers, the electrons, and their mean-free path, leading to the Wiedemann-Franz relation

$$\lambda \propto \kappa = \frac{1}{\rho_e} \tag{4.12}$$

The quantities λ and ρ_e are the axes of Figure 4.10. Data for metals appear at the top left. The broken line shows that the Wiedemann-Franz relation is well obeyed.

But what about the rest of the chart? Electrons do not contribute to thermal conduction in ceramics and polymers. Heat is carried instead by phonons—lattice vibrations of short wavelength. They are scattered by each other and by impurities, lattice defects, and surfaces; it is these that determine the phonon mean-free path, ℓ. The conductivity is still given by equation (4.11), which we write as

$$\lambda = \frac{1}{3}\rho\, C_p\, \bar{c}\, \ell \tag{4.13}$$

but now $\bar{c}$ is the elastic wave speed (around 10^3 m/s—see Figure 4.3), ρ is the density, and C_p is the *specific heat per unit mass* (units: J/kg.K). If the crystal is particularly perfect and the temperature is well below the Debye temperature, as in diamond at room temperature, the phonon conductivity is high: It is for this reason that single crystal silicon carbide and aluminum nitride have thermal conductivities almost as high as copper.

The low conductivity of glass is caused by its irregular amorphous structure; the characteristic length of the molecular linkages (about 10^{-9} m) determines the mean free path. Polymers have low conductivities because the elastic wave speed $\bar{c}$ is low (see Figure 4.3), and the mean free path in the disordered structure is small. Highly porous materials like firebrick, cork, and foams show the lowest thermal conductivities, limited by the thermal conductivity of the gas in their cells.

Graphite and many intermetallic compounds such as C and B_4C, like metals, have free electrons, but the number of carriers is smaller and the resistivity is higher than in metals. Defects such as vacancies and impurity atoms in ionic solids create positive ions that require balancing electrons. These can jump from ion to ion, conducting charge, but slowly because the carrier density is low. Covalent solids and most polymers have no mobile electrons and are insulators ($\rho_e > 10^{12}\,\mu\Omega$.cm)—they lie on the right side of Figure 4.10.

Under a sufficiently high potential gradient, anything will conduct. The gradient tears electrons free from even the most possessive atoms, accelerating them into collision with nearby atoms, knocking out more electrons and creating a cascade. The critical gradient is the *breakdown potential*, V_b (units: MV/m), defined in Chapter 3.

Conducting heat but not electricity

Which materials are both good thermal conductors and good electrical insulators (an unusual combination)? Use Figure 4.10 to find out.

Answer

The chart identifies aluminum nitride, alumina, and silicon nitride as having these properties. They are the ones at the top right.

The thermal conductivity–thermal diffusivity chart

Thermal conductivity, as we have said, governs the flow of heat through a material at steady state. The property governing transient heat flow is the *thermal diffusivity*, a (units: m^2/s). The two are related by

$$a = \frac{\lambda}{\rho C_p} \quad (4.14)$$

where ρ in kg/m^3 is the density. The quantity ρC_p is the *volumetric specific heat* (units: $J/m^3.K$). Figure 4.11 relates thermal conductivity, diffusivity, and volumetric specific heat, at room temperature.

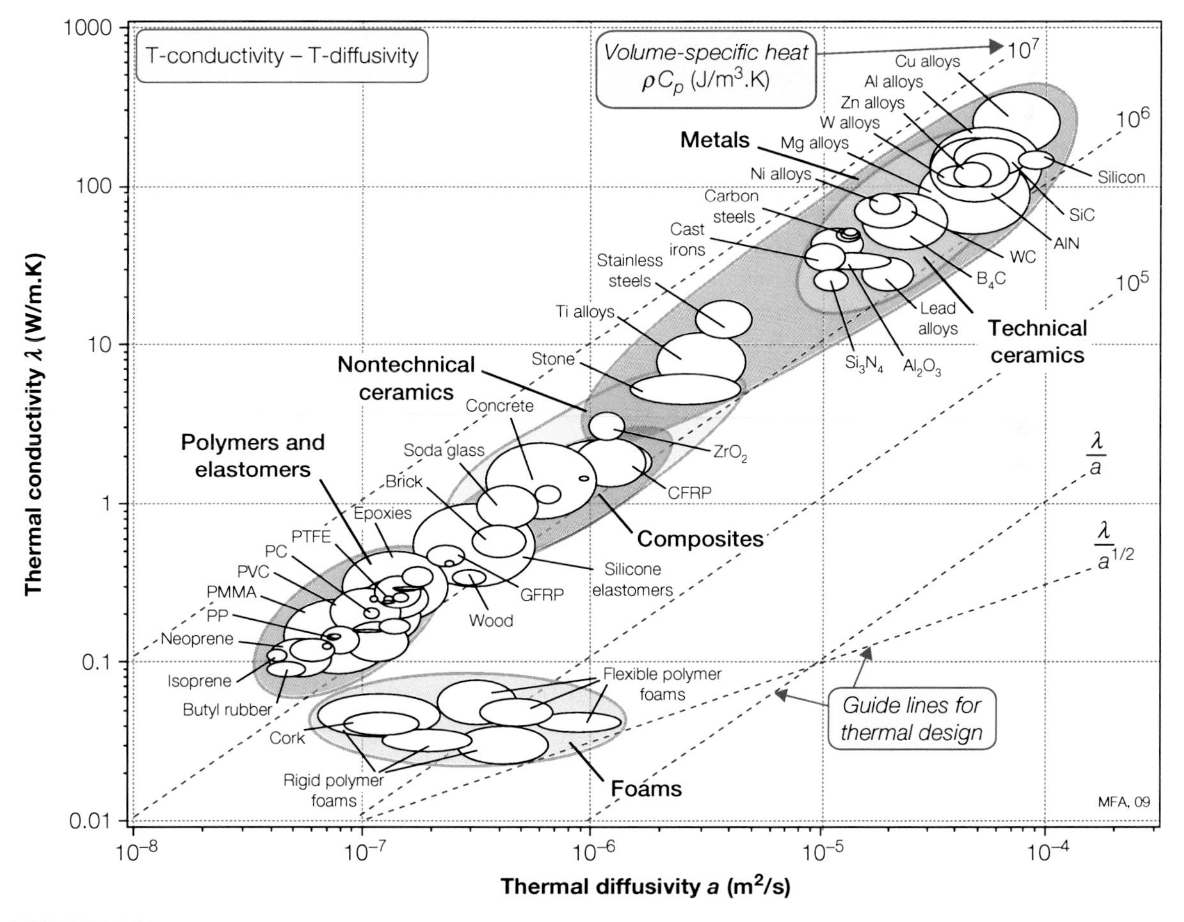

FIGURE 4.11

Thermal conductivity λ plotted against thermal diffusivity a. The contours show the volume-specific heat ρC_V. All three properties vary with temperature; the data here are for room temperature.

Thermal buffers

A good way to protect equipment from sudden temperature change is to encase it in a material with a very low thermal diffusivity, because an external temperature change then takes a long time to reach the inside. Use Figure 4.11 to identify materials that could be good for this.

Answer

The chart identifies isoprene, neoprene, and butyl rubber as potential candidates.

The data span almost five decades in λ and a. Solid materials are strung out along the line[3]

$$\rho C_p \approx 3 \times 10^6 \text{ J/m}^3.\text{K} \tag{4.15}$$

As a general rule, then,

$$\lambda = 3 \times 10^6 a \tag{4.16}$$

(λ in W/m.K and a in m^2/s). Some materials deviate from this rule because they have lower than average volumetric specific heat. The largest deviations are shown by porous solids: foams, low-density firebrick, woods, and the like. Their low density means that they contain fewer atoms per unit volume and, averaged over the volume of the structure, ρC_p is low. The result is that, although foams have low thermal *conductivities* and are widely used for insulation, their thermal *diffusivities* are not necessarily low: They may not transmit much heat, but they reach a steady state quickly. This is important in design, a point illustrated by the case study in Section 6.13.

The thermal expansion–thermal conductivity chart

Almost all solids expand on heating (Figure 4.12). The bond between a pair of atoms behaves like a linear elastic spring when the relative displacement of the atoms is small, but when it is large the spring is nonlinear. Most bonds become stiffer when the atoms are pushed together and less stiff when they are pulled apart. Such bonds are anharmonic. The thermal vibrations of atoms, even at room temperature, involve large displacements; as

---

[3] This can be understood by noting that a solid containing N atoms has $3N$ vibrational modes. Each (in the classical approximation) absorbs thermal energy kT at the absolute temperature T, and the vibrational specific heat is $C_p \approx C_v = 3Nk$ (J/K) where k is Boltzmann's constant (1.34×10^{-23} J/K). The volume per atom Ω for almost all solids lies within a factor of two of 1.4×10^{-29} m^3; thus the volume of N atoms is (NC_p) m^3. The volume-specific heat is then (as the chart shows):

$$\rho C_v \cong 3Nk/N\Omega = \frac{3k}{\Omega} = 3 \times 10^6 \text{ J/m}^3\text{K}$$

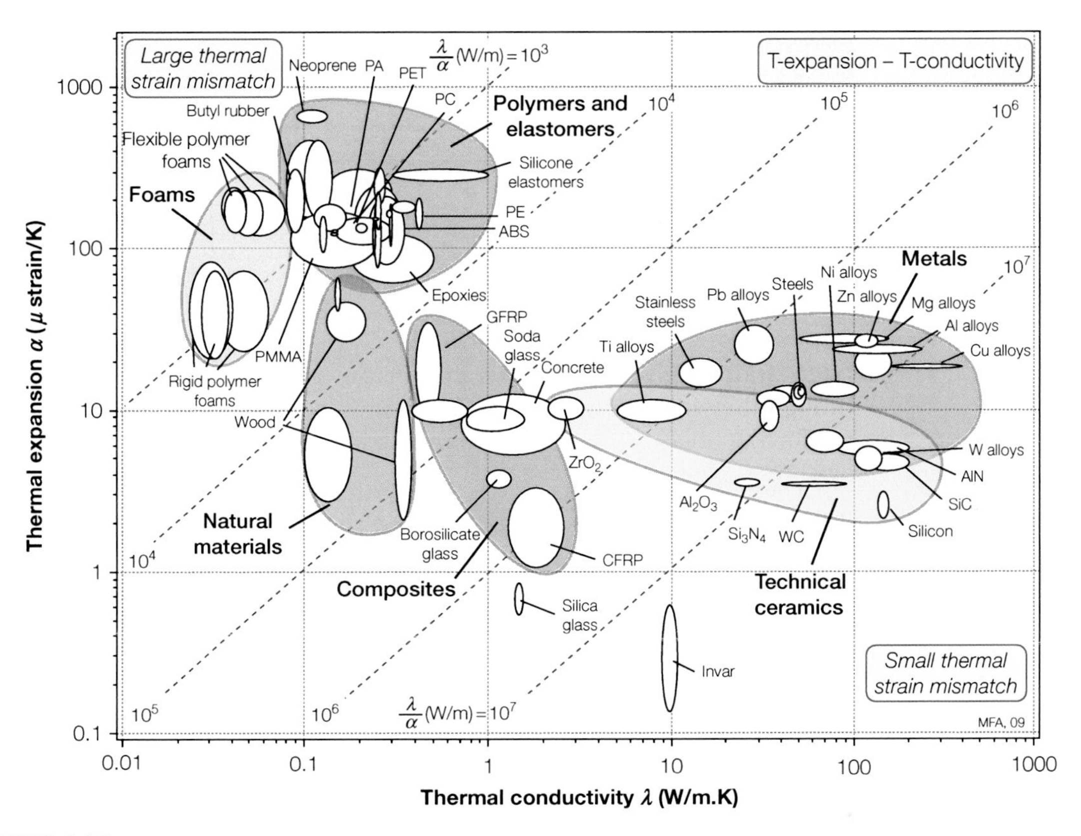

FIGURE 4.12

The linear expansion coefficient α plotted against the thermal conductivity λ. The contours show the thermal distortion parameter λ/α. An extra material, the nickel alloy Invar, has been added to the chart; it is noted for its exceptionally low expansion at and near room temperature, useful in designing precision equipment that must not distort if the temperature changes.

the temperature is raised, the anharmonicity of the bond pushes the atoms apart, increasing their mean spacing. The effect is measured by the linear *expansion coefficient*

$$\alpha = \frac{1}{\ell}\frac{d\ell}{dT} \tag{4.17}$$

where ℓ is a linear dimension of the body.

The expansion coefficient is plotted against the thermal conductivity in Figure 4.12. It shows that polymers have large values of α, roughly 10 times greater than those of metals and almost 100 times greater than those of ceramics. This is because the Van der Waals bonds of the polymer are very anharmonic. Diamond, silicon, and silica glass (SiO_2) have covalent bonds that have low anharmonicity (that is, they are almost linear elastic even at

Thermal actuators

An actuator uses thermal expansion of its active element to generate the actuating force. Use Figure 4.12 to identify the material with the largest expansion coefficient.

Answer

Neoprene, at the upper left of the chart, has a larger value of expansion coefficient than any other on the chart.

large strains), giving them low expansion coefficients. Composites, even though they have polymer matrices, can have low values of α because the reinforcing fibers, particularly carbon, expand very little.

The chart shows contours of λ/α, a quantity important in designing against thermal distortion. An extra material, Invar (a nickel alloy), has been added to the chart because of its uniquely low expansion coefficient at and near room temperature, a consequence of a trade-off between normal expansion and a contraction associated with a magnetic transformation. An application that uses the chart is developed in Chapter 6, Section 6.16.

The thermal expansion–modulus chart

Thermal stress is the stress that appears in a body when it is heated or cooled but prevented from expanding or contracting. It depends on the expansion coefficient, α, of the material and on its modulus, E. A standard development of the theory of thermal expansion leads to the relationship

$$\alpha = \frac{\gamma_G \, \rho \, C_p}{3\,E} \tag{4.18}$$

where γ_G is Gruneisen's constant. It has values between 0.4 and 4, but for most solids it is near 1. Since ρC_p is almost constant (see Equation (4.15)), the equation tells us that α is proportional to $1/E$. Figure 4.13 shows that this is broadly so. Ceramics, with the highest moduli, have the lowest coefficients of expansion; elastomers with the lowest moduli expand the most. Some materials with a low coordination number (silica and some diamond-cubic or zinc-blend structured materials) can absorb energy preferentially in transverse modes, leading to very small or negative values of γ_G and a low expansion coefficient (silica, SiO_2, is an example). Others, like Invar, contract as they lose their ferromagnetism when heated through the Curie temperature; over a narrow range of temperature, they too show near-zero expansion, useful in the manufacture of precision equipment and in glass-metal seals.

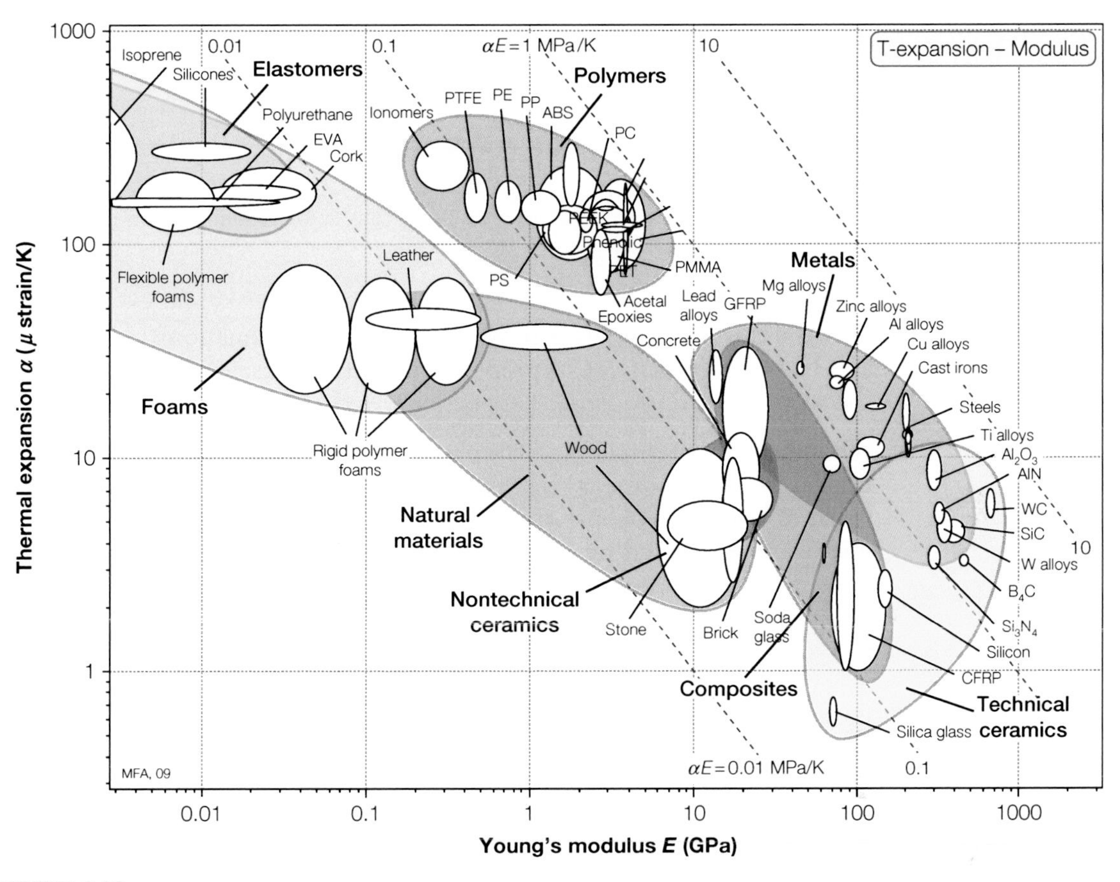

FIGURE 4.13

The linear expansion coefficient α plotted against Young's modulus E. The contours show the thermal stress created by a temperature change of 1°C if the sample is axially constrained. A correction factor C is applied for biaxial or triaxial constraint (see text).

One more useful fact. The moduli of materials scale approximately with their melting point, T_m:

$$E \approx \frac{100\,k\,T_m}{\Omega} \tag{4.19}$$

where k is Boltzmann's constant and Ω is the volume per atom in the structure. Substituting this and equation (4.15) for ρC_p into equation (4.18) for α gives

$$\alpha = \frac{\gamma_G}{100\,T_m} \tag{4.20}$$

—the expansion coefficient varies inversely with the melting point. Equivalently, the thermal strain for all solids, just before they melt, depends only

Thermal stress

Approximately what stress will appear in a rod of steel, rigidly clamped at its ends, if its temperature is changed by 100°C? Use Figure 4.13 to find out.

Answer

The figure shows that, for steel, $\alpha E \approx 3\,\mathrm{MPa/K}$. Thus, a temperature change of 100°C will create a stress of approximately 300 MPa.

on γ_G, and this is roughly a constant at about 1%. Equations (4.18), (4.19), and (4.20) are examples of property correlations, useful for estimating and checking material properties (Appendix A, Section A.12).

Whenever the thermal expansion or contraction of a body is prevented, thermal stresses appear; if large enough, these stresses cause yielding, fracture, or elastic collapse (buckling). It is common to distinguish between thermal stress caused by external constraint (a rod rigidly clamped at both ends, for example) and that which appears without external constraint because of temperature gradients in the body. All scale as the quantity αE, shown as a set of diagonal contours in Figure 4.13. More precisely, the stress $\Delta\sigma$ produced by a temperature change of 1°C in a constrained system, or the stress per °C caused by a sudden change of surface temperature in one that is not constrained, is given by

$$C\,\Delta\sigma = \alpha E \tag{4.21}$$

where $C = 1$ for axial constraint, $(1 - \nu)$ for biaxial constraint or normal quenching, and $(1 - 2\nu)$ for triaxial constraint, where ν is Poisson's ratio. These stresses are large: typically 1 MPa/K. They can cause a material to yield, crack, spall, or buckle when it is suddenly heated or cooled.

The maximum service temperature chart

Temperature affects material performance in many ways. As the temperature is raised, the material may creep, limiting its ability to carry loads. It may degrade or decompose, changing its chemical structure in ways that make it unusable. And it may oxidize or interact in other ways with the environment in which it is used, leaving it unable to perform its function. The approximate temperature at which, for any of these reasons, it is unsafe to use a material is called its *maximum service temperature* T_{max}. Figure 4.14 show this plotted as a bar chart.

The chart gives a birds-eye view of the regimes of temperature in which each material class is usable. Note that few polymers can be used above 200°C, few metals above 800°C, and only ceramics offer strength above 1,500°C.

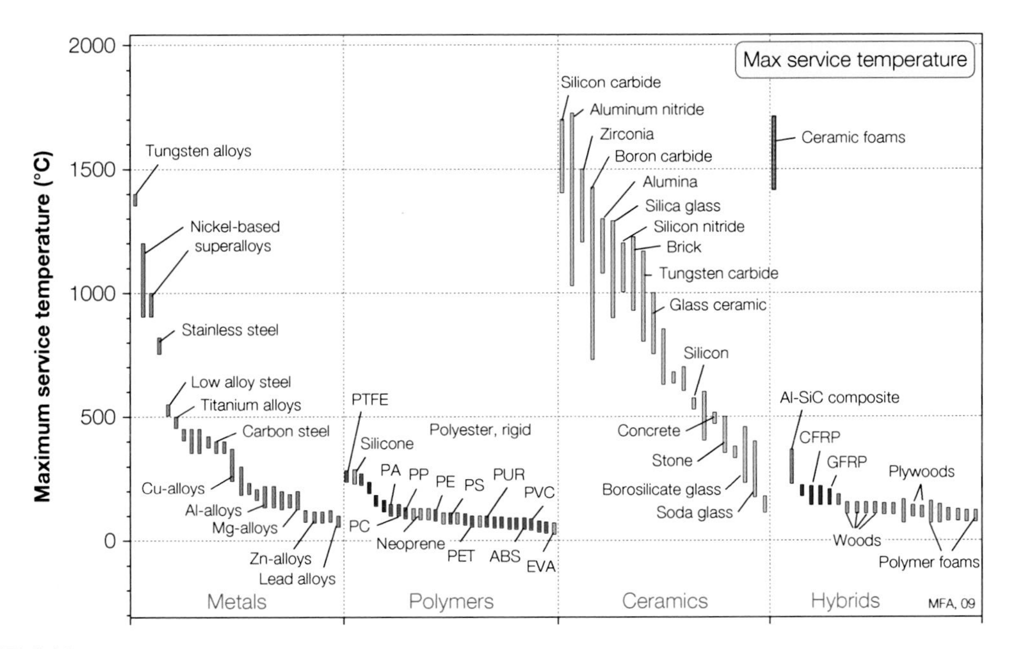

FIGURE 4.14
The maximum service temperature—the temperature above which a material becomes unusable.

Friction and wear

God, it is said, created materials but the devil made surfaces. Surfaces are the source of many problems. When surfaces touch and slide, there is friction; where there is friction, there is wear. Tribologists—those who study friction and wear—are fond of citing the enormous cost, through lost energy and worn equipment, for which these two phenomena are responsible. It is certainly true that, if friction could be eliminated, the efficiency of engines, gear boxes, drive trains, and the like, would increase. If wear could be eradicated, they would also last longer. But before accepting this negative image, one should remember that without wear pencils would not write on paper or chalk on blackboards; without friction, one would slither off the slightest incline.

Use temperature for stainless steel

Stainless steel is proposed for use as part of a structure operating at 500°C. Is it safe to do so?

Answer

Figure 4.14 shows that the maximum use temperature for stainless steel is in the range 700 to 1,100°C. Use at 500°C appears to be practical.

Tribological properties are not attributes of one material alone but of one material sliding on another with, almost always, a third in between. The number of combinations is far too great to allow choice in a simple, systematic way. The selection of materials for bearings, drives, and sliding seals relies heavily on experience. This experience is captured in reference sources (for which see Appendix D). In the end it is these that must be consulted. But it does help to have a feel for the magnitude of friction coefficients and wear rates and to have an idea of how these relate to material class.

When two surfaces are placed in contact under a normal load F_n and one is made to slide over the other, a force F_s opposes the motion. This force is proportional to F_n but does not depend on the area of the surface. This is the single most significant result of studies of friction, since it implies that surfaces do not contact completely but only touch over small patches, the area that is independent of the apparent, nominal area of contact A_n. The *coefficient friction* μ is defined by

$$\mu = \frac{F_s}{F_n} \tag{4.22}$$

Approximate values for μ for dry, unlubricated, sliding of materials on a steel counterface are shown in Figure 4.15. Typically, $\mu \approx 0.5$. Certain materials show much higher values, either because they seize when rubbed together (a soft metal rubbed on itself with no lubrication, for instance) or

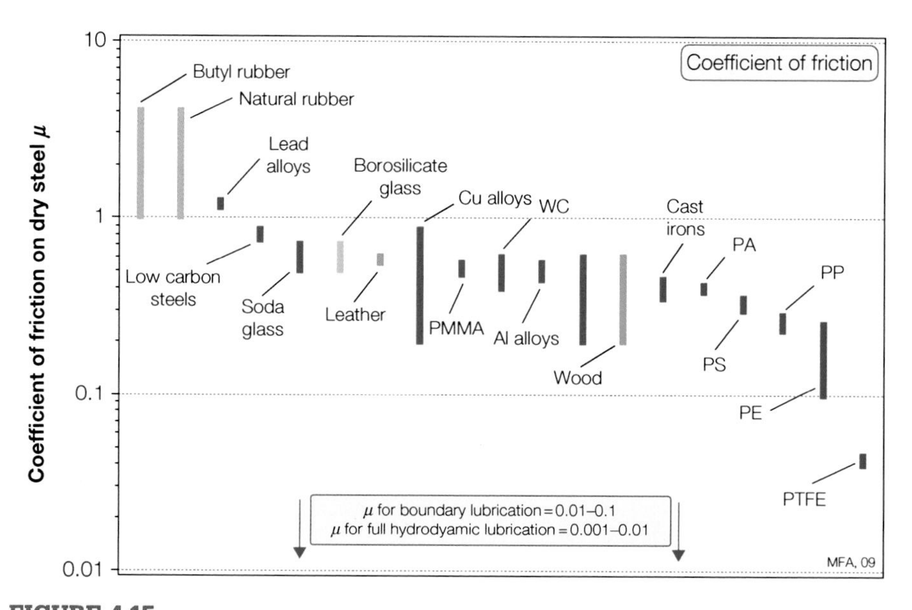

FIGURE 4.15

The friction coefficient μ of materials sliding on an unlubricated steel counterface.

because one surface has a sufficiently low modulus that it conforms to the other (rubber on rough concrete). At the other extreme are sliding combinations with exceptionally low coefficients of friction, such as PTFE or bronze bearings loaded with graphite sliding on polished steel. Here the coefficient of friction falls as low as 0.04, although this is still high compared with friction for lubricated surfaces, as noted at the bottom of the diagram.

When surfaces slide, they wear. Material is lost from both surfaces, even when one is much harder than the other. The *wear rate* W (units: m^2) is conventionally defined as

$$W = \frac{\textit{Volume of material removed from contact surface}}{\textit{Distance slid}} \tag{4.23}$$

A more useful quantity, for our purposes, is the specific wear rate

$$\Omega = \frac{W}{A_n} \tag{4.24}$$

which is dimensionless. It increases with bearing pressure P (which is the normal force F_n divided by the nominal area A_n), such that the ratio

$$k_a = \frac{W}{F_n} = \frac{\Omega}{P} \tag{4.25}$$

is roughly constant. The *wear-rate constant* k_a (units: $(MPa)^{-1}$) is a measure of the propensity of a sliding couple for wear: High k_a means rapid wear at a given bearing pressure.

The bearing pressure P is the quantity specified by the design. The ability of a surface to resist a static contact pressure is measured by its hardness H, so we anticipate that the maximum bearing pressure P_{max} should scale with the hardness of the softer surface:

$$P_{max} = CH$$

where C is a constant. Thus the wear rate of a bearing surface can be written as

$$\Omega = k_a P = C\left(\frac{P}{P_{max}}\right) k_a H \tag{4.26}$$

Two material properties appear in this equation: the wear-rate constant k_a and the hardness, H. They are plotted in Figure 4.16. The dimensionless quantity

$$K = k_a H \tag{4.27}$$

is shown as a set of diagonal contours. Note first that materials of a given class (metals, for instance) tend to lie along a downward-sloping diagonal across the figure, reflecting the fact that low wear rate is associated with high hardness. The best materials for bearings for a given bearing pressure

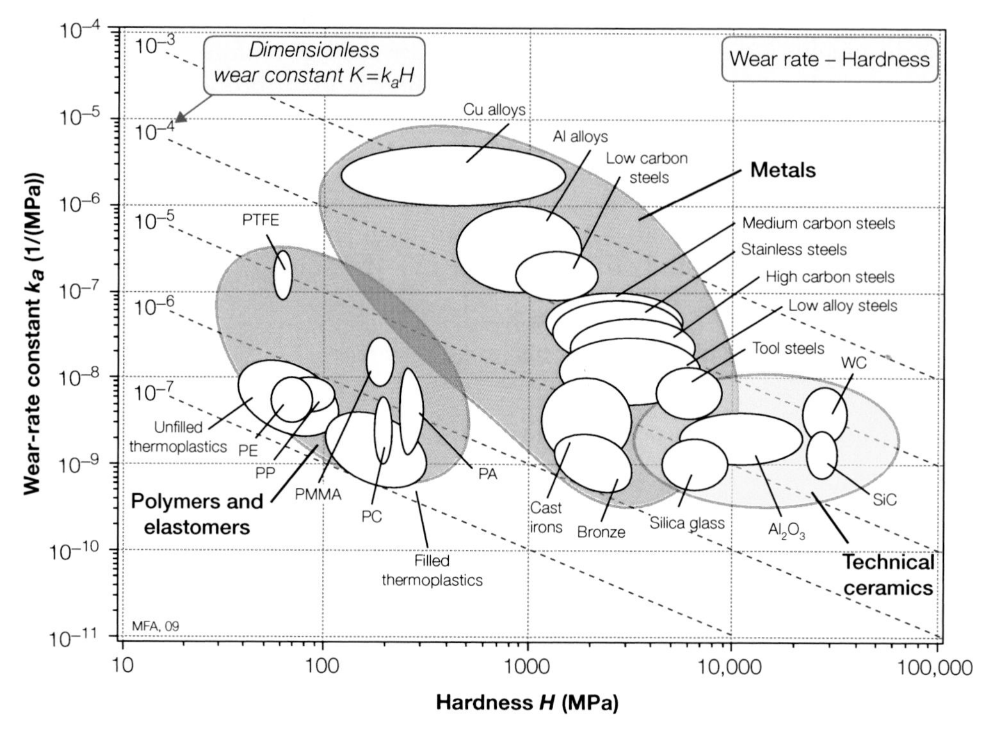

FIGURE 4.16

The normalized wear rate k_A plotted against hardness H, here expressed in MPa rather than Vickers (H in MPa = 10 H_V). The chart gives an overview of the way in which common engineering materials behave.

P are those with the lowest value of k_a, that is, those nearest the bottom of the diagram. On the other hand, an efficient bearing, in terms of size or weight, will be loaded to a safe fraction of its maximum bearing pressure, that is, to a constant value of P/P_{max}; for these, materials with the lowest values of the product $k_a\ H$ are best.

Cost bar charts

Properties like modulus, strength, and conductivity do not change with time. Cost is bothersome because it does. Supply, scarcity, speculation, and

Materials for bearings

Use Figure 4.16 to find two metals and two polymers that offer good wear resistance at constant bearing pressure.

Answer

The figure suggests bronze, cast iron, polycarbonate (PC), and nylon (PA) as good choices.

inflation contribute to considerable fluctuations in the cost per kilogram of a commodity like copper or silver. Data for cost per kg are tabulated for some materials in daily papers and trade journals; those for others are harder to come by. Approximate values for the cost of materials per kg, and their cost per m^3, are plotted in Figures 4.17(a) and (b). Most commodity

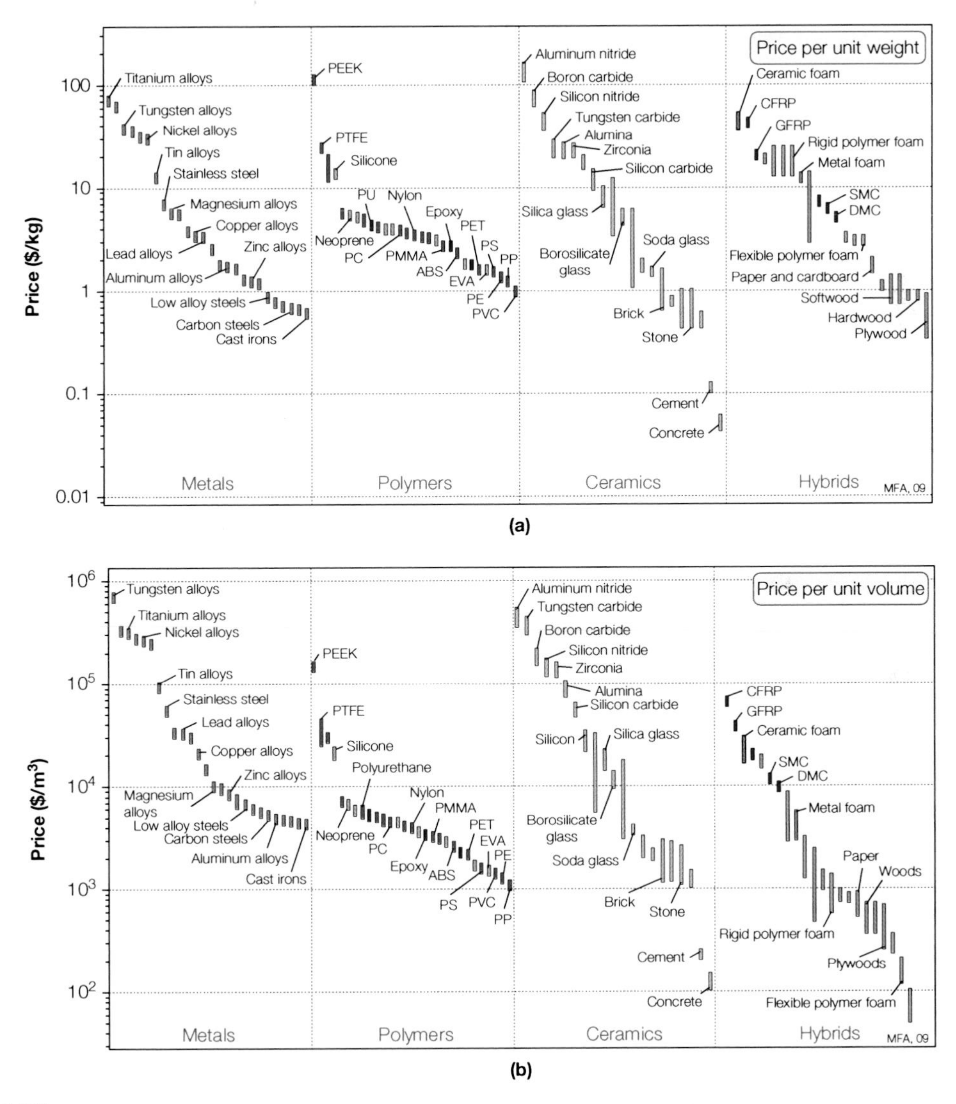

FIGURE 4.17

(a) The approximate price/kg of materials. Commodity materials cost about $1/kg; special materials cost much more. (b) The approximate price/m^3 of materials. Polymers, because they have low densities, cost less per unit volume than most other materials.

Polymer prices

Which four polymers are the least expensive?

Answer

Figure 4.17 shows that the four least expensive polymers, either by weight or by volume, are polypropylene (PP), polyethylene (PE), polyvinylchloride (PVC), and polystyrene (PS). These are the commodity polymers—the ones used in the greatest quantities.

Relative price

Are polypropylene and polyethylene more or less expensive than steel per unit weight? What about per unit volume?

Answer

Figure 4.17 reveals that the two polymers are more expensive than steel per unit weight, but because they have much lower densities, they are cheaper per unit volume. If we want a proper comparison, what should be the basis of comparison? The answer, developed in later chapters, is *per unit function*.

materials (glass, steel, aluminum, and the common polymers) cost between 0.5 and 2.0 $/kg. Because they have low densities, the cost/m^3 of commodity polymers is less than that of metals.

The modulus–relative cost chart

In design for minimum cost, material selection is guided by indices that involve modulus, strength, and cost per unit volume. To make some correction for the influence of inflation and the units of currency in which cost is measured, we define a *relative cost per unit volume* $C_{v,R}$.

$$C_{v,R} = \frac{Cost/kg \times Density\ of\ material}{Cost/kg \times Density\ of\ mild\ steel\ rod} \tag{4.28}$$

At the time of writing, steel reinforcing rod costs about US$0.3/kg.

Figure 4.18 shows the modulus E plotted against relative cost per unit volume $C_{v,R}\rho$ where ρ is the density. Cheap, stiff materials lie toward the top left. Guide lines for selecting materials that are stiff and cheap are plotted in the figure.

The strength–relative cost chart

Cheap strong materials are selected using Figure 4.19. It shows failure strength, defined as before, plotted against relative cost per unit volume. The qualifications for the definition of strength, given earlier, apply here also.

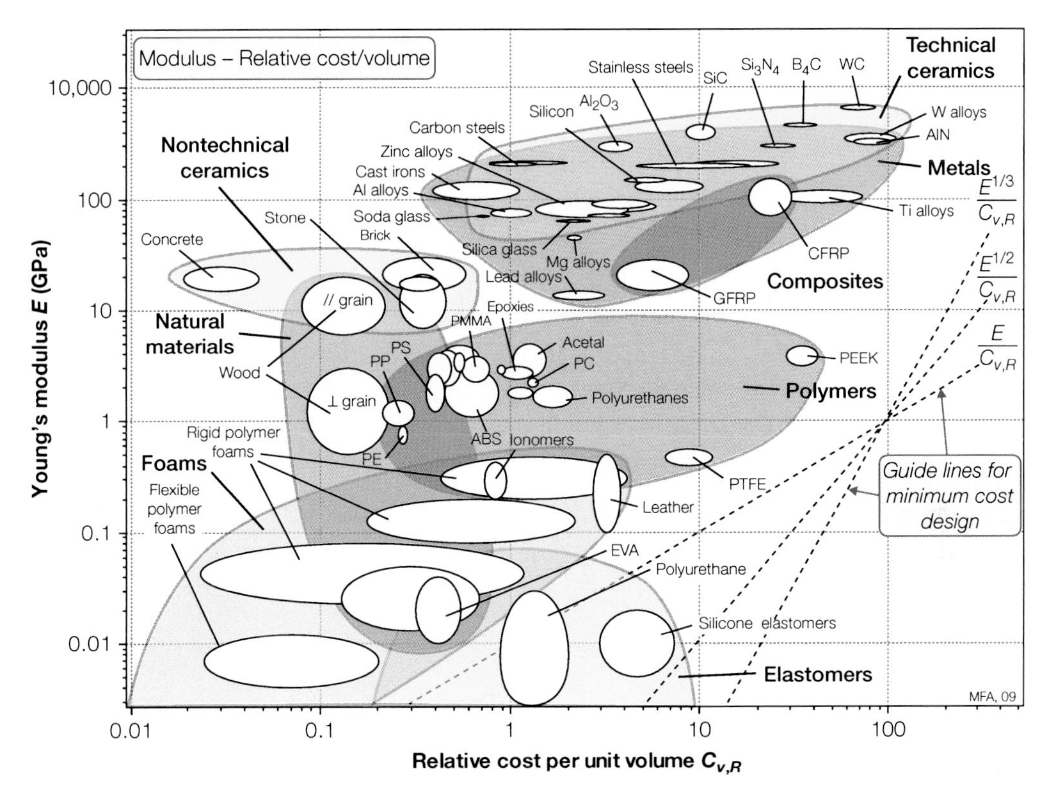

FIGURE 4.18

Young's modulus E plotted against relative cost per unit volume $C_{v,R}$. The design guide lines help selection to maximize stiffness per unit cost.

It must be emphasized that the data plotted here and in Figure 4.18 are less reliable than those in other charts and are subject to unpredictable change. Despite this dire warning, the two charts are genuinely useful. They allow selection of materials using the criterion of "function per unit cost." An example is given in Section 6.5.

4.4 SUMMARY AND CONCLUSIONS

The engineering properties of materials are usefully displayed as material selection charts. Eighteen of them are introduced in this chapter; more appear in later ones. The charts summarize material properties in a compact, easily accessible way, showing the range spanned by each material family and class. By choosing the axes in a sensible way, more information can be displayed. A chart of modulus E against density ρ reveals the longitudinal wave velocity $(E/\rho)^{1/2}$. A chart of fracture toughness K_{1c} against modulus E shows the

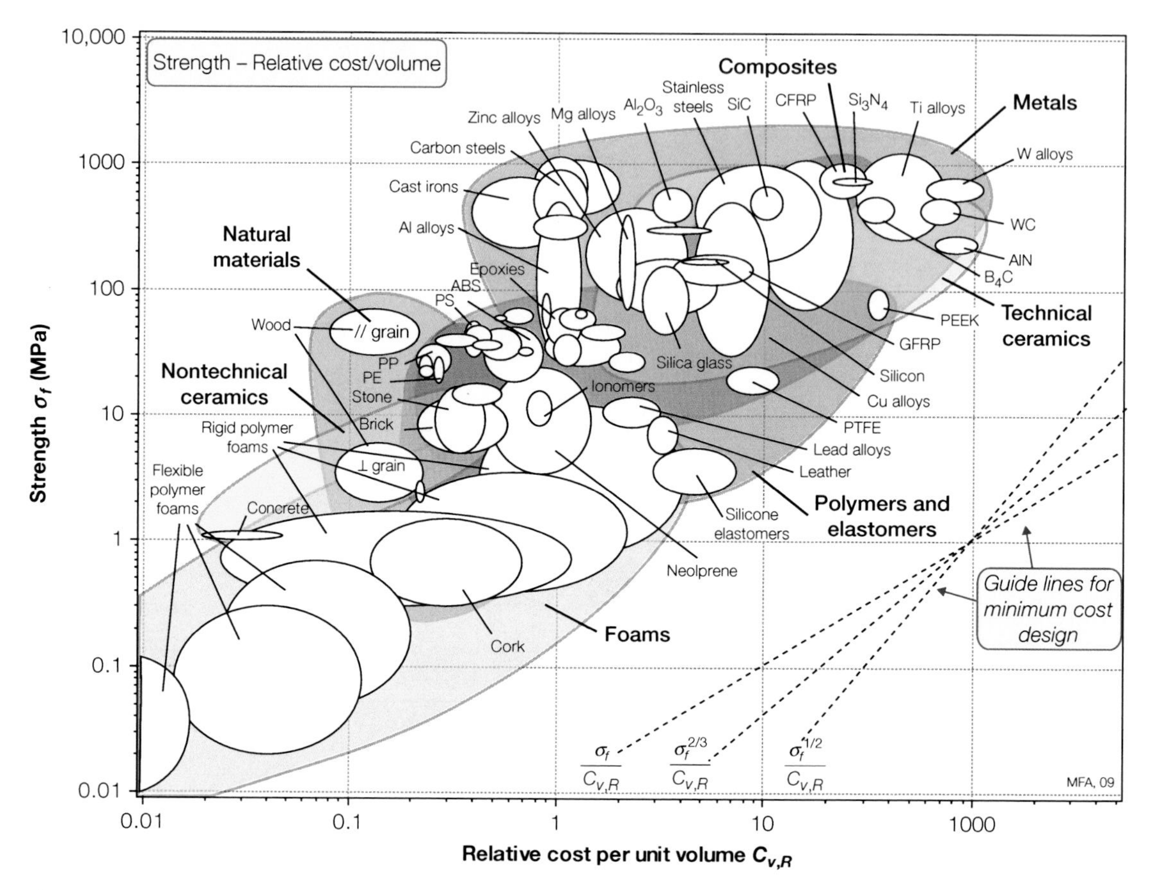

FIGURE 4.19

Strength σ_f plotted against relative cost per unit volume $C_{v,R}$. The design guide lines help selection to maximize strength per unit cost.

toughness G_{1c}. A chart of thermal conductivity λ against diffusivity, a, also gives the volume-specific heat ρC_v. A chart of strength, σ_f, against modulus, E, shows the energy-storing capacity, σ_f^2/E, and there are many more.

The most striking feature of the charts is the way in which members of a material class cluster together. Despite the wide range of modulus and density of metals (as an example), the clusters occupy a field that is distinct from that of polymers or that of ceramic, or that of composites. The same is true of strength, toughness, thermal conductivity, and the rest: The fields frequently overlap, but they always have a characteristic place within the whole picture.

The position of the fields and their relationship can be understood in simple physical terms. The nature of the bonding, the packing density, the lattice resistance, and the vibrational modes of the structure are themselves a function of bonding and packing. It may seem odd that so little mention has been made of microstructure in determining properties. But the charts clearly show that the first-order difference between the properties of materials has its origins in the mass of the atoms, the nature of the inter-atomic forces, and the geometry of packing. Alloying, heat treatment, and mechanical working, to which we return in Chapter 13, all influence microstructure and, through this, properties, giving the elongated bubbles shown on many of the charts; but the magnitude of their effect is less, by factors of 10, than that of bonding and structure.

All the charts have one thing in common: Parts of them are populated with materials and parts are not. Some parts are inaccessible for fundamental reasons that relate to the size of their atoms and the nature of the forces that bind their atoms together. But other parts are empty even though, in principle, they are accessible. If they were accessed, the new materials that lie there could allow novel design possibilities. Ways of doing this are explored further in Chapters 11 and 12.

The charts have numerous applications. One is the checking and validation of data (Appendix A); here use is made of both the range covered by the envelope of material properties and the numerous relations between them (like $E\Omega = 100\,kT_m$), described in Section 4.3. Another concerns the development of, and identification of uses for, new materials; materials that fill gaps in one or more of the charts generally offer some improved design potential. But most important of all, the charts form the basis for a procedure for materials selection. That is developed in the following chapters.

4.5 FURTHER READING

The best general book on the physical origins of the mechanical properties of materials remains that by Cottrell (1964).

Ashby, M.F., Shercliff, H.R., & Cebon, D. (2009). *Materials: Engineering, science, processing and design* (2nd ed.). Butterworth-Heinemann, ISBN 978-1-85617-895-2.

An elementary text that introduces materials through material property charts and develops the selection methods through case studies.

Budinski, K.G., & Budinski, M.K. (2010). *Engineering materials, properties and selection* (9th ed.). Prentice Hall, ISBN 978-0-13-712842-6.

A well-established materials text that deals well with both material properties and processes.

Cottrell, A.H. (1964). *Mechanical properties of matter*. Wiley, Library of Congress Number 65-14262.

An inspirational book, clear, full of insights and simple derivations of the basic equations describing the mechanical behavior of solids, liquids, and gases.

Dieter, G.E. (1999). *Engineering design, a materials and processing approach* (3rd ed.). McGraw-Hill, ISBN 9-780-073-66136-0.

A well-balanced and respected text focusing on the place of materials and processing in technical design.

Farag, M.M. (2008). *Materials and process selection for engineering design* (2nd ed.). CRC Press, Taylor and Francis, ISBN 9-781-420-06308-0.
A materials science approach to the selection of materials.

Shackelford, J.F. (2009). *Introduction to materials science for engineers* (7th ed.). Prentice Hall, ISBN 978-0-13-601260-3.
A well-established materials text with a design slant.

Tabor, D. (1978). *Properties of matter*. Penguin Books.
This text, like that of Cottrell, is notable for its clarity and physical insight.

CHAPTER 5

Materials Selection—The Basics

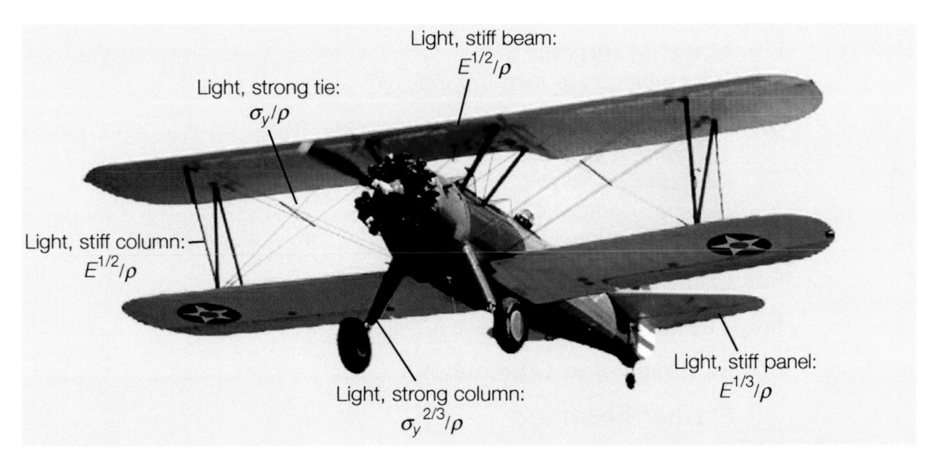

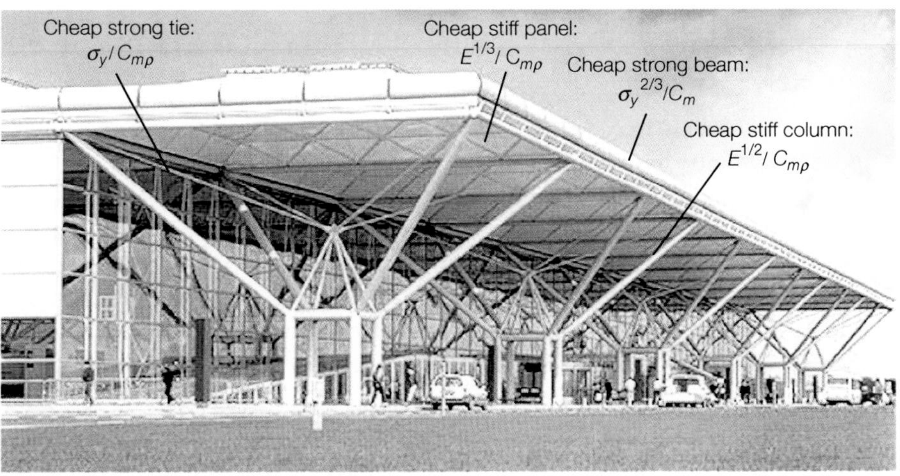

Structures with components loaded in tension, bending, torsion, and compression with material indices. (Image of Stansted Airport courtesy of Norman Foster and Partners, London.)

Materials Selection in Mechanical Design. DOI: 10.1016/B978-1-85617-663-7.00005-9

CONTENTS

5.1 INTRODUCTION AND SYNOPSIS

This chapter sets out the basic procedure for selection, establishing the link between material and function (Figure 5.1). A material has *attributes*: its density, strength, cost, resistance to corrosion, and so forth. A design demands a certain profile of these: a low density, a high strength, a modest cost, and resistance to sea water, perhaps. It is important to start with the full menu of materials as options; failure to do so may mean a missed opportunity. If an innovative choice is to be made, it must be identified early in the design process. Later, too many decisions will have been taken and commitments made to allow radical change: It is now or never.

The task of selection, stated in two lines, is that of

1. identifying the desired attribute profile, and then
2. comparing this with those of real engineering materials to find the best match.

The first step in tackling selection is that of *translation*—examining the design requirements to identify the constraints that they impose on material choice. The immensely wide choice is narrowed, first, by *screening out* the materials that cannot meet the constraints. Further narrowing is achieved by *ranking* the candidates by their ability to maximize performance. This chapter explains how to do both.

The material property charts introduced in Chapter 4 are designed for use with these criteria. Constraints and objectives can be plotted on them, isolating the subset of materials that are the best choice for the design. The whole procedure can be implemented in software as a design tool, allowing computer-aided selection. The procedure is fast and makes for lateral ("what if...?") thinking. Examples of the method are given in Chapter 6.

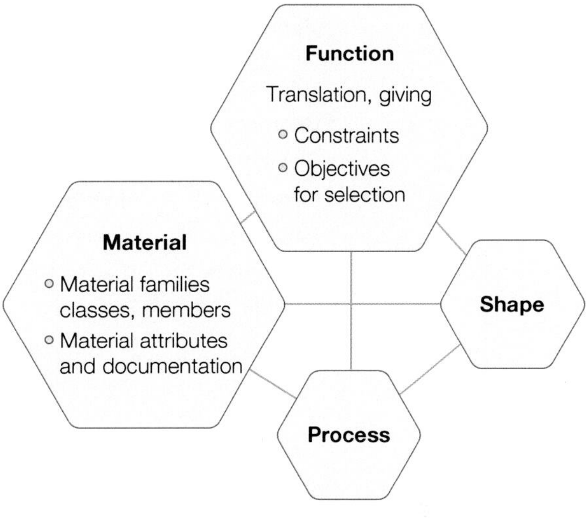

FIGURE 5.1
Material selection is determined by function. Shape sometimes influences selection. This chapter and the next deal with materials selection when this is independent of shape.

5.2 THE SELECTION STRATEGY

Material attributes

Figure 5.2 illustrates how the universe of materials is divided into families, classes, subclasses, and members. Each member is characterized by a set of *attributes*: its properties. As an example, the materials universe contains the family "metals," which in turn contains the class "aluminum alloys," the subclass "6000 series," and finally the particular member "Alloy 6061." It, and every other member of the universe, is characterized by a set of attributes that include its mechanical, thermal, electrical, optical, and chemical properties; its processing characteristics; its cost and availability; and the environmental consequences of its use. We call this its *property profile*. Selection involves seeking the best match between the property profiles of the materials in the universe and the property profile required by the design.

Selection strategies

This chapter is about strategies for selection. It is simpler to start with the selection of a product than with a material; the ideas are the same, but the material has added complications.

You need a new car. To meet your needs it must be a mid-sized four-door family sedan with a gas engine delivering at least 150 horsepower—enough to tow your power boat. Given all of these, you wish it to cost as little to own and run as possible (Figure 5.3, *left*). The three *constraints* are listed on page 101, but they are not all of the same type.

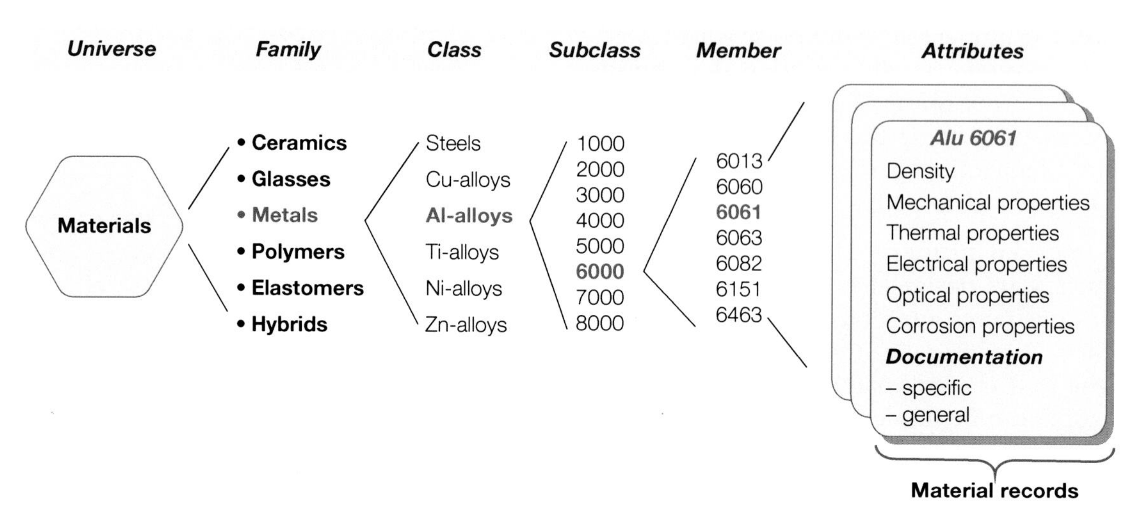

FIGURE 5.2
The taxonomy of the universe of materials and their attributes. Computer-based selection software stores data in a hierarchical structure like this.

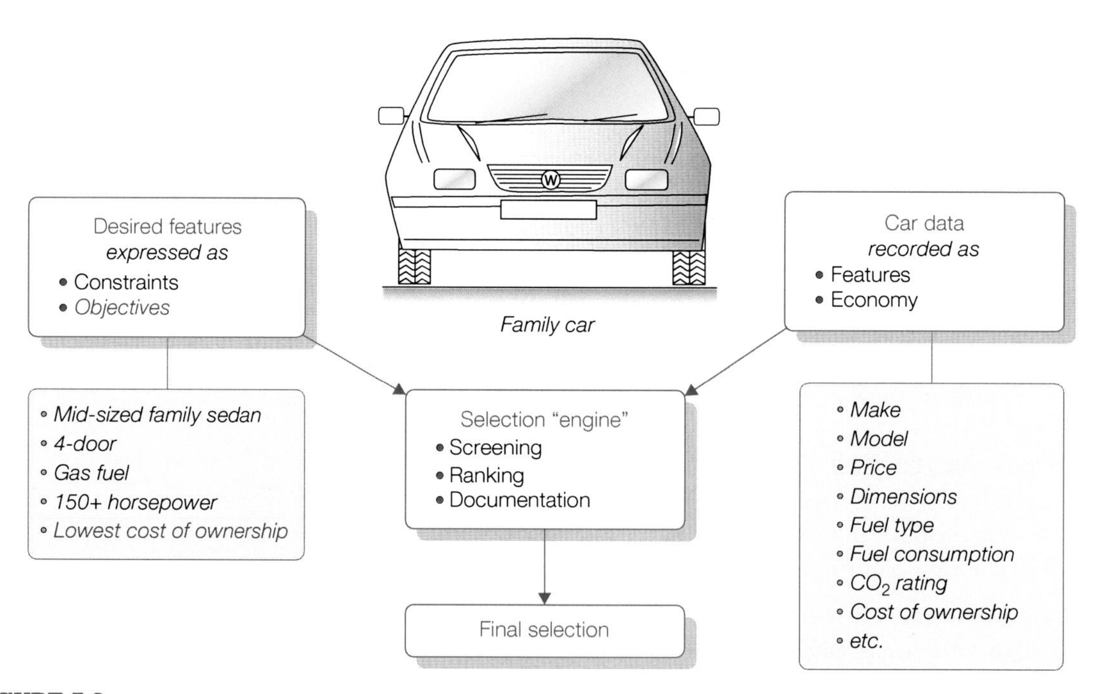

FIGURE 5.3
Choosing a car—an example of the selection strategy. Required features are constraints; they are used to screen out unsuitable cars. The survivors are ranked by cost of ownership.

- The requirements of *four-door family sedan* and *gas power* are *simple constraints;* a car *must* have these to be a candidate.
- The requirement of *at least 150 hp* places a lower limit but no upper limit on power; it is a *limit constraint;* any car with 150 hp or more is acceptable.

The wish for minimum cost of ownership is an *objective,* a criterion of excellence. The most desirable cars, from among those that meet the constraints, are those that minimize this objective.

To proceed you need information about available cars (Figure 5.3, *right*). Car magazines, car makers' websites, and dealers list such information. It includes car type and size, number of doors, fuel type, engine power, and price; car magazines go further and estimate the cost of ownership (meaning the sum of running costs, tax, insurance, servicing, and depreciation), listing it as $/mile or €/km.

Now: decision time (Figure 5.3, *center*). The selection engine (you in this example) uses the constraints to *screen out,* from all the available cars, those that are not four-door gas-powered family sedans with at least 150 hp. Many cars meet these constraints, so the list is still long. You need a way to order it so that the best choices are at the top. That is what the objective is for: It allows you to *rank* the surviving candidates by cost of ownership—those with the lowest values are ranked most highly. Rather than just choosing the one that is cheapest, it is better to keep the top three or four and seek further *documentation,* exploring their other features in depth (delivery time, size of trunk, comfort of seats, guarantee period, and so on) and weighing the small differences in cost against the desirability of these features.

Methods like this are used as a tool for decision making in many fields: in deciding between design options for new products, in optimizing the operating methods for a new plant, in guiding the selection of a site for a new town ... and in selecting materials. We turn to that next.

Selecting materials involves seeking the best match between design requirements and the properties of the materials that might be used to make the design. Figure 5.4 shows the strategy of the last section applied to selecting materials for the protective visor of a safety helmet. On the left are requirements that the material must meet, expressed as constraints and objectives. The constraints: ability to be molded and, of course, transparency. The objective: If the visor is to protect the face, it must be as shatterproof as possible, meaning it must have as high a fracture toughness as possible.

On the right is the database of material attributes, drawn from suppliers' data sheets, handbooks, web-based sources, or software specifically designed

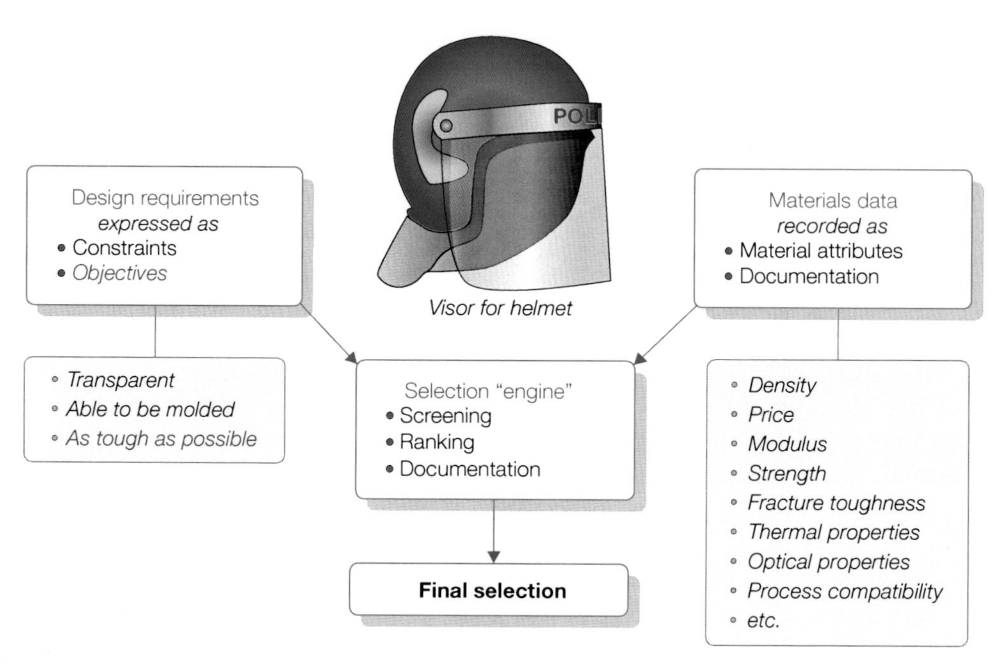

FIGURE 5.4

Choosing a material. Design requirements are first expressed as constraints and objectives. The constraints are used for screening. The survivors are ranked by the objective, expressed as a material index.

for materials selection. The comparison "engine"—you again—applies the constraints on the left to the materials on the right, screening out materials that fail to meet them and delivering a list of viable candidates, just as with cars. The list is then ranked by the fracture toughness. The three or so materials that meet the constraints and have the highest fracture toughness are then explored in depth by seeking documentation for them.

There is, however, a complication. The requirements for the car were straightforward—doors, fuel type, power; all of these are explicitly listed by the manufacturer. The design requirements for a component of a product specify what it should *do* but not what *properties* its materials should have. So the first step in selecting materials is one of *translation*: converting the design requirements (often vague) into constraints and objectives that can be applied to the materials database (Figure 5.5, *top*). The next task is *screening*, as with cars, eliminating the materials that cannot meet the constraints. This is followed by *ranking*, ordering the survivors by their ability to meet a criterion of excellence, such as minimizing cost or maximizing impact resistance. The final task is to explore the most promising candidates in depth, examining how they are used at present, case histories of failures, and how best to design with them; this step is called *documentation*. Now a closer look at each step.

Translation

How are the design requirements for a component (defining what it must do) translated into a prescription for a material? Any engineering component has one or more *functions*: to support a load, to contain a pressure, to transmit heat, and so on. This must be achieved subject to *constraints*: that certain dimensions are fixed, that the component must carry the design loads or pressures without failure, that it insulates or conducts, that it can function in a certain range of temperature and in a given environment, and many more. In designing the component, the designer has an *objective*: to make it as cheap as possible, perhaps, or as light, or as safe, or perhaps some combination of these.

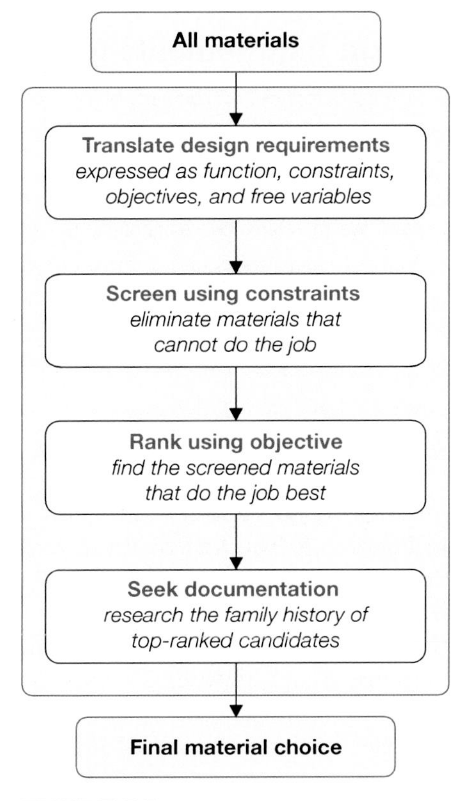

FIGURE 5.5
The strategy for materials selection. The four main steps—translation, screening, ranking, and documentation—are shown here.

Certain parameters can be adjusted to optimize the objective; the designer is free to vary dimensions that have not been constrained by design requirements and, most importanty, free to choose the material for the component. We refer to these as *free variables*. Function, constraints, objectives, and free variables (Table 5.1) define the boundary conditions for selecting a material and—in the case of load-bearing components—a shape for its cross-section. The first step in relating design requirements to material properties is a clear statement of function, constraints, objectives, and free variables.

Table 5.1 Function, Constraints, Objectives, and Free Variables

Function	What does the component do?
Constraints*	What nonnegotiable conditions must be met? What negotiable but desirable conditions must be met?
Objective	What is to be maximized or minimized?
Free variable	Which parameters of the problem is the designer free to change?

**It is sometimes useful to distinguish between "hard" and "soft" constraints. Stiffness and strength might be absolute requirements (hard constraints); cost might be negotiable (soft constraint).*

Design requirements for the helmet visor

A material is required for the visor of a safety helmet to provide maximum facial protection.

Translation

To allow clear vision the visor must be optically transparent. To protect the face from the front, from the sides, and from below it must be doubly curved, requiring that the material can be molded. We thus have two constraints: transparency and ability to be molded.

Fracture of the visor would expose the face to damage; maximizing safety therefore translates into maximizing resistance to fracture. The material property that measures resistance to fracture is the fracture toughness, K_{1c}. The objective is therefore to maximize K_{1c}.

Screening: Attribute limits

Unbiased selection requires that all materials be considered candidates until shown to be otherwise, using the steps in the boxes below "Translate" in Figure 5.5. The first of these, *screening*, eliminates candidates that cannot do the job at all because one or more of their attributes lies outside the limits set by the constraints. As examples, the requirement that "the component must function in boiling water" or that "the component must be transparent" imposes obvious limits on the attributes of *maximum service temperature* and *optical transparency* that successful candidates must meet. We refer to these as *attribute limits*.

Ranking: Material indices

Attribute limits do not, however, help with ordering the candidates that remain. To do this we need optimization criteria. These are found in the material indices, developed next, which measure how well a candidate that has passed the screening step can perform. Performance is sometimes limited by a single property, sometimes by a combination of them. Thus the best materials for buoyancy are those with the lowest density, ρ; those best for thermal insulation are the ones with the smallest values of the thermal conductivity, λ. It is not always the smallest that we want; the best material for a heat exchanger, for instance, is the one with the largest value of λ. Here maximizing or minimizing a single property maximizes performance. But—as we shall see—it is more usual that performance is limited not by one property but by a combination of them. Thus the best materials for a light strong tie-rod or cable are those with the greatest value of the specific strength, σ_f/ρ, where σ_f is failure strength. The best materials for a spring are those with the greatest value of σ_f^2/E, where E is Young's modulus. The property or property group that maximizes performance for a given design is called its *material index*. There are many such indices, each associated with maximizing some aspect of performance.[1] They provide *criteria of excellence* that allow ranking of materials by their ability to perform well in the given application.

To summarize: Screening isolates candidates that are capable of doing the job; ranking identifies those among them that can do the job best.

Screening and ranking for the helmet visor

A search for transparent materials that can be molded delivers the following list. The first four are thermoplastics; the last two, glasses. Fracture toughness values can be found in Appendix A.

Material	Average Fracture Toughness K_{1c} MPa.m$^{1/2}$
Polycarbonate (PC)	3.4
Cellulose acetate (CA)	1.7
Polymethyl methacrylate (acrylic, PMMA)	1.2
Polystyrene (PS)	0.9
Soda-lime glass	0.6
Borosilicate glass	0.6

The constraints have reduced the number of viable materials to six candidates. When ranked by fracture toughness, the top-ranked candidates are PC, CA, and PMMA.

Documentation

The outcome of the steps so far is a ranked short-list of candidates that meet the constraints and that maximize or minimize the criterion of excellence, whichever is required. You could just choose the top-ranked candidate, but what secret vices might it have? What are its strengths and weaknesses? Does it have a good reputation? What, in a word, is its credit rating? To proceed further we seek a detailed profile of each candidate: its *documentation* (Figure 5.5, *bottom*).

Documentation differs greatly from the structured property data used for screening. Typically, it is descriptive, graphical, or pictorial: case studies of previous uses of the material, failure analyses and details of its corrosion, information about availability and pricing, and the like. Such information is found in handbooks, suppliers' data sheets, case studies of use, and failure analyses. Documentation helps narrow the short-list to a final choice, allowing a definitive match to be made between design requirements and material attributes.

Why are all these steps necessary? Without screening and ranking, the candidate pool is enormous and the volume of documentation overwhelming. Dipping

[1] Maximizing performance often means minimizing something: Cost is the obvious example; mass, in transport systems, is another. A low-cost or light component, here, improves performance.

into it, hoping to stumble on a good material, gets you nowhere. But once a small number of potential candidates have been identified by the screening–ranking steps, detailed documentation can be sought for these few alone, and the task becomes viable.

Documentation for materials for the helmet visor

At this point it helps to know how the three top-ranked candidates listed in the last examples box are used. A quick web search reveals the following.

Polycarbonate
Safety shields and goggles; lenses; light fittings; safety helmets; laminated sheet for bullet-proof glazing.

Cellulose Acetate
Spectacle frames; lenses; goggles; tool handles; covers for television screens; decorative trim, steering wheels for cars.

PMMA, Plexiglas
Lenses of all types; cockpit canopies and aircraft windows; containers; tool handles; safety spectacles; lighting, automotive taillights.

This is encouraging: All three materials have a history of use for goggles and protective screening. The one that ranked highest in our list—polycarbonate—has a history of use for protective helmets. We select this material, confident that with its high fracture toughness it is the best choice.

Local conditions

The final choice between competing candidates will often depend on local conditions: in-house expertise or equipment, the availability of local suppliers, and so forth. A systematic procedure cannot help here; the decision must instead be based on local knowledge. This does not mean that the result of the systematic procedure is irrelevant. It is always important to know which material is best, even if for local reasons you decide not to use it.

We explore documentation more fully later. Here we focus on the derivation of property limits and indices.

5.3 MATERIAL INDICES

Constraints set property limits. Objectives define material indices, for which we seek extreme values. When the objective is not coupled to a constraint, the material index is a simple material property. When, instead, the two [better] are coupled, the index becomes a group of properties like those cited above. Where do they come from? This section explains.

Think for a moment of the simplest of mechanical components. The loading on a component can generally be decomposed into some combination of axial tension, bending, torsion, and compression. Almost always, one mode dominates. So common is this that the functional name given to the component describes the way it is loaded: *Ties* carry tensile loads; *beams* and *panels* carry bending moments; *shafts* carry torques; *columns* carry compressive axial loads. The words "tie," "beam," "shaft," and "column" each imply a function. Here we explore constraints, objectives, and resulting material indices for some of these.

The life energy and emissions for transport systems are dominated by the fuel consumed during use. The lighter the system is made, the less fuel it consumes and the less carbon it emits. So a good starting point is *minimum weight design*, subject, of course, to the other necessary constraints, of which the most important here have to do with stiffness and strength. We consider the generic components shown in Figure 5.6: a tie, a panel, and beams, loaded as shown.

Minimizing mass: A light, strong tie A design calls for a tie like those of the biplane in the cover picture. It must carry a tensile force F^* without failure and be as light as possible (Figure 5.6(a)). The length L is specified but the cross-section area A is not. Here, "maximizing performance" means "minimizing the mass while still carrying the load F^* safely." The design requirements, translated, are listed in Table 5.2.

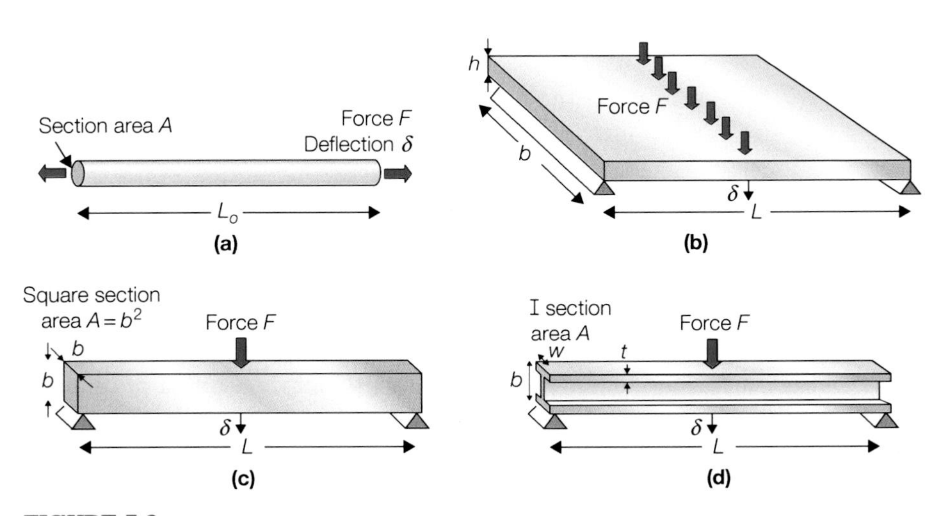

FIGURE 5.6
Generic components: (a) a tie, a tensile component; (b) a panel, loaded in bending; (c) and (d) beams, loaded in bending.

Table 5.2 Design Requirements for the Light, Strong Tie

Function	Tie rod
Constraints	Length L is specified (geometric constraint) Tie must support axial tensile load F^* without failing (functional constraint)
Objective	Minimize the mass m of the tie
Free variables	Cross-section area A Choice of material

We first seek an equation describing the quantity to be maximized or minimized. Here it is the mass m of the tie and it is a minimum that we seek. This equation, called *the objective function,* is

$$m = A\,L\,\rho \tag{5.1}$$

where A is the area of the cross-section and ρ is the density of the material of which it is made. The length L and force F are specified and are therefore fixed; the cross-section A, is free. We can reduce the mass by reducing the cross-section, but there is a constraint: A must be sufficient to carry F^*, requiring that

$$\frac{F^*}{A} \le \sigma_f \tag{5.2}$$

where σ_f is the failure strength. Eliminating A between these two equations gives

$$m \ge (F^*)(L)\left(\frac{\rho}{\sigma_f}\right) \tag{5.3}$$

Functional constraint → (F^*); *Geometric constraint* → (L); *Material properties* → $\left(\frac{\rho}{\sigma_f}\right)$

Note the form of this result. The first bracket contains the specified load F. The second bracket contains the specified geometry (length L of the tie). The last bracket contains the material properties. The lightest tie that will carry F^* safely[2] is that made of the material with the smallest value of ρ/σ_f. We could define this as the material index of the problem, seeking a minimum, but it is more usual when dealing with specific properties to express them in a form for which a maximum is sought. We therefore invert the material

[2] In reality a safety factor, S_f, is always included in such a calculation, so that Equation (5.2) becomes $F/A = \sigma_f/S_f$. If the same safety factor is applied to each material, its value does not influence the choice. We omit it here for simplicity.

properties in Equation (5.3) and define the material index M_t (subscript t for tie) as:

$$M_{t1} = \frac{\sigma_f}{\rho} \tag{5.4}$$

The lightest tie-rod that will carry F^* without failing is that with the largest value of this index, the "specific strength," plotted in Figure 4.6. A similar calculation for a light *stiff* tie (one for which the stiffness S rather then the strength σ_f is specified) leads to the index

$$M_{t2} = \frac{E}{\rho} \tag{5.5}$$

where E is Young's modulus. This time the index is the "specific stiffness," also shown in Figure 4.6. The material group (rather than just a single property) appears as the index in both cases because minimizing the mass m—the objective—was coupled to the constraints of carrying the load F without failing or deflecting too much.

Note the procedure. The length of the rod is specified but we are free to choose the cross-section area A. The objective is to minimize its mass m. We write an equation for m: It is the objective function. But there is a constraint: The rod must carry the load F without yielding (in the first example) or bending too much (in the second). Use this to eliminate the free variable A and read off the combination of properties, M, to be maximized. It sounds easy and it is, so long as you are clear from the start what the constraints are, what you are trying to maximize or minimize, which parameters are specified, and which are free.

That was easy. Now for some slightly more difficult (and important) examples.

Minimizing Mass: A light, stiff panel A *panel* is a flat slab, like a table top. Its length L and width b are specified but its thickness is free. It is loaded in bending by a central load F (see Figure 5.6(b)). The stiffness constraint requires that it must not deflect more than δ . The objective is to achieve this with minimum mass, m. Table 5.3 summarizes the design requirements.

Table 5.3 Design Requirements for a Light, Stiff Panel

Function	Panel
Constraints	Bending stiffness S^* specified (functional constraint) Length L and width b specified (geometric constraints)
Objective	Minimize mass m of the panel
Free variables	Panel thickness h Choice of material

The objective function for the mass of the panel is the same as that for the tie:

$$m = A L \rho = b h L \rho$$

Its bending stiffness S must be at least S^*:

$$S = \frac{C_1 E I}{L^3} \geq S^* \tag{5.6}$$

Here C_1 is a constant that depends only on the distribution of the loads—we don't need its value (you can find it in Appendix B). The second moment of area, I, for a rectangular section is

$$I = \frac{b h^3}{12} \tag{5.7}$$

We can reduce the mass by reducing h, but only so far that the stiffness constraint is still met. Using the last two equations to eliminate h in the objective function gives

$$m = \left(\frac{12 S^*}{C_1 b}\right)^{1/3} (b L^2) \left(\frac{\rho}{E^{1/3}}\right) \tag{5.8}$$

Functional constraint (pointing to $\left(\frac{12 S^*}{C_1 b}\right)^{1/3}$); *Geometric constraints* (pointing to (bL^2)); *Material properties* (pointing to $\left(\frac{\rho}{E^{1/3}}\right)$)

The quantities S^*, L, b, and C_1 are all specified; the only freedom of choice left is that of the material. The index is the group of material properties, which we invert such that a maximum is sought: The best materials for a light, stiff panel are those with the greatest values of

$$M_{p1} = \frac{E^{1/3}}{\rho} \tag{5.9}$$

Repeating the calculation with a constraint of strength rather than stiffness leads to the index

$$M_{p1} = \frac{\sigma_y^{1/2}}{\rho} \tag{5.10}$$

These don't look much different from the previous indices, E/ρ and σ_y/ρ, but they are: They lead to different choices of material, as we shall see in a moment.

Now for another bending problem in which shape plays a role.

Minimizing mass: A light, stiff beam Beams come in many shapes: solid rectangles, cylindrical tubes, I-beams, and more. Some of these have too many free geometric variables to apply the previous method directly. However, if we constrain the shape to be *self-similar* (such that all dimensions of the cross-section change in proportion as we vary the overall size), the problem becomes tractable again. We therefore consider beams in two stages: first, we identify the optimum materials for a light, stiff beam of a prescribed simple shape (a square section); second, we explore how much lighter it could be made for the same stiffness by using a more efficient shape.

Consider a beam of square section $A = b \times b$ that may vary in size but the square shape is retained. It is loaded in bending over a span of fixed length L with a central load F (see Figure 5.6(c)). The stiffness constraint is again that it must not deflect more than δ under F, with the objective that the beam should again be as light as possible. Table 5.4 summarizes the design requirements.

Proceeding as before, the objective function for the mass is

$$m = A L \rho = b^2 L \rho \tag{5.11}$$

The bending stiffness S of the beam must be at least S^*:

$$S = \frac{C_2 EI}{L^3} \geq S^* \tag{5.12}$$

where C_2 is a constant (Appendix B). The second moment of area, I, for a square section beam is

$$I = \frac{b^4}{12} = \frac{A^2}{12} \tag{5.13}$$

For a given L, S^* is adjusted by altering the size of the square section. Now eliminating b (or A) in the objective function for the mass gives

$$m = \left(\frac{12 S^* L^3}{C_2}\right)^{1/2} (L) \left(\frac{\rho}{E^{1/2}}\right) \tag{5.14}$$

Table 5.4 Design Requirements for a Light, Stiff Beam

Function	Beam
Constraints	Length L is specified (geometric constraint) Section shape square (geometric constraint) Beam must support bending load F without deflecting too much, meaning that bending stiffness S is specified as S^* (functional constraint)
Objective	Minimize mass m of the beam
Free variables	Cross-section area A Choice of material

The quantities S^*, L, and C_2 are all specified or constant; the best materials for a light, stiff beam are those with the largest values of index M_b, where

$$M_{b_1} = \frac{E^{1/2}}{\rho} \tag{5.15}$$

Repeating the calculation with a constraint of strength rather than stiffness leads to the index

$$M_{b_2} = \frac{\sigma_y^{2/3}}{\rho} \tag{5.16}$$

This analysis was for a square beam, but the result in fact holds for any shape so long as the shape is held constant. This is a consequence of Equation (5.13); for a given shape, the second moment of area I can always be expressed as a constant times A^2, so changing the shape just changes the constant C_2 in Equation (5.14), not the resulting index.

As noted above, real beams have section shapes that improve their efficiency in bending, requiring less material to get the same stiffness. By shaping the cross-section it is possible to increase I without changing A. This is achieved by locating the material of the beam as far from the neutral axis as possible, as in thin-walled tubes or I-beams (see Figure 5.6 (d)). Some materials are more amenable than others to being made into efficient shapes. Comparing materials on the basis of the index in M_b therefore requires some caution; materials with lower index values may "catch up" by being made into more efficient shapes. We examine this in more detail in Chapter 9.

Minimizing material cost: Cheap ties, panels, and beams When the objective is to minimize cost rather than mass, the indices change again. If the material price is C_m \$/kg, the cost of the material to make a component of mass m is just mC_m. The objective function for the material cost C of the tie, panel or beam then becomes

$$C = m\,C_m = A\,L\,C_m\,\rho \tag{5.17}$$

Proceeding as before leads to indices that have the form of Equations (5.4), (5.5), (5.9), (5.10), (5.15), and (5.16), with ρ replaced by $C_m\,\rho$. Thus the index guiding material choice for a tie of specified strength and minimum material cost is

$$M = \frac{\sigma_f}{C_m\,\rho} \tag{5.18}$$

where C_m is the material price per kg. The index for a cheap stiff panel is

$$M_{p1} = \frac{E^{1/3}}{C_m \rho} \tag{5.19}$$

and so forth (It must be remembered that the material cost is only part of the cost of a shaped component; there is also the manufacturing cost—the cost to shape, join, and finish it.)

Associating material indices with components The components in the cover images of this chapter are labeled with the type of loading support and with the index that guides the choice of material to make them. The biplane typifies lightweight design, meaning that its materials are chosen to carry the design loads at minimum mass. The airport structure uses very large quantities of materials: Here the objective is to carry the design loads safely while minimizing the cost of the material. The guiding indices for each structure are derived from a single objective: minimizing mass in one case, minimizing material cost in the other. Often a design involves more than one objective: in choosing materials for the frame of a bicycle you might wish to minimize both the weight *and* the cost. That requires trade-off methods, the subject of Chapter 7.

How general are material indices?

This is a good moment to describe the method in more general terms. *Structural elements* are components that perform a physical function: They carry loads, transmit heat, store energy and so on. In short, they satisfy *functional requirements*. We have already identified examples: A tie must carry a specified tensile load; a spring must provide a given restoring force or store a given energy; a heat exchanger must transmit heat with a given heat flux, and so on.

The performance of a structural element is determined by three things: the functional requirements, the geometry, and the properties of the material of which it is made.[3] The performance P of the element is described by an equation of the form

$$P = \left[\left(\begin{matrix}\text{Functional}\\ \text{requirements}, F\end{matrix}\right), \left(\begin{matrix}\text{Geometric}\\ \text{parameters}, G\end{matrix}\right), \left(\begin{matrix}\text{Material}\\ \text{properties}, M\end{matrix}\right)\right]$$

or

$$P = f(F, G, M) \tag{5.20}$$

[3] In Chapter 9 we introduce a fourth: section shape.

where P, the *performance metric*, describes some aspect of the performance of the component: its mass, volume, cost, or life, for example; and f means *"a function of." Optimum design* is the selection of the material and geometry that maximize or minimize P, according to its desirability or otherwise.

The three groups of parameters in Equation (5.20) are said to be *separable* when the equation can be written

$$P = f_1(F) \cdot f_2(G) \cdot f_3(M) \tag{5.21}$$

where f_1, f_2, and f_3 are separate functions that are simply multiplied together. It turns out that, commonly, they are. When this is so the optimum choice of material becomes independent of the details of the design; it is the same for *all* geometries, G, and for all values of the function requirement, F. Then the optimum subset of materials can be identified without solving the complete design problem, or even knowing all the details of F and G. This enables enormous simplification: The performance for *all* F and G is maximized by maximizing f_3 (M), which is called the material efficiency coefficient, or *material index* for short. The remaining bit, f_1 $(F) \cdot f_2$ (G), is related to *the structural efficiency coefficient*, or *structural index*. We don't need it now, but we will examine it briefly in Section 5.6.

Each combination of function, objective, and constraint leads to a material index (Figure 5.7); the index is characteristic of the combination and thus of the function the component performs. The method is general and, in later

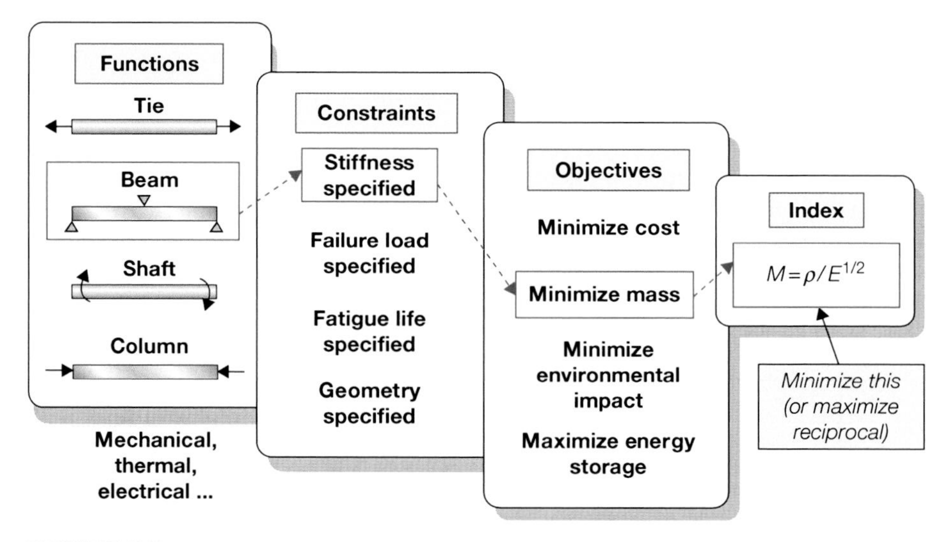

FIGURE 5.7
The specification of function, objective, and constraint leads to a materials index. The combination in the highlighted boxes leads to the index $E^{1/2}/\rho$.

Table 5.5 Examples of Material Indices

Function, Objective, and Constraints	Index
Tie, minimum weight, stiffness prescribed	$\frac{E}{\rho}$
Beam, minimum weight, stiffness prescribed	$\frac{E^{1/2}}{\rho}$
Beam, minimum weight, strength prescribed	$\frac{\sigma_y^{2/3}}{\rho}$
Beam, minimum cost, stiffness prescribed	$\frac{E^{1/2}}{C_m\rho}$
Beam, minimum cost, strength prescribed	$\frac{\sigma_y^{2/3}}{C_m\rho}$
Column, minimum cost, buckling load prescribed	$\frac{E^{1/2}}{C_m\rho}$
Spring, minimum weight for given energy storage	$\frac{\sigma_y^2}{E\rho}$
Thermal insulation, minimum cost, heat flux prescribed	$\frac{1}{\lambda C_p\rho}$
Electromagnet, maximum field, temperature rise prescribed	$\frac{C_p\rho}{\rho_e}$

ρ = density; E = Young's modulus; σ_y = elastic limit; C_m = cost/kg; λ = thermal conductivity; ρ_e = electrical resistivity; C_p = specific heat

chapters, is applied to a wide range of problems. Table 5.5 gives examples of indices and the design problems that they characterize. A fuller catalog of indices is given in Appendix C. New problems throw up new indices, as the case studies of the next chapter will show.

5.4 THE SELECTION PROCEDURE

We can now assemble the four steps into a systematic procedure.

Translation and deriving the index

Table 5.6 lists the steps. Simplified: Identify the material attributes that are constrained by the design, decide what you will use as a criterion of excellence (to be minimized or maximized), substitute for any free variables using one of the constraints, and read off the combination of material properties that optimize the criterion of excellence.

Screening: Applying attribute limits

Any design imposes certain non-negotiable demands ("constraints") on the material of which it is made. We have explained how these are translated

Table 5.6 Translation

Step No.	Action
1	Define the design requirements: *Function:* What does the component do? *Constraints:* Essential requirements that must be met: e.g., stiffness, strength, corrosion resistance, forming characteristics, etc. *Objective:* What is to be maximized or minimized? *Free variables:* Which are the unconstrained variables of the problem?
2	List the constraints (no yield, no fracture, no buckling, etc) and develop an equation for them if necessary.
3	Develop an equation for the objective in terms of the functional requirements, the geometry, and the material properties (*objective function*).
4	Identify the free (unspecified) variables.
5	Substitute the free variables from the constraint equations into the objective function.
6	Group the variables into three groups: functional requirements F, geometry G, and material properties M; thus Performance metric $P \leq f_1(F) \cdot f_2(G) \cdot f_3(M)$ or performance metric $P \leq f_1(F) \cdot f_2(G) \cdot f_3(M)$
7	Read off the material index, expressed as a quantity M that optimizes the performance metric P. M is the criterion of excellence.

into attribute limits. Attribute limits plot as horizontal or vertical lines on material selection charts, illustrated in Figure 5.8. It shows a schematic $E - \rho$ chart, in the manner of Chapter 4. We suppose that the design imposes limits on these of $E > 10$ GPa and $\rho < 3{,}000$ kg/m^3, shown in the figure. The optimizing search is restricted to the window boxed by the limits, labeled "Search region."

Less quantifiable properties such as corrosion resistance, wear resistance, or formability can all appear as limits, which take the form

$$A > A^*$$

or

$$A < A^* \tag{5.22}$$

where A is an attribute (service temperature, for instance) and A^* is a critical value of that attribute, set by the design, that must be exceeded or (in the case of corrosion rate) must *not* be exceeded.

One should not be too hasty in applying attribute limits; it may be possible to engineer a route around them. A component that gets too hot can be cooled; one that corrodes can be coated with a protective film. Many designers apply attribute limits for fracture toughness, K_{1c}, and ductility, ϵ_f, insisting on materials with, as rules of thumb, $K_{1c} > 15$ MPa.m$^{1/2}$ and $\epsilon_f > 2\%$ in order to guarantee adequate tolerance to stress concentrations.

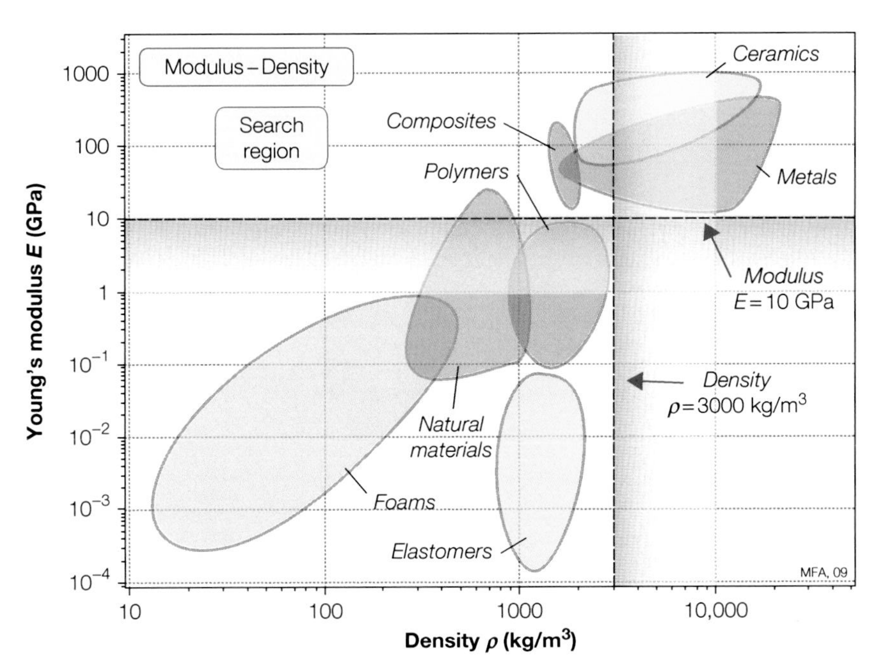

FIGURE 5.8
A schematic $E - \rho$ chart showing a lower limit for E and an upper limit for ρ.

By doing this they eliminate materials that the more innovative designer is able to use to good purpose (the limits just cited for K_{1c} and ϵ_f eliminate all polymers and all ceramics, a rash step too early in the design). At this stage, keep as many options open as possible.

Ranking: Indices on charts

The next step is to seek, from the subset of materials that meet the property limits, those materials that maximize performance. We will use the design of light, stiff components as an example; the other material indices are used in a similar way.

Figure 5.9 shows, as before, the modulus E plotted against density ρ on log scales. The material indices E/ρ, $E^{1/2}/\rho$, and $E^{1/3}/\rho$ can be plotted onto the figure. The condition

$$\frac{E}{\rho} = C$$

or, taking logs,

$$\text{Log}(E) = \text{Log}(\rho) + \text{Log}(C) \tag{5.23}$$

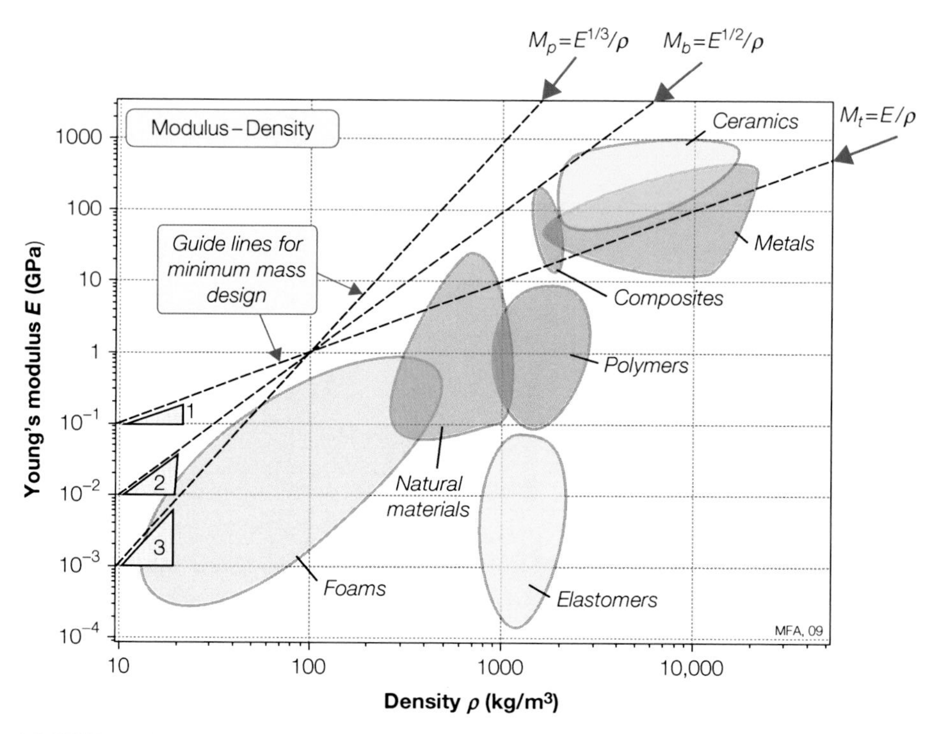

FIGURE 5.9
A schematic $E - \rho$ chart showing guide lines for the three material indices for stiff, lightweight design.

describes a family of straight parallel lines of slope 1 on a plot of Log(E) against Log(ρ), each line corresponding to a value of the constant C. The condition

$$\frac{E^{1/2}}{\rho} = C \tag{5.24}$$

or, taking logs again,

$$\text{Log}(E) = 2\ \text{Log}(\rho) + 2\ \text{Log}(C) \tag{5.25}$$

gives another set, this time with a slope of 2, and

$$\frac{E^{1/3}}{\rho} = C \tag{5.26}$$

gives yet another set, with slope 3. We shall refer to these lines as *selection guide lines*. They give the slope of the family of parallel lines belonging to that index. Where appropriate, the charts of Chapter 4 show guide lines like these.

It is now easy to read off the subset materials that optimally maximize performance for each loading geometry. All the materials that lie on a line of

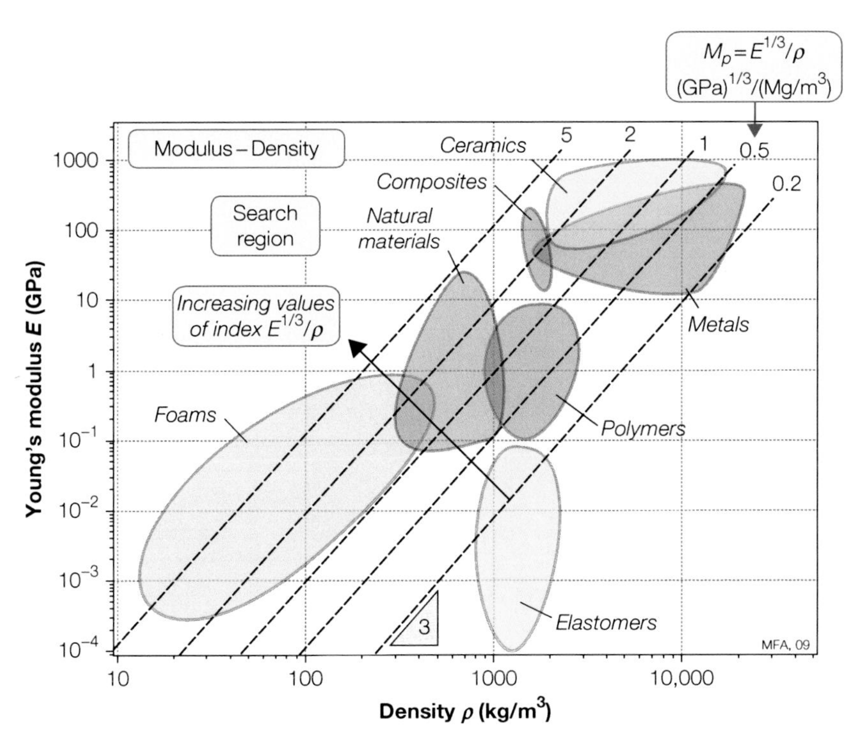

FIGURE 5.10
A schematic $E - \rho$ chart showing a grid of lines for the material index $M = E^{1/3}/\rho$.

constant $E^{1/3}/\rho$ perform equally well as a light, stiff panel; those above the line are better; those below, worse. Figure 5.10 shows a grid of lines corresponding to values of $E^{1/3}/\rho$ from 0.2 to 5 in units of $GPa^{1/3}/(Mg.m^{-3})$. A material with $M = 2$ in these units gives a panel that has one-tenth the weight of one with $M = 0.2$ that has the same stiffness. The subset of materials with particularly good index values is identified by picking a line that isolates a search area containing a reasonably small number of candidates, as shown schematically in Figure 5.11 as a diagonal selection line. Attribute limits can be added, narrowing the search window: That corresponding to $E > 50$ GPa is shown as a horizontal line. The materials lying in the search region meet both criteria. Their number is expanded or contracted by moving the index line down or up.

Documentation

We now have a ranked shortlist of potential candidate materials. The last step is to explore their character in depth. The list of constraints usually contains some that cannot be expressed as simple attribute limits. Many of

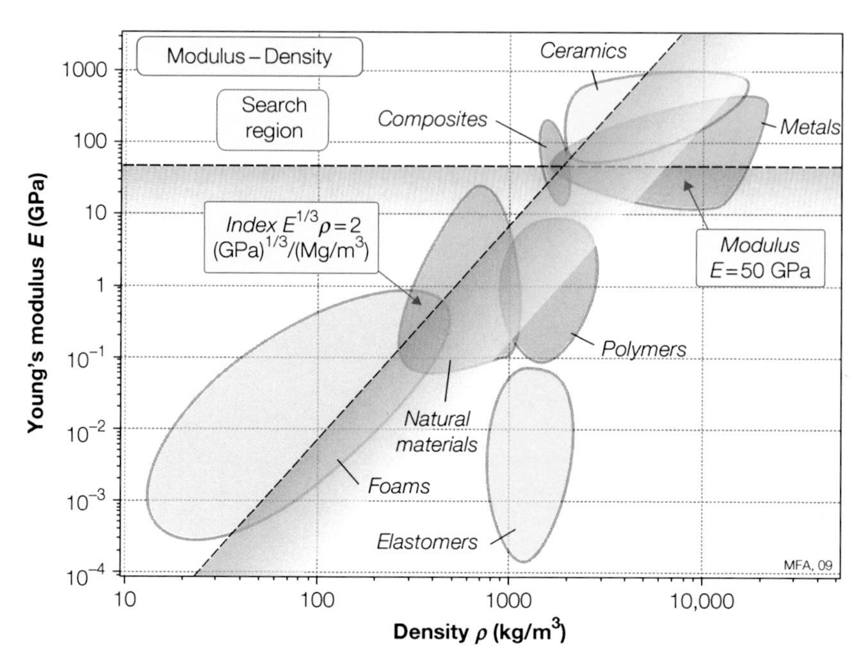

FIGURE 5.11
A selection based on the index $M = E^{1/3}/\rho > 2\ (GPa)^{1/3}\ (Mg/m^3)$ together with the property limit $E > 50$ GPa. The materials contained in the search region become the candidates for the next stage of the selection process.

these relate to the behavior of the material in a given environment or to aspects of the ways in which the material can be shaped, joined, or finished. Such information can be found in handbooks, manufacturers' data sheets and computer-based sources.

And there will be constraints that, at this point, have been overlooked simply because they were not seen as such. Confidence is built by seeking design guidelines, case studies, or failure analyses that document each candidate, building a dossier of its strengths and weaknesses and ways in which these can be overcome. All of these come under the heading of documentation, sources of which are listed in Appendix D.

5.5 COMPUTER-AIDED SELECTION

The charts in Chapter 4 give an overview of material properties, but the number of materials that can be shown on any one of them is obviously limited. Selection using them is practical when there are very few constraints. When there are many—as there usually are—it becomes cumbersome. Both problems are overcome by computer implementation of the methods.

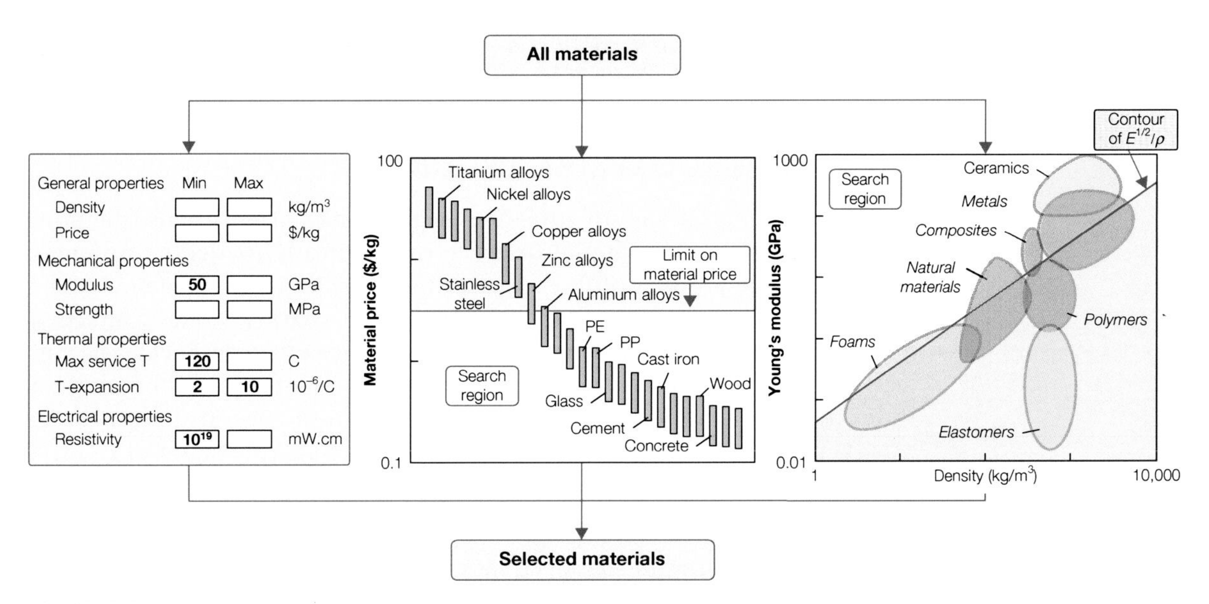

FIGURE 5.12
Computer-aided selection using the CES software. The schematic shows the three types of selection window. They can be used in any order and any combination. The selection engine isolates the subset of materials that passes all the selection stages.

The *CES* material and process selection software[4] is an example of such an implementation. A database contains records for materials, organized in the hierarchical manner shown earlier in Figure 5.2. Each record contains structured attribute data for a material, each attribute stored as a range spanning its typical (or often its permitted) range. It also contains limited documentation in the form of text, images, and references to sources of information about the material. The data are interrogated by a search engine that offers search interfaces shown schematically in Figure 5.12. On the left is a simple query interface for screening of single properties (the cover image of Chapter 3 is an example). The desired upper or lower limits for constrained attributes are entered; the search engine rejects all materials with attributes that lie outside the limits. In the center is shown a second way of interrogating the data: a bar chart like that shown earlier in Figure 4.1.

The bar chart, and the bubble chart in Figure 5.12 (*right*), are ways of both applying constraints and ranking. Used for ranking, a selection line or box is superimposed on the charts with edges that lie at the constrained values of the attributes, eliminating the material in the shaded areas and leaving the materials that meet all the constraints. If ranking is sought instead (having already applied all of the necessary constraints), the line or box is

[4] Granta Design Ltd., Cambridge, U.K. (*www.grantadesign.com*).

positioned so that a few materials—say three—are left in the selected area; these are the top-ranked candidates.

5.6 THE STRUCTURAL INDEX

Books on optimal design of structures (e.g., Shanley, 1960) make the point that the efficiency of material usage in mechanically loaded components depends on the product of three factors: the material index, as defined here; a factor describing section shape, the subject of Chapter 9; and a *structural index*,[5] which contains elements of the G and F of Equation (5.21). The subjects of this book—material and process selection—focus on the material index and on shape; however, we should examine the structural index briefly, partly to make the connection with the classical theory of optimal design and partly because it becomes useful (even to us) when structures are scaled in size.

In design for minimum mass (Equations (5.3), (5.8), and (5.14)), a measure of the efficiency of the design is given by the quantity m/L^3. Equation (5.3) for the light strong tie, when divided by L^3, becomes

$$\frac{m}{L^3} \geq \left(\frac{F^*}{L^2}\right)\left(\frac{\rho}{\sigma_f}\right) \tag{5.27}$$

Equation (5.8) for the light stiff panel becomes

$$\frac{m}{L^3} \geq \left(\frac{12}{C_1}\right)^{1/3}\left(\frac{b^2 S^*}{L^3}\right)^{1/3}\left(\frac{\rho}{E^{1/3}}\right) \tag{5.28}$$

and Equation (5.14) for the light stiff beam becomes

$$\frac{m}{L^3} \geq \left(\frac{12}{C_2}\right)^{1/2}\left(\frac{S^*}{L}\right)^{1/2}\left(\frac{\rho}{E^{1/2}}\right) \tag{5.29}$$

This m/L^3 has the dimensions of density; the lower this pseudodensity the lighter the structure for a given scale and thus the greater the structural efficiency. The first bracketed terms on the right-hand side of Equations (5.28) and (5.29) are constants. The last bracketed term in all three equations is the material index. The remaining term, F^*/L^2 in Equation (5.27), S^*b^2/L^3, in (5.28) and S^*/L in (5.29), is called the *structural index*. It has the dimensions of stress; it is a measure of the intensity of loading. Design proportions that are optimal, minimizing material usage, are optimal for structures of any size provided they all have the same structural index. The performance equations

---

[5] Also called the "structural loading coefficient," the "strain number," or the "strain index."

are written here in a way that isolates the structural index, a convention we shall follow in the case studies of Chapter 6.

The structural index for a component of minimum cost is the same as that for one of minimum mass.

5.7 SUMMARY AND CONCLUSIONS

Material selection is tackled in four steps:

- *Translation*—reinterpreting the design requirements in terms of function, constraints, objectives, and free variables.
- *Screening*—deriving attribute limits from the constraints and applying these to isolate a subset of viable materials.
- *Ranking*—ordering the viable candidates by the value of a material index, the criterion of excellence that maximizes or minimizes some measure of performance.
- *Documentation*—seeking documentation for the top-ranked candidates, exploring aspects of their history, their established uses, their behavior in relevant environments, their availability, and more, until a sufficiently detailed picture is built up that a final choice can be made.

Hard-copy material charts allow a first go at the task, and have the merit of maintaining breadth of vision: All material classes are in the frame, so to speak. But the number of materials is large, they have many properties, and the number of combinations of these appearing in indices is very much larger. It is impractical to print charts to include them all. These problems are overcome by computer implementation, allowing freedom to explore the whole universe of materials and providing detail when required.

The selection procedure described here is extended in Chapter 7 to deal with multiple constraints and objectives, and in Chapter 9 to include section shape. Before moving on to these, it is a good idea to consolidate the ideas so far by applying them to a number of case studies. These follow in Chapter 6.

5.8 FURTHER READING

The following books discuss optimization methods and their application in materials engineering.

Arora, J. S. (1989). *Introduction to optimum design*. McGraw-Hill, ISBN 0-07-002460-X.
An introduction to the terminology and methods of optimization theory.

Ashby, M. F., Shercliff, H. R., & Cebon, D. (2007). *Materials: Engineering, science, processing and design*. Butterworth-Heinemann, ISBN 978-0-7506-8391-3.
An elementary text that introduces materials through material property charts and develops the selection methods through case studies.

Dieter, G. E. (1999). *Engineering design, a materials and processing approach* (3rd ed.). McGraw-Hill, ISBN 9-780-073-66136-0.

A well-balanced and respected text focusing on the place of materials and processing in technical design.

Gordon, J. E. (1976). *The new science of strong materials, or why you don't fall through the floor* (2nd ed.). Penguin Books, ISBN 0-1402-0920-7.

This very readable book presents ideas about plasticity and fracture and describes ways of designing materials to prevent them.

Gordon, J. E. (1978). *Structures, or why things don't fall down*. Penguin Books, ISBN 0-1402-1961-7.

A companion to the other book by Gordon (above), this time introducing structural design.

Shanley, F. R. (1960). *Weight-strength analysis of aircraft structures* (2nd ed.). Dover Publications, Library of Congress Number 60-50107.

A remarkable text, no longer in print, on the design of lightweight structures.

CHAPTER 6

Case Studies: Materials Selection

Image of rower on the Cam in Cambridge, UK, courtesy of Andrew Dunn.

Materials Selection in Mechanical Design. DOI: 10.1016/B978-1-85617-663-7.00006-0

CONTENTS

6.1 INTRODUCTION AND SYNOPSIS

Here we have a collection of case studies illustrating the selection methods of Chapter 5. They are deliberately simplified to avoid obscuring the method under layers of detail. In most cases little is lost by this: The best choice of material for the simple example is the same as that for the more complex, for the reasons given in Section 5.3.

Each case study is laid out in the following way:

- *The problem statement,* setting the scene
- *The translation,* identifying function, constraints, objectives, and free variables, from which emerge the attribute limits and material indices

- *The selection,* in which the full menu of materials is reduced by screening and ranking to a shortlist of viable candidates
- *The postscript,* allowing a commentary on results and philosophy

The first few examples are straightforward, chosen to illustrate the method. Later examples are less obvious and require clear thinking to identify and distinguish objectives and constraints. Confusion here can lead to bizarre and misleading conclusions. Always apply common sense: Does the selection include the traditional materials used for that application? Are some members of the subset obviously unsuitable? If they are, it is usually because a constraint has been overlooked or an objective misapplied. The answer is to rethink them.

Most of the case studies use the hard-copy charts of Chapter 4; those at the end illustrate computer-based methods.

6.2 MATERIALS FOR OARS

Credit for inventing the rowed boat seems to belong to the Egyptians. Boats with oars appear in carved relief on monuments built in Egypt between 3300 and 3000 B.C. Boats, before steam power, could be propelled by poling, by sail, or by oar. Oars gave more control than the other two, the military potential of which was well understood by the Romans, the Vikings, and the Venetians.

Records of rowing races on the Thames in London extend back to 1716. Originally the competitors were watermen, rowing the ferries used to carry people and goods across the river. Gradually gentlemen became involved (notably the young gentlemen of Oxford and Cambridge), sophisticating both the rules and the equipment. The real stimulus for development of boats and oars came in 1900 with the establishment of rowing as an Olympic sport. Since then both have drawn to the fullest on the craftsmanship and materials of their day. Consider, as an example, the oar.

The translation Mechanically speaking, an oar is a beam, loaded in bending. It must be strong enough to carry, without breaking, the bending moment exerted by the oarsman; it must have a stiffness to match the rower's own characteristics; and it must give the right "feel." Meeting the strength constraint is easy. Oars are designed on *stiffness,* that is, to give a specified elastic deflection under a given load.

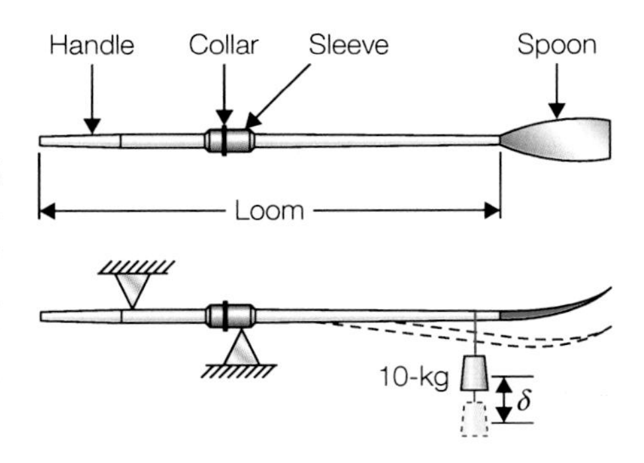

FIGURE 6.1
An oar. Oars are designed on stiffness, measured in the way shown in the *lower* figure, and they must be light.

Figure 6.1 (*top*) shows an oar: A blade or "spoon" is bonded to a shaft or "loom" that carries a sleeve and collar to give positive location in the rowlock. The lower part of the figure shows how the oar stiffness is measured: A 10-kg weight is hung on the oar 2.05 m from the collar and the deflection δ at this point is measured. A soft oar will deflect nearly 50 mm; a hard one only 30. A rower, ordering an oar, will specify how hard it should be.

In addition, the oar must be light; extra weight increases the wetted area of the hull and the drag that goes with it. So there we have it: an oar is a beam of specified stiffness and minimum weight. The material index we want was derived in Chapter 5 as Equation (5.15). It is that for a light, stiff beam:

$$M = \frac{E^{1/2}}{\rho} \tag{6.1}$$

where E is Young's modulus and ρ is the density. There are other obvious constraints. Oars are dropped, and blades sometimes clash. The material must be tough enough to survive this, so brittle materials (those with a toughness G_{1C} less than 1 kJ/m^2) are unacceptable. Given these requirements, summarized in Table 6.1, what materials would you choose to make oars?

The selection Figure 6.2 shows the appropriate chart: that in which Young's modulus is plotted against density ρ. The selection line for the index M has a slope of 2, as explained in Section 5.4; it is positioned so that a small group of materials is left above it. They are the materials with the largest values of M and represent the best choice, provided they satisfy the other constraint (a simple attribute limit on toughness). They contain three classes of material: woods, carbon-reinforced polymers, and certain ceramics (Table 6.2). Ceramics are brittle; the toughness-modulus chart in Figure 4.7 shows that all fail to meet the requirements of the design. The recommendation is clear. Make your oars out of wood or—better—out of CFRP.

Postscript Now we know what oars should be made of. What, in reality, is used? Racing oars and sculls are made either of wood or of a high performance composite: carbon-fiber–reinforced epoxy.

Wooden oars are made today, as they were 100 years ago, by craftsmen working mainly by hand. The shaft and blade are of Sitka spruce from the

Table 6.1 Design Requirements for the Oar

Function	Oar—meaning light, stiff beam
Constraints	Length L specified Bending stiffness S^* specified Toughness $G_{1c} > 1$ kJ/m^2
Objective	Minimize the mass m
Free variables	Shaft diameter Choice of material

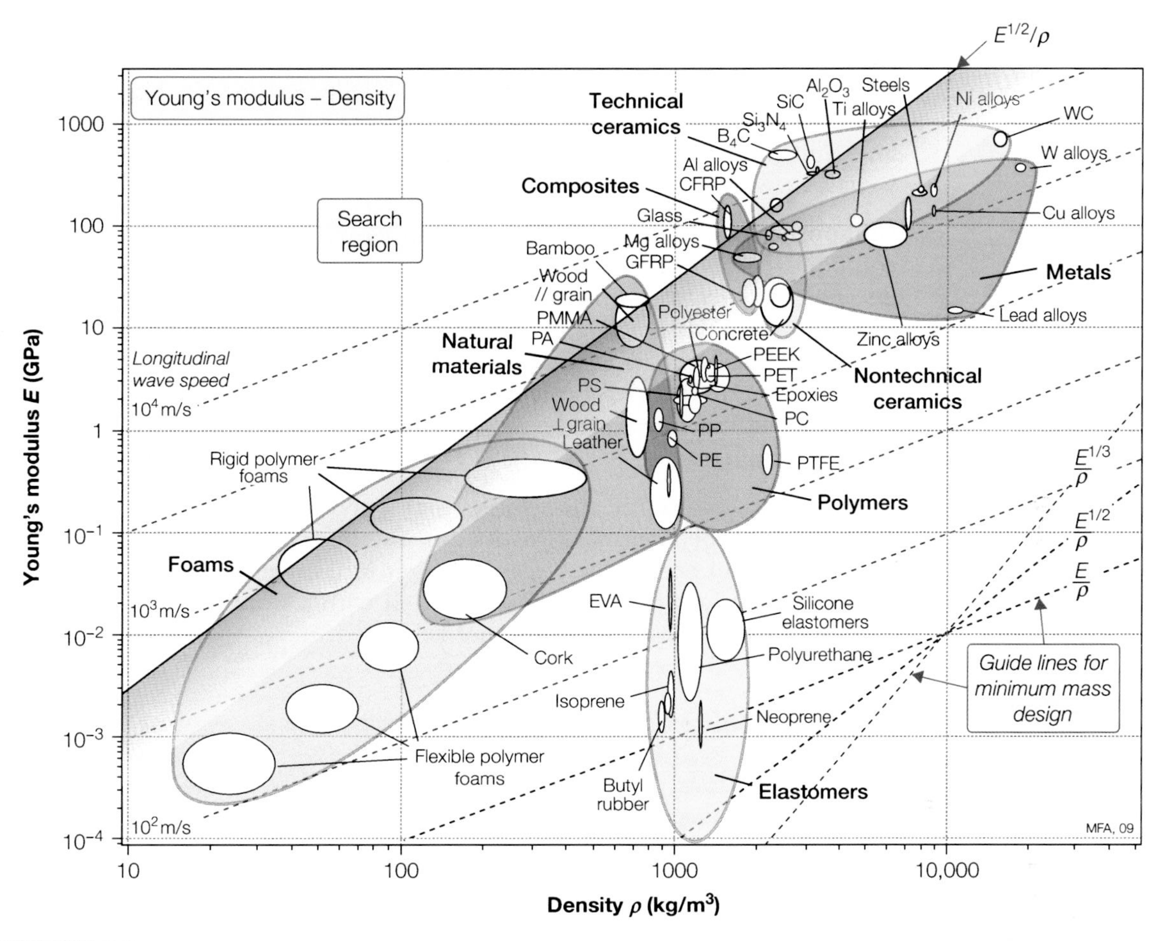

FIGURE 6.2
Materials for oars. CFRP is better than wood because the structure can be controlled.

Table 6.2 Materials for Oars

Material	Index M (GPa)$^{1/2}$/(Mg/m^3)	Comment
Bamboo	4.0–4.5	The traditional material for oars for canoes
Woods	3.4–6.3	Inexpensive, traditional, but with natural variability
CFRP	5.3–7.9	As good as wood, more control of properties
Ceramics	4–8.9	Good M but toughness low and cost high

northern United States or Canada, the further north the better because the short growing season gives a finer grain. The wood is cut into strips, four of which are laminated together to average the stiffness, and the blade is glued to the shaft. The rough oar is then shelved for some weeks to settle down, and finished by hand-cutting and polishing. When finished, a spruce oar weighs between 4 and 4.3 kg.

Composite blades are a little lighter than wood for the same stiffness. The component parts are fabricated from a mixture of carbon and glass fibers in an epoxy matrix, assembled and glued. The advantage of composites lies partly in the saving of weight (typical weight: 3.9 kg) and partly in the greater control of performance: The shaft is molded to give the stiffness specified by the purchaser. Until recently a CFRP oar cost more than a wooden one, but the price of carbon fibers has fallen sufficiently that the two cost about the same.

Could we do better? The chart shows that wood and CFRP offer the lightest oars, at least when normal construction methods are used. Novel composites, not shown on the chart, might permit further weight saving; and functional-grading (a thin, very stiff outer shell with a low-density core) might do it. But both appear, at present, unlikely.

Related reading

Redgrave, S. (1992). *Complete book of rowing.* Partridge Press.

Related case studies

6.3 "Mirrors for large telescopes"
6.4 "Materials for table legs"
10.2 "Spars for human-powered planes"
10.3 "Forks for a racing bicycle"

6.3 MIRRORS FOR LARGE TELESCOPES

There are some very large optical telescopes in the world. The newer ones employ complex and cunning tricks to maintain their precision as they track across the sky—more on that in the Postscript. But if you want a simple telescope, you make the reflector as a single rigid mirror. The largest such telescope is sited on Mount Semivodrike, near Zelenchukskaya in the Caucasus Mountains of Russia. The mirror is 6 m (236 inches) in diameter. To be sufficiently rigid, the mirror is made of glass about 1 m thick and weighs 70 tonnes.

The total cost of a large (236-inch) telescope is, like the telescope itself, astronomical—about US$300 m. The mirror itself accounts for only about 5% of this cost; the rest of the cost is the mechanism that holds, positions, and moves it as it tracks across the sky. This mechanism must be stiff enough to position the mirror relative to the collecting system with a precision about equal to that of the wavelength of light. It might seem, at first sight, that doubling the mass m of the mirror would require that the sections of the support structure be doubled too, so as to keep the stresses (and hence the strains and displacements) the same; but the heavier structure then deflects under its own weight. In practice, the sections have to increase as m^2, and so does the cost.

A century ago, mirrors were made of speculum metal (density: about 8000 kg/m^3). Since then, they have been made of glass (density: 2,300 kg/m^3), silvered on the front surface, so none of the optical properties of the glass are used. Glass is chosen for its mechanical properties only; the 70 tonnes of glass are just a very elaborate support for 100 nm (about 30 grams) of silver. Could one, by taking a radically new look at materials for mirrors, suggest possible routes to the construction of lighter, cheaper telescopes?

The translation At its simplest, the mirror is a circular disk with diameter $2R$ and mean thickness t, simply supported at its periphery (Figure 6.3). When horizontal, it will deflect under its own weight m; when vertical it will not deflect significantly. This distortion (which changes the focal length and introduces aberrations) must be small enough that it does not interfere with performance; in practice, this means that the deflection δ of the midpoint of the mirror must be less than the wavelength of light. Additional requirements are high dimensional stability (no creep) and low thermal expansion (Table 6.3).

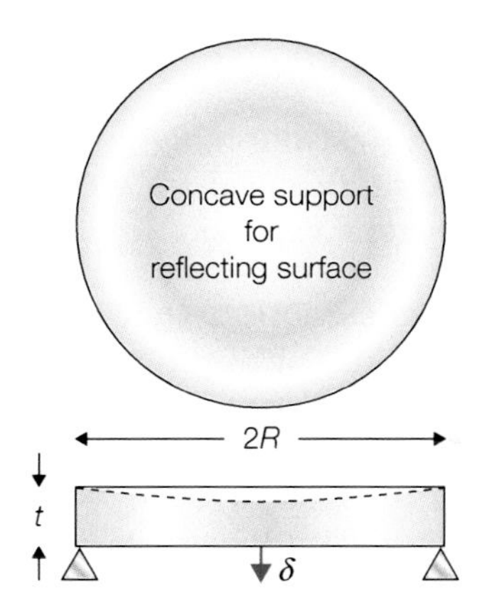

FIGURE 6.3
The mirror of a large optical telescope is modeled as a disk, simply supported at its periphery. It must not sag by more than a wavelength of light at its center.

The mass of the mirror (the property we wish to minimize) is

$$m = \pi R^2 t \rho \tag{6.2}$$

where ρ is the density of the material of the disk. The elastic deflection, δ, of the center of a

Table 6.3 Design Requirements for the Telescope Mirror

Function	Precision mirror
Constraints	Radius R specified Must not distort more than δ under self-weight High dimensional stability: no creep, low thermal expansion
Objective	Minimize the mass, m
Free variables	Thickness of mirror, t Choice of material

horizontal disk due to its own weight is given, for a material with Poisson's ratio of 0.3 (Appendix B), by

$$\delta = \frac{3}{4\pi}\frac{m\,g\,R^2}{E\,t^3} \tag{6.3}$$

The quantity g in this equation is the acceleration due to gravity: 9.81 m/s^2; E, as before, is Young's modulus. We require that this deflection be less than (say) 10 μm. The diameter $2R$ of the disk is specified by the telescope design, but the thickness t is a free variable. Solving for t and substituting this into the first equation gives

$$m = \left(\frac{3\,g}{4\,\delta}\right)^{1/2} \pi\,R^4 \left[\frac{\rho}{E^{1/3}}\right]^{3/2} \tag{6.4}$$

The lightest mirror is the one with the greatest value of the material index

$$M = \frac{E^{1/3}}{\rho} \tag{6.5}$$

We treat the remaining constraints as attribute limits, requiring a melting point greater than 500°C to avoid creep, zero moisture take up, and a low thermal expansion coefficient ($\alpha < 20 \times 10^{-6}$/K).

The selection Here we have another example of elastic design for minimum weight. The appropriate chart is again that relating Young's modulus E and density ρ—but the line we now construct on it has a slope of 3, corresponding to the condition $M = E^{1/3}/\rho$ = constant (Figure 6.4). Glass lies at the value M = 1.7 (GPa)$^{1/3}$.m^3/Mg. Materials that have larger values of M are better; those with lower, worse. Glass is much better than steel or speculum metal (that is why most mirrors are made of glass), but it is less good than magnesium, several ceramics, carbon-fiber– and glass-fiber–reinforced polymers, or—an unexpected finding—stiff foamed polymers. The shortlist before applying the attribute limits is given in Table 6.4.

One must, of course, examine other aspects of this choice. The mass of the mirror, calculated from Equation (6.4), is listed in the table. The CFRP mirror is less than half the weight of the glass one, and its support structure could thus be as much as four times less expensive. The possible saving by using foam is even greater. But could they be made?

Some of the choices—polystyrene foam or CFRP—may at first seem impractical. But the potential cost saving (the factor of 16) is so vast that they are worth examining. There are ways of casting a thin film of silicone rubber or of epoxy onto the surface of the mirror backing (the polystyrene or the CFRP) to give an optically smooth surface that could be silvered. The most obvious obstacle is the lack of stability of polymers—they change dimensions with age, humidity, temperature and so on. But glass itself can be reinforced with carbon

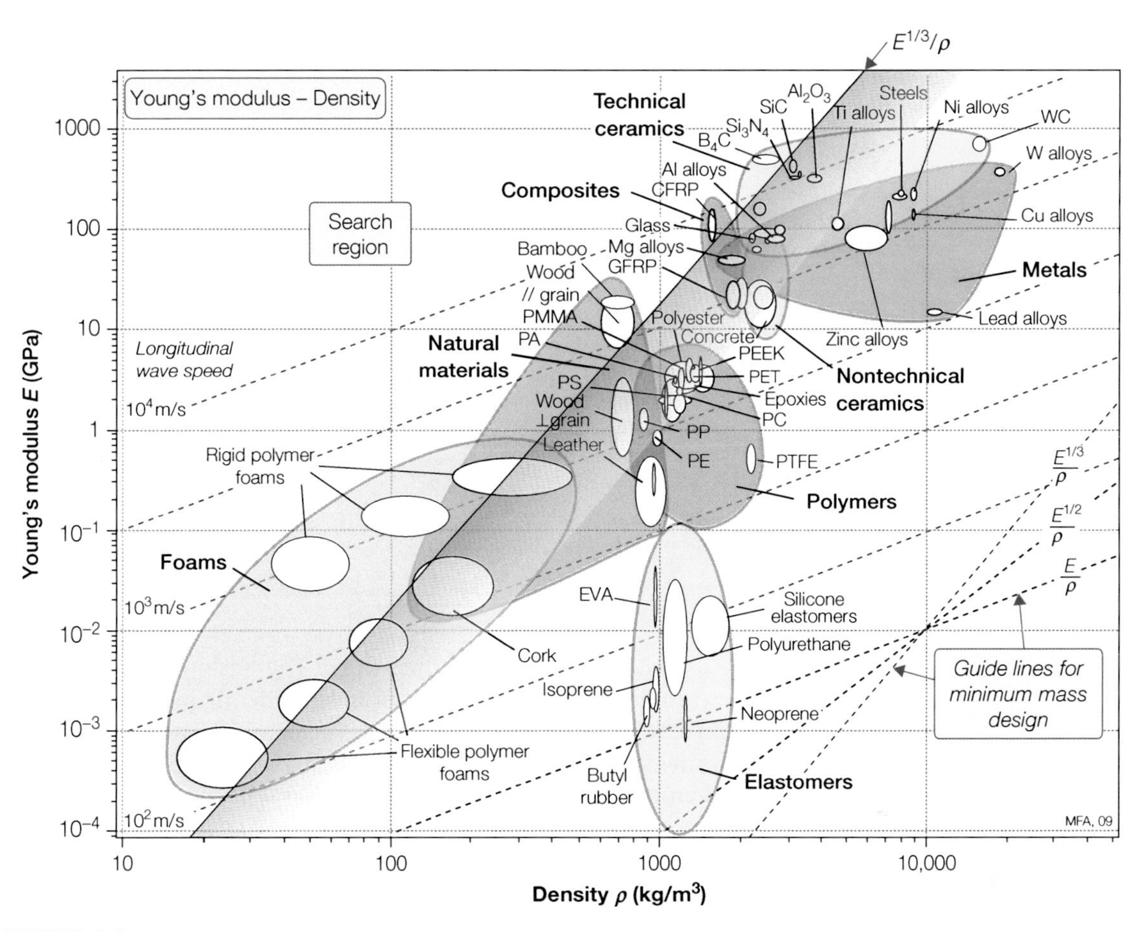

FIGURE 6.4

Materials for telescope mirrors. Glass is better than most metals, among which magnesium is a good choice. Carbon-fiber–reinforced polymers give, potentially, the lowest weight of all, but may lack adequate dimensional stability. Foamed glass is a possible candidate.

fibers; and it can also be foamed to give a material that is denser than polystyrene foam but much lighter than solid glass. Both foamed and carbon-reinforced glass have the same chemical and environmental stability as solid glass. They could provide a route to large cheap mirrors.

Postscript There are, of course, other things you can do. The stringent design criterion ($\delta < 10\,\mu m$) can be partially overcome by engineering design without reference to the material choice. The 8.2-m Japanese telescope on Mauna Kea, Hawaii, and the Very Large Telescope (VLT) at Cerro Paranal Silla in Chile each have a thin glass reflector supported by an array of hydraulic or piezo-electric jacks that exert distributed forces over the back surface,

Table 6.4 Mirror Backing for 200-inch (5.1-m) Telescope

Material	$M = E^{1/3}/\rho$ $(GPa)^{1/3}.m^3/Mg$	m (tonne) $2R = 5.1$ m (from Eq. 6.4)	Comment
Steel (or speculum)	0.74	73.6	Very heavy—the original choice
GFRP	1.5	25.5	Not dimensionally stable enough—use for radio telescope
Al-Alloys	1.6	23.1	Heavier than glass, and with high thermal expansion
Glass	1.7	21.6	The present choice
Mg-Alloys	1.9	17.9	Lighter than glass but high thermal expansion
CFRP	3.0	9	Very light, but not dimensionally stable; use for radio telescopes
Foamed polystyrene	4.5	5	Very light, but dimensionally unstable. Foamed glass?

controlled to vary with the attitude of the mirror. The Keck telescope, also on Mauna Kea, is segmented, with each segment independently positioned to give optical focus. But the limitations of this sort of mechanical system still require that the mirror meet a stiffness target. While stiffness at minimum weight is the design requirement, the material-selection criteria remain unchanged.

Radio telescopes do not have to be quite as precisely dimensioned as optical ones because they detect radiation with a longer wavelength, about 0.25 mm rather than the 0.02 mm of light waves. But they are much bigger (60 m rather than 6) and they suffer from similar distortional problems. A recent 45 m radio telescope built for the University of Tokyo has a parabolic reflector made up of 6000 CFRP panels, each servo-controlled to compensate for macrodistortion. Radio telescopes are now routinely made from CFRP, for exactly the reasons we deduced.

Related case study

6.16 "Minimizing thermal distortion in precision devices"

6.4 MATERIALS FOR TABLE LEGS

Luigi Tavolino, furniture designer, conceives of a lightweight table of daring simplicity: a flat sheet of toughened glass supported on slender, unbraced cylindrical legs (Figure 6.5). The legs must be solid (to make them thin) and as light as possible (to make the table easier to move). They must support the table top and whatever is placed upon it without buckling (Table 6.5). What materials could one recommend?

The translation This is a problem with two objectives[1]: Weight is to be minimized and slenderness maximized. There is one constraint: resistance to buckling. Consider minimizing weight first.

The leg is a slender column of material of density ρ and modulus E. Its length, L, and the maximum load, F, it must carry are determined by the design: They are fixed. The radius r of a leg is a free variable. We wish to minimize the mass m of the leg, given by the objective function

$$m = \pi r^2 L \rho \tag{6.6}$$

subject to the constraint that it supports a load P without buckling. The elastic buckling load F_{crit} of a column of length L and radius r (see Appendix B) is

$$F_{crit} = \frac{\pi^2 E I}{L^2} = \frac{\pi^3 E r^4}{4 L^2} \tag{6.7}$$

using $I = \pi r^4/4$ where I is the second moment of the area of the column. The load F must not exceed F_{crit}. Solving for the free variable, r, and substituting it into the equation for m gives

$$m \geq \left(\frac{4F}{\pi}\right)^{1/2} (L)^2 \left[\frac{\rho}{E^{1/2}}\right] \tag{6.8}$$

The material properties are grouped together in the last pair of brackets. The weight is minimized by selecting the subset of materials with the greatest value of the material index

$$M_1 = \frac{E^{1/2}}{\rho}$$

(a result we could have taken directly from Appendix C).

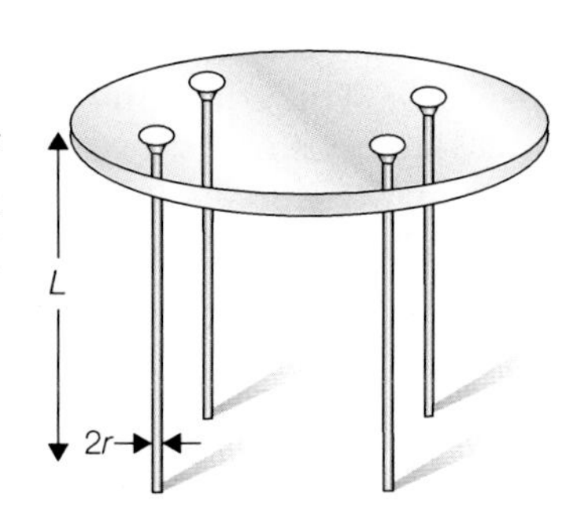

FIGURE 6.5
A lightweight table with slender cylindrical legs. Lightness and slenderness are independent design goals, both constrained by the requirement that the legs must not buckle when the table is loaded. The best choice is a material with high values of both $E^{1/2}/\rho$ and E.

Table 6.5 Design Requirements for Table Legs

Function	Column (supporting compressive loads)
Constraints	Length L specified Must not buckle under design loads Must not fracture if accidentally struck
Objectives	Minimize mass, m Maximize slenderness
Free variables	Diameter of legs, $2r$ Choice of material

[1] Formal methods for dealing with multiple objectives are developed in Chapter 7.

Now slenderness. Inverting Equation (6.7) with F_{crit} set equal to F gives an equation for the thinnest leg that will not buckle:

$$r \geq \left(\frac{4F}{\pi^3}\right)^{1/4} (L)^{1/2} \left[\frac{1}{E}\right]^{1/4} \tag{6.9}$$

The thinnest leg is that made of the material with the largest value of the material index

$$M_2 = E$$

The selection We seek the subset of materials that have high values of $E^{1/2}/\rho$ and E. We need the $E - \rho$ chart again (Figure 6.6). A guideline of slope 2 is drawn on the diagram; it defines the slope of the grid of lines for values of $E^{1/2}/\rho$. The guideline is displaced upward (retaining the slope) until a reasonably small subset of materials is isolated above it; it is shown at the position $M_1 = 5\ \text{GPa}^{1/2}/(\text{Mg/m}^3)$. Materials above this line have higher values of M_1. They are identified in the figure as *woods* (the traditional material for table legs), *composites* (particularly CFRP), and certain *engineering ceramics*. Polymers are out: They are not stiff enough; metals too: They are too heavy (even magnesium alloys, which are the lightest).

The choice is further narrowed by the requirement that, for slenderness, E must be large. A horizontal line on the diagram links materials with equal values of E; those above are stiffer. Figure 6.6 shows that placing this line at $M_1 = 100$ GPa eliminates woods and GFRP. If the legs must be really thin, then the shortlist is reduced to CFRP and ceramics: They give legs that weigh the same as the wooden ones but are barely half as thick. Ceramics, we know, are brittle: They have low values of fracture toughness. Table legs are exposed to abuse—they get knocked and kicked; common sense suggests that an additional constraint is needed, that of adequate toughness. This can be done using Figure 4.7; it eliminates ceramics, leaving CFRP. The cost of CFRP may cause Snr. Tavolino to reconsider his design, but that is another matter: He did not mention cost in his original specification.

It is a good idea to lay out the results as a table, showing not only the materials that are best but those that are second-best—they may, when other considerations are involved, become the best choice. Table 6.6 shows the way to do it.

Postscript Tubular legs, the reader will say, must be lighter than solid ones. True; but they will also be fatter. So it depends on the relative importance Snr. Tavolino attaches to his two objectives—lightness and slenderness—and only he can decide that. If he can be persuaded to live with fat legs, tubing

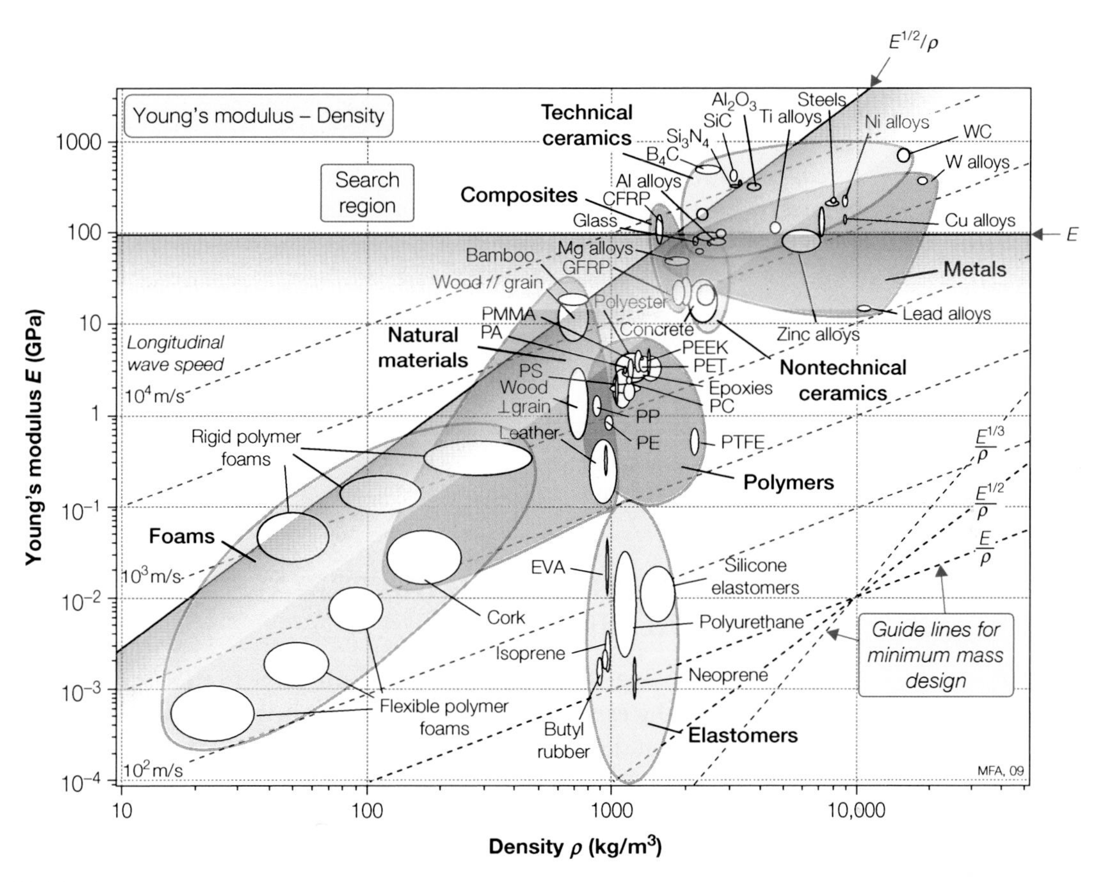

FIGURE 6.6

Materials for light, slender legs. Wood is a good choice; so is a composite such as CFRP, which, having a higher modulus than wood, gives a column that is both light and slender. Ceramics meet the stated design goals, but are brittle.

Table 6.6 Materials for Table Legs

Material	Typical M_1 ($GPa^{1/2}.m^3/Mg$)	Typical M_2 (GPa)	Comment
GFRP	2.5	20	Less expensive than CFRP, but lower M_1 and M_2
Woods	4.5	10	Outstanding M_1; poor M_2 Inexpensive, traditional, reliable
Ceramics	6.3	300	Outstanding M_1 and M_2 Eliminated by brittleness
CFRP	6.6	100	Outstanding M_1 and M_2, but expensive

can be considered—and the material choice may be different. Materials selection when section shape is a variable comes in Chapter 9.

Ceramic legs were eliminated because of low toughness. If (improbably) the goal is to design a light, slender-legged table for use at high temperatures, ceramics should be reconsidered. The brittleness problem can be bypassed by protecting the legs from abuse or by prestressing them in compression.

Related case studies

6.2 "Materials for oars"
6.3 "Mirrors for large telescopes"
8.5 "Conflicting objectives: Table legs again"
10.2 "Spars for human-powered planes"
10.3 "Forks for a racing bicycle"
10.5 "Table legs yet again: Thin or light?"

6.5 COST: STRUCTURAL MATERIALS FOR BUILDINGS

The most expensive thing that most people buy is the house they live in. Roughly half the expense of building a house is the cost of the materials of which it is made, and they are used in large quantities (family house: around 200 tonnes; large apartment block: around 20,000 tonnes). The materials are used in three ways: structurally to hold the building up; as cladding to keep the weather out; and as "internals" to insulate against heat and sound, and to decorate.

Consider the selection of materials for the structure (Figure 6.7). They must be stiff, strong, and cheap. Stiff, so that the building does not flex too much under wind and internal loads; strong, so that there is no risk of it collapsing. And cheap, because such a lot of material is used. The structural frame of a building is rarely exposed to the environment, and is not, in general, visible, so criteria of corrosion resistance or appearance are not important here.

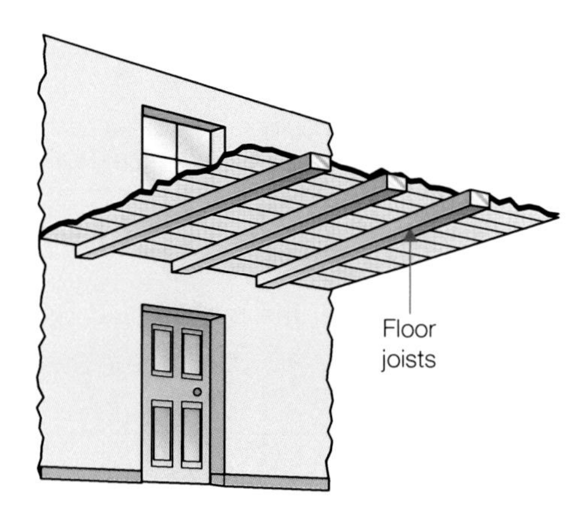

FIGURE 6.7
The materials of a building perform three broad roles. The frame gives mechanical support; the cladding excludes the environment; and the internal surfacing controls heat, light, and sound. The selection criteria depend on the function.

Table 6.7 Design Requirements for Floor Beams	
Function	Floor beam
Constraints	Length L specified Stiffness: must not deflect too much under design loads Strength: must not fail under design loads
Objective	Minimize cost, C
Free variables	Cross-section area of beam, A Choice of material

The design goal is simple: strength and stiffness at minimum cost. To be more specific: Consider the selection of material for floor joists. Table 6.7 summarizes the requirements.

The translation Floor joists are beams; they are loaded in bending. The material index for a stiff beam of minimum mass, m, was developed in Chapter 5 (Equations (5.11) through (5.15)). The cost C of the beam is just its mass, m, times the cost per kg, C_m, of the material of which it is made:

$$C = m\,C_m = A\,L\rho\,C_m \tag{6.10}$$

which becomes the objective function of the problem. Proceeding as in Chapter 5, we find the index for a stiff beam of minimum cost to be

$$M_1 = \frac{E^{1/2}}{\rho\,C_m}$$

The index when strength rather than stiffness is the constraint was not derived earlier. Here it is. The objective function is still Equation (6.10), but the constraint is now that of strength: The beam must support F without failing. The failure load of a beam (Appendix B, Section B.4) is

$$F_f = C_2 \frac{I\,\sigma_f}{y_m\,L} \tag{6.11}$$

where C_2 is a constant, σ_f is the failure strength of the material of the beam, and y_m is the distance between the neutral axis of the beam and its outer filament. We consider a rectangular beam of depth d and width b. We assume the proportions of the beam are fixed so that $d = \alpha b$ where α is the aspect ratio, typically 2 for wood beams. Using this and $I = bd^3/12$ to eliminate A in Equation (6.10) gives the cost of the beam that will just support the load F_f:

$$C = \left(\frac{6\sqrt{\alpha}}{C_2}\,\frac{F_f}{L^2}\right)^{2/3} (L^3) \left[\frac{\rho\,C_m}{\sigma_f^{2/3}}\right] \tag{6.12}$$

The mass is minimized by selecting materials with the largest values of the index

$$M_2 = \frac{\sigma_f^{2/3}}{\rho C_m}$$

The selection Stiffness first. Figure 6.8(a) shows the relevant chart: modulus E against relative cost per unit volume, $C_m\,\rho$ (the chart uses a relative cost C_R, defined in Chapter 4, in place of C_m but this makes no difference to the selection). The shaded band has the appropriate slope for M_1; it isolates concrete, stone, brick, woods, cast irons, and carbon steels. Figure 6.8(b) shows strength against relative cost. The shaded band—M_2 this time—gives almost the same selection. They are listed, with values, in

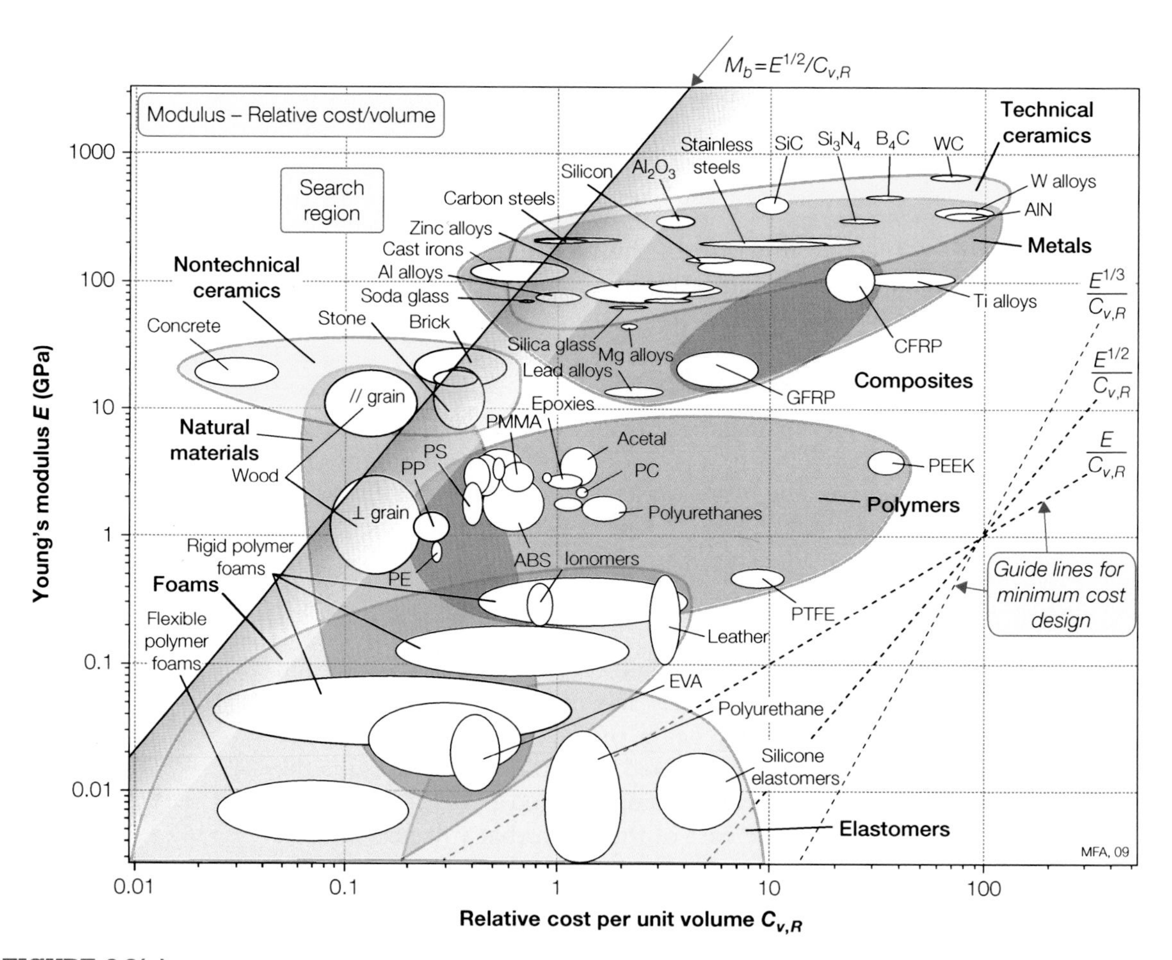

FIGURE 6.8(a)
The selection of cheap, stiff materials for the structural frames of buildings.

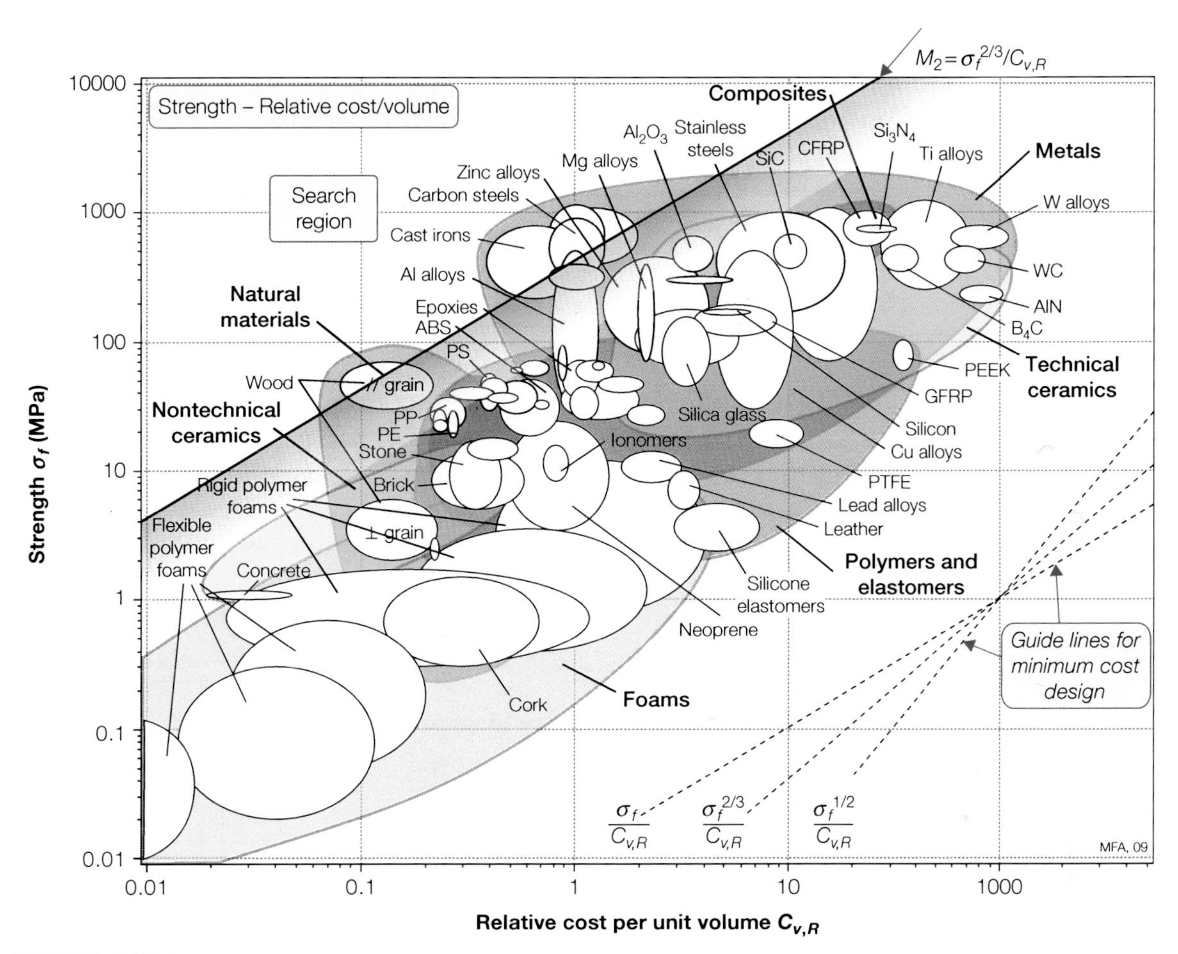

FIGURE 6.8(b)
The selection of cheap, strong materials for the structural frames of buildings.

Table 6.8. They are exactly the materials with which buildings have been, and are, made.

Postscript Concrete, stone, and brick have strength only in compression; the form of the building must use them in this way (columns, arches). Wood, steel, and reinforced concrete have strength in bending and tension as well as compression; steel, additionally, can be given efficient shapes (I-sections, box sections, tubes, discussed in Chapter 9). The form of the building made from these has much greater freedom.

It is sometimes suggested that architects live in the past; that in the twenty-first century they should be building with carbon (CFRP) and fiberglass (GFRP) composites, aluminum and titanium alloys, and stainless steel. Some do, but

Table 6.8 Structural Materials for Buildings

Material	M_1 ($GPa^{1/2}$)/(kg/m^3)	M_2 ($MPa^{2/3}$)/(kg/m^3)	Comment
Concrete	160	14	Use in compression only
Brick	12	12	
Stone	9.3	12	
Woods	21	90	Can support bending and tension as well as compression, allowing greater freedom of shape
Cast iron	17	90	
Steel	14	45	

the last two figures give an idea of the penalty involved: The cost of achieving the same stiffness and strength is between 5 and 50 times greater. Civil construction (buildings, bridges, roads, and the like) is materials-intensive: The cost of the material dominates the product cost, and the quantity used is enormous. Then only the cheapest of materials qualify, and the design must be adapted to use them.

Related reading

Cowan, H.J. & Smith, P.R. (1988). *The science and technology of building materials.* Van Nostrand-Reinhold.

Doran, D.K. (1992). The construction reference book. Butterworth-Heinemann.

Related case studies

6.2 "Materials for oars"
6.4 "Materials for table legs"
8.2 "Floor joists again"
10.4 "Floor joists: wood, bamboo or steel?"

6.6 MATERIALS FOR FLYWHEELS

Flywheels store energy. Small ones—the sort found in children's toys—are made of lead. Old steam engines and modern automobiles have flywheels too; they are made of cast iron. Flywheels have been proposed for power storage and regenerative braking systems for vehicles; a few have been built, some of high-strength steel, some of composites. Lead, cast iron, steel, composites—there is a strange diversity here. What *is* the best choice of material for a flywheel?

An efficient flywheel stores as much *energy per unit weight* as possible. As the flywheel is spun up, increasing its angular velocity ω, it stores more energy. But if the centrifugal stress exceeds the tensile strength of the flywheel, it flies apart. So strength sets an upper limit on the energy that can be stored.

Table 6.9(a) Design Requirements for a Maximum-energy Flywheel	
Function	Flywheel for energy storage
Constraints	Outer radius, R, fixed Must not burst Adequate toughness to give crack tolerance
Objective	Maximize kinetic energy per unit mass
Free variable	Choice of material

Table 6.9(b) Design Requirements for Fixed Velocity	
Function	Flywheel for child's toy
Constraint	Outer radius, R, fixed
Objective	Maximize kinetic energy per unit volume at fixed angular velocity
Free variable	Choice of material

The flywheel of a child's toy is not efficient in this sense. Its angular velocity is limited by the pulling power of the child and never remotely approaches the burst velocity. In this case, and for the flywheel of an automobile engine, we wish to maximize the *energy stored* at a given *angular velocity* in a flywheel with an outer radius, R, that is constrained by the size of the cavity in which it must sit.

The objective and constraints in flywheel design thus depend on its purpose. The two alternative sets of design requirements are listed in Tables 6.9(a) and 6.9(b).

The translation An efficient flywheel of the first type stores as much energy per unit weight as possible, without failing. Think of it as a solid disk of radius R and thickness t, rotating with angular velocity ω (Figure 6.9). The energy U stored in the flywheel (Appendix B) is

$$U = \frac{1}{2} J\omega^2 \qquad (6.13)$$

Here $J = \frac{\pi}{2}\rho R^4 t$ is the polar moment of inertia of the disk and ρ the density of the material of which it is made, giving

$$U = \frac{\pi}{4}\rho R^4 t\omega^2 \qquad (6.14)$$

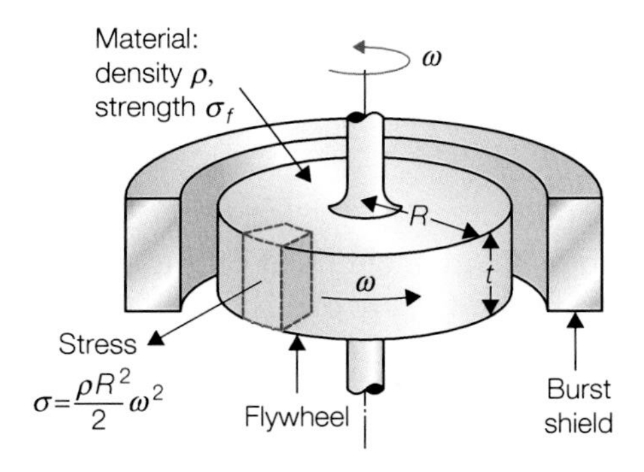

FIGURE 6.9
A flywheel. The maximum kinetic energy it can store is limited by its strength.

The mass of the disk is

$$m = \pi R^2 t \rho \tag{6.15}$$

The quantity to be maximized is the kinetic energy per unit mass, which is the ratio of the last two equations:

$$\frac{U}{m} = \frac{1}{4} R^2 \omega^2 \tag{6.16}$$

As the flywheel is spun up, the energy stored in it increases, but so does the centrifugal stress. The maximum principal stress in a spinning disk of uniform thickness (Appendix B again) is

$$\sigma_{max} = \left(\frac{3+\upsilon}{8}\right) \rho R^2 \omega^2 \approx \frac{1}{2} \rho R^2 \omega^2 \tag{6.17}$$

where υ is Poisson's ratio ($\upsilon \approx 1/3$). This stress must not exceed the failure stress σ_f (with an appropriate factor of safety, here omitted). This sets an upper limit to the product of the angular velocity, ω, and the disk radius, R (the free variables). Eliminating R_ω between the last two equations gives

$$\frac{U}{m} = \frac{1}{2} \left(\frac{\sigma_f}{\rho}\right) \tag{6.18}$$

The best materials for high-performance flywheels are those with high values of the material index

$$M = \frac{\sigma_f}{\rho} \tag{6.19}$$

It has units of kJ/kg.

And now the other sort of flywheel—that of the child's toy. Here we seek the material that stores the most energy per unit volume V at constant velocity, ω. The energy per unit volume at a given ω is (from Equation (6.2)):

$$\frac{U}{V} = \frac{1}{4} \rho R^2 \omega^2.$$

Both R and ω are fixed by the design, so the best material is now that with the greatest value of

$$M_2 = \rho \tag{6.20}$$

The selection Figure 6.10 shows the strength–density chart. Values of M_1 correspond to a grid of lines of slope 1. One such is plotted as a diagonal line at the value M_1 = 200 kJ/kg. Candidate materials with high values of M_1 lie in search region 1, which is toward the top left. The best choices are

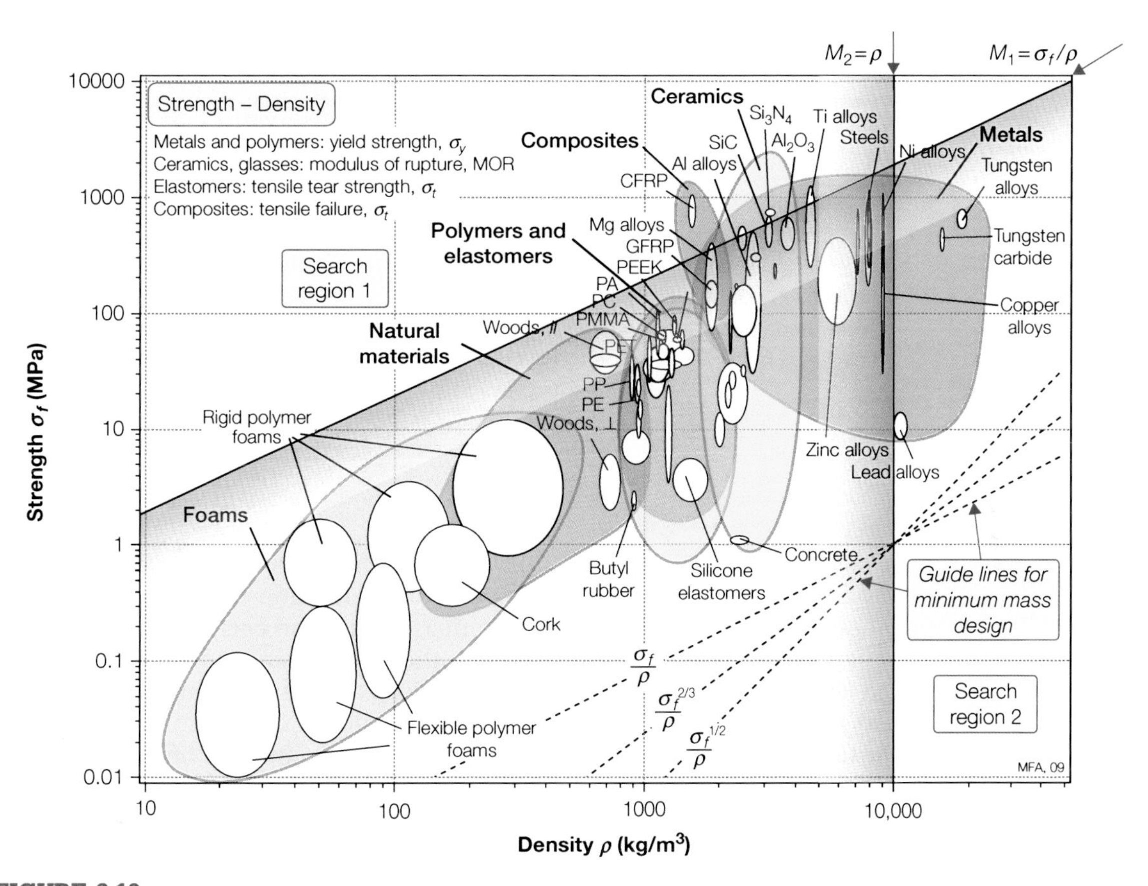

FIGURE 6.10

Materials for flywheels. Composites are the best choices. Lead and cast iron, traditional for flywheels, are good when performance is limited by rotational velocity, not strength.

unexpected ones: composites, particularly CFRP, high-strength titanium alloys, and some ceramics, but these are ruled out by their low toughness.

But what of the lead flywheels of children's toys? There could hardly be two more different materials than CFRP and lead: the one, strong and light; the other, soft and heavy. Why lead? It is because, in the child's toy, the constraint is different. Even a super-child cannot spin the flywheel of his or her toy up to its burst velocity. The angular velocity ω is limited instead by the drive mechanism (pull string, friction drive). Then, as we have seen, the best material is that with the largest density. The second selection line in Figure 6.10 shows the index M_2 at the value 10,000 kg/m^3. We seek materials in search region 2 to the right of this line. Lead is good. Tungsten is better, but more expensive. Cast iron is less good, but cheaper. Gold, platinum, and uranium (not shown on the chart) are the best of all, but may be thought unsuitable for other reasons.

Table 6.10 Energy Density of Power Sources

Source	Energy Density kJ/kg	Comment
Gasoline	20,000	Oxidation of hydrocarbon—mass of oxygen not included
Rocket fuel	5000	Less than hydrocarbons because oxidizing agent forms part of fuel
Flywheels	Up to 400	Attractive, but not yet proved
Lithium-ion battery	Up to 350	Expensive, limited life
Nickel-cadmium battery	170–200	Less expensive than lithium-ion
Lead-acid battery	50–80	Large weight for acceptable range
Springs, rubber bands	Up to 5	Much less efficient method of energy storage than flywheel

Postscript A CFRP rotor is able to store around 400 kJ/kg. A lead flywheel, by contrast, can store only 1 kJ/kg before disintegration; a cast-iron flywheel, about 30. All of these are small compared with the energy density in gasoline: roughly 20,000 kJ/kg. Even so, the energy density in the flywheel is considerable; its sudden release in a failure could be catastrophic. The disk must be surrounded by a burst shield and precise quality control in manufacture is essential to avoid out-of-balance forces. This has been achieved in a number of composite energy-storage flywheels intended for use in trucks and buses, and as an energy reservoir for smoothing wind-power generation.

And now a digression: the electric car. Hybrid gas-electric cars are already on the roads, using advanced battery technology to store energy. But batteries have their problems: The energy density they can contain is low (see Table 6.10); their weight limits both the range and the performance of the car. It is practical to build flywheels with an energy density about equal to that of the best batteries.

Consideration is now being given to a flywheel for electric cars. A pair of counter-rotating CFRP disks are housed in a steel burst shield. Magnets embedded in the disks pass near coils in the housing, inducing a current and allowing power to be drawn to the electric motor that drives the wheels. Such a flywheel could, it is estimated, give an electric car an adequate range at a cost that is competitive with the gasoline engine and with none of the local pollution.

Related reading

Christensen, R.M. (1979). *Mechanics of composite materials* (p. 213 et seq.). Wiley Interscience.

Lewis, G. (1990). *Selection of engineering materials* (Part 1, p. 1). Prentice-Hall.

Medlicott, P.A.C., & Potter, K.D. (1986). The development of a composite flywheel for vehicle applications. In K. Brunsch, H-D. Golden, & C-M. Horkert (Eds.), *High Tech—the way into the nineties* (p. 29). Elsevier.

Related case studies

6.7 "Materials for springs"
6.11 "Safe pressure vessels"
8.3 "Multiple constraints: con-rods for high-performance engines"

6.7 MATERIALS FOR SPRINGS

Springs come in many shapes (Figure 6.11 and Table 6.11) and have many purposes: think of tensile springs (a rubber band, for example), leaf springs, helical springs, spiral springs, torsion bars. Regardless of their shape or use, the best material for a spring of minimum volume is that with the greatest value of σ_f^2/E, and for minimum weight it is that with the greatest value of $\sigma_f^2/\rho E$ (derived in Equation (6.21)). We use springs as a way of introducing two of the most useful of the charts: Young's modulus E plotted against strength σ_f, and specific modulus E/ρ plotted against specific strength σ_f/ρ (Figures 4.5 and 4.6).

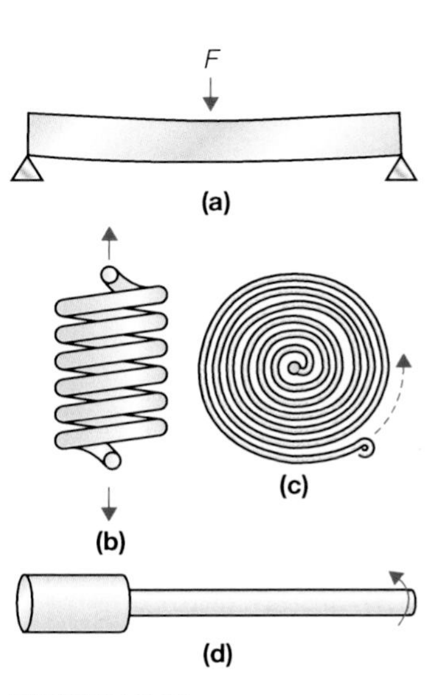

FIGURE 6.11
Springs store energy. The best material for any spring, regardless of its shape or the way in which it is loaded, is that with the highest value of σ_f^2/E or, if weight is important, $\sigma_f^2/\rho E$.

The translation The primary function of a spring is to store elastic energy and—when required—release it again. The elastic energy stored per unit volume in a block of material stressed uniformly to a stress σ is

$$W_v = \frac{1}{2}\frac{\sigma^2}{E} \tag{6.21}$$

where E is Young's modulus. We wish to maximize W_v. The spring will be damaged if the stress σ exceeds the yield stress or failure stress σ_f; the constraint is $\sigma < \sigma_f$. Thus the maximum energy density is

$$W_v = \frac{1}{2}\frac{\sigma_f^2}{E} \tag{6.22}$$

Table 6.11 Design Requirements for Springs

Function	Elastic spring
Constraint	No failure, meaning $\sigma < \sigma_f$ throughout the spring
Objective	Maximum stored elastic energy per unit volume, or maximum stored elastic energy per unit weight
Free variable	Choice of material

Torsion bars and leaf springs are less efficient than axial springs because much of the material is not fully loaded: The material at the neutral axis, for instance, is not loaded at all. For leaf springs

$$W_v = \frac{1}{4}\frac{\sigma_f^2}{E}$$

and for torsion bars

$$W_v = \frac{1}{3}\frac{\sigma_f^2}{E}$$

But—as these results show—this has no influence on the choice of material. The best stuff for a spring regardless of its shape is that with the biggest value of

$$M_1 = \frac{\sigma_f^2}{E} \quad (6.23)$$

If weight, rather than volume, matters, we must divide this by the density ρ (giving energy stored per unit weight), and seek materials with high values of

$$M_2 = \frac{\sigma_f^2}{\rho E} \quad (6.24)$$

The selection The choice of materials for springs of minimum volume is shown in Figure 6.12(a). A family of lines of slope 2 links materials with equal values of $M_1 = \sigma_f^2/E$; those with the highest values of M_1 lie toward the lower right. The heavy line is one of the family; it is positioned at 2.5 MJ/m^3 so that a subset of materials is left exposed. The best choice is a high-strength steel lying near the top end of the line. Other materials are suggested too: CFRP (now used for truck springs); titanium alloys (good but expensive); nylon, PA (children's toys often have nylon springs); and, of course, elastomers. Note how the procedure has identified a candidate from almost every class of materials: metals, polymers, elastomers, and composites. They are listed, with comments, in Table 6.12(a).

Materials selection for light springs is shown in Figure 6.12(b). A family of lines of slope 2 links materials with equal values of

$$M_2 = \left(\frac{\sigma_f}{\rho}\right)^2 \Big/ \left(\frac{E}{\rho}\right) = \frac{\sigma_f^2}{E\rho} \quad (6.25)$$

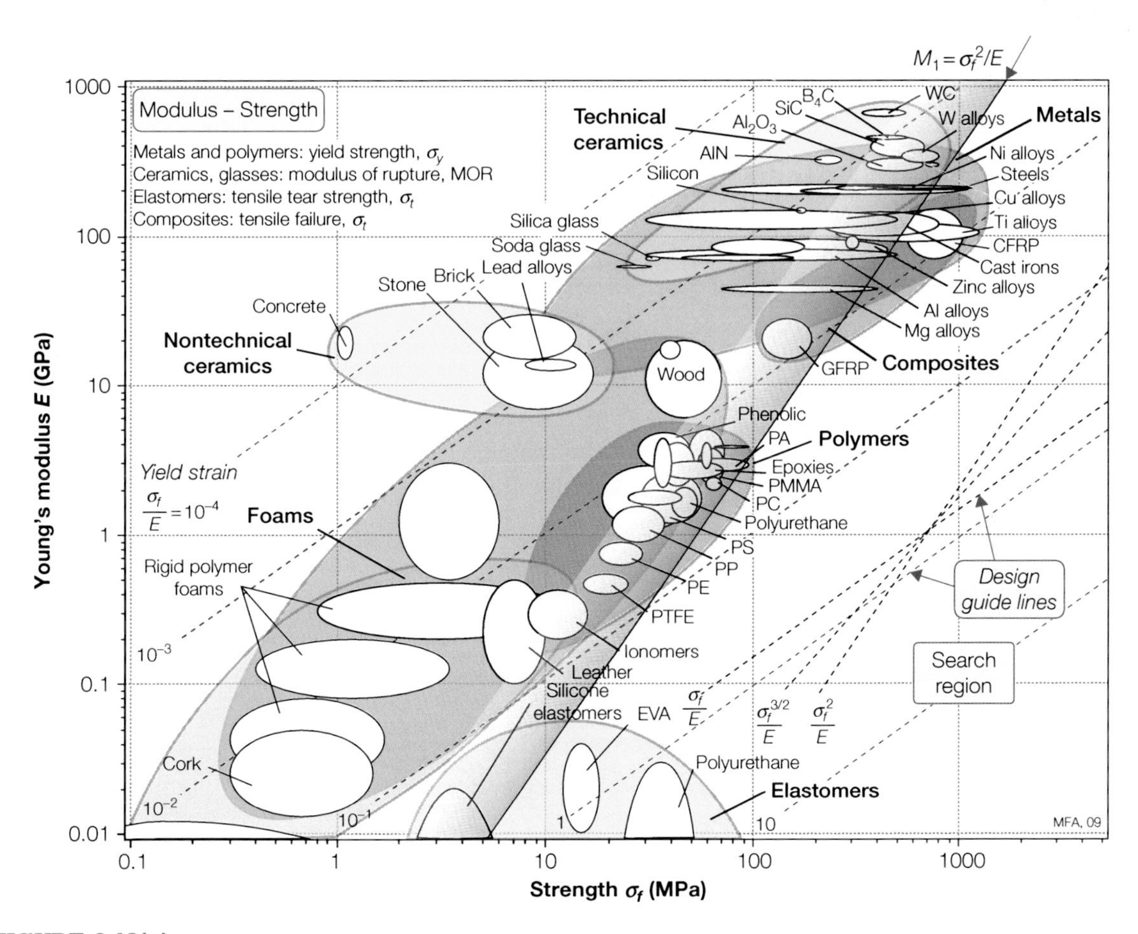

FIGURE 6.12(a)

Materials for small springs. High-strength ("spring") steel is good. Glass, CFRP, and GFRP all, under the right circumstances, make good springs. Elastomers are excellent. Ceramics are eliminated by their low tensile strength.

Table 6.12(a) Materials for Efficient Small Springs

Material	$M_1 = \sigma_f^2/E$ (MJ/m^3)	Comment
Ti alloys	4–12	Expensive, corrosion-resistant
CFRP	6–10	Comparable in performance with steel; expensive
Spring steel	3–7	The traditional choice: easily formed and heat treated
Nylon	1.5–2.5	Inexpensive and easily shaped, but high loss factor
Rubber	20–50	Better than spring steel, but high loss factor

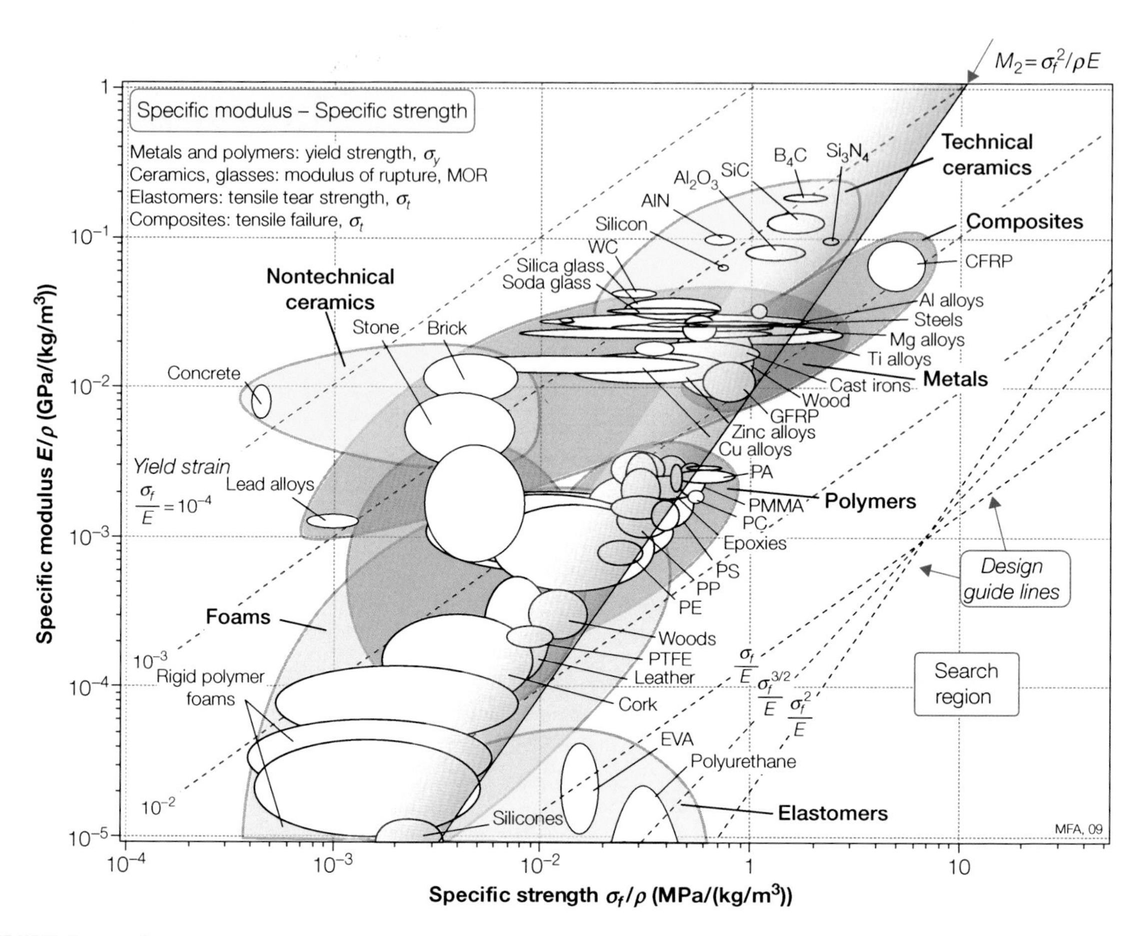

FIGURE 6.12(b)
Materials for light springs. Metals are disadvantaged by their high densities. Composites are good; so is wood. Elastomers are excellent.

One is shown at the value $M_2 = 1$ kJ/kg. Metals, because of their high density, are less good than composites, and much less good than elastomers. (You can store roughly eight times more elastic energy, per unit weight, in a rubber band than in the best spring steel.) Candidates are listed in Table 6.12(b). Wood—the traditional material for archery bows—now appears.

Postscript Many additional considerations enter the choice of a material for a spring. Springs for vehicle suspensions must resist fatigue and corrosion; engine valve springs must cope with elevated temperatures. A subtler property is the loss coefficient, shown in Figure 4.9. Polymers have a relatively high loss factor and dissipate energy when they vibrate; metals, if strongly hardened, do not. Polymers, because they creep, are unsuitable for springs that carry a steady load for long periods of time, though they are

Table 6.12(b) Materials for Efficient Light Springs

Material	$M_1 = \sigma_f^2/\rho E$ (kJ/kg)	Comment
Ti alloys	0.9–2.6	Better than steel; corrosion-resistant; expensive
CFRP	3.9–6.5	Better than steel; expensive
GFRP	1.0–1.8	Better than spring steel; less expensive than CFRP
Spring steel	0.4–0.9	Poor, because of high density
Wood	0.3–0.7	On a weight basis, wood makes good springs
Nylon	1.3–2.1	As good as steel, but with a high loss factor
Rubber	18–45	Outstanding; 20 times better than spring steel; but with high loss factor

still perfectly good for catches and locating springs that spend most of their time unstressed.

Related reading

Boiton, R.G. (1963). The mechanics of instrumentation. *Proc. I. Mech. E.*, *177*(10), 269–288.
Hayes, M. (1990). Materials update 2: springs. *Engineering*, May, p. 42.

Related case studies

6.8 "Elastic hinges and couplings"
10.7 "Shapes that flex: Leaf and strand structures"
10.8 "Ultra-efficient springs"
12.7 "Connectors that don't relax their grip"

6.8 ELASTIC HINGES AND COUPLINGS

Nature makes much use of elastic (or "natural") hinges: Skin, muscle, and cartilage all allow large, recoverable deflections. Humans, too, design with *flexure* and *torsion hinges*: ligaments that connect or transmit a load between components while allowing for limited relative movement between them by deflecting elastically (see Figure 6.13 and Table 6.13). Which materials make good hinges?

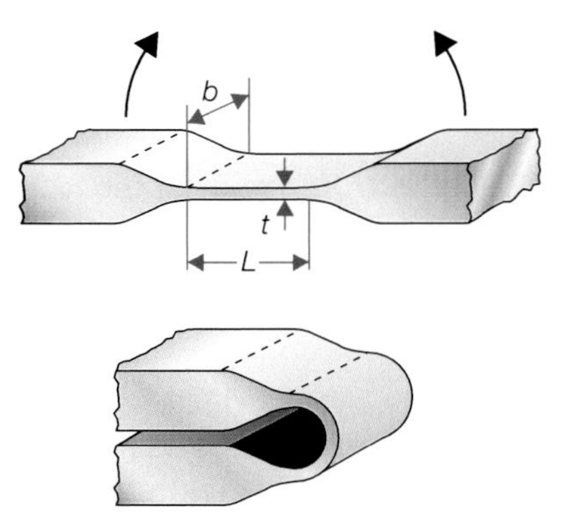

FIGURE 6.13
Elastic or "natural" hinges. The ligaments must bend repeatedly without failing. The cap of a shampoo bottle is an example; elastic hinges are used in high-performance applications too, and are found widely in nature.

The translation Consider the hinge for the lid of a box. The box, lid, and hinge are to be molded as a single unit. The hinge is a thin ligament that flexes elastically as the box is closed, as shown in the figure, but it carries no significant axial

Table 6.13 Design Requirements for Elastic Hinges	
Function	Elastic hinge
Constraint	No failure, meaning $\sigma < \sigma_f$ throughout the hinge
Objective	Maximize elastic flexure
Free variable	Choice of material

loads. Then the best material is the one that (for given ligament dimensions) bends to the smallest radius without yielding or failing. When a ligament of thickness t is bent elastically to a radius R, the surface strain is

$$\varepsilon = \frac{t}{2R} \tag{6.26}$$

and—since the hinge is elastic—the maximum stress is

$$\sigma = E\frac{t}{2R} \tag{6.27}$$

This must not exceed the yield or failure strength σ_f. Thus the minimum radius to which the ligament can be bent without damage is

$$R \geq \frac{t}{2}\left[\frac{E}{\sigma_f}\right] \tag{6.28}$$

The best material is the one that can be bent to the smallest radius, that is, the one with the greatest value of the index

$$M = \frac{\sigma_f}{E} \tag{6.29}$$

The selection We need the $\sigma_f - E$ chart again (Figure 6.14). Candidates are identified by using the guide line of slope 1; a line is shown at the position $M = \sigma_f/E = 2 \times 10^{-2}$. The best choices for the hinge lie to the right of this line: They are all polymeric materials. The shortlist (Table 6.14) includes polyethylene, polypropylene, nylon, and, best of all, elastomers, though these may be too flexible for the body of the box itself. Cheap products with this sort of elastic hinge are generally molded from polyethylene, polypropylene, or nylon. Spring steel and other metallic spring materials (like phosphor bronze) are possibilities: They combine usable σ_f/E with high E, giving flexibility with good positional stability (as in the suspensions of relays). The table gives further details.

Postscript Polymers give more design freedom than metals. The elastic hinge is one example of this, reducing the box, hinge, and lid (three components plus the fasteners needed to join them) to a single box-hinge-lid unit,

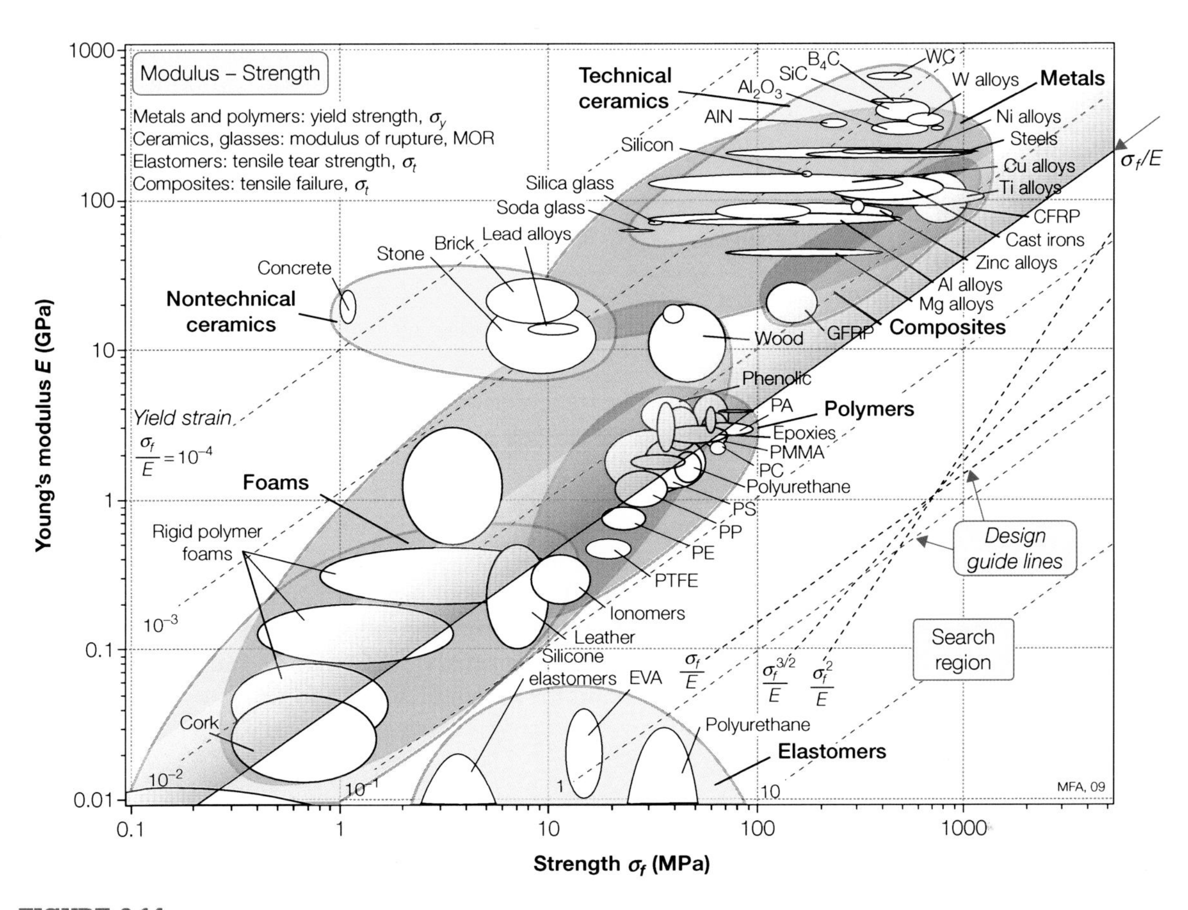

FIGURE 6.14

Materials for elastic hinges. Elastomers are best, but may not be rigid enough to meet other design needs. Polymers such as nylon, PTFE, and PE are better. Spring steel is less good, but much stronger.

Table 6.14 Materials for Elastic Hinges

Material	M ($\times 10^{-3}$)	Comment
Polyethylene	32	Widely used for cheap hinged bottle caps, etc.
Polypropylene	30	Stiffer than polyethylene; easily molded
Nylon	30	Stiffer than polyethylene; easily molded
PTFE	35	Very durable; more expensive than PE, PP, etc.
Elastomers	100–1000	Outstanding, but low modulus
High strength copper alloys	4	M less good than polymers; use when high tensile stiffness is required
Spring steel	6	

molded in one operation. Their spring-like properties allow snap-together, easily joined parts. Another example is the elastomeric coupling—a flexible universal joint, allowing high angular, parallel, and axial flexibility with good shock absorption characteristics. Elastomeric hinges offer many opportunities to be exploited in engineering design.

Related case studies

6.7 "Materials for springs"
6.9 "Materials for seals"
6.10 "Deflection-limited design with brittle polymers"
10.7 "Shapes that flex: Leaf and strand structures"

6.9 MATERIALS FOR SEALS

A reusable elastic seal consists of a cylinder of material compressed between two flat surfaces (Figure 6.15). The seal must form the largest possible contact width, b, while keeping the contact stress, σ, sufficiently low that it does not damage the flat surfaces; and the seal itself must remain elastic so that it can be reused many times. What materials make good seals? Elastomers—everyone knows that. But let us do the job properly; there may be more to be learned. We build the selection around the requirements of Table 6.15.

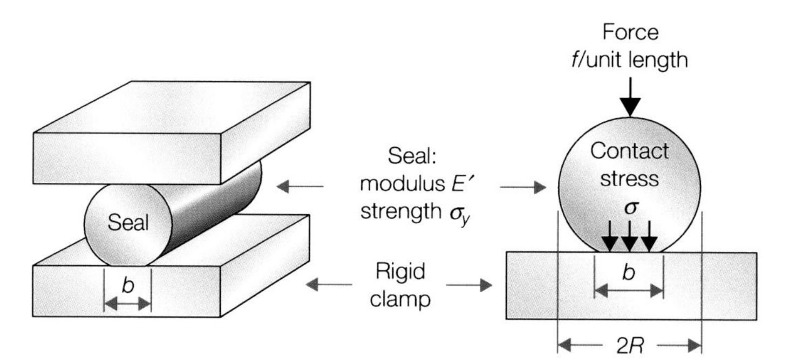

FIGURE 6.15
An elastic seal. A good seal gives a large conforming contact area without imposing damaging loads on itself or on the surfaces with which it mates.

Table 6.15 Design Requirements for Elastic Seals

Function	Elastic seal
Constraints	Limit on contact pressure Low cost
Objective	Maximum conformability to surface
Free variable	Choice of material

The translation A cylinder of diameter $2R$ and modulus E, pressed onto a rigid flat surface by a force f per unit length, forms an elastic contact of width b (Appendix B) where

$$b \approx 2.3\left(\frac{f\,R}{E}\right)^{1/2} \tag{6.30}$$

This is the quantity to be maximized: the objective function. The contact stress, both in the seal and in the surface, is adequately approximated (Appendix A again) by

$$\sigma = 0.57\left(\frac{f\,E}{R}\right)^{1/2} \tag{6.31}$$

The constraint: The seal must remain elastic—that is, σ must be less than the yield or failure strength, σ_f, of the material of which it is made. Combining the last two equations with this condition gives

$$b \leq 4.0\,R\left(\frac{\sigma_f}{E}\right) \tag{6.32}$$

The contact width is maximized by maximizing the index

$$M_1 = \frac{\sigma_f}{E}$$

It is also required that the contact stress σ be kept low to avoid damage to the flat surfaces. Its value when the maximum contact force is applied (to give the biggest width) is simply σ_f, the failure strength of the seal. Suppose the flat surfaces are damaged by a stress of greater than 100 MPa. The contact pressure is kept below this by requiring that

$$M_2 = \sigma_f \leq 100\,\mathrm{MPa}$$

The selection The two indices are plotted on the $\sigma_f - E$ chart in Figure 6.16, isolating elastomers, foams, and cork. The candidates are listed in Table 6.16 with comments. The value of M_2 = 100 MPa admits all elastomers as candidates. If M_2 is reduced to 10 MPa, all but the most compliant elastomers are eliminated, and foamed polymers become the best bet.

Postscript The analysis highlights the functions that seals must perform: large contact area, limited contact pressure, environmental stability. Elastomers maximize the contact area; foams and cork minimize the contact pressure; PTFE and silicone rubbers best resist heat and organic solvents. The final choice depends on the conditions under which the seal will be used.

Related case studies

6.7 "Materials for springs"
6.8 "Elastic hinges and couplings"

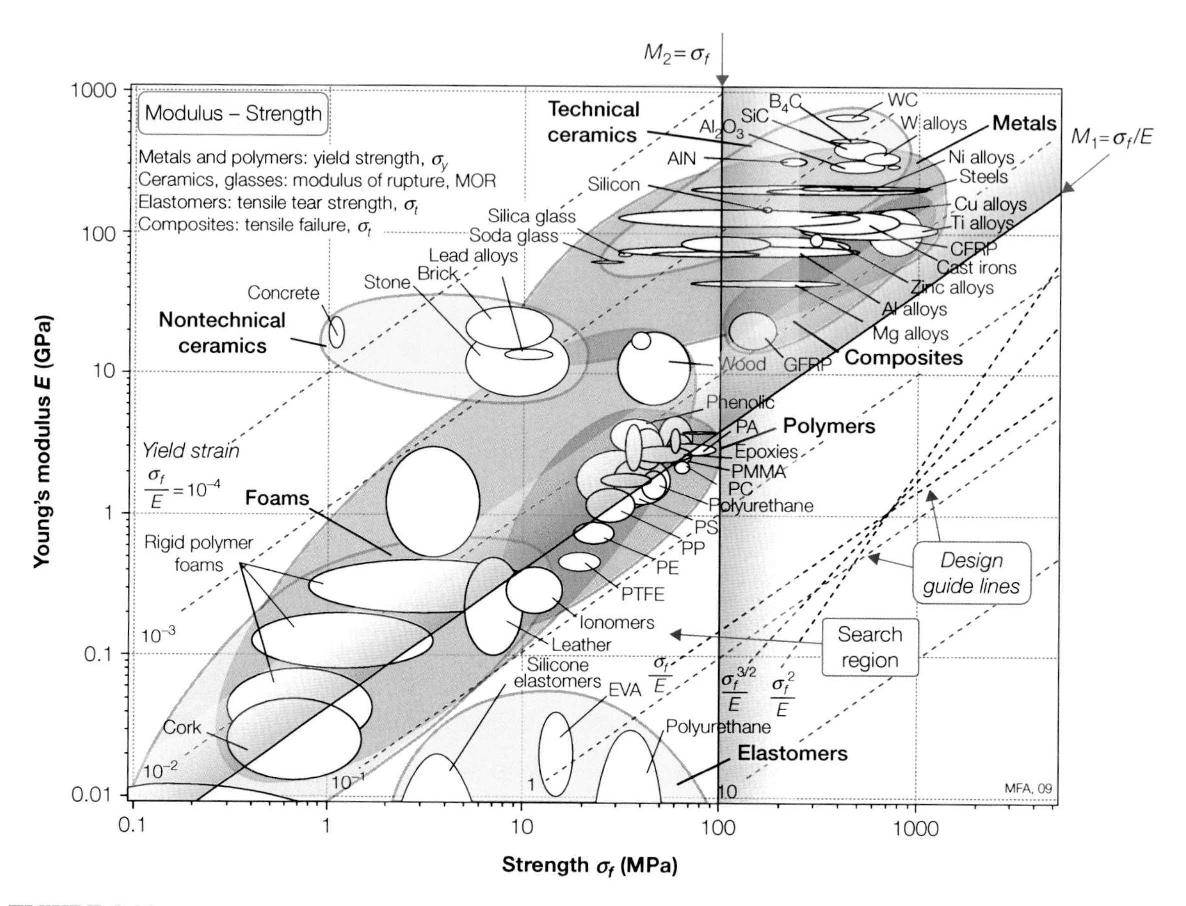

FIGURE 6.16

Materials for elastic seals. Elastomers, compliant polymers, and foams make good seals.

Table 6.16 Materials for Reusable Seals

Material	$M_1 = \frac{\sigma_f}{E}$	Comment
Elastomeric EVA	0.7–1	The natural choice; poor resistance to heat and to some solvents
Polyurethanes	2–5	Widely used for seals
Silicone rubbers	0.2–0.5	Higher temperature capability than carbon-chain elastomers, chemically inert
PTFE	0.05–0.1	Expensive but chemically stable and with high temperature capability
Polyethylenes	0.02–0.05	Inexpensive but liable to take a permanent set
Polypropylenes	0.2–0.04	Inexpensive but liable to take a permanent set
Nylons	0.02–0.03	Near upper limit on contact pressure
Cork	0.03–0.06	Low contact stress, chemically stable
Polymer foams	Up to 0.03	Very low contact pressure; delicate seals

6.10 DEFLECTION-LIMITED DESIGN WITH BRITTLE POLYMERS

The resistance of a material to the propagation of a crack is measured by its plane-strain fracture toughnesses K_{1c}. Among mechanical engineers there is a rule of thumb: avoid materials with $K_{1c} < 15$ MPa.m$^{1/2}$. Almost all metals pass: They have values of K_{1c} in the range of 20 to 100 in these units. White cast iron and some powder-metallurgy products fail; they have values as low as 10 MPa.m$^{1/2}$. Ordinary engineering ceramics have values in the range 1 to 6 MPa.m$^{1/2}$; mechanical engineers view them with deep suspicion. But engineering polymers are even less tough, with K_{1c} in the range 0.5 to 3 MPa.m$^{1/2}$ and yet engineers use them all the time. What is going on here?

When a brittle material is deformed, it deflects elastically until it fractures. The stress at which this happens is

$$\sigma_f = \frac{CK_{1c}}{\sqrt{\pi a_c}} \tag{6.33}$$

where K_c is an appropriate fracture toughness, a_c is the length of the largest crack contained in the material, and C is a constant that depends on geometry but is usually about 1. In a *load-limited* design—a tension member of a bridge, say—the part will fail in a brittle way if the stress exceeds that given by Equation (6.33). Here, obviously, we want materials with high values of K_{1c}.

But not all designs are load-limited; some are *energy-limited*, others are *deflection-limited*. Then the criterion for selection changes. Consider, then, the three scenarios created by the three alternative constraints of Table 6.17.

The translation In load-limited design the component must carry a specified load or pressure without fracturing. It is usual to identify K_c with the plane-strain fracture toughness, K_{1c}, corresponding to the most highly constrained cracking conditions, because this is conservative. Then, as Equation (6.33) shows, the best materials for minimum volume design are those with high values of

$$M_1 = K_{1c} \tag{6.34}$$

Table 6.17 Design Requirements

Function	Resist brittle fracture
Constraints	Design load specified or Design energy specified or Design deflection specified
Objective	Minimize volume (mass, cost)
Free variable	Choice of material

For load-limited design using thin sheet, a plane-stress fracture toughness may be more appropriate; and for multilayer materials, it may be an interface fracture toughness that matters. The point, though, is clear enough: The best materials for load-limited design are those with large values of the appropriate K_c.

But, as we have said, not all design is load-limited. Springs and containment systems for turbines and flywheels are *energy*-limited. Take the spring (refer to Figure 6.11) as an example. The elastic energy per unit volume stored in it is the integral over the volume of

$$U_e = \frac{1}{2}\sigma\varepsilon = \frac{1}{2}\frac{\sigma^2}{E}$$

The stress is limited by the fracture stress of Equation (6.33) so that—if "failure" means "fracture"—the maximum energy the spring can store is

$$U_e^{max} = \frac{C^2}{2\pi a_c}\left(\frac{K_{1c}^2}{E}\right)$$

For a given initial flaw size, energy is maximized by choosing materials with large values of

$$M_2 = \frac{K_{1c}^2}{E} \approx J_c \qquad (6.35)$$

where J_c is the toughness (usual units: kJ/m^2).

There is a third scenario: that of *displacement*-limited design (Figure 6.17). Snap-on bottle tops, snap-together fasteners and the like, are displacement-limited: They must allow sufficient elastic displacement to permit the snap action without failure, requiring a large failure strain ε_f. The strain is related to the stress by Hooke's law $\varepsilon = \sigma/E$, and the stress is limited by the fracture equation (Eq. 6.33). Thus the failure strain is

$$\varepsilon_f = \frac{C}{\sqrt{\pi a_c}}\frac{K_{1c}}{E}$$

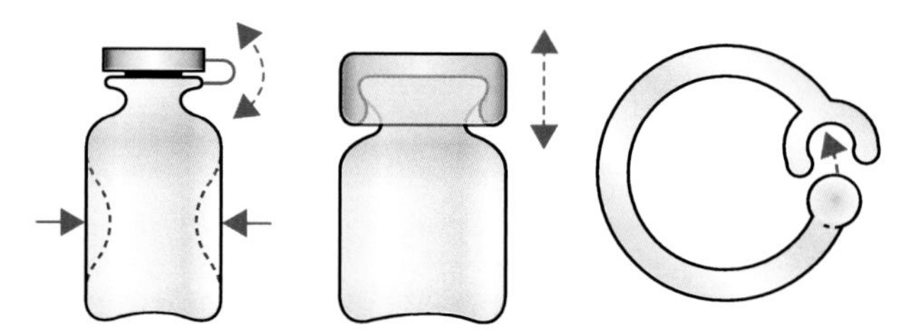

FIGURE 6.17
Load- and deflection-limited design. Polymers, having low moduli, frequently require deflection-limited design methods.

The best materials for displacement-limited design are those with large values of

$$M_3 = \frac{K_{1c}}{E}$$

The selection Figure 6.18 shows the chart of fracture toughness, K_{1c}, plotted against modulus, E. It allows materials to be compared by values of fracture toughness, M_1, by toughness, M_2, and by values of the deflection-limited index M_3. As the engineer's rule of thumb demands, almost all metals have values of K_{1c} that lie above the 15-MPa.m$^{1/2}$ acceptance level for load-limited design, shown as a horizontal selection line in the figure. Polymers and ceramics do not.

The line showing M_2 in Figure 6.18 is placed at the value 1 kJ/m^2. Materials with values of M_2 greater than this have a degree of shock resistance with which engineers feel comfortable (another rule of thumb). Metals,

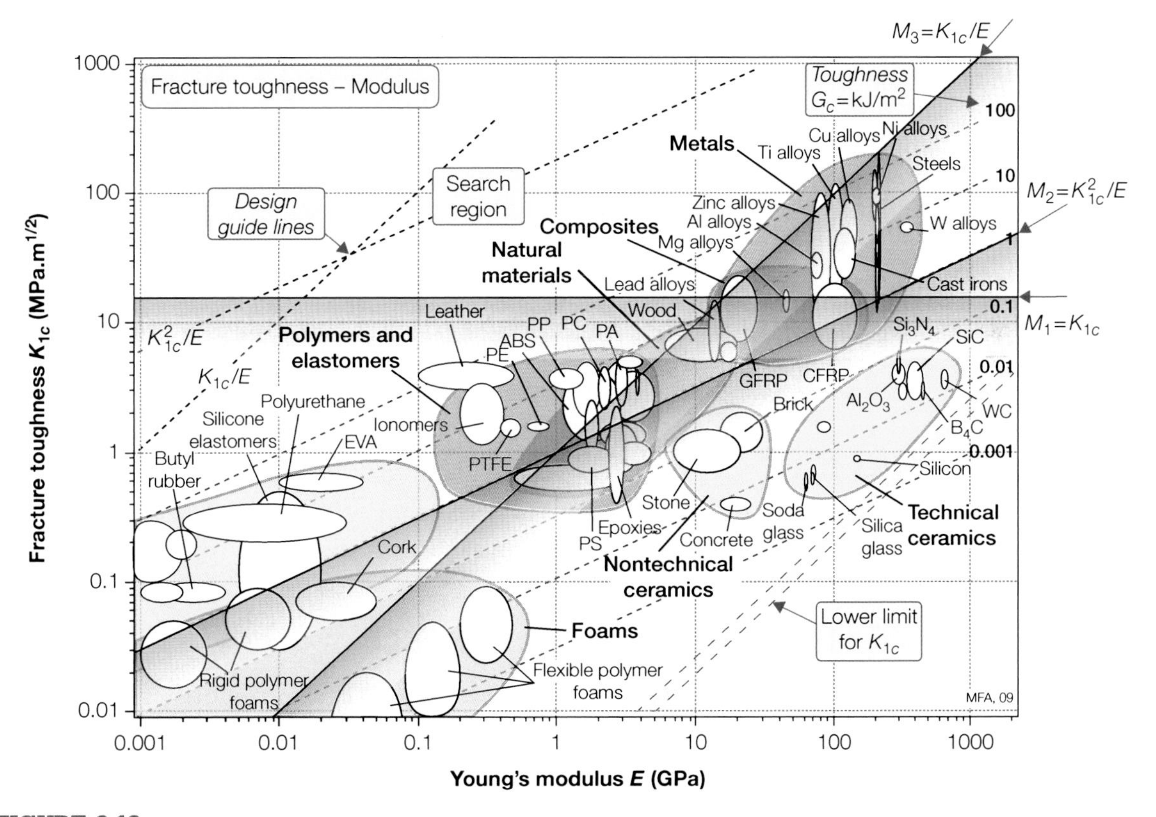

FIGURE 6.18

The selection of materials for load-, deflection-, and energy-limited design. In deflection-limited design, polymers are as good as metals, despite having very low values of fracture toughness.

Table 6.18 Materials for Fracture-Limited Design

Design type and rule of thumb	Material
Load-limited design $K_{1c} > 15$ MPa.m$^{1/2}$	Metals, polymer-matrix composites
Energy-limited design $J_c > 1$ kJ/m^2	Metals, composites, and some polymers
Displacement-limited design $K_{1c}/E > 10^{-3}$m$^{1/2}$	Polymers, elastomers, and the toughest metals

composites, and some polymers qualify; ceramics do not. When we come to deflection-limited design, the picture changes again. The line shows the index $M_3 = K_{1c}/E$ at the value $10^{-3}\text{m}^{1/2}$. It illustrates why polymers find such wide application: When the design is deflection-limited, polymers—particularly polypropylene, ABS, and nylons—are better than the best metals (Table 6.18).

Postscript The figure gives further insights. The mechanical engineers' love of metals (and, more recently, of composites) is inspired not merely by the appeal of their K_{1c} values. They are good by all three criteria (K_{1c}, K_{1c}^2/E, and K_{1c}/E). Polymers have good values of K_{1c}/E and are acceptable by K_{1c}^2/E. Ceramics are poor by all three criteria. Herein lie the deeper roots of the engineers' distrust of ceramics.

Related reading

Background in fracture mechanics and safety criteria can be found in

Brock, D. (1984). *Elementary engineering fracture mechanics.* Martinus Nijoff.
Hellan, K. (1985). *Introduction to fracture mechanics.* McGraw-Hill.
Hertzberg, R.W. (1989). *Deformation and fracture mechanics of engineering materials.* Wiley.

Related case studies

6.7 "Materials for springs"
6.8 "Elastic hinges and couplings"
6.11 "Safe pressure vessels"

6.11 SAFE PRESSURE VESSELS

Pressure vessels, from the simplest aerosol can to the biggest boiler, are designed, for safety, to yield or leak before they break. The details of this design method vary. Small pressure vessels are usually designed to allow general yield at a pressure still too low to cause any crack the vessel may contain to propagate ("yield before break"); the distortion caused by yielding is easy to detect and the pressure can be released safely. With large pressure vessels this may not be possible. Instead, safe design is achieved by ensuring that the

Table 6.19 Design Requirements for Safe Pressure Vessels

Function	Pressure vessel (contain pressure p safely)
Constraint	Radius R specified
Objective	Maximize safety using yield-before-break criterion or leak-before-break criterion
Free variable	Choice of material

smallest crack that will propagate unstably has a length greater than the thickness of the vessel wall ("leak before break"). The leak is easily detected, and it releases pressure gradually and thus safely (Table 6.19). The two criteria lead to different material indices. What are they?

The translation The stress in the wall of a thin-walled spherical pressure vessel of radius R (Figure 6.19) is

$$\sigma = \frac{pR}{2t} \tag{6.36}$$

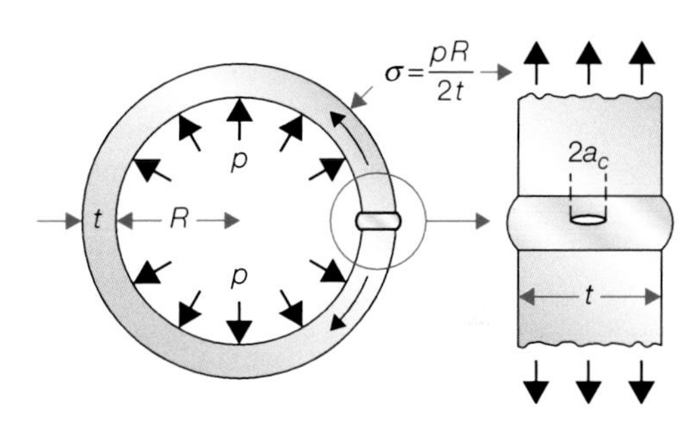

FIGURE 6.19
A pressure vessel containing a flaw. Safe design of small pressure vessels requires that they yield before they break; that of large pressure vessels may require, instead, that they leak before they break.

In pressure vessel design, the wall thickness, t, is chosen so that, at the working pressure p, this stress is less than the yield strength σ_f of the wall (with a safety factor, of course). A small pressure vessel can be examined ultrasonically, or by X-ray methods, or it can be proof-tested, to establish that it contains no crack or flaw of diameter greater than $2a_c^*$; then the stress required to make the crack propagate[2] is

$$\sigma = \frac{CK_{1c}}{\sqrt{\pi a_c^*}} \tag{6.37}$$

where C is a constant near unity and K_{1c} is the plane-strain fracture toughness. Safety can be achieved by ensuring that the working stress is less than this, giving

$$p \leq \frac{2t}{R} \frac{K_{1c}}{\sqrt{\pi a_c^*}} \tag{6.38}$$

The largest pressure (for a given R, t, and a_c^*) is carried by the material with the greatest value of

$$M_1 = K_{1c} \tag{6.39}$$

---

[2] If the wall is sufficiently thin, and close to general yield, it will fail in a plane-stress mode. Then the relevant fracture toughness is that for plane stress, not the smaller value for plane strain.

But this design is not failsafe. If the inspection is faulty or if, for some other reason a crack of length greater than a_c^* appears, catastrophe follows. Greater security is obtained by requiring that the crack will not propagate even if the stress reaches the general yield stress—for then the vessel will deform stably in a way that can be detected. This condition is expressed by setting σ equal to the yield stress σ_y, giving

$$\pi a_c \leq C^2 \left[\frac{K_{1c}}{\sigma_y}\right]^2$$

The tolerable crack size, and thus the integrity of the vessel, is maximized by choosing a material with the largest value of

$$M_2 = \frac{K_{1c}}{\sigma_y} \tag{6.40}$$

Large pressure vessels cannot always be X-rayed or sonically tested; and proof testing them may be impractical. Further, cracks can grow slowly because of corrosion or cyclic loading, so a single examination at the beginning of service life is not sufficient. Then safety can be ensured by arranging that a crack just large enough to penetrate both the inner and the outer surface of the vessel is still stable, because the leak caused by the crack can be detected. This is achieved by setting $a_c^* = t/2$. Safety is ensured if the stress is always less than or equal to

$$\sigma = \frac{CK_{1c}}{\sqrt{\pi t/2}} \tag{6.41}$$

The wall thickness t of the pressure vessel was, of course, designed to contain the pressure p without yielding. From Equation (6.38), this means that

$$t \geq \frac{pR}{2\sigma_y} \tag{6.42}$$

Substituting this into the previous equation (with $\sigma = \sigma_f$) gives

$$p \leq \frac{4\,C^2}{\pi\,R}\left(\frac{K_{1c}^2}{\sigma_y}\right) \tag{6.43}$$

The maximum pressure is carried most safely by the material with the greatest value of

$$M_3 = \frac{K_{1c}^2}{\sigma_y} \tag{6.44}$$

Both M_2 and M_3 could be made large by making the yield strength of the wall, σ_y, very small: Lead, for instance, has high values of both, but you would not choose it for a pressure vessel. That is because the vessel wall must also be kept thin, both for economy of material and to keep it light. The thinnest wall, from Equation (6.42), is that with the largest yield strength, σ_y. Thus we wish also to maximize

$$M_4 = \sigma_y \tag{6.45}$$

further narrowing the choice of material.

The selection These selection criteria are explored using the chart shown in Figure 6.20: the fracture toughness, K_{1c}, plotted against the elastic limit σ_f (meaning yield strength for metals and polymers). The indices M_1, M_2,

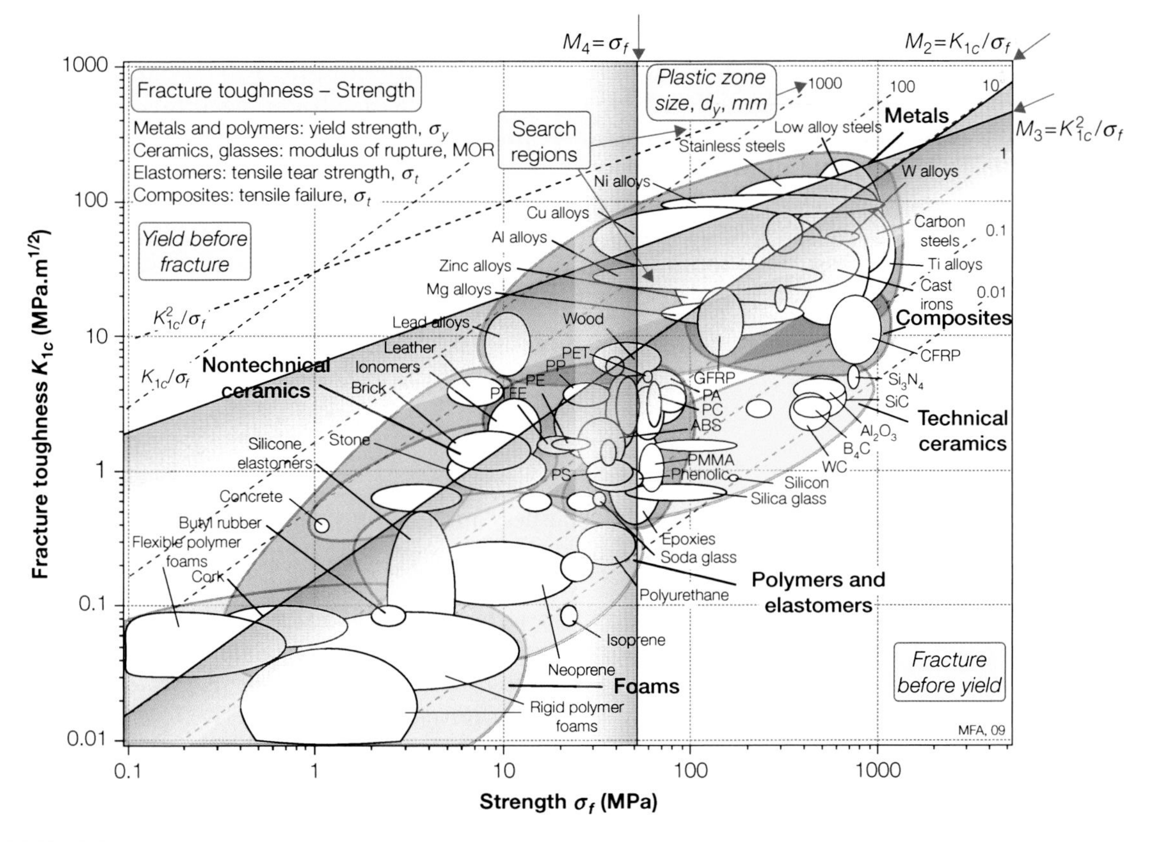

FIGURE 6.20

Materials for pressure vessels. Steel, copper alloys, and aluminum alloys best satisfy the "yield before break" criterion. In addition, a high yield strength allows a high working pressure. The materials in the "search regions" triangle are the best choice. The leak-before-break criterion leads to essentially the same selection.

Table 6.20 Materials for Safe Pressure Vessels

Material	$M_1 = K_{1c}/\sigma_y$ ($m^{1/2}$)	$M_3 = \sigma_y$ (MPa)	Comment
Stainless steels	0.35	300	Nuclear pressure vessels are made of grade 316 stainless steel
Low alloy steels	0.2	800	These are standard in this application
Copper	0.5	200	Hard-drawn copper is used for small boilers and pressure vessels
Aluminum alloys	0.15	200	Pressure tanks of rockets are aluminum
Titanium alloys	0.13	800	Good for light pressure vessels, but expensive

M_3 and M_4 appear as lines of slope 0, 1, and 1/2, and as lines that are vertical. Take "yield before break" as an example. A diagonal line corresponding to a constant value of $M_2 = K_{1c}/\sigma_y$ links materials with equal performance; those above the line are better. The line shown in the figure at a value of M_1, corresponding to a process zone of size 10 mm, excludes everything but the toughest steels, copper, aluminum, and titanium alloys, although some polymers—PP, PE, and PET for instance—nearly make it (pressurized lemonade and beer containers are made of these polymers). Details are given in Table 6.20.

The leak-before-break criterion

$$M_3 = \frac{K_{1c}^2}{\sigma_y}$$

favors low alloy steel and stainless and carbon steels more strongly. Polymers no longer qualify as candidates.

Postscript Large pressure vessels are always made of steel. The ones for models—a model steam engine, for instance—are made of copper. Copper is chosen even though it is more expensive, because of its greater resistance to corrosion. Corrosion rates do not scale with size. The loss of 0.1 mm through corrosion is not serious in a pressure vessel that is 10 mm thick; if it is only 1 mm thick, it becomes a concern.

Boiler failures used to be commonplace—there are even songs about it. Now they are rare, although when safety margins are pared to a minimum (rockets, new aircraft designs) pressure vessels still occasionally fail. This (relative) success is one of the major contributions of fracture mechanics to engineering practice.

Related reading

Background in fracture mechanics and safety criteria can be found in

Brock, D. (1984). *Elementary engineering fracture mechanics.* Martinus Nijoff.
Hellan, K. (1985). *Introduction to fracture mechanics.* McGraw-Hill.
Hertzberg, R.W. (1989). *Deformation and fracture mechanics of engineering materials.* Wiley.

Related case studies

6.6 "Materials for flywheels"
6.10 "Deflection-limited design with brittle polymers"
8.2 "Multiple constraints: Light pressure vessels"

6.12 STIFF, HIGH-DAMPING MATERIALS FOR SHAKER TABLES

Shakers are the members of an obscure and declining New England religious sect, noted for their austere wooden furniture. To those who live elsewhere they are devices for vibration testing (Figure 6.21). This second sort of shaker consists of an electromagnetic actuator driving a table, at frequencies up to 1000 Hz, to which the test object (a space probe, an automobile, an aircraft component, or the like) is clamped. The shaker applies a spectrum of vibration frequencies, f, and amplitudes, A, to the test object to explore its response.

A big table operating at high frequency dissipates a great deal of power. The primary objective is to minimize this, but subject to a number of constraints itemized in Table 6.21. What materials make good shaker tables?

The translation The power p (watts) consumed by a dissipative vibrating system with a sinusoidal input is

$$p = C_1 \, m A^2 \omega^3 \qquad (6.46)$$

where m is the mass of the table, A is the amplitude of vibration, ω is the frequency (rad/s) and C_1 is a constant. Provided the operating frequency ω is significantly less than the resonant frequency of the table, $C_1 \approx 1$. The amplitude A and the frequency ω are prescribed. To minimize the power lost in shaking the table itself, we must minimize its mass m. We idealize the table as a disk of given radius, R. Its thickness, t, is a free variable. Its mass is shown in Equation (6.47).

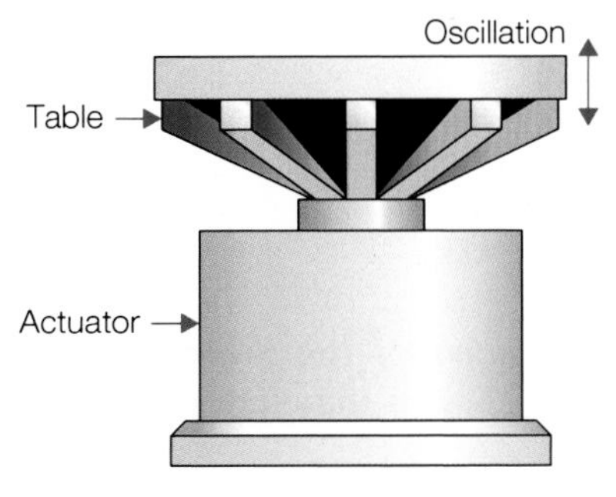

FIGURE 6.21
A shaker table. It is required to be stiff, but have high intrinsic "damping" or loss coefficient.

Table 6.21 Design Requirements for Shaker Tables

Function	Table for vibration tester ("shaker table")
Constraints	Radius, R, specified Must be stiff enough to avoid distortion by clamping forces Natural frequencies above maximum operating frequency (to avoid resonance) High damping to suppress resonance and natural vibrations Tough enough to withstand mishandling and shock
Objective	Minimize power consumption
Free variables	Choice of material Table thickness, t

$$m = \pi R^2 t \rho \tag{6.47}$$

where ρ is the density of the material of which it is made. The thickness influences the bending stiffness of the table—and this is important both to prevent the table flexing too much under clamping loads and because it determines its natural vibration frequencies. The bending stiffness, S, is

$$S = \frac{C_2 E I}{R^3}$$

where C_2 is a constant. The second moment of the section, I, is proportional to $t^3 R$. Thus, for a given stiffness S and radius R,

$$t = C_3 \left(\frac{S R^2}{E}\right)^{1/3}$$

where C_3 is another constant. The thinnest table is the one made from the material with the greatest value of

$$M_1 = E$$

Inserting this expression for t into Equation (6.47), we obtain

$$m = C_3 \pi R^{8/3} S^{1/3} \left(\frac{\rho}{E^{1/3}}\right) \tag{6.48}$$

The mass of the table, for a given stiffness and a minimum vibration frequency, is therefore minimized by selecting materials with high values of

$$M_2 = \frac{E^{1/3}}{\rho}$$

There are three further requirements. The first is high mechanical damping, measured by the loss coefficient, η, to suppress resonance. The second that the yield strength and fracture toughness, K_{1c}, of the table be sufficient to withstand mishandling and clamping forces. And it must not be too thick.

The selection If we had a chart with $E^{1/3}/\rho$ on one axis and η on the other, we could read off the materials with high values of both. Computer-based methods, illustrated in later case studies, allow charts with any desired combination of properties as axes. But for now we will stick with the charts in Chapter 4, requiring a 2-step selection. Figure 6.4 is an $E - \rho$ chart with a contour of $E^{1/3}/\rho$ already plotted on it. Materials with high values lie above or just below it. They are listed in the first column of Table 6.22. We now turn to the $\eta - E$ chart, reproduced as Figure 6.22. Materials with high values of M_1 lie to the right of the vertical line (here set at 30 GPa); those with $\eta > 0.001$ lie above the horizontal line. The search region contains a number of the candidates in the table: CFRP and alloys of magnesium, aluminum and titanium. All are possible candidates. Table 6.22 compares their properties.

Postscript Stiffness, high natural frequencies, and damping are qualities often sought in engineering applications such as engine and machine-tool mountings, supports for precision instruments, and building foundations. The shaker table found its solution (in real life as well as in this case study) in the choice of a cast magnesium alloy.

Sometimes a solution is possible by combining materials (more on this in Chapter 11). The loss coefficient chart shows that polymers and elastomers have high damping. Sheet steel panels, prone to vibration, can be damped by coating one surface with a polymer, a technique exploited in automobiles, typewriters, and machine tools. Aluminum structures can be stiffened (raising natural frequencies) by bonding carbon fiber to them: an approach sometimes use in aircraft design. And structures loaded in bending or torsion can be made lighter, for the same stiffness (again increasing

Table 6.22 Materials for Shaker Tables

Material	$M_1 = E^{1/3}/\rho$ $GPa^{1/3}/(Mg/m^3)$	Loss Coefficient, η	Fr. Toughness K_{1c} $MPa.m^{1/2}$	Comment
Mg-alloys	Up to 2.3	Up to 0.03	15	The best combination of properties
Al-alloys	Up to 1.7	Up to 0.002	30	Less damping than Mg or Ti
Ti-alloys	Up to 1.1	Up to 0.003	60	Good damping but expensive
CFRP	Up to 3.4	Up to 0.003	15	Less damping than Mg-alloys, but possible
Various ceramics	Up to 3.0	About 0.0002	3	Ruled out by low damping and toughness
Various foams	Up to 10	Up to 0.5	0.1	Lacks strength and toughness to carry service loads

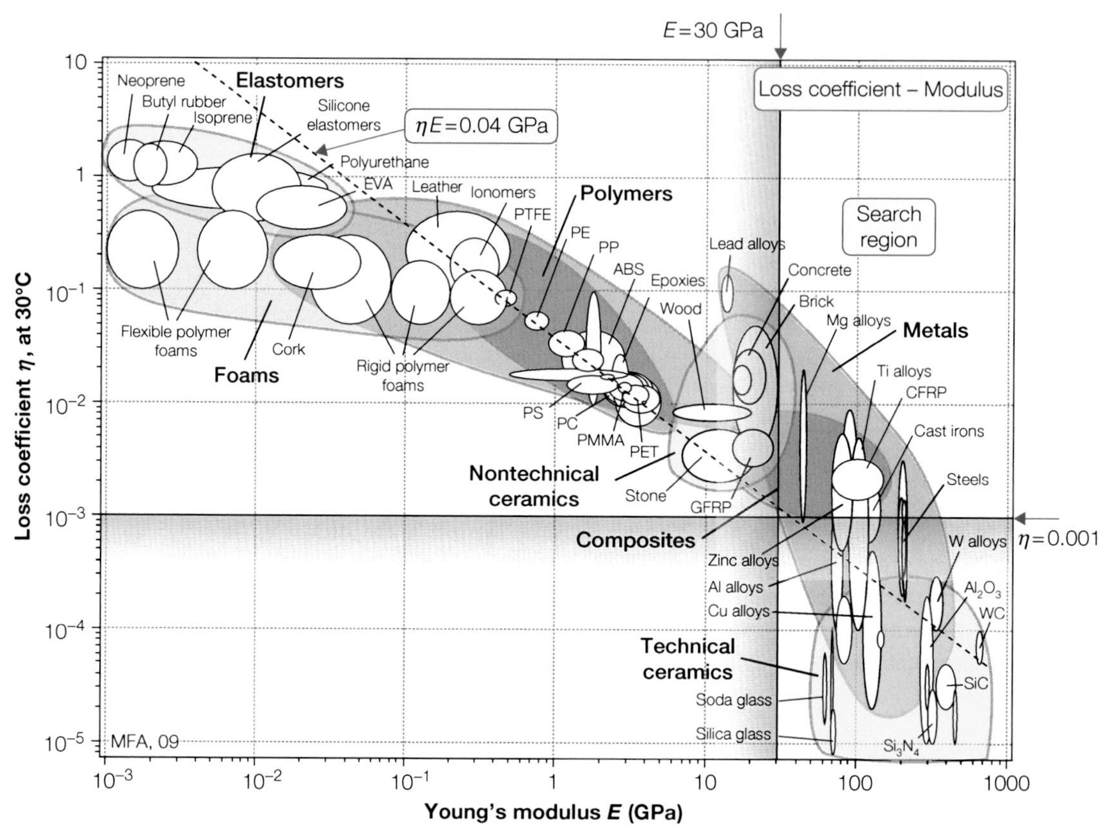

FIGURE 6.22

Selection of materials for the shaker table. Magnesium alloys, cast irons, GFRP, concrete, and the special high-damping Mn-Cu alloys are candidates.

natural frequencies), by shaping them efficiently: by attaching ribs to their underside, for instance. Shaker tables—even the austere wooden tables of the New England Shakers—exploit shape in this way.

Related reading

Tustin, W, & Mercado, R. (1984). *Random vibrations in perspective.* Tustin Institute of Technology Inc., Santa Barbara.

Cebon, D., & Ashby, M.F. (1994). Materials selection for precision instruments. *Meas. Sci. and Technol., 5,* 296–306.

Related case studies

6.4 "Materials for table legs"
6.7 "Materials for springs"
6.16 "Materials to minimize thermal distortion in precision devices"

6.13 INSULATION FOR SHORT-TERM ISOTHERMAL CONTAINERS

Each member of the crew of a military aircraft carries, for emergencies, a radio beacon. If forced to eject, the crew member could find himself in trying circumstances—in water at 4°C, for example (much of the earth's surface is ocean with a mean temperature of roughly this). The beacon guides friendly rescue services, minimizing exposure time.

But microelectronic metabolisms (like those of humans) are upset by low temperatures. In the case of the beacon, its transmission frequency that starts to drift. The design specification for the egg-shaped package containing the electronics (Figure 6.23) requires that, when the temperature of the outer surface is changed by 30°C, the temperature of the inner surface should not change significantly for an hour. To keep the device small, the wall thickness is limited to a w of 20 mm. What is the best material for the package? A dewar system is out—it is too fragile.

A foam of some sort, you might think. But here is a case in which intuition leads you astray. So let us formulate the design requirements (Table 6.23) and do the job properly.

The translation We model the container as a wall of thickness w, thermal conductivity λ. The heat flux q through the wall, once a steady state has been established, is given by Fick's first law:

$$q = -\lambda \frac{dT}{dx} = \lambda \frac{(T_i - T_o)}{w} \qquad (6.49)$$

where T_o is the temperature of the outer surface, T_i is that of the inner one, and dT/dx is the temperature

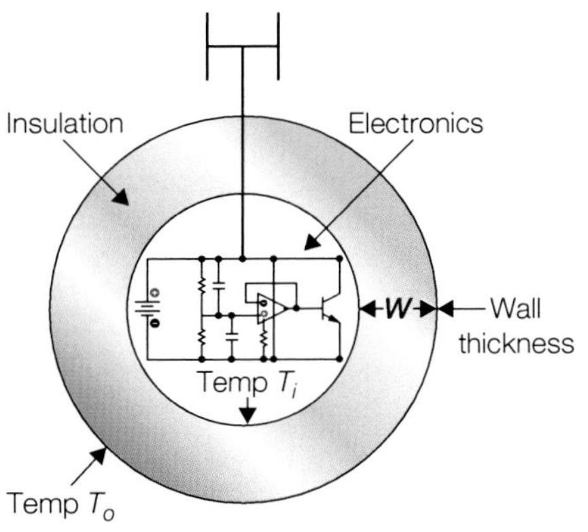

FIGURE 6.23
An isothermal container. It is designed to maximize the time before the inside temperature changes after the outside temperature has suddenly changed.

Table 6.23 Design Requirements for Short-term Insulation

Function	Short-term thermal insulation
Constraint	Wall thickness must not exceed w
Objective	Maximize time t before internal temperature changes when external temperature suddenly drops
Free variable	Choice of material

gradient (Figure 6.23). The only free variable here is the thermal conductivity, λ. The flux is minimized by choosing a wall material with the lowest possible value of λ. The $\lambda - \alpha$ chart (Figure 6.24) shows that this is, indeed, a foam.

But the incorrect question has been answered. The design brief was not to minimize the *heat flux* through the wall, but to maximize the *time* before the temperature of the inner wall changed appreciably. When the surface temperature of a body is suddenly changed, a temperature wave, so to speak, propagates inwards. The distance x it penetrates in time t is approximately $\sqrt{2at}$. Here a is the thermal diffusivity, defined by

$$a = \frac{\lambda}{\rho C_p} \tag{6.50}$$

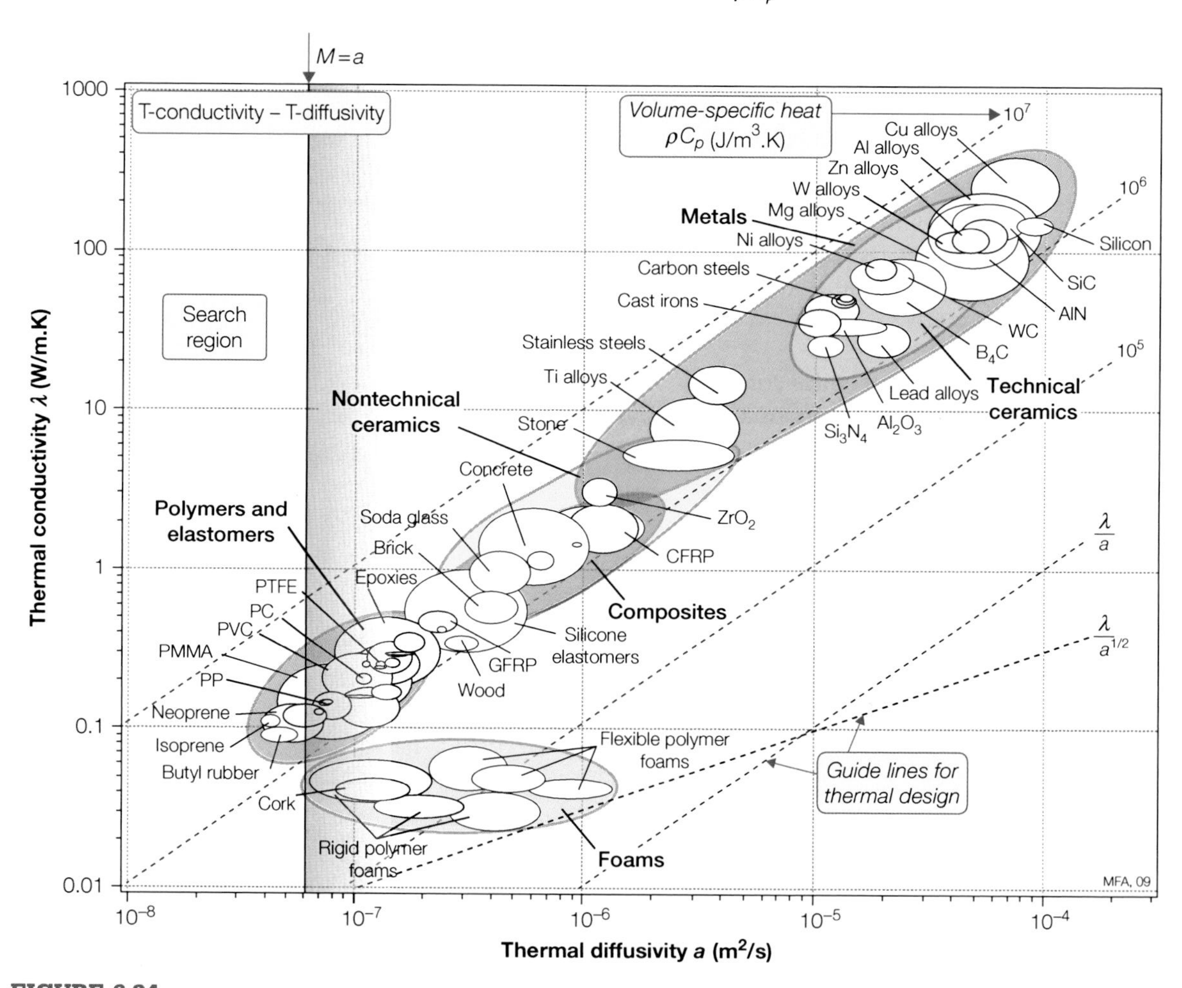

FIGURE 6.24
Materials for short-term isothermal containers. Elastomers are good; foams are not.

where ρ is the density and C_p is the specific heat (see Appendix B). Equating this to the wall thickness w gives

$$t \approx \frac{w^2}{2a} \tag{6.51}$$

The time is maximized by choosing the smallest value of the thermal diffusivity, a, not the conductivity, λ.

The selection Figure 6.24 shows that the thermal diffusivities of foams are not particularly low; it is because they have so little mass and thus heat capacity. The diffusivity of heat in a solid polymer or elastomer is much lower because these materials have specific heats that are exceptionally large. A package made of solid rubber, neoprene, or isoprene would—if of the same thickness—give the beacon a life 10 times greater than one made of (say) a polystyrene foam—though of course it would be heavier. Table 6.24 summarizes the conclusions. The reader can confirm, using Equation (6.51), that 22 mm of neoprene ($a = 5 \times 10^{-8}$ m^2/s, read from Figure 6.24) will allow a time interval of more than 1 hour after an external temperature change before the internal temperature shifts much.

Postscript One can do better than this. The trick is to exploit other ways of absorbing heat. If a liquid—a low-melting wax, for instance—can be found that solidifies at a temperature equal to the minimum desired operating temperature for the transmitter (T_i), it can be used as a "latent-heat sink." Channels in the package are filled with the liquid; the inner temperature can only fall below the desired operating temperature when all the liquid has solidified. The latent heat of solidification must be supplied to do this, giving the package a large (apparent) specific heat and thus an exceptionally low diffusivity for heat at the temperature T_i. The same idea is used, in reverse, in "freezer packs" that solidify when placed in the freezer compartment of a refrigerator and remain cold (by melting, at 4°C) when packed around warm beer cans in a portable cooler.

Table 6.24 Materials for Short-term Thermal Insulation

Material	Comment
Elastomers: Butyl rubber, neoprene and isoprene are examples	Best choice for short-term insulation
Commodity polymers: polyethylenes and polypropylenes	Less expensive than elastomers, but not quite as good for short-term insulation
Polymer foams	Much less good than elastomers for short-term insulation; best choice for long-term insulation at steady state

Related reading

Holman, J.P. (1981). *Heat transfer* (5th ed.). McGraw-Hill.

Related case studies

6.14 "Energy-efficient kiln walls"
6.15 "Materials for passive solar heating"

6.14 ENERGY-EFFICIENT KILN WALLS

The energy cost of one firing cycle of a large pottery kiln (Figure 6.25) is considerable. Part is the cost of the energy that is lost by conduction through the kiln walls; it is reduced by choosing a wall material with a low thermal conductivity, λ, and by making the wall thick. The rest is the cost of the energy used to raise the kiln to its operating temperature; it is reduced by choosing a wall material with a low heat capacity, C_p, and by making the wall thin. Is there a material index that captures these apparently conflicting design goals? And if so, what is a good choice of material for kiln walls? The choice is based on the requirements of Table 6.25.

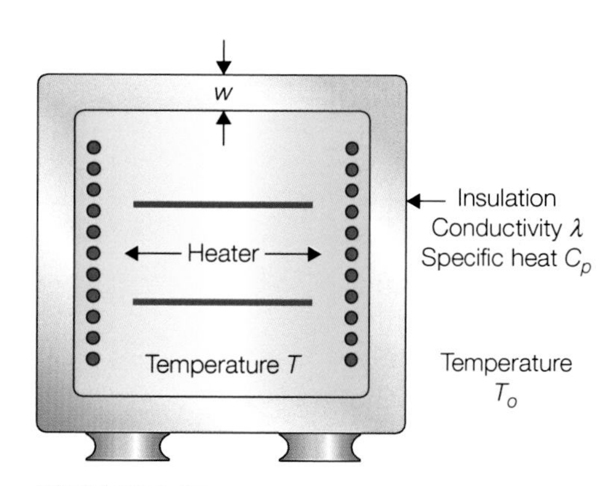

FIGURE 6.25
A kiln. On firing, the kiln wall is first heated to the operating temperature, then held at this temperature. A linear gradient is then expected through the kiln wall.

The translation When a kiln is fired, the internal temperature rises quickly from the ambient, T_o, to the operating temperature, T_i, where it is held for the firing time t. The energy consumed in the firing time has, as we have said, two contributions. The first is the heat conducted out: At steady state the heat loss by conduction, Q_1, per unit area, is given by the first law of heat flow. If held for time t, it is

$$Q_1 = -\lambda \frac{dT}{dx} t = \lambda \frac{(T_i - T_o)}{w} t \tag{6.52}$$

Here dT/dx is the temperature gradient and w is the insulation wall thickness. The second contribution is the heat absorbed by the kiln wall in raising it to T_i, and this can be considerable. Per unit area, it is

$$Q_2 = C_p \rho w \left(\frac{T_i - T_o}{2}\right) \tag{6.53}$$

Table 6.25 Design Requirements for Kiln Walls

Function	Thermal insulation for kiln (cyclic heating and cooling)
Constraints	Maximum operating temperature 1000°C Possible limit on kiln-wall thickness for space reasons
Objective	Minimize energy consumed in firing cycle
Free variables	Kiln wall thickness, w Choice of material

where C_p is the specific heat of the wall material and ρ is its density. The total energy consumed per unit area is the sum of these two:

$$Q = Q_1 + Q_2 = \frac{\lambda (T_i + T_o) t}{w} + \frac{C_p \rho w (T_i - T_o)}{2} \tag{6.54}$$

A wall that is too thin loses much energy by conduction, but absorbs little energy in heating the wall itself. One that is too thick does the opposite. There is an optimum thickness, which we find by differentiating Equation (6.54) with respect to wall thickness w and equating the result to zero, giving

$$w = \left(\frac{2\lambda t}{C_p \rho}\right)^{1/2} = (2at)^{1/2}, \tag{6.55}$$

where $a = \lambda/\rho C_p$ is the thermal diffusivity. The quantity $(2at)^{1/2}$ has dimensions of length and is a measure of the distance heat can diffuse in time t. Equation (6.55) says that the most energy-efficient kiln wall is one that only gets really hot on the outside as the firing cycle approaches completion. Substituting Equation (6.55) back into Equation (6.54) to eliminate w gives

$$Q = (T_i - T_o)(2t)^{1/2}(\lambda C_p \rho)^{1/2}$$

Q is minimized by choosing a material with a low value of the quantity $(\lambda C_p \rho)^{1/2}$, that is, by maximizing

$$M = (\lambda C_p \rho)^{-1/2} = \frac{a^{1/2}}{\lambda} \tag{6.56}$$

By eliminating the wall thickness w, we have lost track of it. It could, for some materials, be excessively large. Before accepting a candidate material, we must check, by evaluating Equation (6.55), how thick the wall made from it will be.

The selection Figure 6.26 shows the $\lambda - a$ chart, here with additional refractories and foams, with a selection line corresponding to $M = a^{1/2}/\lambda$ plotted on it. Polymer foams, cork, and solid polymers are good, but only if the internal temperature is less than 150°C. Real pottery kilns operate near 1000°C, requiring materials with a maximum service temperature above this value. The figure suggests brick (Table 6.26), but here the limitation of the hard-copy charts becomes apparent: There is not enough room to show specialized materials (e.g., refractory bricks and concretes). The limitation is overcome by the computer-based methods mentioned in Chapter 5, allowing a search over a much larger number of materials.

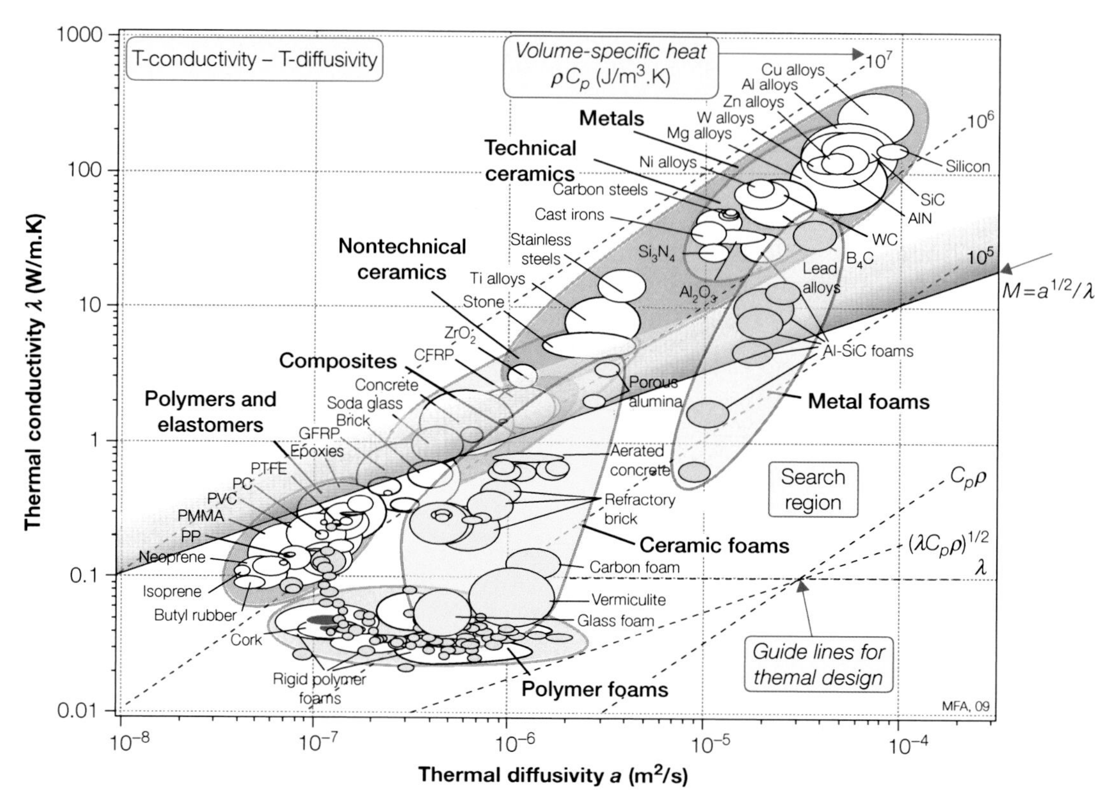

FIGURE 6.26

Materials for kiln walls. Low-density, porous, or foam-like ceramics are the best choice.

Table 6.26 Materials for Energy-efficient Kilns

Material	$M = a^{1/2}/\lambda$ ($m^2K/W.s^{1/2}$)	Thickness w (mm)	Comment
Refractory brick	5×10^{-3}	100	The obvious choice: The lower the density, the better the performance. Special refractory bricks have values of M as high as 3×10^{-3}.
Aerated concrete	2×10^{-3}	110	High-temperature concrete can withstand temperatures up to 1000°C.
Glass and carbon foams	Up to 10^{-2}	140	Both offer exceptional thermal insulation, but are limited to temperatures below 800°C.
Woods	2×10^{-3}	60	The boiler of Stevenson's "Rocket" steam engine was insulated with wood.
Solid elastomers and solid polymers	$2 \times 10^{-3} - 3 \times 10^{-3}$ 2×10^{-3}	50	Good values of material index. Useful if the wall must be very thin. Limited to temperatures below 150°C.
Polymer foam, cork	$3 \times 10^{-3} - 6 \times 10^{-2}$	50–140	The highest value of M—hence their use in house insulation. Limited to temperatures below 150°C.

Having chosen a material, the acceptable wall thickness is calculated from Equation (6.55). It is listed, for a firing time of three hours (approximately 10^4 seconds) in Table 6.26.

Postscript It is not generally appreciated that, in an efficiently designed kiln, as much energy goes in heating up the kiln itself as is lost by thermal conduction to the outside environment. It is a mistake to make kiln walls too thick; a little is saved in reduced conduction loss, but more is lost in the greater heat capacity of the kiln itself.

That, too, is the reason that foams are good: They have a low thermal conductivity *and* a low heat capacity. Centrally heated houses in which the heat is turned off at night suffer a cycle like that of the kiln. Here (because T_i is lower) the best choice is a polymeric foam, cork, or fiberglass (which has thermal properties like those of foams). But, as this case study shows, turning the heat off at night doesn't save you as much as you think because you have to supply the heat capacity of the walls in the morning.

Related reading

Holman, J.P. (1981). *Heat transfer* (5th ed.). McGraw-Hill, ISBN 0-07-029618-9.

Related case studies

6.13 "Insulation for short-term isothermal containers"
6.15 "Materials for passive solar heating"

6.15 MATERIALS FOR PASSIVE SOLAR HEATING

There are a number of schemes for capturing solar energy for home heating: solar cells, liquid-filled heat exchangers, and solid heat reservoirs. The simplest of these is the heat-storing wall: a thick wall, the outer surface of which is heated by exposure to direct sunshine during the day, and from which heat is extracted at night by blowing air over its inner surface (Figure 6.27). An essential of such a scheme is that the time constant for heat flow through the wall be about 12 hours; then the wall first warms on the inner surface roughly 12 hours after the sun first warms the outer surface giving out at night what it took in during the day. We will suppose that, for architectural reasons, the wall must not be more than ½ m thick. What materials maximize the thermal

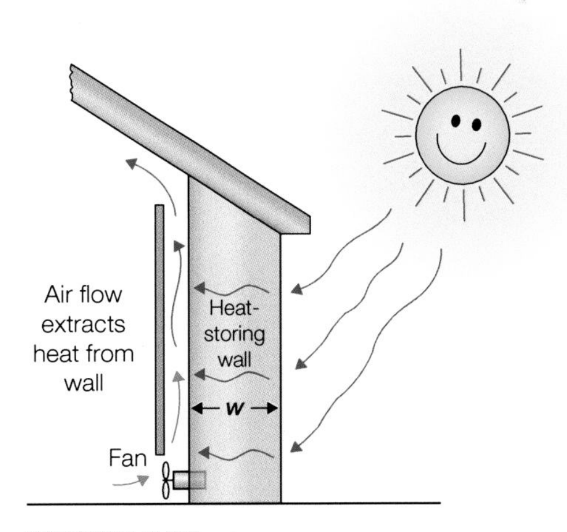

FIGURE 6.27

A heat-storing wall. The sun heats it during the day; heat is extracted from it at night. The heat-diffusion time through the wall must be about 12 hours.

Table 6.27 Design Requirements for Passive Solar Heating

Function	Heat-storing medium
Constraints	Heat diffusion time through wall $t \approx 12$ hours Wall thickness ≤ 0.5 m Adequate working temperature $T_{max} > 100°C$
Objective	Maximize thermal energy stored per unit material cost
Free variables	Wall thickness w Choice of material

energy captured by the wall while retaining a heat-diffusion time of up to 12 hours? Table 6.27 summarizes the requirements.

The translation The heat content, Q, per unit area of wall, when heated through a temperature interval ΔT gives the objective function

$$Q = w \rho C_p \Delta T \tag{6.57}$$

where w is the wall thickness, and ρ C_p is the specific heat per unit volume (the density ρ times the specific heat C_p). The 12-hour time constant is a constraint. It is adequately estimated by the approximation used earlier for the heat-diffusion distance in time t (see Appendix B)

$$w = \sqrt{2at} \tag{6.58}$$

where a is the thermal diffusivity. Eliminating the free variable w gives

$$Q = \sqrt{2t} \Delta T a^{1/2} \rho C_p \tag{6.59}$$

or, using the fact that $a = \lambda/\rho C_p$ where λ is the thermal conductivity,

$$Q = \sqrt{2t} \Delta T \left(\frac{\lambda}{a^{1/2}} \right)$$

The heat capacity of the wall is maximized by choosing material with a high value of

$$M = \frac{\lambda}{a^{1/2}} \tag{6.60}$$

This is the reciprocal of the index of the previous case study. The restriction on thickness w requires (from Equation (6.58)) that

$$a \leq \frac{w^2}{2t}$$

with $w \leq 0.5$ m and $t = 12$ h (4×10^4 s), we obtain an attribute limit

$$a \leq 3 \times 10^{-6} \mathrm{m}^2/\mathrm{s} \tag{6.61}$$

The selection Figure 6.28 shows thermal conductivity λ plotted against thermal diffusivity a with M and the limit on a plotted on it. It identifies the group of materials, listed in Table 6.28: They maximize M_1 while meeting the constraint on wall thickness. Solids are good; porous materials and foams (often used in walls) are not.

Postscript All this is fine, but what of cost? If this scheme is to be used for housing, cost is an important consideration. The approximate costs per unit volume, read from Tables A.3 and A.11 in Appendix A, are listed in the table—it points to the selection of concrete, with stone and brick as alternatives.

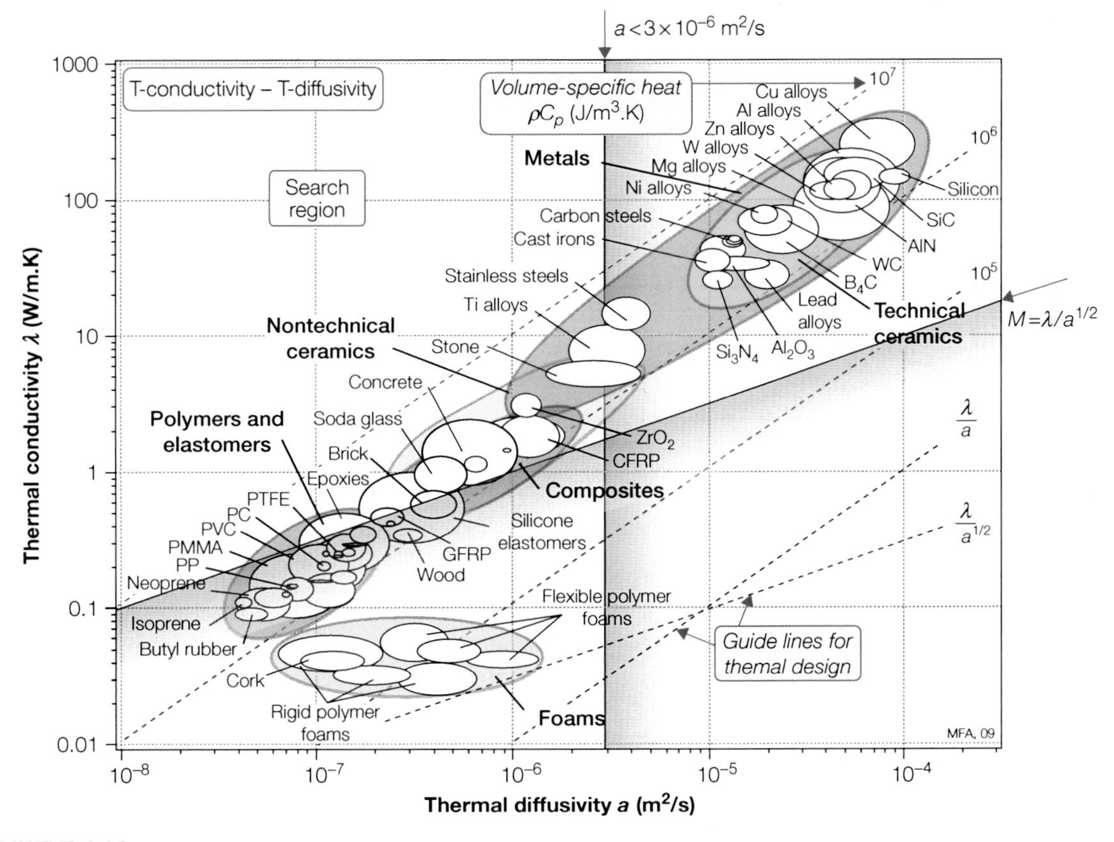

FIGURE 6.28

Materials for heat-storing walls. Cement, concrete, and stone are practical choices; brick is less good.

Table 6.28 Materials for Passive Solar Heat Storage

Material	$M_1 = \lambda/a^{1/2}$ ($W.s^{1/2}/m^2.K$)	Approximate Cost \$/m^3	Comment
Concrete	2.2×10^3	200	The best choice—good performance at minimum cost
Stone	3.5×10^3	1400	Better performance than concrete because specific heat is greater, but more expensive
Brick	10^3	1400	Not as good as concrete
Glass	1.6×10^3	10,000	Useful—part of the wall could be glass
Titanium	4.6×10^3	200,000	An unexpected, but valid, selection; expensive

Related case studies

6.13 "Insulation for short-term isothermal containers"

6.14 "Energy-efficient kiln walls"

6.16 MATERIALS TO MINIMIZE THERMAL DISTORTION IN PRECISION DEVICES

The precision of a measuring device, like a submicrometer displacement gauge, is limited by its stiffness and by the dimensional change caused by temperature gradients. Compensation for elastic deflection can be arranged; and corrections to cope with thermal expansion are possible, too—provided the device is at a uniform temperature. *Thermal gradients* are the real problem: They cause a change of shape—that is, a distortion of the device—for which compensation is not possible. Sensitivity to vibration is also a problem: Natural excitation introduces noise and thus imprecision into the measurement. So it is permissible to allow expansion in precision instrument design, provided distortion does not occur (Chetwynd, 1987). Elastic deflection is allowed, provided natural vibration frequencies are high.

What, then, are good materials for precision devices? Table 6.29 lists the requirements.

The translation Figure 6.29 shows, schematically, such a device: It consists of a probe, a force loop, an actuator, and a sensor. We want a material for the force loop. It will, in general, support heat sources: the fingers of the operator of the device in the figure, or, more usually, electrical or electronic components that generate heat. The relevant material index is found by considering the simple case of one-dimensional heat flow through a rod insulated except at its ends, one of which is at ambient temperature and the other is connected to the heat source. In the steady state, Fourier's law is

$$q = -\lambda \frac{dT}{dx} \tag{6.62}$$

Table 6.29 Design Requirements for Precision Devices	
Function	Force loop (frame) for precision device
Constraints	Must tolerate heat flux Must tolerate vibration
Objective	Maximize positional accuracy (minimize distortion)
Free variable	Choice of material

where q is heat flux per unit area, λ is the thermal conductivity, and $\frac{dT}{dx}$ is the resulting temperature gradient. The strain is related to temperature by

$$\varepsilon = \alpha(T - T_o) \qquad (6.63)$$

where α is the thermal expansion coefficient and T_o is ambient temperature. The distortion is proportional to the gradient of the strain:

$$\frac{d\varepsilon}{dx} = \frac{\alpha\, dT}{dx} = \left(\frac{\alpha}{\lambda}\right)q$$

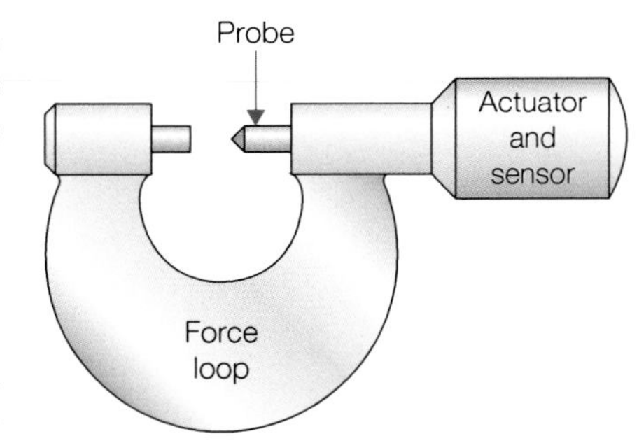

FIGURE 6.29
A schematic of a precision measuring device. Super-accurate dimension-sensing devices include the atomic-force microscope and the scanning-tunneling microscope.

Thus for a given geometry and heat flow, the distortion $d\varepsilon/dx$ is minimized by selecting materials with large values of the index

$$M_1 = \frac{\lambda}{\alpha} \qquad (6.64)$$

The other problem is vibration. The sensitivity to external excitation is minimized by making the natural frequencies of the device as high as possible. The flexural vibrations have the lowest frequencies; they are proportional to

$$M_2 = \frac{E^{1/2}}{\rho}$$

A high value of this index will minimize the problem. Finally, of course, the material must not cost too much.

The selection Figure 6.30 reproduces the chart of expansion coefficient, α, and thermal conductivity, λ. Contours show constant values of the quantity λ/α. A search region is isolated by the line $\lambda/\alpha = 10^7$ W/m, giving the shortlist of Table 6.30. Values of $M_2 = E^{1/2}/\rho$ read from the $E - \rho$ chart in

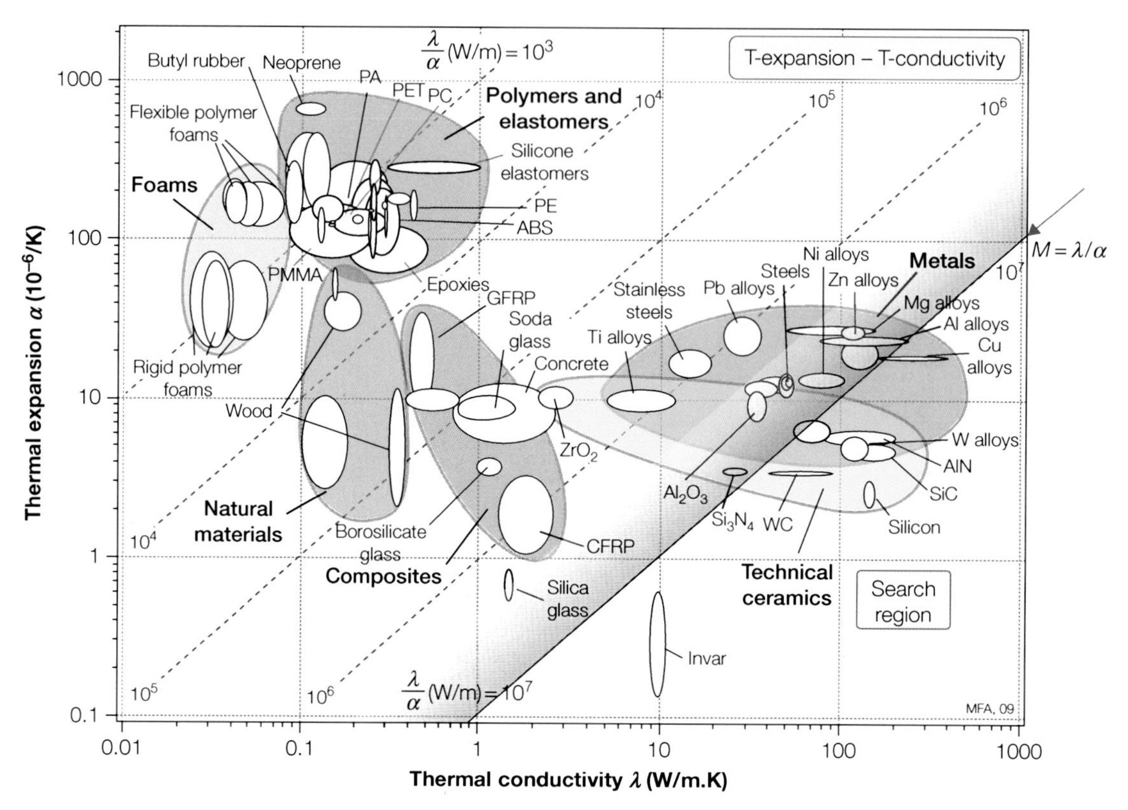

FIGURE 6.30

Materials for precision measuring devices. Metals are less good than ceramics because they have lower vibration frequencies. Silicon may be the best choice.

Table 6.30 Materials to Minimize Thermal Distortion

Material	$M_1 = \lambda/\alpha$ (W/m)	$M_2 = E^{1/2}/\rho$ (GPa$^{1/2}$)/(Mg/m^3)	Comment
Silicon	6×10^7	5.2	Excellent M_1 and M_2
Silicon carbide	3×10^7	6.4	Excellent M_1 and M_2 but more difficult to shape than silicon
Copper	2×10^7	1.3	High density gives poor value of M_2
Tungsten	3×10^7	1.1	Better than copper, silver, or gold, but not as good as silicon or SiC
Aluminum alloys	10^7	3.3	The least expensive and the most easily shaped choice

Figure 4.3 are included in the table. Among metals, copper, tungsten, and the special nickel alloy Invar have the best values of M_1 but are disadvantaged by having high densities and thus poor values of M_2. The best choice is silicon, available in large sections, with high purity. Silicon carbide is an alternative.

Postscript Nanoscale measuring and imaging systems suffer from the problem analyzed here. The atomic-force microscope and the scanning-tunneling microscope both rely on a probe, supported on a force loop, typically with a piezo-electric actuator and electronics to sense the proximity of the probe to the test surface. Closer to home, the mechanism of a video recorder and that of a hard-disk drive qualify as precision instruments; both have a sensor (the read head) attached, with associated electronics, to a force loop. The materials identified in this case study are the best choice for force loop.

Related reading

Chetwynd, D.G. (1987). Selection of structural materials for precision devices. *Precision Engineering, 9*(1), 3.

Cebon, D., & Ashby, M.F. (1994). Materials selection for precision instruments. *Meas. Sci. and Technol., 5*, 296.

Related case studies

6.3 "Mirrors for large telescopes"
6.13 "Insulation for short-term isothermal containers"

6.17 MATERIALS FOR HEAT EXCHANGERS

This and the next two case studies illustrate how the CES software, described in Section 5.5, is used to explore material selection in greater depth.

Heat exchangers take heat from one fluid and pass it to a second (Figure 6.31). The fire-tube array of a steam engine is a heat exchanger, taking heat from the hot combustion gases of the firebox and transmitting it to the water in the boiler. The network of finned tubes in an air conditioner is a heat exchanger, taking heat from the air of the room and dumping it into the working fluid of the conditioner. A key element in all heat exchangers is the tube wall or membrane that separates the two fluids. It is required to

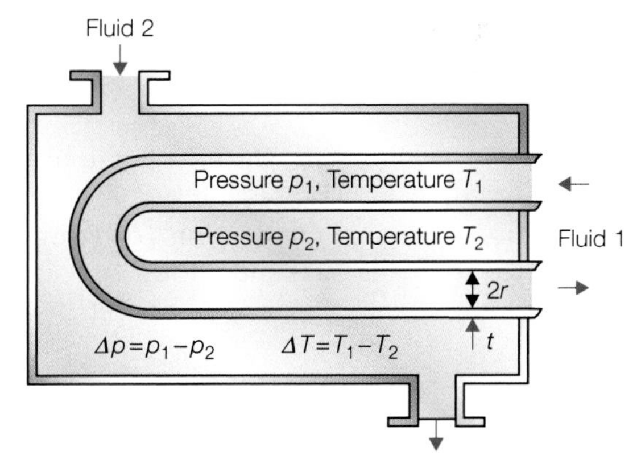

FIGURE 6.31
A heat exchanger. There is a pressure difference Δp and a temperature difference ΔT across the tube wall that also must resist attack by chloride ions.

Table 6.31 Design Requirements for a Heat Exchanger

Function	Heat exchanger
Constraints	Support pressure difference Δp Withstand chloride ions Operating temperature up to 150°C Modest cost
Objectives	Maximize heat flow per unit area (minimum volume exchanger), or maximize heat flow per unit mass (minimum mass exchanger)
Free variables	Tube-wall thickness, t Choice of material

transmit heat, and there is frequently a pressure difference across it, which can be large.

What are the best materials for making heat exchangers? Or, to be specific, what are the best materials for a conduction-limited exchanger with substantial pressure difference between the two fluids, one of them containing chloride ions (seawater)? Table 6.31 contains a summary of these requirements.

The translation First, a little background on heat flow. Heat transfer from one fluid, through a membrane to a second fluid, involves *convective* transfer from fluid 1 into the tube wall, *conduction* through the wall, and *convection* again to transfer it into fluid 2. The heat flux into the tube wall by convection (W/m^2) is described by the heat transfer equation:

$$q = h_1 \Delta T_1 \tag{6.65}$$

in which h_1 is the heat transfer coefficient and ΔT_1 is the temperature drop across the surface from fluid 1 into the wall. Conduction is described by the conduction (or Fourier) equation, which, for one-dimensional heat flow takes the form

$$q = \lambda \frac{\Delta T}{t} \tag{6.66}$$

where λ is the thermal conductivity of the wall (thickness t) and ΔT is the temperature difference across it. It is helpful to think of the *thermal resistance* at surface 1 as $1/h_1$; that of surface 2 as $1/h_2$; and that of the wall itself as t/λ. Then continuity of heat flux requires that the total resistance $1/U$ is

$$\frac{1}{U} = \frac{1}{h_1} + \frac{t}{\lambda} + \frac{1}{h_2} \tag{6.67}$$

where U is called the "total heat transfer coefficient." The heat flux from fluid 1 to fluid 2 is then given by

$$q = U\,(T_1 - T_2) \tag{6.68}$$

where $(T_1 - T_2)$ is the difference in temperature between the two working fluids.

When one of the fluids is a gas—as in an air conditioner—convective heat transfer at the tube surfaces contributes most of the resistance; then fins are used to increase the surface area across which heat can be transferred. But when both working fluids are liquid, convective heat transfer is rapid and conduction through the wall dominates the thermal resistance; $1/h_1$ and $1/h_2$ are negligible compared with t/λ. In this case, simple tube or plate elements are used, making the wall as thin as possible to minimize t/λ. We will consider the second case: conduction-limited heat transfer. Then, the heat flow is adequately described by Equation (6.66).

Consider, then, a heat exchanger with n tubes of length L, each of radius r and wall thickness t. Our aim is to select a material to maximize the total heat flow:

$$Q = qA = \frac{A\lambda}{t}\Delta T \tag{6.69}$$

where $A = 2\,\pi\,r\,L\,n$ is the total surface are of tubing.

This is the objective function. The constraint is that the wall thickness must be sufficient to support the pressure Δp between the inside and outside, as in Figure 6.31. This requires that the stress in the wall remain below the elastic limit, σ_y, of the material of which the tube is made (multiplied by a safety factor—which we can leave out):

$$\sigma = \frac{\Delta pr}{t} < \sigma_y \tag{6.70}$$

This constrains the minimum value of t. Eliminating t between Equations (6.69) and (6.70) gives

$$Q = \frac{A\Delta T}{r\Delta p}(\lambda\sigma_y) \tag{6.71}$$

The heat flow per unit area of the tube wall, Q/A, is maximized by maximizing

$$M_1 = \lambda\sigma_y \tag{6.72}$$

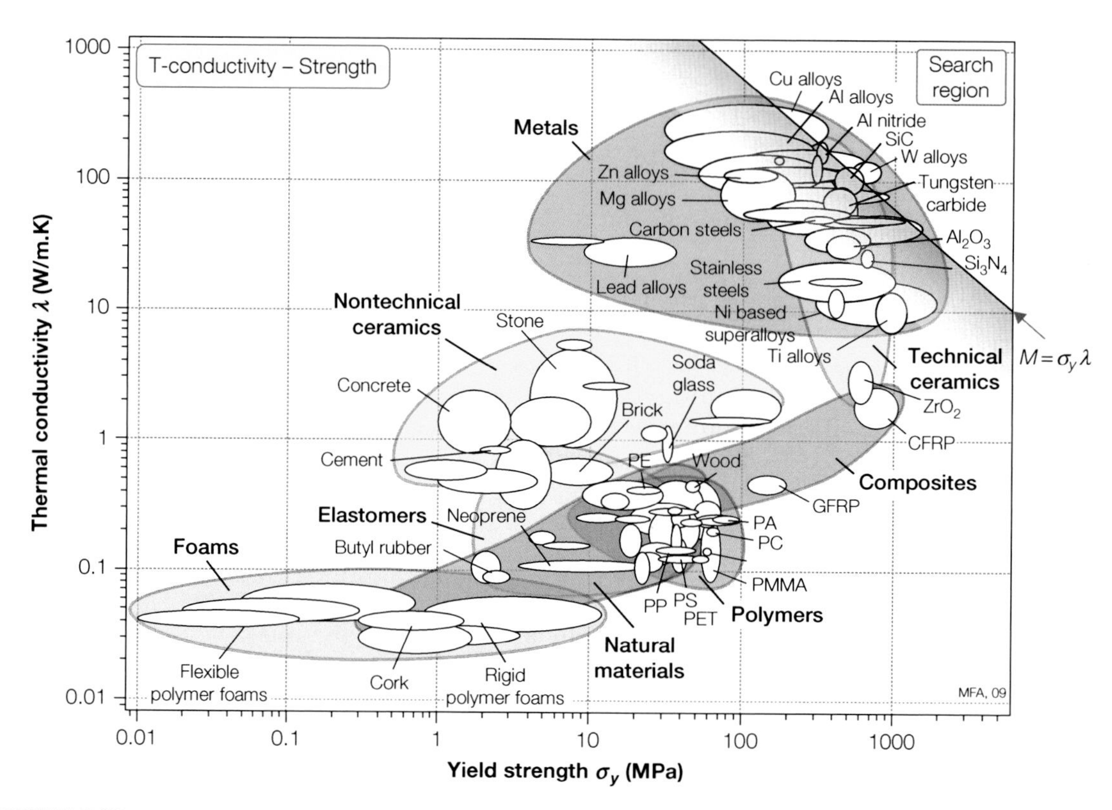

FIGURE 6.32
A chart of yield strength (elastic limit), σ_y, against thermal conductivity, λ, showing the index, M_1.

Four further considerations enter the selection. It is essential to choose a material that can withstand corrosion in the working fluids, which we take to be water containing chloride ions (seawater). Its maximum service temperature must be adequately above the temperature of the hotter working fluid and the material must have sufficient ductility to be drawn into a tube or rolled into a sheet.

The selection A preliminary search for materials with large values of M_1, using the $\lambda - \sigma_f$ chart of Figure 6.32, suggests wrought copper alloys as one possibility. We turn to computer-based methods[3] for more help. Using it we apply limits of 150°C on maximum service temperature, 30% on elongation, a material cost of less than \$6/kg, a rating of "very good" resistance to seawater, and a restriction of the search to copper alloys. With these in place we construct a new chart (Figure 6.33) of σ_y versus λ enabling $M_1 = \sigma_y\lambda$ to be maximized. The materials with large M_1 are listed in Table 6.32.

---

[3] The CES Edu system set at Level 3 (*www.Grantadesign.com*).

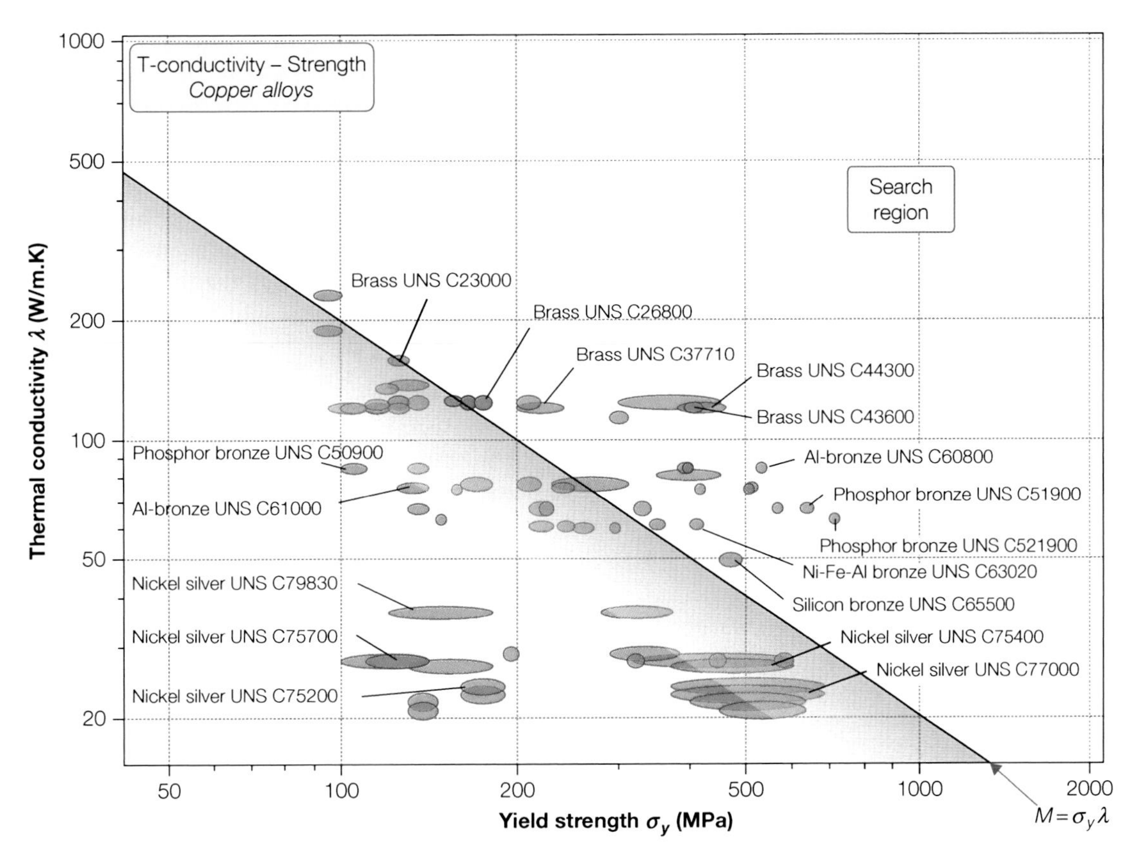

FIGURE 6.33

A more detailed $\lambda - \sigma_y$ for copper alloys, showing the index, M_1.

Table 6.32 Materials for Heat Exchangers

Material	Index M W.MN/m^3.K	Comment
Brasses, naval brass	5×10^4	Liable to dezincification
Phosphor bronzes	4×10^4	Cheap, but not as corrosion resistant as aluminum bronzes
Aluminum-bronzes, wrought	3.8×10^4	An economical and practical choice
Nickel-iron-aluminum-bronzes	2.5×10^4	More corrosion resistant, but more expensive
Silicon bronze	2.2×10^4	Less good than aluminum bronze

Postscript Conduction may limit heat flow in theory, but unspeakable things go on inside heat exchangers. Seawater—often one of the working fluids—seethes with biofouling organisms that attach themselves to tube walls and thrive there, like barnacles on a boat, creating a layer of high thermal resistance impeding fluid flow. A search for documentation reveals that some materials are more resistant to biofouling than others; copper-nickel alloys are particularly good, probably because the organisms dislike copper salts, even in very low concentrations. Otherwise, the problem must be tackled by adding chemical inhibitors to the fluids or by scraping—the traditional winter pastime of boat owners.

It is sometimes important to minimize the weight of heat exchangers. Repeating the calculation to seek materials, the maximum value of Q/m (where m is the mass of the tubes) gives, instead of M_1, the index

$$M_2 = \frac{\lambda \sigma_y^2}{\rho} \quad (6.73)$$

where ρ is the density of the material of which the tubes are made. (The strength σ_y is now raised to the power of 2 because the weight depends on wall thickness as well as density, and wall thickness varies as $1/\sigma_y$—see Equation (6.67)). Similarly, the cheapest heat exchangers are those made of the material with the greatest value of

$$M_3 = \frac{\lambda \sigma_y^2}{C_m \rho} \quad (6.74)$$

where C_m is the cost per kg of the material. In both cases aluminum alloys score highly because they are both light and cheap. The selections are not shown but can readily be explored using the CES system.

Related reading

Holman, J.P. (1981). *Heat transfer* (5th ed.). McGraw-Hill.

Related case studies

6.11 "Safe pressure vessels"
6.16 "Materials to minimize thermal distortion in precision devices"
6.18 "Heat sinks for hot microchips"

6.18 HEAT SINKS FOR HOT MICROCHIPS

A microchip may only consume milliwatts, but it is dissipated in a tiny volume. The power is low but the *power density* is high. As chips shrink and clock speeds grow, heating becomes a problem. The Pentium chip of today's PCs already reaches 85°C, requiring forced cooling. Multiple-chip modules (MCMs) pack as many as 130 chips onto a single substrate. Heating is kept under control by attaching the chip to a heat sink (Figure 6.34), taking pains to ensure good thermal contact between the chip and the sink. The heat sink now becomes a critical component, limiting further development of the electronics. How can its performance be maximized?

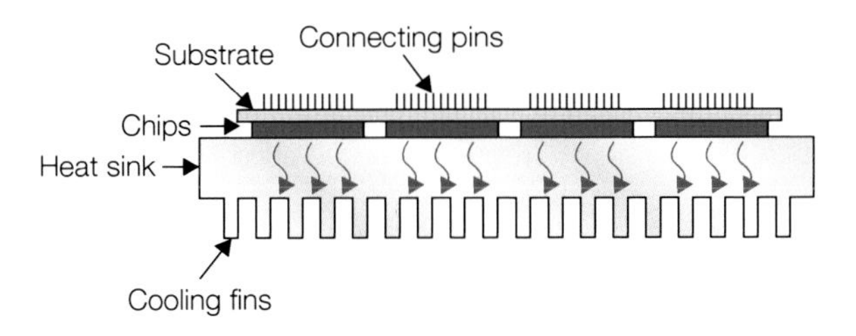

FIGURE 6.34
A heat sink for power microelectronics. The material must insulate electrically but conduct heat as well as possible.

Table 6.33 Design Requirements for Heat Sinks

Function	Heat sink
Constraints	Material must be "good insulator," or $\rho_e > 10^{18}\ \mu\Omega$.cm Maximum service temperature > 150°C All dimensions are specified
Objective	Maximize thermal conductivity, λ
Free variable	Choice of material

To prevent electrical coupling and stray capacitance between the chip and the heat sink, the heat sink must be a good electrical insulator, meaning a resistivity, $\rho_e > 10^{18}\ \mu\Omega$.cm. But to drain heat away from the chip as fast as possible, it must also have the highest possible thermal conductivity, λ. The translation step is summarized in Table 6.33, where we assume that all dimensions are constrained by other aspects of the design.

The translation Resistivity is treated as a *constraint*, a go/no go criterion. Materials that fail to qualify as a "good insulator" or that have a resistivity greater than the value listed in the table are screened out. The thermal conductivity is treated as an *objective*: Of the materials that meet the constraint, we seek those with the largest values of λ and rank them by this—it becomes the material index for the design. If we assume that all dimensions are fixed by the design, there remains only one free variable in seeking to maximize heat flow: the choice of material. The procedure is to *screen* on resistivity, then *rank* on conductivity.

The selection The steps can be implemented using the $\lambda - \rho_e$ chart in Figure 4.10, reproduced as Figure 6.35. Draw a vertical line at $\rho_e = 10^{18}\ \mu\Omega$.cm, then pick off the materials that lie above this line, and have the highest λ. The initial result: aluminum nitride, AlN, alumina, Al_2O_3, or silicon nitride, Si_3N_4. Of the three, aluminum nitride has the highest thermal conductivity.

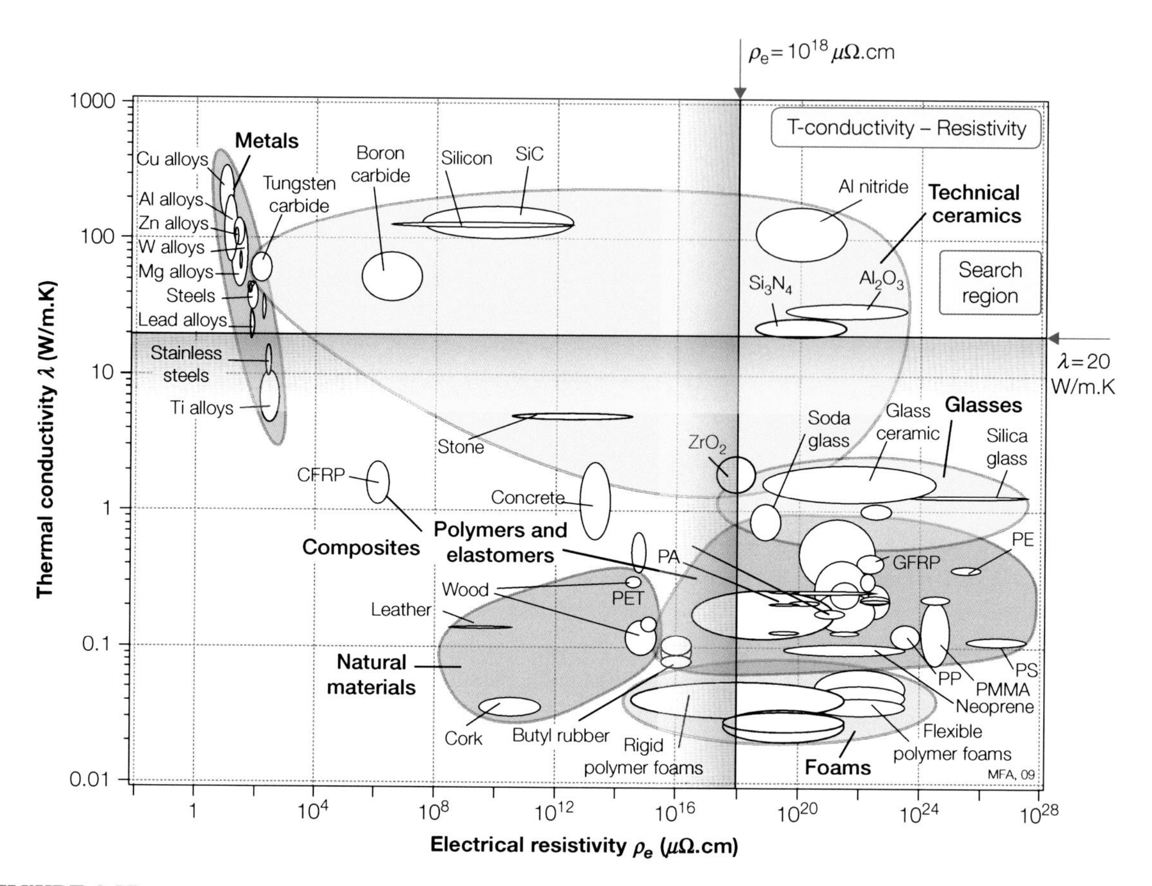

FIGURE 6.35

The $\lambda - \rho_e$ chart with the attribute limit $\rho_e > 10^{18}\mu\Omega$.cm and the index λ plotted on it. The selection is refined by raising the position of the λ selection line.

The $\lambda - \rho$ chart in Figure 6.35 includes all material classes and, of necessity, can show only a limited number of each. To get further we turn to the computer-based CES system, applying the constraints $\rho_e \geq 10^{18}\,\mu\Omega$.cm, $\lambda \geq 20$ W/m.K and a *maximum service temperature* $\geq 120°$C. The result (Figure 6.36) confirms and extends the earlier finding, suggesting the additional materials beryllia, BeO, and magnesia, MgO and diamond, C (with a thermal conductivity three times larger than any of the other candidates), as options. Diamond is outstanding but is probably impracticable for reasons of cost. Compounds of beryllium are toxic and for this reason perhaps undesirable. That leaves aluminum nitride, the earlier choice.

Typical uses of aluminum nitride

Substrates and heat sinks for microcircuits, chip carriers, electronic components; windows, heaters, chucks, clamp rings, gas distribution plates.

Postscript A quick search for documentation, seeking "Applications of aluminum nitride," yields the text at the right. Heat sinks are specifically mentioned. The method has led us quickly to a reliable choice.

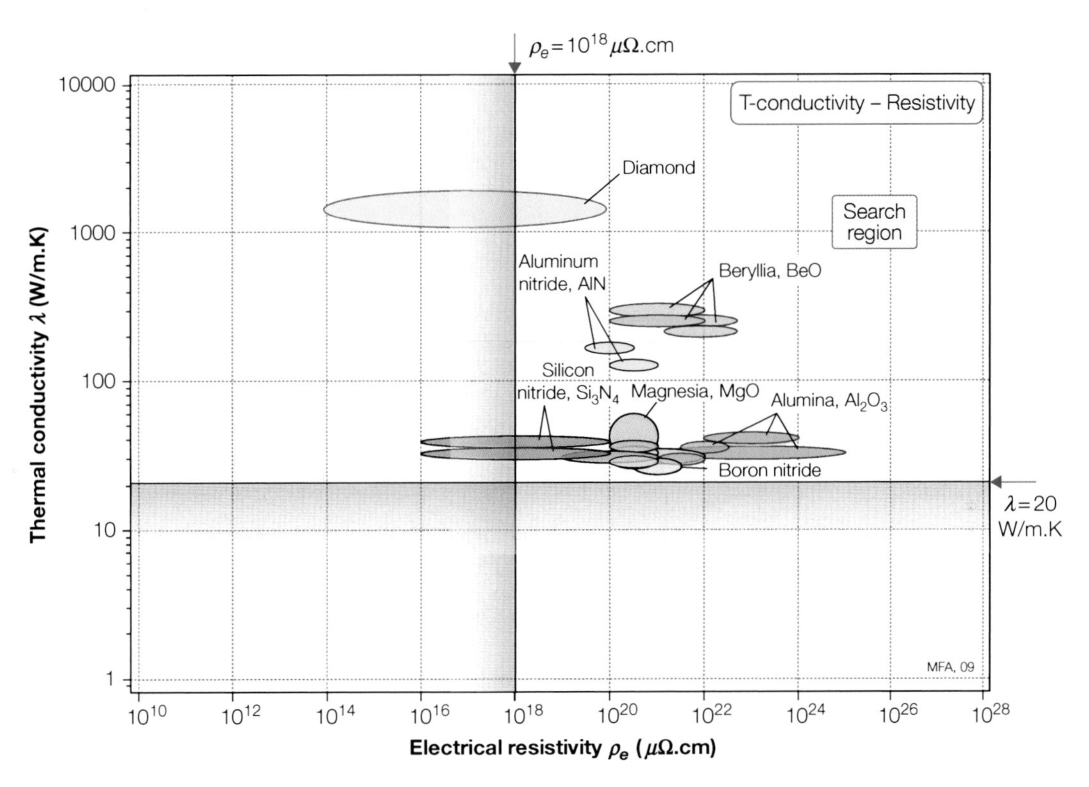

FIGURE 6.36

The $\lambda - \rho_e$ chart at greater resolution with the same attribute limit $\rho_e > 10^{18} \mu\Omega.\text{cm}$ and the index λ plotted on it. The selection is refined by raising the position of the λ selection line. The greater resolution allows more refined choice.

6.19 MATERIALS FOR RADOMES

When the BBC[4] wants to catch you watching television without a license, it parks outside your house an unmarked vehicle equipped to detect high-frequency radiation. The vehicle looks normal enough, but it differs from the norm in one important respect: The body skin is not made of pressed steel but of a material transparent to microwaves. The requirements of the body are much the same as those of the protective dome enclosing the delicate detectors that pick up high-frequency signals from space or those that protect the radar equipment on ships, aircraft, and spacecraft. What are the best materials to make them?

---

[4] The British Broadcasting Corporation derives its income from the license fee paid by owners of television receivers. Failure to pay the fee deprives the BBC of its income; hence the sophisticated detection scheme.

Table 6.34 Design Requirements for a Radome

Function	Radome
Constraints	Support pressure difference Δp Tolerate temperature up to T_{max}
Objective	Minimize dielectric loss in transmission of microwaves
Free variables	Thickness of skin, t Choice of material

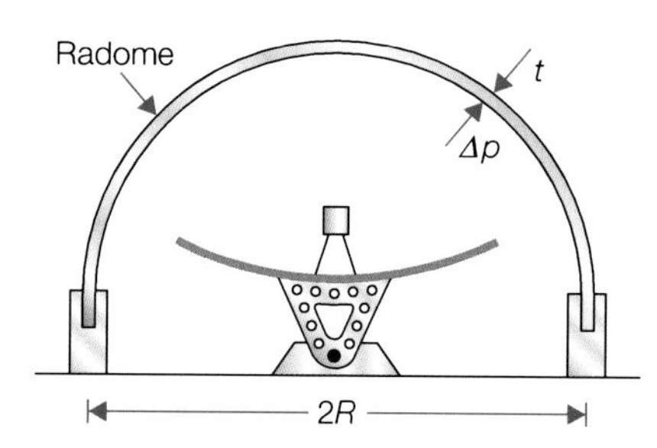

FIGURE 6.37
A radome. It must be transparent to microwaves yet support wind loads and, in many applications, a pressure difference.

The function of a radome is to shield a microwave antenna from the adverse effects of the environment while having as little effect as possible on electrical performance. When detecting incoming signals that are weak to begin with, even a small attenuation of the signal as it passes through the radome decreases the sensitivity of the system. Yet the radome must withstand structural loads, loads caused by pressure difference between the inside and outside of the dome, and—in the case of supersonic flight—high temperatures. Table 6.34 summarizes the design requirements.

The translation Figure 6.37 shows an idealized radome. It is a hemispherical skin of microwave-transparent material of radius R and thickness t, supporting a pressure difference, Δp between its inner and outer surfaces. The two critical material properties in determining radome performance are the dielectric constant, ε_r, and the electric loss tangent $\tan \delta$. Losses are of two types: *reflection* and *absorption*. The fraction of the signal that is reflected is related to the dielectric constant ε_r; the higher the frequency, the higher the reflected fraction. Air has a dielectric constant of 1; a radome with the same dielectric constant, if it were possible, would not reflect any radiation ("stealth" technology seeks to achieve this).

The second, and often more important, loss is that due to absorption as the signal passes through the skin of the radome. When an electromagnetic wave of frequency f (cycles/sec) passes through a dielectric with loss tangent $\tan \delta$, the fractional *power loss* in passing through a thickness dt is

$$\left|\frac{dU}{U_o}\right| = \frac{fA^2\varepsilon_o}{2}(\varepsilon_r \tan \delta)dt \tag{6.75}$$

where A is the electric amplitude of the wave and ε_o the permittivity of vacuum. For a thin shell (thickness t), the loss per unit area is thus

$$\left|\frac{\Delta U}{U_o}\right| = \frac{fA^2\varepsilon_o t}{2}(\varepsilon_r \tan\delta) \tag{6.76}$$

This is the quantity we wish to minimize—the objective function—and this is achieved by making the skin as thin as possible. But the need to support a pressure difference Δp imposes a constraint. The pressure difference creates a stress

$$\sigma = \frac{\Delta p R}{2t} \tag{6.77}$$

in the skin. If it is to support Δp, this stress must be less than the failure stress σ_f of the material of which it is made, imposing a constraint on the thickness:

$$t \geq \frac{\Delta p R}{2\sigma_f}$$

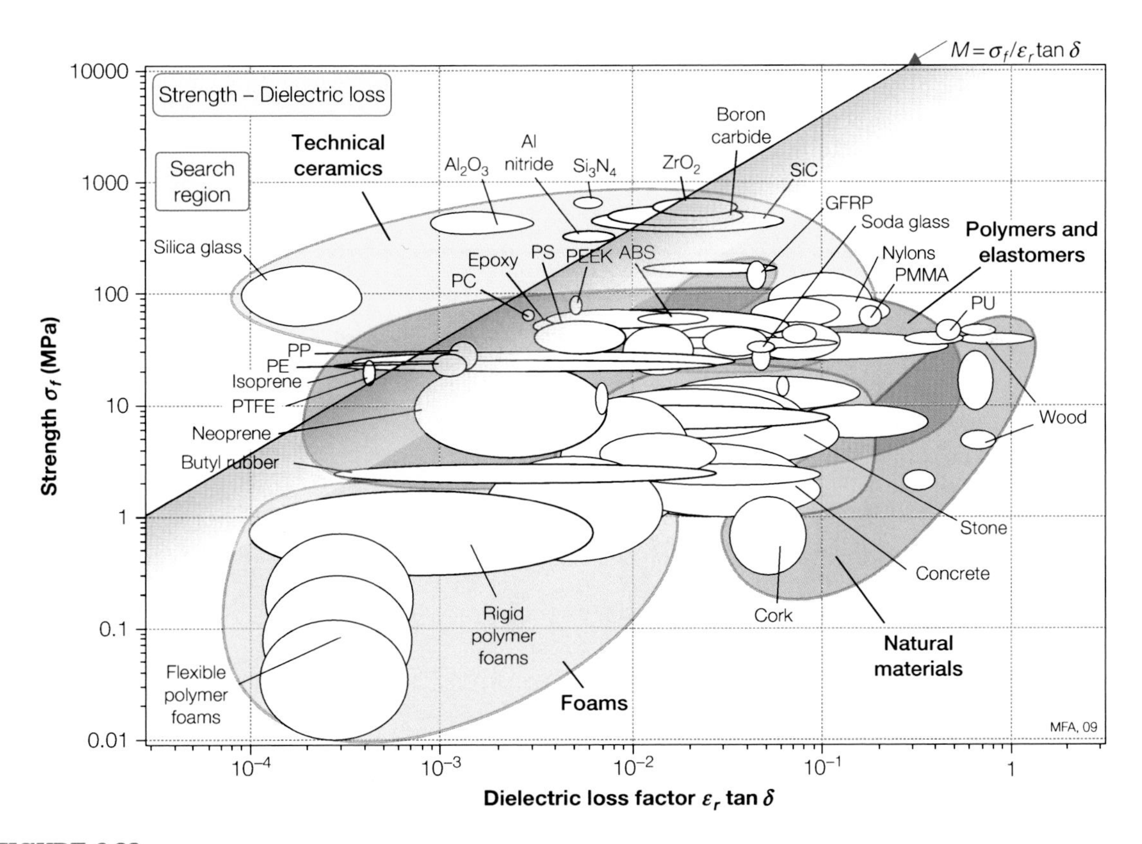

FIGURE 6.38

The elastic limit, σ_f, plotted against the loss factor, $\varepsilon_r \tan\delta$, showing the index, M.

Substituting this into the Equation (6.76) gives

$$\left|\frac{\Delta U}{U}\right| = \frac{fA^2\varepsilon_o \Delta pR}{4}\left(\frac{\varepsilon_r \tan\delta}{\sigma_f}\right) \quad (6.78)$$

The power loss is minimized by maximizing the index

$$M = \frac{\sigma_f}{\varepsilon_r \tan\delta} \quad (6.79)$$

The selection A preliminary survey using the strength–dielectric loss chart in Figure 6.38 shows that polymers have attractive values of M but have poor strength. Some ceramics have excellent values of M and are stable to high temperatures. We need to explore these two classes of material in more depth. Appropriate charts are shown in Figures 6.39(a) and (b), the first for polymers, both plain and glass-reinforced, the second for ceramics. The axes are σ_f and $\varepsilon \tan\delta$. Both have a selection line of slope 1 showing the index M. The selection is summarized in Table 6.35. The materials of

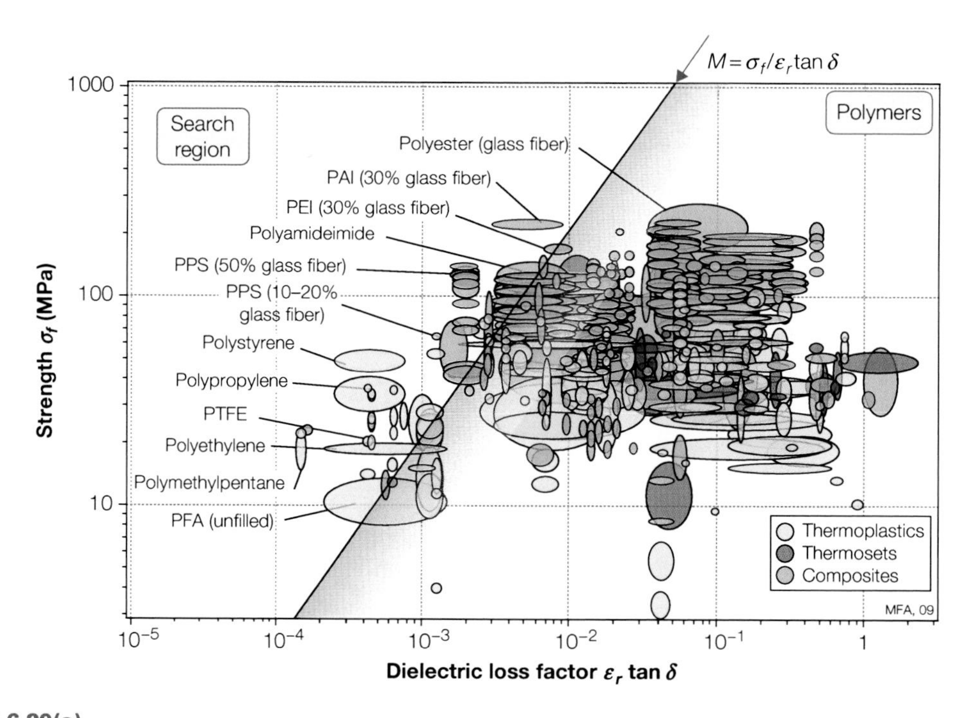

FIGURE 6.39(a)

The elastic limit, σ_f, plotted against the power factor, $\varepsilon_r \tan\delta$, in detail, for polymers, filled polymers, and composites.

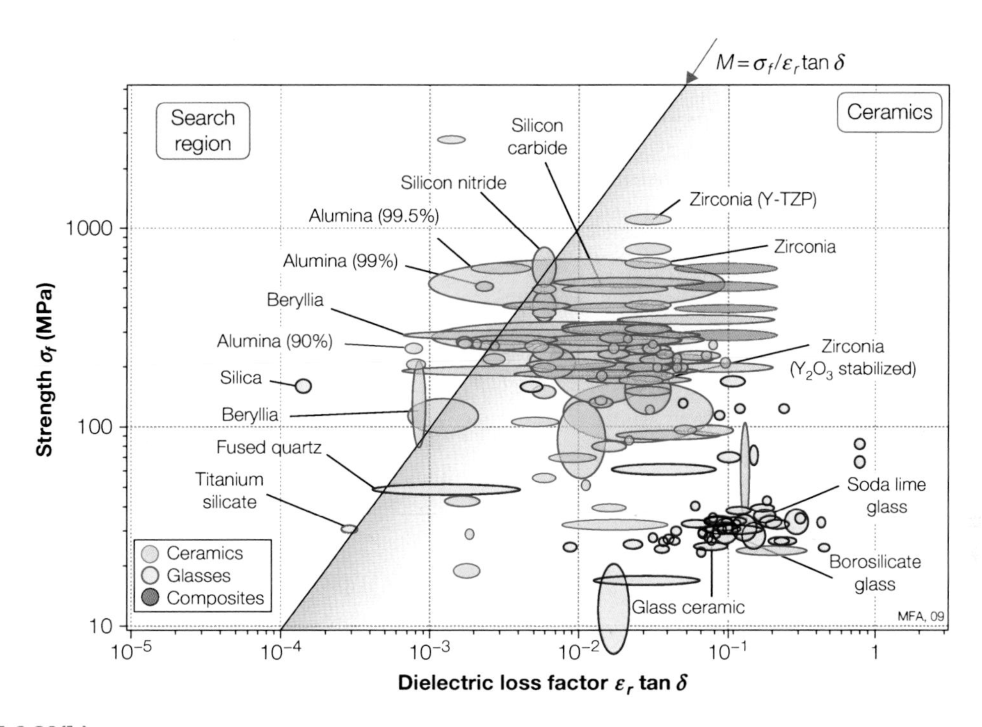

FIGURE 6.39(b)

The elastic limit, σ_f, plotted against the power factor, $\varepsilon_r \tan\delta$, in detail, for ceramics, and glasses.

Table 6.35 Materials for Radomes

Material	Comment
PTFE, polyethylene, polypropylene, polystyrene, and polyphenylene sulphide (PPS)	Minimum dielectric loss, but limited to near room temperature
Glass-reinforced polyester, PTFE, polyethylenes and polypropylenes, polyamideimide	Slightly greater loss, but greater strength and temperature resistance
Silica, alumina, beryllia, silicon carbide	The choice for re-entry vehicles and rockets where heating is great

the first row—PTFE, polyethylene, and polypropylene—maximize M. If greater strength or impact resistance is required, the fiber-reinforced polymers of the second row are the best choices. When, additionally, high temperatures are involved, the ceramics listed in the third row become candidates.

Postscript What are real radomes made of? Among polymers, PTFE and polycarbonate are the commonest materials. Both are very flexible. Where structural rigidity is required (as in the BBC van), GFRP (epoxy or polyester reinforced with woven glass cloth) is used, although with some loss of performance. When performance is at a premium, glass-reinforced PTFE is used instead. For skin heating up to 300°C, polyimides meet the requirements; beyond that temperature, the choice has to be ceramics. Silica (SiO_2), alumina (Al_2O_3), beryllia (BeO) and silicon nitride (Si_3N_4) are all employed. The choices we have identified are all there.

Related reading

Huddleston, G.K., & Bassett, H.L. (1984). Radomes. In R. C. Johnson, & H. Jasik (Eds.). *Antenna engineering handbook* (2nd ed.), chapter 44. McGraw-Hill.

Lewis, C.F. (1988). Materials keep a low profile. *Mechanical engineer*, June, 37–41.

Related case studies

6.11 "Safe pressure vessels"

8.2 "Multiple constraints: Light pressure vessels"

6.20 SUMMARY AND CONCLUSIONS

The case studies in this chapter illustrate how the choice of material is narrowed from the initial broad menu to a small subset that can examined in depth. Most designs make certain nonnegotiable demands on a material: It must withstand a temperature greater than T, it must resist corrosive fluid, F, and so forth. These constraints narrow the choice to a few broad classes of material. The choice is narrowed further by seeking the combination of properties that maximize performance (combinations like $E^{1/2}/\rho$) or maximize safety (combinations like K_{1c}/σ_f) or conduction or insulation (like $a^{1/2}/\lambda$). These, plus economics, isolate a small subset of materials for further consideration.

The final choice between these will depend on more detailed information on their properties, considerations of manufacture, economics, and aesthetics. These are discussed in the chapters that follow.

6.21 FURTHER READING

The texts listed next contain detailed case studies about materials selection. They generally assume that a short list of candidates is already known and argue their relative merits, rather than start with a clean slate, as we do here.

Callister, W.D. (2003). *Materials science and engineering, an introduction* (6th ed.). John Wiley, ISBN 0-471-13576-3.

Charles, J.A., Crane, F.A.A., & Furness, J.A.G. (1997). *Selection and use of engineering materials* (3rd ed.). Butterworth-Heinemann, ISBN 0-7506-3277-1.
A materials science approach to the selection of materials.

Dieter, G.E. (1999). *Engineering design, a materials and processing approach* (3rd ed.). McGraw-Hill, ISBN 9-780-073-66136-0.

A well-balanced and respected text focusing on the place of materials and processing in technical design.

Farag, M.M. (2008). *Materials and process selection for engineering design* (2nd ed.). CRC Press, Taylor and Francis, ISBN 9-781-420-06308-0.

A materials science approach to the selection of materials.

Lewis, G. (1990). *Selection of engineering materials.* Prentice-Hall, ISBN 0-13-802190-2.

A collection of case studies illustrating the choice of materials for a range of engineering applications.

Shackelford, J.F. (2009). *Introduction to materials science for engineers* (7th ed.). Prentice Hall, ISBN 978-0-13-601260-3.

A well-established materials text with a design slant.

CHAPTER 7

Multiple Constraints and Conflicting Objectives

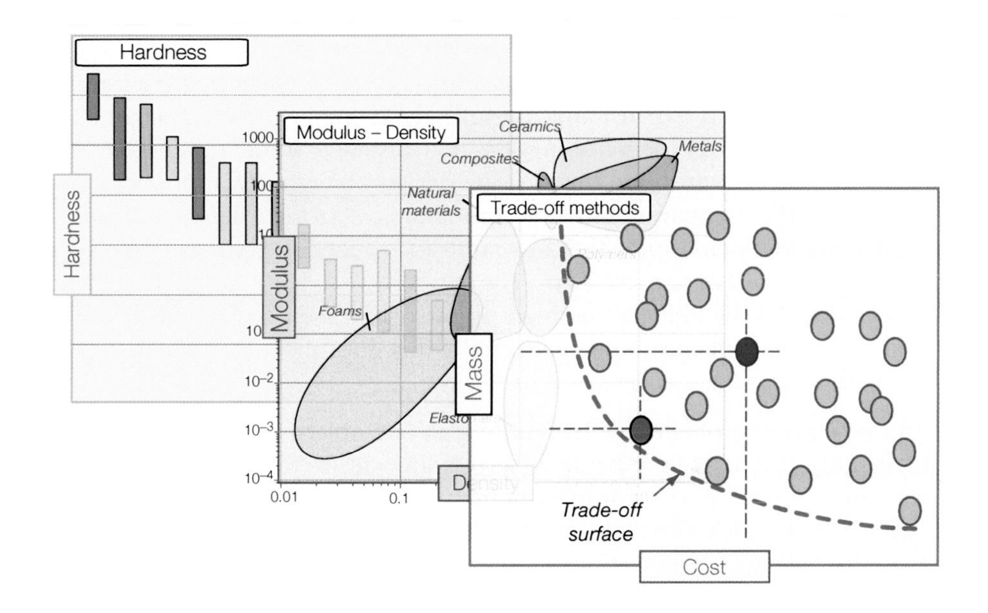

CONTENTS

Materials Selection in Mechanical Design. DOI: 10.1016/B978-1-85617-663-7.00007-2

7.1 INTRODUCTION AND SYNOPSIS

Most decisions in life involve trade-offs. Sometimes the trade-off is to cope with conflicting constraints: I must pay this bill but I must also pay that one—you pay the one that is most pressing. At other times the trade-off is to balance divergent objectives: I want to be rich but I also want to be happy—and resolving this is more dificult since you must balance the two, and wealth is rarely measured in the same units as happiness.

So it is with selecting materials and processes. The selection must satisfy several often conflicting constraints. In the design of an aircraft wing spar, weight must be minimized, with constraints on stiffness, fatigue strength, toughness and geometry. In the design of a disposable hot-drink cup, cost is what matters; it must be minimized subject to constraints on stiffness, strength, and thermal conductivity, though painful experience suggests that designers sometimes neglect the last. In this class of problem there is one design objective (minimization of weight or cost) with many constraints, a situation we met in Chapter 5. Its solution is straightforward: Apply the constraints in sequence, rejecting at each step the materials that fail to meet them. The survivors are viable candidates. Rank them by their ability to meet the single objective and then explore documentation for the top-ranked candidates. Usually this does the job, but sometimes there is an extra twist. It is described in Section 7.2.

A second class of problem involves more than one objective, and here the conflict is more severe. Nature being what it is, the choice of materials that best meets one objective will not usually be that which best meets others. The designer charged with selecting a material for a wing spar that must be both light and cheap faces an obvious difficulty: The lightest materials are not always the least expensive, and vice versa. To make any progress, the designer needs a way of trading weight against cost—a problem we have not encountered until now.

There are a number of quick though subjective ways of dealing with multiple constraints and conflicting objectives: the method of weight factors and methods employing fuzzy logic. They are discussed in the appendix at the end of this chapter. They are a good way of getting into the problem, so to speak, but they rely heavily on personal judgment; their subjective nature must be recognized. Subjectivity is eliminated by employing the active constraint method to resolve multiple constraints (Section 7.2) and by combining conflicting objectives into a single penalty function (Section 7.3). These are standard tools of multicriteria optimization. To use them, we must adopt, in this chapter, the convention that all objectives are expressed as quantities to be minimized; without it the penalty function method does not work.

So now the important stuff. Figure 7.1 is the road map. We start on the top path and work down.

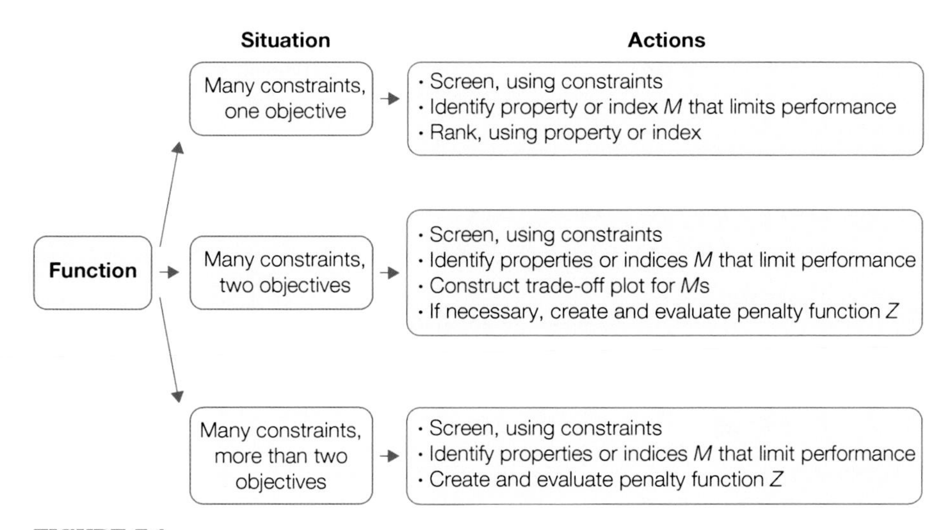

FIGURE 7.1
Strategies for tackling selection with multiple constraints and conflicting objectives.

7.2 SELECTION WITH MULTIPLE CONSTRAINTS

Nearly all material selection problems are overconstrained, meaning that there are more constraints than free variables. We saw multiple constraints in Chapters 5 and 6. Recapitulating, we identify the constraints and the objective imposed by the design requirements, and apply the following steps.

- Screen, using each constraint in turn.
- Rank, using the performance metric describing the objective (often mass, volume, or cost) or simply by the value of the material index that appears in the equation for the metric.
- Seek documentation for the top-ranked candidates and use this to make the final choice.

Steps 1 and 2 are illustrated in Figure 7.2, which we think of as the central methodology. The box on the left represents screening by imposing constraints on properties, on requirements such as corrosion resistance, or on the ability to be processed in a certain way. That on the right—here a bar chart for cost for the surviving candidates—indicates how they are ranked. All very simple.

But not so fast. There is one little twist. It concerns the special case of a single objective that can be limited by more than one constraint. As an example, the requirements for a tie-rod of minimum mass might specify both stiffness and strength, leading to two independent equations for the

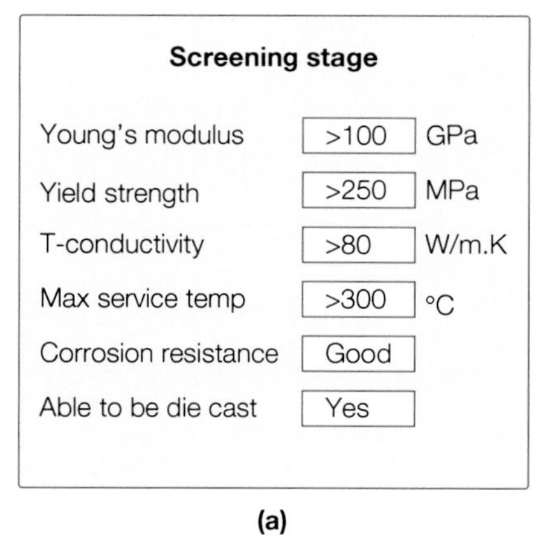

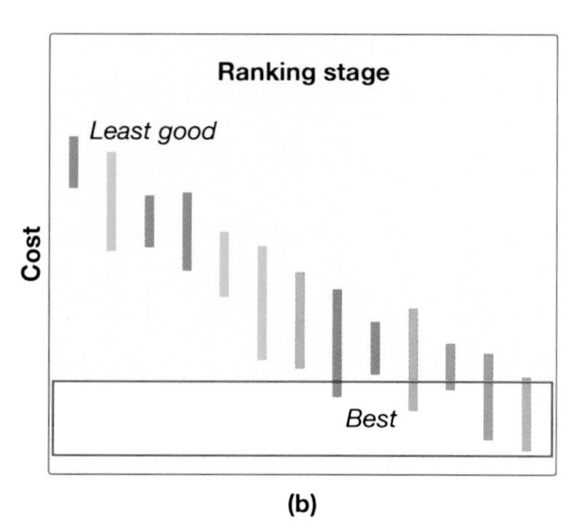

FIGURE 7.2
Selection with multiple constraints (a) and a single objective (b). Screen using the constraints; rank using the objective.

mass. Following exactly the steps of Chapter 5, Equation (5.3), the situation is described by the chain of reasoning shown in Figure 7.3.

If stiffness is the dominant constraint, the mass of the rod is m_1; if it is strength, the mass is m_2. If the tie is to meet the requirements on both, its mass has to be the greater of m_1 and m_2. Writing

$$\tilde{m} = \max(m_1, m_2) \tag{7.1}$$

we search for the material that offers the smallest value of $\tilde{m}$. This is an example of a "min–max" problem, not uncommon in the world of

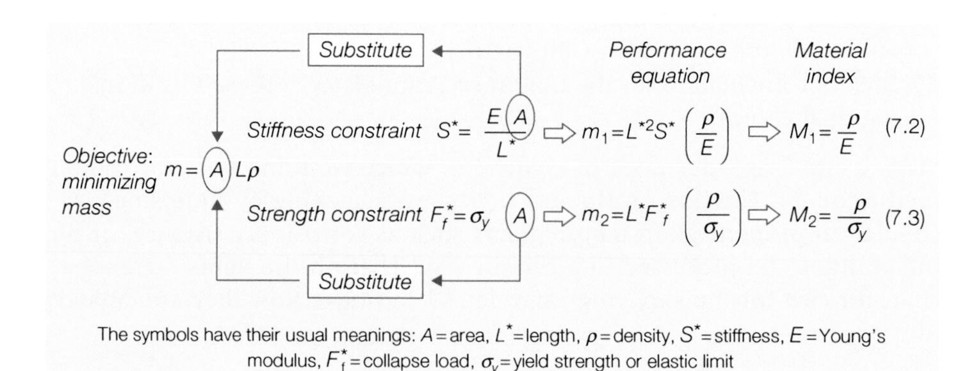

FIGURE 7.3
One objective (here, minimizing mass) with two constraints leads to two performance equations, each with its own value of M.

optimization. We seek the smallest value (min) of a metric that is the larger (max) of two or more alternatives.

The analytical method Powerful methods exist for solving min–max problems when the metric (here, mass) is a continuous function of the control variables (the things on the right of the two performance equations shown in Figure 7.3). But here one of the control variables is the material, and we are dealing with a population of materials, each of which has its own unique values of material properties. The problem is discreet, not continuous.

One way to tackle the problem is to evaluate both m_1 and m_2 for each member of the population, assign the larger of the two to each member, and then rank the members by the assigned value, seeking a minimum. Here is an example.

An example of multiple constraints

A material is required for a light tie of specified length L, stiffness S, and collapse load F_f with the values

$$L^* = 1\,\text{m} \quad S^* = 3 \times 10^7\ \text{N/m} \quad F_f^* = 10^5\,\text{N}$$

Answer

Substituting these values and the material properties shown in the table into Equations (7.2) and (7.3) in Figure 7.3 gives the values for m_1 and m_2 shown in the table. The last column shows $\tilde{m}$ calculated from Equation (7.1). For these design requirements Ti-6-4 is emphatically the best choice: It allows the lightest tie that satisfies both constraints.

Selection of a material for a light, stiff, strong tie

Material	ρ **Kg/m^3**	***E*** **GPa**	σ_y **MPa**	m_1 **kg**	m_2 **kg**	$\tilde{m}$ **kg**
1020 Steel	7,850	200	320	1.12	2.45	2.45
6061 Al	2,700	70	120	1.16	2.25	2.25
Ti-6-4	4,400	115	950	1.15	0.46	1.15

If the constraints are now changed to

$$L^* = 3\,\text{m} \quad S^* = 10^8\text{N/m} \quad F_f^* = 3 \times 10^4\,\text{N}$$

the selection changes. Now steel is the best choice: It gives the lightest tie that satisfies all the constraints. Try it.

When there are 3,000 rather than 3 materials to choose from, simple computer codes can be used to sort and rank them. But this numerical approach lacks visual immediacy and the stimulus for creative thinking that a more graphical method allows. We describe this next.

The graphical method Suppose, for a population of materials, that we plot m_1 against m_2 as suggested by Figure 7.4(a). Each bubble represents a material. (All the variables in both equations for m_1 and m_2 are specified except the material, so the only difference between one bubble and another is the material.) We wish to minimize mass, so the best choices lie somewhere near the bottom left. But where, exactly? The choice if stiffness is paramount and strength is unimportant must surely differ from that if the opposite were true. The line $m_1 = m_2$ separates the chart into two regions. In one, $m_1 > m_2$ and constraint 1 (stiffness) is dominant. In the other, $m_2 > m_1$ and constraint 2 (strength) dominates. In region 1 our objective is to minimize m_1, since it is the larger of the two; in region 2 the opposite is true. This defines a box-shaped selection envelope with its corner on the $m_1 = m_2$ line. The nearer the box is pulled to the bottom left, the smaller is $\tilde{m}$. The best choice is the last material left in the box.

This explains the idea, but there is a better way to implement it. Figure 7.4(a), with m_1 and m_2 as axes, is specific to single values of L^*, S^*, and F_f^*; if these change we need a new chart. Suppose, instead, that we plot the material indices $M_1 = \rho/E$ and $M_2 = \rho/\sigma_y$ that are contained in the performance equations, as shown in Figure 7.4(b). Each bubble still represents a material, but now its position depends only on material properties, not on the values

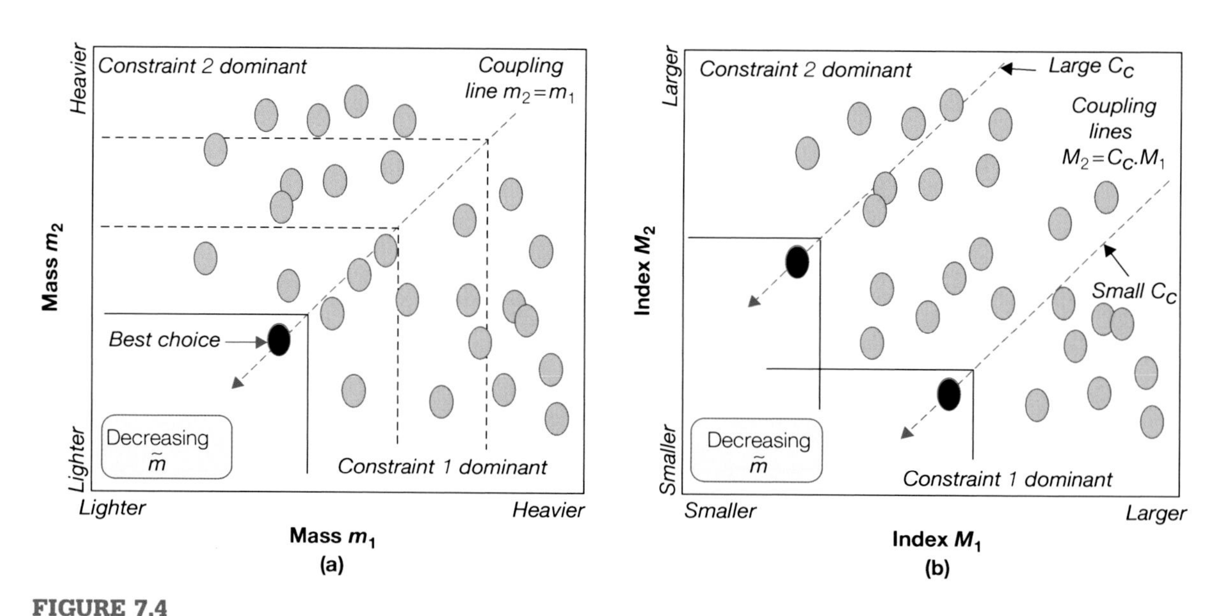

FIGURE 7.4
The graphical approach to min–max problems. (a) Coupled selection using performance metrics (here, mass m). (b) A more general approach: coupled selection using material indices M and a coupling constant C_C.

of L^*, S^*, and F_f^*. The condition $m_1 = m_2$, substituting from Equations (7.2) and (7.3) in Figure 7.3, yields the relationship

$$M_2 = \left(\frac{L^* S^*}{F_f^*}\right) M_1 \tag{7.4}$$

or, on logarithmic scales

$$Log\,(M_2) = Log\,(M_1) + \log\left(\frac{L^* S^*}{F_f^*}\right) \tag{7.5}$$

This describes a line of slope 1, in a position that depends on the value of L^*S^*/F_f^*. We refer to this line as the coupling line and to L^*S^*/F_f^* as the coupling constant, symbol C_c. The selection strategy remains the same: A box, with its corner on the coupling line, is pulled down toward the bottom left. But the chart is now more general, covering all values of L^*, S^*, and F_f^*. Changing either one of these, or the geometry of the component (here described by L^*) moves the coupling line and changes the selections.

Worked examples are given in Chapter 8.

7.3 CONFLICTING OBJECTIVES

Real-life materials selection almost always requires that a compromise be reached between conflicting objectives. Three crop up all the time:

- *Minimizing mass*—a common target in designing things that move or have to be moved, that oscillate, or that must response quickly to a limited force (think of aerospace and ground transport systems).
- *Minimizing volume*—because less material is used, and because space is increasingly precious (think of the drive for ever thinner, smaller mobile phones, portable computers, MP3 players, etc., and the need to pack more and more functionality into a fixed volume).
- *Minimizing cost*—profitability depends on the difference between cost and value (more on this in Chapters 13 and 16); the most obvious way to increase this difference is to reduce cost.

To these we must now add a fourth objective:

- *Minimizing environmental impact*—the damage to our surroundings caused by material production, product manufacture and product use (Chapter 15).

There are, of course, other objectives specific to particular applications. Some are just one of the four expressed previously but in different words.

The objective of maximizing power-to-weight ratio translates into minimizing mass for a given power output. Maximizing energy storage in a spring, battery, or flywheel means minimizing the volume for a given stored energy. Some can be quantified in engineering terms, such as maximizing reliability (although this can translate into achieving a given wear resistance or corrosion resistance at minimum cost). Others cannot, such as maximizing consumer appeal—an amalgam of performance, styling, image, and marketing.

So we have four common objectives, each characterized by a performance metric P_i. At least two are involved in the design of almost any product. Conflict arises because the choice that optimizes one objective will not, in general, do the same for the others; the best choice is thus a compromise, optimizing none but pushing all as close to their optima as their interdependence allows. And this highlights the central problem: How is mass to be compared with cost, or volume with environmental impact? Unlike the performance equations shown in Figure 7.3, each is measured in different units; they are incommensurate. We need strategies to deal with this. They come in a moment. First, some definitions.

Trade-off strategies Consider the choice of material to minimize both cost (performance metric P_1) and mass (performance metric P_2) while also meeting a set of constraints such as a required maximum service temperature, or corrosion resistance in a certain environment. Following the standard terminology of optimizations theory, we define a solution as a viable choice of material, meeting all the constraints but not necessarily optimal by either of the objectives. Figure 7.5 is a plot of P_1 against P_2 for alternative solutions, each bubble describing one of them. The solutions that minimize P_1 do not minimize P_2, and vice versa. Some solutions, such as that at *A*, are far from optimal—all the solutions in the box attached to *A* have lower values of both P_1 and P_2. Solutions like *A* are said to be dominated by others. Solutions like those at *B* have the characteristic that no other solutions exists with lower values of both P_1 and P_2. These are said to be nondominated solutions. The line or surface on which they lie is called the nondominated or optimal trade-off surface. The values of P_1 and P_2 corresponding to the nondominated set of solutions are called the Pareto set.

There are three strategies for progressing further. The solutions on or near the trade-off surface offer the best compromise; the rest can be rejected. Often, this is enough to identify a shortlist, using intuition to rank them, for which documentation can now be sought (Strategy 1). Alternatively (Strategy 2), one objective can be reformulated as a constraint, as illustrated in Figure 7.6. Here an upper limit is set on cost; the solution that minimizes the other constraint can then be read off. But this is cheating: It is not a true optimization. To achieve that, we need Strategy 3: penalty functions.

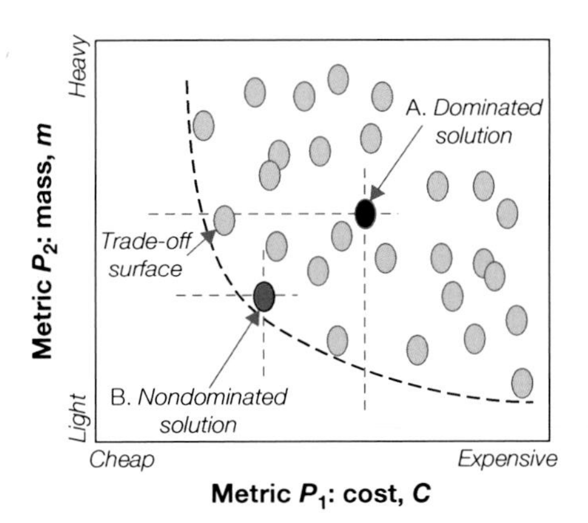

FIGURE 7.5
Multiple objectives: We seek the material that cominimizes mass and cost. Each bubble is a solution—a material choice that meets all constraints. The trade-off surface links nondominated solutions.

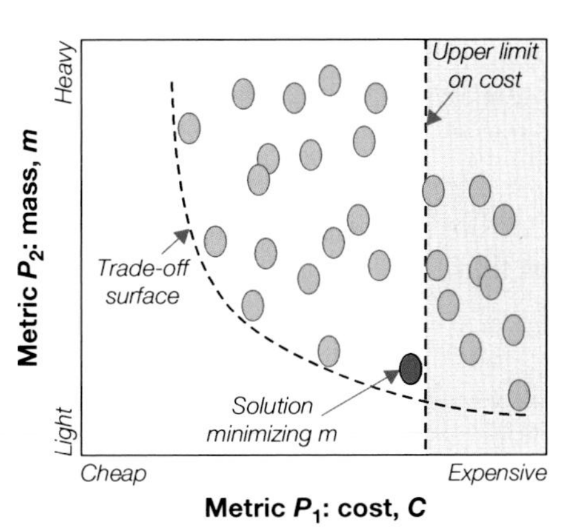

FIGURE 7.6
The trade-off plot with a simple constraint imposed on cost. The solution with the lowest mass can now be read off, but it is not a true optimization.

Penalty functions The trade-off surface identifies the subset of solutions that offer the best compromises between the objectives. Ultimately, though, we want a single solution. One way to do this is to aggregate the various objectives into a single objective function, formulated such that its minimum defines the most preferable solution. To do this we define a locally linear penalty function[1] Z:

$$Z = \alpha_1 P_1 + \alpha_2 P_2 + \alpha_3 P_3 \ldots. \quad (7.6)$$

The best choice is the material with the smallest value of Z. The αs are called exchange constants (or, equivalently, utility constants or scaling constants); they convert the units of performance into the units of Z, which are usually currency ($). The exchange constants are defined by

$$\alpha_i = \left(\frac{\partial Z}{\partial P_i}\right)_{P_j,\, j \neq i} \quad (7.7)$$

They measure the increment in penalty for a unit increment in a given performance metric, all others being held constant. If, for example, the metric P_2 is mass m, then α_2 is the change in Z associated with a unit increase in m.

Frequently one of the objectives to be minimized is cost, C, so that $P_1 = C$. Then it makes sense to measure Z in units of currency. With this choice, a

[1] Also called a *value function* or *utility function*. The method allows a local minimum to be found. When the search space is large, it is necessary to recognize that the values of the exchange constants α_i may themselves depend on the values of the performance metrics P_i.

unit change in C gives a unit change in Z, with the result that $\alpha_1 = 1$ and Equation (7.6) becomes

$$Z = C + \alpha_2 P_2 + \alpha_3 P_3 \ldots. \tag{7.8}$$

Consider now the earlier example in which P_1 = cost, C, and P_2 = mass. m, so that

$$Z = C + \alpha m \tag{7.9}$$

or

$$m = -\frac{1}{\alpha}C + \frac{1}{\alpha}Z \tag{7.10}$$

Then α is the change in Z associated with a unit increase in m. Equation 7.10 defines linear a relationship between m and C. This plots as a family of parallel penalty lines, each for a given value of Z, as shown in Figure 7.7(a). The slope of the lines is the reciprocal of the exchange constant, $-1/\alpha$. The value of Z decreases toward the *bottom left*: The best choices lie there. The optimum solution is the one nearest the point at which a penalty line is tangential to the trade-off surface since it is the one with the smallest value of Z. Narrowing the choice to just one candidate at this stage is not sensible—we do not yet know what the search for documentation will reveal. Instead, we choose the subset of solutions that lie closest to the tangent point.

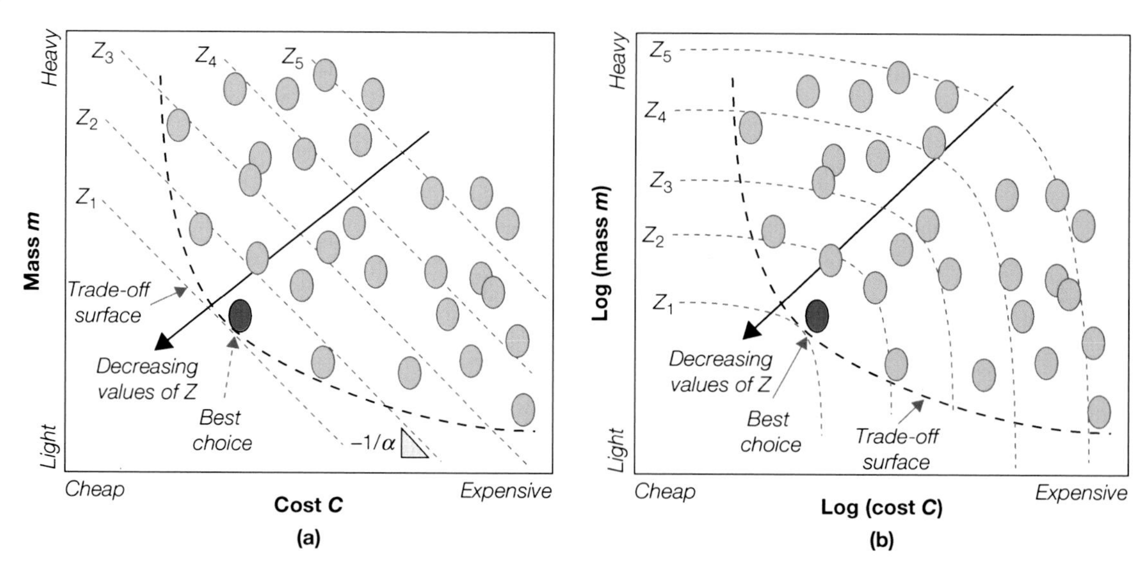

FIGURE 7.7

(a) The penalty function Z superimposed on the trade-off plot. The contours of Z have a slope of $-1/\alpha$. The contour that is tangent to the trade-off surface identifies the optimum solution. (b) The same, plotted on logarithmic scales; the linear relation now appears as curved lines.

Using penalty functions

The exchange constant for weight saving in light trucks is α = \$12/kg, meaning that the value of weight reduction over the life of the vehicle is \$12 for each kilogram saved. A maker of such vehicles offers three models. The first uses steel panels for the body work. The second uses aluminum, costs \$2,500 more but weighs 300 kg less. The third offers carbon-fiber paneling, costs \$8,000 more, and weighs 500 kg less. Which is the best buy?

Answer

The penalty functions for the steel (1) and aluminum (2) vehicles are

$$Z_1 = C_1 + \alpha m_1$$

and

$$Z_2 = C_2 + \alpha m_2$$

The aluminum vehicle is attractive only if its Z is lower than the Z of the steel one. Writing

$$\Delta Z = Z_2 - Z_1 = C_2 - C_1 + \alpha (m_2 - m_1) = 2500 - 12 \times 300 = -1100$$

The aluminum-paneled vehicle offers a life saving of \$1,100—it is a good buy. Repeating the comparison for the composite-paneled vehicle gives a value of ΔZ = +\$2,000. It is not a good buy.

There is one little quirk. Almost all the material selection charts use logarithmic scales, for very good reasons (Chapter 4). A linear relationship, plotted on log scales, appears as a curve, as shown in Figure 7.7(b). But the procedure remains the same: The best candidates are those nearest the point at which one of these curves just touches the trade-off surface.

Relative penalty functions When, as is often the case, we seek a better material for an existing application, it is more helpful to compare the new material choice with the existing one. To do this we define a relative penalty function:

$$Z^* = \frac{C}{C_o} + \alpha^* \frac{m}{m_o} \tag{7.11}$$

in which the subscript *o* means properties of the existing material and the asterisk * on Z^* and α^* is a reminder that both are now dimensionless. The relative exchange constant α^* measures the fractional gain in value for a given fractional gain in performance. Thus $\alpha^* = 1$ means that, at constant Z,

$$\frac{\Delta C}{C_o} = -\frac{\Delta m}{m_o}$$

and that halving the mass is perceived to be worth twice the cost.

Figure 7.8 shows the relative trade-off plot, here on linear scales. The axes are C/C_o and m/m_o. The material currently used in the application appears at the

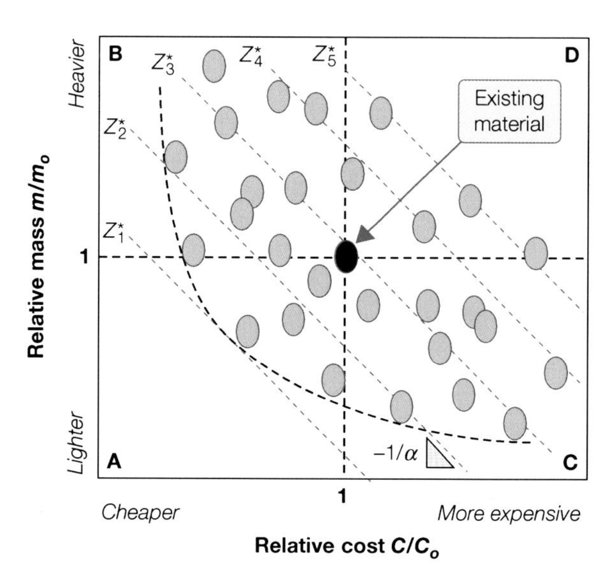

FIGURE 7.8
A relative trade-off plot, useful when exploring substitution of an existing material with the aim of reducing mass or cost or both. The existing material sits at the coordinates (1,1). Solutions in sector A are both lighter and cheaper.

coordinates (1,1). Solutions in sector A are both lighter and cheaper than the existing material, those in sector B are cheaper but heavier, those in sector C are lighter but more expensive, and those in sector D are uninteresting. Contours of Z^* can be plotted onto the figure. The contour that is tangent to the relative trade-off surface again identifies the optimum search area. As before, when logarithmic scales are used, the contours of Z^* become curves. The case studies of Chapter 8 make use of relative penalty functions.

So if values for the exchange constants are known, a completely systematic selection is possible. But that is a big "if." It is discussed next.

Values for the exchange constants, α An exchange constant is a measure of the penalty of unit increase in a performance metric, or—more easily understood—it is the value or "utility" of a unit decrease in the metric. Its magnitude and sign depend on the application. Thus the utility of weight saving in a family car is smal, though significant; in aerospace it is much larger. The utility of heat transfer in house insulation is directly related to the cost of the energy used to heat the house; that in a heat exchanger for power electronics can be much higher because it increases electrical performance. The utility can be real, meaning that it measures a true saving of cost. But it can, sometimes, be perceived, meaning that the consumer, influenced by scarcity, advertising, or fashion, will pay more or less than the true value of these metrics.

In many engineering applications the exchange constants can be derived approximately from technical models for the lifetime cost of a system. Thus the utility of weight saving in transport systems is derived from the value of

Table 7.1 Exchange Constants α for the Mass–Cost Trade-off for Transport Systems

Sector: Transport systems	Basis of estimate	Exchange constant, α (US$/kg)
Family car	Fuel saving	1–2
Truck	Payload	5–20
Civil aircraft	Payload	100–500
Military aircraft	Payload, performance	500–1,000
Space vehicle	Payload	3,000–10,000

the fuel saved or of the increased payload, evaluated over the life of the system. Table 7.1 gives approximate values for α. The most striking thing about them is the enormous range. The exchange constant depends in a dramatic way on the application in which the material will be used. It is this that lies behind the difficulty in adopting aluminum alloys for cars despite their universal use in aircraft, the much greater use of titanium alloys in military than in civil aircraft, and the restriction of beryllium for use in space vehicles.

Exchange constants can be estimated in various ways. The cost of launching a payload into space lies in the range \$3,000 to \$10,000/kg; a reduction of 1 kg in the weight of the launch structure would allow a corresponding increase in payload, giving the ranges of α shown in the table. Similar arguments based on increased payload or decreased fuel consumption give the values shown for civil aircraft, commercial trucks, and automobiles. The values change with time, reflecting changes in fuel costs, legislation to increase fuel economy and the like. Special circumstances can change them dramatically—a jet engine builder who has guaranteed a certain power/weight ratio for his engine may be willing to pay more than \$1,000 to save a kilogram if it is the only way the guarantee can be met, or (expressed as a penalty) he will be penalized \$1,000/kg if it is not.

These values for the exchange constant are based on engineering criteria. More difficult to assess are values based on perceived worth. The value for the weight/cost trade-off for a bicycle is an example. To the enthusiast, a lighter bike is a better bike. Figure 7.9 shows just how much cyclists value reduction in weight. It is a trade-off plot of mass and cost of bicycles, using data from bike magazines. The tangent to the trade-off line at any point gives a measure of the exchange constant: It ranges from \$20/kg to \$2,000/kg, depending on the mass. Does it make sense for the ordinary cyclist to pay \$2,000 to reduce the mass of the bike by 1kg when, by dieting, he could reduce the mass of the system (himself plus the bike) by more without spending a penny? Possibly. But mostly it is perceived value. One aim of advertising is to increase the perceived worth of a product, thereby increasing its value without increasing its cost. It influences the exchange constants for family cars, and it is the motive behind the use of titanium for watches, carbon fiber for frames of glasses, and exotic materials in much sports equipment. For these, the value of α is harder to pin down.

There are other circumstances in which establishing the exchange constant can be difficult. One example is environmental impact—the damage to the environment caused by manufacture, use, or disposal of a product. Minimizing environmental impact has now become an important objective, almost as important as minimizing cost. Ingenious design can reduce the first without driving up the second too much. But how much is unit decrease in eco-impact worth? Until an exchange constant is agreed on or imposed, it is difficult for the designer to respond.

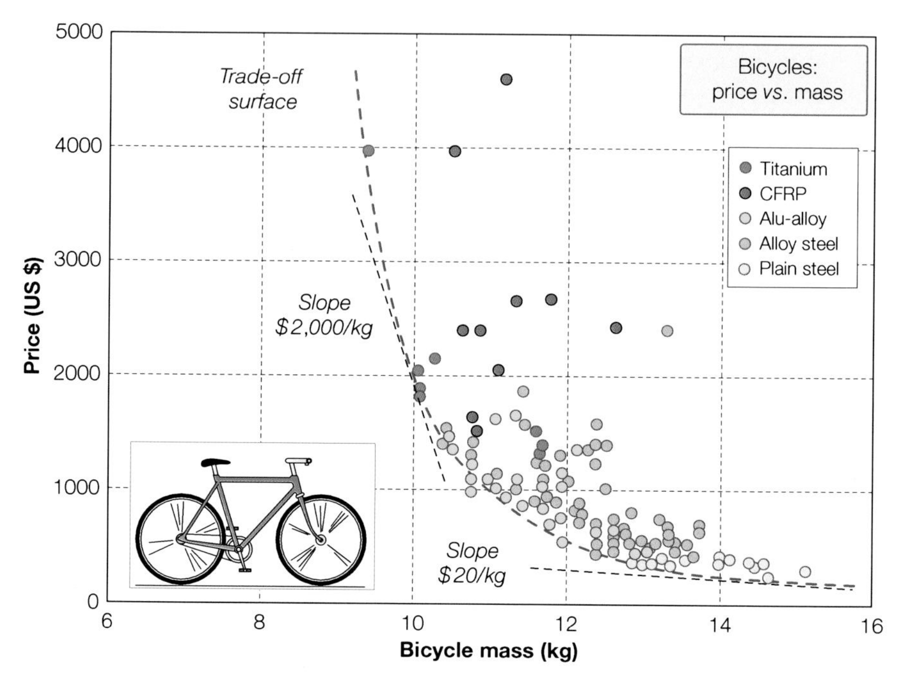

FIGURE 7.9

A cost–mass trade-off plot for bicycles using data from a bike magazine. Solutions are color-coded with the material of which the frame is made. The tangent to the trade-off surface at any point gives an estimate of the exchange constant. It depends on the application. To a consumer seeking a cheap bike for shopping, the value of weight saving is low ($20/kg). To an enthusiast who wants performance, it can be high ($2,000/kg).

However, things are not quite as difficult as they at first appear. Useful engineering decisions can be reached even when exchange constants are imprecisely known, as explained in the next section.

How do exchange constants influence choice? The discreteness of the search space for material selection means that a given solution on the trade-off surface is optimal for a certain range of values of α; outside this range another solution becomes the optimal choice. The range can be large, so any value of the exchange constant within the range leads to the same choice of material. This is illustrated in Figure 7.10. For simplicity, the solutions have been moved so that, in this figure, only three are potentially optimal. For $\alpha \leq 0.1$ (so that $1/\alpha \geq 10$), solution A is the optimum; for $0.1 < \alpha < 10$, solution B is the best choice; and for $\alpha \geq 10$, it is solution C. This information is captured in the bar on the *right* of the figure, showing the range of values of α, subdivided at the points at which a change of optimum occurs and labeled with the solution that is optimal in each range.

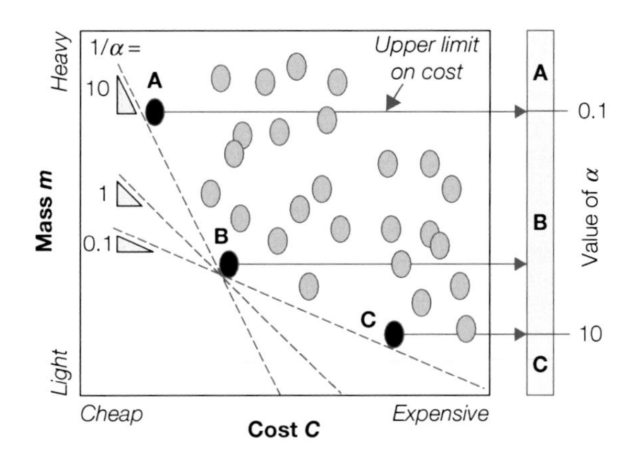

FIGURE 7.10
It is often the case that a single material (or subset of materials) is optimal over a wide range of values of the exchange constant. Then approximate values for exchange constants are sufficient to reach precise conclusions about the choice of materials.

7.4 SUMMARY AND CONCLUSIONS

The method of material indices allows a simple, transparent procedure for selecting materials to minimize a single objective while meeting a set of simple constraints. But things are rarely that simple—different measures of performance compete, and a compromise has to be found between them.

Judgment can be used to rank the importance of the competing constraints and objectives. Weight factors or fuzzy logic, described in Section 7.6, put the judgment on a more formal footing, but can also obscure its consequences. When possible, judgment should be replaced by analysis. For multiple constraints, this is done by identifying the active constraint and basing the design on this. The procedure can be made graphical by deriving coupling equations that link the material indices; then simple constructions on material selection charts with the indices as axes unambiguously identify the subset of materials that maximize performance while meeting all the constraints. Compound objectives require the formulation of a penalty function, Z, containing one or more exchange constants, α_i; it allows all objectives to be expressed in the same units (usually cost). Minimizing Z identifies the optimum choice.

When multiple constraints operate, or a compound objective is involved, the best choice of material is far from obvious. It is here that the methods developed have real power. Chapter 8 gives examples.

7.5 FURTHER READING

Ashby, M. F. (2002). Multi-objective optimization in material design and selection. *Acta Mater, 48*, 359–369.

An exploration of the use of trade-off surfaces and utility functions for material selection.

Bader, M.G. (1977). Composites applications and design. In. *Proc ICCM-11, Gold Coast, Australia Vol. 1*. ICCM: London.
An example of trade-off methods applied to the choice of composite systems.

Bourell, D.L. (1997). Decision matrices in materials selection. In G.E. Dieter (Ed.), *ASM Handbook*, Vol. 20. *Materials selection and design* (pp. 291–296). ASM International, ISBN 0-87170-386-6291–296.
An introduction to the use of weight factors and decision matrices.

Clark, J.P., Roth, R., & Field, F.R. (1997). Techno-economic issues in material science. In G.E. Dieter (Ed.), *ASM Handbook*, Vol. 20. *Materials selection and design* (pp. 255–265). ASM International, ISBN 0-87170-386.
The authors explore methods of cost and utility analysis, and of environmental issues, in materials selection.

Dieter, G.E. (2000). *Engineering design—A materials and processing approach* (3rd ed.). 150–153, 255–257. McGraw-Hill, ISBN 0-07-366136-8.
A well-balanced and respected text, now in its 3rd edition, focusing on the role of materials and processing in technical design.

Field, F.R., & de Neufville, R. (1988). Material selection—maximizing overall utility. *Metals and Materials*, June, 378–382.
A summary of utility analysis applied to material selection in the automobile industry.

Goicoechea, A., Hansen, D.R., & Duckstein, L. (1982). *Multi-objective decision analysis with engineering and business applications*. Wiley, ISBN 0-4710-64017.
A good starting point for the theory of multi-objective decision making.

Keeney, R.L., & Raiffa, H. (1993). *Decisions with multiple objectives: preferences and value trade-offs* (2nd ed.). Cambridge University Press, ISBN 0-521-43883-7.
A notably readable introduction to methods of decision making with multiple, competing objectives.

Papalambros, P.Y., & Wilde, D.J. (2000). *Principles of optimal design, modeling and computation* (2nd ed.). Cambridge University Press, ISBN 0-521-62727-3.
An introduction to methods of optimal engineering design.

Pareto, V. (1906). Manuale di economica politica. Societa Editrice Libraria, Milano, Italy. Translated into English by Schwier, A. S. (1971) as *Manual of Political Economics*, Macmillan.
A book that is much quoted but little read; the origin of the concept of a trade-off surface as an approach to multi-objective optimization.

Sawaragi, Y., Nakayama, H., & Tanino, T. (1985). *Theory of multi-objective optimization*. Academic Press, ISBN 0-12-620370-9.
Multi-objective optimization in all its gruesome detail. Exhaustive but not the best place to start.

7.6 APPENDIX: WEIGHT FACTORS AND FUZZY METHODS

Suppose you want a component with a required stiffness (constraint 1) and strength (constraint 2) and it must be as light as possible (one objective). You could choose materials with high modulus E for stiffness, and then the subset of these that have high elastic limits σ_y for strength, and the subset of those that have low density ρ for light weight. Then again, if you wanted a material with a required stiffness (one constraint) that was simultaneously as light (objective 1) and as cheap (objective 2) as possible, you could apply the constraint and then find the subset of survivors that are light and the subset of those survivors that are inexpensive. Some selection systems work that way, but it is not a good idea because there is no guidance in deciding relative importance of the limits on stiffness, strength, weight, and cost. This is not a trivial difficulty: It is exactly this relative importance that makes aluminum the prime structural material for aerospace and steel that for ground-based structures.

These problems of relative importance are old; engineers have sought methods to overcome them for at least a century. The traditional approach is that of assigning weight factors to each constraint and objective and using them to guide choice in ways that are summarized in the following. The upside: Experienced engineers can be good at assessing relative weights. The downside: The method relies on judgment. In assessing weights, judgments can differ, and there are subtler problems; one of them is discussed next. For this reason, this chapter has focused on systematic methods. But one should know about the traditional methods because they are still widely used.

The method of weight factors

Weight factors seek to quantify judgment. The method works like this. The key properties or indices are identified and their values M_i are tabulated for promising candidates. Since their absolute values can differ widely and depend on the units in which they are measured, each is first scaled by dividing it by the largest index of its group, $(M_i)_{max}$, so that the largest after scaling has the value 1. Each is then multiplied by a weight factor, w_i, with a value between 0 and 1, expressing its relative importance for the performance of the component. This gives a weighted index W_i:

$$W_i = w_i \frac{M_i}{(M_i)_{max}} \tag{7.13}$$

For properties that are not readily expressed as numerical values, such as weldability or wear resistance, rankings such as A to E are expressed instead by a numeric rating—A = 5 (very good) to E = 1 (very bad)—then divided by the highest rating value as before. For properties that are to be minimized, like corrosion rate, the scaling uses the minimum value $(M_i)_{min}$, expressed in the form

$$W_i = w_i \frac{(M_i)_{min}}{M_i} \tag{7.14}$$

The weight factors w_i are chosen such that they add up to 1, that is: $w_i < 1$ and $\Sigma\, w_i = 1$. There are numerous schemes for assigning their values (see

Section 7.5); all require, in varying degrees, judgment. The property judged to be most important is given the largest w; the second most important, the second largest; and so on. The w_i s are calculated from Equations (7.13) and (7.14) and summed. The best choice is the material with the largest value of the sum

$$W = \Sigma_i W_i \tag{7.15}$$

Sounds simple, but there are problems, some obvious (like subjectivity in assigning the weights) and some more subtle. Here is an example.

A material is sought to make a light component (low ρ) that must be strong (high σ^y). Table 7.2 gives values for four possible candidates. Weight, in our judgment, is more important than strength, so we assign it the weight factor

$$w_1 = 0.75$$

That for strength is then

$$w_2 = 0.25$$

Normalize the index values (as in Equations (7.13) and (7.14)) and sum them (Equation (7.15)) to give W. The second to last column of the table shows the result: Beryllium wins easily; Ti-6Al-4V comes second; 6061 aluminum, third. But observe what happens if beryllium (which is very expensive and can be toxic) is omitted from the selection, leaving only the first three materials. The same procedure now leads to the values of W in the last column: 6061 aluminum wins; Ti-6Al-4V is second. Removing one, nonviable, material from the selection has reversed the ranking of those that remain. Even if the weight factors could be chosen with accuracy, this dependence of the outcome on the population from which the choice comes is disturbing. The method is inherently unstable and sensitive to irrelevant alternatives.

Table 7.2 Example of Use of Weight Factors

Material	ρ Mg/m^3	σ_y MPa	*W* (inc. Be)	*W* (excl. Be)
1020 steel	7.85	320	0.27	0.34
6061 Al (T4)	2.7	120	0.55	0.78
Ti-6Al-4V	4.4	950	0.57	0.71
Beryllium	1.86	170	0.79	—

Fuzzy logic

Fuzzy logic takes weight factors one step further. Figure 7.11 (*upper left*) shows the probability $P(R)$ of a material having a property with a value R in a given range. Here the property has a well-defined range for each of the four materials A, B, C, and D (the values are crisp in the terminology of the

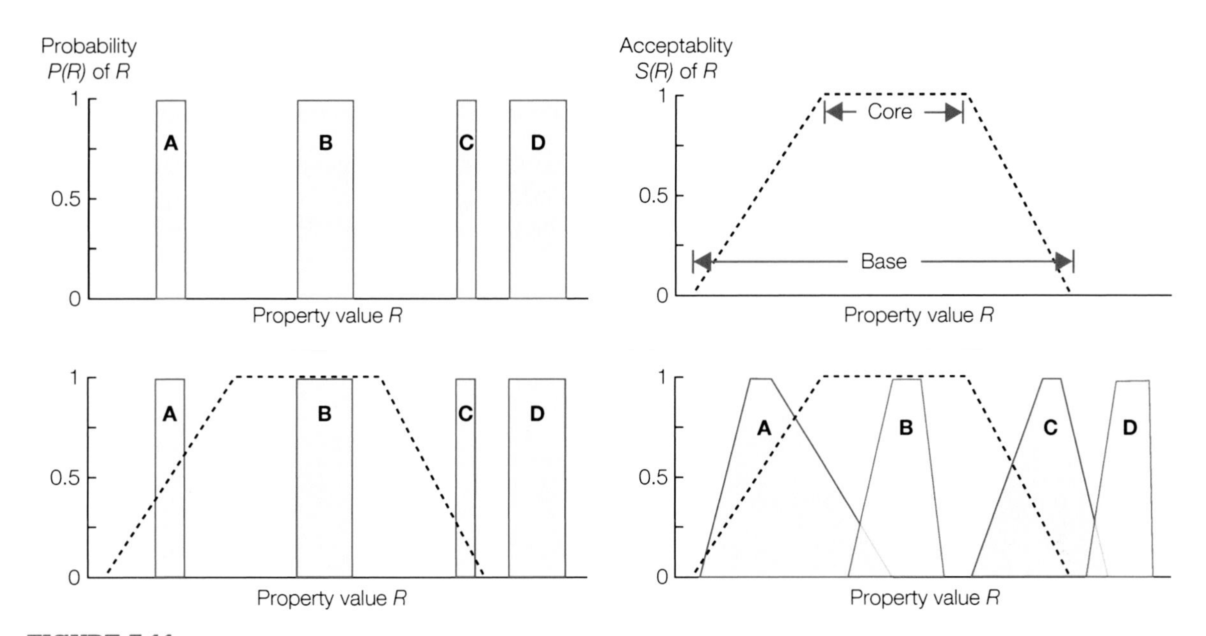

FIGURE 7.11

Fuzzy selection methods. Sharply defined properties and a fuzzy selection criterion (*top row*) are combined to give weight factors for each material (*center*). The properties themselves can be given fuzzy ranges (*lower right*).

field). The selection criterion, shown at the top right, identifies the range of R that is sought for the properties, and it is fuzzy. In other words, it has a well-defined core defining the ideal range sought for the property, with a wider base, extending the range to include boundary regions in which the value of the property is allowable but with decreasing acceptability as the edges of the base are approached. This defines the probability $S(R)$ of a choice being a successful one.

The superposition of the two figures, shown at the lower left in Figure 7.11, illustrates a single selection stage. Desirability is measured by the product $P(R).S(R)$. Here material B is fully acceptable—it acquires a weight of 1. Material A is acceptable but with a lower weight, here 0.5; C is acceptable with a weight of roughly 0.25, and D is unacceptable—it has a weight of 0. At the end of the first selection stage, each material in the database has one weight factor associated with it. The procedure is repeated for successive stages, which could include indices derived from other constraints and objectives. The weights for each material are aggregated—by multiplying them together, for instance—to give each a super-weight with a value between 0 (totally unacceptable) and 1 (fully acceptable by all criteria). The method can be refined further by giving fuzzy boundaries to the material properties or indices as well as to the selection criteria, as illustrated at the lower right in Figure 7.11. Techniques exist to choose the positions of the cores and the bases; despite the sophistication, however, the basic problem remains: The selection of the ranges $S(R)$ is a matter of judgment.

Weight factors and fuzzy methods all have merit when more rigorous analysis is impractical. They can be a good first step. But if you really want to identify the best material for a complex design, you need the methods of Sections 7.2 and 7.3.

CHAPTER 8

Case Studies: Multiple Constraints and Conflicting Objectives

A disk-brake and caliper.

Materials Selection in Mechanical Design. DOI: 10.1016/B978-1-85617-663-7.00008-4

CONTENTS

8.1 INTRODUCTION AND SYNOPSIS

These case studies illustrate how the techniques described in the last chapter work.[1] They are deliberately simplified to avoid clouding the illustration with unnecessary detail. The simplification is rarely as critical as it may at first appear: The choice of material is determined primarily by the physical principles of the problem, not by details of geometry. The principles remain the same when much of the detail is removed so that the selection is largely independent of these. The methods developed in Chapter 7 are so widely useful that they appear in the case studies in later chapters as well as this one. A reference is made to related case studies at the end of each section.

We start with three examples of coupled constraints using the methods in Section 7.2. We then explore three examples of conflicting objectives with the methods in Section 7.3.

8.2 MULTIPLE CONSTRAINTS: LIGHT PRESSURE VESSELS

What is the best material to make a light pressure vessel? Its radius R is prescribed. It must contain a pressure p without failing by yield or by fast fracture. And it must be as light as possible (Table 8.1).

---

[1] The charts now become more complicated. They have, as axes, combinations of properties. Some are the indices derived in earlier chapters; some are new. All are made with the same CES computer-based system (see *www.grantadesign.com*) that was used to make all the other charts in this book.

Table 8.1 Design Requirements for Safe Pressure Vessels

Function	Light pressure vessel (contain pressure p safely)
Constraints	Radius R specified Must not fail by yield Must not fail by fast fracture
Objective	Minimize mass
Free variables	Wall thickness t Choice of material

This is a "min–max" problem of the sort described in Section 7.2. Pressure vessels were the subject of Section 6.11. We draw on some of the results developed there to tackle this new problem.

The translation The mass of a thin-walled spherical pressure vessel (Figure 6.19) is

$$m = 4\pi R^2 t\rho$$

where t is the wall thickness and ρ the density of the material of which it is made. This is the objective function, the quantity we wish to minimize.

The stress in the wall of the vessel is

$$\sigma = \frac{pR}{2t} \tag{8.1}$$

The condition that the vessel must not yield requires that $\sigma < \sigma_y$, where σ_y is the yield strength of the wall material. This requires a wall thickness of

$$t \geq \frac{pR}{2\sigma_y}$$

resulting in a mass of

$$m_1 \geq 2\pi R^3 p\left(\frac{\rho}{\sigma_y}\right) \tag{8.2}$$

containing the index

$$M_1 = \frac{\rho}{\sigma_y} \tag{8.3}$$

The condition of no fast fracture requires that $\sigma < \sigma_f$, where

$$\sigma_f = \frac{K_{Ic}}{C\sqrt{\pi a_c^*}}$$

Here $2a_c^*$ is the diameter of the largest crack or flaw contained in the wall, C is a constant which we take to be unity, and K_{1c} is the plane-strain fracture toughness. This requires a wall thickness of

$$t \geq \frac{pR\sqrt{\pi a_c^*}}{2K_{1c}}$$

resulting in a mass of

$$m_2 \geq 2\pi R^3 p\sqrt{\pi a_c^*}\left(\frac{\rho}{K_{1c}}\right) \tag{8.4}$$

containing the index

$$M_2 = \frac{\rho}{K_{1c}} \tag{8.5}$$

Equating the two masses gives the equation of the coupling line linking M_1 and M_2:

$$M_2 = \frac{1}{\sqrt{\pi a_c^*}} M_1 \tag{8.6}$$

with the coupling constant

$$C_c = 1/\sqrt{\pi a_c^*}$$

The selection The two indices, M_1 and M_2, are the axes of Figures 8.1 and 8.2. Both show the position of the coupling lines for different values of the crack length a_c^*. Select the one that describes the longest crack the material could contain—usually identified with the resolution limit of the nondestructive tests used to detect cracks. Above it the fracture constraint determines the mass; below it the yield constraint does so.

The first chart covers all classes of materials, but with low resolution. The selection is made by placing a box with its corner on the appropriate coupling line and sliding it down the line, reducing both M_1 and M_2 until it contains only a small number of candidates for which documentation can be sought. The chart suggests that aluminum alloys and steels are the best choices. To get further we need more detail. Figure 8.2 provides it.[2] The axes are the same, but the materials are limited to high-strength low alloy steels, stainless steels, aluminum alloys, and polymer-matrix composites. The chart confirms the earlier finding and identifies the specific alloys that offer the lightest vessel for a given a_c^*. Here the procedure is taken to an extreme, leaving just one material identified in each of the selection boxes. The results are summarized in Table 8.2.

---

[2] Constructed with the CES software.

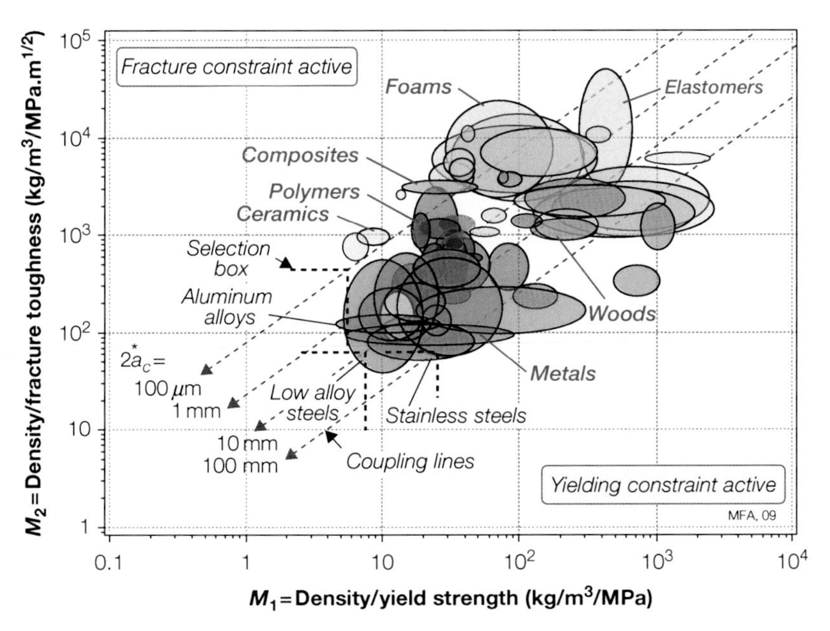

FIGURE 8.1

Over-constrained design leads to two or more performance indices linked by coupling equations. The diagonal broken lines show the coupling equation for four values of crack length $2a_c^*$. The selection lines intersect on the appropriate coupling line given the box-shaped search areas.

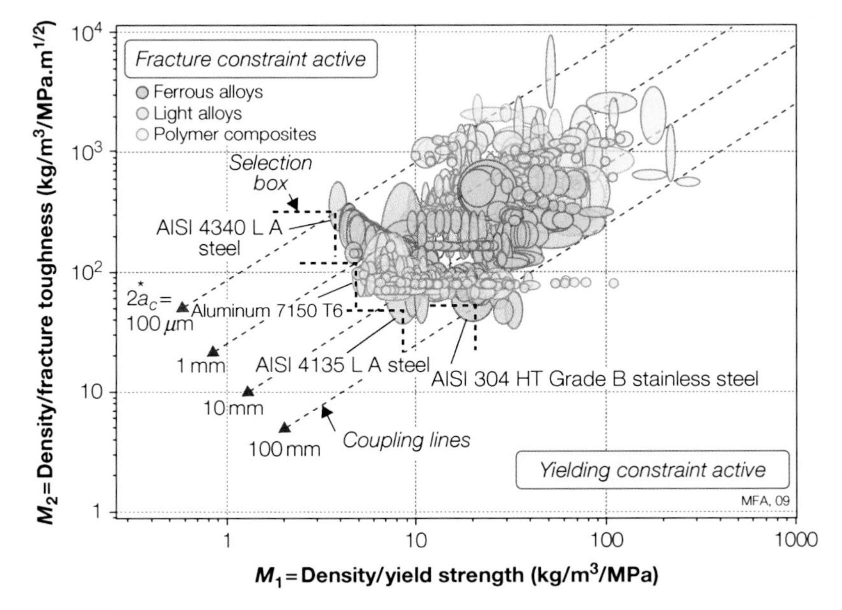

FIGURE 8.2

The same chart as in Figure 8.1, but with higher resolution. The colored ovals show data for 960 steels, aluminum alloys, and polymer-matrix composites.

Table 8.2 Materials for Light Pressure Vessels

Crack Size $2a_c^*$, mm	Selected Material
0.1	Low alloy steel AISI 4340 tempered 206°C
1	7150 aluminum alloy, T6 temper
10	Low alloy steel AISI 4135 tempered
100	Stainless steel AISI 304, HT grade B

Related case study

6.11 "Safe pressure vessels"

8.3 MULTIPLE CONSTRAINTS: CON-RODS FOR HIGH-PERFORMANCE ENGINES

A connecting rod in a high-performance engine, compressor or pump is a critical component: If it fails, catastrophe follows. Yet to minimize inertial forces and bearing loads, it must weigh as little as possible, implying the use of light, strong materials, stressed near their limits. When minimizing cost is the objective, con-rods are frequently made of cast iron because it is so cheap. But what are the best materials for con-rods when the objective is to maximize performance? Table 8.3 summarizes the design requirements for a connecting rod of minimum weight with two constraints: that it must carry a peak load F without failing either by fatigue or by buckling elastically. For simplicity, we assume that the shaft has a rectangular section $A = bw$ (Figure 8.3).

The translation This, like the last, is a "min–max" problem. The objective function is an equation for the mass that we approximate as

$$m = \beta A L \rho \tag{8.7}$$

where L is the length of the con-rod, ρ the density of the material of which it is made, A the cross-section of the shaft, and β a constant multiplier to allow for the mass of the bearing housings.

Table 8.3 The Design Requirements: Connecting Rods

Function	Connecting rod for reciprocating engine or pump
Constraints	Must not fail by high-cycle fatigue Must not fail by elastic buckling Stroke, and thus con-rod length L, specified
Objective	Minimize mass
Free variables	Cross section A Choice of material

The fatigue constraint requires that

$$\frac{F}{A} \leq \sigma_e \tag{8.8}$$

where σ_e is the endurance limit of the material of which the con-rod is made. (Here and elsewhere, we omit the safety factor that would normally enter an equation of this sort since it does not influence the selection.) Using Equation (8.8) to eliminate A in (8.7) gives the mass of a con-rod that will just meet the fatigue constraint:

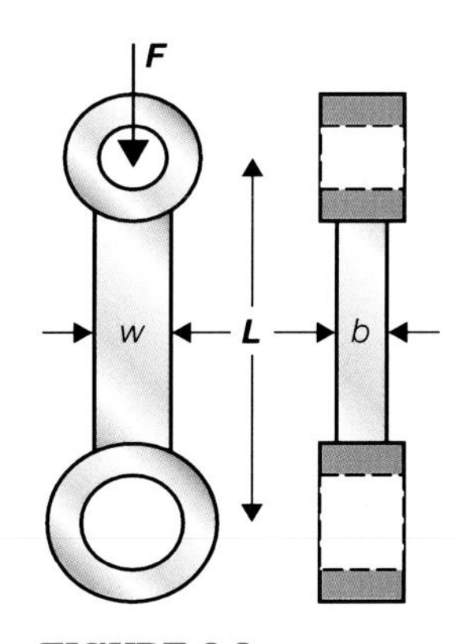

FIGURE 8.3
A connecting rod. It must not buckle or fail by fatigue—an example of multiple constraints.

$$m_1 = \beta F L \left(\frac{\rho}{\sigma_e}\right) \tag{8.9}$$

containing the material index

$$M_1 = \frac{\rho}{\sigma_e} \tag{8.10}$$

The buckling constraint requires that the peak compressive load F does not exceed the Euler buckling load:

$$F \leq \frac{\pi^2 E I}{L^2} \tag{8.11}$$

with

$$I = \frac{b^3 w}{12}$$

(Appendix B). Writing $b = \alpha w$, where α is a dimensionless "shape constant" characterizing the proportions of the cross-section, and eliminating A from Equation (8.7) gives a second equation for the mass:

$$m_2 = \beta \left(\frac{12 F}{\alpha \pi^2}\right)^{1/2} L^2 \left(\frac{\rho}{E^{1/2}}\right) \tag{8.12}$$

containing the second material index:

$$M_2 = \frac{\rho}{E^{1/2}} \tag{8.13}$$

The con-rod, to be safe, must meet both constraints. For a given length, L, the active constraint is the one leading to the largest value of the mass, m. Figure 8.4 shows the way in which m varies with L (a sketch of

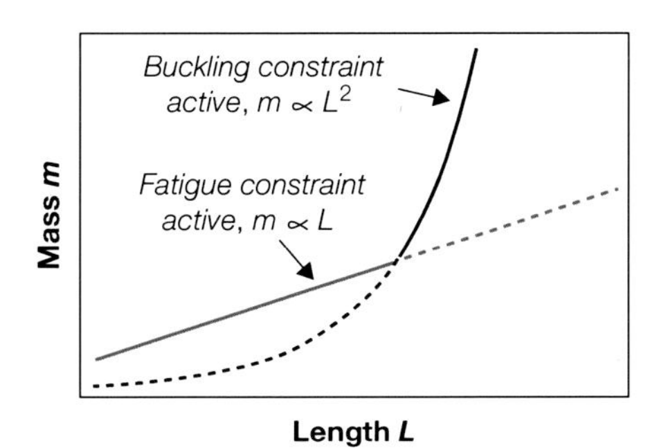

FIGURE 8.4
The equations for the mass of the con-rod are shown schematically as a function of L.

Equations (8.9) and (8.11)) for a single material. Short con-rods are liable to fatigue failure; long ones are prone to buckle.

The selection Consider first the selection of a material for the con-rod from among the limited list in Table 8.4. The specifications are

$$L = 200\,\text{mm} \quad F = 50\,\text{kN}$$
$$\alpha = 0.8 \quad \beta = 1.5$$

The table lists the mass m_1 of a rod that will just meet the fatigue constraint, and the mass m_2 that will just meet that on buckling (Equations (8.9) and (8.12)). For three of the materials, the active constraint is fatigue; for two, it is buckling. The quantity $\tilde{m}$ in the last column of the table is the larger of m_1 and m_2 for each material; it is the lowest mass that meets both constraints. The material offering the lightest rod is that with the smallest value of $\tilde{m}$. Here it is the titanium alloy Ti-6Al-4V. The metal-matrix composite Duralcan 6061–20% SiC is a close second. Both weigh less than half as much as a cast-iron rod.

Well, that is one way to use the method, but it is not the best. First, it assumes some "preselection" procedure has been used to obtain the materials listed in the table, but does not explain how this is to be done; second, the results apply only to the values of F and L listed previously—change these, and the selection changes. If we wish to escape these restrictions, the graphical method is the way to do it.

The mass of the rod that will survive both fatigue and buckling is the larger of the two masses m_1 and m_2 (Equations (8.9) and (8.12)). Setting them equal gives the equation of the coupling line (defined in Section 7.2):

$$M_2 = \left[\left(\frac{\alpha\pi^2}{12}\cdot\frac{F}{L^2}\right)^{1/2}\right]\cdot M_1 \tag{8.14}$$

The quantity in square brackets is the coupling constant C_c. It contains the quantity F/L^2—the "structural loading coefficient" of Section 5.6.

Table 8.4 Selection of a Material for the Con-rod

Material	ρ kg/m^3	E GPa	σ_e MPa	m_1 kg	m_2 kg	$\tilde{m} = \max(m_1, m_2)$ kg
Nodular cast iron	7150	178	250	**0.43**	0.22	0.43
HSLA steel 4140 (o.q. T-315)	7850	210	590	0.20	**0.28**	0.28
Al S355.0 casting alloy	2700	70	95	**0.39**	0.14	0.39
Duralcan Al-SiC(p) composite	2880	110	230	**0.18**	0.12	0.18
Titanium 6Al 4V	4400	115	530	0.12	**0.17**	**0.17**

Materials with the optimum combination of M_1 and M_2 are identified by creating a chart with these indices as axes. Figure 8.5 illustrates this, using a database of light alloys, but including cast iron for comparison. Coupling lines for two values of F/L^2 are plotted on it, taking $\alpha = 0.8$. Two extreme selections are shown, one isolating the best subset when the structural loading coefficient F/L^2 is high, the other when it is low. Beryllium and its alloys emerge as the best choice for all values of C_c within this range. Leaving them aside, the best choices when F/L^2 is large ($F/L^2 = 5$ MPa) are titanium alloys such as Ti-6Al-4V. For the low value ($F/L^2 = 0.05$ MPa), magnesium alloys such as AZ61 offer lighter solutions than aluminum or titanium. Table 8.5 lists the conclusions.

Postscript Con-rods have been made from all the materials in the table: aluminum and magnesium in road cars, titanium and (rarely) beryllium in racing engines. Had we included CFRP in the selection, we would have found that it, too, performs well by the criteria we have used. This conclusion has been reached by others, who have tried to do something about it: At least three designs of CFRP con-rods have been prototyped. It is not easy to design a CFRP con-rod. It is essential to use continuous fibers, which must be wound in such a way as to enclose both the shaft and the bearing housings; and the shaft

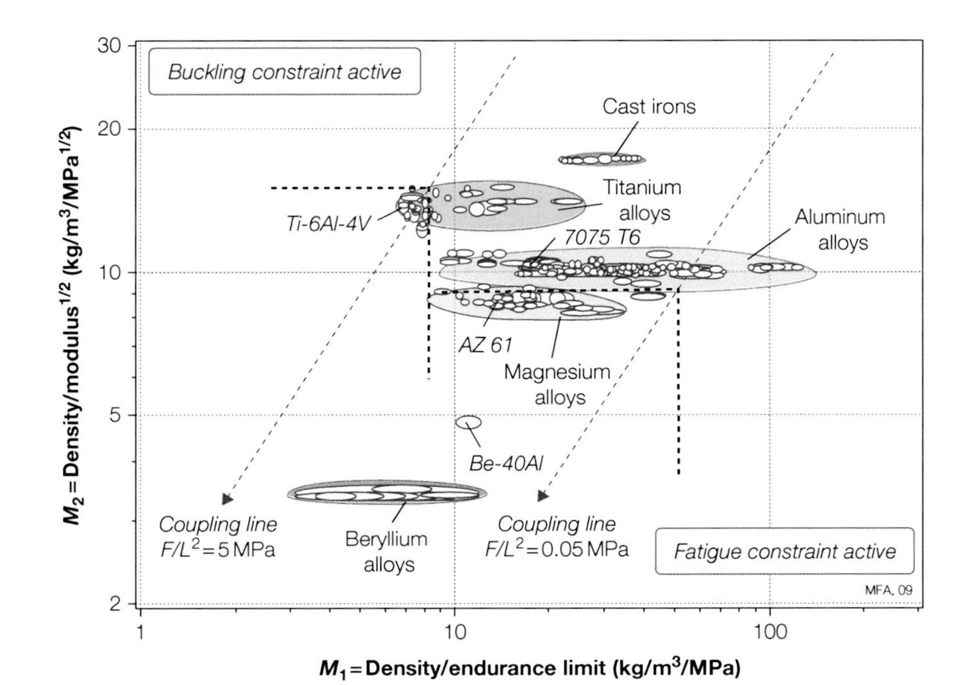

FIGURE 8.5
The coupled-constraint construction for the con-rod. The diagonal broken lines show the coupling equation for two extreme values of F/L^2. The box-shaped search areas are shown.

Table 8.5 Materials for High-performance Con-rods

Material	Comment
Magnesium alloys	AZ61 and related alloys offer good all-around performance
Titanium alloys	Ti-6-4 is the best choice for high F/L^2
Beryllium alloys	The ultimate choice, but difficult to process and very expensive
Aluminum alloys	Cheaper than titanium or magnesium, but lower performance

must have a high proportion of fibers that lie parallel to the direction in which F acts. You might, as a challenge, think how you would do it.

Related case studies

6.4 "Materials for table legs"
8.4 "Multiple constraints: windings for high-field magnets"
10.3 "Forks for a racing bicycle"
10.5 "Table legs yet again: Thin or light?"

8.4 MULTIPLE CONSTRAINTS: WINDINGS FOR HIGH-FIELD MAGNETS

Physicists, for reasons of their own, like to see what happens to things in high magnetic fields. "High" means 50 Tesla or more. The only way to get such fields is the old-fashioned one: Dump a huge current through a wire-wound coil like that shown schematically in Figure 8.6; neither permanent magnets (practical limit: 1.5 T) nor super-conducting coils (present limit: 25 T) can achieve such high fields. The current generates a field pulse that lasts as long as the current flows. The upper limits on the field and its duration are set by the material of the coil itself: If the field is too high, magnetic forces blow the coil apart; if too long, it melts. So choosing the right material for the coil is critical. What should it be? The answer depends on the pulse length.

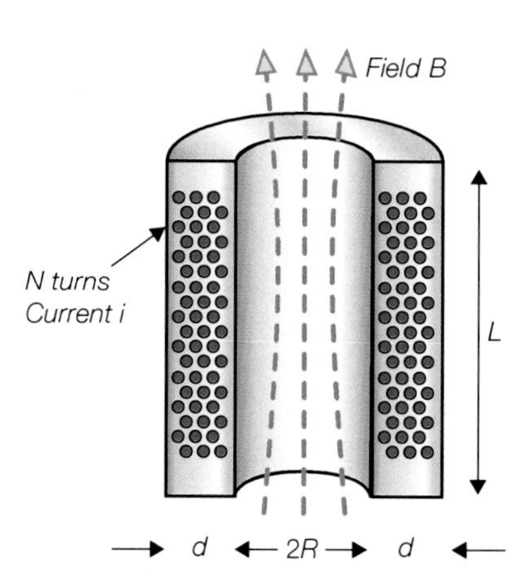

FIGURE 8.6
Windings for high-powered magnets. There are two constraints: The magnet must not overheat, and it must not fail under the radial magnetic forces.

Pulsed fields are classified according to their duration and strength, as in Table 8.6. The requirements for the survival of the magnet producing them are summarized in Table 8.7. There is one objective—to maximize the field—with two constraints

Table 8.6 Duration and Strength of Pulsed Fields

Classification	Duration	Field Strength
Continuous	1 s–∞	< 30 T
Long	100 ms–1 s	30–60 T
Standard	10–100 ms	40–70 T
Short	10–1000 μs	70–80 T
Ultrashort	0.1–10 μs	> 100 T

Table 8.7 The Design Requirements: High-field Magnet Windings

Function	Magnet windings
Constraints	No mechanical failure Temperature rise $< 100°C$ Radius R and length L of coil specified
Objective	Maximize magnetic field
Free variable	Choice of material for the winding

deriving from the requirement of survivability: that the windings are strong enough to withstand the radial force on them caused by the field and that they do not heat up too much.

The translation Detailed modeling gets a little complicated, so let us start with some intelligent guesses (*IGs*). First, if the windings must carry load (the first constraint) they must be strong—the higher the strength, the greater the field they can tolerate. So (*IG 1*) we want materials with a high elastic limit σ_y. Second, a current i flowing for a time t_p through a coil of resistance R_e dissipates $i^2 R_e t_p$ joules of energy, and if this takes place in a volume V, the temperature rise is

$$\Delta T = \frac{i^2 R_e t_p}{V C_p \rho}$$

where C_p is the specific heat of the material and ρ is its density. So (*IG 2*) to maximize the current (and thus the field B) we need materials with low values of $R_e/C_p\rho$ or, since resistance R_e is proportional to resistivity ρ_e for a fixed coil geometry, materials with low $\rho_e/C_p\rho$.

Both guesses are correct. This has gotten us a long way; a simple search for material with high σ_y—or rather low $M_1 = 1/\sigma_y$ (since we must express objectives in a form to be minimized)—and low $M_2 = \rho_e/C_p\rho$ will find a sensible subset. The two are plotted in Figure 8.7 for some 1200 metals and alloys (ignore for the moment the black selection boxes). The materials with the best combination of indices lie along the lower envelope of the populated region. Strength is the dominant constraint when pulses are short, requiring materials

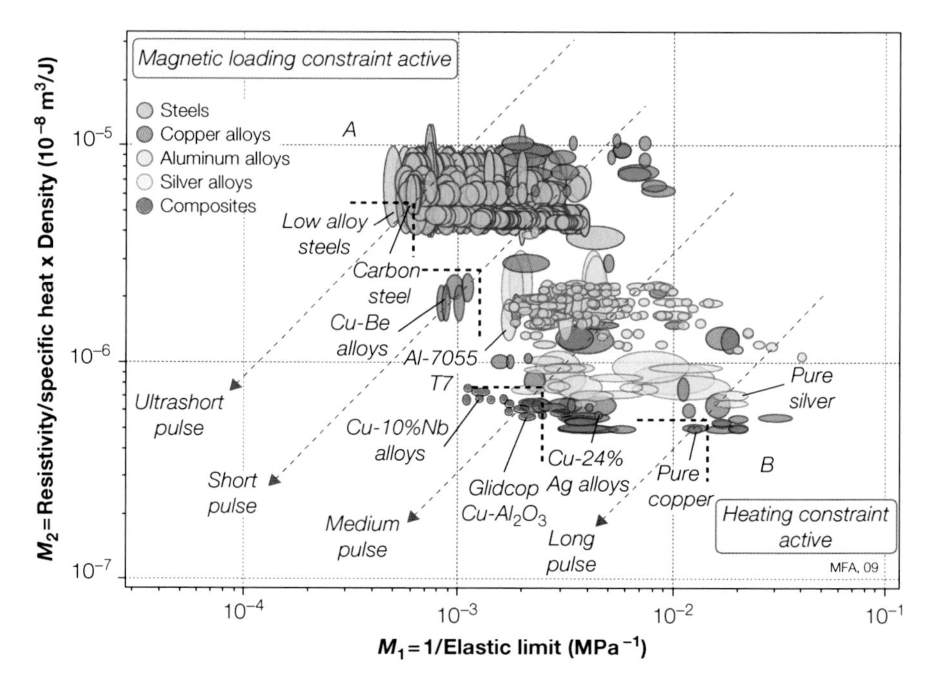

FIGURE 8.7
The two material groups that determine the choice of material for winding of high-powered magnets or electric motors. The axes are the two "guesses" made in the text—the modeling confirms the choice and allows precise positioning of the selection lines for a given pulse duration.

with low M_1; those near **A** are the best choice. Heating is the dominant constraint when pulses are long, and materials near **B**, with low M_2, are the answer.

This is progress, and it may be enough. If we want greater resolution, we must abandon guesswork (intelligent though it was) and apply min–max methods, requiring more detailed modeling. It gets a bit involved—if it looks too grim, skip to the next section, "The selection."

Consider first destruction by magnetic loading. The field, B (units: Weber/m^2), in a long solenoid like that of Figure 8.6 is

$$B = \frac{\mu_o N i}{L} \cdot \lambda_f \cdot F(\alpha, \beta) \tag{8.15}$$

where μ_o is the permeability of air ($4\pi \times 10^{-7}$ Wb/A.m), N is the number of turns, i is the current, L is the length of the coil, λ_f is the fill factor that accounts for the thickness of insulation (λ_f = cross-section of conductor/cross-section of coil), and $F(\alpha, \beta)$ is a geometric constant (the "shape factor") that depends on the proportions of the magnet, the value of which need not concern us.

The field creates a force on the current-carrying coil. It acts radially outward, like the pressure in a pressure vessel, even though it is actually a body force, not a surface force. Its magnitude, p per unit area, is

$$p = \frac{B^2}{2\mu_o \cdot F(\alpha, \beta)}$$

The pressure generates a stress σ in the windings and their casing:

$$\sigma = \frac{pR}{d} = \frac{B^2}{2\mu_o \cdot F(\alpha, \beta)} \cdot \frac{R}{d} \tag{8.16}$$

This must not exceed the yield strength σ_y of the windings, giving the first limit on B:

$$B_1 \leq \left(\frac{2\mu_o d\sigma_y \cdot F(\alpha, \beta)}{R} \right)^{1/2} \tag{8.17}$$

The field is maximized by maximizing σ_y, that is, by minimizing

$$M_1 = \frac{1}{\sigma_y} \tag{8.18}$$

vindicating *IG 1*.

Now consider destruction by overheating. High-powered magnets are initially cooled in liquid nitrogen to –196°C in order to reduce the resistance of the windings; if the windings warm above room temperature, the resistance, R_e, in general becomes too large. The entire energy of the pulse, $\int i^2 R_e dt \approx i^2 \overline{R}_e t_p$ is converted into heat (here $\overline{R}_e$ is the average of the resistance over the heating cycle and t_p is the length of the pulse); and since there is insufficient time for the heat to be conducted away, this energy causes the temperature of the coil to rise by ΔT, where

$$\Delta T = \frac{i^2 \overline{R}_e t_p}{C_p \rho V} = \frac{B^2}{\mu_o^2} \cdot \frac{\rho_e t_p}{d^2 C_p \rho} \tag{8.19}$$

Here ρ_e is the resistivity of the material of the windings, V its volume, C_p its specific heat (J/kg.K), and ρ its density. If the upper limit for the temperature is 200 K, $\Delta T_{max} \leq 100$ K, giving the second limit on B:

$$B_2 \leq \left(\frac{\mu_o^2 d^2 C_p \rho \lambda_f \Delta T_{max}}{t_p \rho_e} \right)^{1/2} F(\alpha, \beta) \tag{8.20}$$

The field is maximized by minimizing

$$M_2 = \frac{\rho_e}{C_p \rho} \tag{8.21}$$

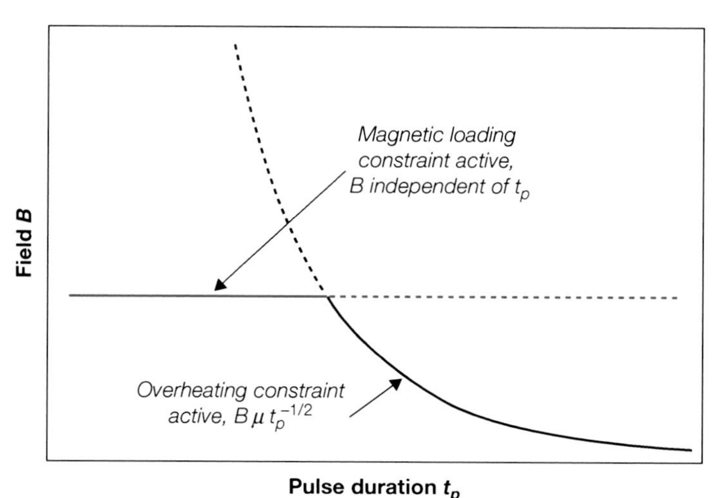

FIGURE 8.8
The two equations for B are sketched, indicating the active constraint.

in accord with *IG 2*. The two equations for B are sketched as a function of pulse time t_p in Figure 8.8. For short pulses, the strength constraint is active; for long ones, the heating constraint is dominant.

The selection Table 8.8 lists material properties for three alternative windings. The sixth column gives the strength-limited field strength B_1; the seventh column, the heat-limited field B_2 evaluated for the following values of the design requirements:

$$t_p = 10\,\text{ms} \quad \lambda_f = 0.5 \quad \Delta T_{\max} = 100\,\text{K}$$
$$F(\alpha, \beta) = 1 \quad R = 0.05\,\text{m} \quad d = 0.1\,\text{m}$$

Strength is the active constraint for the copper-based alloys; heating for the steels. The last column lists the limiting field $\tilde{B}$ for the active constraint. The Cu-Nb composites offer the largest $\tilde{B}$.

Table 8.8 Selection of a Material for a High-field Magnet, Pulse Length 10 ms

Material	ρ kg/m^3	σ_y MPa	C_p J/kg.K	ρ_e 10^{-8} Ω.m	B_1 Wb/m^2	B_2 Wb/m^2	$\tilde{B}$ Wb/m^2
High-conductivity copper	8940	250	385	1.7	**35**	113	35
Cu-15%Nb composite	8900	780	368	2.4	**62**	92	**62**
HSLA steel	7850	1600	450	25	89	**30**	30

So far, so good. But we have the same problem that appeared in the last case study—someone preselected the three materials in the table; surely there must be others? And the choice we reached is specific to a magnet with the dimensions listed above and a pulse time t_p of 10 ms. What happens if we change these? We need the graphical method.

The cross-over point in Figure 8.6 is that at which Equations (8.17) and (8.20) are equal, giving the coupling line

$$M_2 = \left[\frac{\mu_o R d \lambda_f F(\alpha, \beta) \Delta T_{max}}{2t_p}\right] \cdot M_1 \tag{8.22}$$

The quantity in square brackets is the coupling constant C_c; it depends on the pulse length t_p.

Now back to Figure 8.7. The axes, as already said, are the two indices M_1 and M_2. Three selections are shown, one for ultrashort-pulse magnets, the other two for longer pulses. Each selection box is a contour of constant field B; its corner lies on the coupling line for the appropriate pulse duration. The best choice, for a given pulse length, is that contained in the box that lies farthest down its coupling line. The results are summarized in Table 8.9.

Postscript The case study, as developed here, is an oversimplification. Magnet design today is very sophisticated, involving nested sets of electro- and super-conducting magnets (up to nine deep), with geometry the most important variable. But the selection scheme for coil materials has validity: When pulses are long, resistivity is the primary consideration; when they are

Table 8.9 Materials for High-field Magnet Windings

Material	Comment
Continuous and long pulse	
High-conductivity coppers Pure silver	Best choice for low-field, long-pulse magnets (heat-limited)
Short pulse	
Copper-AL_2O_3 composites (Glidcop) H-C copper cadmium alloys H-C copper zirconium alloys H-C copper chromium alloys Drawn copper-niobium composites	Best choice for high-field, short-pulse magnets (heat- and strength-limited)
Ultrashort pulse, ultrahigh field	
Copper-beryllium-cobalt alloys High-strength, low alloy steels	Best choice for high-field, short-pulse magnets (strength-limited)

very short, it is strength, and the best choice for each is that developed here. Similar considerations enter the selection of materials for very-high-speed motors, for bus bars, and for relays.

Related reading

Herlach, F. (1988). The technology of pulsed high-field magnets. *IEEE Transactions on Magnetics, 24*, 1049.

Wood, J.T., Embury, J.D., & Ashby, M.F. (1995). An approach to material selection for high field magnet design. *Acta Metal. et Mater., 43*, 212.

Related case study

8.3 "Multiple constraints: con-rods for high-performance engines"

8.5 CONFLICTING OBJECTIVES: TABLE LEGS AGAIN

We now turn from coupled constraints to conflicting objectives, applying the methods of Section 7.3. We start with a simple example, revisiting the selection of materials for slender table legs.

The translation The design requirements set down by Snr. Tavolino for his table (refer to Table 6.5) involved two objectives: that the leg be as light *and* as thin as possible. The mass m of a leg (Equation (6.8)) is proportional to

$$M_1 = \frac{\rho}{E^{1/2}}$$

The thickness $2r$ (Equation (6.9)) scales with

$$M_2 = \frac{1}{E}$$

Snr. Tavolino wishes to minimize both.

The selection The proper way to tackle multiobjective problems like this one is to make a trade-off plot. Figure 8.9 is an example: M_1 on the vertical axis, M_2 on the horizontal one. For clarity, only ferrous alloys, light alloys, composites, and woods appear on the plot—that set includes almost all materials that could be thought of as candidates. The classes cluster very tightly because both modulus and density have narrow ranges.

Legs made of spruce or fir are potentially lighter than those made of any other material. Composites offer legs that are almost as light and much thinner. But, interestingly, they do not offer the thinnest of all—steel does

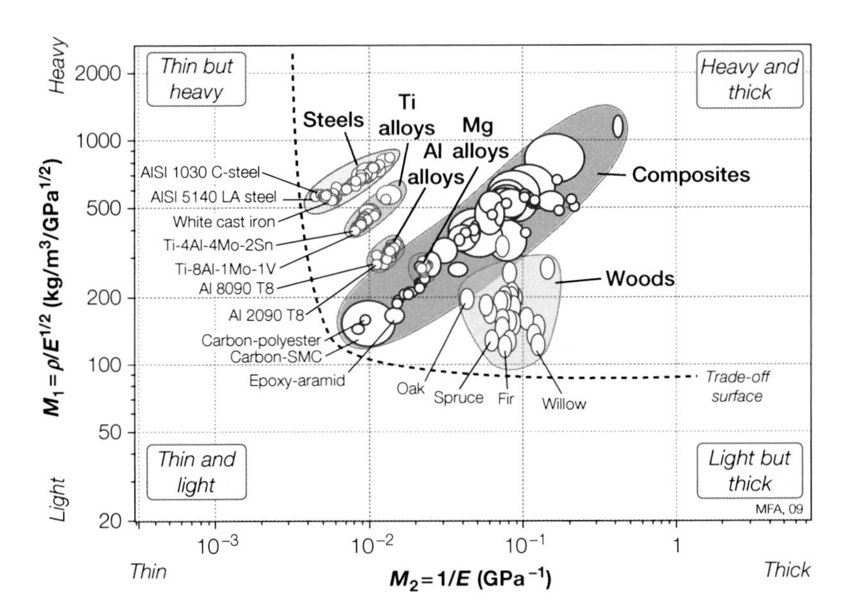

FIGURE 8.9
The trade-off plot for the table leg. Materials that lie near the trade-off surface are identified.

better. Light alloys allow legs that are lighter than steel but not nearly as thin. Overall, composites offer the best compromise—very low weight and attractive slenderness.

Postscript But (as we asked in Section 6.4), wouldn't a tubular leg be lighter? Fatter, but how much fatter? Wouldn't the tubular solution be the best compromise? The answer will have to wait a little longer—until Section 10.5.

Related case study

6.4 "Materials for table legs"

8.6 CONFLICTING OBJECTIVES: WAFER-THIN CASINGS FOR MUST-HAVE ELECTRONICS

Slenderness in consumer electronics—portable computers, mobile phones, PDAs, and MP3 players—is a major design driver. The ideal is a device that you can slip into a shirt pocket and not even know that it is there. The casing has to be stiff and strong enough to protect the electronics—the display, particularly—from damage. Casings used to be made of molded ABS or polycarbonate. To be stiff enough an ABS case has to be at least 2 mm thick, too much for today's designs in which thinness and lightness are very highly valued. But the consequences of making the casing too thin are severe: Mobile phones get sat on, and portable computers end up under piles of books. If the casing is insufficiently stiff, it flexes and damages the screen. The challenge: to identify materials for casings that are at least as stiff as a 2-mm ABS case but thinner and lighter. We must recognize that the thinnest may not be the lightest, and vice versa. A trade-off will be needed. Table 8.10 summarizes the requirements.

The translation We idealize the loading on a panel of the casing in the way shown in the Figure 8.10. External loads cause it to bend. The bending stiffness is

$$S = \frac{48\,EI}{L^3}$$

with

$$I = \frac{Wt^3}{12} \tag{8.23}$$

where E is Young's modulus, I is the second moment of the area of the panel, and the dimensions L, W, and t are shown in the figure. The stiffness S must equal or exceed a design requirement S^* if the panel is to perform its function properly. Combining the two equations and solving for the thickness t gives Equation (8.24).

Table 8.10 The Design Requirements: Casing for Portable Electronics	
Function	Light, thin (cheap) casing
Constraints	Bending stiffness S^* specified Dimensions L and W specified
Objectives	Minimize thickness of casing Minimize mass of casing (Minimize material cost)
Free variables	Thickness t of casing wall Choice of material

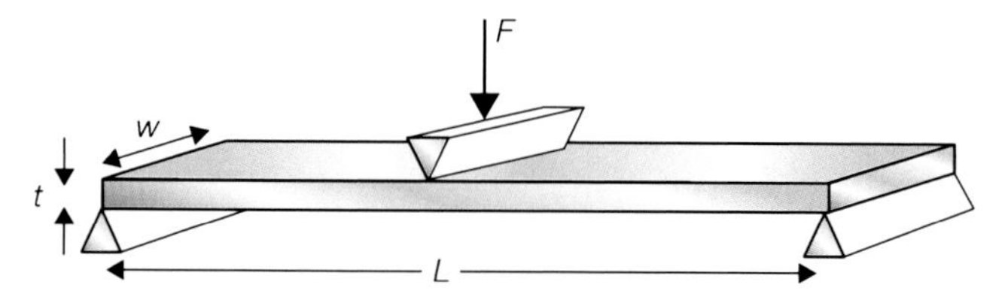

FIGURE 8.10
The casing can be idealized as a panel of dimensions $L \times W$ and thickness t, loaded in bending.

$$t \geq \left(\frac{S^* L^3}{4 EW} \right)^{1/3} \quad (8.24)$$

The thinnest panel is that made from the material with the smallest value of the index:

$$M_1 = \frac{1}{E^{1/3}}$$

The mass of the panel per unit area, m_a, is just ρt, where ρ is its density—the lightest panel is that made from the material with the smallest value of

$$M_2 = \frac{\rho}{E^{1/3}} \quad (8.25)$$

We use the existing ABS panel, stiffness S^*, as the standard for comparison. If ABS has a modulus E_o and a density ρ_o, then a panel made from any other material (modulus E, density ρ) will, according to Equation (8.24), have a thickness t relative to that of the ABS panel t_o given by

$$\frac{t}{t_o} = \left(\frac{E_o}{E} \right)^{1/3} \quad (8.26)$$

and a relative mass per unit area of

$$\frac{m_a}{m_{a,o}} = \left(\frac{\rho}{E^{1/3}} \right) \left(\frac{E_o^{1/3}}{\rho_o} \right) \quad (8.27)$$

We wish to explore the trade-off between t/t_o and $m_a/m_{a,o}$ for possible solutions.

The selection Figure 8.11 shows the necessary plot, here limited to a few material classes for simplicity. It is divided into four sectors with ABS at the center at the coordinates (1,1). The solutions in sector **A** are both thinner and lighter than ABS, some by a factor of 2. Those in sector **B** and **C** are better by one metric but worse by the other. Those in sector **D** are worse by both. To narrow in on an optimal choice we sketch in a trade-off surface, shown as the broken line. The solutions nearest to this surface are, in terms of one metric or the other, good choices. Intuition guides us to those in or near Sector **A**.

This is already enough to suggest choices that offer savings in thickness and in weight. If we want to go further we must formulate a *relative penalty function*. Define Z^*, measured in units of currency, as

$$Z^* = \alpha_t^* \frac{t}{t_o} + \alpha_m^* \frac{m_a}{m_{a,o}} \quad (8.28)$$

The exchange constant α_t^* measures the decrease in penalty—or gain in value—for a fractional decrease in thickness; α_m^*, for a fractional decrease in mass.

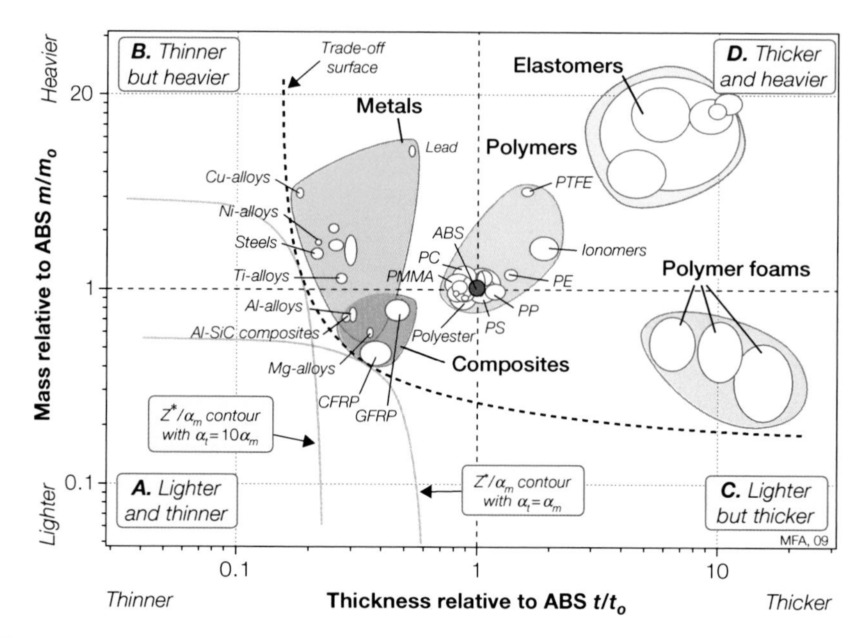

FIGURE 8.11

The relative thickness and mass of casings made from alternative materials. Those near the trade-off surface are identified.

As an example, set $\alpha_t^* = \alpha_m^*$, meaning that we value both equally. Then solutions with equal penalty Z^* are those on the contour

$$\frac{Z^*}{\alpha_m^*} = \frac{t}{t_o} + \frac{m_a}{m_{a,o}} \tag{8.29}$$

where the first term on the right is given by Equation (8.26) and the second by Equation (8.27). This is plotted for a selection of metals, polymers, and composites in Figure 8.12. ABS lies near the middle of the polymer group. CFRP, GFRP, titanium, aluminum, and magnesium all offer casings that have lower (better) values of Z^*.

The problem with this plot is that it is specific to a single value of the ratio α_t^*/α_m^*. If the relative importance of thinness and lightness are changed, the ranking changes too. We need a more general method. It is provided by constructing penalty contours on the trade-off plot. Two are shown as blue lines in Figure 8.11. The linear relationship of Equation (8.29) plots as a family of curves (not straight lines because of the log scales), with Z^*/α_m^* decreasing toward the bottom left. The absolute value of Z^*/α_m^* does not matter—all we need it for is to identify the point at which a contour is tangent to the trade-off surface as shown in Figure 8.11. The solutions nearest this point are the optimum choices: CFRP, magnesium alloys, and Al-SiC composites.

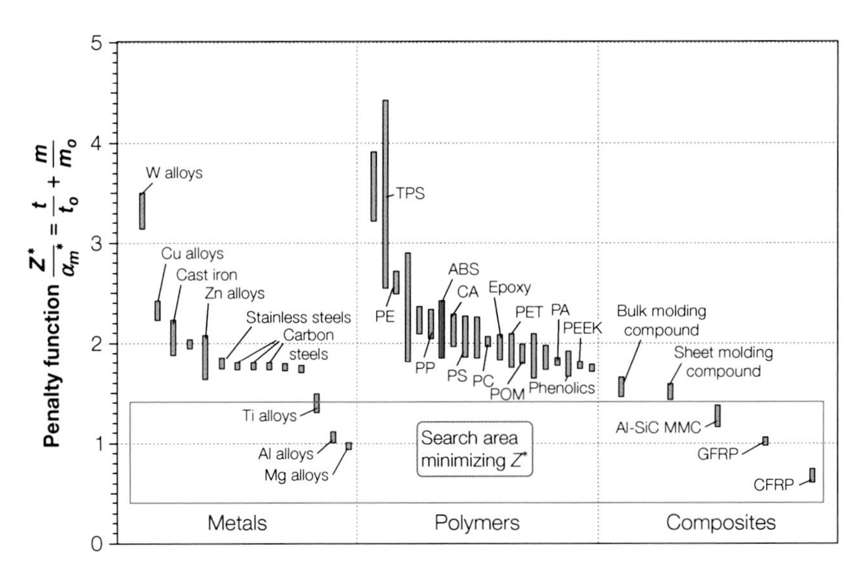

FIGURE 8.12

The penalty function Z^*/α_m^* when $\alpha_t^* = \alpha_m^*$. ABS lies near the middle of the polymer column. Materials below it have a lower penalty—they are better choices.

If, instead, we set $\alpha_t^* = 10\alpha_m^*$, meaning that thinness is much more highly valued than lightness, the contour moves to the second position shown on Figure 8.11. Now titanium and even steel become attractive candidates.

Postscript Back in 1997, when extreme thinness and lightness first became major design drivers, the conclusions reached here were new. At that time almost all casings for handheld electronics were made of ABS, polycarbonate, or, occasionally, steel. Now, 12 or more years later, examples of aluminum, magnesium, titanium, and even CFRP casings can be found in currently marketed products. The value of the case study (which dates from 1997) is as an illustration of the way in which systematic methods can be applied to multiobjective selection.

Related case study

8.7 "Conflicting objectives: Materials for a disk-brake caliper"

8.7 CONFLICTING OBJECTIVES: MATERIALS FOR A DISK-BRAKE CALIPER

It is unusual—very unusual—to ask whether cost is important in selecting a material and to get the answer "No." But it does sometimes happen, notably when the material is to perform a critical function in space (beryllium for structural components, iridium for radiation screening), in medical procedures (think of gold tooth fillings), and in equipment for highly competitive sports (one racing motorcycle had a cylinder head made of solid silver for its high thermal conductivity). Here is another example—materials for the brake calipers of a Formula 1 racing car.

The translation The brake caliper can be idealized as two beams of length L, depth b, and thickness h, locked together at their ends (see this chapter's cover picture and Figure 8.13). Each beam is loaded in bending when the brake is applied, and because braking generates heat, it gets hot. The lower schematic represents one of the beams. Its length L and depth b are given. The beam stiffness S is critical: If it is inadequate the caliper will flex, impairing braking efficiency and allowing vibration. Its ability to transmit heat, too, is critical since part of the heat generated in braking must be conducted out through the caliper. Table 8.11 summarizes the requirements.

The mass of the caliper scales with that of one of the beams. Its mass per unit area is simply

$$m_a = h\rho \text{ (units: kg/m}^2\text{)} \tag{8.30}$$

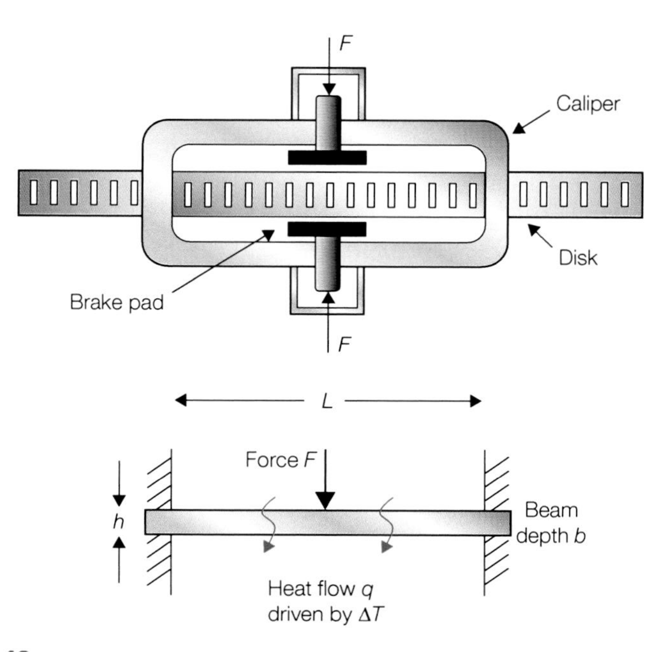

FIGURE 8.13
A schematic of a brake caliper. The long arms of the caliper are loaded in bending and must conduct heat well to prevent overheating.

Table 8.11 The Design Requirements: Brake Caliper

Function	Brake caliper
Constraints	Bending stiffness S^* specified Dimensions L and b specified
Objectives	Minimize mass of caliper Maximize heat transfer through caliper
Free variables	Thickness h of caliper wall Choice of material

where ρ is the density of the material of which it is made. Heat transfer q depends on the thermal conductivity λ of the material of the beam; the heat flux per unit area is

$$q_a = \lambda \frac{\Delta T}{h} \text{ (units: Watts/m}^2\text{)} \tag{8.31}$$

where ΔT is the temperature difference between the surfaces.

The quantities L, b, and ΔT are specified. The only free variable is the thickness h. But there is a constraint: The caliper must be stiff enough to ensure that it does not flex or vibrate excessively. To achieve this we require that

$$S = \frac{C_1 EI}{L^3} = \frac{C_1 Ebh^3}{12L^3} \geq S^* \text{ (units: N/m)} \tag{8.32}$$

where S^* is the desired stiffness, E is Young's modulus, C_1 is a constant that depends on the distribution of load, and $I = bh^3/12$ is the second moment of the area of the beam. Thus

$$h \geq \left(\frac{12S^*}{C_1 bE}\right)^{1/3} L \tag{8.33}$$

Inserting this into Equations (8.30) and (8.31) gives equations for the mass m_a of the arm and the heat q_a transferred through it, per unit area:

$$m_a \geq \left(\frac{12S^*}{C_1 b}\right)^{1/3} L \left(\frac{\rho}{E^{1/3}}\right) \text{ (units: kg/m}^2\text{)} \tag{8.34}$$

$$q_a = \frac{\Delta T}{L} \left(\frac{C_1 b}{12S^*}\right)^{1/3} (\lambda E^{1/3}) \text{ (units: W/m}^2\text{)} \tag{8.35}$$

The first equation contains the material index:

$$M_1 = \frac{\rho}{E^{1/3}}$$

the second (expressed such that a minimum is sought) contains the index

$$M_2 = \frac{1}{\lambda E^{1/3}}$$

The standard material for a brake caliper is nodular cast iron—it is cheap and stiff, but it is also heavy and a relatively poor conductor. We use this as a standard for comparison, normalizing Equations (8.34) and (8.35) by the values for cast iron (density ρ_o, modulus E_o, and conductivity λ_o), giving

$$\frac{m_a}{m_{a,o}} = \left(\frac{\rho}{E^{1/2}}\right)\left(\frac{E_o^{1/3}}{\rho_o}\right) \tag{8.36}$$

and

$$\frac{q_{a,o}}{q_a} = \frac{\lambda_o E_o^{1/3}}{\lambda E^{1/3}} \tag{8.37}$$

The equation for q_a has been inverted so that the best choice of material is that which minimizes both of these. Figure 8.14 shows a chart with these as axes. It is divided into four quadrants, centered on cast iron at the point (1,1). Each bubble describes a material. Those in the lower left are better than cast iron by both objectives; an aluminum caliper, for example, has half the weight and offers twice the heat transfer. The ultimate choice is beryllium or its alloy Be 40%Al.

To go further we formulate the relative penalty function

$$Z^* = \alpha_m^* \left(\frac{m_a}{m_{a,o}}\right) + \alpha_q^* \left(\frac{q_{a,o}}{q_a}\right) \tag{8.38}$$

in which the terms in brackets are given by Equations (8.36) and (8.37) and the exchange constants α_m^* and α_q^* measure the relative value of a fractional saving of weight or increase in heat transfer relative to cast iron. The penalty function is plotted in Figure 8.14 for three values of the ratio α_q^*/α_m^* of the exchange constants. Each is tangent to a trade-off surface that excludes the "exotic" beryllium alloys, which otherwise dominate the selection for all values. For $\alpha_q^*/\alpha_m^* = 0.1$, meaning mass reduction is of prime importance, magnesium alloys are the best choice. If mass reduction and heat transfer are given equal weight ($\alpha_q^*/\alpha_m^* = 1$), aluminum alloys become a good choice. If heat transfer is the overriding consideration ($\alpha_q^*/\alpha_m^* = 10$), alloys based on copper win. But if you really want the best, it has to be beryllium.

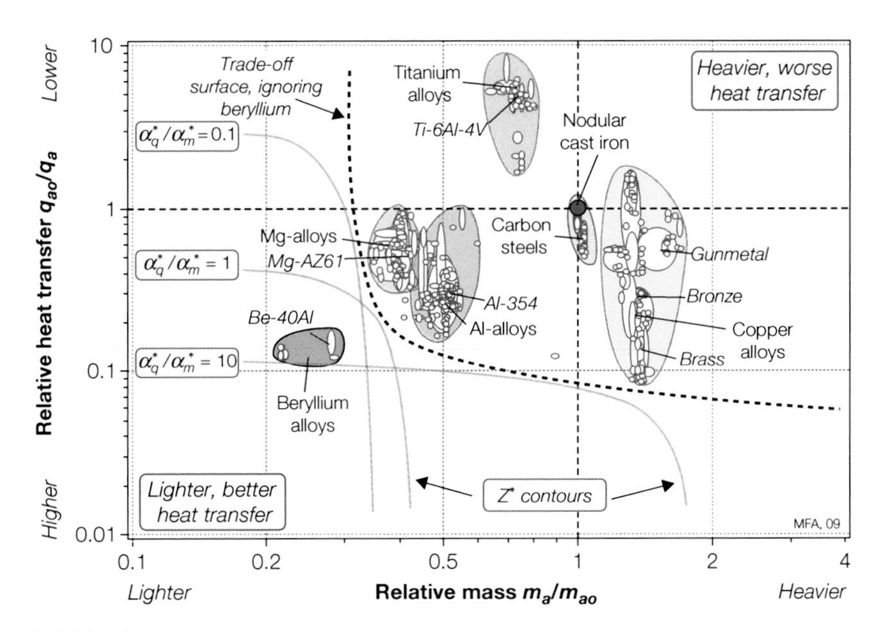

FIGURE 8.14

A chart with Equations (8.27) and (8.28) as axes. Beryllium and its alloys are the preferred choice, minimizing mass and maximizing heat transfer. But if these exotics are excluded, the choice becomes dependent on the ratio α_q^*/α_m^*. The trade-off surface and penalty contours for three values of α_q^*/α_m^* are shown.

Postscript Ferrari Racing did, at one time, commission beryllium brake calipers. Today restrictions on materials, imposed to make Formula 1 more competitive, outlaw their use.

Related case studies

6.16 "Materials to minimize thermal distortion in precision devices"
8.6 "Wafer-thin casings for must-have electronics"

8.8 SUMMARY AND CONCLUSIONS

Most designs are overconstrained: They must simultaneously meet several competing, and often conflicting, requirements. But although they conflict, an optimum selection is still possible. The "active constraint" method, developed in Chapter 7, allows the selection of materials that optimally meet two or more constraints. It is illustrated here by three case studies, two of them mechanical, one electro-mechanical.

Greater challenges arise when the design must meet two or more conflicting objectives (such as minimizing mass, volume, cost, and environmental impact). Here we need a way to express all the objectives in the same units, a "common currency," so to speak. The conversion factors are called the "exchange constants." Establishing the value of the exchange constant is an important step in solving the problem. With it, a penalty function Z can be constructed that combines the objectives. Materials that minimize Z meet all the objectives in a properly balanced way. The most obvious common currency is cost itself, requiring an "exchange rate" to be established between cost and the other objectives. This can be done for mass and—at least in principle—for other objectives too. The method is illustrated by three further case studies.

CHAPTER 9

Selection of Material and Shape

Extruded shapes. (Images courtesy of Thomas Publishing, www.Thomasnet.com—*www.thomasnet.com/articles/image/plastic-extrusions.jpg.*)

Materials Selection in Mechanical Design. DOI: 10.1016/B978-1-85617-663-7.00009-6

CONTENTS

9.1 INTRODUCTION AND SYNOPSIS

Pause for a moment and reflect on how shape is used to modify the ways in which materials behave. A material has a modulus and a strength, but it can be made stiffer and stronger when loaded in bending or twisting by shaping it into an I-beam or a hollow tube. It can be made less stiff by flattening it into a flat leaf or winding it, in the form of wire, into a helix. Thinned shapes help dissipate heat; cellular shapes help conserve it. There are shapes to maximize electrical capacitance and to conserve magnetic field; shapes that control optical reflection, diffraction, and refraction; shapes to reflect a sound, and shapes to absorb it. Shape is even used to change the way a material feels, making it smooth or rough, slippery or grippy. And of course, it is shape that distinguishes the Venus de Milo from the marble block from which she was carved. It is a rich subject.

Here we explore one part of it—the way shape can be used to increase the mechanical efficiency of a material. Shaped sections carry bending, torsional, and axial-compressive loads more efficiently than do solid sections. By "shaped" we mean that the cross-section is formed as a tube, a box section, an I-section, or the like. By "efficient" we mean that, for given loading conditions, the section uses as little material as possible. Tubes, boxes, and I-sections will be referred to as "simple shapes." Even greater efficiencies are possible with sandwich panels (thin load-bearing skins bonded to a foam or honeycomb interior) and with more elaborate structures (the Warren truss, for instance).

This chapter extends selection methods so as to include shape (Figure 9.1). Often this is not needed: In the Case Studies of Chapter 6, shape either did not enter at all, or, when it did, it was not a variable (that is, we compared different materials with the same shape). But when two different materials are available,

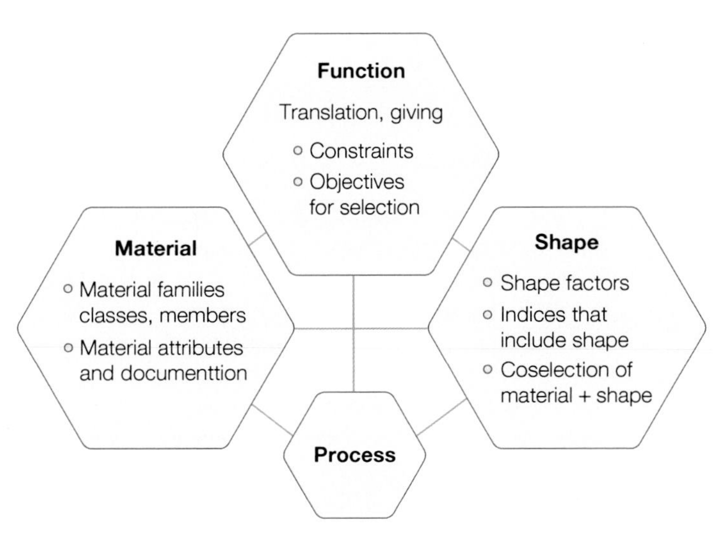

FIGURE 9.1
Section shape is important for certain modes of loading. When shape is a variable, a new term—the shape factor ϕ—appears in some of the material indices.

each with its own section shape, the more general problem arises: how to choose the best combination from among the vast range of materials and the section shapes that are available or could potentially be made. Take the example of a bicycle: Its forks are loaded in bending. They could, say, be made of steel or of wood—early bikes *were* made of wood. But steel is available as a thin-walled tube whereas wood is not; wood components, usually, are solid. A solid wood bicycle is certainly lighter for the same stiffness than a solid steel one, but is it lighter than one made of steel tubing? Might a magnesium I-section be lighter still? How, in short, is one to choose the best combination of material and shape?

A procedure for answering these and related questions is developed in this chapter. It involves the definition of *shape factors*. A *material* can be thought of as having properties but no shape. A *structure* is a material made into a shape (Figure 9.2). Shape factors are measures of the efficiency of material usage. Further, they enable the definition of material indices, such as those of Chapter 5, but now include shape. When shape is constant, the indices reduce exactly to those of Chapter 5; when shape is a variable, however, the shape factor appears in the expressions for the indices. They let you compare shaped materials and guide the choice of the best combination of material and shape. The symbols used in the development are listed for convenience in Table 9.1. Don't be put off by them; the ideas are not difficult.

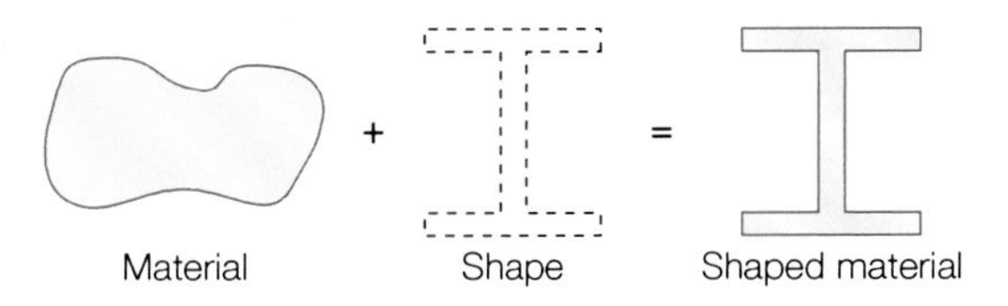

FIGURE 9.2
Mechanical efficiency is obtained by combining material with macroscopic shape. The shape is characterized by a dimensionless shape factor ϕ.

Table 9.1 Definition of Symbols

Symbol	Definition
M	Moment (Nm)
F	Force (N)
E	Young's modulus of the material of the section (GPa)
G	Shear modulus of the material of the section (GPa)
σ_f	Yield or failure strength of the material of the section (MPa)
ρ	Density of the material of the section (kg/m^3)
m_l	Mass per unit length of the section (kg/m)
A	Cross-sectional area of the section (m^2)
I	Second moment of area of the section (m^4)
I_o	Second moment of area of the square reference section (m^4)
Z	Section modulus of the section (m^3)
Z_o	Section modulus of the square reference section (m^3)
K	Torsional moment of area (m^4)
K_o	Torsional moment of area for the square reference section (m^4)
Q	Torsional section modulus (m^3)
Q_o	Torsional section modulus for the square reference section (m^3)
ϕ_B^e	Macro shape factor for elastic bending deflection (–)
ϕ_B^f	Macro shape factor for onset of plasticity or failure in bending (–)
ϕ_T^e	Macro shape factor for elastic torsional deflection (–)
ϕ_T^f	Macro shape factor for onset of plasticity or failure in torsion (–)
ψ_B^e	Micro shape factor for elastic bending deflection (–)
ψ_B^f	Micro shape factor for onset of plasticity or failure in bending (–)
ψ_T^e	Micro shape factor for elastic torsional deflection (–)
ψ_T^f	Micro shape factor for onset of plasticity or failure in torsion (–)
S_B	Bending stiffness (N/m)
S_T	Torsional stiffness (N.m)
(EI)	Essential term in bending stiffness (N.m^2)
$(Z\sigma_f)$	Essential term in bending strength (N.m)
t	Web and flange thickness (m)
c	Web height (m)
d	Section height ($2t + c$) of sandwich (m)
b	Section (flange) width (m)
L	Section length (m)

9.2 SHAPE FACTORS

The loads on a component can be decomposed into those that are axial, those that exert bending moments, and those that exert torques. One of these usually dominates to such an extent that structural elements are specially designed to carry it, and these have common names. Thus *ties* carry tensile loads; *beams* carry bending moments; *shafts* carry torques; *columns* carry compressive axial loads. Figure 9.3 shows these modes of loading applied to shapes that resist them well. The point it makes is that the best material-shape combination depends on the mode of loading. In what follows, we separate the modes, dealing with each separately.

In axial tension, the area of the cross-section is important but its shape is not: All sections with the same area will carry the same load. Not so in bending: Beams with hollow-box sections or I-sections are better than solid

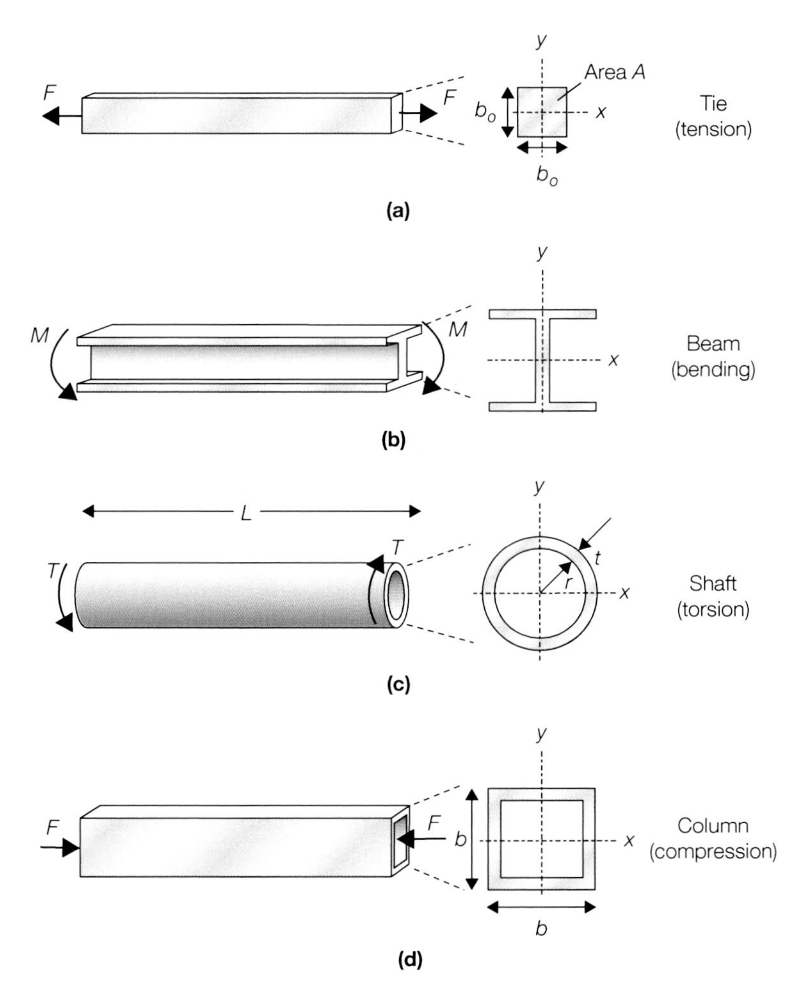

FIGURE 9.3
Common modes of loading and the section shapes that are chosen to support them: (a) axial tension, (b) bending, (c) torsion, and (d) axial compression, which can lead to buckling.

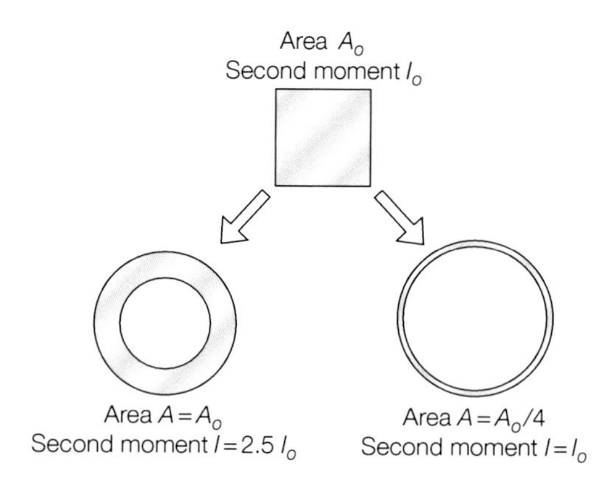

FIGURE 9.4
The effect of section shape on bending stiffness EI: a square-section beam compared: *left*, with a tube of the same area (but 2.5 times stiffer); *right*, with a tube with the same stiffness (but 4 times lighter).

sections of the same cross-sectional area. Torsion, too, has its efficient shapes: Circular tubes, for instance, are more efficient than either solid sections or I-sections. To characterize this we need a metric—a way of measuring the structural efficiency of a section shape, independent of the material of which it is made. An obvious metric is that given by the ratio ϕ (phi) of the stiffness or strength of the shaped section to that of a "neutral" reference shape, which we take to be that of a solid square section with the same cross-sectional area A, and thus the same mass per unit length m_l, as the shaped section (Figure 9.4).

Elastic bending of beams The bending stiffness S of a beam (Figure 9.3b) is proportional to the product EI

$$S \propto \frac{EI}{L^3}$$

Here E is Young's modulus and I is the second moment of area of the beam of length L about the axis of bending (the x axis):

$$I = \int_{section} y^2 dA \tag{9.1}$$

where y is measured normal to the bending axis and dA is the differential element of area at y. Values of the moment I and of the area A for common sections are listed in the first two columns of Table 9.2. Those for the more complex shapes are approximate, but completely adequate for present needs. The second moment of area, I_o, for a reference beam of square section with edge length b_o and section area $A = b_o^2$ is simply

$$I_o = \frac{b_o^4}{12} = \frac{A^2}{12} \tag{9.2}$$

(Here and elsewhere the subscript o refers to the solid square reference section.) The bending stiffness of the shaped section differs from that of a square one with the same area A by the factor ϕ_B^e where

$$\phi_B^e = \frac{S}{S_o} = \frac{EI}{EI_o} = \frac{12\, I}{A^2} \tag{9.3}$$

We call ϕ_B^e the *shape factor for elastic bending*. Note that it is dimensionless—I has dimensions of $(\text{length})^4$ and so does A^2. It depends only on shape, not on scale: Big and small beams have the same value of ϕ_B^e if their section

Table 9.2 Moments of Sections (with units)

Section Shape	Area A (m^2)	Moment I (m^4)	Moment K (m^4)	Moment Z (m^3)	Moment Q (m^3)
h, b	bh	$\frac{bh^3}{12}$	$\frac{bh^3}{3}\left(1-0.58\frac{b}{h}\right)$ $(h>b)$	$\frac{bh^2}{6}$	$\frac{b^2h^2}{(3h+1.8b)}$ $(h>b)$
a, a	$\frac{\sqrt{3}}{4}a^2$	$\frac{a^4}{32\sqrt{3}}$	$\frac{\sqrt{3}a^4}{80}$	$\frac{a^3}{32}$	$\frac{a^3}{20}$
$2r$	πr^2	$\frac{\pi}{4}r^4$	$\frac{\pi}{2}r^4$	$\frac{\pi}{4}r^3$	$\frac{\pi}{2}r^3$
$2a$, $2b$	πab	$\frac{\pi}{4}a^3b$	$\frac{\pi a^3b^3}{(a^2+b^2)}$	$\frac{\pi}{4}a^2b$	$\frac{\pi}{2}a^2b$ $(a<b)$
t, $2r_i$, $2r_o$	$\pi(r_o^2-r_i^2)$ $\approx 2\pi rt$	$\frac{\pi}{4}(r_o^4-r_i^4)$ $\approx \pi r^3t$	$\frac{\pi}{2}(r_o^4-r_i^4)$ $\approx 2\pi r^3t$	$\frac{\pi}{4r_o}(r_o^4-r_i^4)$ $\approx \pi r^2t$	$\frac{\pi}{2r_o}(r_o^4-r_i^4)$ $\approx 2\pi r^2t$
t, h, b	$2t(h+b)$ $(h,b>>t)$	$\frac{1}{6}h^3t\left(1+3\frac{b}{h}\right)$	$\frac{2tb^2h^2}{(h+b)}\left(1-\frac{t}{h}\right)^4$	$\frac{1}{3}h^2t\left(1+3\frac{b}{h}\right)$	$2tbh\left(1-\frac{t}{h}\right)^2$
t, $2a$, $2b$	$\pi(a+b)t$ $(a,b>>t)$	$\frac{\pi}{4}a^3t\left(1+\frac{3b}{a}\right)$	$\frac{4\pi(ab)^{5/2}t}{(a^2+b^2)}$	$\frac{\pi}{4}a^2t\left(1+\frac{3b}{a}\right)$	$2\pi t(a^3b)^{1/2}$ $(b>a)$
t, h_i, h_o, b	$b(h_o-h_i)$ $\approx 2bt$ $(h,b>>t)$	$\frac{b}{12}(h_o^3-h_i^3)$ $\approx \frac{1}{2}bth_o^2$	–	$\frac{b}{6h_o}(h_o^3-h_i^3)$ $\approx bth_o$	–

Continued

Table 9.2 Moments of Sections (with units) *continued*

Section Shape	Area A (m^2)	Moment I (m^4)	Moment K (m^4)	Moment Z (m^3)	Moment Q (m^3)
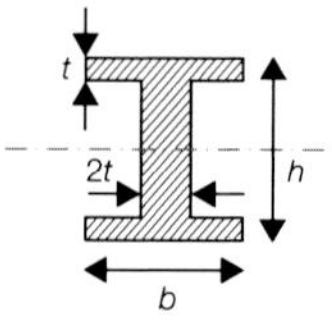	$2t(h+b)$ $(h,b>>t)$	$\frac{1}{6}h^3t\left(1+3\frac{b}{h}\right)$	$\frac{2}{3}bt^3\left(1+4\frac{h}{b}\right)$	$\frac{1}{3}h^2t\left(1+3\frac{b}{h}\right)$	$\frac{2}{3}bt^2\left(1+4\frac{h}{b}\right)$
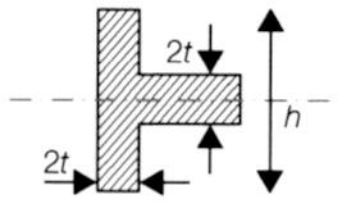	$2t(h+b)$ $(h,b>>t)$	$\frac{t}{6}(h^3+4bt^2)$	$\frac{t^3}{3}(8b+h)$	$\frac{t}{3h}(h^3+4bt^2)$	$\frac{t^2}{3}(8b+h)$
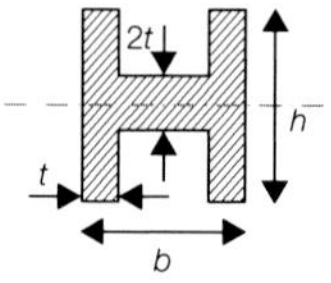	$2t(h+b)$ $(h,b>>t)$	$\frac{t}{6}(h^3+4bt^2)$	$\frac{2}{3}ht^3\left(1+4\frac{b}{h}\right)$	$\frac{t}{3h}(h^3+4bt^2)$	$\frac{2}{3}ht^2\left(1+4\frac{b}{h}\right)$

shapes are the same.[1] This is shown in Figure 9.5. The three members of each horizontal group differ in scale but have the same shape factor—each member is a magnified or shrunken version of its neighbors. Shape-efficiency factors ϕ_B^e for common shapes in bending, calculated from the expressions for A and I in Table 9.2, are listed in the first column of Table 9.3. Solid equi-axed sections (circles, squares, hexagons, octagons) all have values very close to 1—for practical purposes they can be set equal to 1. But if the section is elongated, hollow, or of I-section, things change; a thin-walled tube or a slender I-beam can have a value of 50 or more. A beam with $\phi_B^e = 50$ is 50 times stiffer than a solid beam of the same weight.

Figure 9.6 is a plot of I against A for values of ϕ_B^e (Equation (9.3)). The contour for $\phi_B^e = 1$ describes the square-section reference beam. Those for $\phi_B^e = 10$ and $\phi_B^e = 100$ describe more efficient shapes, as suggested by the icons at the bottom left, in each of which the axis of bending is horizontal. But it is not always high stiffness that is wanted. Springs, cradles, suspensions, cables, and other structures that must flex yet have high tensile strength rely on having a low bending stiffness. Then we want low shape efficiency. It is achieved by spreading the material in a plane containing the axis of bending to form sheets or wires, as suggested by the contours for $\phi_B^e = 0.1$ and 0.01.

[1] This elastic shape-efficiency factor is related to the radius of gyration, R_g, by $\phi_B^e = 12R_g^2/A$. It is related to the "shape parameter," k_1, of Shanley (1960) by $\phi_B^e = 12k_1$.

Calculating shape factors

A tube has a radius $r = 10$ mm and a wall thickness $t = 1$ mm. How much stiffer is it in bending than a solid cylinder of the same mass per unit length m_l?

Answer

The difference is the ratio of the two shape factors. The shape factor for the tube, from Table 9.3, is $\phi_B^e = \frac{3}{\pi}\left(\frac{r}{t}\right) = 9.55$. That for a solid circular section is $\phi_B^e = \frac{3}{\pi} = 0.955$. The tube is stiffer by a factor of 10.

Elastic twisting of shafts (Figure 9.3(c)) Shapes that resist bending well may not be so good when twisted. The stiffness of a shaft—the torque T divided by the angle of twist θ—is proportional to GK, where G is its shear modulus and K is its torsional moment of area. For circular sections, K is identical with the polar moment of area J:

$$J = \int_{\text{section}} r^2 dA \tag{9.4}$$

where dA is the differential element of area at the radial distance r, measured from the center of the section. For noncircular sections, K is less than J; it is defined such that the angle of twist θ is related to the torque T by

$$S_T = \frac{T}{\theta} = \frac{KG}{L} \tag{9.5}$$

where L is the length of the shaft and G is the shear modulus of the material of which it is made. Approximate expressions for K are listed in Table 9.2.

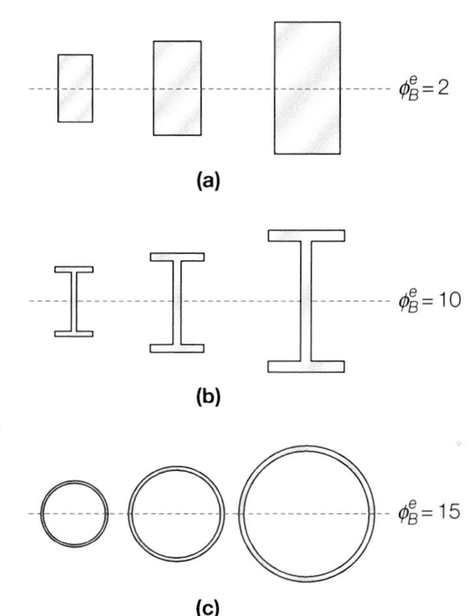

FIGURE 9.5
(a) A set of rectangular sections with $\phi_B^e = 2$. (b) A set of I-sections with $\phi_B^e = 10$. (c) A set of tubes with $\phi_B^e = 15$. Members of a set differ in size but not in shape.

The shape factor for elastic twisting is defined, as before, by the ratio of the torsional stiffness of the shaped section S_T to that of a solid square shaft S_{To} of the same length L and cross-section A, which, using Equation (9.5), is

$$\phi_T^e = \frac{S_T}{S_{To}} = \frac{K}{K_o} \tag{9.6}$$

Table 9.3 Shape-efficiency Factors

Section Shape	Bending Factor ϕ_B^e	Torsional Factor ϕ_T^e	Bending Factor ϕ_B^f	Torsional Factor ϕ_T^f
h, b	$\frac{h}{b}$	$2.38\frac{h}{b}\left(1-0.58\frac{b}{h}\right)(h>b)$	$\left(\frac{h}{b}\right)^{0.5}$	$1.6\sqrt{\frac{b}{h}}\frac{1}{\left(1+0.6\frac{b}{h}\right)}$ $(h>b)$
a, a	$\frac{2}{\sqrt{3}}=1.15$	0.832	$\frac{3^{1/4}}{2}=0.658$	0.83
2r	$\frac{3}{\pi}=0.955$	1.14	$\frac{3}{2\sqrt{\pi}}=0.846$	1.35
2a, 2b	$\frac{3}{\pi}\frac{a}{b}$	$\frac{2.28\,ab}{(a^2+b^2)}$	$\frac{3}{2\sqrt{\pi}}\sqrt{\frac{a}{b}}$	$1.35\sqrt{\frac{a}{b}}(a<b)$
t, $2r_i$, $2r_o$	$\frac{3}{\pi}\left(\frac{r}{t}\right)(r>>t)$	$1.14\left(\frac{r}{t}\right)$	$\frac{3}{\sqrt{2\pi}}\sqrt{\frac{r}{t}}$	$1.91\sqrt{\frac{r}{t}}$
t, h, b	$\frac{1}{2}\frac{h}{t}\frac{(1+3b/h)}{(1+b/h)^2}$ $(h,b>>t)$	$\frac{3.57b^2\left(1-\frac{t}{h}\right)^4}{th\left(1+\frac{b}{h}\right)^3}$	$\frac{1}{\sqrt{2}}\sqrt{\frac{h}{t}}\frac{\left(1+\frac{3b}{h}\right)}{\left(1+\frac{b}{h}\right)^{3/2}}$	$3.39\sqrt{\frac{h^2}{bt}}\frac{1}{\left(1+\frac{h}{b}\right)^{3/2}}$
t, 2a, 2b	$\frac{3}{\pi}\frac{a}{t}\frac{(1+3b/a)}{(1+b/a)^2}$ $(a,b>>t)$	$\frac{9.12\,(ab)^{5/2}}{t(a^2+b^2)(a+b)^2}$	$\frac{3}{2\sqrt{\pi}}\sqrt{\frac{a}{t}}\frac{\left(1+\frac{3b}{a}\right)}{\left(1+\frac{b}{a}\right)^{3/2}}$	$5.41\sqrt{\frac{a}{t}}\frac{1}{\left(1+\frac{a}{b}\right)^{3/2}}$
t, h_i, h_o, b	$\frac{3}{2}\frac{h_o^2}{bt}(h,b>>t)$	–	$\frac{3}{\sqrt{2}}\frac{h_o}{\sqrt{bt}}$	–

Table 9.3 *continued*

Section Shape	Bending Factor ϕ_B^e	Torsional Factor ϕ_T^e	Bending Factor ϕ_B^f	Torsional Factor ϕ_T^f
	$\frac{1}{2}\frac{h}{t}\frac{(1+3b/h)}{(1+b/h)^2}$ $(h,b>>t)$	$1.19\left(\frac{t}{b}\right)\frac{\left(1+\frac{4h}{b}\right)}{\left(1+\frac{h}{b}\right)^2}$	$\frac{1}{\sqrt{2}}\sqrt{\frac{h}{t}}\frac{\left(1+\frac{3b}{h}\right)}{\left(1+\frac{b}{h}\right)^{3/2}}$	$1.13\sqrt{\frac{t}{b}}\frac{\left(1+\frac{4h}{b}\right)}{\left(1+\frac{h}{b}\right)^{3/2}}$
	$\frac{1}{2}\frac{h}{t}\frac{(1+4bt^2/h^3)}{(1+b/h)^2}$ $(h,b\ t)$	$0.595\left(\frac{t}{h}\right)\frac{\left(1+\frac{8b}{h}\right)}{\left(1+\frac{b}{h}\right)^2}$	$\frac{3}{4}\sqrt{\frac{h}{t}}\frac{\left(1+\frac{4bt^2}{h^3}\right)}{\left(1+\frac{b}{h}\right)^{3/2}}$	$0.565\sqrt{\frac{t}{h}}\frac{\left(1+\frac{8b}{h}\right)}{\left(1+\frac{b}{h}\right)^{3/2}}$
	$\frac{1}{2}\frac{h}{t}\frac{(1+4bt^2/h^3)}{(1+b/h)^2}$ $(h,b>>t)$	$1.19\left(\frac{t}{h}\right)\frac{\left(1+\frac{4b}{h}\right)}{\left(1+\frac{b}{h}\right)^2}$	$\frac{3}{4}\sqrt{\frac{h}{t}}\frac{\left(1+\frac{4bt^2}{h^3}\right)}{\left(1+\frac{b}{h}\right)^{3/2}}$	$1.13\sqrt{\frac{t}{h}}\frac{\left(1+\frac{4b}{h}\right)}{\left(1+\frac{b}{h}\right)^{3/2}}$

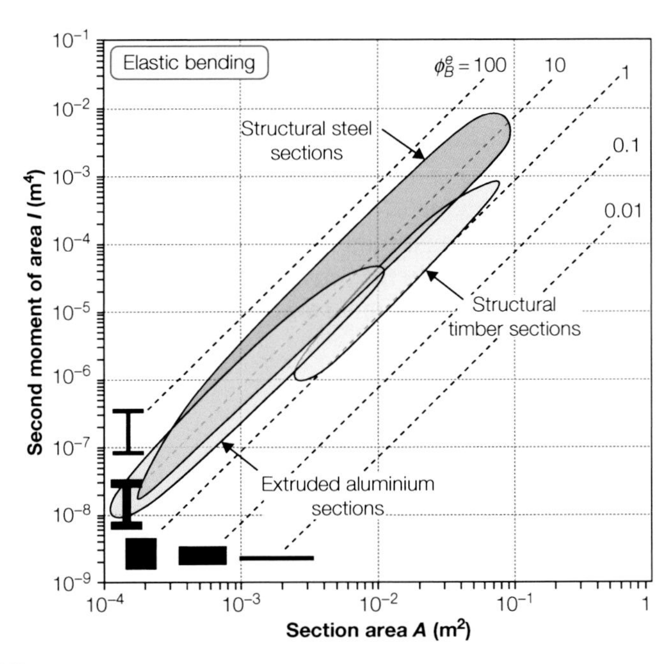

FIGURE 9.6
The second moment of area I plotted against section area A. Efficient structures have high values of the ratio I/A^2; inefficient structures (ones that bend easily) have low values. Real structural sections have values of I and A that lie in the shaded zones. Note that there are limits on A and on the maximum shape efficiency ϕ_B^e that depend on material.

The torsional constant K_o for a solid square section (Table 9.2, top row with $b = h$) is

$$k_o = 0.14\,A^2$$

giving

$$\phi_T^e = 7.14\frac{K}{A^2} \tag{9.7}$$

It, too, has the value 1 for a solid square section, and has values near 1 for any solid, equi-axed section; but for thin-walled shapes, particularly tubes, it can be large. As before, sections with the same value of ϕ_T^e differ in size but not shape. Values derived from the expressions for K and A in Table 9.2 are listed in Table 9.3.

Failure in bending Plasticity starts when the stress, somewhere, first reaches the yield strength σ_y; fracture occurs when this stress first exceeds the fracture strength σ_{fr}; fatigue failure, if it exceeds the endurance limit σ_e. Any one of these constitutes failure. As in earlier chapters, we use the symbol σ_f for the failure stress, meaning "the local stress that will first cause yielding or fracture or fatigue failure."

In bending, the stress σ is largest at the point y_m on the surface of the beam that lies furthest from the neutral axis. Its value is

$$\sigma = \frac{M\,y_m}{I} = \frac{M}{Z} \tag{9.8}$$

where M is the bending moment. Failure occurs when this stress first exceeds σ_f. Thus, in problems of beam failure, shape enters through the *section modulus*, $Z = I/y_m$. The strength efficiency of the shaped beam ϕ_B^f is measured by the ratio Z/Z_o, where Z_o is the section modulus of a reference beam of square section with the same cross-sectional area, A:

$$Z_o = \frac{b_o^3}{6} = \frac{A^{3/2}}{6} \tag{9.9}$$

Thus,

$$\phi_B^f = \frac{Z}{Z_o} = \frac{6\,Z}{A^{3/2}} \tag{9.10}$$

Like the other shape-efficiency factor, it is dimensionless and therefore independent of scale. As before, $\phi_B^f = 1$ describes the square-section reference beam. Table 9.3 gives expressions for ϕ_B^f for other shapes derived from the values of the section modulus, Z, in Table 9.2. A beam with a failure shape-efficiency factor of 10 is 10 times stronger in bending than a solid square section of the same weight. Figure 9.7 is a plot of Z against A for values of ϕ_B^f (Equation (9.10)). The other contours describe shapes that are more or less efficient, as suggested by the icons.

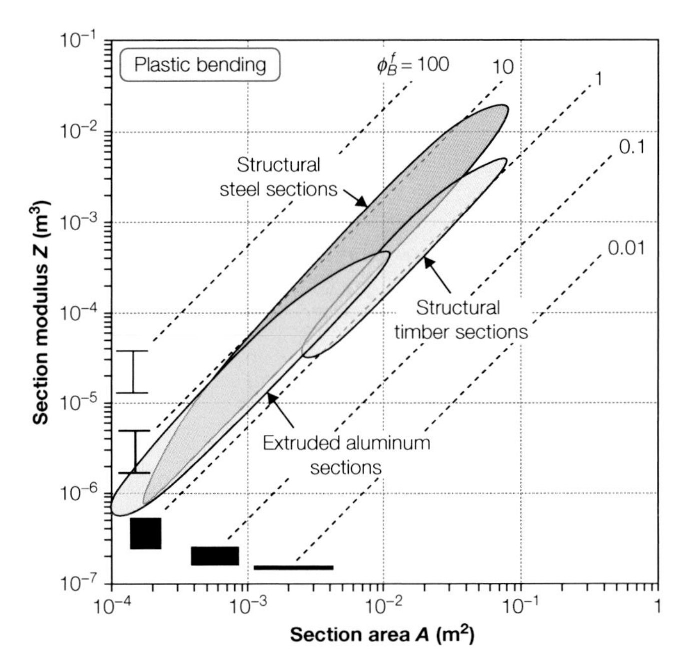

FIGURE 9.7
The section modulus Z plotted against section area A. Efficient structures have high values of the ratio $Z/A^{3/2}$; inefficient structures (ones that bend easily) have low values. Real structural sections have values of Z and A that lie in the shaded zones. Note that there are limits on A and on the maximum shape efficiency ϕ_B^f that depend on material.

Evaluating shape factors

A beam has a square-box section with a height h = 100 mm, a width b = 100 mm, and a wall thickness t = 5 mm. What is the value of its shape factor ϕ_B^f?

Answer

The shape factor for the box section, from Table 9.3, is $\phi_B^f = \frac{1}{\sqrt{2}}\sqrt{\frac{h}{t}}\frac{\left(1+\frac{3b}{h}\right)}{\left(1+\frac{b}{h}\right)^{3/2}} = 4.47$. The box section is stronger than a solid square-section beam of the same mass per unit length by a factor of 4.5.

Failure in torsion In torsion the problem is more complicated. For circular rods or tubes subjected to a torque T (as in Figure 9.3(c)) the shear stress τ is a maximum at the outer surface, at the radial distance r_m from the axis of bending:

$$\tau = \frac{Tr_m}{J} \qquad (9.11)$$

The quantity J/r_m, in twisting has the same character as I/y_m in bending. For noncircular sections with ends that are free to warp, the maximum surface stress is given instead by

$$\tau = \frac{T}{Q} \tag{9.12}$$

where Q, with units of m^3, now plays the role in torsion that Z plays in bending. This allows the definition of a shape factor, ϕ_T^f, for failure in torsion, following the same pattern as before:

$$\phi_T^f = \frac{Q}{Q_o} = 4.8\frac{Q}{A^{3/2}} \tag{9.13}$$

Values of Q and ϕ_T^f are listed in Tables 9.2 and 9.3. Shafts with solid equiaxed sections all have values of ϕ_T^f close to 1.

Fully plastic bending or twisting (such that the yield strength is exceeded throughout the section) involves a further pair of shape factors. Generally speaking, shapes that resist the onset of plasticity well are resistant to full plasticity also, so ϕ_B^{pl} does not differ much from ϕ_B^f. New shape factors for these are not, at this stage, necessary.

Axial loading: Column buckling A column of length L, loaded in compression, buckles elastically when the load exceeds the Euler load

$$F_c = \frac{n^2 \pi^2 E I_{min}}{L^2} \tag{9.14}$$

Inreasing strength by shaping

A slender, solid cylindrical column of height L supports a load F. If overloaded, the column fails by elastic buckling. By how much is the load-bearing capacity increased if the solid cylinder is replaced by a hollow circular tube of the same cross-section A?

Answer

Replacing I_{min} in Equation 9.14 by $\phi_B^e A^2/12$ from Equation 9.3 gives

$$F_c = \frac{n^2 \pi^2 A^2}{12 L^2} E \phi_B^e$$

The failure load is increased by the ratio of the shape factor for the tube to the shape factor of the solid cylinder. The shape factor for a thin-walled tube is $\phi_B^e = 3r/\pi t$; for the solid cylinder it is $\phi_B^e = 3/\pi$ (Table 9.3). The ratio is r/t, where r is the radius and t is the wall thickness of the tube.

where n is a constant that depends on the end-constraints. The resistance to buckling, then, depends on the smallest second moment of area, I_{min}, and the appropriate shape factor (ϕ_B^e) is the same as that for elastic bending (Equation (9.3)) with I replaced by I_{min}.

9.3 LIMITS TO SHAPE EFFICIENCY

The conclusions so far: If you wish to make stiff, strong structures that are efficient (using as little material as possible), make the shape-efficiency factors as large as possible. It would seem, then, that the bigger the value of ϕ the better. True, but there are limits. We examine these next.

Empirical limits There are practical limits for the slenderness of sections, and these determine, for a given material, the maximum attainable efficiencies. These limits may be imposed by manufacturing constraints: The difficulty or expense of making an efficient shape may simply be too great. More often they are imposed by the properties of the material itself because these determine the failure mode of the section. We explore these limits in two ways. The first is empirical: by examining the shapes in which real materials—steel, aluminum, and so on—are actually made, recording the limiting efficiency of available sections. The second is by the analysis of the mechanical stability of shaped sections.

Standard sections for beams, shafts, and columns are generally prismatic. Prismatic shapes are easily made by rolling, extrusion, drawing, pultrusion, or sawing (see the chapter opening picture). The section may be solid, closed-hollow (like a tube or box), or open-hollow (an I-, U-, or L-section, for instance). Each class of shape can be made in a range of materials. Some are available as standard, off-the-shelf sections, notably structural steel, extruded aluminum alloy, pultruded GFRP (glass fiber reinforced polyester, or epoxy), and structural timber. Figure 9.8 shows values for I and A (the same axes as in Figure 9.6) for 1,880 standard sections made from these four materials, with contours of the shape factor ϕ_B^e superimposed. Some of the sections have $\phi_B^e \approx 1$; they are the ones with solid cylindrical or square sections. More interesting is that none have values of ϕ_B^e greater than about 65; there is an *upper limit* for the shape. A similar plot for Z and A (the axes of Figure 9.7) indicates an upper limit for ϕ_B^f of about 15. When these data are segregated by material,[2] it is found that each has its own upper limit of shape and that they differ greatly. Similar limits hold for the torsional shape factors. They are listed in Table 9.4 and were plotted as shaded bands on Figures 9.6 and 9.7.

[2] Birmingham & Jobling (1996); Weaver & Ashby (1997).

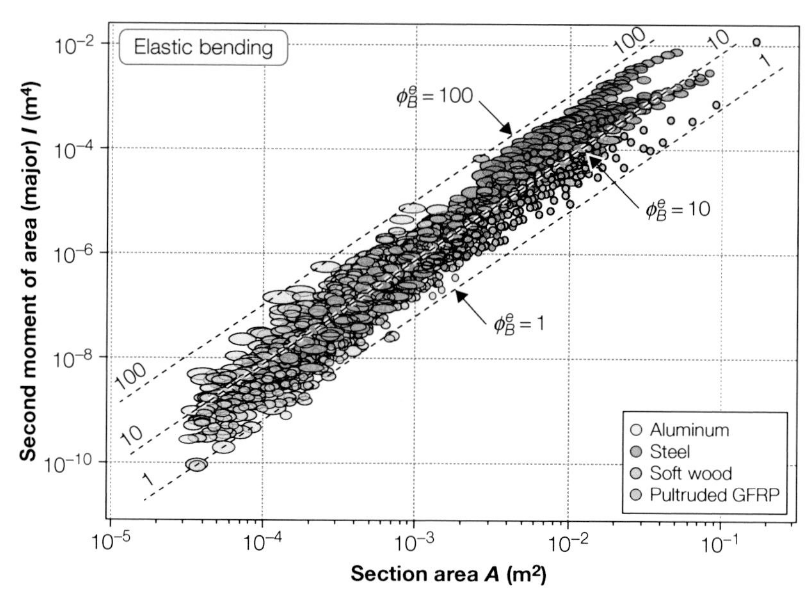

FIGURE 9.8

Log (I) plotted against log (A) for standard sections of steel, aluminum, pultruded GFRP, and wood. Contours of ϕ_B^e are shown, illustrating that there is an upper limit. A similar plot for log (Z) against log (A) reveals an upper limit for ϕ_B^f.

Table 9.4 Empirical Upper Limits for the Shape Factors ϕ_B^e, ϕ_T^e, ϕ_B^f, and ϕ_T^f

Material	$(\phi_B^e)_{max}$	$(\phi_T^e)_{max}$	$(\phi_B^f)_{max}$	$(\phi_T^f)_{max}$
Structural steel	65	25	13	7
6061 aluminum alloy	44	31	10	8
GFRP and CFRP	39	26	9	7
Polymers (e.g., nylons)	12	8	5	4
Woods (solid sections)	5	1	3	1
Elastomers	< 6	3	–	–

The upper limits for shape efficiency are important. They are central to the design of structures that are light or for which, for other reasons (cost, perhaps), the material content should be minimized. Two questions then arise. What sets the upper limit on shape efficiency? And why does the limit depend on material? One explanation is simply the difficulty of making them—a manufacturing constraint. Steel, for example, can be drawn to thin-walled tubes or formed (by rolling, folding, or welding) into efficient I-sections; shape factors as high as 50 are common. Wood cannot be so easily shaped; plywood technology could, in principle, be used to make thin tubes or I-sections, but in practice shapes with values of ϕ_B^e greater than 5 are uncommon. Composites, too, can be limited by the present difficulty in making them into thin-walled prismatic shapes, though the technology for doing this now exists.

But there is a more fundamental constraint on shape efficiency. It has to do with local buckling.

Limits imposed by local buckling When efficient shapes *can* be fabricated, the limits of the efficiency are set by the competition between failure modes. Inefficient sections fail in a simple way: They yield, they fracture, or they suffer large-scale buckling. In seeking greater efficiency, a shape is chosen that raises the load required for the simple failure modes, but in doing so the structure is pushed nearer the load at which new modes—particularly those involving *local* buckling—become dominant.

It is a characteristic of shapes that approach their limiting efficiency that two or more failure modes occur at almost the same load. Why? Here is a simple-minded explanation. If failure by one mechanism occurs at a lower load than all others, the section shape can be adjusted to suppress it; but this pushes the load upward until another mechanism becomes dominant. If the shape is described by a single variable (ϕ), when two mechanisms occur at the same load you have to stop—no further shape adjustment can improve things. Adding webs, ribs, or other stiffeners, gives additional variables, allowing shape to be optimized further, but we shall not pursue that here.

The best way to illustrate the preceding is with a simple example. Think of a drinking straw—it is a thin-walled hollow tube about 5 mm in diameter. It is made of polystyrene, but there is not much of it. If the straw were collapsed into a solid cylinder, the cylinder would have a diameter of less than 1 mm, and because polystyrene has a low modulus, it would have a low bending stiffness. If bent sufficiently it would fail by plastic yielding; if bent further, by fracturing. Now restore it to its straw shape and bend it. It is much stiffer than before, but as it bends it ovalizes and then fails suddenly and unpredictably by kinking—a form of local buckling (try it).

A fuller analysis[3] indicates that the maximum practical shape efficiency—when not limited by manufacturing constraints—is indeed dictated by the onset of local buckling. A thick-walled section, loaded in bending, yields before it buckles locally. It can be made more efficient by increasing its slenderness in ways that increase I and Z, thereby increasing both its stiffness and the load it can carry before it yields, but reducing the load at which the increasingly slender walls of the section start to buckle. When the load for

[3] See Gerard (1956) and Weaver & Ashby (1997) in Section 9.9.

local buckling falls below that for yield, it fails by buckling—and this is undesirable because buckling is defect-dependent and can lead to sudden, unpredictable collapse. The implication drawn from detailed versions of plots like that of Figure 9.8 is that real sections are designed to avoid local buckling, setting the upper limit on shape efficiency. Not surprisingly, the limit depends on the material—those with low strength and high modulus yield easily but do not easily buckle, and vice versa. A rough rule of thumb follows from this:

$$(\phi_B^e)_{\max} \approx 2.3\left(\frac{E}{\sigma_f}\right)^{1/2} \tag{9.15a}$$

and[4]

$$(\phi_B^f)_{\max} \approx \sqrt{(\phi_B^e)_{\max}} \tag{9.15b}$$

allowing approximate estimates for the maximum shape efficiency of materials. Wood, according to these equations, is capable of far greater shape efficiency than that of standard timber sections. This high efficiency can be realized by plywood technology, but such sections are not standard.

Much higher efficiencies are possible when precise loading conditions are known, allowing customized application of stiffeners and webs to suppress local buckling. This allows a further increase in the ϕ's until failure or new, localized buckling modes appear. These, too, can be suppressed by a further hierarchy of structuring; ultimately, the ϕ's are limited only by manufacturing constraints. But this is getting more sophisticated than we need for a general selection of material and shape. Equation (9.15) will do all we need.

9.4 EXPLORING MATERIAL-SHAPE COMBINATIONS

Stiffness-limited design The $E - \rho$ material property chart displays properties of materials. The shape-efficiency chart of Figure 9.6 captures information about the influence of shape on bending stiffness. By linking them,[5] the performance of the section can be explored. In Figure 9.9 the two charts are placed in opposite corners of a square. The material property chart (here much simplified, with only a few materials) is at the top left. The shape-efficiency chart is at the lower right, with its axes interchanged so

---

[4] A consequence of the fact that $I/Zh \approx 0.5$ where h is the depth of the section.

[5] Birmingham (1996).

that I is along the bottom and A is up the side; shaded bands on it show the populated areas, derived from plots like that in Figure 9.8. The remaining two quadrants automatically form two more charts, each sharing axes with the first two. That at the upper right has axes of E and I; the diagonal lines show the bending stiffness of the section EI. That at the lower left has axes of A and ρ; the diagonal contours show the metric of performance: the mass per unit length, $m_l = \rho A$, of the section.

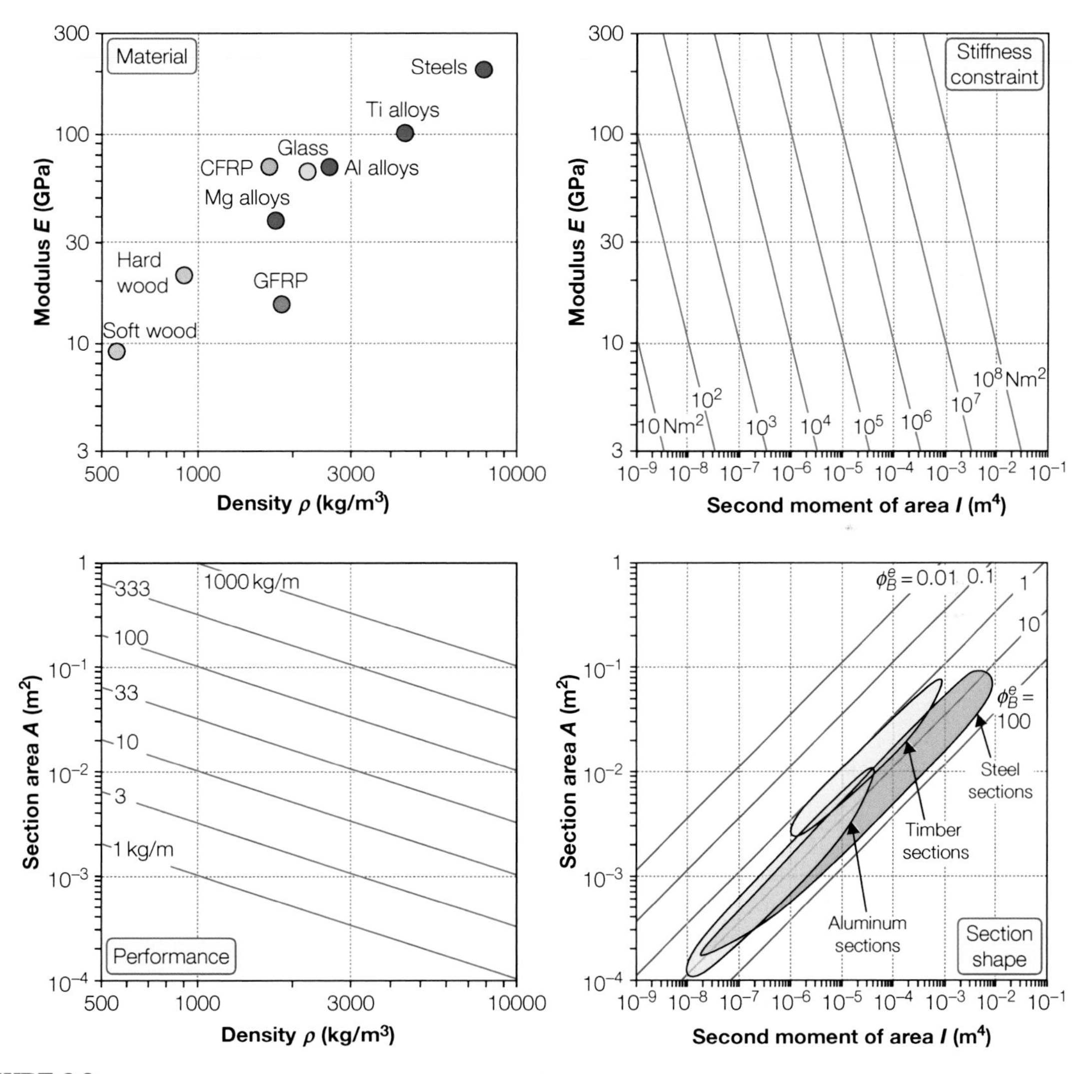

FIGURE 9.9

The 4-quadrant chart assembly for exploring structural sections for stiffness-limited design during bending. Each chart shares its axes with its neighbors.

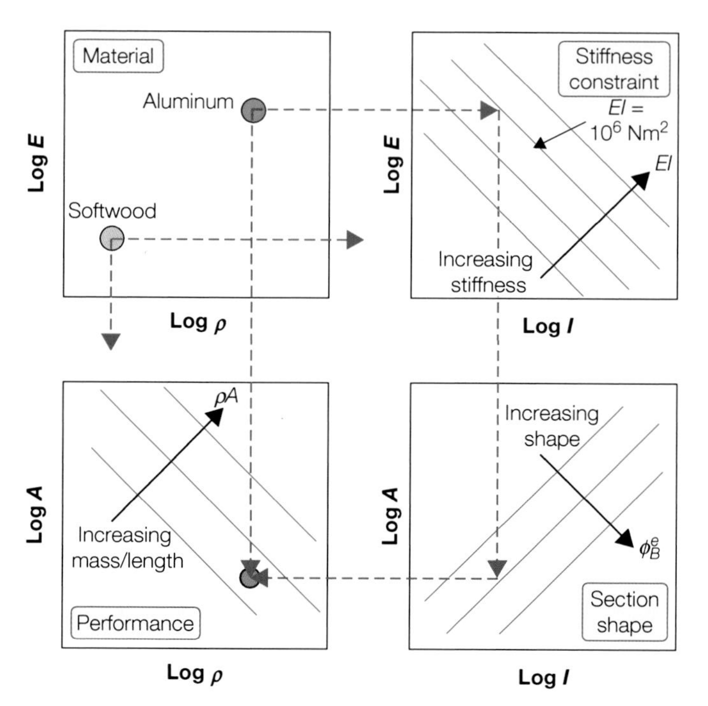

FIGURE 9.10
A schematic showing how the 4-quadrant chart is used.

This set of charts enables the assessment and comparison of stiffness-limited sections. It can be used in several ways, of which the following is typical. It is illustrated in Figure 9.10.

- Choose a material for the section and mark its modulus E and density ρ onto the **material property chart** in the first quadrant of the figure.
- Choose the desired section stiffness (EI); it is a constraint that must be met by the section. Extend a horizontal line from the value of E for the material to the appropriate contour in the **stiffness constraint chart** in the second quadrant.
- Drop a vertical from this point onto the **section shape chart** in the third quadrant to the line describing the shape factor ϕ_B^e for the section. Values of I and A outside the shaded bands are forbidden.
- Extend a horizontal line from this point to the **performance chart** in the final quadrant. Drop a vertical from the material density ρ in the material chart. The intersection shows the mass per unit length of the section.

The example of Figure 9.11 compares the mass of rolled steel and extruded aluminum sections both with $\phi_B^e = 10$, and timber sections with $\phi_B^e = 2$,

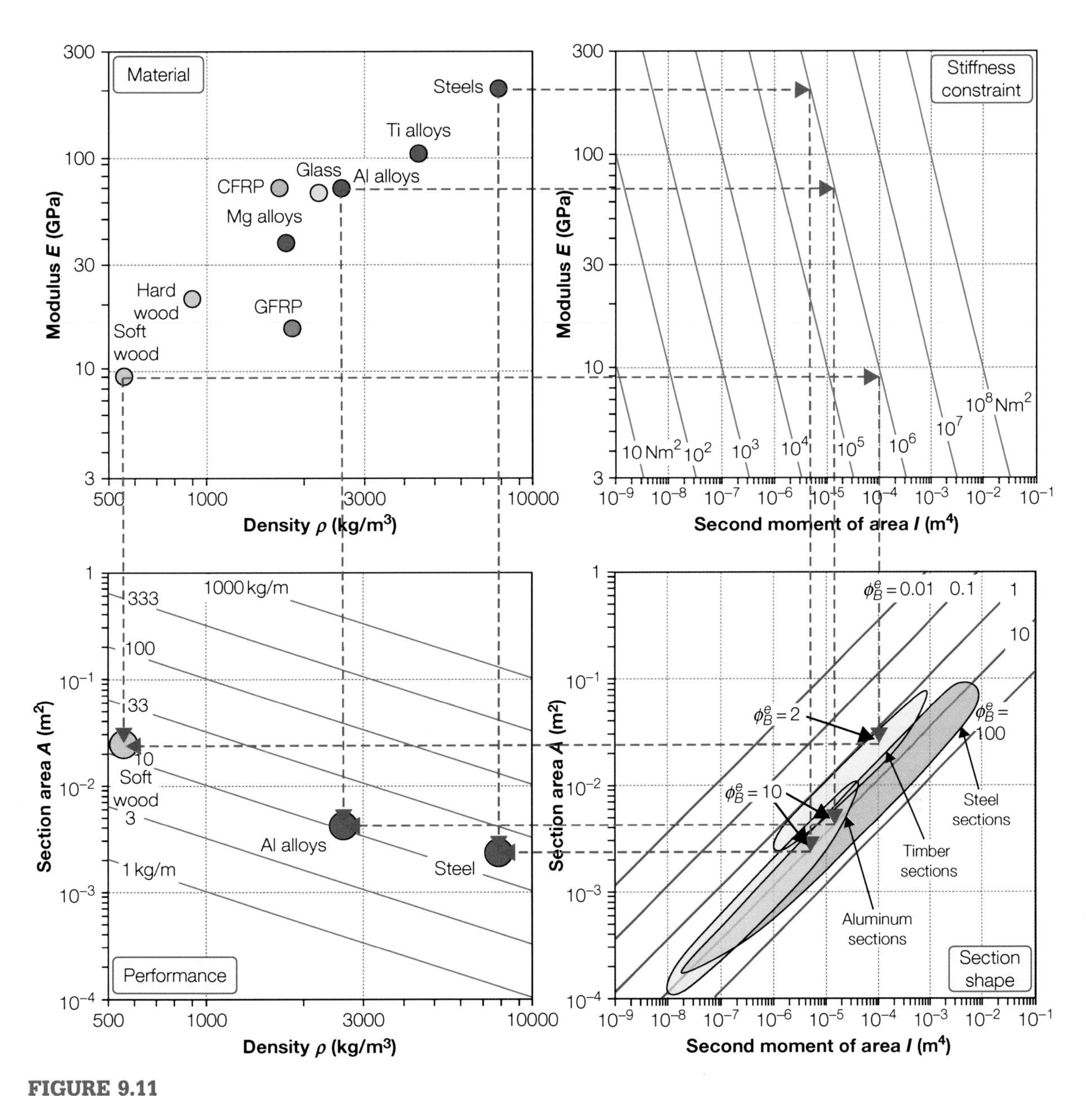

FIGURE 9.11

A comparison of steel, aluminum, and wood sections for a stiffness-limited design with $EI = 10^6$ Nm^2. Aluminum gives a section with a mass of 10 kg/m; steel is almost three times heavier.

with the constraint of a bending stiffness of 10^6 N.m^2. The extruded aluminum section gives the lightest beam. Remarkably, an efficiently shaped steel beam is nearly as light—for a given bending stiffness—as is one made of timber, even though the density of steel is 12 times greater than that of wood. It is because of the higher shape factor possible with steel.

Strength-limited design The reasoning for this follows a similar path. In Figure 9.12 the strength-density ($\sigma_f - \rho$) **material property chart** is placed at the top left. The *Z*-*A* **shape chart** (refer to Figure 9.7 with the axes interchanged) is at the bottom right. As before, the remaining two quadrants generate two more charts. The **strength constraint chart** at the upper right has axes of σ_f and Z; the diagonal lines show the bending strength of the section, $Z\sigma_f$. The **performance chart** at the lower left has the same axes as

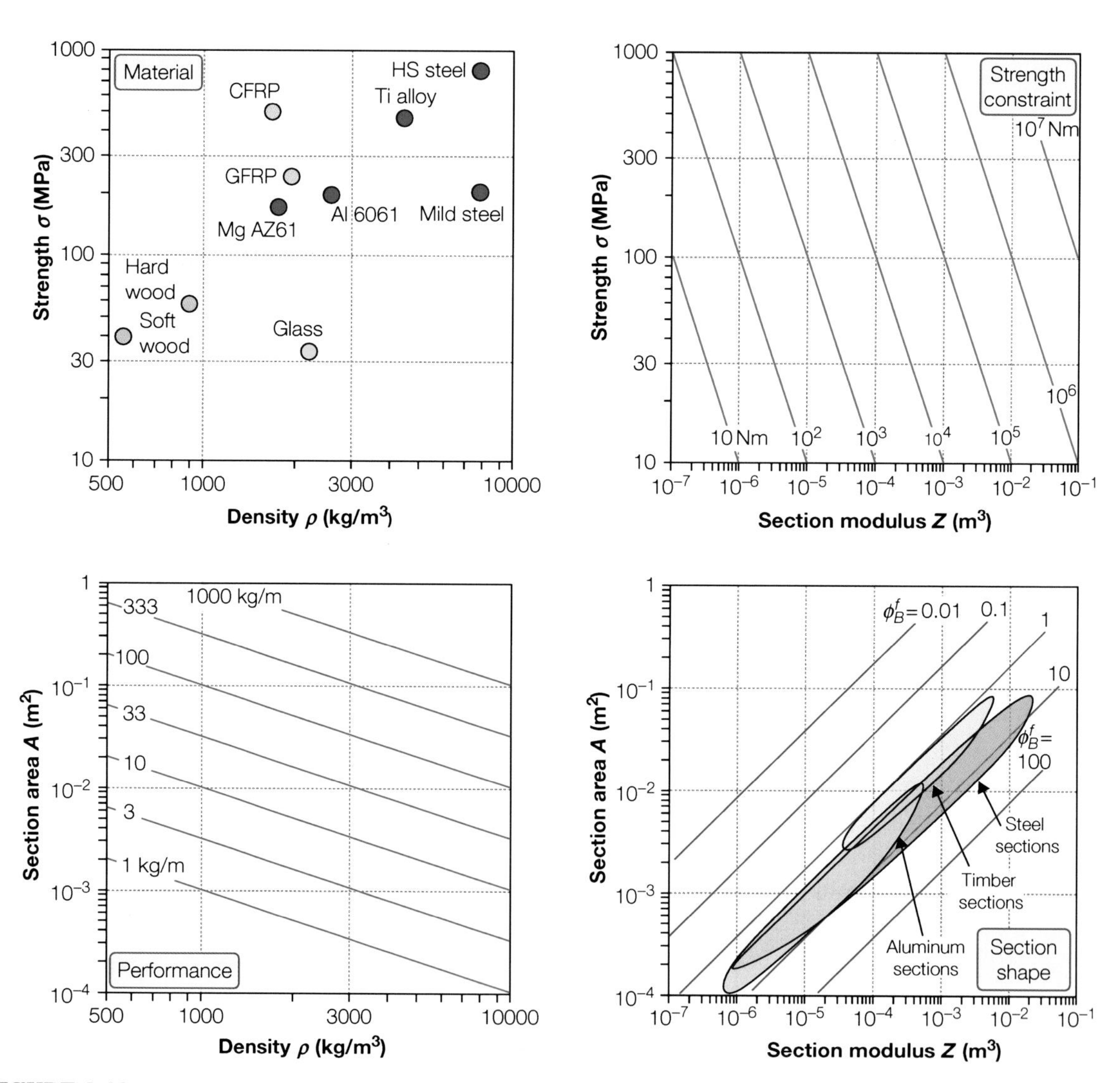

FIGURE 9.12

The 4-quadrant chart assembly for exploring structural sections for strength-limited design. Like those for stiffness, each chart shares its axes with its neighbors.

before—A and ρ—and the diagonal contours again show the metric of performance: the mass per unit length, $ml = \rho A$, of the section.

It is used in the same way as that for stiffness-limited design. Try it for steel with $\phi_B^f = 15$, aluminum with $\phi_B^f = 10$, and GFRP with $\phi_B^f = 5$ for a required failure moment of $Z\sigma_f = 10^4$ N.m. You will find that GFRP offers the lightest solution.

9.5 MATERIAL INDICES THAT INCLUDE SHAPE

The chart-arrays of Figures 9.9 and 9.12 link material, shape, constraint, and performance objective in a graphical, though rather clumsy, way. There is a neater way—build them into the material indices of Chapter 5. Remember that most indices don't need this refinement—the performance they characterize does not depend on shape. But stiffness and strength-limited design do. The indices for these can be adapted to include the relevant shape factor, such that they characterize material-shape combinations.

The method is illustrated in the following for minimum weight design. It can be adapted to other objectives in obvious ways. It follows the derivations of Chapter 5, with one extra step to bring in the shape.

Elastic bending of beams Consider the selection of a material for a beam of specified bending stiffness S_B^* and length L (the constraints), to have minimum mass m (the objective). The mass m of a beam of length L and section area A is given, as before, by

$$m = A\,L\,\rho \tag{9.16}$$

Its bending stiffness is

$$S_B = C_1\,\frac{EI}{L^3} \tag{9.17}$$

where C_1 is a constant that depends only on the way the loads are distributed on the beam. Replacing I by $\phi_B^e A^2/12$ (Equation (9.3)) gives

$$S_B = \frac{C_1}{12}\,\frac{E}{L^3}\,\phi_B^e\,A^2 \tag{9.18}$$

Using this to eliminate A in Equation (9.16) and inserting the target stiffness S_B^* gives the mass of the beam.

$$m = \left(\frac{12\,S_B^*}{C_1}\right)^{1/2} L^{5/2}\left[\frac{\rho}{(\phi_B^e E)^{1/2}}\right] \tag{9.19}$$

Everything in this equation is specified except the term in square brackets, and this depends only on material and shape. For beams with the *same* shape (and thus the same value of ϕ_B^e), the best choice is the material with the greatest value of $E^{1/2}/\rho$—the result derived in Chapter 5. But if we want the lightest material-shape combination, it is the one with the greatest value of the index

$$M_1 = \frac{(\phi_B^e E)^{1/2}}{\rho} \tag{9.20}$$

This index allows material-shape combinations to be ranked. Here is an example.

Elastic twisting of shafts The procedure for elastic twisting of shafts is similar. A shaft of section A and length L is subjected to a torque T. It twists through an angle θ. We want the torsional stiffness, $S_T = T/\theta$, to meet a specified target, S_T^*, at minimum mass. The torsional stiffness is

$$S_T = \frac{KG}{L} \tag{9.21}$$

where G is the shear modulus. Replacing K by ϕ_T^e using Equation (9.7) gives

$$S_T = \frac{G}{7.14\ L}\ \phi_T^e A^2 \tag{9.22}$$

Efficiency gain through shape

A shaped material is required for a stiff beam of minimum mass. Four materials are available with the properties and typical shapes listed in Table 9.5. Which material-shape combination has the lowest mass for a given stiffness?

Answer

The second to last column of the table shows the simple "fixed shape" index $E^{1/2}/\rho$: Wood has the greatest value, more than twice that of steel. But when each material is shaped efficiently (last column), wood has the *lowest* value of M_1—even steel is better; the aluminum alloy wins, surpassing steel and GFRP.

Table 9.5 Selection of Material and Shape for a Light, Stiff Beam

Material	ρ (Mg/m^3)	E (GPa)	ϕ_B^e	$E^{1/2}/\rho$	$(\phi_B^e E)^{1/2}/\rho$
1020 steel	7.85	205	20	1.8	8.2
6061-T4 Al	2.7	70	15	3.1	**12.0**
GFRP (isotropic)	1.75	28	8	2.9	8.5
Wood (oak)	0.9	13.5	2	**4.1**	5.8

Using this to eliminate A in Equation (9.16) and inserting the target stiffness S_T^* gives

$$m = \left(7.14\,\frac{S_T^*}{L^3}\right)^{1/2} L^{3/2}\left[\frac{\rho}{(\phi_T^e G)^{1/2}}\right] \tag{9.23}$$

The best material-shape combination is the one with the greatest value of

$$\frac{(\phi_T^e\, G)^{1/2}}{\rho}$$

The shear modulus G is closely related to Young's modulus E. For practical purposes we approximate G by $3/8E$; when the index becomes

$$M_2 = \frac{(\phi_T^e\, E)^{1/2}}{\rho} \tag{9.24}$$

For shafts of the same shape, this reduces to $E^{1/2}/\rho$ again. When shafts differ in both material and shape, the material index (Equation 9.24) is the one to use.

Equations (9.19) and (9.23) give a way of calculating shape factors for complex structures like bridges and trusses. Inverting Equation (9.19), for example, gives

$$\phi_B^e = \frac{12\; S_B}{C_1}\,\frac{L^5}{m^2}\left[\frac{\rho^2}{E}\right] \tag{9.25}$$

Thus if the mass of the structure, its length, and its bending stiffness are known (as they are for large bridge spans) and the density and modulus of the material of which it is made are known, the shape factor can be calculated. For existing bridge spans, this gives values between 50 and 200. These are larger than the maximum values in Table 9.4 because the bridges are "structured structures" with two or more levels of structure. The high values of ϕ are examples of the way in which shape efficiency can be increased by a hierarchy of structuring.

Shape factors from experimental data

A hollow tubular aluminum extrusion of complex ribbed shape has a mass per unit length m_l = 0.3 kg/m. A length $L = 1$ m of the extrusion, loaded in 3-point bending by a central load of $W = 10$ kg suffers a mid-point deflection $\delta = 2$ mm. What is the shape factor ϕ_B^e of the section? (For aluminum $E = 70$ GPa and $\rho = 2{,}700$ kg/m^3. For 3-point bending, $C_1 = 48$; see Appendix A.)

Answer

The force exerted by the load W is $F = Wg = 98.1$ N. The stiffness of the beam is $S_B = F/\delta = 4.9 \times 10^4$ N/m. Inserting the data into Equation 9.25 gives $\phi_B^e = 13.2$.

Failure of beams and shafts The procedure is the same. A beam of length L, loaded in bending, must support a specified load F without failing and be as light as possible. When section shape is a variable the best choice is found as follows. Failure occurs if the moment exceeds

$$M = Z\sigma_f$$

where Z is the section modulus and σ_f is the stress at which failure occurs. Replacing Z by the shape factor ϕ_B^f of Equation (9.10) gives

$$M = \frac{\sigma_f}{6}\,\phi_B^f\,A^{3/2} \tag{9.26}$$

Substituting this into Equation (9.16) for the mass of the beam gives

$$m = (6\,M)^{2/3}L\left[\frac{\rho^{3/2}}{\phi_B^f\,\sigma_f}\right]^{2/3} \tag{9.27}$$

The best material-shape combination is that with the greatest value of the index

$$M_3 = \frac{(\phi_B^f\,\sigma_f)^{2/3}}{\rho} \tag{9.28}$$

A similar analysis for torsional failure give:

$$M_4 = \frac{(\phi_T^f\,\sigma_f)^{2/3}}{\rho} \tag{9.29}$$

Choosing material and shape combinations

A shaped material is required for a strong beam of minimum mass. Four materials are available with the properties and shapes listed in Table 9.6. Which combination has the lowest mass for a given bending strength?

Answer

The second to last column of the table shows the simple "fixed-shape" index $\sigma_f^{2/3}/\rho$: Wood has the greatest value, more than three times that of steel. But when each material is shaped (last column) the aluminum alloy wins, surpassing steel and GFRP.

Table 9.6 Selection of Material and Shape for a Light, Strong Beam

Material	ρ (Mg/m^3)	σ_f (MPa)	ϕ_B^f	$\sigma_f^{2/3}/\rho$	$(\phi_B^f\sigma_f)^{2/3}/\rho$
1020 steel, normalized	7.85	330	5	6.1	17.8
6061-T4 Al	2.7	110	4	8.5	**21.4**
GFRP SMC (isotropic)	2.0	80	3	9.3	19.3
Wood (oak), along grain	0.9	50	1.5	**15**	19.7

At constant shape both indices reduce to the familiar ${\sigma_f}^{2/3}/\rho$ of Chapter 5; but when shape as well as material are to be compared, the full index must be used.

Selection for strength follows a similar routine, using the index M_3 of Equation (9.28).

9.6 GRAPHICAL COSELECTING USING INDICES

Shaped materials can be plotted onto material property charts. All the selection criteria still apply. Here is how it works.

The material index for elastic bending (Equation (9.28)) can be rewritten as

$$M_1 = \frac{(\phi_B^e E)^{1/2}}{\rho} = \frac{(E/\phi_B^e)^{1/2}}{\rho/\phi_B^e} = \frac{E^{*\,1/2}}{\rho^*} \tag{9.30}$$

The equation says: a material with modulus E and density ρ, when structured, can be thought of as a new material with a modulus and density of

$$E^* = \frac{E}{\phi_B^e} \text{ and } \rho^* = \frac{\rho}{\phi_B^e} \tag{9.31}$$

The $E - \rho$ chart is shown schematically in Figure 9.13. The "new" material properties E^* and ρ^* can be plotted onto it. Introducing shape ($\phi_B^e = 10$, for example) moves the material M to the lower left along a line of slope 1, from position E, ρ to position $E/10$, $\rho/10$, as shown in the figure. The selection criteria are plotted onto the figure as before: A constant value of the index of $E^{1/2}/\rho$ for instance, plots as a straight line of slope 2; it is shown for one value of $E^{1/2}/\rho$ as a heavy broken line. The introduction of shape has moved the material from a position below this line to one above; its performance has improved. Elastic twisting of shafts is treated in the same way.

Materials selection based on strength (rather than stiffness) at minimum weight uses a similar procedure. The material index for failure in bending (Equation (9.28)) can be rewritten as follows

$$M_3 = \frac{(\phi_B^f \sigma_f)^{\frac{2}{3}}}{\rho} = \frac{\left(\sigma_f/(\phi_B^f)^2\right)^{\frac{2}{3}}}{\rho/(\phi_B^f)^2} = \frac{\sigma_f^{*\,2/3}}{\rho^*} \tag{9.32}$$

The material with strength σ_f and density ρ, when shaped, behaves in bending like a new material of strength and density

$$\sigma_f^* = \frac{\sigma_f}{(\phi_B^f)^2} \text{ and } \rho^* = \frac{\rho}{(\phi_B^f)^2} \tag{9.33}$$

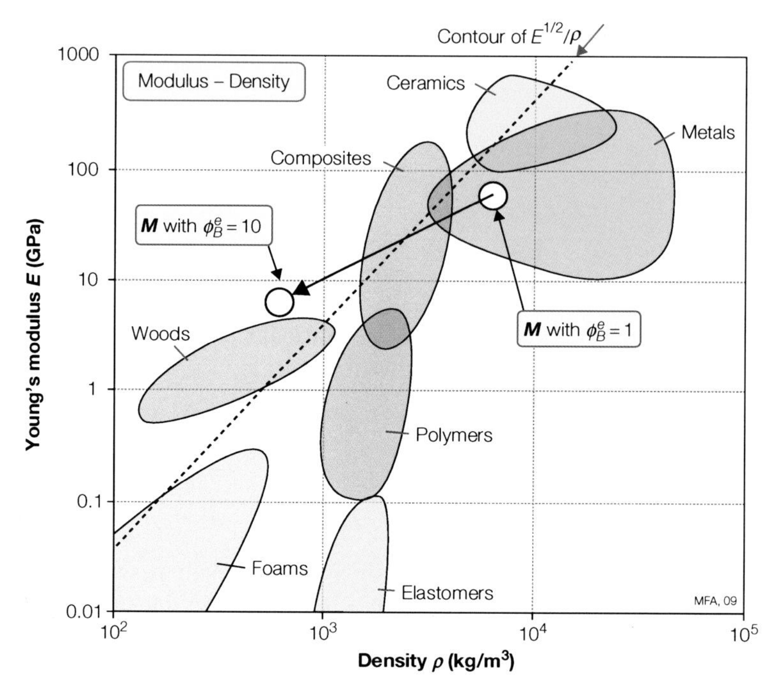

FIGURE 9.13
The structured material behaves like a new material with a modulus $E^* = E/\phi_B^e$ and a density $\rho^* = \rho/\phi_B^e$, moving it from below the broken selection line to above it. A similar procedure can be applied for bending strength, as described in the text.

The rest will be obvious. Introducing shape ($\phi_B^f = 3$, say) moves a material M along a line of slope 1, taking it, in the schematic, from position σ_f, ρ below the material index line (the broken line) to position $\sigma_f/9$, $\rho/9$, which lies above it. The performance has again improved. Torsional failure is analyzed by using ϕ_T^f in place of ϕ_B^f.

The value of this approach is that the plots have retained their generality. It allows selection by any of the earlier criteria, correctly identifying materials for the lightest ties, beams, or panels.

9.7 ARCHITECTURED MATERIALS: MICROSCOPIC SHAPE

Survival in nature is closely linked to structural efficiency. The tree that, with a given resource of cellulose, grows the tallest captures the most sunlight. The creature that, with a given allocation of hydroxyapatite, develops the strongest bone structure and wins the most fights or—if it is prey rather than a predator—runs the fastest. Structural efficiency means survival. It is worth asking how nature does it.

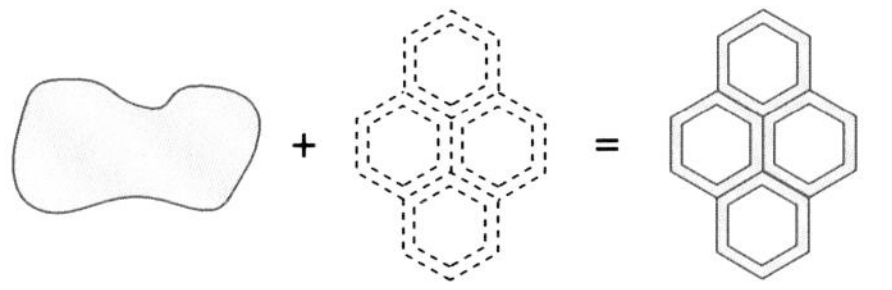

Material Shaped material Macro-shape

FIGURE 9.14
Mechanical efficiency can be obtained by combining material with microscopic, or internal, shape, which repeats itself to give an extensive structure. The shape is characterized by microscopic shape factors, ψ.

Microscopic shape The shapes that are listed earlier in Tables 9.2 and 9.3 achieve efficiency through their *macroscopic* shape. Structural efficiency can be achieved in another way: through shape on a small scale—*microscopic* or *microstructural* (Figure 9.14). Wood is an example. The solid component of wood (a composite of cellulose, lignin, and other polymers) is shaped into small prismatic cells, dispersing the solid further from the axis of bending or twisting of the branch or trunk of the tree, increasing both stiffness and strength. The added efficiency is characterized by a set of *microscopic shape factors,* ψ *(Psi)*, with definitions exactly like those of ϕ. The characteristic of microscopic shape is that the structure repeats itself: It is *extensive*. The microstructured solid can be thought of as a "material" in its own right: It has a modulus, a density, a strength, and so forth. Shapes can be cut from it that—provided they are large compared with the size of the cells—inherit its properties. It is possible, for instance, to fabricate an I-section from wood, and such a section has macroscopic shape (as defined earlier) as well as microscopic shape as suggested by Figure 9.15. It is shown in a moment that the total shape factor for a wooden I-beam is the *product* of the shape factor for the wood structure and that for the I-beam, and this can be large.

Many natural materials have microscopic shape. Wood is just one example. Bone, the stalks and leaves of plants, and the cuttlebone of a squid all have structures that give high stiffness at low weight. It is more difficult to think of human-made examples, though it would appear possible to make them. Figure 9.16 shows four extensive structures with microscopic shape, all of them found in nature. The first is a wood-like structure of hexagonal-prismatic cells; it is isotropic in the plane of the section when the cells are regular hexagons. The second is an array of fibers separated by a foamed matrix typical of palm wood; it too is isotropic in-plane. The third is an axi-symmetric structure of concentric cylindrical shells separated by a foamed matrix, like the stem of some plants. The fourth is a layered structure, a sort of multiple sandwich panel, like the shell of the cuttlefish.

Microscopic shape factors Consider the gain in bending stiffness when a solid square beam like that shown as a solid square of side b_o in Figure 9.16 is expanded, at constant mass, to a larger square section with any one of the structures that surround it in the figure. The bending stiffness S_s of the original

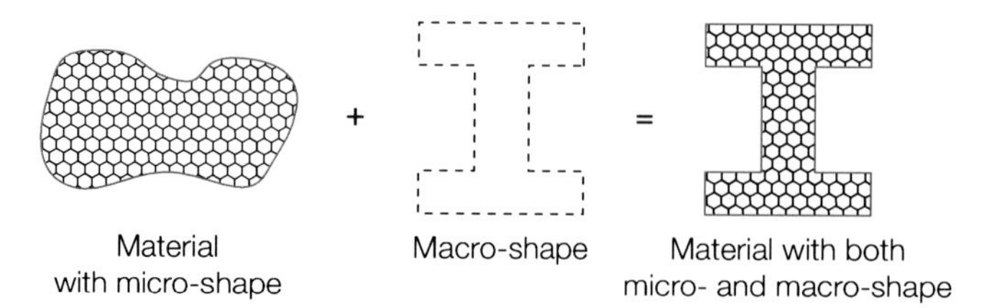

FIGURE 9.15
Microstructural shape can be combined with macroscopic shape to give efficient structures. The overall shape factor is the product of the microscopic and macroscopic shape factors.

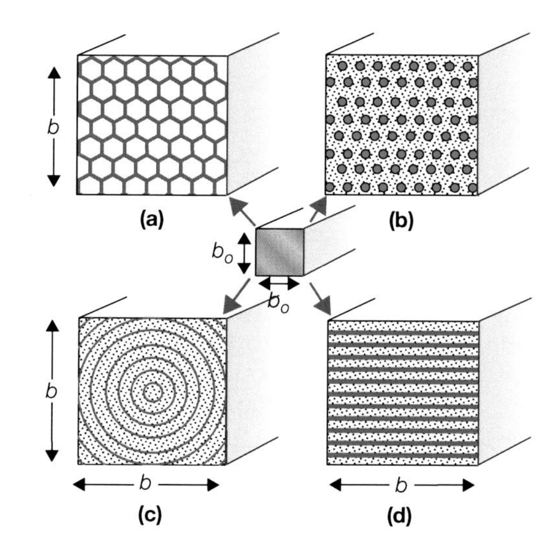

FIGURE 9.16
Four extensive microstructured materials that are mechanically efficient: (a) prismatic cells, (b) fibers embedded in a foamed matrix, (c) concentric cylindrical shells with foam between, and (d) parallel plates separated by foamed spacers.

solid beam is proportional to the product of its modulus E_s and its second moment of areas I_s:

$$S_s \propto E_s I_s \tag{9.34}$$

where the subscript s means "a property of the solid beam" and $I_s = b_o^4/12$. When the beam is expanded at constant mass its density falls from ρ_s to ρ and its edge length increases from b_o to b where

$$b = \left(\frac{\rho_s}{\rho}\right)^{1/2} b_o \tag{9.35}$$

with the result that its second moment of area increases from I_s to

$$I = \frac{b^4}{12} = \frac{1}{12}\left(\frac{\rho_s}{\rho}\right)^2 b_o^4 = \left(\frac{\rho_s}{\rho}\right)^2 I_s \tag{9.36}$$

If the cells, fibers, or rings in Figure 9.16(a), (b), or (c) extend parallel to the axis of the beam, the modulus parallel to this axis falls from that of the solid, E_s, to

$$E = \left(\frac{\rho}{\rho_s}\right) E_s \tag{9.37}$$

The bending stiffness of the expanded beam scales as EI, so that it is stiffer than the original solid beam by the factor

$$\psi_B^e = \frac{S}{S_s} = \frac{EI}{E_s I_s} = \frac{\rho_s}{\rho} \tag{9.38}$$

We refer to ψ_B^e as the *microscopic shape factor for elastic bending*. That for prismatic structures like those in Figure 9.16(a) is simply the reciprocal of the relative density, ρ/ρ_s. Note that, in the limit of a solid (when $\rho = \rho_s$), ψ_B^e takes the value 1, as it obviously should. A similar analysis for failure in bending gives the shape factor

$$\psi_B^f = \left(\frac{\rho_s}{\rho}\right)^{1/2} \tag{9.39}$$

Torsion, as always, is more difficult. When the structure of Figure 9.16(c), which has circular symmetry, is twisted, its rings act like concentric tubes. For these,

$$\psi_T^e = \frac{\rho_s}{\rho} \quad \text{and} \quad \psi_T^f = \left(\frac{\rho_s}{\rho}\right)^{1/2} \tag{9.40}$$

The other structures have lower torsion stiffness and strength (and thus lower microshape factors) for the same reason that I-sections, good in bending, perform poorly in torsion.

Structuring, then, converts a solid with modulus E_s and strength $\sigma_{f,s}$ to a new solid with properties E and σ_f. If this new solid is formed to an efficient macroscopic shape (a tube, say, or an I-section) its bending stiffness, to take an example, increases by a further factor of ϕ_B^e. Then the stiffness of the beam, expressed in terms of that of the solid of which it is made, is

$$S = \psi_B^e \, \phi_B^e \, S_s \tag{9.41}$$

that is, the shape factors simply multiply. The same is true for strength.

Stiffness from microscopic shape

What is the gain in bending stiffness EI if a beam, initially with a solid cross-section, is expanded to create a prismatic structure like those in Figure 9.16? If, instead, it is expanded to create a foam-like structure for which $E = (\frac{\rho}{\rho_s})^2 E_s$, what is the gain in bending stiffness?

Answer

The shape factor ψ_B^e in Equation (9.38) measures the ratio of the stiffness of a prismatic microstructured beam to that of a solid beam of the same mass. The gain in EI scales as $\psi_B^e = \rho_s/\rho$. Repeating the derivation using the expression for foam modulus in terms of relative density given above, we find $\psi_B^e = 1$. Foaming gives no gain in bending stiffness.

This is an example of structural hierarchy and the benefits it brings. It is possible to extend it further: The individual cell walls or layers could, for instance, be structured, giving a third multiplier to the overall shape factor, and these units, too, could be structured. Nature does this to good effect, but for man-made structures there are difficulties. There is the obvious difficulty of manufacture, imposing economic limits on the levels of structuring. And there is the less obvious difference of reliability. If the structure is optimized at all levels of structure, a failure of a member at any one level can trigger failure at the level above, causing a cascade that ends with the failure of the structure as a whole. The more complex the structure, the harder it becomes to ensure the integrity at all levels. This might be overcome by incorporating redundancy (or a safety factor) at each level, but this implies a cumulative loss of efficiency. Two levels of structure are usually as far as it is practical to go.

As pointed out earlier, a microstructured material can be thought of as a new material. It has a density, a strength, a thermal conductivity, and so on; difficulties arise only if the sample size is comparable to the cell size, when "properties" become size-dependent. This means that microstructured materials can be plotted on the material charts—indeed, wood appears on them already—and that all the selection criteria developed in Chapter 5 apply, unchanged, to the microstructured materials. This line of thinking is developed further in Chapter 12, which includes material charts for a range of natural materials.

9.8 SUMMARY AND CONCLUSIONS

The designer has two groups of variables with which to optimize the performance of a load-bearing component: the material properties and the shape of the section. They are not independent. The best material, in a given application, depends on the shapes in which it is available or to which it could potentially be formed.

Table 9.7 Definitions of Shape Factors*

Design Constraint	Bending	Torsion
Stiffness	$\phi_B^e = \frac{12I}{A^2}$	$\phi_T^e = \frac{7.14K}{A^2}$
Strength	$\phi_B^f = \frac{6Z}{A^{3/2}}$	$\phi_T^f = \frac{4.8Q}{A^{3/2}}$

* *A; I, K, Z, and Q are defined in the text and tabulated in Table 9.2.*

The contribution of shape is isolated by defining four shape factors. The first, ϕ_B^e, is for the elastic bending and buckling of beams; the second, ϕ_T^e, is for the elastic twisting of shafts; the third, ϕ_B^f, is for the plastic failure of beams loading in bending; the last, ϕ_T^f, is for the plastic failure of twisted shafts (Table 9.7). The shape factors are dimensionless numbers that characterize the efficiency of use of the material in each mode of loading. They are defined such that all four have the value 1 for a solid square section. With this definition, all equiaxed solid sections (solid cylinders and hexagonal and other polygonal sections) have shape factors of close to 1. Efficient shapes that disperse the material far from the axis of bending or twisting (I-beams, hollow tubes, box sections, etc.) have values that are much larger. They are assembled for common shapes in Table 9.3 earlier in the chapter.

The shapes into which a material can, in practice, be made are limited by manufacturing constraints and by the restriction that the section should yield before it suffers local buckling. These limits can be plotted on a "shape chart" that when combined with a material property chart in a four-chart array (refer to Figures 9.9 and 9.12), allow the potential of alternative material-shape combinations to be explored. While this is educational, there is a more efficient alternative: developing indices that include the shape factors. The best material-shape combination for a light beam with a prescribed bending stiffness is the one that maximizes the material index

$$M_1 = \frac{(E\,\phi_B^e)^{1/2}}{\rho}$$

The material-shape combination for a light beam with a prescribed strength is that which maximizes the material index

$$M_3 = \frac{(\phi_B^f\,\sigma_f)^{2/3}}{\rho}$$

These allow shaped sections to be plotted on property charts. They are used for selection in exactly the same way as the indices of Chapter 5.

Similar combinations involving ϕ_T^e *and* ϕ_T^f give the lightest stiff or strong shaft. Here, the criterion of "performance" was meeting a design specification at minimum weight. Other such material-shape combinations maximize other performance criteria: minimizing cost rather than weight, for example, or maximizing energy storage.

The procedure for selecting material-shape combinations is best illustrated by examples. These, and exercises in shape selection and the use of shape factors, can be found in the next chapter.

9.9 FURTHER READING

Ashby, M.F. (1991). Material and shape. *Acta Metall. Mater., 39,* 1025–1039.
The paper in which the ideas in this chapter were first developed.

Birmingham, R.W., & Jobling, B. (1996). *Material selection: Comparative procedures and the significance of form. International Conference on Lightweight Materials in Naval Architecture,* The Royal Institution of Naval Architects, London.
The article in which the 4-quadrant charts like those in Figures 9.12 and 9.15 were introduced.

Gerard, G. (1956). *Minimum weight analysis of compression structures.* New York University Press, Library of Congress Catalog Number 55-10052.
This and the book by Shanley, cited below, establish the principles of minimum weight design. Both, unfortunately, are out of print but can be found in libraries.

Gere, J.M., & Timoshenko, S.P. (1985). *Mechanics of materials.* Wadsworth International.
An introduction to the mechanics of elastic solids.

Gibson, L.J., & Ashby, M.F. (1997). *Cellular solids* (2nd ed.). Cambridge University Press, ISBN 0-521-49560-1.
A broad-based introduction to the structure and properties of foams and cellular solids of all types.

Gibson, L.J., Ashby, M.F., & Hurley, B. (2010). *Cellular solids in nature.* Cambridge University Press, ISBN 978-0-521-19544-7.
An exploration of architectured materials in nature.

Parkhouse, J.G. (1984). Structuring: a process of material dilution. In H. Nooshin (Ed.), *Proceedings of the Third International Conference on Space Structures* (pp. 367). Elsevier.
Parkhouse develops an unusual approach to analyzing material efficiency in structures.

Shanley, F.R. (1960). *Weight-strength analysis of aircraft structures.* (2nd ed.). Dover Publications, Library of Congress Catalog Number 60-501011.
This and the book by Gerard, cited above, establish the principles of minimum weight design. Both, unfortunately, are out of print but can be found in libraries.

Timoshenko, S.P., & Gere, J.M. (1961). *Theory of elastic stability.* McGraw-Hill, Koga Kusha Ltd., Library of Congress Catalog Number 59-8568.
The definitive text on elastic buckling.

Weaver, P.M., & Ashby, M.F. (1998). Material limits for shape efficiency. *Prog. Mat. Sci., 41,* 61–128.
A review of the shape-efficiency of standard sections, and the analyses leading to the results used in this chapter for the maximum practical shape factors for bending and twisting.

Young, W.C. (1989). *Roark's formulas for stress and strain* (6th ed.). McGraw-Hill, ISBN 0-07-100373-8.
A sort of Yellow Pages of formulae for stress and strain, cataloging the solutions to thousands of standard mechanics problems.

CHAPTER 10

Case Studies: Material and Shape

A pair of electric aircraft. The lightweight frames are pin-jointed aluminum tubing. (Image courtesy of David Bremner, editor of *Microlight Flying* magazine.)

CONTENTS

Materials Selection in Mechanical Design. DOI: 10.1016/B978-1-85617-663-7.00010-2

10.1 INTRODUCTION AND SYNOPSIS

This chapter, like Chapters 6 and 8, is a collection of case studies. They illustrate the use of shape factors, of the 4-quadrant chart construction, and of material indices that include shape. They are necessary for the restricted class of problems in which section shape directly influences performance—that is, when the prime function of a component is to carry loads that can cause it to bend, twist, or buckle.

Indices that include shape provide a tool for optimizing the coselection of material and shape. The important ones are summarized in Table 10.1. The selection procedure is, first, to identify candidate materials and the section shapes in which each is available or could be made. The relevant material properties and shape factors for each are tabulated and the relevant index is evaluated. The best combination of material and shape is that with the greatest value of the index. The same information can be plotted on material selection charts, allowing a graphical solution to the problem—one that often suggests further possibilities.

The method has other uses. It gives insight into the way in which natural materials—many of which are very efficient—have evolved. Bamboo is an example: It has both internal or microscopic shape and a tubular, macroscopic shape, giving it very attractive properties. This and other aspects are brought out in the case studies that follow.

Table 10.1 Indices with Shape: Stiffness- and Strength-limited Design at Minimum Weight

Component Shape, Loading, and Constraints	Stiffness-limited Design*	Strength-limited Design*
Tie (tensile member) Load, stiffness, and length specified, section-area free	$\frac{E}{\rho}$	$\frac{\sigma_f}{\rho}$
Beam (loaded in bending) Loaded externally or by self-weight, stiffness, strength and length specified, section area and shape free	$\frac{(\phi_B^e E)^{1/2}}{\rho}$	$\frac{(\phi_B^f \sigma_f)^{2/3}}{\rho}$
Torsion bar or tube Loaded externally, stiffness, strength, and length specified, section area and shape free	$\frac{(\phi_T^e E)^{1/2}}{\rho}$	$\frac{(\phi_T^f \sigma_f)^{2/3}}{\rho}$
Column (compression strut) Collapse load by buckling or plastic crushing, strength and length specified, section area and shape free	$\frac{(\phi_B^e E)^{1/2}}{\rho}$	$\frac{\sigma_f}{\rho}$

* *The shape factors ϕ_B^e and ϕ_B^f are for bending; ϕ_T^e and ϕ_T^f are for torsion. For minimum cost design, replace ρ by $C_m\rho$ in the indices.*

10.2 SPARS FOR HUMAN-POWERED PLANES

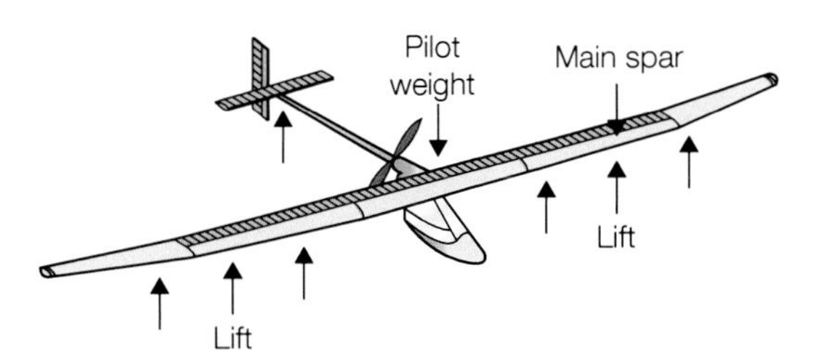

FIGURE 10.1
The loading on a human-powered plane is carried by two spars, one spanning the wings and the other linking the wings to the tail. Both are designed for stiffness at minimum weight.

In most engineering design the objectives are complex, often requiring a trade-off between performance and cost. But in designing a spar for a human-powered plane the objective is simple: The spar must be as light as possible and yet stiff enough to maintain the aerodynamic efficiency of the wings. Strength, reliability, even cost, hardly matter when records are about to be broken. The plane (Figure 10.1) has two main spars: the transverse spar supporting the wings, and the longitudinal spar carrying the tail assembly. Both are loaded primarily in bending (torsion cannot, in reality, be neglected, although we shall do so here).

Some 60 human-powered planes have flown successfully. Planes of the first generation were built of balsa wood, spruce, and silk. The second generation relied on aluminum tubing for the load-bearing structure.[1] The present, third, generation uses carbon-fiber/epoxy spars, molded to appropriate shapes. How has this evolution come about? And how much further can it go?

The translation and the selection We seek a material-and-shape combination that minimizes mass for a given bending stiffness (Table 10.2). The measure of performance, read from Table 10.1, is

$$M_1 = \frac{(\phi_B^e E)^{1/2}}{\rho} \tag{10.1}$$

Data for five materials are assembled in the upper part of Table 10.3. If all have the same shape, M_1 reduces to the familiar $E^{1/2}/\rho$ and the ranking is that of the fourth column. Balsa and spruce are extraordinarily efficient; that is why model aircraft builders use them now and the builders of real aircraft relied so heavily on them in the past. Solid CFRP comes close. Steel and aluminum are far behind.

Now add shape. Achievable shape factors for the five materials appear in column 5 of the table; they are typical of commercially available sections and are well below the maximum for each material. The effect of shaping the section, to a rectangle for the woods and to a box section for aluminum and CFRP, gives the results in the last column. Aluminum is now marginally better than the woods; CFRP is best of all.

---

[1] The electric-powered microlight planes in the chapter cover picture have tubular aluminum frames.

Table 10.2 Design Requirements for Wing Spars

Function	Wing spar
Constraints	Specified stiffness Length specified
Objective	Minimum mass
Free variables	Choice of material Section shape and scale

Table 10.3 Materials for Wing Spars

Material	Modulus E (GPa)	Density ρ (kg/m^3)	Index* $E^{1/2}/\rho$ (GPa$^{1/2}$/Mg/m^3)	Shape Factor ϕ_B^e	Index* M_1 $(\phi_B^e E)^{1/2}/\rho$ (GPa$^{1/2}$/Mg/m^3)
Balsa	4.6	210	10	2	14
Spruce	10.3	450	8	2	11
Steel	205	7,850	1.8	25	9
Al 7075 T6	70	2,700	3	25	15
CFRP	115	1,550	7	10	22
Beryllium	300	1,840	9.3	15	36
Be 38%Al (AlBeMet 162)	185	2,100	6.5	15	25
Borosilicate glass	63	2,200	3.6	10	11

* *The values of the index are based on mean values of the material properties.*

The same information is shown graphically in Figure 10.2, using the method of Section 9.6. Each shape is treated as a new material with modulus $E^* = E/\phi_B^e$ and $\rho^* = \rho/\phi_B^e$. The values of E^* and ρ^* are plotted on the chart. The superiority of both the aluminum tubing with $\phi_B^e = 25$ and the CFRP box sections with $\phi_B^e = 10$ are clearly demonstrated.

Postscript Why is wood so good? With no shape it does as well as heavily shaped steel. It is because wood *is* shaped: Its cellular structure gives it internal microshape, increasing the performance of the material in bending; it is nature's answer to the I-beam.. Advances in the technology of drawing thin-walled aluminum tubes allowed a shape factor that cannot be reproduced in wood, giving aluminum a performance edge—a fact that did not escape the designers of the second generation of human-powered planes. There is a limit of course: Tubes that are too thin will kink (as described in Chapter 9), setting an upper limit to the shape factor for aluminum at about 40. Further advances required a new material with lower density and higher modulus, conditions met by CFRP. Add shape to it and CFRP outperforms them all.

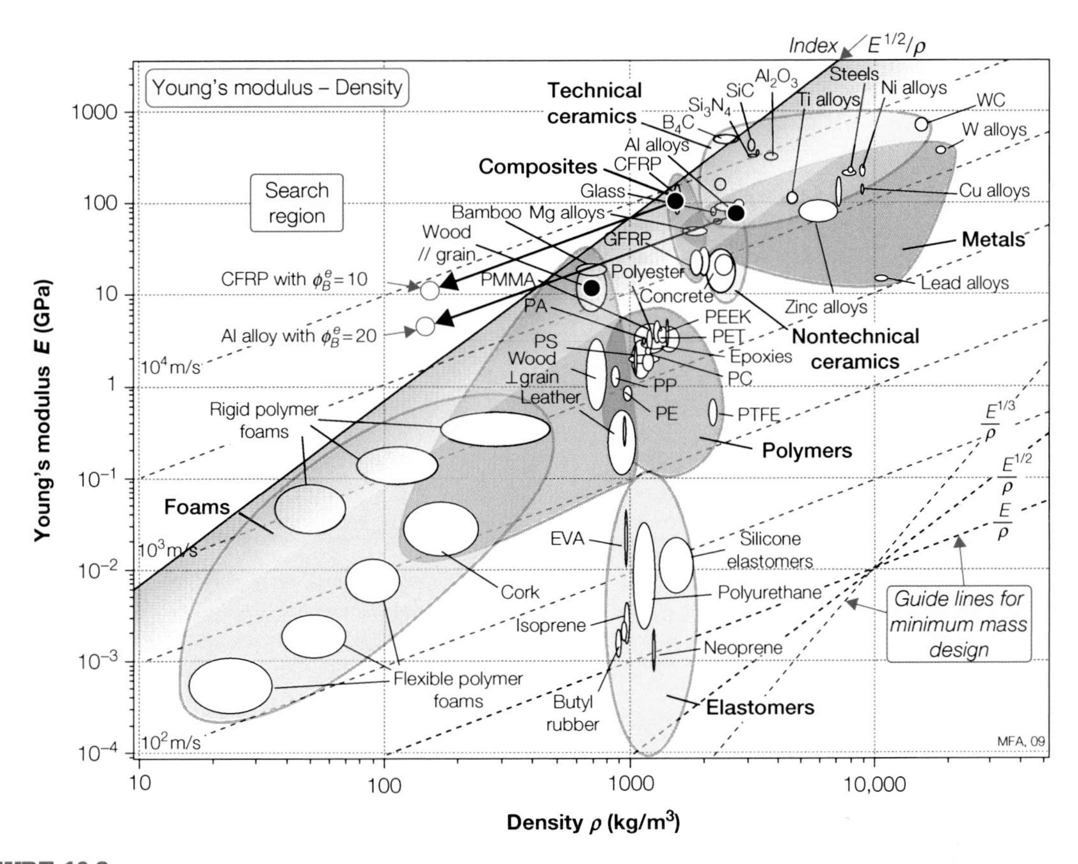

FIGURE 10.2
The materials and shapes for wing spars, plotted on the modulus-density chart. A spar made of CFRP with a shape factor of 10 outperforms spars made of aluminum with a shape factor of 20 and wood with shape factor close to 1.

Can we do better? Not easily, but if it is really worth it, maybe. Chapter 11 develops methods for designing material combinations that outperform anything that one material could do by itself, but we will leave these for later. What can be done with a single material? If we rank materials by $E^{1/2}/\rho$, we get a list headed by diamond, boron, and—ah!—beryllium, a metal used for space structures that can be shaped. After that come more ceramics, not viable because of their brittleness and the difficulty of forming them into any useful shape. And, far below, magnesium and aluminum alloys. But above these, in the middle of the ceramics, is...glass. A glider with glass wing spars? Sounds crazy, but think for a moment. Toughened glass, bulletproof glass, glass floors; glass can be used as a structural material. And it is easy to shape. It is not even that expensive.

So suppose beryllium or glass were given shape—could either one outperform CFRP? Again, not easily, but maybe. Here we have to guess. In theory

(Equation (9.15a)) beryllium could be given a shape factor of 60, glass of 30. Being more realistic, factors of 15 and 10 are possible. The lower part of the table shows what all this means. Beryllium out-performs CFRP, by a large margin. Glass doesn't. Well, it was a thought. For more, see Chapter 12.

Related reading: Human-powered flight

There is a large body of literature. Try these:

Bliesner, W. (1991). The design and construction details of the Marathon Eagle. In: Technology for human powered aircraft, *Proceedings of the Human-powered Aircraft Group Half Day Conference.* The Royal Aeronautical Society, London.

Details of another attempt to build a record-breaking plane.

Drela, M. & Langford, J.D. (1985). Man-powered flight. *Scientific American*, p. 122.

A concise history of human-powered flight up to 1985.

Grosser, M (1981). *Gossamer odyssey.* Dover Publications, ISBN 0-486-26645-1.

Accounts of the Gossamer Condor and Albatross, which were attempts to capture the world record for human-powered flight.

Nadel, E.R., & Bussolari, S.R., (1988). The Daedalus project: physiological problems and solutions. *American Scientist*, July–August.

An account of the Daedalus project, a competition for human-powered flight from Crete and mainland Greece—the mythical route of Daedelus and his father.

Sherwin, K. (1971). *Man powered flight.* Model & Allied Publications, Argus Books Ltd.

Sherwin, K. (1976). *To fly like a bird.* Bailey Brothers & Swinfen Ltd., ISBN 561-00283-5.

Related case studies

6.5 "Cost: Structural materials for buildings"
10.3 "Forks for a racing bicycle"
10.4 "Floor joists: Wood, bamboo, or steel?"

10.3 FORKS FOR A RACING BICYCLE

The first consideration in bicycle design (Figure 10.3) is strength. Stiffness matters, of course, but the initial design criterion is that the frame and forks should not collapse in normal use. The loading on the forks is predominantly *bending*. If the bicycle is for racing, mass is a primary consideration: The forks should be as light as possible. What is the best choice of material and shape?

The translation and the selection We model the forks as beams of length L that must carry a maximum load P (both fixed by the design) without plastic collapse or fracture (Table 10.4). The forks are tubular, of radius r and fixed wall thickness t. The mass is to be minimized. The fork is a light, strong beam. Further details of load and geometry are unnecessary: The best material and shape, from Table 10.1, is that with the greatest value of

$$M_2 = \frac{(\phi_B^f \sigma_f)^{2/3}}{\rho} \qquad (10.2)$$

Table 10.5 lists seven candidate materials with their properties. If the forks are solid, meaning that $\phi_B^f = 1$, spruce wins (see second to last column of table). Bamboo is special because it grows as a hollow tube with a macroscopic shape factor ϕ_B^f of about 2.2, giving it a bending strength that is much higher than solid spruce (last column). When shape is added to the other materials, however, the ranking changes. The shape factors listed in the table are achievable using normal production methods. Steel is good. Titanium 6-4 is better. Best of all is CFRP. Magnesium, despite its low density, is poor in strength-limited applications.

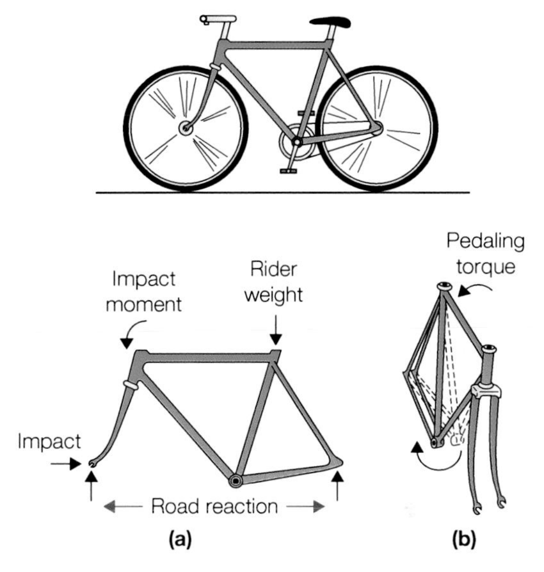

FIGURE 10.3

The bicycle. The forks are loaded in bending. The lightest forks that will not collapse plastically under a specified design load are those made of the material and shape with the greatest value of $(\phi_B^f \sigma_f)^{2/3}/\rho$.

Postscript Bicycles have been made of all seven of the materials listed in the table—you can still buy bicycles made of six of them. Early bicycles were made of wood; present-day racing bicycles, of steel, aluminum, or CFRP; sometimes they are interleaved with carbon fibers that have layers of glass or Kevlar to improve fracture resistance. Mountain bicycles, for which strength and impact resistance are particularly important, have steel or titanium forks.

The reader may be perturbed by the cavalier manner in which the theory for a straight beam with an end load acting normal to it is applied to a curved beam loaded at an acute angle. No alarm is necessary. When (as explained in Section 5.4) the variables describing the functional requirements (F), the geometry (G), and the materials (M) in the performance equation are separable, the details of loading and geometry affect the terms F and G but not M. This is an example: Beam curvature and angle of application of load do not change the material index, which depends only on the design requirement of strength in bending at minimum weight.

Table 10.4 Design Requirements for Bicycle Forks

Function	Bicycle forks
Constraints	Must not fail under design loads—a strength constraint Length specified
Objective	Minimize mass
Free variables	Choice of material Section shape

Table 10.5 Material for Bicycle Forks

Material	Strength σ_f (MPa)	Density ρ (kg/m^3)	Shape Factor ϕ_B^f	Index* $\sigma_f^{2/3}/\rho$ (MPa)$^{2/3}$/(Mg/m^3)	Index M_2* $(\phi_B^f \sigma_f)^{2/3}/\rho$ (MPa)$^{2/3}$/(Mg/m^3)
Spruce (Norwegian)	75	450	1.5	39	51
Bamboo	70	700	2.2	24	41
Steel (Reynolds 531)	880	7,850	7.5	12	46
Alu (6061-T6)	250	2,700	5.9	15	49
Titanium 6%Al-4%V	950	4,420	5.9	22	72
Magnesium AZ 61	165	1,810	4.25	17	45
CFRP	375	1,550	4.25	33	87

* *The values of the indices are based on mean values of the material properties.*

Related reading: Bicycle design

Oliver, T. (1992). *Touring bikes, a practical guide.* Crowood Press, ISBN 1-85223-339-7.

A good source of information about bike materials and construction, with tables of data for the steels used in tube sets.

Sharp, A. (1979). *Bicycles and tricycles, an elementary treatise on their design and construction.* MIT Press, ISBN 0-262-69066-7.

A far-from-elementary treatise, despite its title, first published in 1977. It is the place to look if you need the mechanics of bicycles.

Watson R., & Gray, M. (1978). *The Penguin book of the bicycle.* Penguin, ISBN 0-1400-4297-0.

Watson and Gray describe the history and use of bicycles. Not much on design, mechanics, or materials.

Whitt, F.R., & Wilson, D.G. (1982). *Bicycling science* (2nd ed.). The MIT Press, ISBN 0-262-73060-X.

A book by two MIT professorial bike enthusiasts, more easily digestible than Sharp, and with a good chapter on materials.

Wilson, D.G. (1986). A short history of human powered vehicles. *American Scientist, 74,* p. 350.

Typical *Scientific American* article: *good content, balance, and presentation. A good starting point.*

Bike magazines such as Mountain Bike, Which Bike? *and* Cycling and Mountain Biking *carry extensive tables of available bikes and their characteristics—type, maker, cost, weight, gear set, etc.*

Related case studies

6.5 "Cost: Structural materials for buildings"
10.2 "Spars for human-powered planes"
10.4 "Floor joists: Wood, bamboo, or steel?"

10.4 FLOOR JOISTS: WOOD, BAMBOO, OR STEEL?

Floors are supported on joists: beams that span the space between the walls (Figures 6.7 and 10.4). Let us suppose that a joist is required to support a specified bending load (the "floor loading") without sagging excessively or failing; it must be cheap. Traditionally, joists in the United States and Europe are made of wood with a rectangular section of aspect ratio 2:1, giving an elastic shape factor (refer to Table 9.3) of $\phi_B^e = 2$. In Asian countries bamboo, with a "natural" shape factor of about $\phi_B^e = 3.2$, is a replacement for wood in smaller buildings. But as wood becomes scarcer and buildings need to be larger, steel replaces wood and bamboo as the primary structural material. Standard steel I-section joists have shape factors in the range $5 < \phi_B^e < 25$ (special I-sections can have much larger values). Are steel I-joists a better choice than wooden ones? Table 10.6 contains a summary of the design requirements.

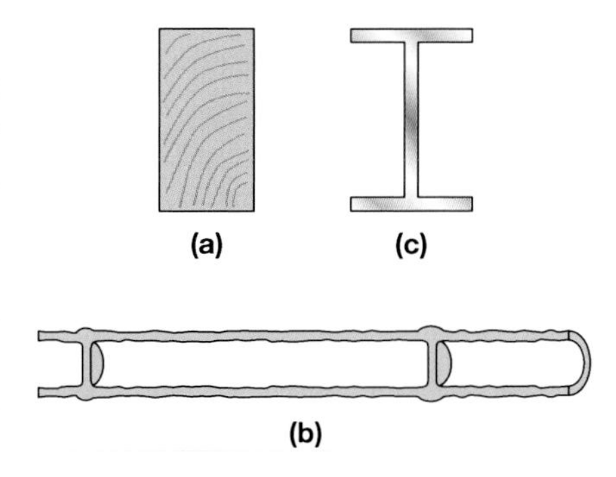

FIGURE 10.4
The cross-sections of a wooden beam ($\phi_B^e = 2$), a steel I-beam ($\phi_B^e = 10$), and bamboo ("natural" shape factors of $\phi_B^e = 3.2$). The values of ϕ_B^e are calculated from the ratios of dimensions of each beam, using the formulae of Table 9.3.

The translation and the selection Consider stiffness first. The cheapest beam, for a given stiffness, is that with the largest value of the index (from Table 10.1 with ρ replaced by $C_m \rho$ to minimize cost):

$$M_1 = \frac{(\phi_B^e E)^{1/2}}{C_m \rho} \tag{10.3}$$

Data for the modulus E, the density ρ, the material cost C_m, and the shape factor ϕ_B^e are listed in Table 10.7, together with the values of the index M_1 with and without shape. The steel beam with $\phi_B^e = 25$ has a slightly larger value of M_1 than wood, meaning that it is a little cheaper for the same stiffness.

But what about strength? The best choice for a light beam of specified strength is one that maximizes the material index:

$$M_2 = \frac{(\phi_B^f \sigma_f)^{2/3}}{C_m \rho} \tag{10.4}$$

The quantities of failure strength σ_f shape factor ϕ_B^f, and index M_2 are also given in the table. Wood performs better than even the most efficient steel I-beam.

Postscript So the conclusion: As far as performance per unit material cost is concerned, there is not much to choose between the standard wood and

Table 10.6 Design Requirements for Floor Joists

Function	Floor joist
Constraints	Length specified Stiffness specified Strength specified
Objective	Minimum material cost
Free variables	Choice of material Section shape

Table 10.7(a) Materials for Floor Joists

Material	Density ρ (kg/m³)	Cost C_m US$/kg	Flexural Modulus E (GPa)	Flexural Strength σ_f (MPa)
Wood (pine)	490	1.0	9.5	41
Bamboo	700	1.9	17	42
Steel (1020)	7850	0.65	205	355

Table 10.7(b) Indices for Floor Joists

Material	Shape Factor ϕ_B^e	Shape Factor ϕ_B^f	Index $E^{1/2}/C_m\rho$	Index* M_1 $(\phi_B^e E)^{1/2}/C_m\rho$	Index $\sigma_f^{2/3}/C_m\rho$	Index* M_2 $(\phi_B^f \sigma_f)^{2/3}/C_m\rho$
Wood (pine)	2	1.4	6.3	8.9	24	30
Bamboo	3.2	2	3.1	5.5	9	14
Steel (1020)	15	4	2.8	11	9.8	25

* *The values of the indices are based on means of the material properties. The units of the indices for elastic deflection are* $(GPa)^{1/2}/kg/m^3$; those for failure are $(MPa)^{2/3}/kg/m^3$.

the standard steel sections used for joists. As a general statement, this is no surprise—if one were much better than the other, the other would no longer exist. But—looking a little deeper—wood dominates in certain market sectors, bamboo in others, and, in others still, steel. Why?

Wood and bamboo are indigenous to some countries and grow locally; steel has to come further, with associated transport costs. Assembling wood structures is easier than assembling those of steel; wood is more forgiving of mismatches of dimensions, it can be trimmed on site, you can hammer nails into it anywhere. It is a user-friendly material.

But wood is a variable material, and, like us, is vulnerable to the ravages of time, prey to fungi, insects, and small mammals with sharp teeth. The problems they create in a small building—a family home, say—are easily overcome, but in a large commercial building—an office block, for instance—they create greater risks and are harder to fix. Here, steel wins.

Related reading

Cowan, H.J., & Smith, P.R. (1988). *The science and technology of building materials.* Van Nostrand, ISBN 0-442-21799-4.

A broad survey of materials for the structure, cladding, insulation, and interior surfacing of buildings—an excellent introduction to the subject.

Farrelly, D. (1984). *The book of bamboo.* Sierra Club Books, ISBN 0-87156-825-X.

An introduction to bamboo and its many varieties.

Janssen, J.J. (1995). *Building with bamboo: A handbook.* Practical Action, ISBN 1-85339-203-0.

Bamboo remains a building material of great importance, as well as the material of flooring, matting, and basketmaking. The techniques of building with bamboo have a long history, documented here.

Related case studies

6.5 "Cost: Structural materials for buildings"
10.2 "Spars for human-powered planes"
10.3 "Forks for a racing bicycle"

10.5 TABLE LEGS YET AGAIN: THIN OR LIGHT?

Luigi Tavolino's table (Section 6.4 and Figure 6.5) is a great success. He decides to develop a range of less expensive tubular-legged furniture. Some are to have thin legs, others have to be light. He needs a more general way of exploring material and shape.

The translation and the selection Tubes can be made with almost any radius r and wall thickness t, though (as we know from Chapter 9) it is impractical to make them with r/t too large because they buckle locally, and that is bad. Luigi chooses GFRP as the material for his legs, and for this material the maximum available shape factor is $\phi_B^e = 10$ corresponding to a value of $r/t = 10.5$ (refer to Table 9.3). The cross-sectional area A and the second moment of area I of a thin-walled tube are

$$A = 2\pi r t$$

and

$$I = \pi r^3 t$$

from which

$$r = \left(\frac{2I}{A}\right)^{1/2} \tag{10.5}$$

This allows contours of constant r to be plotted on the A—I quadrant of the four quadrant diagram shown in Figure 10.5 (they form a family of lines of slope 1), which can now be used to select either for minimizing mass or for minimizing tube radius.

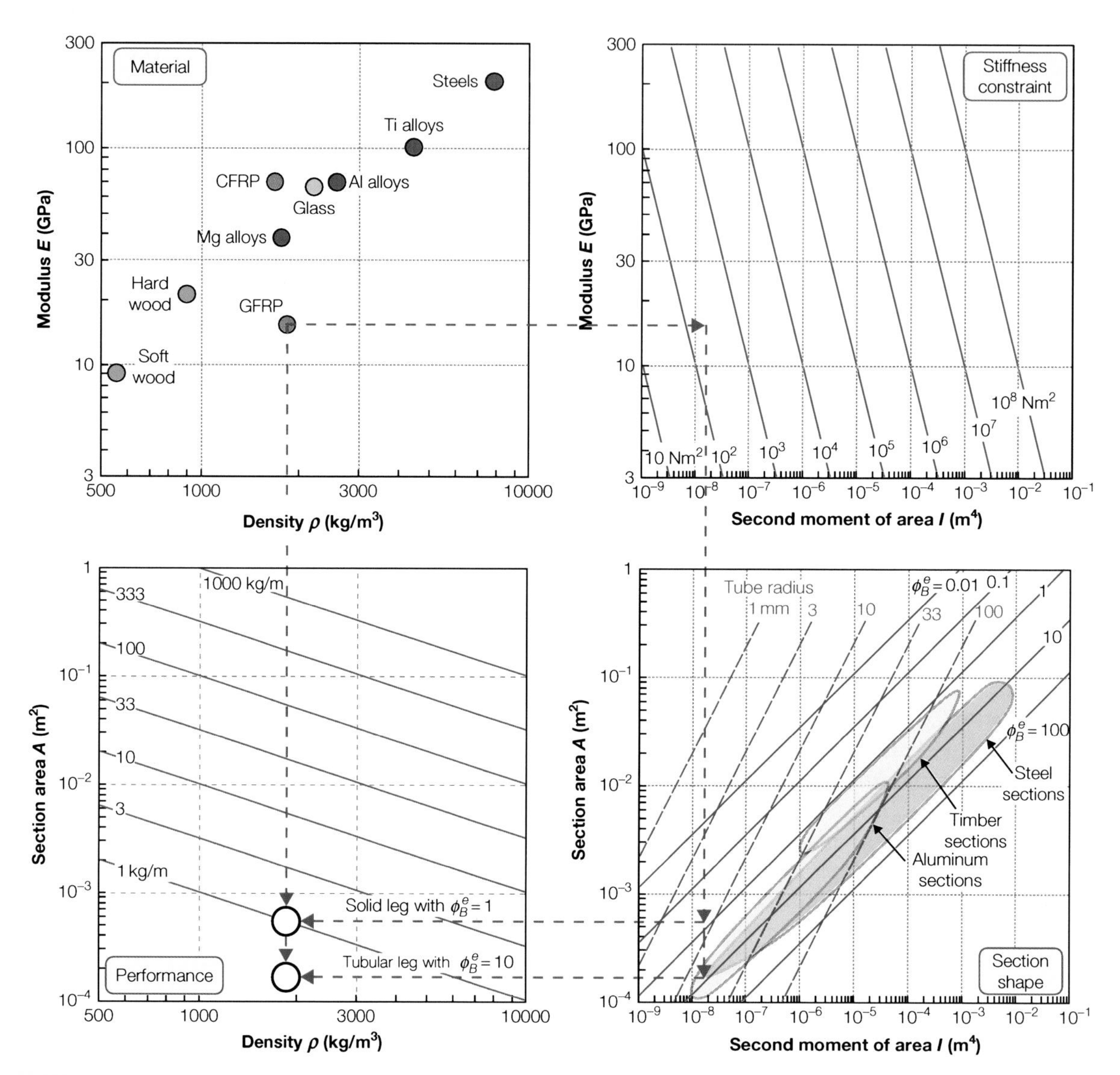

FIGURE 10.5

The chart assembly for exploring stiffness-limited design with tubes. The *I-A* quadrant now has contours of tube radius *r*. The thinnest GFRP leg is the one with $\phi_B^e = 1$. The lightest one is that with $\phi_B^e = 10$, the maximum value for GFRP.

Luigi might use it in the following way. He designs a table with cylindrical, unbraced legs, each with length $L = 1$ m. For safety, each leg must support 50 kg without buckling, requiring that

$$\frac{\pi^2}{4}\frac{E\,I}{\ell^3} \geq 500\ \text{N}$$

from which

$$E\,I \geq 200\ \mathrm{N \cdot m^2}$$

Luigi now plots a rectangular selection path on the four-quadrant diagram in the way shown in Figure 10.5. The lightest leg is that with the largest allowable value of ϕ_B^e (10 in the case of GFRP), but this gives a fat leg, one nearly 20 mm in radius, as read from the radius contours. The thinnest leg is a solid one, and that has $\phi_B^e = 1$. It is much thinner—only 5 mm in radius—but it is also almost three times heavier.

Postscript So Luigi will have to compromise: fat and light or thin and heavy—or somewhere in between. Or he could use another material. The strength of the four-quadrant method is that he can explore other materials with ease. Repeating the construction for aluminum, putting an upper limit on ϕ_B^e of, say, 15, or for CFRP with the same upper limit as GFRP is only a moment's work.

Related case studies

6.4 "Materials for table legs"
8.5 "Conflicting objectives: table legs again"

10.6 INCREASING THE STIFFNESS OF STEEL SHEET

How can you make steel sheet stiffer? There are many reasons you might wish to do so. The most obvious: to enable stiffness-limited sheet structures to be lighter than they are now; to allow panels to carry larger in-plane compressive loads without buckling; and to raise the natural vibration frequencies of sheet structures. Bending stiffness is proportional to $E.I$ (E is Young's modulus, I is the second moment of area of the sheet, equal to $t^3/12$ per unit width where t is the sheet thickness). There is nothing much you can do to change the modulus of steel, which is always close to 210 GPa. The answer is to add a bit of shape, increasing I. So consider the design brief of Table 10.8.

The translation and the selection The age-old way to make sheet steel stiffer is to corrugate it, giving it a roughly sinusoidal profile. The corrugations increase the second moment of area of the sheet about an axis normal to the corrugations themselves. The resistance to bending in one direction is thereby increased, but in the cross-direction it is not changed at all.

Corrugations are the clue, but—to be useful—they must stiffen the sheet in all directions, not just one. A hexagonal grid of dimples (Figure 10.6)

Table 10.8 Design Requirements for Stiffened Steel Sheet

Function	Steel sheet for stiffness-limited structures
Constraints	Profile limited to a maximum deviation ± 5 times the sheet thickness from flatness Cheap to manufacture
Objective	Maximize bending stiffness of sheet
Free variables	Section profile

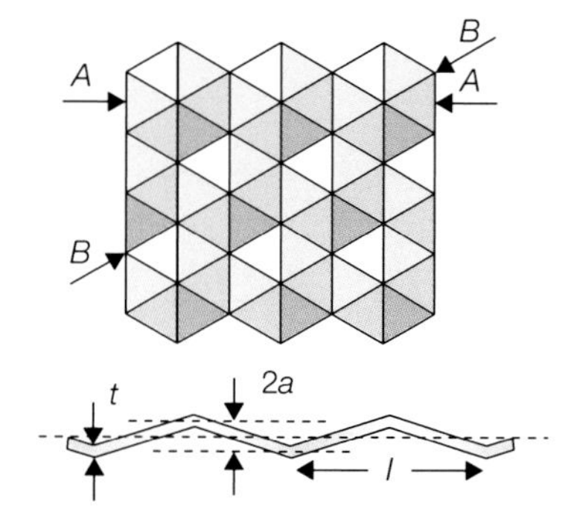

FIGURE 10.6
A sheet with a profile of adjacent hexagonal dimples that increases its bending stiffness and strength. Shape factors for the section *A–A* are calculated in the text. Those along other trajectories are lower but still significantly greater than 1.

achieves this. There is now no direction of bending that is not dimpled. The dimples need not be hexagons; any pattern arranged in such a way that you cannot draw a straight line across it without intersecting dimples will do. Hexagons are probably about the best.

Consider an idealized cross-section as in the lower part of Figure 10.6, which shows the section *A–A* enlarged. As before, we define the shape factor as the ratio of the stiffness of the dimpled sheet to that of the flat sheet from which it originated. The second moment of area of the flat sheet per unit width is

$$I_o = \frac{t^3}{12} \tag{10.6}$$

That of the dimpled sheet with amplitude a is

$$I \approx \frac{1}{12}(2a+t)^2\, t \tag{10.7}$$

giving a shape factor, defined as before as the ratio of the stiffness of the sheet before and after corrugating, of

$$\phi_B^e = \frac{I}{I_o} \approx \frac{(2a+t)^2}{t^2} \tag{10.8}$$

Note that the shape factor has the value unity when the amplitude is zero, but increases as the amplitude increases. The equivalent shape factor for failure in bending is

$$\phi_B^f = \frac{Z}{Z_o} \approx \frac{(2a+t)}{t} \tag{10.9}$$

These equations predict large gains in stiffness and strength. The reality is a little less rosy. This is because, while all cross-sections of the sheet are dimpled, only those that cut through the peaks of the dimples have an amplitude equal to the peak height (all others have less) and, even among these, only some have adjacent dimples; the section B–B, for example, does not. Despite this, and limits set by the onset of local buckling, the gain is real.

Postscript Dimpling can be applied to most rolled-sheet products. It is done by making the final roll pass through mating rolls with meshing dimples, adding little to the cost. It is most commonly applied to sheet steel. Here it finds applications in the automobile industry for bumper armatures, seat frames, and side impact bars, allowing weight saving without compromising mechanical performance. Stiffening sheet also raises its natural vibration frequencies, making them harder to excite, thus helping to suppress vibration in panels.

But a final word of warning: Stiffening the sheet may change its failure mechanism. Flat sheet yields when bent; dimpled sheet, if thin, can fail by a local buckling mode. It is this that ultimately limits the useful extent of dimpling.

Related reading: stiffening sheet by dimpling

Fletcher, M. (1998). Cold-rolled dimples gauge strength. *Eureka* (May).

A brief account of the shaping of steel panels to improve bending stiffness and strength.

Related case study

10.7 "Shapes that flex: Leaf and strand structures"

10.7 SHAPES THAT FLEX: LEAF AND STRAND STRUCTURES

Flexible cables, leaf and helical springs, and flexural hinges require low, not high, structural efficiency. Here the requirement is for low flexural or torsional stiffness about one or two axes while retaining high stiffness and strength in other directions. The simplest example is the torsion bar (Figure 10.7(a)), with a shape factor $\phi_T^e \approx 1$. The single-leaf spring (Figure 10.7(b)) allows much lower values of ϕ_B^e ($\phi_B^e = h/w$; the dimensions are defined in the figure).

Multistrand cables and multileaf assemblies do much better. Consider the change in efficiency when a solid square beam of section $A = b^2$ is subdivided into n cylindrical strands, each of radius r (Figures 10.7(c)(d)), such that

$$n\pi r^2 = b^2 \tag{10.10}$$

The axial stiffness of the original bar is proportional to EA, and its bending stiffness is S_o to EI_o, where E is the modulus of the material and I_o is

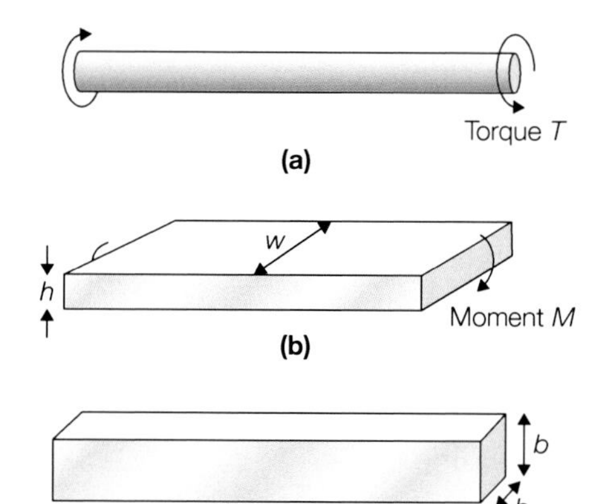

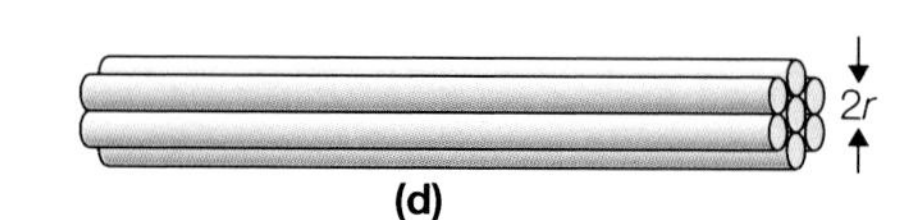

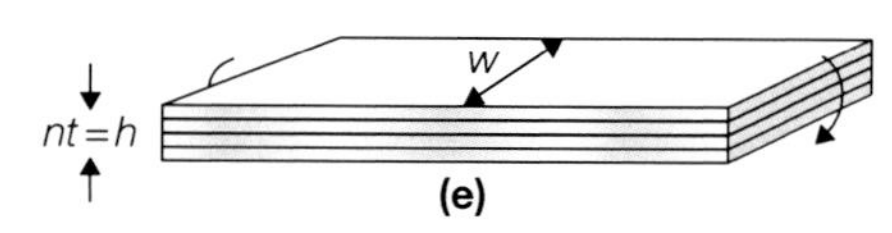

FIGURE 10.7
A simple torsion bar (a); a flat panel of width w used in the definition of ϕ_B^e for the single-leaf structure (b); the standard shape used in the definition of ϕ_B^e for the cable (c); a cable with the same section area as the standard shape of (c) but made up of n cylindrical strands (d); the multileaf structure (e) with the same width and cross-sectional area as the panel in (b).

the second moments of area. When made into a cable of n parallel cylindrical strands, the axial stiffness remains unchanged, but the bending stiffness S falls to a value proportional to nEI, where I is the second moment of a single strand. Thus

$$S \propto n E \frac{\pi r^4}{4} = \frac{1}{4\pi} E \frac{b^4}{n} \tag{10.11}$$

The structural efficiency of the cable is therefore

$$\phi_B^e = \frac{S}{S_o} = \frac{n E I}{E I_o} = \frac{3}{n \pi} \tag{10.12}$$

The number of strands, n, can be very large, allowing the flexural stiffness to be adjusted over a large range while leaving the axial stiffness unchanged.

Multileaf assemblies allow even more dramatic anisotropy. If the thick leaf shown at (b) in the figure is divided into the stack of thin leaves of the same width shown at (e), the bending stiffness changes from

$$S_o = E w h^3/12$$

to

$$S = n E w t^3/12$$

where $t = h/n$. The shape efficiency of the stack is

$$\phi_B^e = \frac{S}{S_o} = \frac{n t^3}{h^3} = \frac{1}{n^2} \tag{10.13}$$

Thus a stack of 10 layers is 100 times less stiff than a simple leaf of the same total thickness, while the in-plane stiffness is unchanged.

Postscript When a segmented structure is bent the segments slide against each other. The analysis we have just done assumed that there was no friction opposing the sliding. A large friction prevents the sliding, and, until sliding starts, the bending stiffness is the same as that of an unsegmented bar or plate. Leaf springs are lubricated or are interleaved with Teflon shims to make sliding easy.

Related case studies

10.6 "Increasing the stiffness of steel sheet"
10.8 "Ultra-efficient springs"

10.8 ULTRA-EFFICIENT SPRINGS

Springs, we deduced in Section 6.7, store energy. They are best made of a material with a high value of σ_f^2/E, or, if mass is more important than volume, of $\sigma_f^2/\rho E$. Springs can be made more efficient still by shaping their section. Just how much more is revealed below.

We take as a measure of performance the energy stored per unit volume of solid of which the spring is made; we wish to maximize this energy. Energy per unit weight and per unit cost are maximized by similar procedures (Table 10.9).

The translation and the selection Consider a single-leaf spring first (Figure 10.8(a)). A leaf spring is an elastically bent beam. The energy stored in a bent beam, loaded by a force F, is

$$U = \frac{1}{2}\frac{F^2}{S_B} \tag{10.14}$$

where S_B, the bending stiffness of the spring, is given by Equation (9.17) or, after replacing I by ϕ_B^e, by Equation (9.18), which, repeated, is

$$S_B = \frac{C_1}{12\,L^3}\;E\,\phi_B^e A^2 \tag{10.15}$$

The force F in Equation (10.14) is limited by the onset of yield; its maximum value is

$$F_f = C_2\,Z\,\frac{\sigma_f}{L} = \frac{C_2}{6\,L}\;\sigma_f\,\phi_B^f\;A^{3/2} \tag{10.16}$$

Table 10.9 Design Requirements for Ultra-efficient Springs

Function	Material-efficient spring
Constraint	Must remain elastic under design loads
Objective	Maximum stored energy per unit volume (or mass or cost)
Free variables	Choice of material Section shape

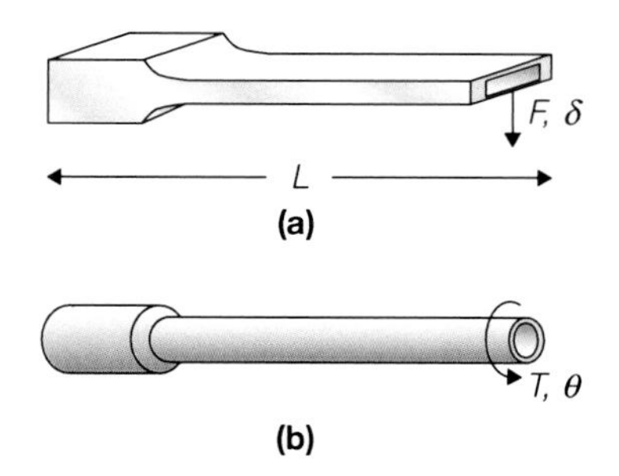

FIGURE 10.8
Hollow springs use material more efficiently than solid springs (a). Best in bending is the hollow rectangular or elliptical section; best in torsion is the tube (b).

(The constants C_1 and C_2 are tabulated in Appendix B, Tables B.3 and B.4.) Assembling these gives the maximum energy the spring can store:

$$\frac{U_{max}}{V} = \frac{C_2{}^2}{6\,C_1}\left(\frac{(\phi_B^f\,\sigma_f)^2}{\phi_B^e\,E}\right) \tag{10.17}$$

where $V = A\,L$ is the volume of solid in the spring. The best material and shape for the spring—the one that uses the least material—is that with the greatest value of the quantity

$$M_1 = \frac{(\phi_B^f\,\sigma_f)^2}{\phi_B^e\,E} \tag{10.18}$$

For a fixed section shape, the ratio involving the two ϕs is a constant: The best choice of material is that with the greatest value of σ_f^2/E —the same result as before. When shape is a variable, the most efficient shapes are those with large $(\phi_B^f)^2/\phi_B^e$. Values for these ratios are tabulated for common section shapes in Table 10.10; hollow-box and elliptical sections are up to three times more efficient than solid shapes.

Torsion bars and helical springs are loaded in torsion (Figure 10.3(b)). A similar calculation gives

$$\frac{U_{max}}{V} = \frac{1}{6.5}\,\frac{(\phi_T^f\,\sigma_f)^2}{\phi_T^e\,G} \tag{10.19}$$

The most efficient material and shape for a torsional spring is that with the largest value of

$$M_2 = \frac{(\phi_T^f\,\sigma_f)^2}{\phi_T^e\,E} \tag{10.20}$$

(where G has been replaced by $3E/8$). The criteria are the same: When shape is not a variable, the best torsion bar materials are those with high values of σ_f^2/E. Table 10.10 shows that the best shapes are hollow tubes, with a ratio of $(\phi_T^f)^2/\phi_T^e$ that is twice that of a solid cylinder; all other shapes are less efficient. Springs that store the maximum energy per unit weight (instead of unit volume) are selected with indices given by replacing E by $E\rho$ in Equations (10.18) and (10.20). For maximum energy per unit cost, replace E by $E\,C_m\,\rho$ where C_m is the material cost per kg.

Table 10.10 Shape Factors for the Efficiency of Springs

Section Shape	$(\phi_B^f)^2/\phi_B^e$	$(\phi_T^f)^2/\phi_T^e$
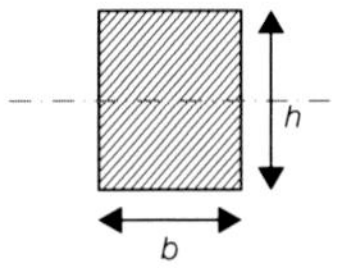 Rectangle, $b \times h$	1	$1.08\frac{b^2}{h^2}\frac{1}{\left(1+0.6\frac{b}{h}\right)^2\left(1-0.58\frac{b}{h}\right)}$
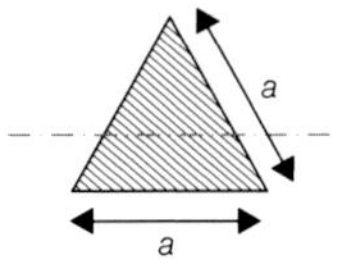 Equilateral triangle, side a	0.38	0.83
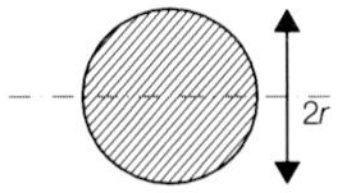 Circle, diameter $2r$	0.75	1.6
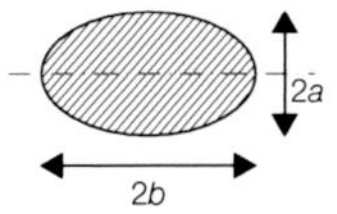 Ellipse, $2a \times 2b$	0.75	$0.8\left(1+\frac{a^2}{b^2}\right)$ $(a < b)$
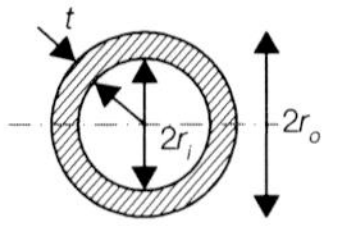 Tube, t, $2r_i$, $2r_o$	1.5	3.2
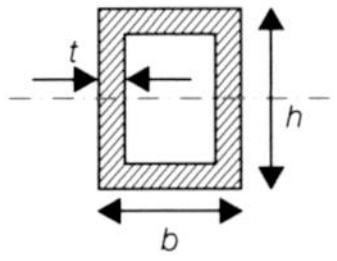 Box section, t, $b \times h$	$\frac{(1+3b/h)}{(1+b/h)}$ $(h,b >> t)$	$3.32\frac{1}{(1-t/h)^4}$........$(h,b >> t)$
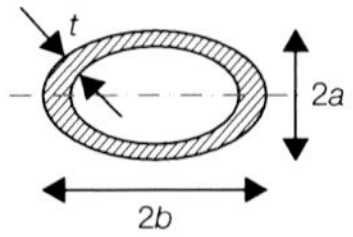 Elliptical tube, t, $2a \times 2b$	$\frac{3}{4}\frac{(1+3b/a)}{(1+b/a)}$ $(a,b >> t)$	$3.2\frac{(1+a^2/b^2)}{(1+a/b)}\left(\frac{b}{a}\right)^{3/2}$ $(a,b >> t)$
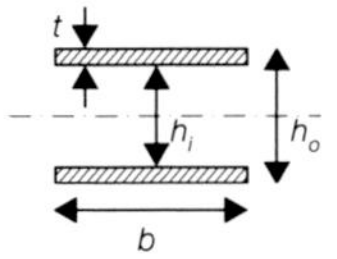 Sandwich, t, h_i, h_o, b	3	–

Continued

Table 10.10 Shape Factors for the Efficiency of Springs *continued*

Section Shape	$(\phi_B^f)^2/\phi_B^e$	$(\phi_T^f)^2/\phi_T^e$
I-section (t, 2t, h, b)	$\frac{(1+3b/h)}{(1+b/h)}$ $(h,b >> t)$	$1.07\,\frac{(1+4h/b)}{(1+h/b)}$ $(h,b >> t)$
T-section (2t, 2t, h)	$1.13\,\frac{(1+4bt^2/h^3)}{(1+b/h)}$ $(h >> t)$	$0.54\,\frac{(1+8b/h)}{(1+b/h)}$ $(h,b >> t)$
H-section (2t, t, h, b)	$1.13\,\frac{(1+4bt^2/h^3)}{(1+b/h)}$ $(h >> t)$	$1.07\,\frac{(1+4h/b)}{(1+h/b)}$ $(h,b >> t)$

Postscript Hollow springs are common in vibrating and oscillating devices and for instruments in which inertial forces must be minimized. The hollow elliptical section is widely used for springs loaded in bending; the hollow tube, for those loaded in torsion. More about this problem can be found in the classic paper by Boiten (1963).

Related reading: Design of efficient springs

Boiten, R.G. (1963). Mechanics of instrumentation. *Proc. I. Mech. E., 177*, 269.
A definitive analysis of the mechanical design of precision instruments.

Related case studies

6.7 "Materials for springs"
10.7 "Shapes that flex: Leaf and strand structures"

10.9 SUMMARY AND CONCLUSIONS

In designing components that are loaded such that they bend, twist, or buckle, the designer has two groups of variables with which to optimize performance: *material* and *section shape*. The best choice of material depends on the shapes in which it is available or to which it could potentially be formed. The procedure of Chapter 9 gives a method for optimizing the coupled choice of material and shape.

The procedure is illustrated in this chapter. Often the designer has available certain stock materials in certain shapes. If so, the one with the greatest value of the appropriate material index (of which a number were listed in Table 10.1) maximizes performance. Sometimes sections can be specially designed; then material properties and design loads determine a maximum practical value for the shape factor above which local buckling leads to failure; again, the procedure gives an optimal choice of material and shape.

Further gains in efficiency are possible by combining microscopic with macroscopic shape—something we return to later.

CHAPTER 11

Designing Hybrid Materials

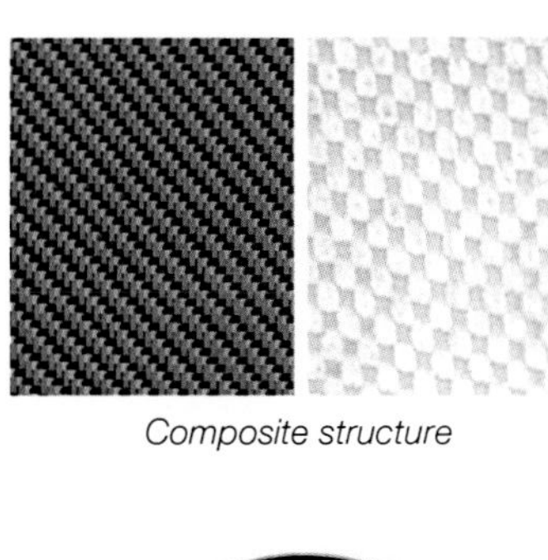

Composite structure

Sandwich structure

Segmented structure

Cellular structure

Hybrid materials.

Materials Selection in Mechanical Design. DOI: 10.1016/B978-1-85617-663-7.00011-4

CONTENTS

11.1 INTRODUCTION AND SYNOPSIS

Why do horse breeders cross a horse with a donkey, delivering a mule? Why do farmers prefer hybrid corn to the natural strain? Mules, after all, are best known for their stubbornness, and—like hybrid corn—they cannot reproduce, so you have to start again for each generation. So—why? Because, although they have some attributes that are not as good as those of their forbears, they have others—hardiness, strength, resistance to disease—that are better. The botanical phrase "hybrid vigor" sums it up.

So let us explore the idea of hybrid materials—combinations of two or more materials assembled in such a way as to have attributes not offered by either one alone (Figure 11.1, central circle). Like the mule, we may find that some attributes are less good (the cost, for example), but if the ones we want are better, something is achieved. Particulate and fibrous composites are examples of one type of hybrid, but there are many others: sandwich structures, lattice structures, segmented structures, and more. Here we explore ways of designing hybrid materials, emphasizing the choice of components, their configuration, their relative volume fraction, and their scale (Table 11.1). The new variables expand design space, allowing the creation of new "materials" with specific property profiles.

And that highlights one of the challenges. How are we to compare a hybrid like a sandwich with monolithic materials like, say, polycarbonate or titanium? To do this we must think of the sandwich not only as a hybrid

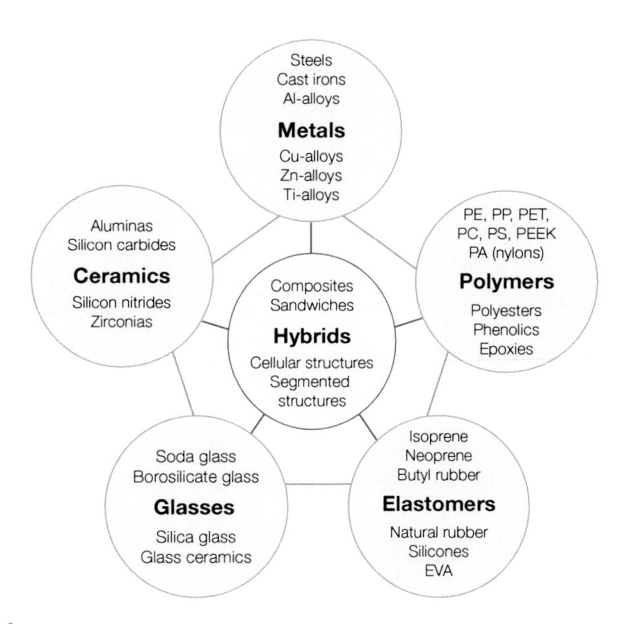

FIGURE 11.1
Hybrid materials combine the properties of two (or more) monolithic materials or of one material and space. They include fibrous and particulate composites, foams and lattices, sandwiches, and almost all natural materials.

Table 11.1 Ingredients of Hybrid Design

Components	The choice of materials to be combined
Configuration	The shape and connectivity of the components
Relative volumes	The volume fraction of each component
Scale	The length scale of the structural unit

with faces on one material bonded to a core of another, but as a "material" in its own right, with its own set of effective properties; it is these that allow the comparison.

The approach adopted here is one of breadth rather than precision. The aim is to assemble methods to allow the properties of alternative hybrids to be scanned and compared, seeking those that best meet a given set of design requirements. Once materials and configuration have been chosen, standard methods such as optimization routines and finite element analyses can be used to refine them. But what the standard methods are *not* good at is the quick scan of alternative combinations. That is where the approximate methods developed below, in which material and configuration become the variables, pay off.

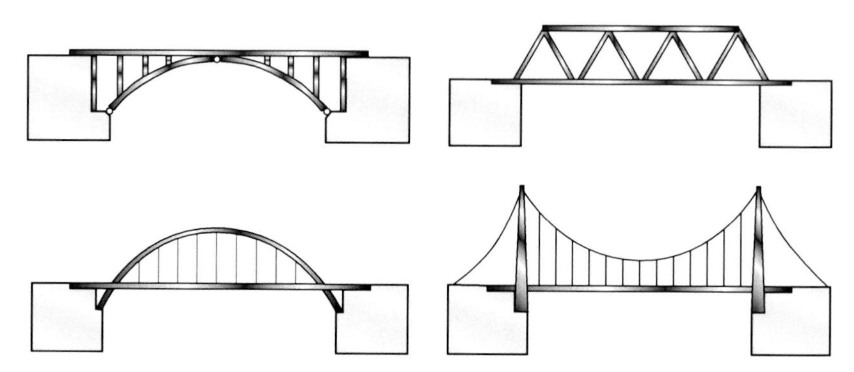

FIGURE 11.2
Four configurations for a bridge. The design variables describing the performance of each differ. Optimization of performance becomes possible only when a configuration has been chosen.

The word "configuration" requires elaboration. Figure 11.2 shows four different configurations of bridge. In the first all members are loaded in compression. In the second, members carry both tension and compression, depending on how the bridge is loaded. In the third and fourth, the suspension cables are loaded purely in tension. Any one of these can be optimized, but no amount of optimization will cause one to evolve into another because this involves a discrete jump in configuration, each characterized by its own set of variables.[1]

Hybrid design has the same feature: The classes of hybrid are distinguished by their configuration. Here we focus on four classes, each with a number of discrete members. The images on the first page of this chapter suggest what they look like. To avoid a mouthful of words every time we refer to one, we use the shorthand: *composite, sandwich, cellular*, and *segmented structures*. *Composites* combine two solid components, one (the reinforcement) as fibers or particles, contained in the other (the matrix). Their properties are some average of those of the components, and, on a scale large compared to that of the reinforcement, they behave as if they were homogeneous materials. *Sandwiches* have outer faces of one material supported by a core of another, usually a low-density material—a configuration that gives a flexural stiffness per unit weight that is greater than that offered by either component alone.

---

[1] Numerical tools are emerging that allow a degree of *topological optimization*, that is, the development of a configuration. They work like this. Start with an envelope—a set of boundaries—and fill it with a homogenous "material" with a relative density initially set at 0.5, with properties that depend linearly on relative density. Impose constraints, meaning the mechanical, thermal, and other loads the structure must support, give it a criterion of excellence, and let it condense out into regions of relative density 1 and regions where it is 0 retaining only changes that increase the measure of excellence. The method is computationally intensive but has had some success in suggesting configurations that use a material efficiently (see "Further reading" for more information).

Cellular structures are combinations of material and space (which can, of course, be filled with another material). We distinguish two types: In the first, the low connectivity of the struts allows them to bend when the lattice is loaded; in the other, the higher connectivity suppresses bending, forcing the struts to stretch. *Segmented structures* are materials subdivided in one, two, or three dimensions; the subdivisions both lower the stiffness and, by dividing the material into discrete units, impart damage tolerance.

The approach we adopt is to use bounding methods to estimate the properties of each configuration. With these, the properties of a given pair of materials in a given configuration can be calculated. These can then be plotted on material selection charts, which become tools for selecting both configuration and material.

11.2 HOLES IN MATERIAL-PROPERTY SPACE

All the charts of Chapter 4 have one thing in common: Parts of them are populated with materials and parts are not—there are *holes* (Figure 11.3). Some parts are inaccessible for fundamental reasons that relate to the size of atoms and the nature of the forces that bind them together. But other parts are empty even though, in principle, they could be filled.

Is anything to be gained by developing materials (or material combinations) that lie in these holes? The material indices show where this is profitable. A grid of lines of one index—E/ρ—is plotted in Figure 11.3. If the filled areas can be expanded in the direction of the arrow (i.e., to greater values of E/ρ) it will enable lighter, stiffer structures to be made. The arrow lies normal to the index lines. It defines a *vector for material development*.

One approach to filling holes—the long-established one—is that of developing new metal alloys, new polymer chemistries, and new compositions of glass and ceramic so as to extend the populated areas of the property charts; Figure 1.2 illustrated how this has evolved over time. But developing new materials can be an expensive and uncertain process, and the gains tend to be incremental rather than step-like. An alternative is to combine two or more existing materials so as to allow a superposition of their properties—in short, to create hybrids. The spectacular success of carbon and glass-fiber–reinforced composites at one extreme, and of foamed materials at another (hybrids of material and space) in filling previously empty areas of the property charts is encouragement enough to explore ways in which such hybrids can be designed.

What might we hope to achieve? Figure 11.4 shows two materials, M_1 and M_2, plotted on a chart with properties P_1 and P_2 as axes. The figure shows four scenarios, each typical of a certain class of hybrid. Depending on the

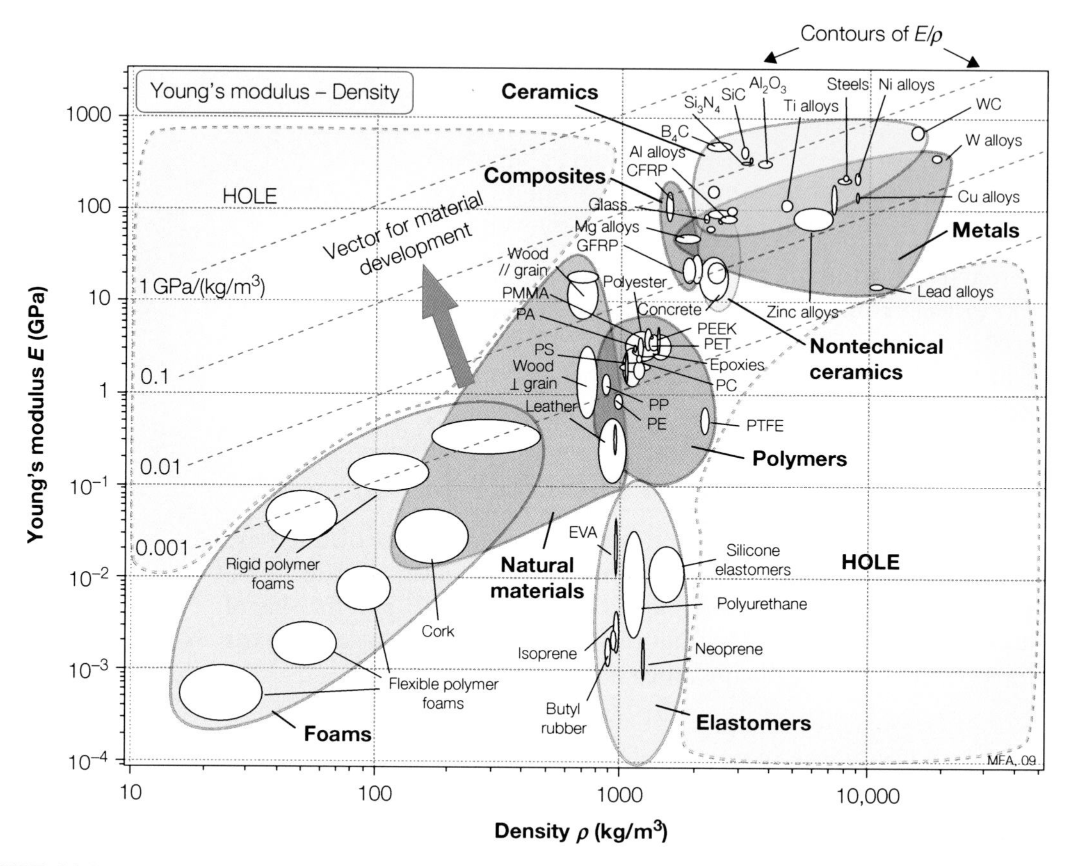

FIGURE 11.3

Holes in modulus-density space, with contours of specific modulus E/ρ. Material development that extended the occupied territory in the direction of the arrow (the "vector for development") allows components with greater stiffness to weight than any current material allows.

configuration of the materials and the way they are combined, we may find any one of the following.

- *"The best of both" scenario.* The ideal is often the creation of a hybrid with the best properties of both components. There are examples, most commonly when a bulk property of one material is combined with the surface properties of another. Zinc-coated steel has the strength and toughness of steel with the corrosion resistance of zinc. Glazed pottery exploits the formability and low cost of clay with the impermeability and durability of glass.
- *"The rule of mixtures" scenario.* When bulk properties are combined in a hybrid, as in structural composites, the best that can be obtained is often the arithmetic average of the properties of the components, weighted by

their volume fractions. Thus unidirectional fiber composites have an axial modulus (the one parallel to the fibers) that lies close to the rule of mixtures.

- *"The weaker link dominates" scenario.* Sometimes we have to live with a lesser compromise, typified by the stiffness of particulate composites, in which the hybrid properties fall below those of a rule of mixtures, lying closer to the harmonic than the arithmetic mean of the properties. Although the gains are less spectacular, they are still useful.
- *"The least of both" or weakest-link scenario.* Sometimes it is not the greatest but the least of the properties that we seek. Fire sprinkler systems, for example, use a wax-metal hybrid designed to fail, releasing the spray when the melting point of the lower-melting material (the wax) is exceeded.

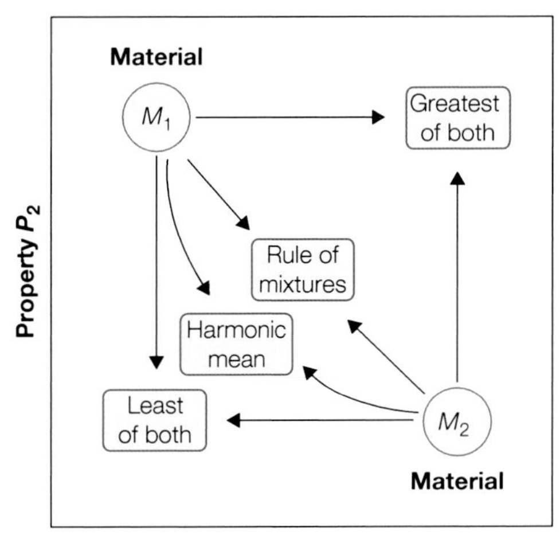

FIGURE 11.4
The possibilities of hybridization. The properties of the hybrid reflect those of its component materials, combined in one of several possible ways.

These set certain fixed points, but the list is not exhaustive. Other combinations are possible. These will emerge in the following.

When is a hybrid a "material"? There is a certain duality in how hybrids are considered and discussed. Some, like filled polymers, composites, or wood are treated as materials in their own right, each characterized by its own set of material properties. Others, such as galvanized steel, are seen as one material (steel) to which a coating of a second (zinc) has been applied, even though it could be regarded as a new material with the strength of steel but the surface properties of zinc ("stinc," perhaps?). Sandwich panels illustrate the duality, sometimes viewed as two sheets of face material separated by a core material, and sometimes—to allow comparison with bulk materials—as "materials" with their own density, flexural stiffness, and strength. To call any one of these a "material" and to characterize it as such is a useful shorthand, allowing designers to use existing methods when designing with them. But if we are to design the hybrid itself, we must deconstruct it and think of it as a combination of materials (or of material and space) in a chosen configuration.

11.3 THE METHOD: "A + B + CONFIGURATION + SCALE"

First, a working definition: A hybrid material is a combination of two or more materials in a predetermined configuration, relative volume, and scale, optimally serving a specific engineering purpose," which we paraphrase as "A + B + configuration + scale." Here we allow for the widest possible choice of A and

B, including the possibility that one of them is a gas or simply space. These new design variables expand design space, allowing an optimization of properties that is not possible if the choice is limited to single, monolithic materials.

The basic idea, illustrated in Figure 11.5, is this. Monolithic materials offer a certain portfolio of properties on which much engineering design is based. Design requirements isolate a sector of material-property space. If this sector contains materials, the requirements can be met by choosing one of them. But if the design requirements are sufficiently demanding, no single material may be found that can meet them all: The requirements lie in a hole in property space. Then the way forward is to identify and separate the conflicting requirements, seeking optimal material solutions for each, and then combine them in ways that retain the desirable attributes of both. The best choice is the one that ranks most highly when measured by the performance metrics that motivate the design: minimizing mass or cost or maximizing some aspect of performance (criterion of excellence). The alternative combinations are examined and assessed, using the criterion of excellence to rank them. The output is a specification of a hybrid in terms of its component materials and their configurations.

The four classes of hybrid structures—composites, sandwiches, cellular, and segmented—are explored in the next four sections, applying the method.

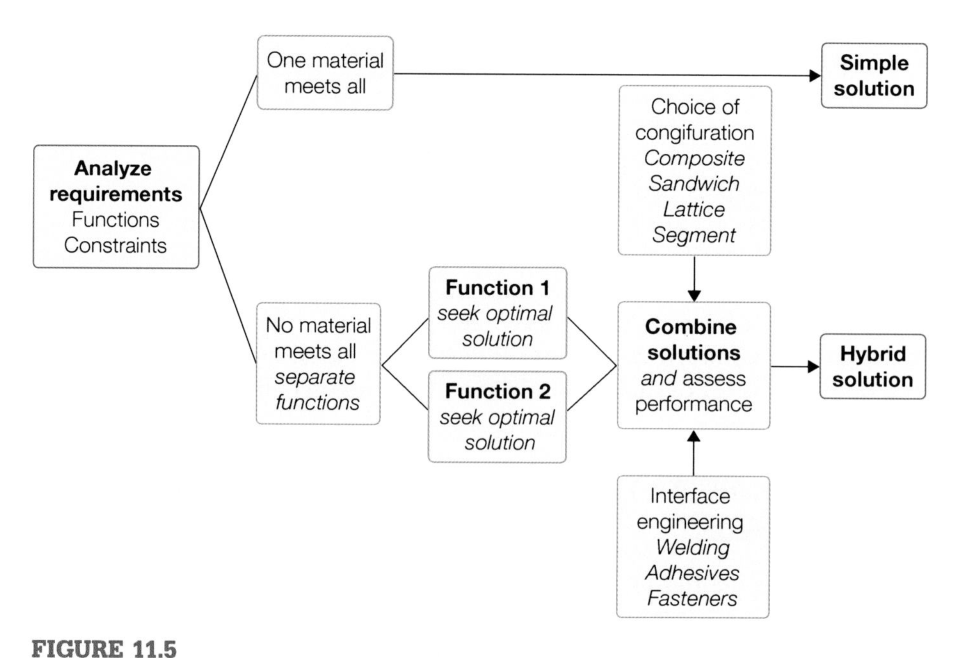

FIGURE 11.5
The steps in designing a hybrid to meet given design requirements.

11.4 COMPOSITES

Aircraft engineers, automobile makers, and designers of sports equipment all have one thing in common: They all want materials that are stiff, strong, tough, and light. The single-material choices that best achieve this are the *light alloys*: alloys that are based on magnesium, aluminum, and titanium. Much research aims at improving their properties. But they are not all that light—polymers have much lower densities. Nor are they all that stiff—ceramics are much stiffer and, particularly in the form of small particles or thin fibers, much stronger. These facts are exploited in the family of structural hybrids that we usually refer to as *particulate and fibrous composites*.

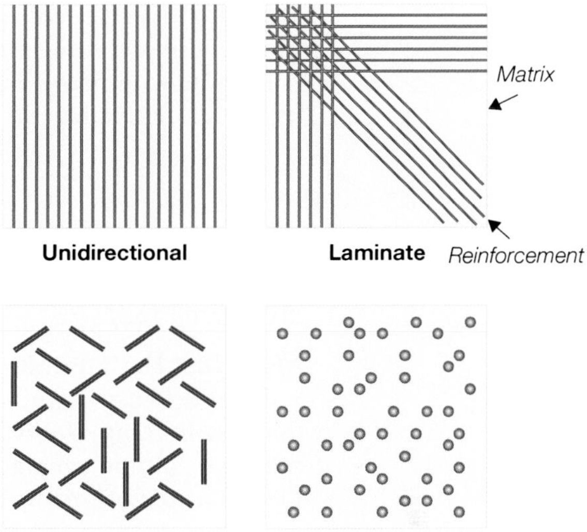

Unidirectional Laminate Matrix Reinforcement Chopped fiber Particulate

FIGURE 11.6
Schematic of hybrids of the composite type: unidirectional fibrous, laminated fiber, chopped fiber, and particulate composites. Bounds and limits, described in the text, bracket the properties of all of these.

Any two materials can, in principle, be combined to make a composite, and they can be mixed in many geometries (Figure 11.6). In this section we restrict the discussion to *fully dense, strongly bonded composites* such that there is no tendency for the components to separate at their interfaces when the composite is loaded, and to those in which the scale of the reinforcement is large compared to that of the atom or molecule size and the dislocation spacing, allowing the use of continuum methods.

On a macroscopic scale—one that is large compared to that of the components—a composite behaves like a homogeneous solid with its own set of mechanical, thermal, and electrical properties. Calculating these precisely can be done, but is difficult. It is much easier to bracket them by *bounds* or *limits*: upper and lower values between which the properties lie. The term "bound" will be used to describe a rigorous boundary, one that the value of the property *cannot*—subject to certain assumptions—exceed or fall below. It is not always possible to derive bounds; the best that can be done is to derive "limits" outside which it is *unlikely* that the value of the property lies. The important point is that the bounds orlimits bracket the properties of *all* the configurations of matrix and reinforcement shown in Figure 11.6; by using them we escape from the need to model individual geometries.

Criteria of excellence We need criteria of excellence to assess any given hybrid's merit. These are provided by the material indices of Chapter 5. If a possible hybrid has a value of any one of these that exceeds those of existing materials, it achieves our goal.

Density When a volume fraction f of a reinforcement r (density ρ_r) is mixed with a volume fraction $(1-f)$ of a matrix m (density ρ_m) to form a composite with no residual porosity, the composite density $\widetilde{\rho}$ is given exactly by a rule of mixtures (an arithmetic mean weighted by volume fraction):

$$\widetilde{\rho} = f\rho_r + (1-f)\rho_m. \tag{11.1}$$

The geometry or shape of the reinforcement does not matter except in determining the maximum packing fraction of reinforcement and thus the upper limit for f.

Modulus The modulus of a composite is bracketed by the well-known Voigt and Reuss bounds. The upper bound, $\widetilde{E}_u$, is obtained by postulating that, on loading, the two components suffer the same strain; the stress is then the volume average of the local stresses, and the composite modulus follows a rule of mixtures:

$$\widetilde{E}_u = f E_r + (1-f)E_m. \tag{11.2a}$$

Here E_r is the Young's modulus of the reinforcement and E_m that of the matrix. The lower bound, $\widetilde{E}_L$, is found by postulating instead that the two components carry the same stress; the strain is the volume average of the local strains, and the composite modulus is

$$\widetilde{E}_L = \frac{E_m E_r}{f E_m + (1-f)E_r} \tag{11.2b}$$

More precise bounds are possible, but the simple ones are adequate to illustrate the method.

To see how the bounds are used, consider the following example.

Composite design for stiffness at minimum mass

A beam is at present made of an aluminum alloy. Beryllium is both lighter and less dense than aluminum; the ceramic alumina (Al_2O_3) is also stiffer but denser. Can hybrids of aluminum with either of these offer improved performance, measured by the criterion of excellence $E^{1/2}/\rho$ derived in Chapter 5?

The three materials are plotted on a segment of the $E-\rho$ property space in Figure 11.7. Composites made by mixing them have densities given exactly by Equation (11.1) and moduli that are bracketed by the bounds of Equations (11.2a) and (11.2b). Both of these moduli depend on the volume fraction of reinforcement and, through this, on the density. Upper and lower bounds for the modulus-density relationship can thus be plotted on the $E-\rho$ chart using volume fraction f as a parameter, as shown in Figure 11.7. Any composite made by combining aluminum

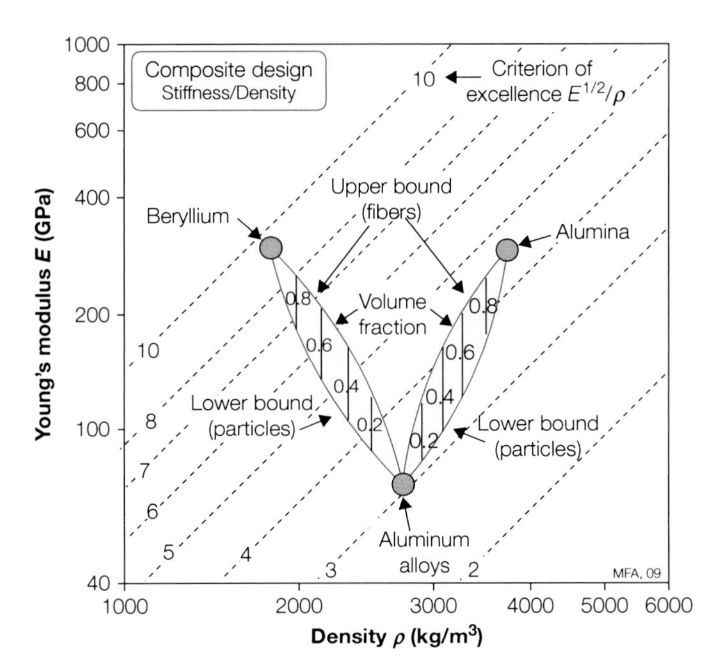

FIGURE 11.7
Part of the $E - \rho$ property chart, showing aluminum alloys, beryllium, and alumina (Al_2O_3). Bounds for the moduli of hybrids made by mixing them are shown. The diagonal contours plot the criterion of excellence $E^{1/2}/\rho$.

with alumina will have a modulus contained in the envelope for Al–Al_2O_3; the same for Al-Be. Fibrous reinforcement gives a modulus in a direction parallel to the fibers near the upper bound; particulate reinforcement or transversely loaded fibers give moduli near the lower one.

The criterion of excellence, $E^{1/2}/\rho$, is plotted as a grid in the Figure 11.7. The bound-envelope for Al–Be composites extends almost normal to the grid, while that for Al–Al_2O_3 lies at a shallow angle to it. Beryllium fibers improve performance (as measured by $E^{1/2}/\rho$) roughly four times as much as alumina fibers do, for the same volume fraction. The difference for particulate reinforcement is even more dramatic. The lower bound for Al-Be lies normal to the contours: 30% of particulate beryllium increases $E^{1/2}/\rho$ by a factor of 1.5. The lower bound for Al–Al_2O_3 is, initially, parallel to the $E^{1/2}/\rho$ grid: 30% of particulate Al_2O_3 gives almost no gain. The underlying reason is that both beryllium and Al_2O_3 increase the modulus, but only beryllium decreases the density; the criterion of excellence is more sensitive to density than to modulus.

The commercial alloy AlBeMet (62% Be, 38% Al) exploits this idea. The two metals are mutually insoluble, creating a 2-phase composite of Al and Be with $E^{1/2}/\rho = 6.5$ compared with 3.1 for Al alone.

Strength Of all the bounds and limits described in this chapter, those for strength are the least satisfactory. The nonlinearity of the problem, the

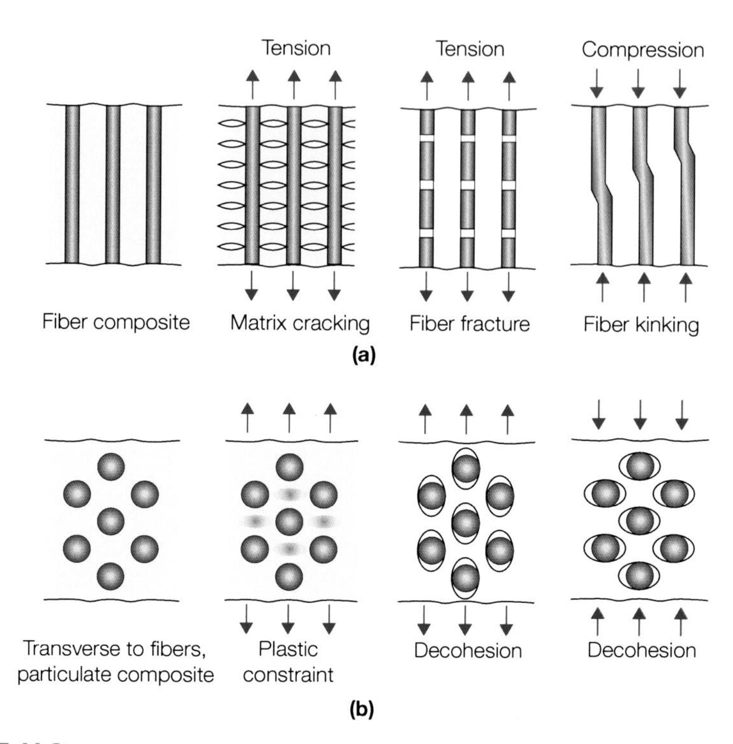

FIGURE 11.8
Failure modes in composites.

multitude of failure mechanisms[2] and the sensitivity of strength and toughness against impurities and processing defects makes accurate modeling difficult. The literature contains many calculations for special cases: reinforcement by unidirectional fibers or by a dilute dispersion of spheres. We wish to avoid models that require detailed knowledge of how a particular architecture behaves, and we seek less restrictive limits.

As the load on a continuous fiber composite is increased, it is redistributed between the components until one suffers general yield or fracture (Figure 11.8(a)). Beyond this point the composite has suffered permanent deformation or damage but can still carry load; final failure requires yielding or fracture of both. The composite is strongest if both reach their failure state simultaneously. Thus the upper bound for a continuous fiber ply like that labeled "Unidirectional" in Figure 11.6 loaded parallel to the fibers (the axial strength in tension, subscript a) is a rule of mixtures:

$$(\tilde{\sigma}_f)_{u,a} = f(\sigma_f)_r + (1-f)(\sigma_f)_m \tag{11.3a}$$

[2] The treatment here is the simplest that allows the method to be demonstrated. The full complexity of the modeling of composite failure is documented in the texts in "Further reading."

where $(\sigma_f)_m$ is the strength of the matrix and $(\sigma_f)_r$ is that of the reinforcement. If one fails before the other, the load is carried by the survivor. Thus a lower bound for strength in tension is given by

$$(\tilde{\sigma}_f)_{L,a} = \text{Greater of } (f(\sigma_f)_r, (1-f)(\sigma_f)_m) \tag{11.3b}$$

Creating anisotropy

The elastic and plastic properties of bulk monolithic solids are frequently anisotropic, but weakly so—the properties do not depend strongly on direction. Hybridization gives a way of creating controlled anisotropy, and it can be large. We have already seen an example in Figure 11.7, which shows the upper and lower bounds for the moduli of composites. The longitudinal properties of unidirectional long-fiber composites lie near the upper bound; the transverse properties, near the lower one. The vertical width of the band between them measures the anisotropy. A unidirectional continuous-fiber composite has a maximum anisotropy ratio R_a given by the ratio of the bounds—in this example

$$R_a = \frac{\tilde{E}_u}{\tilde{E}_L} = f^2 + (1-f)^2 + f(1-f)\left(\frac{E_m}{E_r} + \frac{E_r}{E_m}\right)$$

In Figure 11.7 the maximum R_a is only 1.5. A more dramatic example involving thermal properties is given in Chapter 12.

Determining the transverse strength (Figure 11.8(b)) is more difficult. It depends on interface bond strength, fiber distribution, stress concentrations, and voids. In general the transverse strength is less than that of the unreinforced matrix, and the strain to failure, too, is less. In a continuous ductile matrix containing strongly bonded, nondeforming, particles or fibers, the flow in the matrix is constrained. The constraint increases the stress required for flow in the matrix, giving an upper-limiting tensile strength of

$$\text{Lesser of} \begin{cases} (\tilde{\sigma}_f)_{u,t} \approx (\sigma_f)_m \left(\dfrac{1}{1-f^{1/2}}\right) \\ (\tilde{\sigma}_f)_{u,t} \approx (\sigma_f)_r \end{cases} \tag{11.4a}$$

More usually the transverse strength is lower than that of the matrix alone because of stress concentration and debonding at the fiber-matrix interface. Hull gives the approximate lower limit for tensile strength as

$$(\tilde{\sigma}_f)_{L,t} \approx (\sigma_f)_m (1-f^{1/2}) \tag{11.4b}$$

The two pairs of limits allow the potential of a given choice of reinforcement and matrix to be explored. Figure 11.9 shows the limits for axial and transverse strength of an epoxy-glass composite ply.

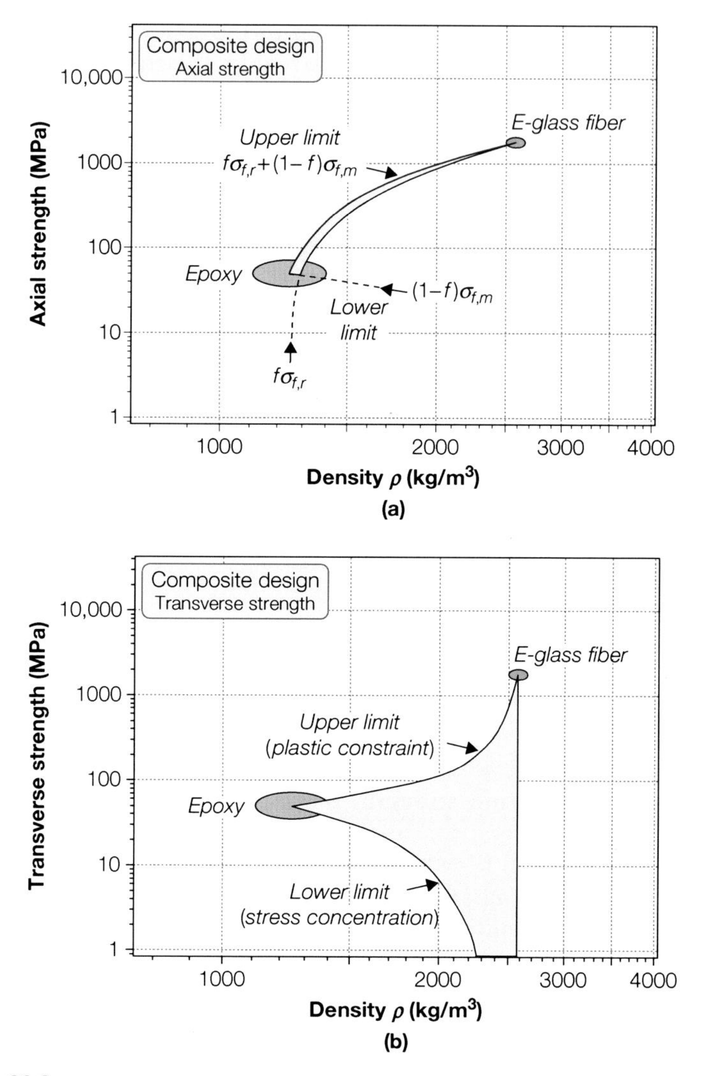

FIGURE 11.9

The limits for axial (a) and transverse (b) strength of a composite ply.

Continuous fiber composites can fail in compression by fiber kinking (Figure 11.8(a), extreme right). The kinking is resisted by the shear strength of the matrix, approximately $(\sigma_f)_m/2$. This leads an axial compressive stress for buckling of fibers of

$$(\tilde{\sigma}_c)_{u,a} = \frac{1}{\vartheta}\frac{(\sigma_f)_m}{2} \approx 14(\sigma_f)_m \tag{11.5}$$

Here ϑ is the initial misalignment of the fibers from the axis of compression, in radians. Experiments show that a typical value, in carefully aligned composites, is $\vartheta \approx 0.035$, giving the final value shown on the right of the equation. We identify the upper bound with the lesser of this and Equation (11.3a). When, instead, misalignment is severe, meaning $\vartheta \approx 1$, the strength falls greatly. We identify the lower bound for compressive failure with that of the matrix, $(\sigma_f)_m$.

Specific heat The specific heats of solids at constant pressure, C_p, are almost the same as those at constant volume, C_v. If they were identical, the heat capacity per unit volume of a composite would, like the density, be given exactly by a rule-of-mixtures:

$$\widetilde{\rho}\,\widetilde{C}_p = f\,\rho_r(C_p)_r + (1-f)\rho_m(Cp)m \tag{11.6}$$

where $(C_p)_r$ is the specific heat of the reinforcement and $(C_p)_m$ is that of the matrix (the densities enter because the units of C_p are J/kg.K). A slight difference appears because thermal expansion generates a misfit between the components when the composite is heated; the misfit creates local pressures on the components and thus changes the specific heat. The effect is very small and need not concern us further.

Thermal expansion coefficient The thermal expansion of a composite can, in some directions, be greater than that of either component; in others, less. This is because an elastic constant—Poisson's ratio—couples the principal elastic strains; if the matrix is prevented from expanding in one direction (by embedded fibers, for instance), then it expands more in the transverse directions. For simplicity we shall use the approximate lower bound:

$$\widetilde{\alpha}_L = \frac{E_r\alpha_r f + E_m\alpha_m(1-f)}{E_r f + E_m(1-f)} \tag{11.7}$$

(it reduces to the rule of mixtures when the moduli are the same) and the upper bound:

$$\widetilde{\alpha}_u = f\alpha_r(1+v_r) + (1-f)\alpha_m(1+v_m) - \alpha_L[fv_r + (1-f)v_m] \tag{11.8}$$

where α_r and α_m are the two expansion coefficients and v_r and v_m are the Poisson's ratios.

Thermal conductivity Thermal conductivity determines heat flow at a steady rate. A composite of two materials, bonded to give good thermal contact, has a thermal conductivity λ that lies between those of the individual components, λ_m and λ_r. Not surprisingly, a composite containing parallel continuous fibers has a conductivity, parallel to the fibers, given by a rule of mixtures:

$$\widetilde{\lambda}_u = f\lambda_r + (1-f)\lambda_m \tag{11.9}$$

This is an upper bound: In any other direction the conductivity is lower. The transverse conductivity of a parallel-fiber composite (again assuming good bonding and thermal contact) lies near the lower bound first derived by Maxwell:

$$\tilde{\lambda}_L = \lambda_m \left(\frac{\lambda_r + 2\lambda_m - 2f(\lambda_m - \lambda_r)}{\lambda_r + 2\lambda_m + f(\lambda_m - \lambda_r)} \right) \tag{11.10}$$

Particulate composites, too, have a conductivity near this lower bound. Poor interface conductivity can make λ drop below it. Debonding or an interfacial layer between reinforcement and matrix can cause this; so, too, can a large difference of modulus between reinforcement and matrix (because this reflects phonons, creating an interface impedance) or a structural scale that is shorter than the phonon wavelengths.

Thermal diffusivity The thermal diffusivity

$$a = \frac{\lambda}{\rho C_p}$$

determines heat flow when conditions are transient, that is, when the temperature field changes with time. It is formed from three of the earlier properties: λ, ρ, and C_p. The second and third of these are given exactly by Equations (11.1) and (11.6), allowing the diffusivity to be expressed as

$$\tilde{a} = \frac{\tilde{\lambda}}{f\rho_r (C_p)_r + (1-f)\rho_m (C_p)_m} \tag{11.11}$$

Its upper and lower bounds are found by substituting those for $\tilde{\lambda}$ (Equations (11.9) and (11.10)) into this equation.

Composite design for controlled thermal response

Thermo-mechanical design involves the specific heat, C_p, the thermal expansion, α, the conductivity, λ, and the diffusivity, a. These composite properties are bounded by Equations (11.6) through (11.11). They are involved in a number of indices. One is the criterion for minimizing thermal distortion derived in Section 6.16: maximizing the index λ/α.

Figure 11.10 shows a small part of the $\alpha - \lambda$ materials selection chart, with a grid showing the criterion of excellence, λ/α, superimposed on it. Three materials are shown: aluminum, boron nitride (BN), and silicon carbide (SiC). The thermal properties of Al–BN and Al–SiC composites are bracketed by envelopes calculated from the bounding equations. (Both α and λ have upper and lower bounds so there are four possible combinations for each material pair. Those shown in the figure are the outermost pair of the four.) The plot reveals immediately that SiC reinforcement in aluminum increases performance (as measured by λ/α); reinforcement with BN decreases it.

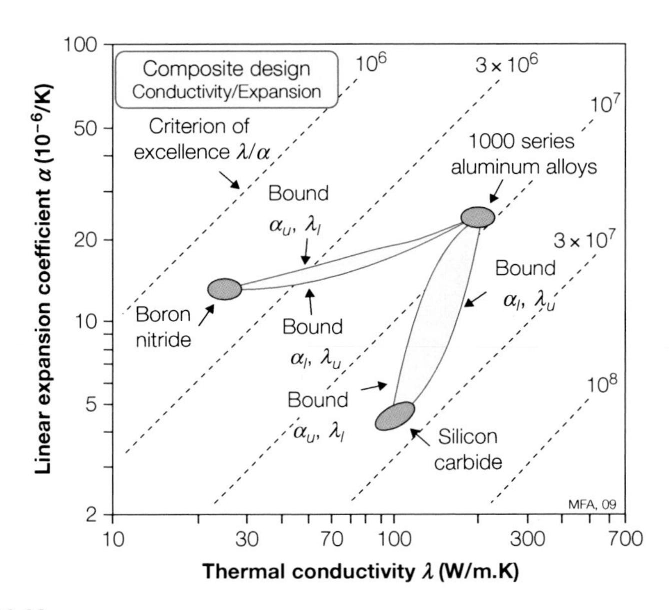

FIGURE 11.10

A part of expansion-coefficient/conductivity space showing aluminum alloys, boron nitride, and silicon carbide. The properties of Al-BN and Al-SiC composites are bracketed by the bounds of Equations (11.7)–(11.10). The Al-SiC composites enhance performance; the Al-BN composites reduce it.

Dielectric constant The dielectric constant $\tilde{\varepsilon}_d$ is given by a rule of mixtures:

$$\tilde{\varepsilon}_d = f\varepsilon_{d,r} + (1-f)\varepsilon_{d,m} \tag{11.12}$$

where $\varepsilon_{d,r}$ is the dielectric constant of the reinforcement and $\varepsilon_{d,m}$ that of the matrix.

Electrical conductivity and percolation When the electrical conductivities κ of the components of a composite are of comparable magnitude, bounds for the electrical conductivity are given by those for thermal conductivity with λ replaced by κ. When, instead, they differ by many orders of magnitude (a metallic powder dispersed in an insulating polymer, for instance), questions of *percolation* arise. Percolation is discussed in Section 12.3.

Filling property space with composites We end this section with two figures illustrating how the development of composites has filled out the occupancy of material-property space. The first, Figure 11.11, is a modulus-density $(E-\rho)$ section. The areas filled by unreinforced metals and polymers appear as pale red and pale blue envelopes; members are labeled in gray (they are taken from the original $E-\rho$ chart in Figure 4.3). Polymer matrix composites occupy the bolder purple zone; metal matrix composites, the

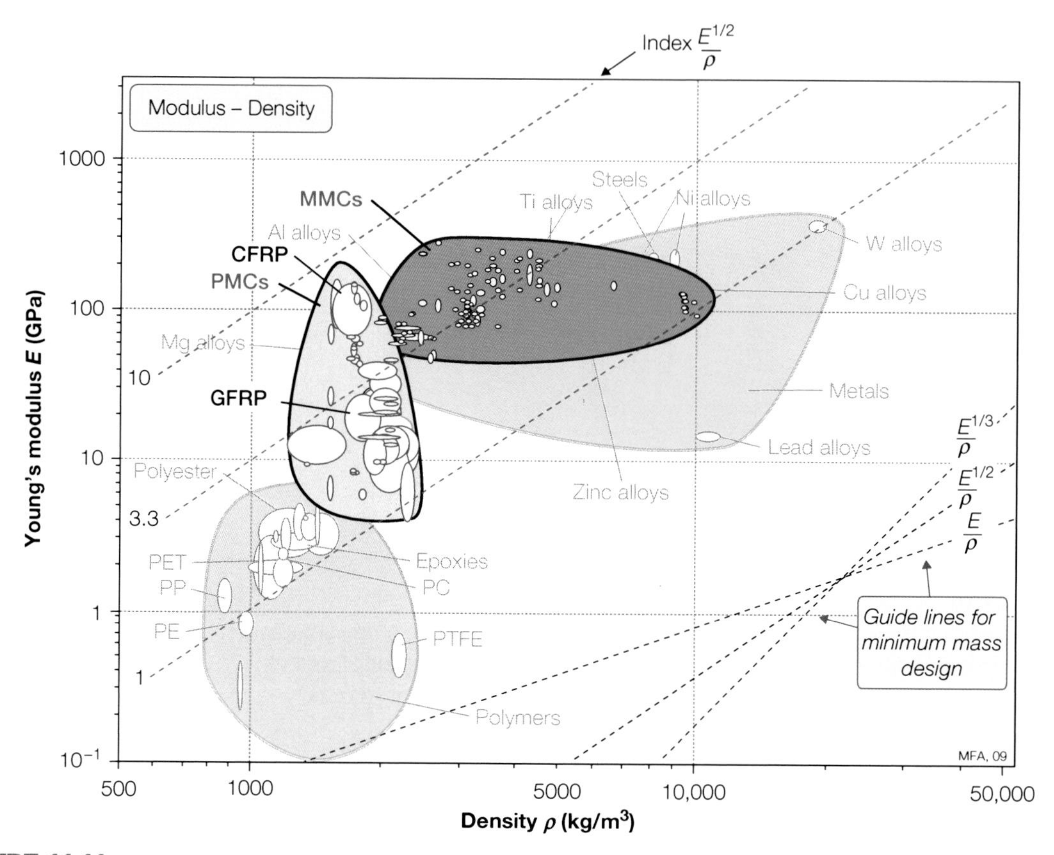

FIGURE 11.11
Polymer (PMC) and metal (MMC) matrix composites expand the occupied area of modulus-density space. (Each of the small bubbles in the bold envelopes labeled PMCs and MMCs describe a composite. Data from the CES Edu '09 database.)

darker red one. Both extend into areas that were previously empty. Using any one of the indices for light, stiff structures (E/ρ, $E^{1/2}/\rho$ and $E^{1/3}/\rho$) as a criterion of excellence, we find that composites offer performance that was not previously attainable.

Figure 11.12 paints a similar picture for the strength-density ($\sigma_y - \rho$) section. The color coding is the same as that in Figure 11.11. Composites again expand the populated area in a direction that, using the indices for light strong structures (σ_y/ρ, $\sigma_f^{2/3}/\rho$, and $\sigma_f^{1/2}/\rho$) as a criterion, offers enhanced performance.

11.5 SANDWICH STRUCTURES

A sandwich panel epitomizes the concept of a hybrid. It combines two materials in a specified geometry and scale, configured such that one forms the faces and the other the core to give a structure of high bending stiffness

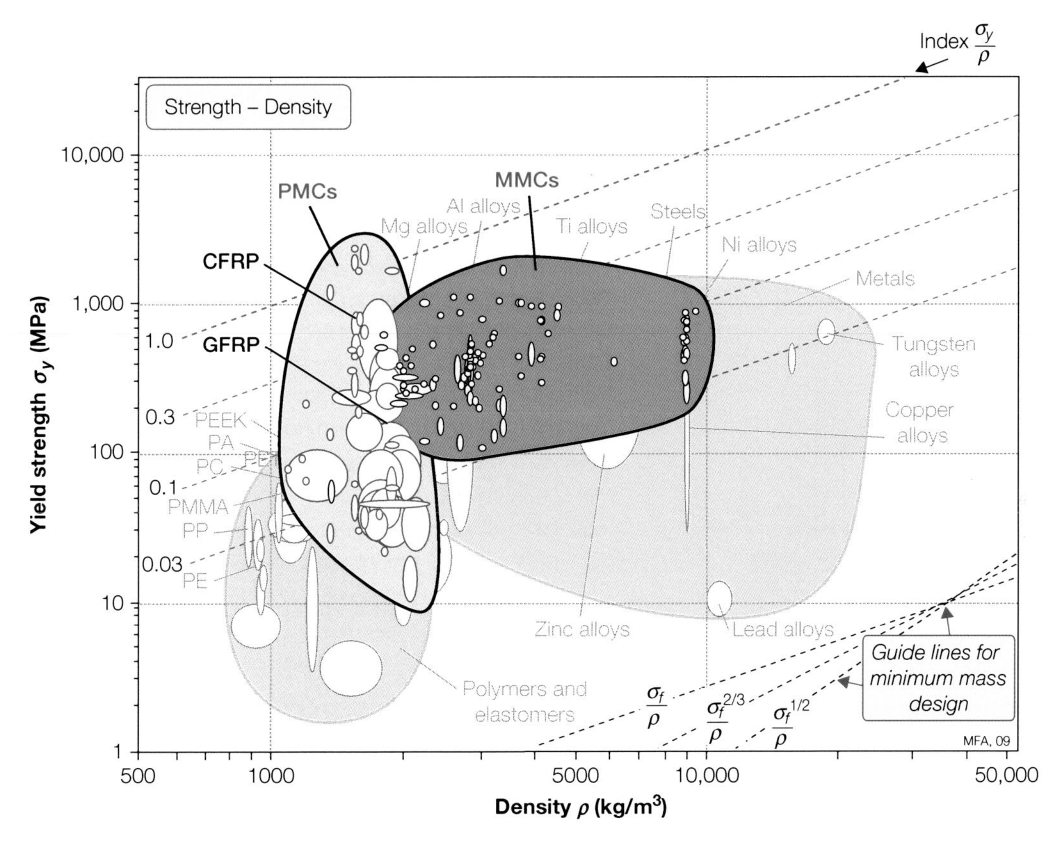

FIGURE 11.12

Polymer (PMC) and metal (MMC) matrix composites also expand the occupied area of strength-density space. (Each of the small bubbles in the bold envelopes labeled PMCs and MMCs describe a composite. Data from the CES Edu '09 database.)

and strength at low weight Figure 11.13). The separation of the faces by the core increases the moment of inertia of the section, I, and its section modulus, Z, producing a structure that resists bending and buckling loads well. Sandwiches are used where weight saving is critical: in aircraft, trains, motor vehicles, portable structures, and sports equipment. Nature, too, makes use of sandwich designs: Sections through the human skull, the wing of a bird, and the stalk and leaves of many plants show a low-density foam-like core separating solid faces.[3]

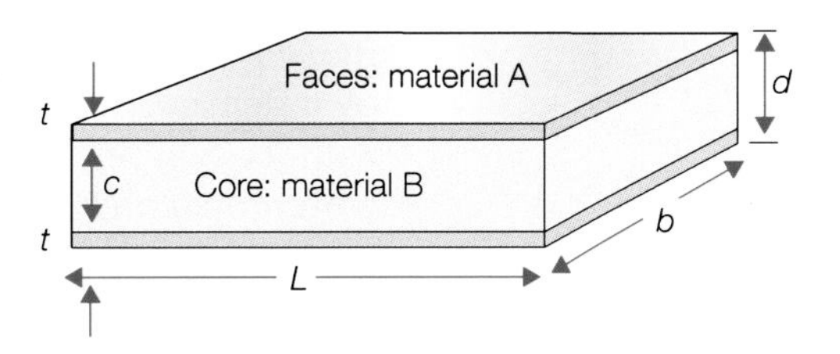

FIGURE 11.13

The sandwich. The face thickness is t, the core thickness is c, and the panel thickness is d.

[3] Allen (1969) and Zenkert (1995) give good introductions to the design of sandwich panels for engineering applications. Gibson et al. (2010) do the same for their use in nature.

Table 11.2 The Symbols

Symbol	Meaning and Usual Units
t, c, d	Face thickness, core thickness, and overall panel thickness (m)
L, b	Panel length and width (m)
m_a	Mass per unit area of panel (kg/m^2)
$f = 2t/d$	Relative volumes occupied by faces
$(1 - f) = c/d$	Relative volume occupied by core
I	Second moment of area (m^4)
ρ_f, ρ_c	Densities of face and core material (kg/m^3)
$\tilde{\rho}$	Equivalent density of panel (kg/m^3)
E_f	Young's modulus of faces (GN/m^2)
E_c, G_c	Young's modulus and shear modulus of core (GN/m^2)
$\tilde{E}_{in\text{-}plane}, \tilde{E}_{flex}$	Equivalent in-plane and flexural modulus of panel (GN/m^2)
σ_f	Yield strength of faces (MN/m^2)
σ_c, τ_c	Yield strength and shear-yield strength of core (MN/m^2)
$\tilde{\sigma}_{in\text{-}plane}$	Equivalent in-plane strength of panel (MN/m^2)
$\tilde{\sigma}_{flex1}, \tilde{\sigma}_{flex2}, \tilde{\sigma}_{flex3}$	Equivalent flexural strength of panel, depending on mechanism of failure (MN/m^2)

The faces, each of thickness t, carry most of the load, so they must be stiff and strong; because they form the exterior surfaces of the panel they must also tolerate the environment in which it operates. The core, of thickness c, occupies most of the volume; it must be light and stiff and strong enough to carry the shear stresses necessary to make the whole panel behave as a load-bearing unit, (if the core is much thicker than the faces, these stresses are small).

A sandwich as a "material" So far we have spoken of the sandwich as a *structure*: faces of material A supported on a core of material B, each with its own density, modulus, and strength. But we can also think of it as a *material* with its own set of properties, and this is useful because it allows comparison with more conventional materials. To do this we calculate *equivalent material properties* for the sandwich, and we identify them, like composites, by a tilde (e.g., $\tilde{\rho}$, $\tilde{E}$). The quantities $\tilde{\rho}$ and $\tilde{E}$ can be plotted on the modulus-density chart, allowing a direct comparison with all the other materials on the chart. All the constructions using material indices apply unchanged. The symbols that appear in this section are defined in Table 11.2.

Finding equivalent properties of structured materials by experiment

Consider a sandwich structure with solid skins separated by a cellular core. The panel has an equivalent density equal to its mass divided by its volume,

m_a/d, where m_a is its mass per unit area and $d = 2t + c$ is its overall thickness. It has a flexural stiffness EI, measured by loading the panel in bending and recording the deflection. We define an equivalent homogeneous material with $\tilde{\rho} = \rho$ and $\widetilde{EI} = EI$, where $\tilde{I} = bd^3/12$ is the second moment of area for a *homogeneous* panel with the same dimensions as the real one. The equivalent density and modulus are then

$$\tilde{\rho} = \frac{m_a}{d} \tag{11.13}$$

$$\tilde{E} = \frac{12\,EI}{b\,d^3} \tag{11.14}$$

Loading the panel to failure allows the failure moment, M_f, to be measured experimentally. It is then possible to define an equivalent flexural strength via $\tilde{Z}\tilde{\sigma}_{flex} = M_f$, where $\tilde{Z} = bd^2/4$ is the (fully plastic) section modulus of the panel. The equivalent flexural strength is then

$$\tilde{\sigma}_{flex} = \frac{4\,M_f}{b\,d^2} \tag{11.15}$$

A brief example will illustrate the method.

Tests on a carbon-aramid sandwich panel used as flooring in Boeing aircraft gave the results in the following table.

Face material	0.25 mm carbon/phenolic
Core material	3.2 mm cell, 147 kg/m^3, aramid honeycomb
Panel weight per unit area, m_a	2.69 kg/m^2
Panel length, L	510 mm
Panel width, b	51 mm
Panel thickness, d	10.0 mm
Flexural stiffness, EI	122 Nm^2
Failure moment, M_f	196 Nm

The equivalent density from Equation (11.13) is

$$\tilde{\rho} = \frac{m_a}{d} = 269\,\text{kg/m}^3$$

The equivalent modulus $\tilde{E}$ from Equation (11.14) is

$$\tilde{E} = \frac{12\,EI}{b d^3} = 28.8\,\text{GPa}$$

The equivalent flexural strength $\tilde{\sigma}_{flex}$ from Equation (11.15) is

$$\tilde{\sigma}_{flex} = \frac{4\,M_f}{b\,d^2} = 154\,\text{MPa}$$

Equivalent properties of sandwich structures by analysis

In this section we develop equations for the stiffness and strength of sandwich panels and express them as properties of an equivalent homogeneous material. The symbols are defined earlier in Table 11.2.

Equivalent density The equivalent density of the sandwich (its mass divided by its volume) is

$$\tilde{\rho} = f\rho_f + (1-f)\rho_c \tag{11.16}$$

Here f is the volume fraction occupied by the faces: $f = 2t/d$.

Mechanical properties Sandwich panels are designed to be stiff and strong during bending. In thinking of the panel as a "material," we must therefore distinguish the in-plane modulus and strength from those in bending. The effective in-plane modulus $\tilde{E}_{in\text{-}plane}$ and strength $\tilde{\sigma}_{in\text{-}plane}$ are given, to an adequate approximation, by the rule of mixtures.

Equivalent flexural modulus Flexural properties are quite different. The flexural compliance (the reciprocal of the stiffness) has two contributions: one from the bending of the panel as a whole, the other from the shear of the core (Figure 11.14). They add. The bending stiffness is

$$EI = \frac{b}{12}(d^3 - c^3)E_f + \frac{bc^3}{12}E_c$$

The shear stiffness is

$$AG = \frac{bd^2}{c}G_c$$

Here the dimensions, d, c, t, and L are identified in Figure 11.13, E_f is Young's modulus of the face sheets, G_c is the shear modulus of the core, and A is the area of its cross-section.

Summing deflections gives

$$\delta = \frac{12PL^3}{B_1 b\{(d^3 - c^3)E_f + c^3E_c\}} + \frac{PLc}{B_2 d^2 bG_c} \tag{11.17}$$

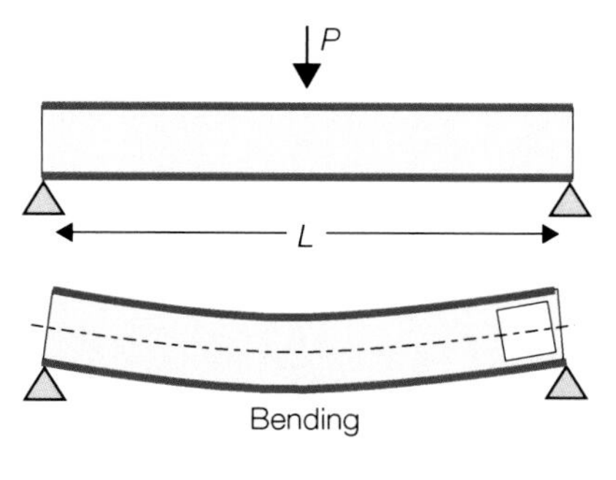

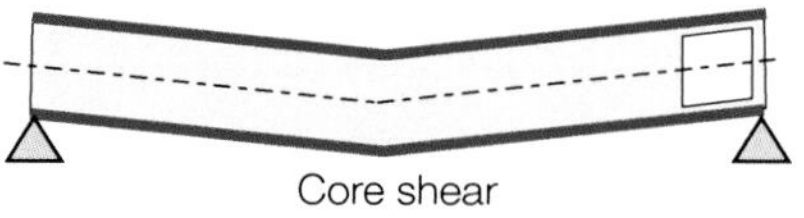

FIGURE 11.14
Sandwich panel flexural stiffness. There are contributions from bending and core shear.

The load configuration determines the values of the constants B_1 and B_2, as

Table 11.3 Constants to Describe Modes of Loading

Mode of Loading	Description	B_1	B_2	B_3	B_4
	Cantilever, end load	3	1	1	1
	Cantilever, uniformly distributed load	8	2	2	1
	Three-point bend, central load	48	4	4	2
	Three-point bend, uniformly distributed load	384/5	8	8	2
	Ends built in, central load	192	4	8	2
	Ends built in, uniformly distributed load	384	8	12	2

summarized in Table 11.3. Comparison with $\delta = \frac{12PL^3}{\tilde{E}d^3b}$ for the "equivalent" material gives

$$\frac{1}{\tilde{E}_{flex}} = \frac{1}{12}\frac{1}{E_f\left\{\left(1-(1-f)^3\right)+\frac{E_c}{E_f}(1-f)^3\right\}} + \frac{B_1}{B_2}\left(\frac{d}{L}\right)^2\frac{(1-f)}{G_c} \tag{11.18}$$

Note that, except for the bending-to-shear balancing term $(d/L)^2$, the equivalent property is scale-independent (as a material property should be); the only variable is the relative thickness of faces and core, f. The bending stiffness (EI) is recovered by forming the $\tilde{E}\tilde{I}$ where $\tilde{I}$ is the second moment of a homogeneous panel ($\tilde{I} = bd^3/12$).

Equivalent flexural strength Sandwich panels can fail in many different ways (Figure 11.15). The failure mechanisms compete, meaning that the one at the lowest load dominates. We calculate an *equivalent flexural strength* for each mode, then seek the lowest.

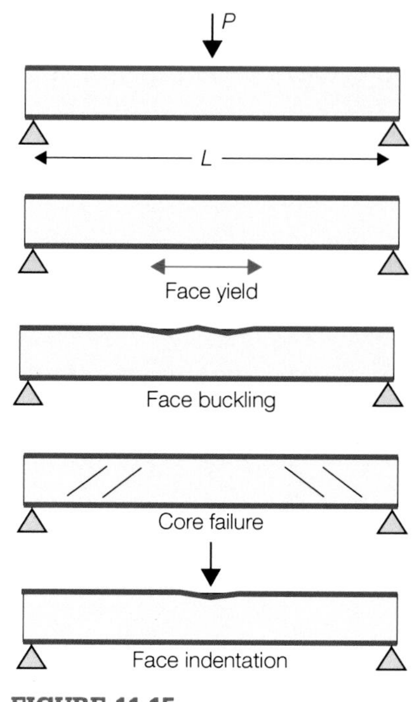

FIGURE 11.15
Failure modes of sandwich panels in flexure.

Face yield The fully plastic moment of the sandwich is

$$M_f = \frac{b}{4}\{(d^2 - c^2)\,\sigma_f + c^2\sigma_c\}$$

Using the fact that $c/d = (1-f)$, Equation (11.15) gives the following equivalent failure strength when face yielding is the dominant failure mode:

$$\tilde{\sigma}_{flex\,1} = \left(1 - (1-f)^2\right)\,\sigma_f + (1-f)^2\sigma_c \tag{11.19}$$

which, again, is independent of scale.

Face buckling In flexure, one face of the sandwich is in compression (Figure 11.16). If it buckles, the sandwich fails. The face stress at which this happens[4] is

$$\sigma_b = 0.57\,(E_f E_c^2)^{1/3} \tag{11.20}$$

Buckling is a problem only when faces are thin and the core offers little support. The failure moment M_f is then well approximated by

$$M_f = 2\,\sigma_b\,t\,b\,c = 1.14\,(E_f E_c^2)^{1/3}\,t\,b\,c$$

Which, via the previous Equation (9.2), gives

$$\tilde{\sigma}_{flex\,2} = 1.14 f (E_f E_c^2)^{1/3} \tag{11.21}$$

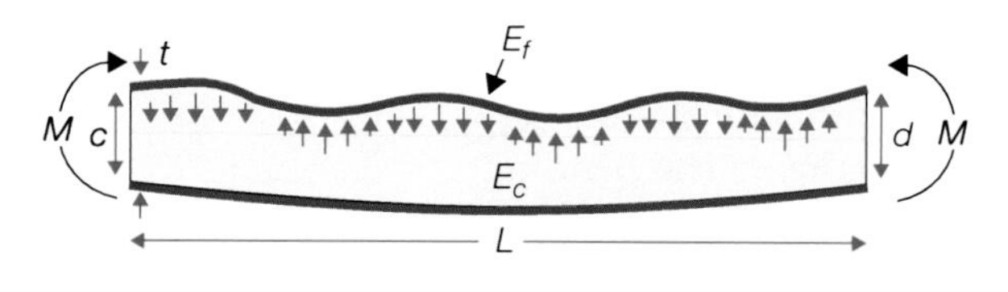

FIGURE 11.16
Face buckling.

Core shear Failure by core shear (Figure 11.17) occurs at the load

$$P_f = B_4 bc\,(\tau_c + \frac{t^2}{cL}\sigma_f)$$

[4] Derivations of this and the other equations cited here can be found in Ashby et al. (2000) and Gibson and Ashby (1997) in "Further reading" at the end of this chapter.

Here the first term results from shear in the core, the second from the formation of plastic hinges in the faces. Equating to

$$P_f = \frac{B_3 b d^2}{4L}\widetilde{\sigma}_3$$

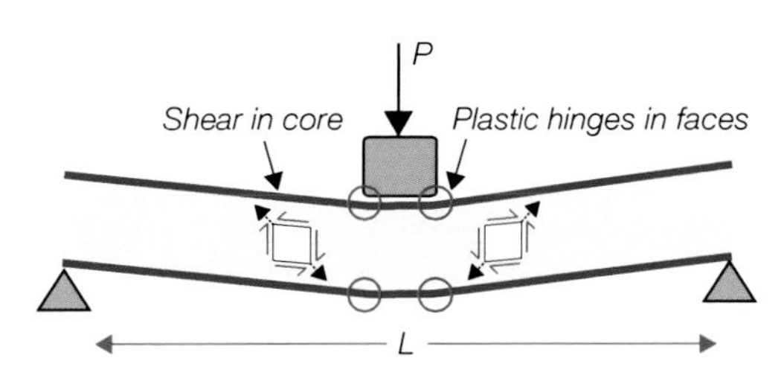

FIGURE 11.17
Core shear.

gives the equivalent strength when failure is by shear:

$$\widetilde{\sigma}_{flex\,3} = \frac{B_4}{B_3}\left\{4\frac{L}{d}(1-f)\,\tau_c + f^2\sigma_f\right\} \qquad (11.22)$$

(The load configuration determines the constant values B_3 and B_4, as summarized earlier in Table 11.3.) When the core material is roughly isotropic (as foams are), τ_c can be replaced by $\sigma_c/2$. When it is not (an example is a honeycomb core), τ_c must be retained.

Indentation The indentation pressure $P_{ind} = P/a$ is

$$\frac{F}{ab} = p_{ind} = \frac{2t}{a}\left(\sigma_y^f\sigma_y^c\right)^{1/2} + \sigma_y^c \qquad (11.23)$$

from which we find the minimum face thickness to avoid indentation (refer to Ashby et al., 2000, in Section 11.9).

The efficiency of sandwich structures Sandwiches are compared with monolithic materials as illustrated in Figures 11.18 and 11.19. In the first of these shows the equivalent density $\widetilde{\rho}$ and equivalent flexural modulus, $\widetilde{E}$ (Equations (11.16) and (11.18)) for sandwiches, using the data in Table 11.4. Here CFRP face sheets are combined with a high-performance foam core in different ratios, stepping through values of $2t/d$ for a chosen ratio of d/L to give the trajectory shown. Its doubly curved shape arises because of the interplay of bending and shear modes of deformation. Contours show values of the index for a light, stiff panel:

$$M_3 = \frac{E^{1/3}}{\rho}$$

The optimum panel, from a stiffness per unit weight perspective, is the one at which a contour is tangent to the trajectory. The figure shows that this occurs at $f \approx 0.04$, giving a panel that is 2.8 times lighter than a solid CFRP panel of the same stiffness (or $(2.8)^3 = 22$ times stiffer for the same mass).

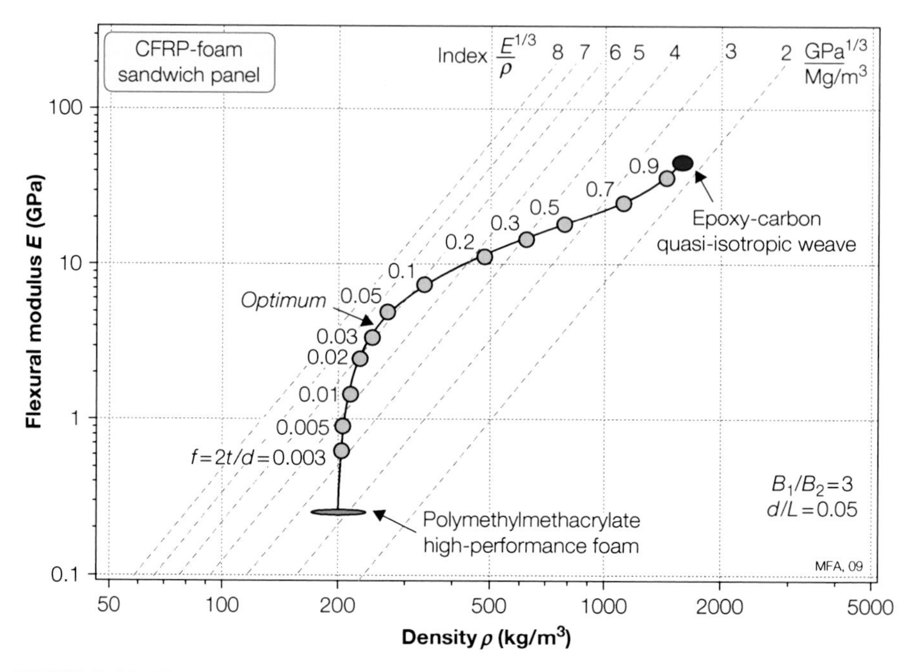

FIGURE 11.18

The equivalent modulus and density of a CFRP-foam sandwich are compared with those of monolithic materials. The contours of the index $E^{1/3}/\rho$ allow optimization of the proportions of the sandwich.

Strength (Figure 11.19) is handled in a similar way, but here there is the problem of competing mechanisms. We take the equivalent failure strength to be the least of $\tilde{\sigma}_{flex1}$, $\tilde{\sigma}_{flex2}$, and $\tilde{\sigma}_{flex3}$ (Equations (11.19), (11.21), and (11.22)), thereby properly accounting for the competition between them. For the conditions chosen here, face buckling dominates for $f < 0.025$; face yield dominates from $f = 0.025$ to $f = 0.1$, when a switch to core shear takes place. The envelope shows the achievable strength of CFRP–foam sandwich structures, and allows direct comparison with monolithic materials. Contours show the index for light strong structures:

$$M_6 = \frac{\sigma_f^{1/2}}{\rho}$$

which measures material efficiency when flexural strength is the main requirement. The optimum lies just below $f = 0.1$, at which the panel is 2.0 times lighter than a solid CFRP panel of the same strength (or $2.0^2 = 4.0$ times stronger for the same mass).

Table 11.4 Data for Face and Core of Sandwich

Face and Core Material	Density ρ (kg/m^3)	Modulus E (GPa)	Strength σ_f (MPa)
Carbon-epoxy quasi-isotropic weave	1570	46	550
Polymethacrylimide high-performance foam	200	0.255	6.8

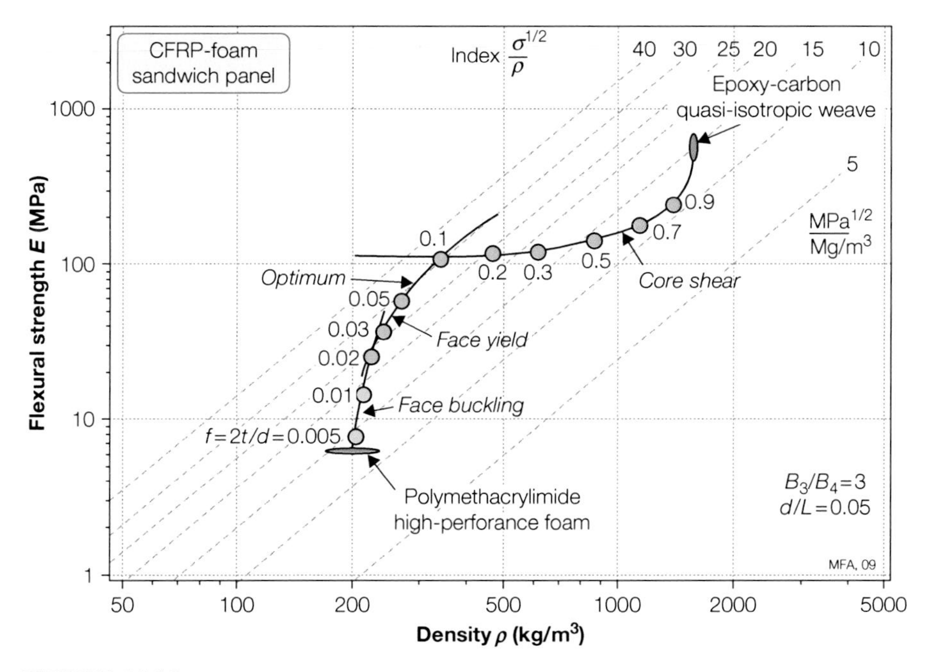

FIGURE 11.19

The equivalent strength and density of a CFRP-foam sandwich are compared with those of monolithic materials. The envelope shows the least strong of the competing failure modes. The contours of the index $\sigma^{1/2}/\rho$ allow optimization of the proportions of the sandwich. Indentation is included by imposing a minimum on the thickness ratio $2t/d$.

Indentation was not included in this competition because it is a local mechanism—it depends on the area of contact (or of impact) with the indenter, often an event for which the panel was not primarily designed. Protection is possible by estimating a "worst case" for the indentation load and area and calculating the value t/d required to withstand it. This is done using Equation (11.23) to calculate a lower safe limit for t/d, which is then applied as a constraint to the trajectory.

Thermal properties Thermal properties are treated in a similar way. The specific heat C_p follows a rule of mixtures (Equation (11.6)). The in-plane thermal conductivity $\lambda_{//}$ also follows such a rule (Equation (11.9)). The through-thickness conductivity, $\lambda_{\perp}$, is given by the harmonic mean

$$\widetilde{\lambda}_{\perp} = \left(\frac{f}{\lambda_f} + \frac{(1-f)}{\lambda_c} \right)^{-1} \tag{11.24}$$

Thermal expansion in-plane is complicated by the fact that the faces and core have different expansion coefficients, but being bonded together they are forced to suffer the same strain. This constraint leads to an in-plane expansion coefficient of

$$\widetilde{\alpha}_{//} = \frac{f\,E_f\,\alpha_f + (1-f)E_c\,\alpha_c}{f\,E_f + (1-f)E_c} \tag{11.25}$$

The through-thickness coefficient is simpler; it is given by the weighted mean

$$\widetilde{\alpha}_{\perp} = f\,\alpha_f + (1-f)\,\alpha_c \tag{11.26}$$

Through-thickness thermal diffusivity is not a single-valued quantity, but depends on time. At short times heat does not penetrate the core and the diffusivity is that of the face, but at longer times the diffusivity tends to the value given by the ratio $\widetilde{\lambda}/\widetilde{\rho}\widetilde{C}_p$.

Electrical properties The dielectric constant of a sandwich, as with composites, is given by a rule of mixtures—that is, Equation (11.12)—with $f = 2t/d$. Polymer foams have very low dielectric constants, so sandwiches with GFRP faces and polymer foam cores allow the construction of stiff, strong shells with exceptionally low dielectric loss. In-plane electrical conductivity, too, follows a rule of mixtures. Through-thickness electrical conductivity, like that of heat, is described by the harmonic mean—that is, the equivalent of Equation (11.24).

Filling property space with sandwich structures We end this section, as we did the last, with two figures illustrating how sandwich structures can expand the occupancy of material-property space. The first, Figure 11.20, is a flexural modulus-density ($E - \rho$) section. The areas filled by metals, polymers, ceramics, composites and foams appear as pale envelopes; members are labeled in gray. The stiffness and density of the CFRP–foam sandwiches of

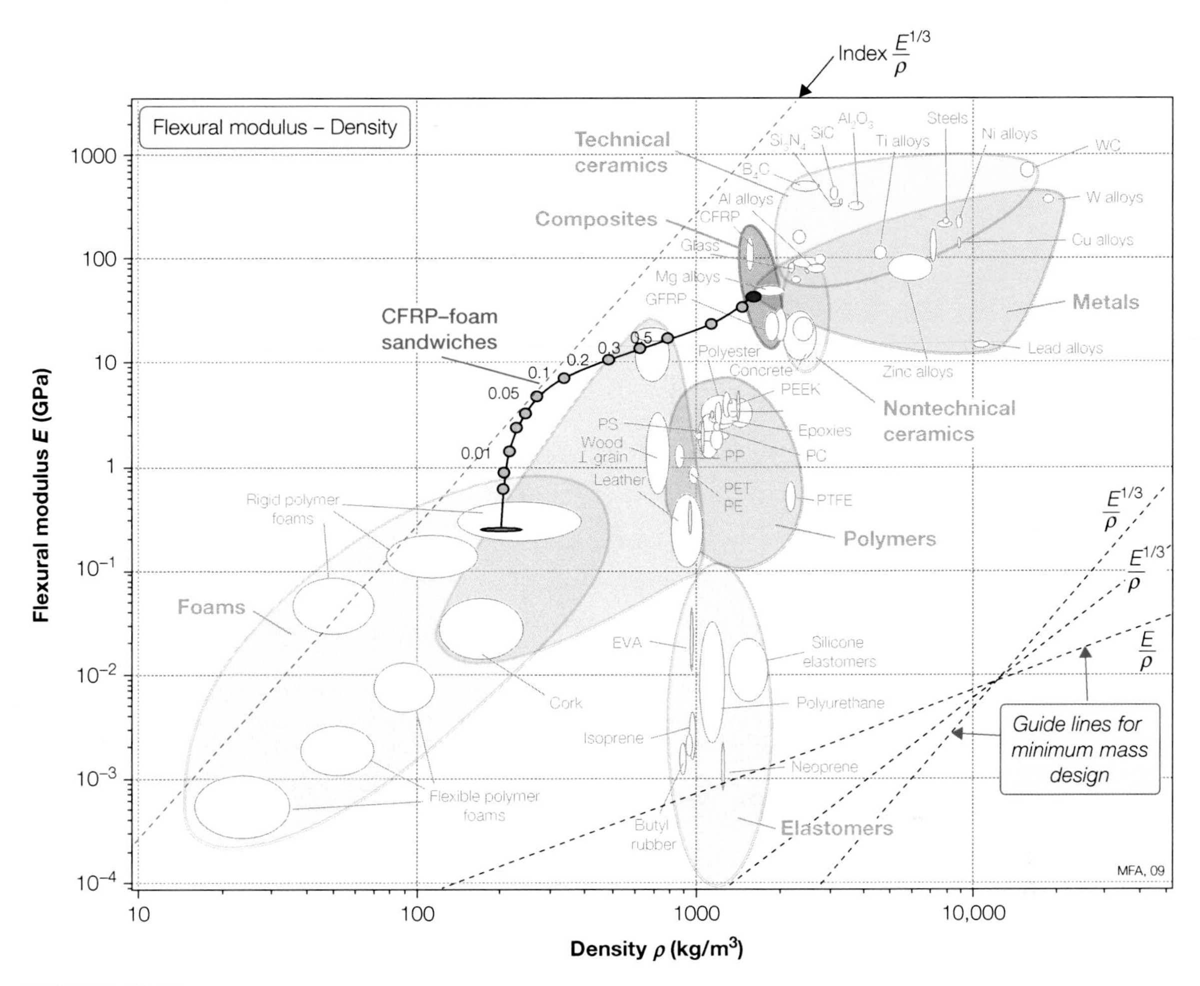

FIGURE 11.20

The sandwich data of Figure 11.18 superimposed on a modulus-density chart, showing the exceptional value of the flexural stiffness index $E^{1/3}/\rho$.

Figure 11.18 are superimposed. Those with $0.01 < f < 0.2$ extend into an area that was previously empty. Using the index $E^{1/3}/\rho$ for a light, stiff panel as a criterion of excellence, we find that sandwiches offer performance not attainable before.

Figure 11.21 tells a similar story for the strength-density $(\sigma_f - \rho)$ section. The color coding is the same as that in the previous figure. The strength-density trajectory from Figure 11.19 is overlaid. Sandwiches again expand the populated area in a direction that, using the index for a light, strong panel $(\sigma_f^{1/2}/\rho)$ as a criterion, offers enhanced performance.

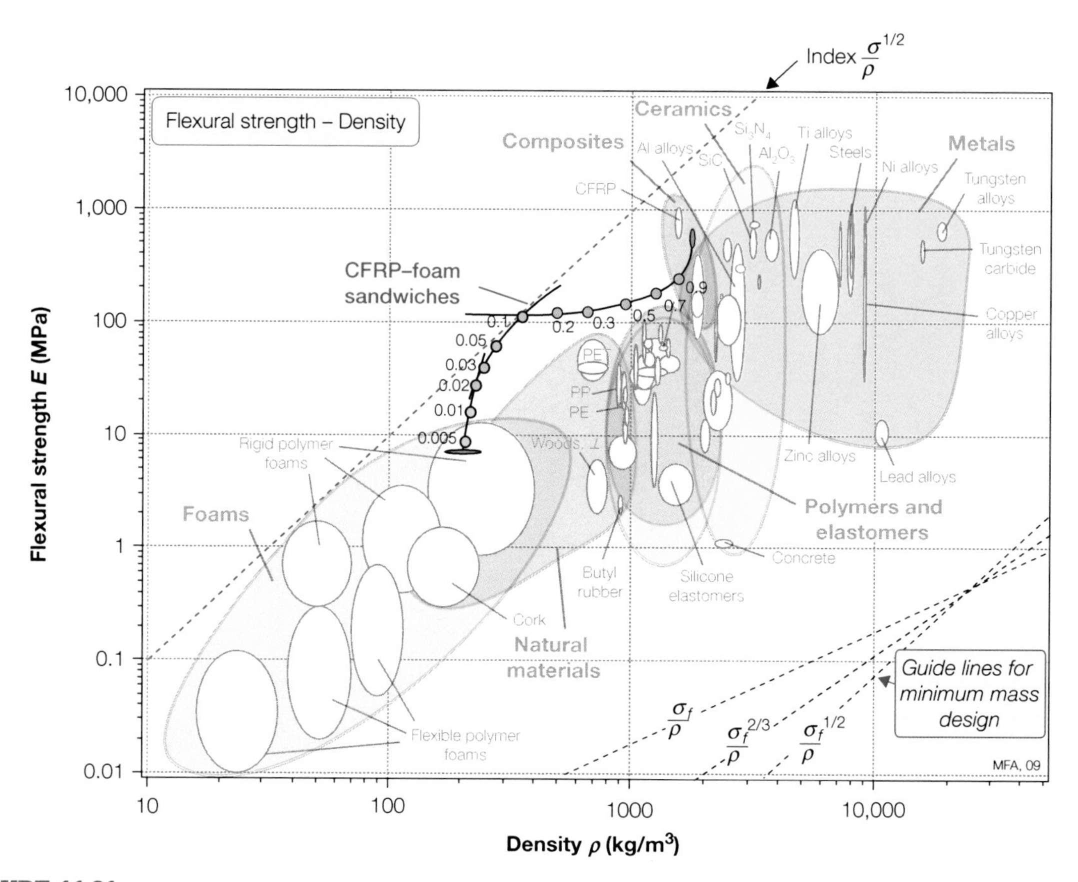

FIGURE 11.21

The sandwich data of Figure 11.19 superimposed on a strength-density chart, showing the exceptional value of the flexural stiffness index $\sigma^{1/2}/\rho$.

11.6 CELLULAR STRUCTURES: FOAMS AND LATTICES

Cellular structures—foams and lattices—are hybrids of a solid and a gas. The properties of the gas might at first seem irrelevant, but this is not so. The thermal conductivity of low-density foams of the sort used for insulation is determined by the conductivity of the gas contained in its pores; and the dielectric constant and breakdown potential, and even the compressibility, can depend on the gas properties.

There are two distinct species of cellular solid. The distinction is most obvious in their mechanical properties. The first, typified by foams, are *bending-dominated structures*; the second, typified by triangulated lattice structures, are *stretch dominated*—a distinction explained more fully next. To give an idea of the difference: A foam with a relative density of 0.1 (meaning that the solid cell walls occupy 10% of the volume) is less stiff by a factor of 10 than a triangulated lattice of the same relative density. The word "configuration" has special relevance here.

Foams: Bending-dominated structures Foams are cellular solids made by expanding polymers, metals, ceramics, or glasses with a foaming agent—a generic term for one of many ways of introducing gas, much as yeast does in breadmaking. Figure 11.22 shows an idealized cell of a low-density foam. It consists of solid cell walls or edges surrounding a void space containing a gas or fluid. Cellular solids are characterized by their *relative density*, which for the structure shown here (with $t << L$) is

$$\frac{\widetilde{\rho}}{\rho_s} \approx \left(\frac{t}{L}\right)^2 \qquad (11.27)$$

where $\widetilde{\rho}$ is the density of the foam, ρ_s is the density of the solid of which it is made, L is the cell size, and t is the thickness of the cell edges.

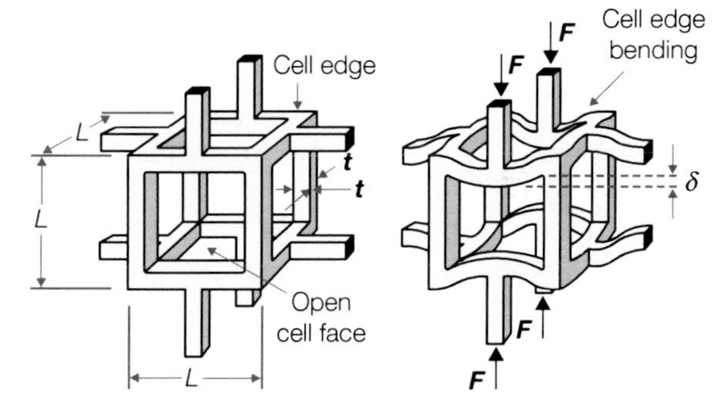

FIGURE 11.22
A cell in a low-density foam. When the foam is loaded, the cell edges bend, giving a low-modulus structure.

Foams have the characteristic that, when they are loaded, their cell walls *bend*, with consequences that we now analyze.

Mechanical properties The compressive stress-strain curve of bending-dominated foams looks like those in Figure 11.23. The material is linear elastic, with modulus $\widetilde{E}$ up to its elastic limit, at which point the cell edges yield, buckle, or fracture. The foam continues to collapse at a nearly constant stress (the "plateau stress" $\widetilde{\sigma}_{pl}$) until opposite sides of the cells impinge (the "densification strain" $\widetilde{\varepsilon}_d$), when the stress rises rapidly. The mechanical properties are calculated in the ways developed next (details are in the texts listed in "Further reading").

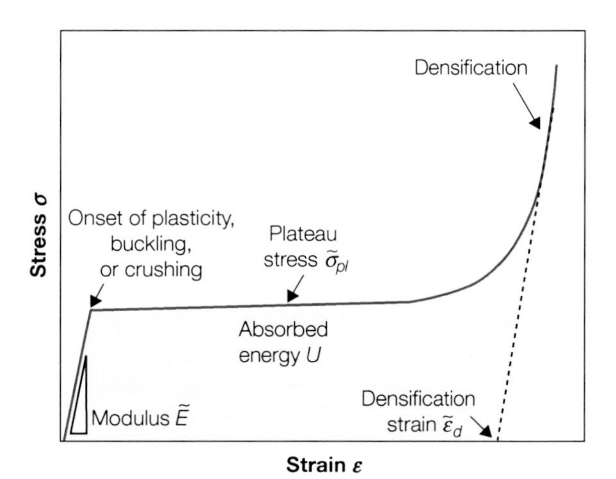

FIGURE 11.23
The plateau stress is determined by buckling, plastic bending, or fracturing of the cell walls.

A remote compressive stress σ exerts a force $F \propto \sigma L^2$ on the cell edges, causing them to bend and leading to a bending deflection δ, as shown in Figure 11.22. For the open-celled structure shown in the figure, the bending deflection (Appendix B.3) scales as

$$\delta \propto \frac{FL^3}{E_s I} \qquad (11.28)$$

where E_s is the modulus of the solid of which the foam is made and $I = \frac{t^4}{12}$ is the second moment of area of the cell edge of square cross-section, $t \times t$. The compressive strain suffered by the cell as a whole is then $\varepsilon = 2\delta/L$. Assembling these results gives the modulus $\widetilde{E} = \sigma/\varepsilon$ of the foam as

$$\frac{\widetilde{E}}{E_s} \propto \left(\frac{\widetilde{\rho}}{\rho_s}\right)^2 \quad \text{(bending – dominated behavior)} \tag{11.29}$$

Since $\widetilde{E} = E_s$ when $\widetilde{\rho} = \rho_s$, we expect the constant of proportionality to be close to unity—a speculation confirmed both by experiment and by numerical simulation. The quadratic dependence means that a small decrease in relative density causes a large drop in modulus (Figure 11.24(a)).

A similar approach can be used to model the plateau stress of the foam. The cell walls yield, as shown in Figure 11.25(a), when the force exerted on them exceeds their fully plastic moment (refer to Appendix A, Equation (A.4)),

$$M_f = \frac{\sigma_s t^3}{4} \tag{11.30}$$

where σ_s is the yield strength of the solid of which the foam is made. This moment is related to the remote stress by $M \propto FL \propto \sigma L^3$. Assembling these results gives the failure strength $\widetilde{\sigma}_{pl}$:

$$\frac{\widetilde{\sigma}_{pl}}{\sigma_{f,s}} = C\left(\frac{\widetilde{\rho}}{\rho_s}\right)^{3/2} \quad \text{(bending-dominated behavior)} \tag{11.31}$$

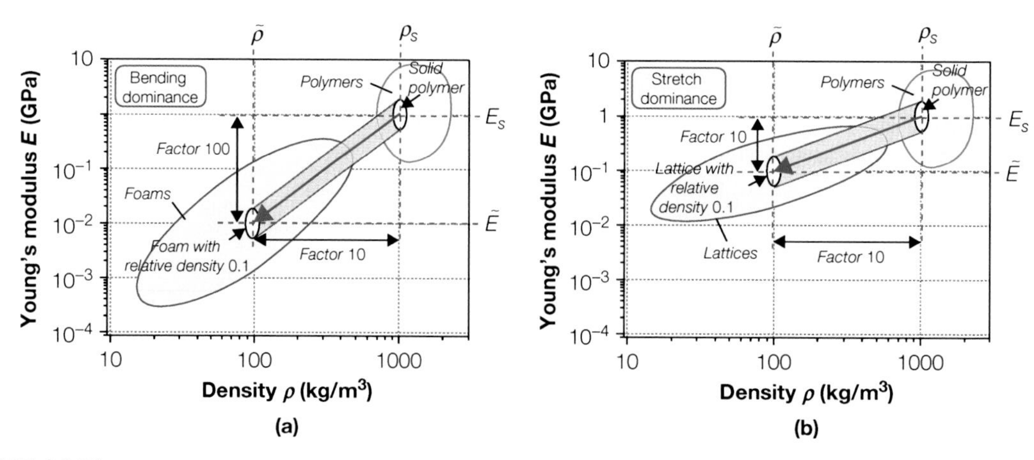

FIGURE 11.24
Foaming creates bending-dominated structures with lower modulus and density (a). Lattices that are stretch-dominated have moduli that are much greater than those of foams of the same density (b).

where the constant of proportionality, $C \approx 0.3$, has been established both by experiment and by numerical computation.

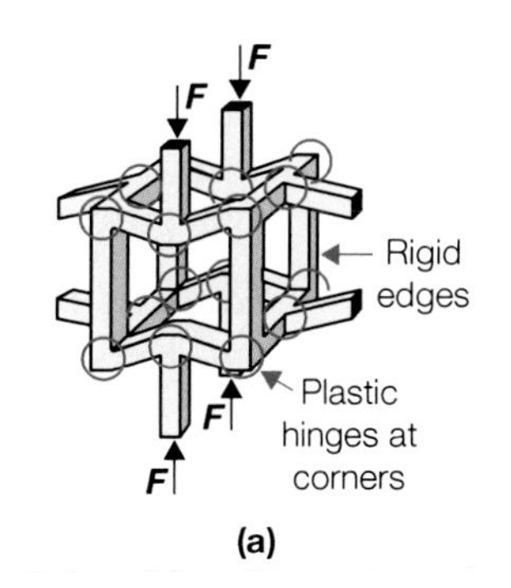

Elastomeric foams collapse not by yielding but by elastic bucking; brittle foams, by cell-wall fracture (Figures 11.25(b) and (c)). As with plastic collapse, simple scaling laws describe this behavior well. Collapse by buckling (refer to Appendix B, Table B.5) occurs when the stress exceeds $\widetilde{\sigma}_{el}$, given by

$$\frac{\widetilde{\sigma}_{el}}{E_s} \approx 0.05\left(\frac{\widetilde{\rho}}{\rho_s}\right)^2 \tag{11.32}$$

and by cell-wall fracture (Appendix B.4 again) when it exceeds $\widetilde{\sigma}_{cr}$:

$$\frac{\widetilde{\sigma}_{cr}}{\sigma_{cr,s}} \approx 0.3\left(\frac{\widetilde{\rho}}{\rho_s}\right)^{3/2} \tag{11.33}$$

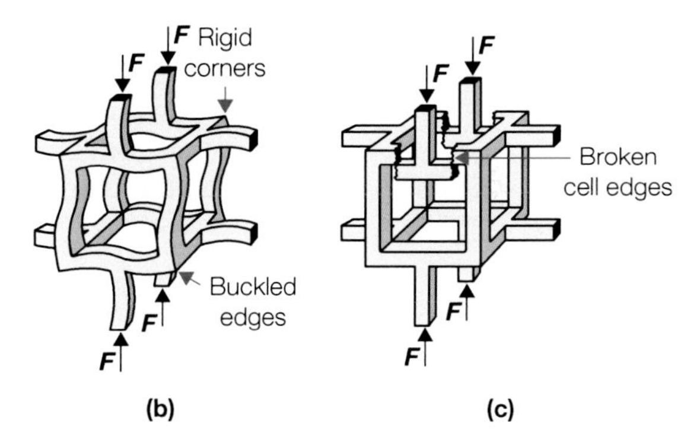

FIGURE 11.25

Collapse of foams. (a) When a foam made of a plastic materials is loaded beyond its elastic limit, the cell edges bend plastically. (b) An elastomeric foam, by contrast, collapses by the elastic buckling of its cell edges. (c) A brittle foam collapses by the successive fracturing of cell edges.

where $\sigma_{cr,s}$ is the flexural strength of the cell-wall material. Densification, when the stress rises rapidly, is a purely geometric effect: The opposite sides of the cells are forced into contact and further bending or buckling is not possible. It is found to occur at a strain $\widetilde{\varepsilon}_d$ (the densification strain) of

$$\widetilde{\varepsilon}_d \approx 1 - 1.4\left(\frac{\widetilde{\rho}}{\rho_s}\right) \tag{11.34}$$

Foams are often used for cushioning and packaging and to protect against impact. The useful energy that a foam can absorb per unit volume is approximated by

$$\widetilde{U} \approx \widetilde{\sigma}_{pl}\,\widetilde{\varepsilon}_d \tag{11.35}$$

where $\widetilde{\sigma}_{pl}$ is the plateau stress—the yield, buckling, or fracturing strength of the foam, whichever is least.

This behavior is not confined to open-cell foams with the structure idealized, as shown earlier in Figure 11.22. Most closed-cell foams also follow these scaling laws, at first sight an unexpected result because the cell faces must carry membrane stresses when the foam is loaded, and these should lead to a linear dependence of both stiffness and strength on relative density. The explanation lies in the fact that the cell faces are very thin; they buckle or rupture at stresses so low that their contribution to stiffness and strength is small, leaving the cell edges to carry most of the load.

Thermal properties The specific heat of foams, when expressed in units of $J/m^3.K$, is given by a rule of mixtures, summing the contributions from the solid and the gas. The thermal expansion coefficient of an open cell foam is the same as that of the solid from which it is made. The same is true of rigid closed-cell foams but not necessarily of low-density elastomeric foams because the expansion of the gas within the cells can expand the foam itself, giving an apparently higher coefficient.

The cells in most foams are sufficiently small that convection of the gas within them is completely suppressed. The thermal conductivity of the foam is thus the sum of that conducted through the cell walls and that through the still air (or other gas) they contain. To an adequate approximation,

$$\widetilde{\lambda} = \frac{1}{3}\left(\left(\frac{\widetilde{\rho}}{\rho_s}\right) + 2\left(\frac{\widetilde{\rho}}{\rho_s}\right)^{3/2}\right)\lambda_s + \left(1 - \left(\frac{\widetilde{\rho}}{\rho_s}\right)\right)\lambda_g \qquad (11.36)$$

where λ_s is the conductivity of the solid and λ_g that of the gas (for dry air it is 0.025 W/m.K). The term associated with the gas is important: Blowing agents for foams intended for thermal insulation are chosen to have a low value of λ_g.

Electrical properties Insulating foams are attractive for their low dielectric constant, $\widetilde{\varepsilon}_r$, falling toward 1 (the value for air or vacuum) as the relative density decreases:

$$\widetilde{\varepsilon}_r = 1 + (\varepsilon_{r,s} - 1)\left(\frac{\widetilde{\rho}}{\rho_s}\right) \qquad (11.37)$$

where $\varepsilon_{r,s}$ is the dielectric constant of the solid of which the foam is made. The electrical conductivity follows the same scaling law as the thermal conductivity.

Lattice: Stretch-dominated structures

If conventional foams have low stiffness because the configuration of their cell edges allows them to bend, might it not be possible to devise other configurations in which the cell edges are made to stretch instead? This thinking leads to the idea of *micro-truss lattice structures*. To understand these we need one of those simple yet profound fundamental laws: the Maxwell stability criterion.

Equation (11.38) is the condition that a pin-jointed frame (meaning one that is hinged at its corners) made up of b struts and j frictionless joints, like those shown in Figure 11.26, is to be both *statically* and *kinematically determinate* (meaning that it is rigid and does not fold up when loaded) in two dimensions:

$$M = b - 2j + 3 = 0 \qquad (11.38)$$

In three dimensions the equivalent equation is

$$M = b - 3j + 6 = 0 \qquad (11.39)$$

If $M < 0$, the frame is a *mechanism*. It has no stiffness or strength; it collapses if loaded. If its joints are locked, preventing rotation (as they are in a lattice), the bars of the frame *bend* when the structure is loaded, just as those in Figure 11.22 do. If, instead, $M \geq 0$ the frame ceases to be a mechanism. If it is loaded, its members carry tension or compression (even when pin-jointed), and it becomes a *stretch-dominated* structure. Locking the hinges now makes little difference because slender structures are much stiffer when stretched than when bent. There is an underlying

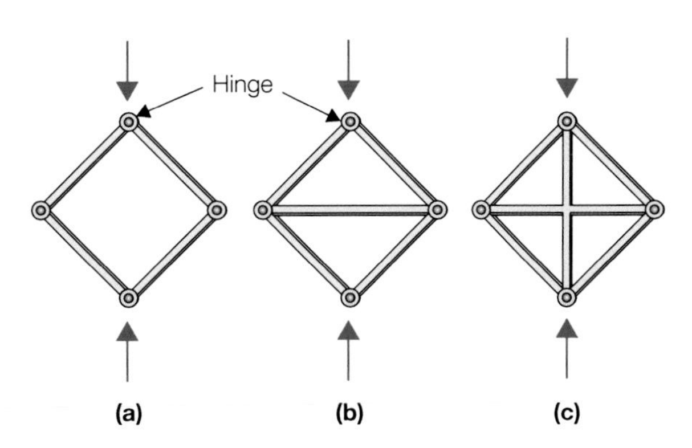

FIGURE 11.26
The pin-jointed frame at (a) is a mechanism. If its joints are welded together, the cell edges bend. The pin-jointed, triangulated frame at (b) is stiff when loaded because the transverse bar carries tension, preventing collapse. When the frame's joints are welded, its stiffness and strength hardly change. The frame at (c) is over-constrained. If the horizontal bar is tightened, the vertical bar is put in tension even when there are no external loads.

Estimating foam properties

A polyethylene foam has a density ρ of 150 kg/m^3. The density, modulus, strength, and thermal conductivity of the polyethylene are listed in the table. What would you expect the same properties of the foam to be?

	Density ρ_s (kg/m^3)	Young's Modulus E_s (GPa)	Flexural Strength $\sigma_{f,s}$ (MPa)	Thermal Conductivity λ_s (W/m.K)
High-molecular-weight polyethylene	950	0.94	33	0.195

Answer

The relative density of the foam is $\tilde{\rho}/\rho_s = 0.16$. Using Equations (11.29), (11.31), and (11.36), we find

	Young's Modulus $\tilde{E}$ (GPa)	Flexural Strength $\tilde{\sigma}_{pl}$ (MPa)	Thermal Conductivity $\tilde{\lambda}$ (W/m.K)
HMW polyethylene foam	0.024	0.63	0.04

Estimating lattice properties

A stretch-dominated polyethylene lattice has a density ρ of 150 kg/m^3. The density, modulus, strength, and thermal conductivity of the polyethylene are the same as those listed for the last example. What would you expect the properties of the lattice to be?

Answer

The relative density of the lattice is $\tilde{\rho}/\rho_s = 0.16$. Using Equations (11.40), (11.41), and (11.36), we find

	Young's Modulus $\tilde{E}$ (GPa)	Flexural Strength $\tilde{\sigma}$ (MPa)	Thermal Conductivity $\tilde{\lambda}$ (W/m.K)
HMW polyethylene lattice	0.05	1.8	0.04

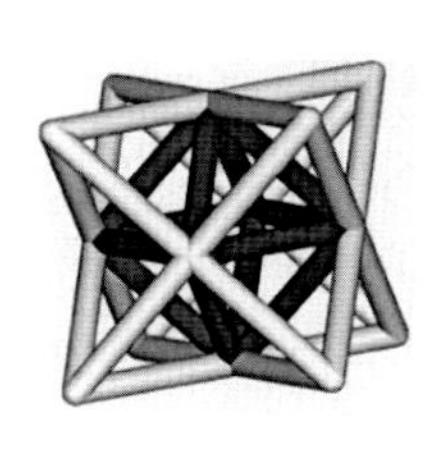
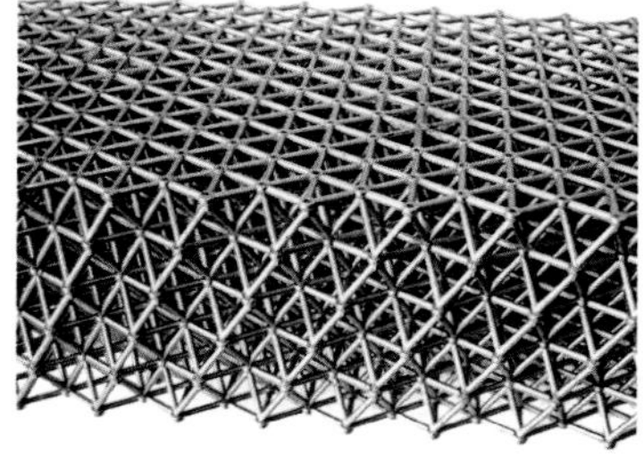

FIGURE 11.27
A micro-truss structure and its unit cell. This is a stretch-dominated structure, and it is over-constrained, meaning that it is possible for it to be in a state of self-stress.

principle here: *Stretch-dominated structures have high structural efficiency; bending-dominated structures have low.*

Mechanical properties These criteria provide a basis for the design of efficient microlattice structures. For the cellular structure of Figure 11.22, Maxwell's $M < 0$ and bending dominates. For the structure shown in Figure 11.27, however, $M > 0$ and it behaves as an almost isotropic, stretch-dominated structure. On average one-third of its bars carry tension when the structure is loaded in simple tension. Thus

$$\frac{\tilde{E}}{E_s} \approx \frac{1}{3}\left(\frac{\tilde{\rho}}{\rho_s}\right) \text{ (isotropic stretch-dominated behavior)} \quad (11.40)$$

The modulus is linear, not quadratic, in density (Figure 11.24b), giving a much stiffer structure for the same density. Collapse occurs when the cell edges yield, giving the collapse stress

$$\frac{\tilde{\sigma}}{\sigma_{f,s}} \approx \frac{1}{3}\left(\frac{\tilde{\rho}}{\rho_s}\right) \text{ (isotropic stretch-dominated behavior)} \quad (11.41)$$

This is an upper bound since it assumes that the struts yield in tension or compression when the structure is loaded. If the struts are slender, they may buckle before they yield. Then the "strength," like that of a buckling foam (Equation (11.32)), is

$$\frac{\tilde{\sigma}_{el}}{E_s} \approx 0.2\left(\frac{\tilde{\rho}}{\rho_s}\right)^2 \quad (11.42)$$

Thermal and electrical properties The bending/stretching distinction influences mechanical properties profoundly, but has no effect on thermal or electrical properties, which are adequately described by the equations listed above for foams.

Filling property space with cellular structures All of the charts in Chapter 4 have an envelope labeled "Foams," indicating where the properties' commercial polymer foams lie. It is highlighted in Figure 11.28 which again

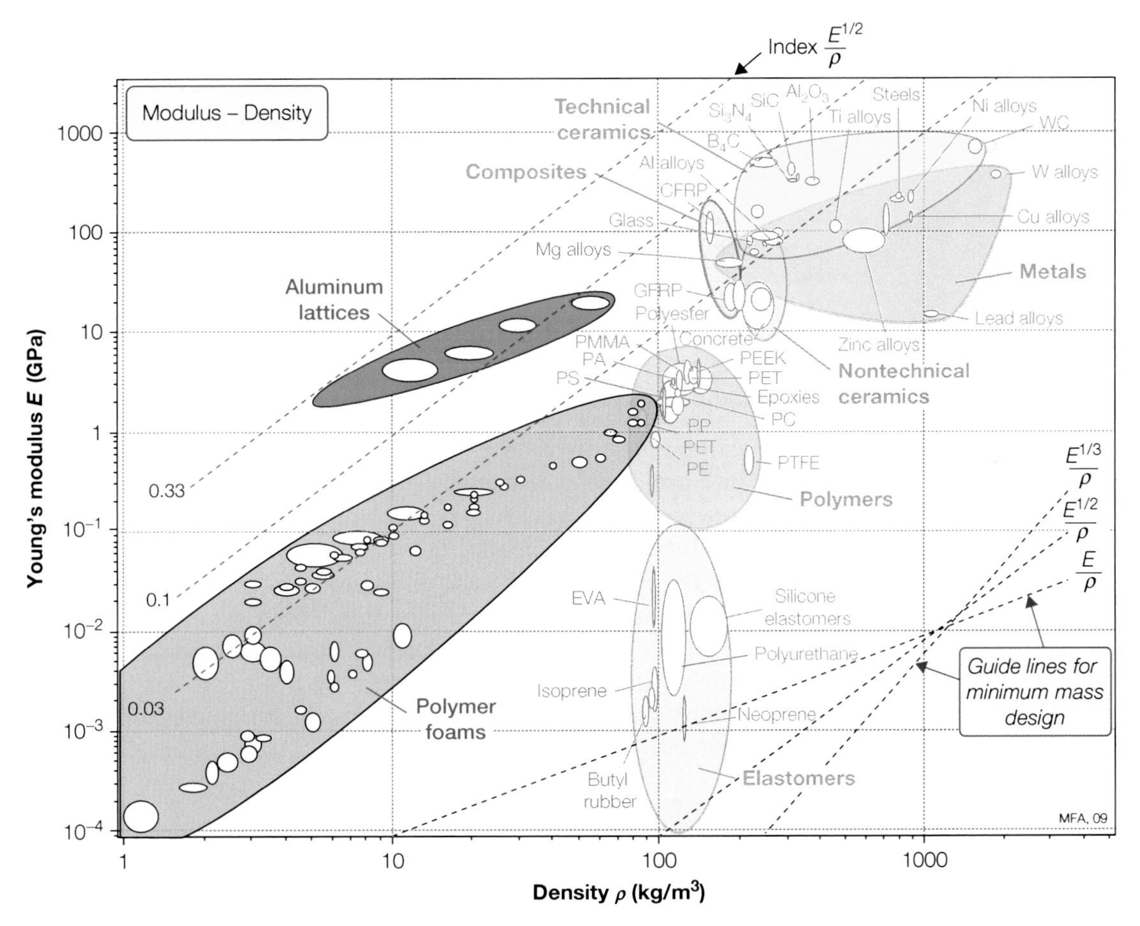

FIGURE 11.28

Foams and micro-truss structures are hybrids of material and space. Their mechanical response depends on their structure. Foams are usually bending-dominated and lie along a line of slope 2 on this chart. Micro-truss structures are stretch-dominated and lie on a line of slope 1. Both extend the occupied area of this chart by many decades.

shows the modulus-density ($E - \rho$) section through material-property space. The polymer foam envelope is extended along a line of slope 2, as predicted by Equation (11.29). Lattice structures, by contrast, extend along a line of slope 1, as Equation (11.40) predicts. Both fill areas of $E - \rho$ space that are not filled by solid materials. Lattice structures push the filled area to higher values of the indices

$$E^{1/2}/\rho \quad \text{and} \quad E^{1/3}/\rho$$

11.7 SEGMENTED STRUCTURES

Subdivision as a design variable

Shape can be used to reduce flexural stiffness and strength as well as to increase them. Springs, suspensions, flexible cables, and other structures that must flex yet have high tensile strength use shape to give a low bending stiffness. This is achieved by shaping the material into strands or leaves, as explained in Section 10.7. The slender strands or leaves bend easily but do not stretch when the section is bent: an n-strand cable is less stiff by a factor of $3/\pi n$ than the solid reference section; an n-leaf panel by a factor $1/n^2$.

Subdivision allows mixing. If one or more of the segments is replaced by a second material, combinations can be created that have property combinations not found in monolithic materials. The method is best illustrated by the example that follows.

Subdivision can be used in another way: to impart *damage tolerance*. A glass window, hit by a projectile, will shatter. One made of small glass bricks, laid as bricks usually are, will lose a brick or two but not shatter totally; it is damage-tolerant. By subdividing and separating the material, a crack in one segment does not penetrate into its neighbors, allowing local but not global failure. This is the principle of "topological toughening." Builders in stone and brick have exploited the idea for thousands of years: Stone and brick are brittle, but buildings made of them—even those made without cement ("dry-stone building")—survive ground movement, even small earthquakes, through their ability to deform with some local failure but without total collapse.

Materials for long-span power cables

In the design of suspended power cables, the objectives are to minimize electrical resistance and at the same time maximize the strength since this allows the greatest span. This is an example of multiobjective optimization, discussed in Chapter 7. As explained there, each objective is expressed, by convention, such that a minimum for it is sought. We thus seek materials with the lowest values of resistivity ρ_e and the reciprocal of yield strength, $1/\sigma_y$.

Figure 11.29 shows the result: Materials that best meet the design requirements lie toward the bottom left. But here there is a hole: All 1700 metals and alloys plotted here have properties that lie above the broken red line, none below. Those with the lowest resistance—copper, aluminum, and certain of their alloys—are not very strong; those that are strongest—drawn carbon and low-alloy steel—do not conduct very well.

Now consider a cable made by interleaving strands of copper and steel such that each occupies half the cross-section. If the steel carries no current and the copper no load (the most pessimistic scenario), the performance of the cable will lie at the point shown on the figure—it has twice the resistivity of the copper and half the strength of the steel. It lies in a part of property space that was empty, offering performance not previously possible. Other ratios of copper to steel fill other parts of the space; by varying the ratio, the shaded envelope in the figure is covered. Similar hybrids of aluminum and steel fill a different area, as is easily seen by repeating the construction using "1000 series Al alloys" in place of "OFHC copper, hard" in the combination. Their combinations of ρ_e and σ_y are less good, but they are lighter and cheaper and for this reason are widely used.

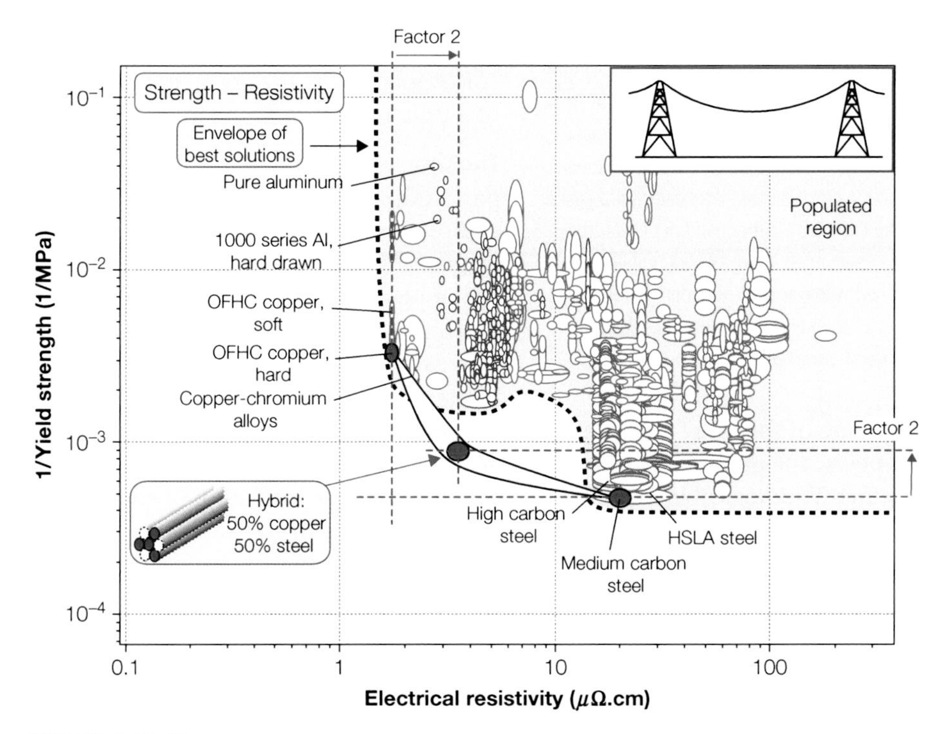

FIGURE 11.29

Designing a hybrid–here one with high strength and high electrical conductivity. The figure shows the resistivity and reciprocal of tensile strength for 1,700 metals and alloys. The construction is for a hybrid of hard-drawn OFHC copper and drawn medium carbon steel, but the figure itself allows many hybrids to be investigated.

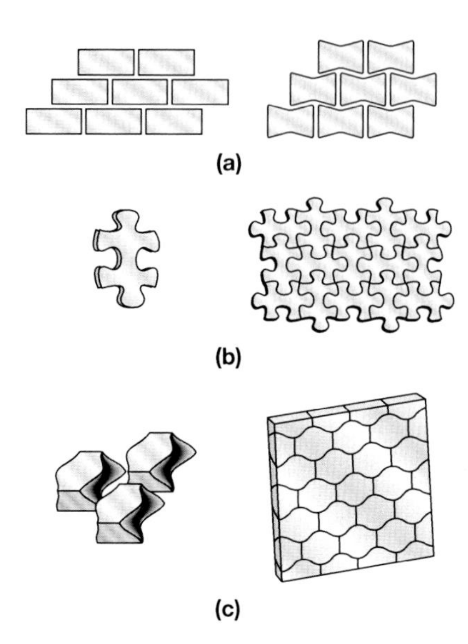

FIGURE 11.30
Examples of topological interlocking: discrete, unbonded structures that carry load. (a) Brick-like assemblies of rectangular blocks carry axial compression, but not tension or shear. (b) The 2-dimensional interlocking of a jigsaw puzzle carries in-plane loads. (c) Units that, when assembled into a continuous layer and clamped within a rigid boundary around the edges of the layer, can carry out-of-plane loads and bending moments.

Taking the simplest view, two things are necessary in order to have topological damage tolerance: discreteness of the structural units and an interlocking of the units in such a way that the array as a whole can carry a load. Brick-like arrangements (Figure 11.30(a) carry large loads and are damage-tolerant in compression and shear, but disintegrate under tension. Strand- and layer-like structures are damage-tolerant in tension because if one strand fails, the crack does not penetrate its neighbors—the principle of multistrand ropes and cables. The jigsaw puzzle configuration (Figure 11.30(b)) carries in-plane tension, compression and shear, but at the cost of introducing a stress concentration factor of about $\sqrt{R/r}$, where R is the approximate radius of a unit and r that of the interlock.

The sources listed in Section 11.9 explore a particular set of topologies that rely on compressive or rigid boundary conditions to create continuous layers that tolerate out-of-plane forces and bending moments, illustrated in Figure 11.30(c). This is done by creating interlocking units with nonplanar surfaces that have curvature both in the plane of the array and normal to it. Provided the array is constrained at its periphery, the nesting shapes limit the relative motion of the units, locking them together. Topological interlocking of this sort allows the formation of continuous layers that can be used for ceramic claddings or linings to give surface protection. And of course the units need not be made of one material alone. Since the only requirement is that of interlocking shape, the segments can be made of different materials. Just as a builder constructing a brick wall can use porous bricks for ventilation and transparent bricks to admit light, the designer of a segmented hybrid can add functionality through the choice of material for the units.

11.8 SUMMARY AND CONCLUSIONS

The properties of engineering materials can be thought of as defining the axes of a multidimensional space, with each property as a dimension. Chapter 4 showed how this space can be mapped. The maps reveal that some areas of property space are occupied and others are empty—there are holes. The holes can sometimes be filled by making hybrids: combinations of two (or more) materials in a chosen configuration and scale. Design requirements isolate a small box in a multidimensional material-property space. If it is occupied by materials, the requirements can be met. However, if the box hits a hole in any one of the dimensions, we need a hybrid. A number of families of configuration exist, each offering different combinations of functionality. Individual configurations are characterized by a set of bounds, bracketing their effective properties. The methods developed in this chapter provide tools for exploring alternative material configuration combinations. Chapter 12 gives examples of their use.

11.9 FURTHER READING

Hybrid materials—general

Bendsoe, M.P., & Sigmund, O. (2003). *Topology optimization, theory, methods, and applications.* Springer-Verlag, ISBN 3-540-42992-1.
The first comprehensive treatment of emerging methods for optimizing configuration and scale.

Kromm, F.X., Quenisset, J.M, Harry, R., & Lorriot, T. (2001). An example of multimaterial design. *Proc Euromat '01.* Rimini, Italy.
One of the first papers to address hybrid design.

McDowell, D.L., Allen, J., Mistree, F., Panchal, J., & Choi, H-J (2009). *Integrated design of multiscale materials and products.* Elsevier, ISBN 978-1-85617-662-0.
McDowell et al. seek to integrate the "materials by design approach" of Olson/Ques Tek LLC with the "materials selection" approach developed in this book and implemented in the CES software.

Composites

Ashby, M.F. (1993). Criteria for selecting the components of composites. *Acta Mater. 41,* 1313–1335.
A compilation of models for composite properties, introducing the methods developed here.

Budiansky, B., & Fleck, N.A. (1993). Compressive failure of fibre composite. *J. Mech. Phys. Solids, 41,* 183–211.
The definitive analysis of fiber kinking in compression of composites.

Chamis, C.C. (1987). Engineers guide to composite materials. *Am. Soc. Metals,* pp. 3–24.
A compilation of models for composite properties.

Clyne, T.W., & Withers, P.J. (1993). *An introduction to metal matrix composites.* Cambridge University Press, ISBN 0-521-41808-9.
A broad introduction to the modeling of metal-matrix composites—a companion volume to Hull and Clyne.

Hull, D., & Clyne, T.W. (1996). *An introduction to composite materials.* Cambridge University Press, ISBN 0-521-38855-4.
A broad introduction to the modeling of polymer-matrix composites—a companion volume to Clyne and Withers.

Schoutens, J.E., & Zarate, D.A. (1986). Structural indices in design optimization with metal matrix composites. *Composites, 17,* 188.
A compilation of models for composite properties.

Watt, J.P., Davies, G.F., & O'Connell, R.J. (1976). Reviews of geophysics and space. *Physics, 14,* p. 541.
A compilation of models for composite properties.

Sandwich structures

Allen, H.G. (1969). *Analysis and design of structural sandwich panels.* Pergamon Press.
The bible: the book that established the principles of sandwich design.

Ashby, M.F., Evans, A.G., Fleck, N.A., Gibson, L.J., Hutchinson, J.W., & Wadley, H.N.G. (2000). *Metal foams: A design guide.* Butterworth-Heinemann, ISBN 0-7506-7219-6.
A text establishing the experimental and theoretical basis of the properties of metal foams, with data for real foams and examples of their applications.

Gill, M.C. (2009). *Simplified sandwich panel design. www.mcgillcorp.com/doorway/pdf/97_Summer.pdf.*

Pflug, J., & Verpoest, I. (2006). Sandwich materials selection charts. *Journal of Sandwich Structures and Materials, 8*(5), 407–421.

Pflug, J., Vangrimde, B., & Verpoest, I. (2003). *Material efficiency and cost effectiveness of sandwich materials.* SAMPE US Proceedings.

Pflug, J., Verpoest, I., & Vandepitte, D. (2004). SAND.CORE Workshop, Brussels, December.

Zenkert, D. (1995). *An introduction to sandwich construction.* Engineering Advisory Services Ltd., Solihull, U.K., Chameleon Press Ltd., ISBN 0-947-81777-8.
A primer on the basic analysis of sandwich structures.

Cellular structures

Deshpande, V.S., Ashby, M.F., & Fleck, N.A. (2001). Foam topology: Bending versus stretching dominated architectures. *Acta Mater., 49,* 1035–1040.
A discussion of the bending vs. stretching dominated topologies.

Gibson, L.J., & Ashby, M.F. (1997). *Cellular solids, structure and properties* (2nd ed.). Cambridge University Press, ISBN 0-521-49560-1.
A monograph analyzing the properties, performance and uses of foams and giving the derivations and experimental verification of the equations used in Section 11.6.

Gibson, L.J., Ashby, M.F., & Harley, B. (2010). *Cellular bio-materials.* Cambridge University Press, ISBN 9-7805-2119-5447.
An analysis of the functions of cellular solids in nature.

Segmented structures

Dyskin, A.V., Estrin, Y., Kanel-Belov, A.J., & Pasternak, E. (2001). Toughening by fragmentation: How topology helps. *Advanced Engineering Materials, 3,* 885–888.

Dyskin, A.V., Estrin, Y., Kanel-Belov, A.J., & Pasternak, E. (2003). Topological interlocking of platonic solids: A way to novel materials and structures. *Phil. Mag., 83,* 197–203.
The above two papers introduce interlocking configurations that carry bending loads yet provide damage tolerance.

Autruffe, A., Pelloux, F., Brugger, C., Duval, P., Brechet, Y., & Fivel, M. (2007). Indentation behaviour of interlocked structures made of ice: Influence of the friction coefficient. *Advanced Engineering Materials, 9*(8), 664–666.

Stauffer, D., & Aharony, A. (1994). *Introduction to percolation theory* (2nd ed.). Taylor and Francis, ISBN 0-7484-0253-5.
A personal but very readable introduction to percolation theory.

Weibull, W. (1951). A statistical distribution function of wide applicability. *J. Appl. Mech., 18,* p. 293.
The originator or the "weakest-link" model of a brittle solid.

CHAPTER 12

Case Studies: Hybrids

Racing yachts used hybrids throughout—mast and boom: epoxy-carbon fiber; hull: GRFP and CFRP skinned polymethyl-methyl-acrylamid foamed core sandwich; sails: Kevlar-nylon mix weave with thermally bonded PET skin.

Materials Selection in Mechanical Design. DOI: 10.1016/B978-1-85617-663-7.00012-6

CONTENTS

12.1 INTRODUCTION AND SYNOPSIS

Chapter 11 explored hybrids of four types: composites, sandwiches, lattices, and segmented structures. Each is associated with a set of models that allow its properties to be estimated. In this chapter we illustrate the models' use to design hybrids to fill specified needs—needs that cannot be filled by single material choices. The cover picture is a reminder of the widespread use of hybrids to maximize performance: The yacht is almost entirely made of hybrid materials.

12.2 DESIGNING METAL MATRIX COMPOSITES

The most common state of loading in structures is that of bending. One measure of excellence in designing materials to carry bending moments at minimum weight is the index $E^{1/2}/\rho$, where E is Young's modulus and is the ρ the density. Alloys of aluminum and of magnesium rank highly by this criterion; titanium alloys and steels are not as good. How could the performance of magnesium (the best of the lot) be enhanced further? Table 12.1 summarizes the challenge.

Table 12.1 Design Requirements for the Material of the Panel

Function	Light, stiff beam
Constraint	Magnesium matrix
Objective	Maximize stiffness to weight in bending (index $E^{1/2}/\rho$)
Free variable	Choice of reinforcement and volume fraction

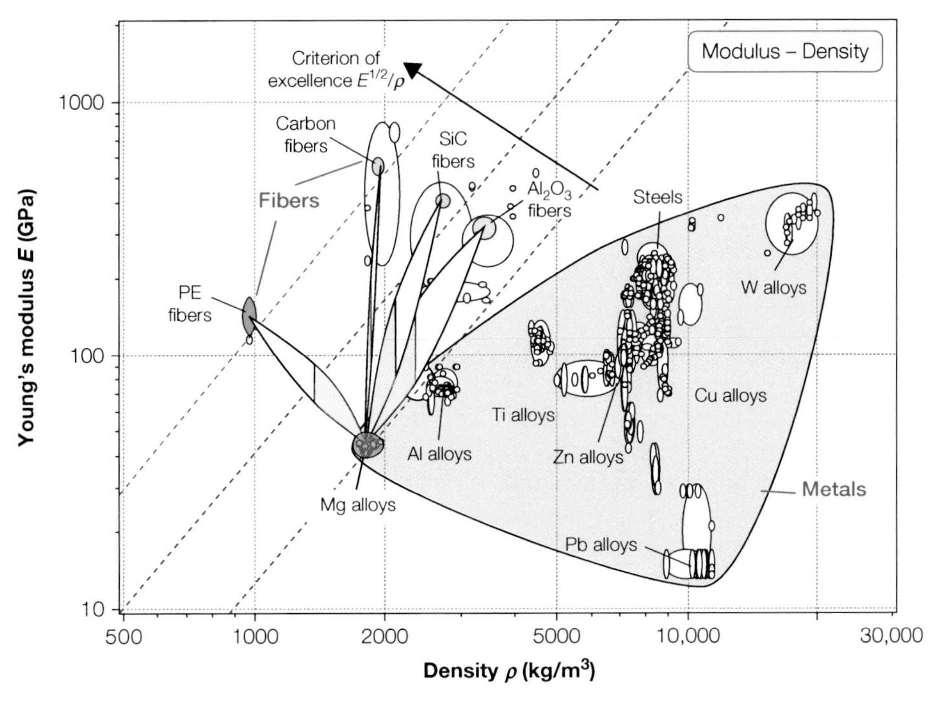

FIGURE 12.1

Possible magnesium-matrix composites. The lozenges show the areas bracketed by the upper and lower bounds of Table 12.2. The green shaded areas within them extend up to a volume fraction of 0.5.

Table 12.2 Superposition Rules for Composite Density and Modulus*

Property	Lower Bound	Upper Bound
Density	$\tilde{\rho} = f\rho_r + (1-f)\rho_m$ (exact)	
Modulus	$\tilde{E}_L = \dfrac{E_m E_r}{f E_m + (1-f)E_r}$	$\tilde{E}_u = f E_r + (1-f)E_m$

* *Subscripts m and r mean "matrix" and "reinforcement"; f = volume fraction.*

The method Figure 12.1 shows a chart of E and ρ for metals and fibers. Magnesium alloys appear at the extreme left of the red "Metals" envelope. The criterion of excellence is shown as a set of diagonal contours, increasing toward the top left. Magnesium ranks slightly higher than aluminum and much higher than titanium and steel.

Table 12.2 summarizes the superposition rules for density and modulus. They are plotted as envelopes of attainable performance for four magnesium-based composites; the upper edge of each envelope is the upper bound; the lower edge, the lower bound. There is an upper limit for the volume fraction, which we will set at 0.5. It is shown as a vertical bar within each envelope. Only the shaded part of the envelope below the bar is accessible. The diagonal lines plot the criterion of excellence, $E^{1/2}/\rho$. The combinations with the highest values of this quantity offer the greatest gain in stiffness per unit weight.

The results It is immediately apparent that the greatest promise is shown by the composites of magnesium with drawn polyethylene (PE) or carbon fibers; magnesium-SiC is less good. Magnesium-Al_2O_3 offers almost no gain at all. The method allows the potential of rapidly exploring alternative choices.

Postscript Magnesium-PE composites look good and are promising, but there remains the challenge of actually making them. Polyethylene fibers are already used in ropes and cables because of their high stiffness, strength, and low weight. But they are destroyed by temperatures much above 120°C, so casting or sintering of the magnesium around the fibers is not an option. One possibility is to use drawn PE sheets rather than fibers, and fabricate a multilayer laminate by bonding the PE sheets between sheets of magnesium. A second possibility is to explore ternary composites: dispersing magnesium powder in an epoxy and using this mix as the matrix to contain the PE fibers, for example. Otherwise we must fall back on magnesium-carbon, an attractive option.

Related case study

6.12 "Stiff, high damping materials for shaker tables"

12.3 FLEXIBLE CONDUCTORS AND PERCOLATION

A material is required for demountable seals in specialized equipment. The material must conform to the curved surfaces between which it is clamped, it must be electrically conducting to prevent charge build-up, and it should be moldable. Table 12.3 summarizes the requirements.

The method and results Metals, carbon, and some carbides and intermetallics are good conductors, but they are stiff and cannot be molded (Figure 12.2). Thermoplastic and thermosetting elastomers can be molded and are flexible, but they do not conduct. How are they to be combined? Metal coating of polymers is workable if the product will be used in a protected environment, but coatings are easily damaged. If a robust, flexible

Table 12.3 Materials for Flexible Conductors

Function	Flexible conducting solid
Constraints	Low Young's modulus to allow conformation Low resistivity to permit conduction ($\rho_e < 1000$ $\mu\Omega$.cm)
Objective	Able to be molded
Free variable	Choice of matrix, reinforcement, configuration, and volume fraction

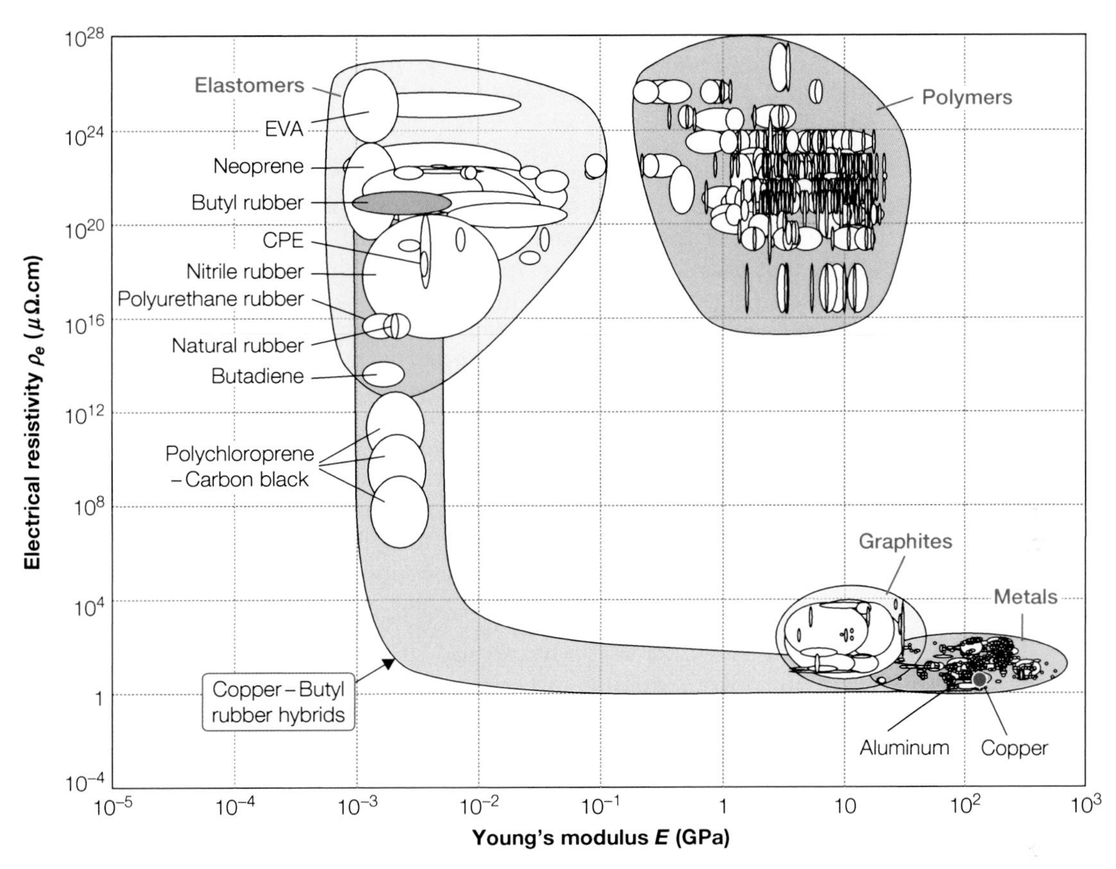

FIGURE 12.2
When conducting particles or fibers are mixed into an insulating elastomer, a hole in the material-property space is filled. Carbon-filled butyl rubbers lie in this part of the space.

product is needed, bulk rather than surface conduction is essential. This can be achieved by dispersing conducting material in a polymer. But how much? And in what shape?

Think of mixing conducting and insulating spheres of the same size by shaking them up in a jar. If there are only a few conducting spheres, they do not touch and the array as a whole is an insulator. If each conducting sphere contacts just one other, there is still no connecting path. If, on average, each touches two, there is still no path. Adding more conducting spheres gives larger clusters, but they can be large yet still discrete. For bulk conduction we need *connectivity*: The array first becomes a conductor when a single trail of contacts links one surface to the other, that is, when the volume fraction f of the conducting spheres reaches the *percolation threshold* f_c. Percolation problems are easy to describe but difficult to solve. Research since 1960 has provided approximate solutions to most of the percolation problems associated with the design of hybrids (see "Further reading" for a review). For simple cubic packing $f_c = 0.248$; for close packing $f_c = 0.180$. For a random array (the sphere shaken in a jar) it is somewhere in between—approximately 0.2.[1]

Making the spheres smaller smears the transition out. The percolation threshold is still 0.2, but the first connecting path is now thin and extremely devious—it is the only one out of the vast number of almost complete paths that actually connects. Increase the volume fraction and the number of conducting paths increases initially as $(f - f_c)^2$, then linearly, reverting to a rule of mixtures. If the particles are very small, as much as 40% may be needed to give good conduction. But a loading of 40% seriously degrades the moldability and flexibility of the polymer.

Shape gives a way out. If the spheres are replaced by slender fibers, they touch more easily and the percolation threshold falls. If their aspect ratio is $\beta = L/d$ (where L is the fiber length, d the diameter), then, empirically, the percolation threshold falls from f_c to roughly $f_c/\beta^{1/2}$, so an aspect ratio of 9 reduces p_c, the volume fraction of the conducting phase, by a factor of 3.

The concept of percolation is a necessary tool in designing hybrids. Electrical conductivity works that way; so too does the passage of liquids through foams or porous media—no connected path and no fluid flows; just one (out of a million possibilities) and there is a leak. Add a few more connections and there is a flood. Percolation ideas are particularly important in understanding the transport properties of hybrids: Properties that determine the flow of electricity or heat, of fluid, or of flow by diffusion, especially when the differences in properties of the components are extreme, because it is then that connected paths matter.

Figure 12.2 is a chart of electrical resistivity and modulus with metals, polymers, and composites plotted. There is a hole at the lower left where materials with the property combination we want would lie. The pale-green L-shaped envelope encloses the properties of hybrids of rubber with chopped copper wires. The resistivity drops sharply at the percolation threshold, which, for wires, falls to a few percent. The material retains all the flexibility of the elastomer but behaves as a bulk conductor.

Postscript Conducting elastomers exploiting these ideas are widely available. Such materials find application in antistatic clothing and mats, as pressure sensing elements, and even as solderless connections.

Related case study

12.4 "Extreme combinations of thermal and electrical conduction"

---

[1] These results are for infinite, or at least very large, arrays. Experiments generally give values in the range 0.19 to 0.22, with some variability because of the finite size of the samples.

12.4 EXTREME COMBINATIONS OF THERMAL AND ELECTRICAL CONDUCTION

Materials that are good electrical conductors are always good thermal conductors too. Copper, for example, excels at both. Most polymers, by contrast, are electrical insulators (meaning that their conductivity is so low that for practical purposes they do not conduct at all), and as solids go they are also poor thermal conductors—polyethylene is an example. Thus the "high-high" and "low-low" combinations of conduction can be met by monolithic materials, and there are plenty of them. The "high-low" and "low-high" combinations are a different matter: Nature gives us very few of either. The challenge is summarized in Table 12.4: Using only copper and polyethylene, devise hybrid materials that achieve both of these combinations (data for both are in Table 12.5).

The method and results Figure 12.3 shows two possible hybrid configurations. Both are of type "composite" but with very different configurations and volume fractions. The first is a tangle of fine copper wires embedded in a PE matrix. To describe its performance we draw on the bounds of Section 11.4. Electrical conductivity requires percolation; as explained in Section 12.3, the percolation threshold is minimized by using conducting wires of high aspect ratio. Above the percolation threshold the conductivity tends toward a rule of mixtures (Equation (11.9), with thermal conductivity replaced by electrical conductivity):

$$\widetilde{\kappa}_1 = f\,\kappa_{cu} + (1-f)\,\kappa_{PE} \tag{12.1}$$

Table 12.4 Requirements for the Conducting Hybrids of Copper and Polyethylene

Function	Extreme conduction combinations
Constraint	Materials: copper and polyethylene
Objective	Maximize difference between electrical and thermal conductivities
Configuration	Free choice
Free variable	Configuration and relative volume fractions of the two materials

Table 12.5 Data for Copper and HDPE

Material	Electrical Conductivity ($1/\mu\Omega$.cm)	Thermal Conductivity (W/m.K)
High conductivity copper	0.6	395
High density polyethylene	1×10^{-25}	0.16

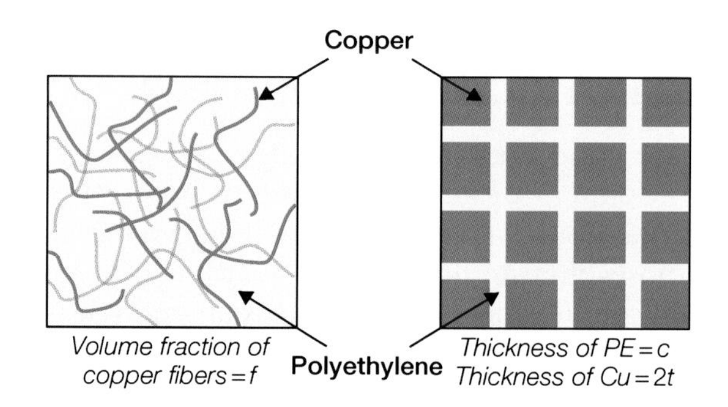

FIGURE 12.3
Two alternative configurations of copper and polyethylene, shown here in two dimensions but easily generalized to three. The one on the left has high electrical conductivity but low thermal conductivity; the one on the right has the opposite.

The thermal conductivity for a random array like this will lie near the lower bound (Equation (11.10)):

$$\tilde{\lambda}_1 = \lambda_{PE}\left(\frac{\lambda_{Cu} + 2\lambda_{PE} - 2f(\lambda_{PE} - \lambda_{Cu})}{\lambda_{Cu} + 2\lambda_{PE} + f(\lambda_{PE} - \lambda_{Cu})}\right) \tag{12.2}$$

The composite properties $\tilde{\kappa}_1$ and $\tilde{\lambda}_1$ are plotted in Figure 12.4, stepping through values of f, giving the upper curve.

The second hybrid is a multilayered composite with three orthogonal families of PE sheet separating blocks of copper. We draw again on the bounds of Section 11.4. When the PE layers are thin, the through-thickness thermal resistances add; the same is true of the electrical resistances. This means that both the electrical and thermal conductivities are given by the harmonic means (Equation (11.24)):

$$\tilde{\kappa}_2 = \left(\frac{f}{\kappa_{Cu}} + \frac{(1-f)}{\kappa_{PE}}\right)^{-1} \tag{12.3}$$

$$\tilde{\lambda}_2 = \left(\frac{f}{\lambda_{Cu}} + \frac{(1-f)}{\lambda_{PE}}\right)^{-1} \tag{12.4}$$

The composite properties $\tilde{\kappa}_2$ and $\tilde{\lambda}_2$ are also plotted in Figure 12.4, giving the lower curve. The shape of the curve is a consequence of the very different ranges of values of the two properties: a factor of 1000 for λ, a factor of 10^{25} for κ. The two hybrids differ dramatically in their behavior. The hybridization has allowed the creation of "materials" with extreme combinations of conductivities.

Postscript Hybrids of the first type are widely used to give electrical screening to computer and TV cabinets. Those of the second type are less common, but

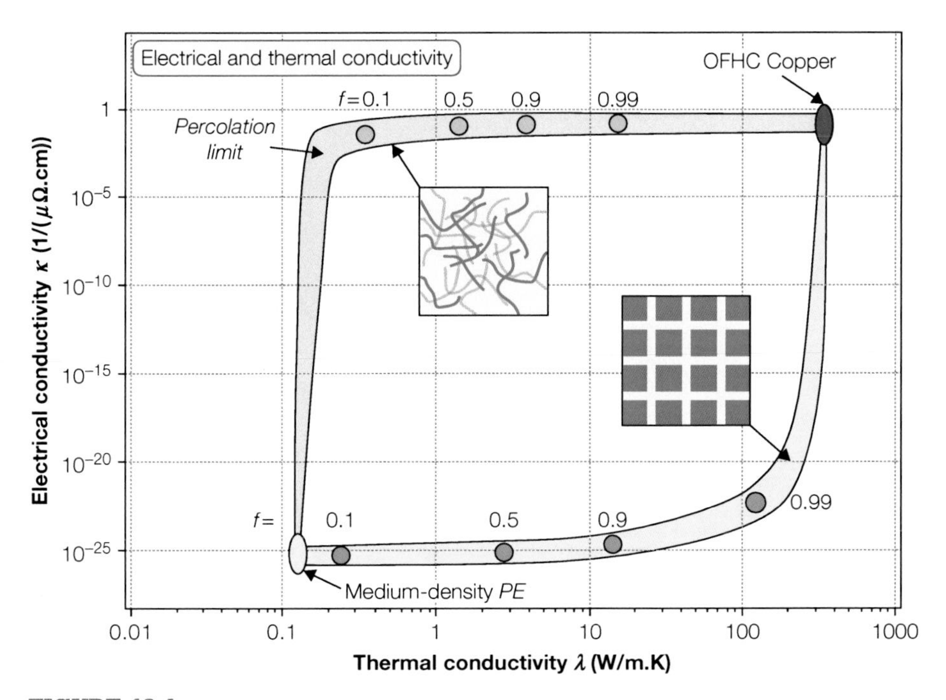

FIGURE 12.4

Two alternative hybrid configurations of copper and polyethylene give very different combinations of thermal and electrical conductivity, and create new "materials" with properties that are not found in homogeneous materials.

could find application as heat sinks for power electronics in which large-scale electrical conduction would lead to coupling and eddy current losses. How would you make the second type of hybrid? Perhaps by thermally bonding a stack of copper sheets interleaved with polyethylene film, dicing the stack normal to the layers and restacking the slices with PE film interlayers, and finally dicing and restacking a third time to give the last set of layers.

Related case studies

6.17 "Materials for heat exchangers"
12.3 "Flexible conductors and percolation"
12.7 "Connectors that don't relax their grip"

12.5 REFRIGERATOR WALLS

The panels of a refrigerator or freezer like that in Figure 12.5 perform two primary functions. The first is to insulate, and for this the through-thickness thermal conductivity must be minimized. The second is mechanical: The walls provide stiffness and strength, and support the shelves on which the contents rest. For a given thickness of panel, the first is achieved by

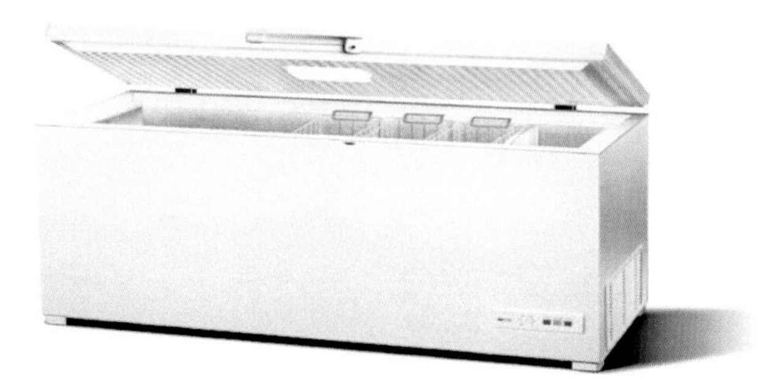

FIGURE 12.5
A refrigerator. The panels of the container unit must insulate, protect against the external environment, and be stiff and strong in bending.

Table 12.6 Design Requirements for the Insulating Panel

Function	Insulating panel
Constraints	Sufficient stiffness to suppress vibration and support internal loads Low cost Protect against environment Not too thick
Objective	Minimize heat transfer through thickness
Free variable	Material for faces and core; their relative thicknesses

Table 12.7 Superposition Rules for Sandwich Stiffness and Conductivity

Property	Lower Bound	Upper Bound
Flexural modulus	$\widetilde{E}_{flex} = \left(1 - \left(1 - \frac{2t}{d}\right)^3\right) E_f K_s$	$\widetilde{E}_{flex} = \left(1 - \left(1 - \frac{2t}{d}\right)^3\right) E_f$
Through-thickness conductivity	$\widetilde{\lambda}_{\perp} = \left(\frac{2t/d}{\lambda_f} + \frac{(1-2t/d)}{\lambda_c}\right)^{-1}$ (exact)	

minimizing $\widetilde{\lambda}$, where $\widetilde{\lambda}$ is the appropriate thermal conductivity. The second is achieved by seeking materials or hybrids that maximize $\widetilde{E}_{flex}$, where $\widetilde{E}_{flex}$ is the flexural modulus. (See Table 12.6.)

The method and results We select a hybrid of type "sandwich" and explore how the performance of various combinations of face and core compare with each other and with monolithic materials. The quantity $\widetilde{E}_{flex}$ for the sandwich is given by Equation (11.18) and its through-thickness thermal conductivity is given by Equation (11.24). Simplified equations describing them are assembled in Table 12.7, in which t is the face sheet thickness, d the panel thickness, E_f the modulus of the face material, λ_f and λ_c the conductivities of face and core, and K_s the knock-down factor to allow for core shear, ideally equal to 1 (no shear) but potentially as low as 0.5.

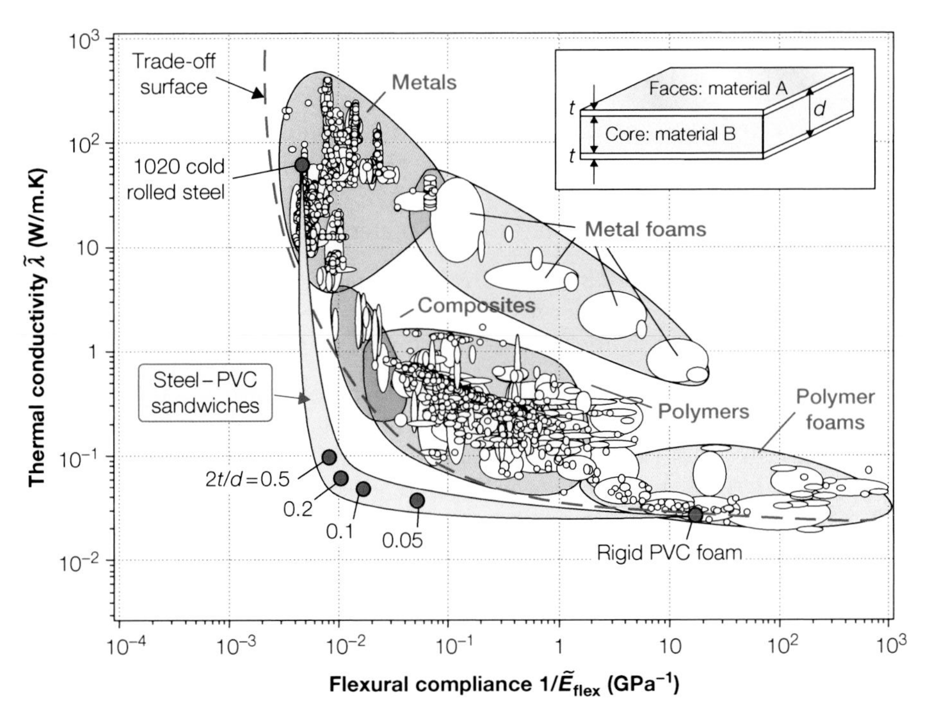

FIGURE 12.6

The blue L-shaped trajectory plots the performance of steel-faced, PVC foam cored sandwiches. The thermal performance is plotted on the vertical axis, the mechanical performance on the horizontal one. Both are to be minimized.

Figure 12.6 shows the appropriate plot, using the thermal conductivity $\tilde{\lambda}_{\perp}$ and the flexural compliance $1/\tilde{E}_{flex}$ instead of its inverse so that a minimum is sought for both quantities. The approximate performance of a sandwich with mild steel faces and a core of rigid foamed PVC is plotted using the equations in the table for four values of skin-to-core thickness; all other values are contained in the blue band. The panel offers combinations of stiffness and insulation that cannot be matched by monolithic metals, composites, polymers, or foams. Other combinations of face and core (aluminum or SMC faces with foamed polystyrene core, for example) can be evaluated efficiently using the same chart. Neither combination performs as well as the steel-polystyrene combination, though both come close.

Postscript Adhesive technology has advanced rapidly over the last two decades. Adhesives are now available to bond almost any two materials, and to do so with high bond strength (though some adhesives are expensive). Fabricating the sandwich should not be a problem.

Related case studies

6.14 "Energy-efficient kiln walls"

12.6 "Materials for microwave-transparent enclosures"

12.6 MATERIALS FOR MICROWAVE-TRANSPARENT ENCLOSURES

Microwave-transparent radomes were introduced in Section 6.19. The radome is a thin panel or shell, requiring flexural stiffness and strength but also requiring as low a dielectric constant, ε_r, as possible. Could hybrids offer better performance than monolithic materials?

The method and results Sandwich structures offer flexural stiffness and strength, and allow some control of electrical properties. We therefore explore these, seeking to meet the requirements of Table 12.8.

Figure 12.7 shows the flexural strength, σ_{flex}, of foams, polymers, and ceramics, plotted against the dielectric constant, ε_r. Many polymers have dielectric constants between 2 and 5. Dielectric response is an extensive property—if these polymers are foamed, the dielectric constant of the foam falls linearly with the relative density, approaching 1 at low densities (Equation (11.37)):

$$\tilde{\varepsilon}_r = 1 + (\varepsilon_{r,s} - 1)\left(\frac{\tilde{\rho}}{\rho_s}\right) \tag{12.5}$$

where $\varepsilon_{r,s}$ is the dielectric constant of the solid of which the foam is made and $\tilde{\rho}/\rho_s$ is its relative density. Foams, however, are not very strong. GFRP, with a dielectric constant of 5, is much stronger. Based on a survey of possible faces and cores from among those plotted in Figure 12.7, we choose to explore a sandwich with faces of GFRP and a core of a low-/medium-density expanded polymer foam. The flexural strength of the sandwich, provided it is properly manufactured and the core material has sufficient strength, is (Equation (11.19))

$$(\tilde{\sigma}_{flex})_U = (1 - (1-f)^2)\,\sigma_f + (1-f)^2 \sigma_c \tag{12.6}$$

The dielectric constant for the panel (Equation (11.12) with $f = 2t/d$) is

$$\tilde{\varepsilon}_r = f\,\varepsilon_f + (1-f)\varepsilon_c \tag{12.7}$$

Table 12.8 Requirements for Low Dielectric Constant Radome Skin

Function	Material for protection of microwave detector
Constraint	Must meet constraints on flexural strength σ_{flex}
Objective	Minimize dielectric constant ε_r
Free variables	Choice of material for face and core Relative thickness of the two materials

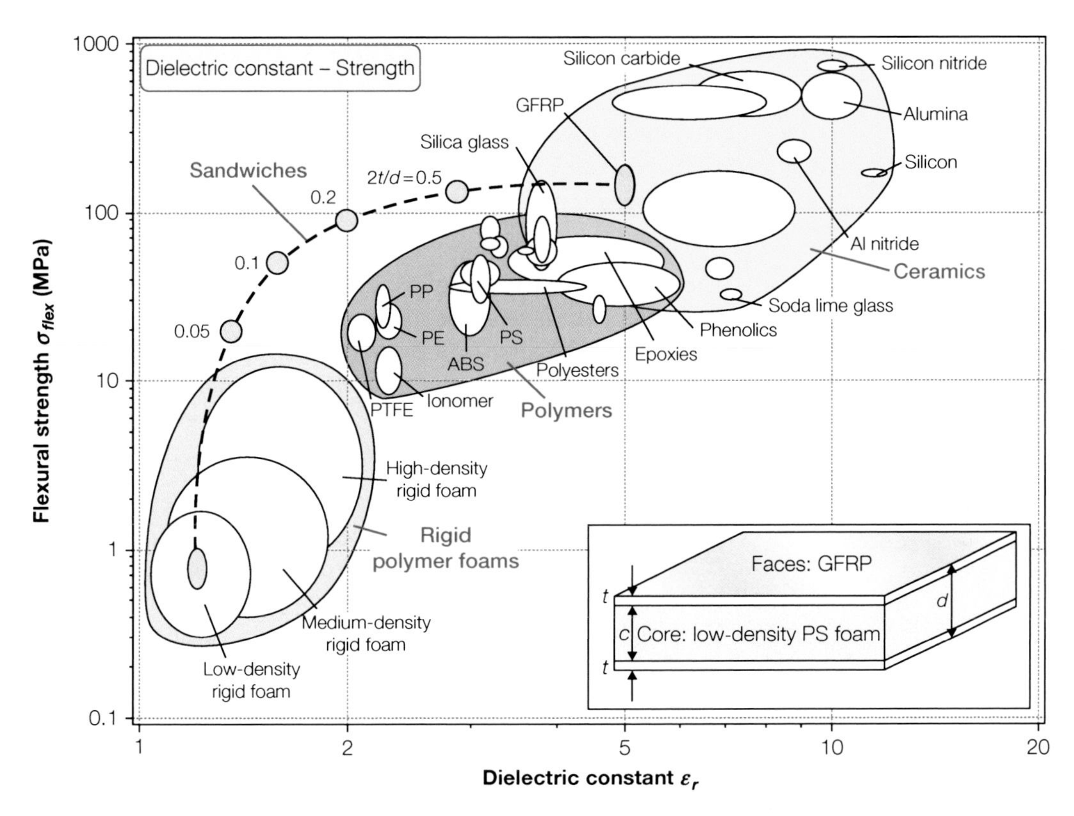

FIGURE 12.7

A plot of flexural modulus and dielectric constant for low dielectric constant materials. The trajectory shows the possibilities offered by hybrids of GFRP and polymer foam.

The simplest way to explore dialectric constants is to plot the two using $f = 2t/d$ as a parameter to link them. Figure 12.7 shows the results. The sandwich allows the creation of a set of materials with combinations of flexural strength and dielectric constant that outperform all the homogeneous materials that appear in the figure—indeed it outperforms even the best composites (not shown here). The figure works by identifying the desired flexural strength, σ_{flex}, and reading off the values of $2t/d$ and dielectric constant.

Related reading

Huddleston, G.K., & Bassett, H.L. (1984). Radomes. In R.C. Johnson, & H. Jasik (Eds.). *Antenna engineering handbook* (2nd ed.), chapter 44. McGraw-Hill, ISBN 0071475745.

Lewis, C.F. (1998). Materials keep a low profile. *Mechanical Engineer* (June), 37–41.

Related case study

6.19 "Materials for radomes"

12.7 CONNECTORS THAT DON'T RELAX THEIR GRIP

There are kilometers of wiring in a car. The transition to drive-by-wire control systems will increase this further. Wires have ends; they don't do much unless the ends are connected to something. The connectors are the problem: They loosen with time until, eventually, the connection is lost.

Car makers, responding to market forces, now design cars to run for at least 300,000 kilometers and last, on average, 10 years. The electrical system is expected to operate *without servicing* for the lifetime of the car. Its integrity is vital: You would not be happy in a drive-by-wire car with loose connectors. With increasing instrumentation on engine and exhaust systems, many of the connectors get hot; some have to maintain good electrical contact at temperatures up to 200°C (see Table 12.9).

The primary material choice for connectors today is a copper-beryllium alloy, Cu 2% Be: It has excellent conductivity and the high strength needed to act as a spring to give the required clamping force at the connection. But the maximum long-term service temperature of copper-beryllium alloys is only about 130°C; at higher temperatures, creep relaxation causes the connector to loosen its grip. The challenge: to suggest a way to solve this problem.

The method and results The answer is to separate the functions, select the best material for each, check for compatibility, and combine the materials to make a hybrid. So here goes. Function 1: Conduct electricity. Copper excels at this; no other affordable material is as good. Its alloys (among them copper-beryllium) are stronger, but at the cost of some loss of conductivity and a great increase in price. We choose copper to provide the conduction. Function 2: Provide clamping force for the lifetime of the vehicle. The material chosen to fill function 2 will have to be bonded to the copper, and if the combination is not to distort when heated, it must have the same expansion coefficient.

Figures 12.8 and 12.9 guide the choice. The first shows the maximum service temperature and expansion coefficient for copper, Cu-2% Be, and a range of steels. The box encloses materials with the same expansion coefficient as copper. Type 302 and 304 austenitic stainless steels match copper in expansion coefficient and can be used at much higher temperatures. But do they

Table 12.9 Design Requirements for the Connector

Function	Hybrid conductor for electrical connector
Constraints	Provide good electrical connection Maintain clamping force at 200°C for life of vehicle
Objective	Minimize cost
Free variable	Material 1 and 2; their relative thicknesses

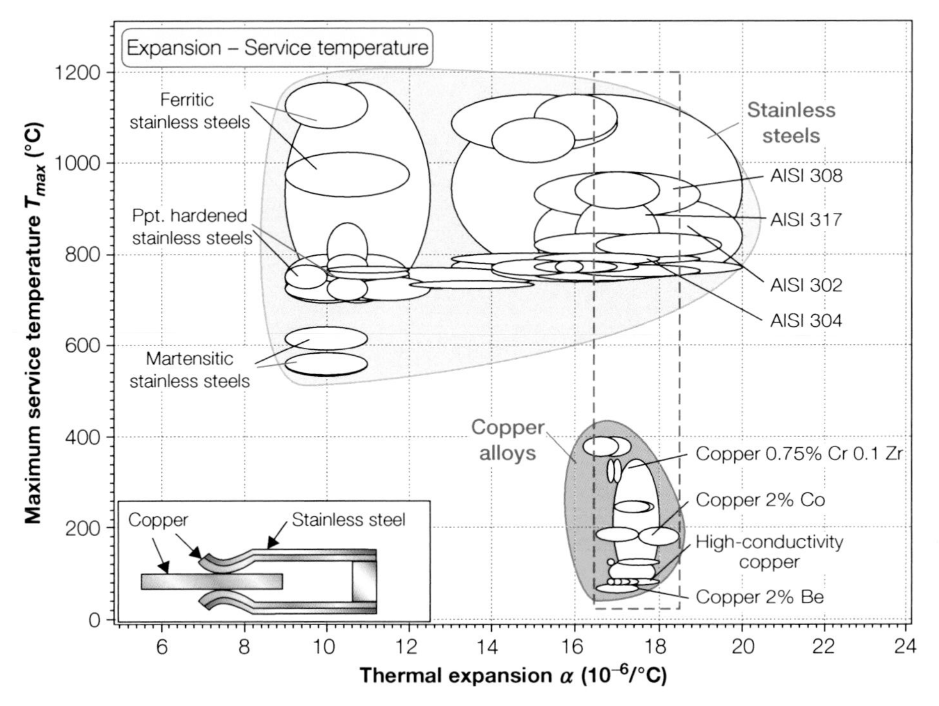

FIGURE 12.8

A hybrid connector. We seek materials with matching thermal expansion, one of which retains strength and stiffness well above 200°C. Copper is chosen for Material 1 because of its excellent electrical conductivity. Type 302 or 304 stainless steel is a good choice for Material 2.

make good springs? And are they affordable? Figure 12.9 answers these questions. Good materials for springs (Section 6.7) are those with high values of σ_f^2/E—this appears as one axis of the chart. The other axis is approximate price/kg. The chart shows that both 302 and 304 stainless steels, in the wrought condition, are almost as good as Cu 2% Be as a spring and considerably cheaper.

Postscript The proposed solution, then, is a hybrid of copper and type 302 stainless steel, roll-bonded to form a bilayer like that shown in the inset on both figures. Further detailed design will, of course, be needed to establish the thicknesses of each layer, the best degree of cold work, the formability, and the resistance to the environment in which it will be used. But the method has guided us to a sensible concept, quickly and efficiently.

Related case studies

6.7 "Materials for springs"
10.8 "Ultra-efficient springs"

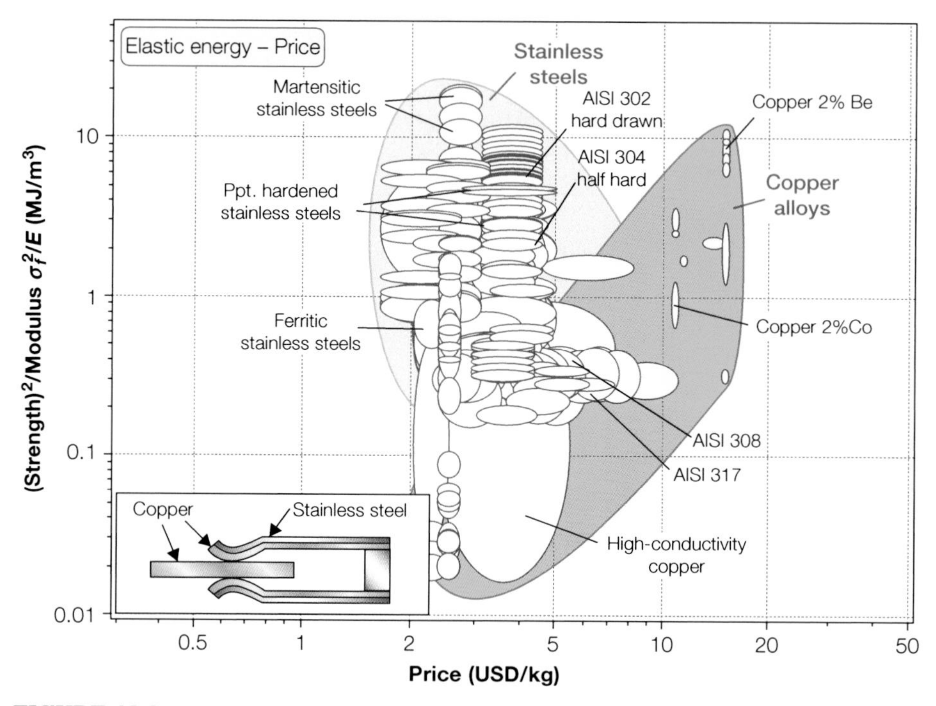

FIGURE 12.9

The connector has two jobs—to conduct and to exert a clamping force that does not relax. High-conductivity copper and 304 stainless steel are both much cheaper that Cu 2% Be. Roll-bonding them will of course add cost, but in large volume production it could be competitive—and it solves the relaxation problem.

12.8 EXPLOITING ANISOTROPY: HEAT-SPREADING SURFACES

A saucepan made from a single material, when heated on an open flame, develops hot spots that can burn its contents at those spots. That is because the saucepan is thin; heat is transmitted through the thickness more quickly than it can be spread transversely to bring the entire pan surface to a uniform temperature. The metals of which saucepans are usually made—cast iron, aluminum, stainless steel, or copper—have thermal conductivities that are isotropic, the same in all directions. What we clearly want is a thermal conductivity that is higher in the transverse direction than in the through-thickness direction. A bilayer (or multilayer) hybrid can achieve this. Table 12.10 summarizes the situation.

The method and results Heat transmitted in the plane of a bilayer (Figure 12.10, *inset*) has two parallel paths; the total heat flux is a sum of

Table 12.10 Design Requirements for the Panel

Function	Heat-spreading surface
Constraint	Temperature up to 200°C without distortion
Objective	Maximize thermal anisotropy while maintaining good conduction
Free variable	Choice of materials and their relative thicknesses

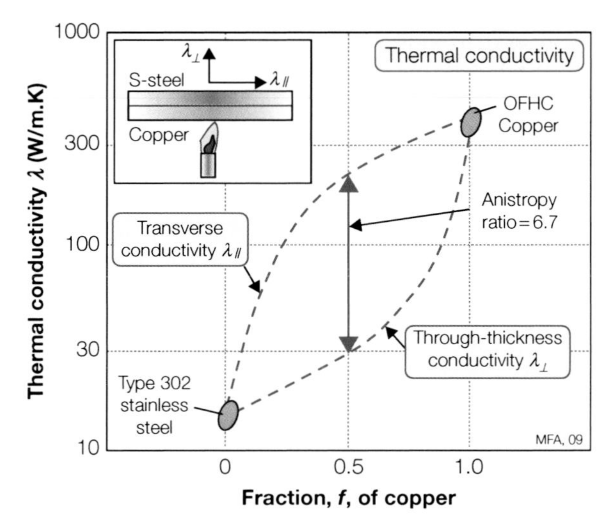

FIGURE 12.10

Creating anisotropy. A bilayer of copper and stainless steel creates a "material" with good conductivity and an anisotropy ratio greater than 6.

the heat flux in each of the paths. If it is made of a layer of material 1 with thickness t_1 and conductivity λ_1, bonded to a layer of material 2 with thickness t_2 and conductivity λ_2, the conductivity parallel to the layers is

$$\tilde{\lambda}_{//} = f\,\lambda_1 + (1-f)\lambda_2 \tag{12.8}$$

(Equation (11.9) with $f = t_1/(t_1 + t_2)$. Perpendicular to the layers the conductivity is

$$\tilde{\lambda}_{\perp} = \left(\frac{f}{\lambda_1} + \frac{(1-f)}{\lambda_2}\right)^{-1} \tag{12.9}$$

(the harmonic mean of Equation (11.24)).

If the bilayer is made of materials that differ greatly in thermal expansion coefficient, the pan will distort when heated. We therefore seek a pair of materials with almost the same expansion coefficient but that differ as much as possible in conductivity in order to maximize the difference between Equations (12.10) and (12.11).

The results Copper is an excellent thermal conductor. We saw, in Case Study 12.7, that the thermal expansion of copper matches that of 302 or 304 stainless steel. The thermal conductivities of the two, however, differ. Figure 12.10 shows the conductivities $\widetilde{\lambda}_{\perp}$ and $\widetilde{\lambda}_{//}$ of the bilayer as a function of the relative thickness f for a bilayer of copper (λ_1 = 390 W/m.K) and austenitic stainless steel (λ_2 = 16 W/m.K). The maximum separation between $\widetilde{\lambda}_{\perp}$ and $\widetilde{\lambda}_{//}$ occurs broadly at f = 0.5 (each material occupies about half the thickness) when the ratio of the two conductivities (the anisotropy ratio) is 6.7. The hybrid has extended the occupied area of property space along an unusual dimension—thermal anisotropy.

Related case studies

6.15 "Materials for passive solar heating"
12.7 "Connectors that don't relax their grip"

12.9 THE MECHANICAL EFFICIENCY OF NATURAL MATERIALS

"As a general principle natural selection is continually trying to economize every part of the organization."

—Charles Darwin, writing 150 years ago

Natural materials are remarkably efficient. By efficient we mean that they fulfill the complex requirements posed by the way plants and animals function and that they do so using as little material as possible. Many of these requirements are mechanical in nature: the need to support static and dynamic loads created by the mass of the organism or by wind loading, the need to store and release elastic energy, the need to flex through large angles, and the need to resist buckling and fracture.

Virtually all natural materials are hybrids. They consist of a relatively small number of polymeric and ceramic components or building blocks that are often composites themselves. Plant cell walls, for instance, combine cellulose, hemicellulose, and pectin, and can be lignified; animal tissue consists largely of collagen, elastin, keratin, chitin, and minerals such as salts of calcium or silica. From these limited "ingredients" nature fabricates a remarkable range of structured hybrids. Wood, bamboo, and palm consist of cellulose fibers in a lignin/hemicellulose matrix, shaped to hollow prismatic cells of varying wall thickness. Hair, nail, horn, wool, reptilian scales, and hooves are made of keratin while insect cuticle contains chitin in a matrix of protein. The dominant ingredient of mollusk shell is calcium carbonate, bonded with a few percent of protein. Dentine, bone, and antler are formed of "bricks" of hydroxyapatite cemented together with collagen. Collagen is the basic structural element for soft and hard tissues in animals, such as tendon, ligament, skin, blood vessels, muscle, and cartilage, often used in ways that exploit shape.

From a mechanical point of view, there is nothing very special about the individual building blocks. Cellulose fibers have Young's moduli that are about the same as those of nylon fishing line, but a lot less than those of steel; and the lignin-hemicellulose matrix in which they are embedded has properties very similar to those of epoxy. Hydroxyapatite has a fracture toughness comparable with that of human-made ceramics; that is to say it is brittle. It is thus the *configuration* of the components that give rise to the striking efficiency of natural materials.

The complex hierarchies of configurations that make up wood and bone are illustrated in Figures 12.11 and 12.12. The stiffness, strength, and toughness of wood derive largely from the stiffness, strength, and toughness of the cellulose molecule, shown on the left of Figure 12.11. Crystalline microfibrils are built up of aligned molecules some 30 to 60 nm in length. These form the reinforcing fibers of the lamellae, the matrix of which is amorphous lignin and hemicellulose. Stacks of lamellae in the four-layer pattern and fiber orientation (*right*) become the structural material of the cell wall. The fibrils of the primary cell wall interweave randomly, as in cotton wool. In

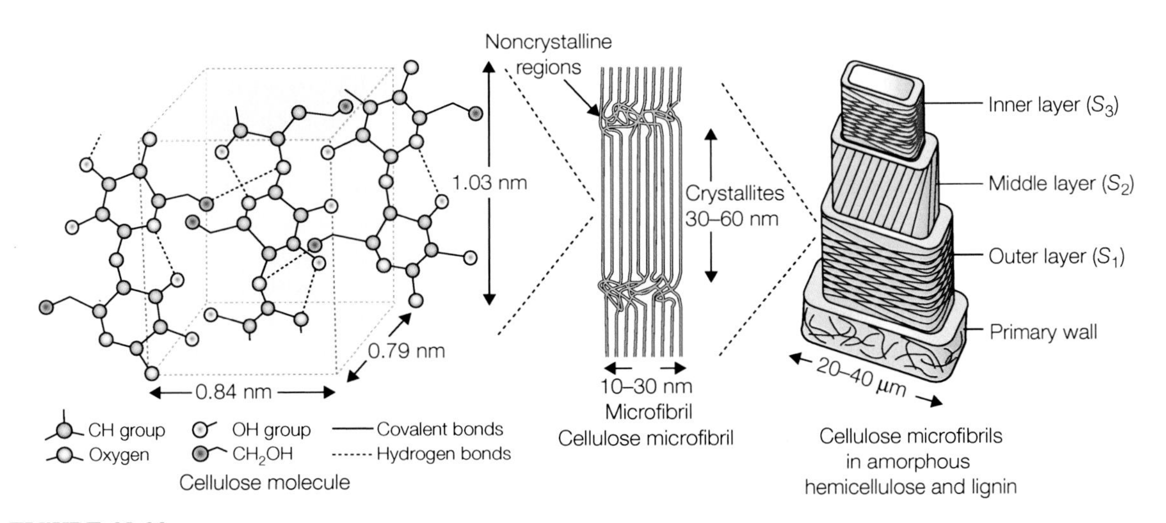

FIGURE 12.11
The hierarchical structure of wood.

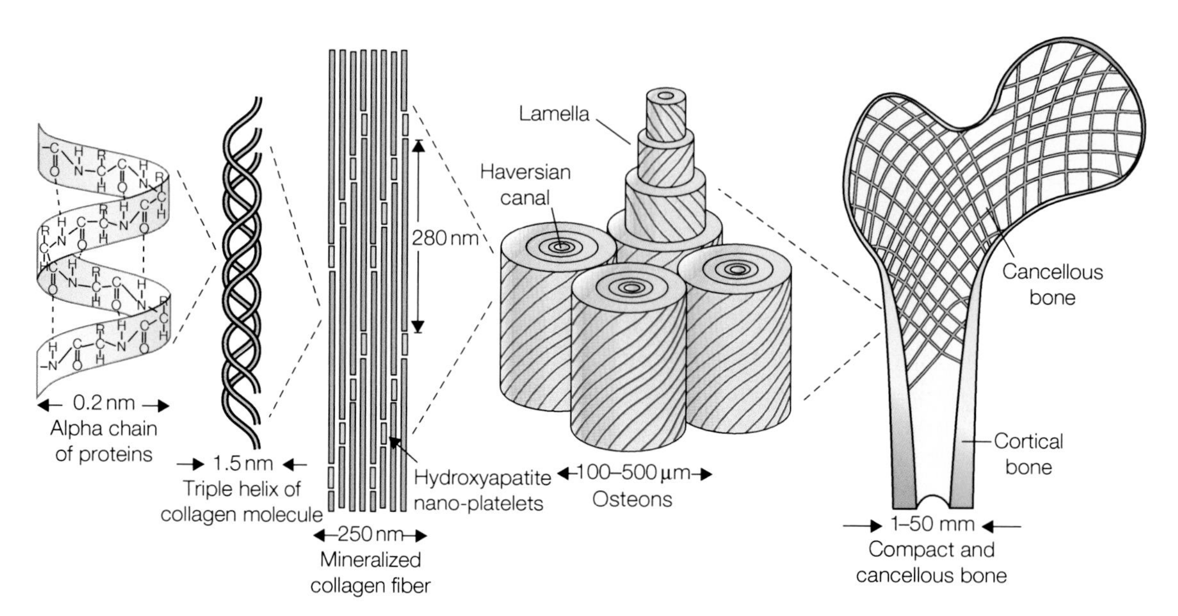

FIGURE 12.12
The hierarchical structure of bone.

subsequent layers the fibrils are parallel and closely packed. The outer S_1 layer has lamellae with alternate right- and left-handed spiral winding of fibrils. Beneath it, the thicker S_2 layer has fibrils that are more nearly oriented along the axis of the cell. The innermost S_3 layer has a lay-up like that of S_1. The cell as a whole is bonded to its neighbors by the middle lamella (not shown), a lignin-pectin complex devoid of cellulose.

There is an interesting parallel between the hierarchical structure of bone and that of wood, despite the great differences in their molecular chemistry (Figure 12.12). The starting point here is the triple-helical structure of the collagen molecule, shown on the left. But—unlike wood—this becomes the matrix, not the reinforcement of the mineralized tissue of bone. Hydroxyapatite nanoplatelets deposit in the nascent tissue and increase in volume fraction over time to produce the mature osteons, with an ordered lay-up of highly mineralized fibers with the strength and stiffness to support structural loads that bone must carry in a mature organism. At the most macro scale, nearly fully dense compact bone provides the outer structure of whole bone while highly porous trabecular bone fills the vertebrae, shell-like bones such as the skull, and the ends of the long bones such as the femur.

The same as with engineering materials, the building blocks of natural materials can be grouped into classes: *bio-ceramics* (calcite, aragonite, hydroxyapatite), *biopolymers* (the organic building blocks: polysaccharides; cellulose;

and the proteins chitin, collagen, silk, and keratin), and *natural elastomers* (elastin, resilin, abductin, skin, artery, and cartilage). These combine to give a range of hybrids, among them composites and sandwiches (bone, antler, enamel, dentine, shell, cuttle, and coral), cellular structures (natural cellular materials such as wood, cork, palm, bamboo, trabecular bone), and segmented structures (scales, hair). Their properties, like those of engineering materials, can be explored and compared using material property charts. We conclude this chapter by examining charts for natural hybrids and the building blocks from which they are made.

The Young's modulus–density chart Figure 12.13 shows data for Young's modulus, E, and density, ρ. Those for the classes of natural materials are circumscribed by large envelopes; class members are shown as smaller bubbles within them. Data for woods, palms, corks, trabecular (foam-like)

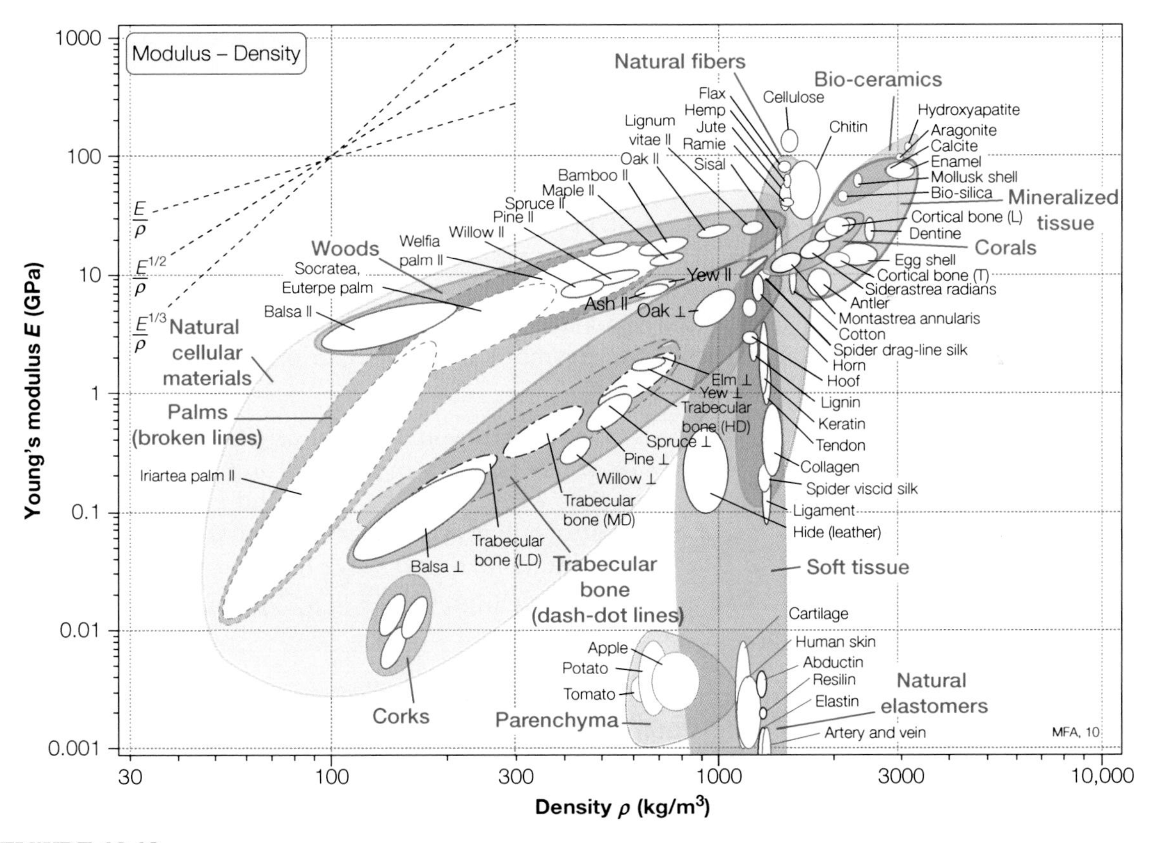

FIGURE 12.13

A material-property chart for natural materials, plotting Young's modulus against density. The guide lines identify structurally efficient materials that are light and stiff.

bone, and corals are enclosed in subsidiary envelopes. Natural fibers (silk, flax, jute, and the like) have their own envelope, as does mineralized tissue (bone, shell, etc.) and soft tissue (ligament, cartilage, and so forth). Where they differ, the moduli parallel (symbol ||) and perpendicular (symbol ⊥) to the fiber orientation or grain have been plotted separately. Trabecular bone exhibits a particularly wide range of densities; for these, three bubbles for high density (HD), medium density (MD), and low density (LD) are plotted. The familiar stiffness guidelines E/ρ, $E/p^{1/2}/\rho$, and $E^{1/3}/\rho$ are shown, each representing the material index for a particular mode of loading.

The natural polymer with the highest efficiency in tension, measured by the index E/ρ, is cellulose; it exceeds that of steel by a factor of about 2.6. The high values for the fibers flax, hemp, and cotton derive from this. Wood, palm, and bamboo are particularly efficient in bending and resistant to buckling, as indicated by the high values of the flexure index $E^{1/2}/\rho$ when loaded parallel to the grain. That for balsa wood, for example, can be five times greater than that for steel.

The tensile strength–density chart Figure 12.14 shows data for the strength, σ_f, and density, ρ, of natural materials. The color coding and envelope scheme parallels that of Figure 12.13. For natural ceramics the tensile strength is identified with the flexural strength (modulus of rupture) in the beam-bending symbol (T). For natural polymers and elastomers, the strengths are tensile strengths. And for natural cellular materials, the tensile stress is either the plateau stress, or flexural strength, symbol (T), depending on the nature of the material. Where they differ, the strengths parallel (symbol ||) and perpendicular (symbol ⊥) to the fiber orientation or grain have been plotted separately.

Evolution to provide tensile strength would—we anticipate—lead to materials with high values of σ_f/ρ, where strength in bending or buckling is required we expect to find materials with high $\sigma_f^{2/3}/\rho$. Silk and cellulose have the highest values of σ_f/ρ; that of silk is even higher than that of carbon fibers. The fibers of flax, hemp, and cotton, too, have high values of this index. Bamboo, palm, and wood have high values of $\sigma_f^{2/3}/\rho$, giving resistance to flexural failure.

The Young's modulus–strength chart Data for the strength, σ_f, and modulus, E, of natural materials are shown in Figure 12.15. Two of the combinations are significant. Materials with large values of σ_f^2/E store elastic energy and make good springs; those with large values of σ_f/E have exceptional resilience. Both are plotted in the figure. Silks (including the silks of spider webs) stand out as exceptionally efficient, having values of σ_f^2/E that exceed those of spring steel or rubber. High values of the other index, σ_f/E,

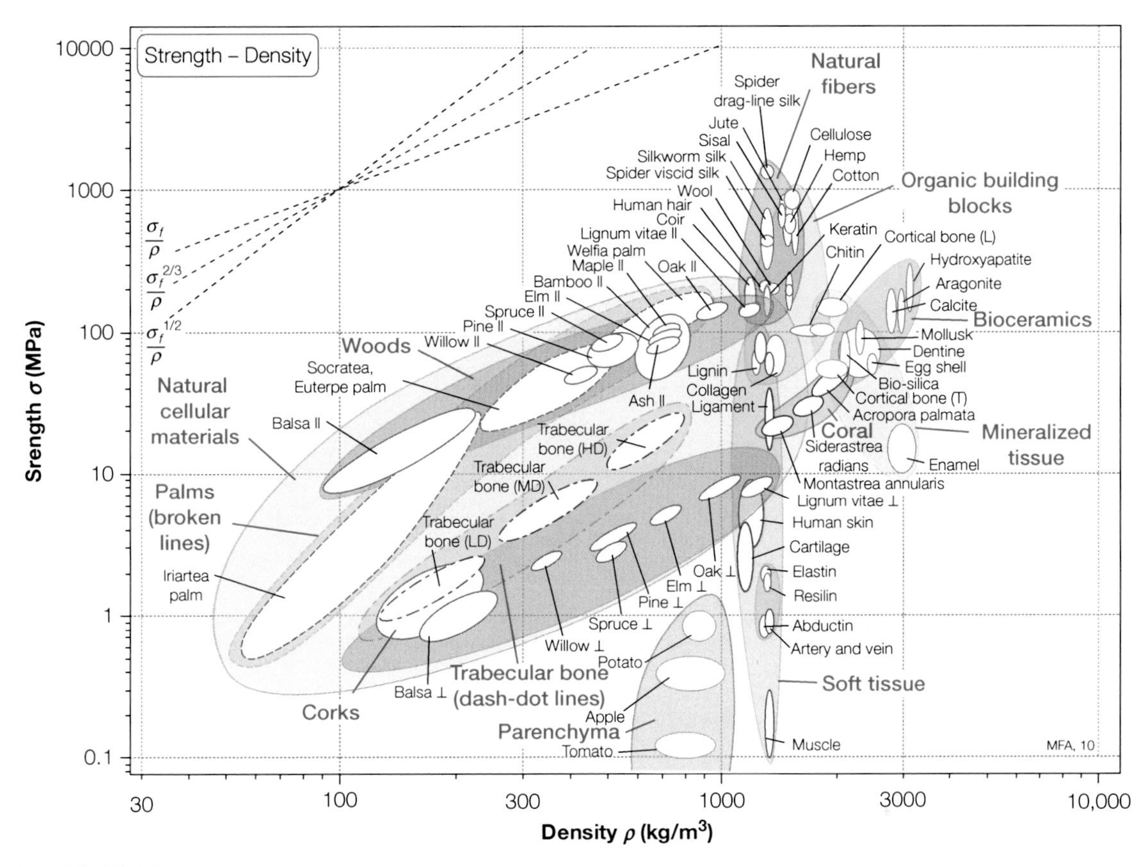

FIGURE 12.14

A material-property chart for natural materials, plotting strength against density. Guide lines identify structurally efficient materials that are light and strong.

mean that a material allows large, recoverable deflections and, for this reason, makes good elastic hinges. Palm (coconut timber) has a higher value of this index than wood, allowing palms to flex in a high wind. Nature makes much use of these: Skin, leather, and cartilage are all required to act as flexural and torsional hinges.

The toughness–Young's modulus chart The toughness of a material measures its resistance to the propagation of a crack. The limited data for the toughness, J_c, and Young's modulus, E, of natural materials are shown in Figure 12.16. When the component is required to absorb a given *impact energy* without failing, the best material will have the largest value of J_c. These materials lie at the top of Figure 12.16: Antler, hoof, horn, bamboo, and woods stand out. When instead a component containing a crack must

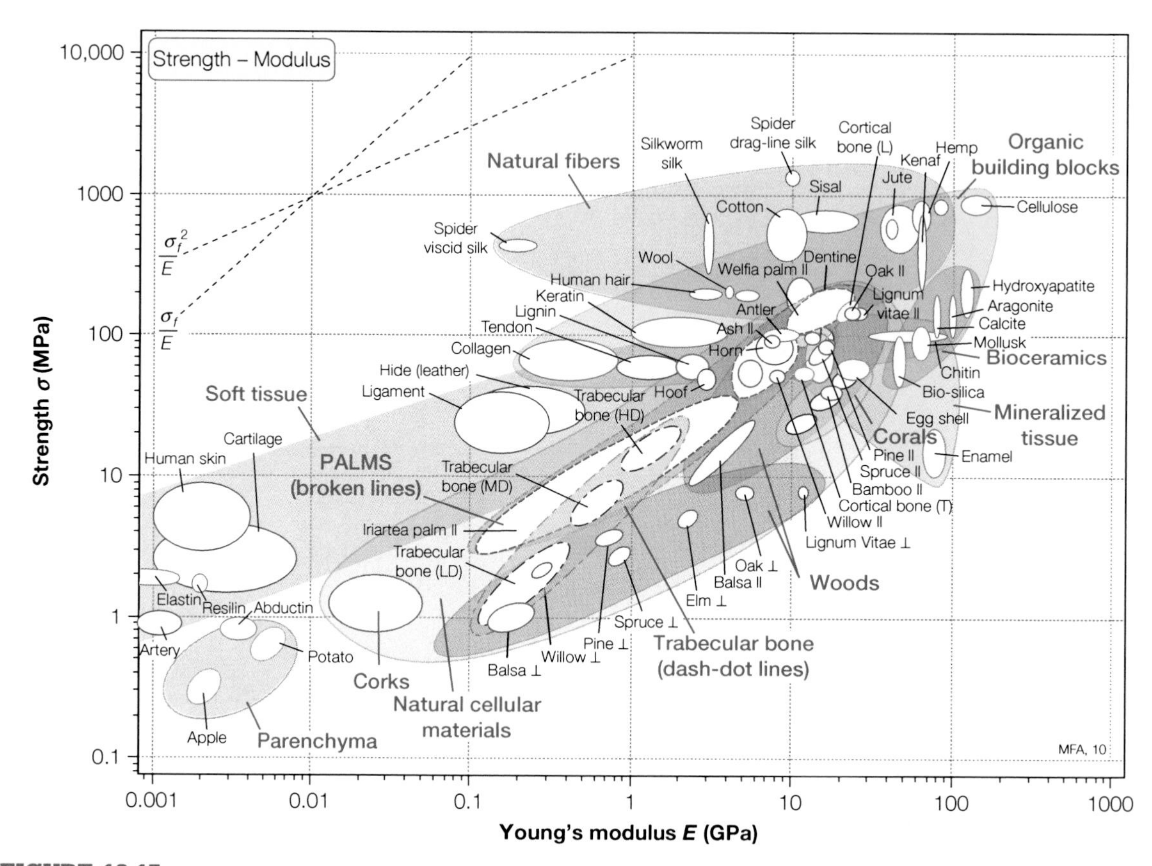

FIGURE 12.15

A material-property chart for natural materials, plotting Young's modulus against strength. The guide lines identify materials that store the most elastic energy per unit volume and that make good elastic hinges.

carry a given *load* without failing, the safest choice of material is that with the largest values of the fracture toughness $K_{1c} \approx (EJ_c)^{1/2}$. Diagonal contours sloping from the upper left to the lower right in Figure 12.16 show values of this index. Mollusk shell and tooth enamel stand out.

Many engineering materials (e.g., steels, aluminum, alloys) have values of J_c and K_{1c} that are much higher than those of the best natural materials. However, the toughness of natural ceramics like nacre, dentine, cortical (dense) bone, and enamel is an order of magnitude higher than that of conventional engineering ceramics like alumina. Their toughness derives from their segmented structure: platelets of ceramics such as calcite, hydroxyapatite or aragonite, bonded by a small volume fraction of polymer, usually collagen; it increases with decreasing mineral content and increasing collagen content.

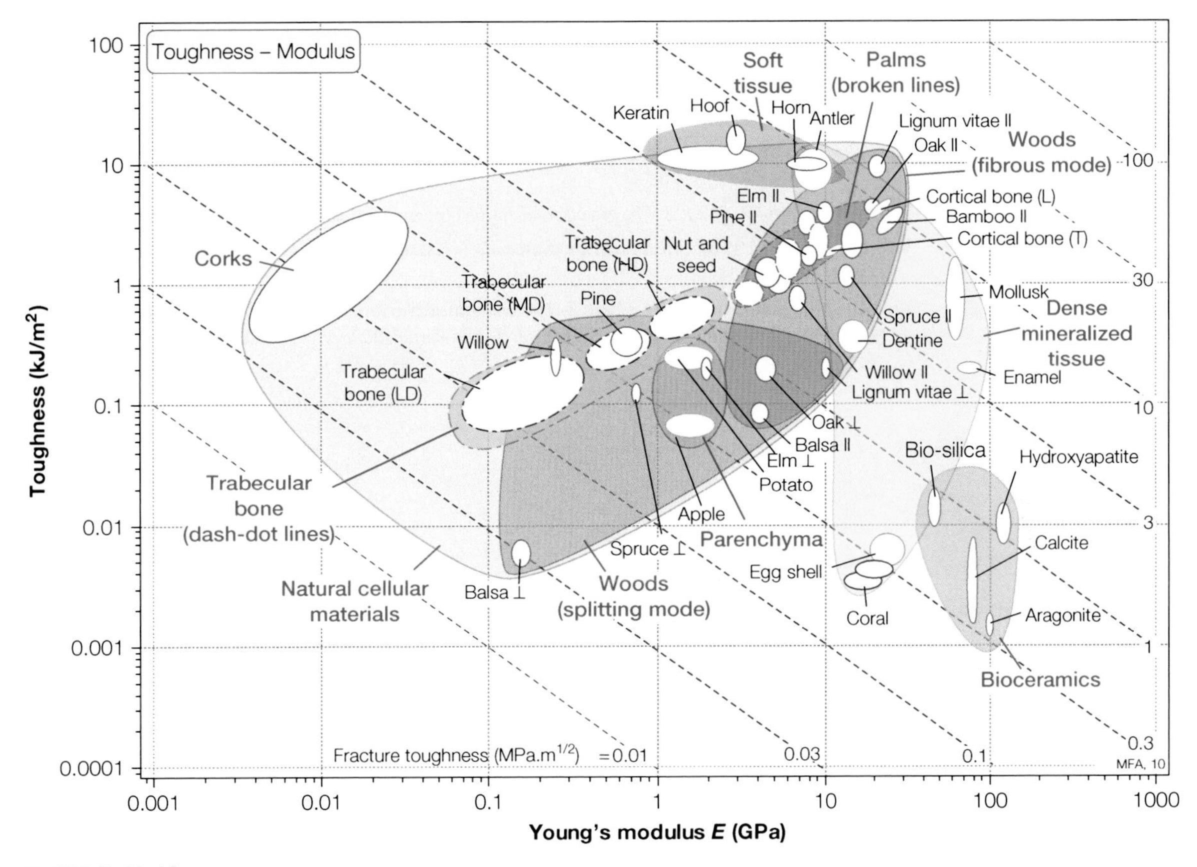

FIGURE 12.16

A material-property chart for natural materials, plotting toughness against Young's modulus. The guide lines show fracture toughness $(EJ_c)^{1/2}$ $(\text{MPa})^{1/2}$.

Postscript The charts bear out Darwin's idea that natural materials evolve to make the most of what is available to them—they are efficient, in the sense used earlier. But there is more to it than that. For deeper insight, see the books listed in "Further reading."

12.10 FURTHER READING: NATURAL MATERIALS[2]

Beukers, A., & van Hinte, E. (1998). *Lightness. The inevitable renaissance of minimum energy structures.* 010 Publishers, ISBN 9064503346.

Currey, J.D., Wainwright, S.A., & Biggs, W.D. (1982). *Mechanical design in organisms.* Princeton University Press, ISBN 0691083088.

[2] See also Appendix D.

Gibson, L.J., Ashby, M.F., & Harley, B.A. (2010). *Cellular materials in nature and in medicine.* Cambridge University Press, ISBN 9-7805-2119-5447.
A monograph exploring the structure, mechanics, and use of cellular materials in nature.

McMahon, T., & Bonner, J. (1983). *On size and life.* American Books, ISBN 0716750007.

Sarikaya, M., & Aksay, I.A. (Eds.) (1995). *Biomimetics: Design and processing of materials.* AIP Press, ISBN 1563961962.

Thompson, D'A.W. (1992). *On growth and form.* Dover Publications, ISBN 0486671356.

Vincent, J.F.V. (1990). *Structural biomaterials (revised ed.).* Princeton University Press, ISBN 0691025134.

Vincent, J.F.V., & Currey, J.D. (1980). The mechanical properties of biological materials. In *Proceedings of the Symposia of the Society for Experimental Biology,* 34. Cambridge University Press for the Society for Experimental Biology.

CHAPTER 13

Processes and Process Selection

Die casting. Image courtesy of Thomas Publishing, New York.

CONTENTS

Materials Selection in Mechanical Design. DOI: 10.1016/B978-1-85617-663-7.00013-8

13.1 INTRODUCTION AND SYNOPSIS

A *process* is a method of shaping, joining, or finishing a material. *Sand casting, injection molding, fusion welding,* and *electro-polishing* are all processes; there are hundreds of them. The choice, for a given component, depends on the material of which it is to be made; on its size, shape, and required precision; and on how many are to be made—in short, on the *design requirements.*

To select processes we must first classify them. Section 13.2 develops the classification. It is used to structure Section 13.3, in which process *families* and their *attributes* are described: the materials they can handle, the shapes they can make, and the precision with which they can do it.

Processing has dual functions. The obvious one is that of shaping, joining, and finishing. The less obvious one is that of property control. Metals are strengthened by rolling and forging; steels are heat-treated to enhance hardness and toughness; polymers are drawn to increase modulus and strength; and ceramics are hot-pressed, again to increase strength. Process-property relationships are explored more closely in Section 13.4.

Process selection—finding the best match between process attributes and design requirements—is the subject of Sections 13.5 and 13.6. In using the methods developed there, one should not forget that material, shape and processing interact (Figure 13.1). Material properties and shape limit the choice of

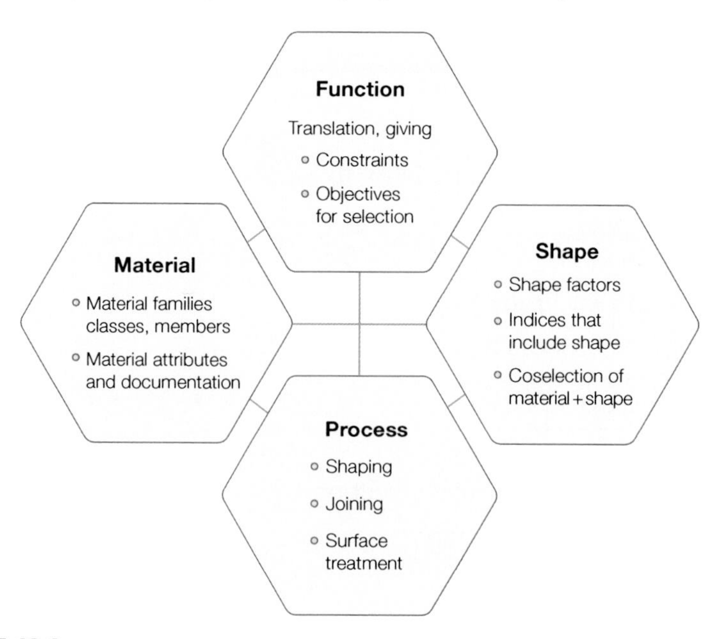

FIGURE 13.1
Processing selection depends on material and shape. The "process attributes" are used as criteria for selection.

process: Ductile materials can be forged, rolled, and drawn; those that are brittle may have to be shaped using powder methods. Materials that melt at modest temperatures to low-viscosity liquids can be cast; those that do not have to be processed by other routes. Shape, too, can influence the choice of process. Slender shapes can be made easily by rolling or drawing but not by casting. Hollow shapes cannot be made by forging, but they can by casting or molding. Conversely, processing affects properties. Rolling and forging change the hardness and texture of metals and align the inclusions they contain, enhancing strength and ductility. Heat treatment allows manipulation of strength, ductility, and toughness. Composites do not exist until they are processed; before processing, they are just a soup of polymer and a sheaf of fibers.

Like the other aspects of design, process selection is an iterative procedure. The first iteration gives one or more possible process routes. The design must then be rethought to adapt it, as far as possible, to ease manufacture by the most promising route. The final choice is based on a comparison of *process cost*, requiring the use of cost models developed in Section 13.6, and on *documentation*: guidelines, case histories, and examples of process routes used for related products. Documentation also helps in dealing with the coupling between process and material properties.

The chapter ends, as always, with a summary and annotated recommendations for further reading.

13.2 CLASSIFYING PROCESSES

Manufacturing processes can be classified under the headings shown in Figure 13.2. *Primary shaping* create shapes. The first row lists six classes of primary forming processes: casting, molding, deformation, powder methods, methods for forming composites, and special methods such as rapid prototyping. *Secondary processes* modify shapes or properties; here they are shown as "machining," which adds features to an already shaped body, and "heat treatment," which enhances surface or bulk properties. Below these come *joining* an, *surface treatment* or *finishing*.

Figure 13.2 has merit as a flow chart; it shows a progression through the manufacturing route. It should not be treated too literally, however; the order of the steps can be varied to suit the needs of the design. The point it makes is that there are three broad process families: shaping, joining, and finishing. The attributes of one family differ so greatly from those of another that, in assembling and structuring data for them, they must be treated separately.

To organize processes in more detail, we need a hierarchical classification like that used for materials in Chapter 5. Figure 13.3 shows part of it. The process universe has three families: shaping, joining, and finishing. In this figure, the shaping family is expanded to show classes: casting, deformation,

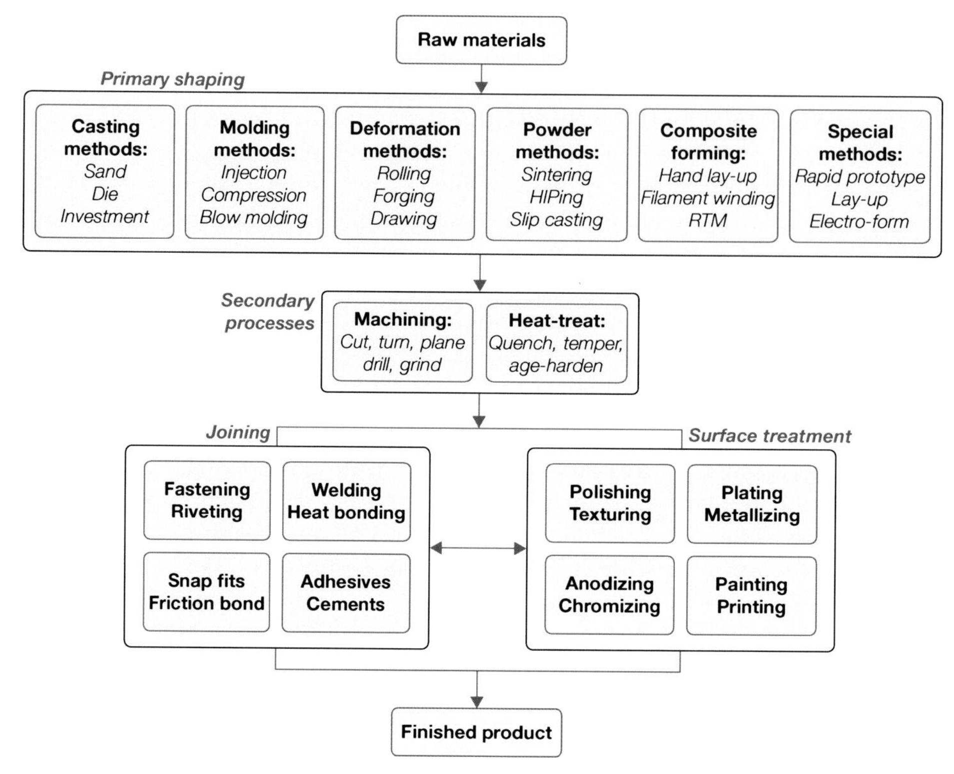

FIGURE 13.2
The classes of process. The first row contains the family of shaping processes; below lie the secondary processes of machining and heat treatment, followed by the families of joining and finishing (surface treatment) processes.

molding, and so on. One of these—molding—is again expanded to show its members: rotation molding, blow molding, injection molding, and so forth. Each of these has certain attributes: the materials it can handle, the shapes it can makes, their size, precision, and an optimum batch size (the number of units that molding can make economically).

Expanding members of the shaping family

Expanding the casting family at about the same level of detail as that used for molding in the figure gives the result:

1. Sand casting
2. Shaping–casting
3. Die casting
4. Investment casting

The other two families are partly expanded in Figure 13.4. The joining family contains three broad classes: adhesives, welding, and fasteners. In this

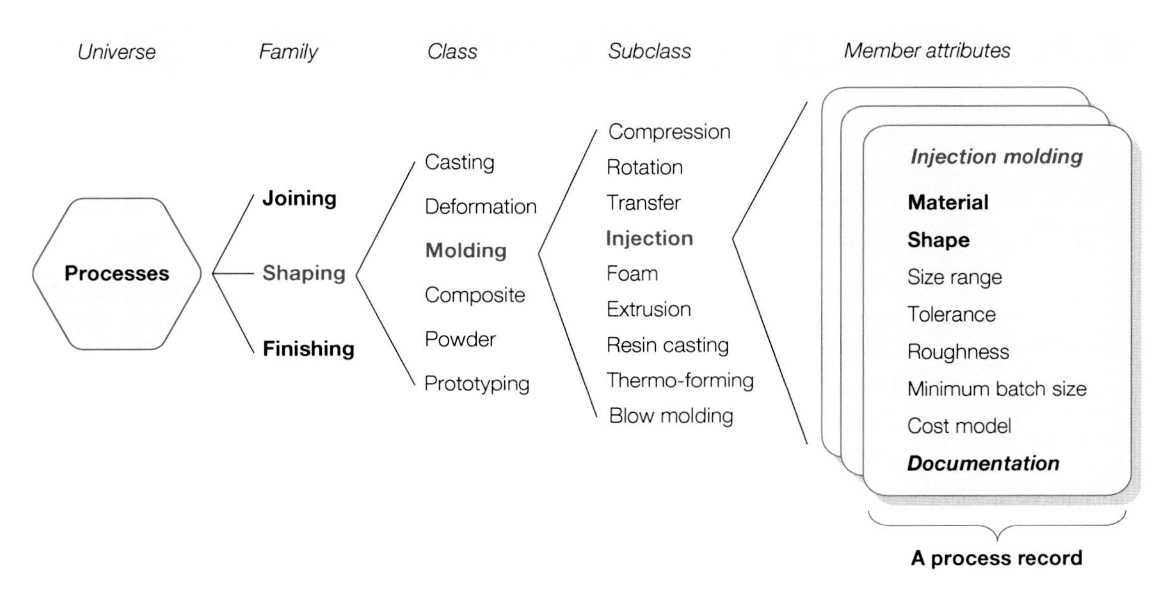

FIGURE 13.3
The taxonomy of the kingdom of process with part of the *shaping* family expanded. Each member is characterized by a set of attributes. Process selection involves matching these to the requirements of the design.

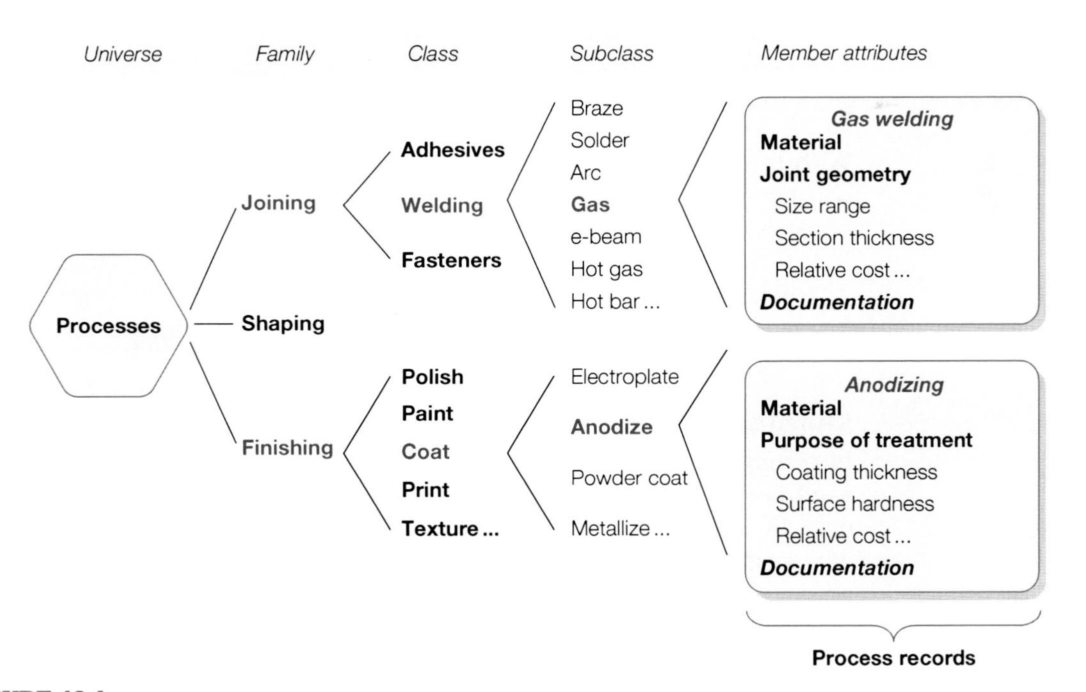

FIGURE 13.4
The taxonomy of the process kingdom again, with the families of *joining* and *finishing* partly expanded.

figure one of them—welding—is expanded to show its members. As before, each member has attributes. The first is the materials it can join. After that the attribute list differs from that for shaping. Here the geometry of the joint and the way it will be loaded are important, as are requirements that the joint can or cannot be disassembled, that it be watertight, that it be electrically conducting, and the like.

The lower part of the figure expands the family of finishing. Some of the classes it contains are shown; one—coating—is expanded to show some of its members. As with shaping and joining, the material to be coated is an important attribute but the others differ. Most important is the purpose of the treatment (protection, surface hardening, decoration, etc.), followed by properties of the coating itself.

Expanding the joining family

Expanding the fasteners family at about the same level of detail as that used for molding in the figure, gives the result:

1. Rivets and staples
2. Joining—fasteners
3. Threaded fasteners
4. Sewing
5. Snap fits

With this background we can embark on a lightning tour of processes. It will be kept as concise as possible; details can be found in the numerous books listed in "Further reading" (Section 13.9).

13.3 THE PROCESSES: SHAPING, JOINING, FINISHING

Shaping processes

In ***casting*** (Figure 13.5), a liquid is poured or forced into a mold where it solidifies by cooling. Casting is distinguished from molding, which comes next, by the low viscosity of the liquid: It fills the mold by flow under its own weight (as in gravity sand and investment casting) or under a modest pressure (as in die casting and pressure sand casting). Sand molds for one-off castings are cheap; metal dies for die casting large batches can be expensive. Between these extremes lie a number of other casting methods: shell, investment, plaster mold, and so forth.

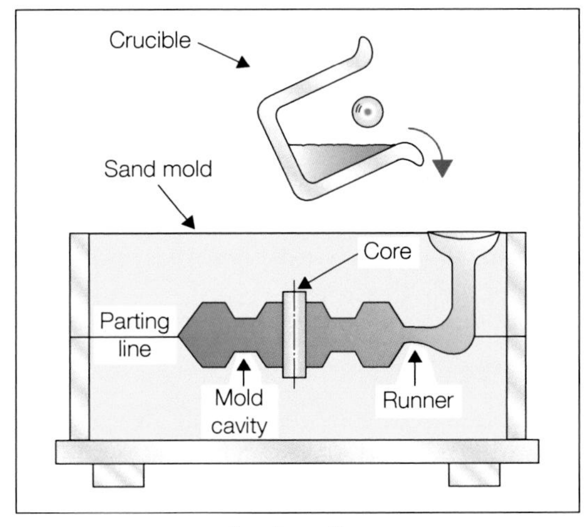

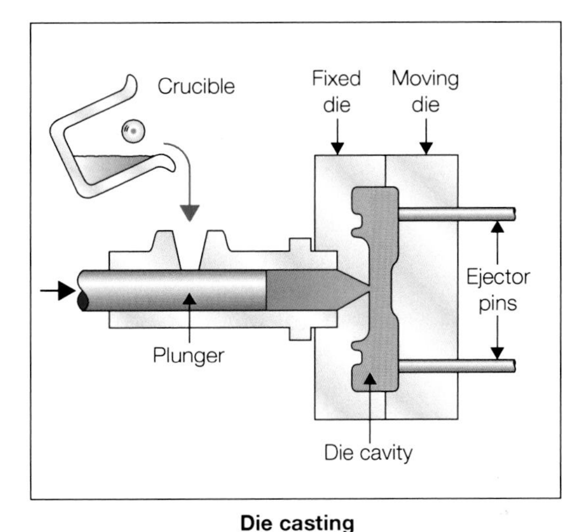

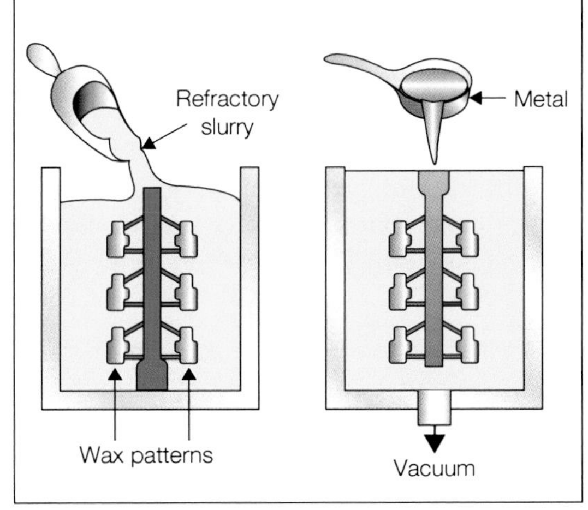

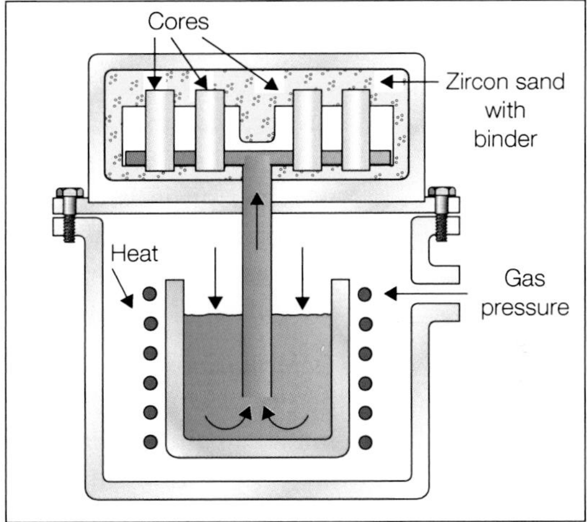

Low-pressure casting

FIGURE 13.5

Casting processes. In *sand casting*, liquid metal is poured into a split sand mold. In *die casting*, liquid is forced under pressure into a metal mold. In *investment casting*, a wax pattern is embedded in a refractory, melted out, and the cavity filled with metal. In *pressure casting*, a die is filled from below, giving control of atmosphere and of the flow of metal into the die.

Cast shapes must be designed for easy flow of liquid to all parts of the mold, and for progressive solidification that does not trap pockets of liquid in a solid shell, giving shrinkage cavities. Whenever possible, section thicknesses are made uniform (the thickness of adjoining sections should not differ by more than a factor of 2). The shape is designed so that the pattern and the finished casting can be removed from the mold. Keyed-in shapes are avoided because they lead to "hot tearing" (a tensile creep fracture) as the solid cools and shrinks. The tolerance and surface finish of a casting vary from poor for sand casting to excellent for precision die castings; they are quantified in Section 13.6.

When metal is poured into a mold, the flow is turbulent, trapping surface oxide and debris within the casting, giving casting defects. These are avoided by filling the mold from below in such a way that flow is laminar, driven by a vacuum or gas pressure, as shown in Figure 13.5.

Molding Figure 13.6 shows casting, adapted to materials that are very viscous when molten, particularly thermoplastics and glasses. The hot, viscous fluid is pressed or injected into a die under considerable pressure, where it cools and solidifies. The die must withstand repeated application of pressure, temperature, and the wear involved in separating and removing the part, and therefore it is expensive. Elaborate shapes can be molded, but at the penalty of complexity in die shape and in the way it separates to allow removal. The molds for thermo-forming, by contrast, are cheap. Variants of the process use gas pressure or vacuum to press a heated polymer sheet onto a single-part mold. Blow molding, too, uses a gas pressure to expand a polymer or glass blank into a split outer die. It is a rapid, low-cost process well suited for mass production of cheap parts like milk bottles. Polymers, such as metals, can be extruded; virtually all rods, tubes, and other prismatic sections are made in this way.

Deformation processing Figure 13.7 shows that the deformation process can be hot, warm, or cold—cold, that is, relative to the melting point T_m of the material being processed. Extrusion, hot forging, and hot rolling ($T > 0.55\ T_m$) have much in common with molding, though the material is a true solid, not a viscous liquid. The high temperature lowers the yield strength and allows simultaneous recrystallization, both of which lower the forming pressures. Warm working ($0.35\ T_m < T < 0.55\ T_m$) allows recovery but not recrystallization. Cold forging, rolling, and drawing ($T < 0.35\ T_m$) exploit work hardening to increase the strength of the final product, but at the penalty of higher forming pressures.

Forged parts are designed to avoid sudden changes in thickness and sharp radii of curvature since both require large local strains that can cause the material to tear or to fold back on itself ("lapping"). Hot forging of metals allows larger changes of shape but generally gives a poor surface and tolerance because of oxidation and warpage. Cold forging gives greater precision and finish, but forging pressures are higher and the deformations are limited by work hardening.

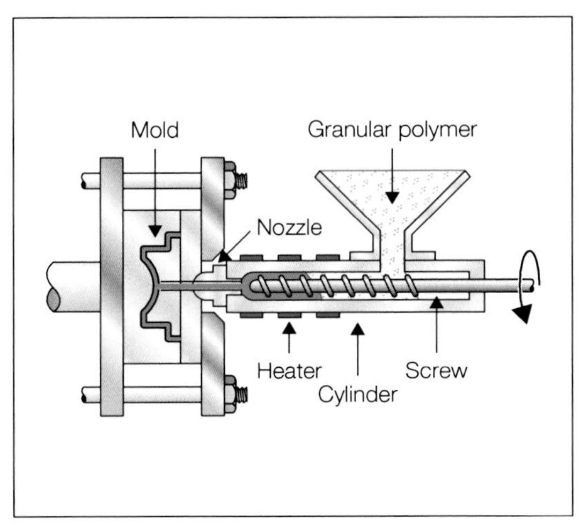

Injection molding

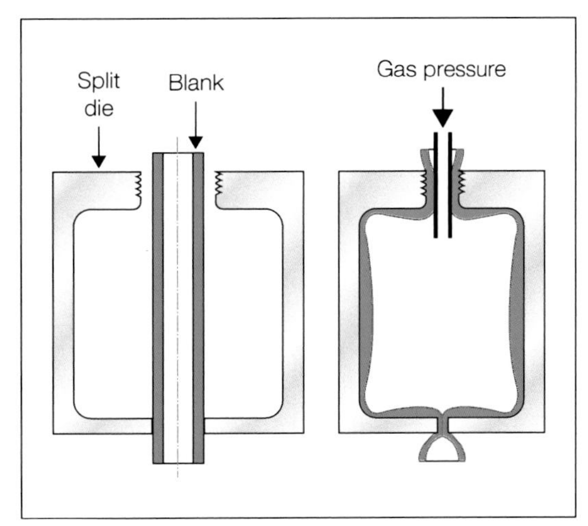

Blow molding

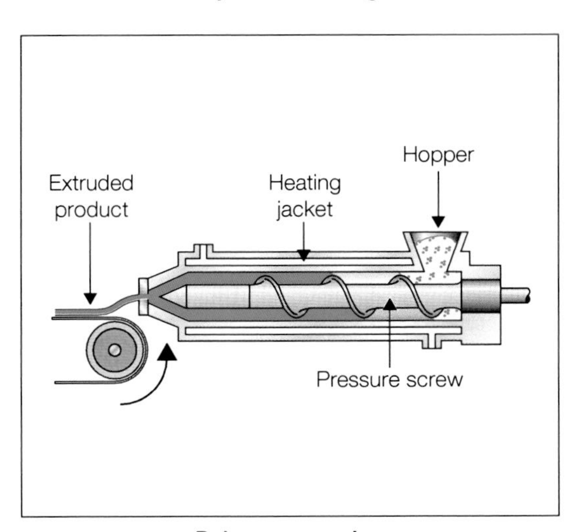

Polymer extrusion

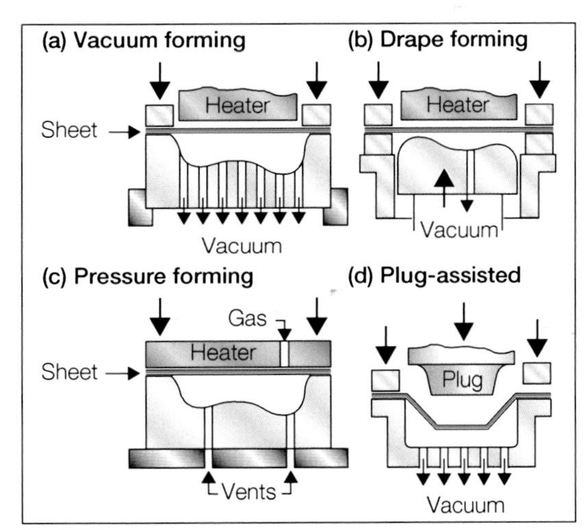

Thermo-forming

FIGURE 13.6

Molding processes. In *injection molding*, a granular polymer (or filled polymer) is heated, compressed, and sheared by a screw feeder, forcing it into the mold cavity. In *blow molding*, a tubular blank of hot polymer or glass is expanded by gas pressure against the inner wall of a split die. In *polymer extrusion*, shaped sections are formed by extrusion through a shaped die. In *thermo-forming*, a sheet of thermoplastic is heated and deformed into a female die by vacuum or gas pressure.

Powder methods Figure 13.8 shows that powder methods create the shape by pressing and then sintering fine particles of the material. The powder can be cold-pressed and then sintered (heated at up to 0.8 T_m to give diffusion bonding); it can be pressed in a heated die ("die pressing"); or, contained in a thin preform, it can be heated under a hydrostatic pressure ("hot isostatic pressing"

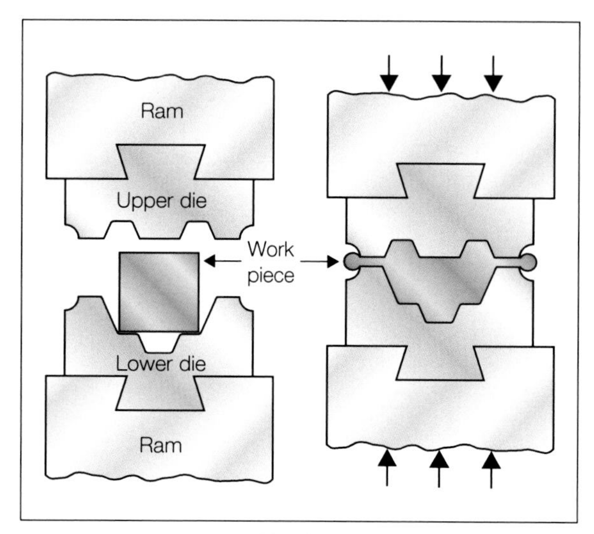

Forging

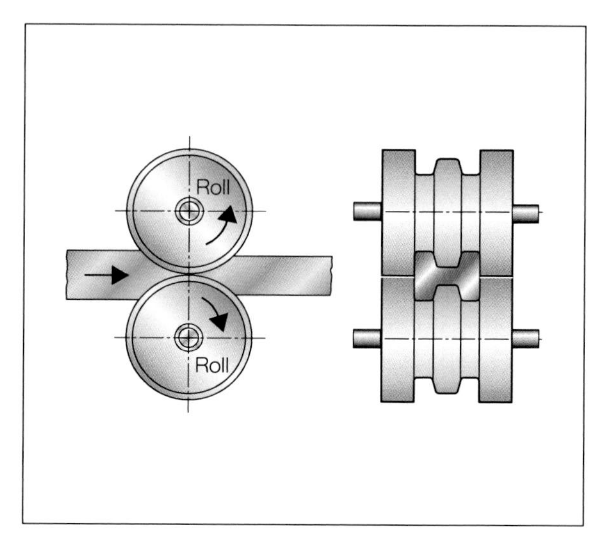

Rolling

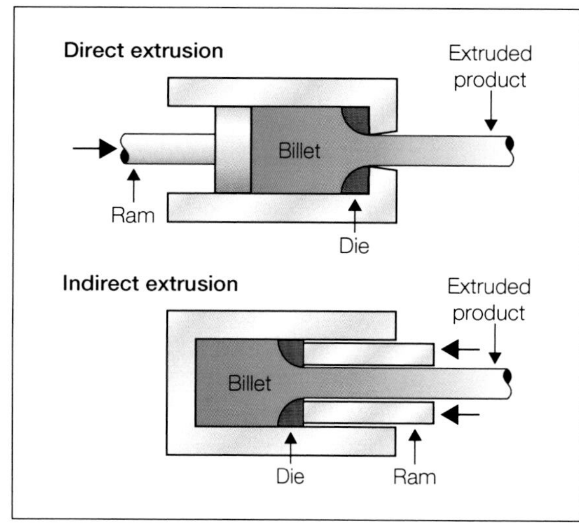

Extrusion

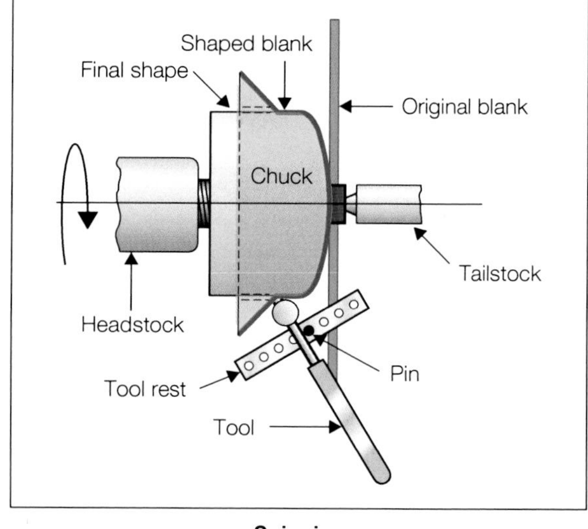

Spinning

FIGURE 13.7

Deformation processes. In *forging*, a slug of metal is shaped between two dies held in the jaws of a press. In *rolling*, a billet or bar is reduced in section by compressive deformation between the rolls. In *extrusion*, metal is forced to flow through a die aperture to give a continuous prismatic shape. All three processes can be hot ($T > 0.85\ T_m$), warm ($0.55\ T_m < T < 0.85\ T_m$), or cold ($T < 0.35\ T_m$). In *spinning*, a spinning disc of ductile metal is shaped over a wooden pattern by repeated sweeps of the smooth, rounded tool.

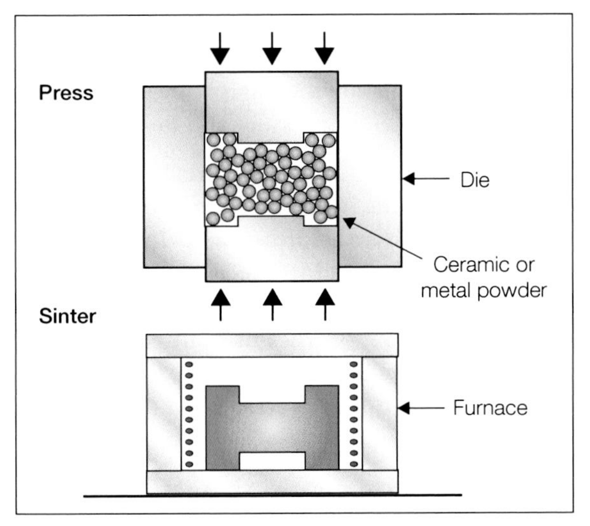

Die-pressing and sintering

Hot isostatic pressing

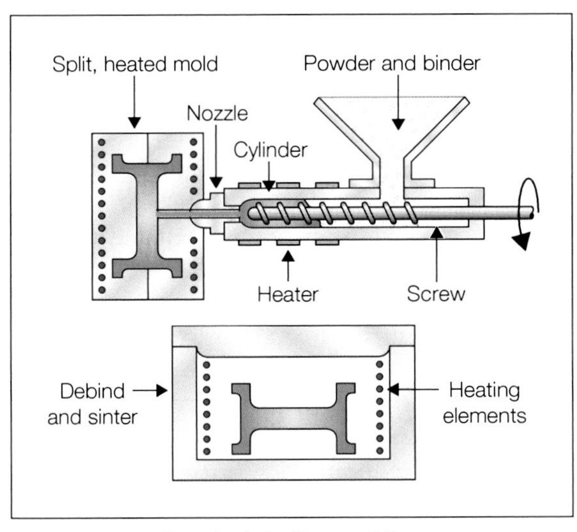

Powder injection molding

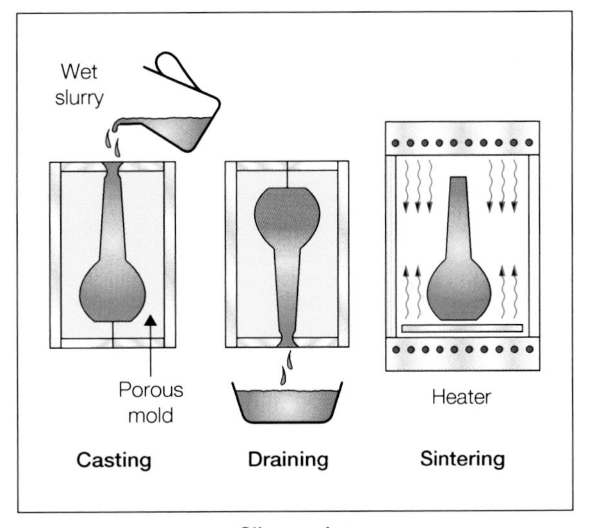

Slip casting

FIGURE 13.8

Powder processing. In *die-pressing* and *sintering* the powder is compacted in a die, often with a binder, and the green compact is then fired to give a more or less dense product. In *hot isostatic pressing*, powder in a thin, shaped shell or preform is heated and compressed by an external gas pressure. In *powder injection molding*, powder and binder are forced into a die to give a green blank that is then fired. In *slip casting*, a water-based powder slurry is poured into a porous plaster mold that absorbs the water, leaving a powder shell that is subsequently fired.

or "HIPing"). Metals that are too high-melting to cast and too strong to deform can be made (by chemical methods) into powders and then shaped in this way. But the processes are not limited to "difficult" materials; almost any material can be shaped by subjecting it, as a powder, to pressure and heat.

Powder processing is most widely used for small metallic parts like gears and bearings for cars and appliances. It is economic in its use of material; it allows parts to be fabricated from materials that cannot be cast, deformed, or machined; and it can give a product that requires little or no finishing. Since pressure is not transmitted uniformly through a bed of powder, the length of a die-pressed powder part should not exceed 2.5 times its diameter. Sections must be near uniform because the powder will not flow easily around corners. And the shape must be simple and easily extracted from the die.

Ceramics, difficult to cast and impossible to deform, are routinely shaped by powder methods. In slip casting, a water-based powder slurry is poured into a plaster mold. The mold wall absorbs water, leaving a semi-dry skin of slurry over its inner wall. The remaining liquid is drained out, and the dired slurry shell is fired to give a ceramic body. In powder injection molding (the way spark-plug insulators are made) a ceramic powder in a polymer binder is molded in the conventional way; the molded part is fired, burning the binder and sintering the powder.

Composite forming methods, as shown in Figure 13.9, make polymer-matrix composites reinforced with continuous or chopped fibers. Large components are fabricated by filament winding or by laying up preimpregnated mats of carbon, glass, or Kevlar fiber ("pre-preg") to the required thickness, pressing, and curing. Parts of the process can be automated, but it remains a slow manufacturing route; and, if the component is a critical one, extensive ultrasonic testing may be necessary to confirm its integrity. Higher integrity is given by vacuum-bag or pressure-bag molding, which squeezes bubbles out of the matrix before it polymerizes. Lay-up methods are best suited to a small number of high-performance, tailor-made components.

More routine components (car bumpers, tennis racquets) are made from chopped-fiber composites by pressing and heating a "dough" of resin containing the fibers, known as bulk molding compound (BMC) or sheet molding compound (SMC), in a mold, or by injection molding a rather more fluid mixture into a die. The flow pattern is critical in aligning the fibers, so the designer must work closely with the manufacturer to exploit the composite properties fully.

Rapid prototyping systems (RPS) allow single examples of complex shapes to be made from numerical data generated by CAD solid-modeling software (see Figure 13.10). The motive may be that of visualization: The aesthetics of an object may be evident only when viewed as a prototype. It may be

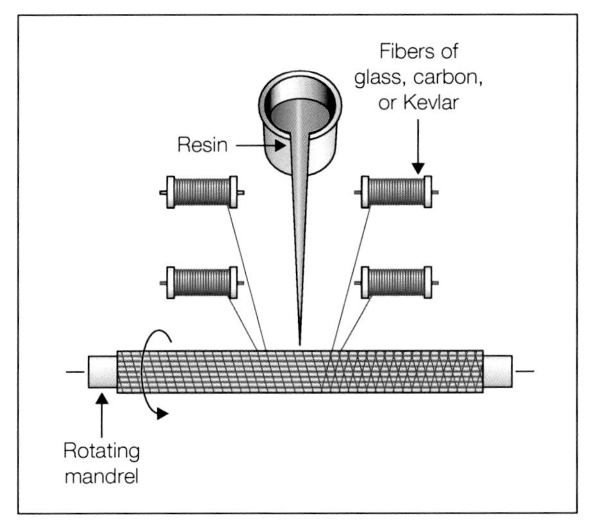

Filament winding

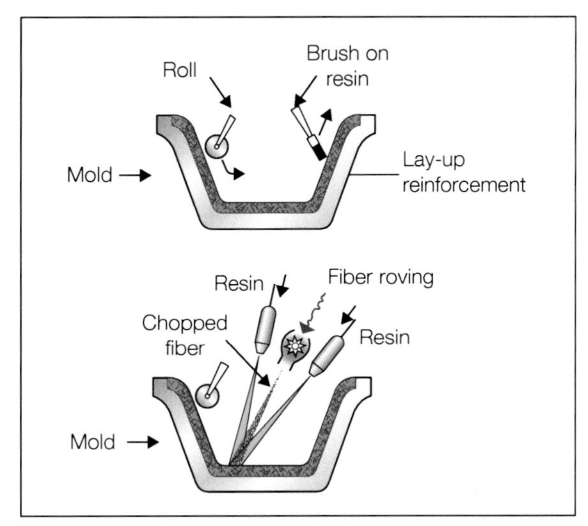

Hand and spray lay-up

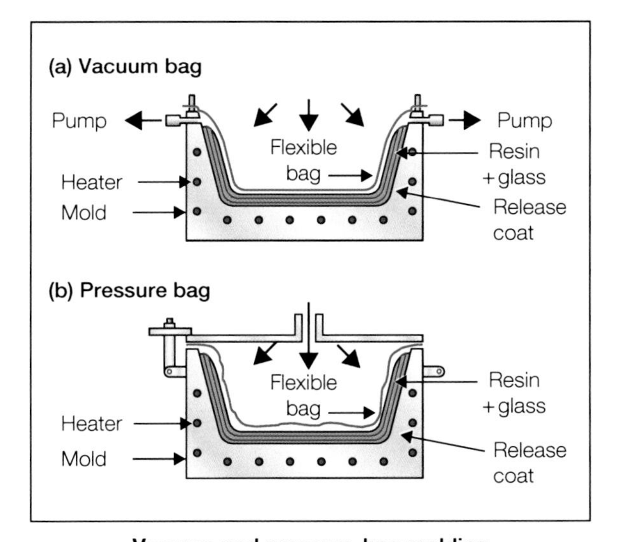

Vacuum and pressure-bag molding

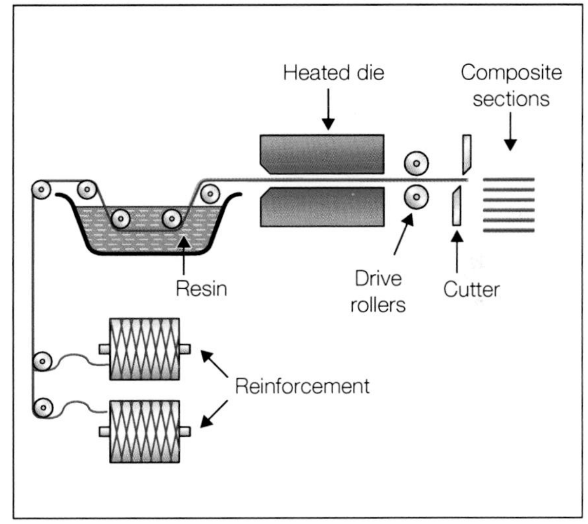

Pultrusion

FIGURE 13.9

Composite forming methods. In *filament winding*, fibers of glass, Kevlar, or carbon are wound onto a former and impregnated with a resin-hardener mix. In *roll* and *spray lay-up*, fiber reinforcement is laid up in a mold onto which the resin-hardener mix is rolled or sprayed. In *vacuum-bag* and *pressure-bag molding*, laid-up fiber reinforcement, impregnated with resin-hardener mix, is compressed and heated to cause polymerization. In *pultrusion*, fibers are fed through a resin bath into a heated die to form continuous prismatic sections.

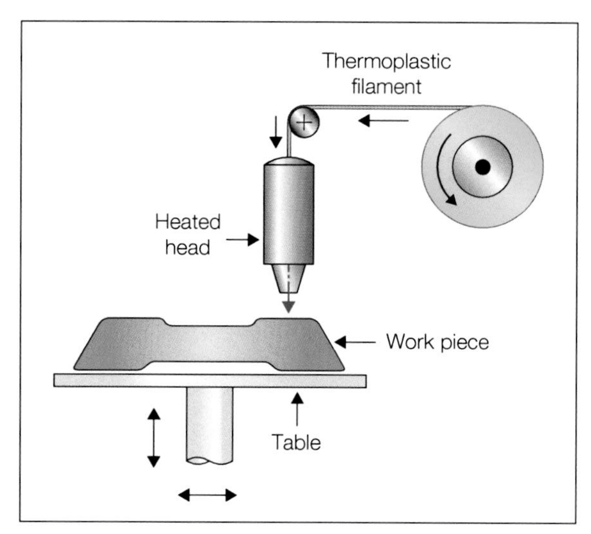

Deposition modeling

Stero-lithography (SLA)

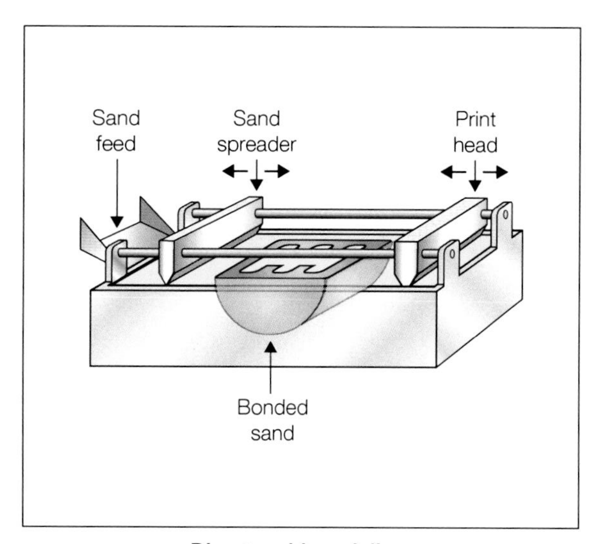

Direct mold modeling

Laminated object manufacture (LOM)

FIGURE 13.10

Rapid prototyping. In *deposition modeling* and *ballistic particle manufacture* (BPM), a solid body is created by the layer-by-layer deposition of polymer droplets. In *stereo-lithography* (SLA), a solid shape is created layer by layer by laser-induced polymerization of a resin. In *direct mold modeling*, a sand mold is built up layer by layer by selective spraying of a binder from a scanning print-head. In *laminated object manufacture* (LOM), a solid body is created from layers of paper, cut by a scanning laser beam and bonded with a heat-sensitive polymer.

that of pattern-making: The prototype becomes the master from which molds for conventional processing, such as casting, can be made. Or—in complex assemblies—it may be that of validating intricate geometry, ensuring that parts fit, can be assembled, and are accessible.

All RPSs can create shapes of great complexity with internal cavities, overhangs, and transverse features, though the precision, at present, is limited to $\pm$ 0.3 mm at best. All RP methods build shapes layer by layer, rather like three-dimensional printing, and are slow (typically 4–40 hours per unit). There are at least six broad classes.

1. The shape is built up from a thermoplastic fed to a single scanning head that extrudes it like a thin layer of toothpaste (*fused deposition modeling* or FDM), exudes it as tiny droplets (*ballistic particle manufacture*, BPM), or ejects it in a patterned array like a bubble-jet printer ("3D printing").
2. Scanned-laser–induced polymerization of a photo-sensitive monomer (*stereo-lithography* or SLA). After each scan, the work piece is incrementally lowered, allowing fresh monomer to cover the surface.
3. Scanned laser cutting of bondable paper elements. Each paper-thin layer is cut by a laser beam and heat-bonded to the next one.
4. Screen-based technology like that used to produce microcircuits (*solid ground curing* or SGC). A succession of screens admits UV light to polymerize a photo-sensitive monomer, building shapes layer by layer.
5. *Selected laser sintering* (SLS) allows components to be fabricated directly in thermoplastic, metal, or ceramic. A laser, as in SLA, scans a bed of particles, sintering a thin surface layer where the beam strikes. A new layer of particles is swept across the surface and the laser-sintering step is repeated, building up a three-dimensional body.
6. *Bonded sand molding* offers the ability to make large complex metal parts easily. Here a multi-jet print-head squirts a binder onto a bed of loose casting sand, building up the mold shape much as selected laser sintering does, but more quickly. When complete, the mold is lifted from the remaining loose sand and used in a conventional casting process.

To be useful, prototypes made by RPS are used as silicone molding masters, allowing replicas to be cast using high-temperature resins or metals.

Almost all engineering components, whether made of metal, polymer, or ceramic, are subjected to some kind of ***machining*** (Figure 13.11) during manufacture. To make this possible they should be designed to make gripping and jigging easy and to keep the symmetry high: Symmetric shapes need fewer operations. Metals differ greatly in their *machinability*, a measure of the ease of chip formation, the ability to give a smooth surface, and the ability to give economical tool life (evaluated in a standard test). Poor machinability means higher cost.

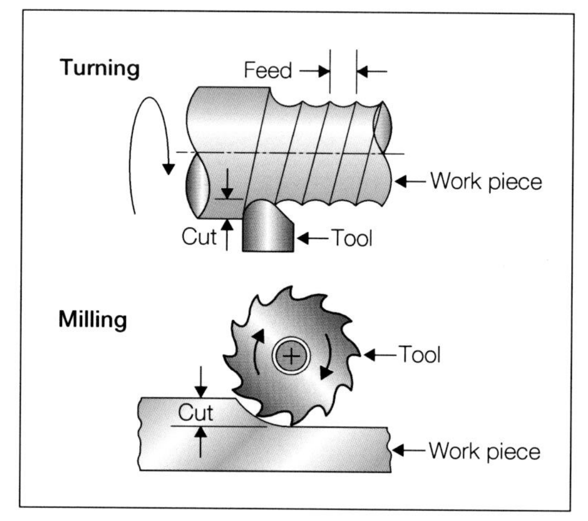

Turning and milling

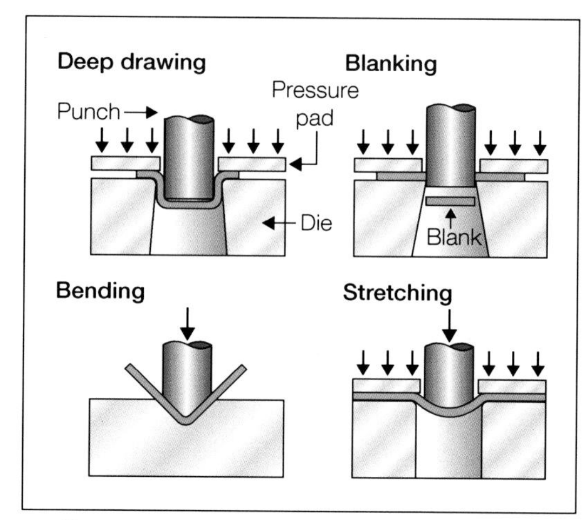

Drawing, blanking, bending, and stretching

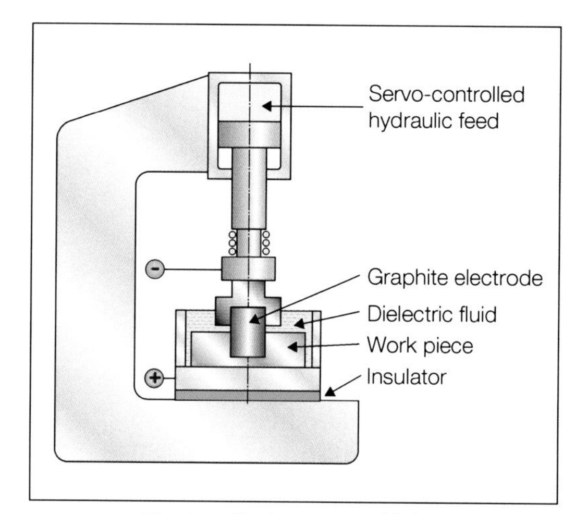

Electro-discharge machining

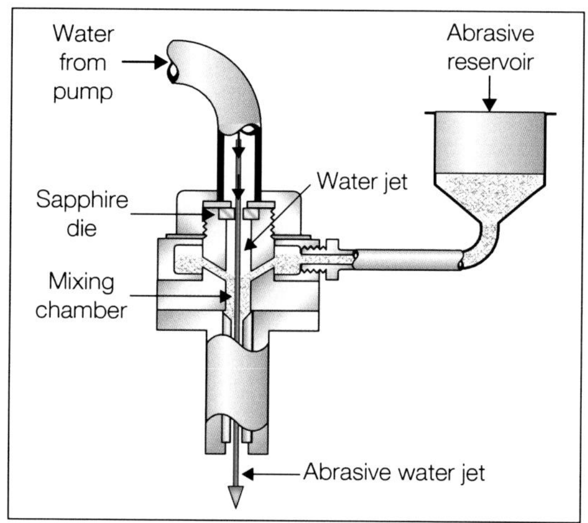

Water-jet cutting

FIGURE 13.11

Machining operations. In *turning* and *milling*, the sharp, hardened tip of a tool cuts a chip from the work piece surface. In *drawing*, *blanking*, and *stretching*, sheet is shaped and cut to give flat and dished shapes. In *electro-discharge machining*, electric discharge between a graphite electrode and the work piece, submerged in a dielectric such as paraffin, erodes the work piece to the desired shape. In *water-jet cutting*, an abrasive entrained in a high-speed water jet erodes the material in its path.

Most polymers are molded into a final shape. When necessary they can be machined but their low moduli mean that they deflect elastically during the machining operation, limiting tolerance. Ceramics and glasses can be ground and lapped to high tolerance and finish (think of the mirrors of telescopes). There are many "special" machining techniques with particular applications including: electro-discharge machining (EDM), ultrasonic cutting, chemical milling, and cutting by water and sand jets and electron and laser beams.

Sheet metal forming involves punching, bending, and stretching. Holes cannot be punched to a diameter less than the thickness of the sheet. The minimum radius to which a sheet can be bent, its *formability*, is sometimes expressed in multiples of the sheet thickness t: A value of 1 is good; one of 4 is average. Radii are best made as large as possible, and never less than t. The formability also determines the amount the sheet can be stretched or drawn without necking and failing. The *forming limit diagram* gives more precise information: It shows the combination of principal strains in the plane of the sheet that will cause failure. The part is designed so that the strains do not exceed this limit.

Machining is often a secondary operation applied to castings, moldings, or powder products to increase finish and tolerance. Higher finish and tolerance mean higher cost; over-specifying either is a mistake.

Joining processes

Joining Figure 13.12 shows that joining is made possible by a number of techniques. Almost any material can be joined with adhesives, though ensuring a sound, durable bond can be difficult. Bolting, riveting, stapling, and snap fitting are commonly used to join polymers and metals, and have the feature that they can be disassembled if need be. Welding, the largest class of joining processes, is widely used to bond metals and polymers; specialized techniques have evolved to deal with each class. Friction welding and friction-stir welding rely on the heat and deformation generated by friction to create a bond between different metals. Ceramics can be diffusion-bonded to themselves, to glasses, and to metals.

If components are to be welded, the material of which they are made must be characterized by a high *weldability*. Like *machinability*, it measures a combination of basic properties. A low thermal conductivity allows welding with a low rate of heat input, but can lead to greater distortion on cooling. Low thermal expansion gives small thermal strains with less risk of distortion. A solid solution is better than an age-hardened alloy because, in the heat-affected zone on either side of the weld, over-aging and softening can occur.

Welding always leaves internal stresses that are roughly equal to the yield strength of the parent material. They can be relaxed by heat treatment but this is expensive, so it is better to minimize their effect by good design. To achieve this, parts to be welded are made of equal thickness whenever possible, the welds are located where stress or deflection is least critical, and the total number of welds is minimized.

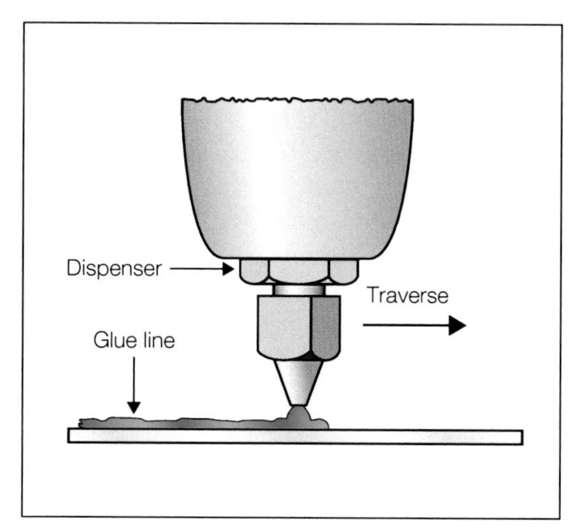

Adhesives

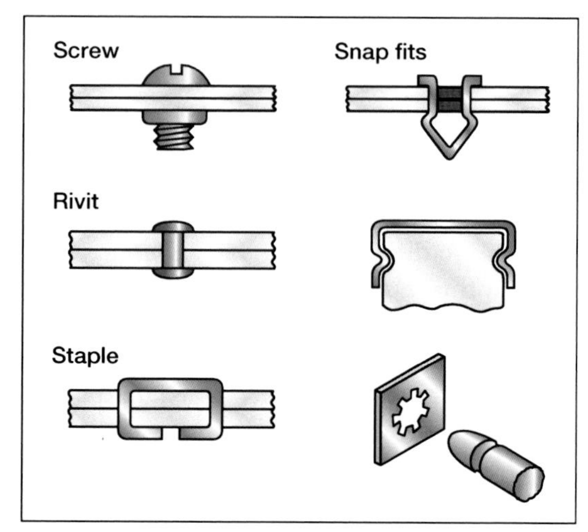

Fasteners

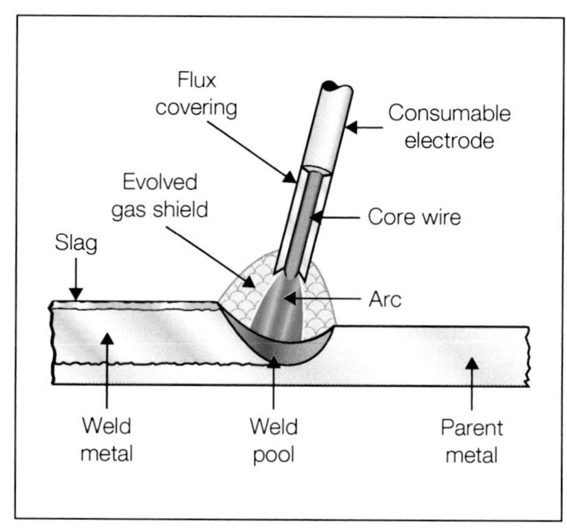

Welding: manual metal arc

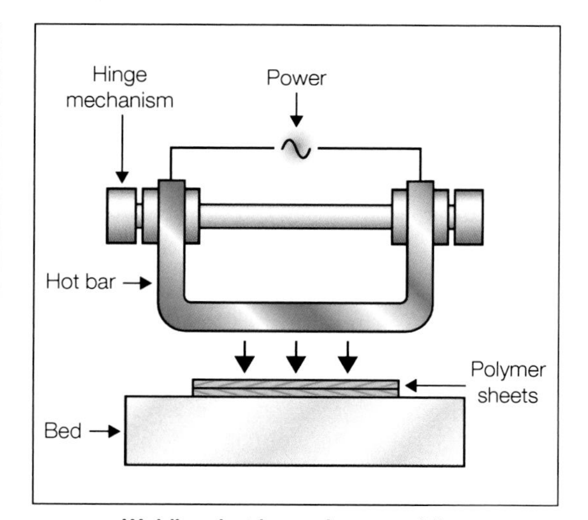

Welding: hot bar polymer welding

FIGURE 13.12

Joining operations. In *adhesive bonding*, a film of adhesive is applied to one surface, which is then pressed onto the mating one. *Fastening* is achieved by bolting; riveting; stapling; push-through snap fastener; push-on snap fastener; or rod-to-sheet snap fastener. In *metal fusion welding*, metal is melted, and more is added from a filler rod to give a bond or coating. In *thermoplastic polymer welding*, heat is applied to the polymer components, which are simultaneously pressed together to form a bond.

The large-volume use of fasteners is costly; welding, crimping, or the use of adhesives can be more economical. Design for assembly (DFA) methods provide a checklist to guide minimizing assembly time.

Finishing processes

Finishing involves treatments applied to the surface of the component or assembly. Some aim to improve mechanical and other engineering properties, others to enhance appearance.

Finishing treatments to improve engineering properties (Figure 13.13) Grinding, lapping, and polishing increase precision and smoothness, particularly important for bearing surfaces. Electroplating deposits a thin metal layer onto the surface of a component to give resistance to corrosion and abrasion. Plating and painting are both made easier by a simple part shape with largely convex surfaces: Channels, crevices, and slots are difficult to reach. Anodizing, phosphating, and chromating create a thin layer of oxide, phosphate, or chromate on the surface, imparting corrosion resistance.

Heat treatment is a necessary part of the processing of many materials. Age-hardening alloys of aluminum, titanium, and nickel derive their strength from a precipitate produced by a controlled heat treatment: quenching from a high temperature followed by aging at a lower one. The hardness and toughness of steels are controlled in a similar way: by quenching from the "austenitizing" temperature (about 800°C) and tempering. The treatment can be applied to the entire component, as in bulk carburizing, or just to a surface layer, as in flame hardening, induction hardening, and laser surface hardening.

Quenching—very rapid cooling—is a savage procedure; the sudden thermal contraction associated with it can produce stresses large enough to distort or crack the component. The stresses are caused by a nonuniform temperature distribution, and this, in turn, is related to the geometry of the component. To avoid damaging stresses, the section thickness should be as uniform as possible, and nowhere so large that the quench rate falls below the critical value required for successful heat treatment. Stress concentrations should be avoided because they are a source of quench cracks. Materials that have been molded or deformed may contain internal stresses that can be removed, at least partially, by stress-relief anneals—another sort of heat treatment.

Finishing treatments that enhance aesthetics (Figure 13.14) The processes just described can be used to enhance the visual and tactile attributes of a material: Electroplating and anodizing are examples. There are many more, of which painting is the most widely used. Organic-solvent–based paints give durable coatings with high finish, but the solvent poses environmental problems. Water-based paints overcome these, but dry more slowly and the resulting paint film is less perfect. In polymer powder-coating and polymer powder-spraying a film of thermoplastic—nylon, polypropylene, or polyethylene—is deposited on the surface, giving a protective layer that can be brightly colored.

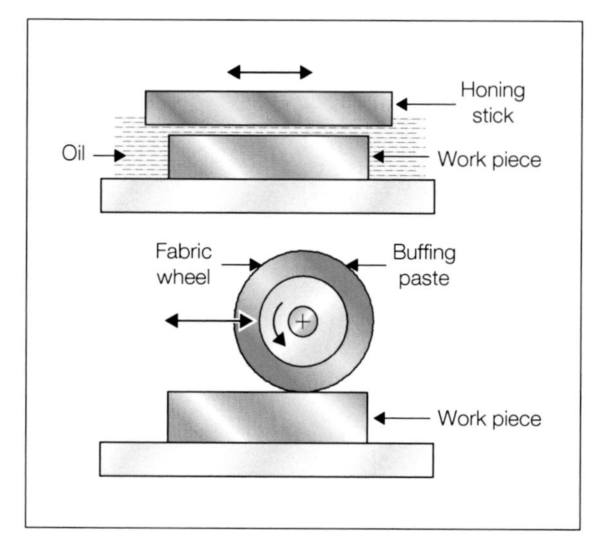

Mechanical polishing

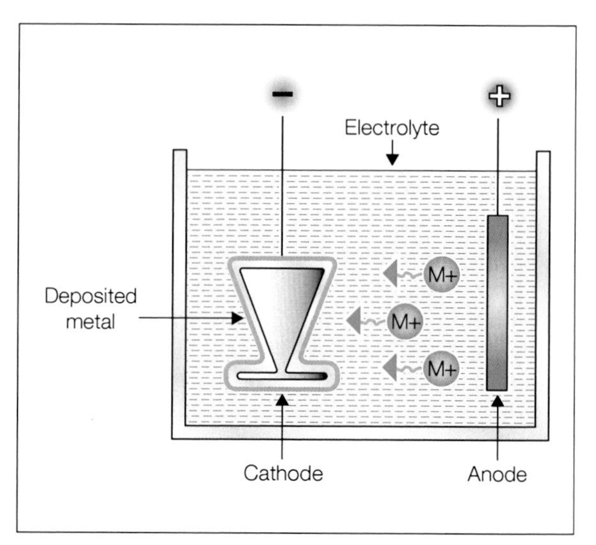

Electroplating

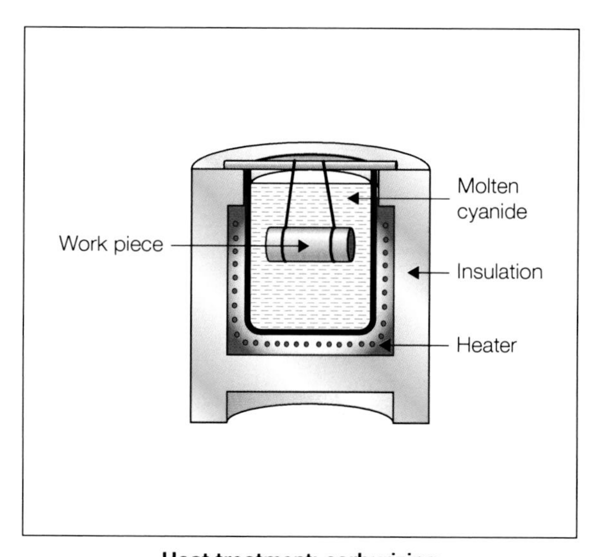

Heat treatment: carburizing

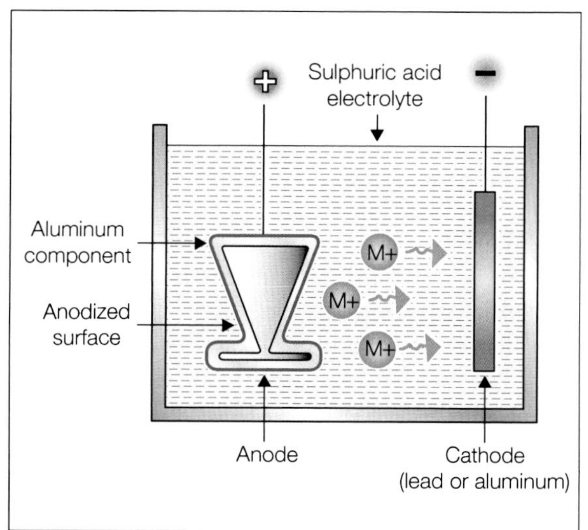

Anodizing

FIGURE 13.13

Finishing processes to protect and enhance properties. In *mechanical polishing*, the roughness of a surface is reduced and its precision increase, by material removal using finely ground abrasives. In *electroplating*, metal is plated onto a conducting work piece by electro-deposition in a plating bath. In *heat treatment*, a surface layer of the work piece is hardened and made more corrosion resistant by the inward diffusion of carbon, nitrogen, phosphorous, or aluminum from a powder bed or molten bath. In *anodizing*, a surface oxide layer is built up on the work piece (which must be aluminum, magnesium, titanium, or zinc) by a potential gradient in an oxidizing bath.

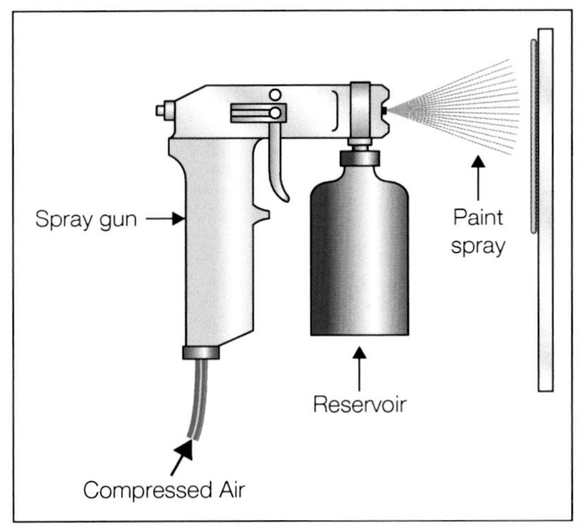

Paint spraying

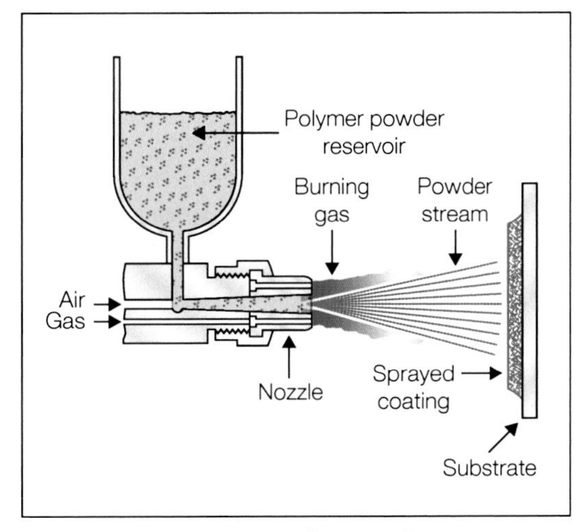

Polymer powder spraying

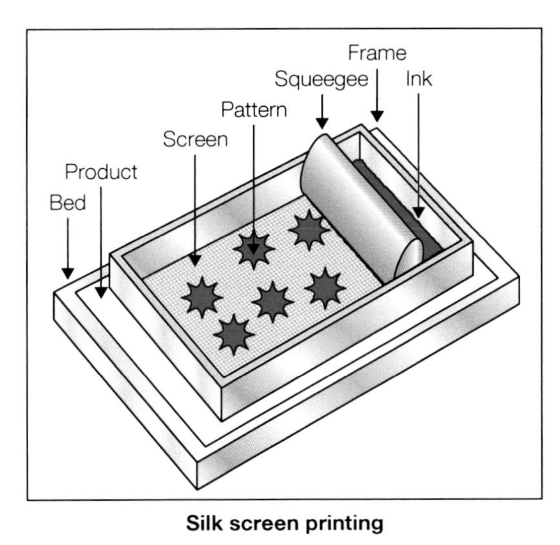

Silk screen printing

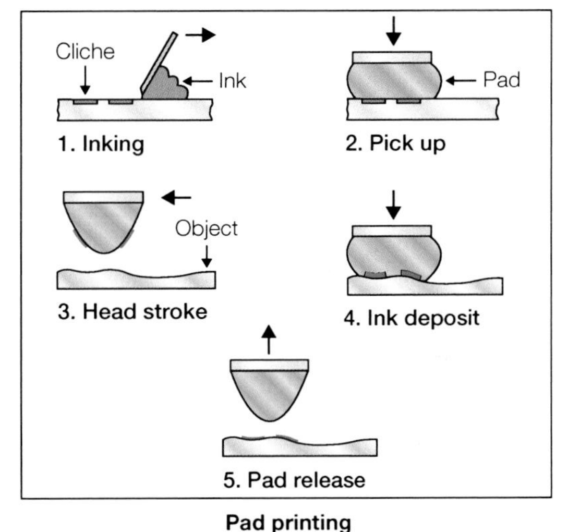

Pad printing

FIGURE 13.14

Finishing processes to enhance appearance. In *paint spraying*, a pigment in an organic- or water-based solvent is sprayed onto the surface to be decorated. In *polymer powder-coating* a layer of thermoplastic is deposited on the surface by direct spraying in a gas flame, or by immersing the hot work piece in a bed of powder. In *silk-screen printing*, ink is wiped onto the surface through a screen onto which a blocking pattern has been deposited, allowing ink to pass in selected areas only. In *pad-printing*, an inked pattern is picked up on a rubber pad and applied to the surface, which can be curved or irregular.

In screen printing an oil-based ink is squeegeed through a mesh on which a blocking film holds back the ink where it is not wanted; full-color printing requires the successive use of up to four screens. Screen printing is widely used to apply designs to flat surfaces. Curved surfaces require the use of pad printing, in which a pattern, etched onto a metal "cliche," is inked and picked up on a soft rubber pad. The pad is pressed onto the product, depositing the pattern on its surface; the compliant rubber conforms to the curvature of the surface.

Enough of the processes themselves; for more detail the reader will have to consult "Further reading" in Section 13.9. Now a brief look at what processing does to properties.

13.4 PROCESSING FOR PROPERTIES

The extent of the material bubbles in the property charts gives an idea of the degree to which properties can be manipulated by processing. The annotations in Figure 13.15, a copy of the modulus-strength chart, illustrates the degree to which these two properties for metals can (or cannot) be controlled by alloying, heat treatment, and cold work. The rather different

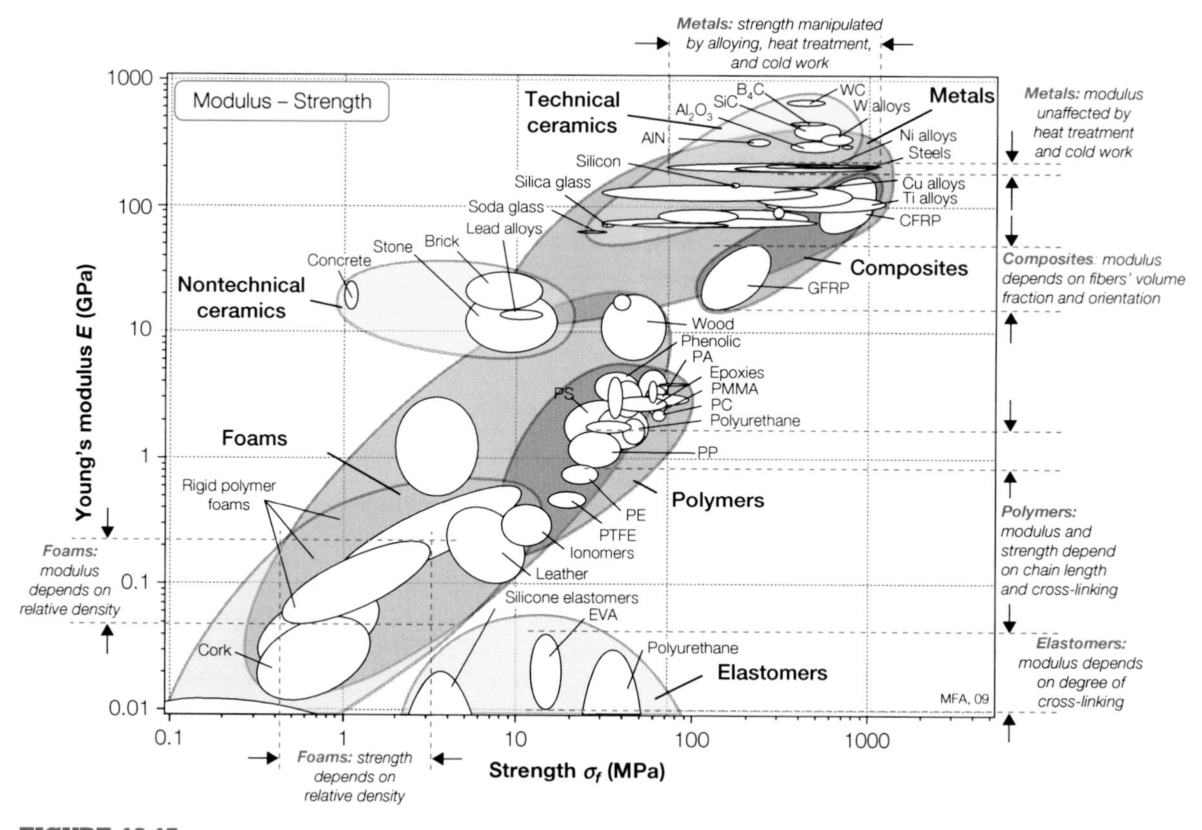

FIGURE 13.15

The extent of the material bubbles on the property charts gives an idea of the degree to which properties can be manipulated by processing.

shapes of the bubbles for composites and for foams reflect the way the properties of the first depend on fiber content and orientation; those of the second, on the extent of foaming, measured by the porosity or relative density (Chapter 11). The modulus and strength of polymers and elastomers depend on the chain length and degree of cross-linking, aspects of structure directly controlled by processing. The strength, particularly, of ceramics depends on porosity, another aspect of microstructure that is directly influenced by processing.

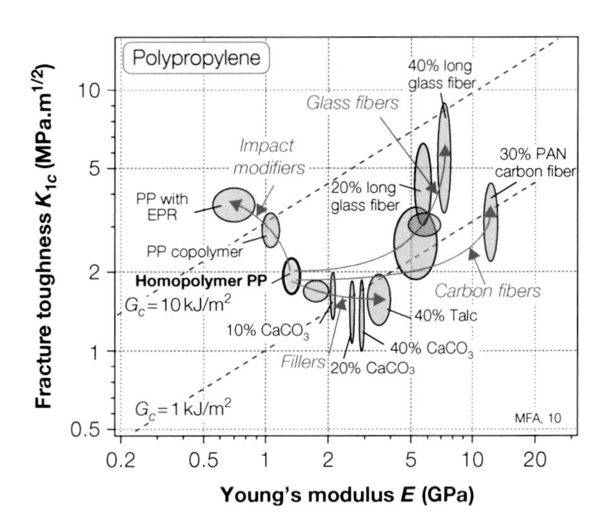

FIGURE 13.16

The strength and toughness of polypropylene, showing the effect of fillers, impact modifiers, and fibers.

We have already seen, in Figures 11.7 and 11.10, how fiber content, shape, and orientation control composite properties, and in Figure 11.24 how the foam density and connectivity dramatically influence the properties of cellular solids. The strengths of polymers span around a factor of 5 and their fracture toughnesses a factor of 20, depending on chemistry, chain length, and degree of cross-linking. More dramatic changes are possible by blending, filling, reinforcing, or plasticizing. Figure 13.16 shows how these influence the modulus E and fracture toughness K_{Ic} of polypropylene, PP. Blending or copolymerization with elastomers, such as EPR or EDPM ("impact modifiers"), reduces the modulus but increases the fracture toughness K_{Ic} and toughness G_c. Filling with inexpensive powdered glass, talc, or calcium carbonate more than doubles the modulus; however, that is at the expense of some loss of toughness. Plasticizing (blending with low-molecular-weight polymers) lowers the modulus even more dramatically. Between them, these processes can change the polymer modulus by a factor of 100 and the toughness by a factor of 10.

Of all the properties that materials scientists and engineers have sought to manipulate, the strength of metals and alloys is probably the most explored. It is easy to see why—Table 13.1 lists some of their applications. Solution hardening, precipitation hardening, and work hardening are combined to give desired combinations of strength and toughness. Figure 13.17 illustrates the large gains in strength of copper alloys that these mechanisms allow. Good things, however, have to be paid for. Here the payment required for increased strength is loss of ductility, in this case measured by the elongation to fracture ε_f.

Strength and ductility are structure-sensitive properties—they depend on composition and microstructure, and these, in turn, are controlled by

Table 13.1 Metal Alloys with Typical Applications, Indicating the Strengthening Mechanisms Used

Alloy	Typical Uses	Solution Hardening	Precipitate Hardening	Work Hardening
Pure Al	Kitchen foil			✓✓✓
Pure Cu	Wire			✓✓✓
Cast Al, Mg	Automotive parts	✓✓✓	✓	
Bronze (Cu-Sn), Brass (Cu-Zn)	Marine components	✓✓✓	✓	✓
Non-heat-treatable wrought Al	Ships, cans, structures	✓✓✓		✓✓✓
Heat-treatable wrought Al	Aircraft, structures	✓	✓✓✓	✓
Low carbon steels	Car bodies, structures, ships, cans	✓✓✓		✓✓✓
Low alloy steels	Automotive parts, tools	✓	✓✓✓	✓
Stainless steels	Pressure vessels	✓✓✓	✓	✓✓✓
Cast Ni alloys	Jet engine turbines	✓✓✓	✓✓✓	

Symbols: ✓✓✓ = routinely used; ✓ = sometimes used.

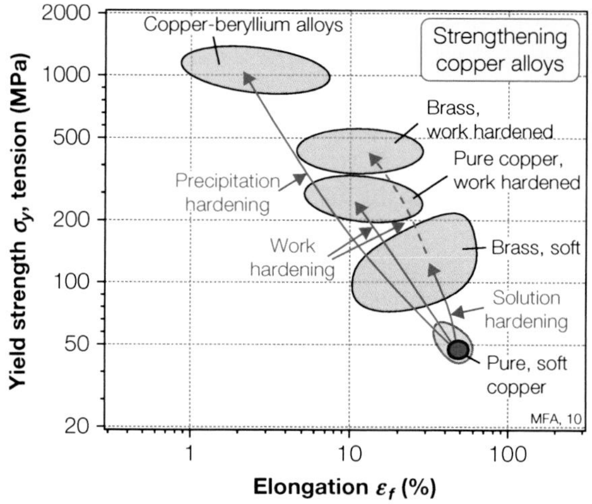

FIGURE 13.17
Strengthening mechanisms and the consequent drop in ductility, here shown for copper alloys. The mechanisms are frequently combined. The greater the strength, the lower the ductility (the elongation to fracture ε_f).

processing. Hardness, fatigue strength, fracture toughness, and thermal and electrical conductivities, too, are structure-sensitive properties. Much processing is fine-tuned to produce particular combinations of these. They are best illustrated as mini property charts, like Figures 13.16 and 13.17. Materials for heat transfer—heat exchangers, chemical engineering equipment—require good thermal conductivity with high strength. Figure 13.18 shows how processing changes these two properties for alloys based on aluminum hardened by each of the three main mechanisms: solid solution, work hardening, and precipitation hardening. Work hardening strengthens significantly without changing the conductivity much, while solid solution and precipitation hardening introduce more scattering centers, giving a drop in conductivity. Many electrical applications—high-speed motors and power transmission, for instance—require materials with good conductivity and high strength.

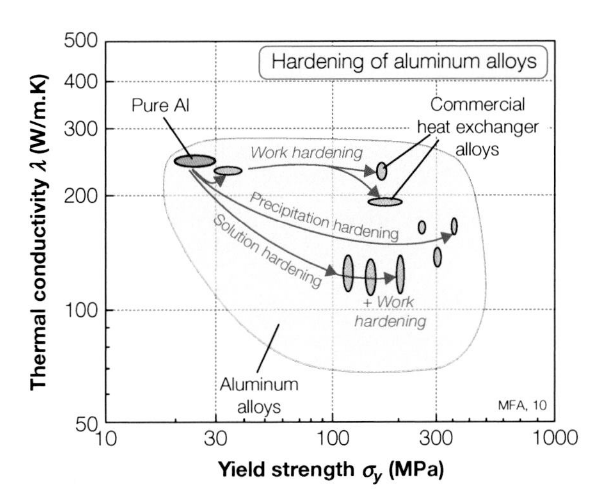

FIGURE 13.18
Thermal conductivity and strength for aluminum alloys.

Figure 13.19 illustrates this for one of the best conductors that we have: copper. Adding solute increases its strength, but the solute atoms also act as scattering centers, increasing the electrical resistivity too. Dislocations add strength (by what we called *work hardening*) and they too scatter electrons a little, though not as much as solute. Precipitates offer the greatest gain in strength with only a slight loss of conductivity. Precipitation hardening (with low residual solute) and work hardening are therefore the best ways to strengthen conductors.

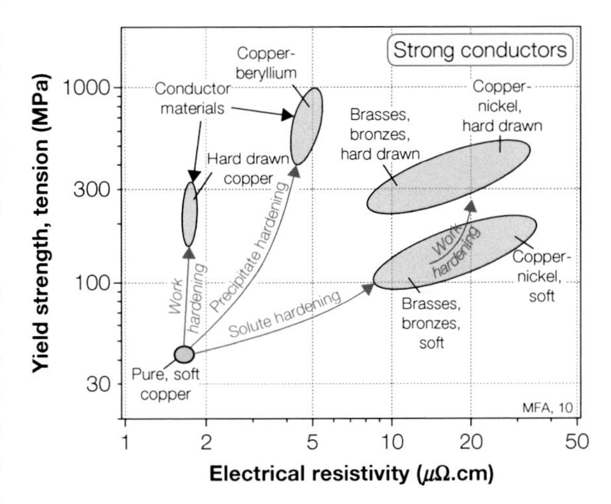

FIGURE 13.19
The best choice of material for a cable is one with high strength and low resistivity, but strengthening mechanisms increase resistivity. Work hardening and precipitation hardening do so less than solute hardening.

Nowhere is processing for properties more important than in the heat treatment of steels. Figure 13.20 shows the sequence for a medium-carbon steel. The steel is solutionized (heated into the FCC austenite field) to dissolve all of the carbon it then quenched in water or oil, which causes it to transform to hard, brittle martensite. Martensite has high hardness but is so brittle—its fracture toughness is so low that it is almost useless as a structural material. Tempering reduces the hardness and yield strength but restores toughness to a degree that depends on the tempering temperature and time. The desired property set is obtained by controlling these.

Thus processing plays a central role in manipulating material properties. Processes are chosen for their ability to create shapes and to create properties.

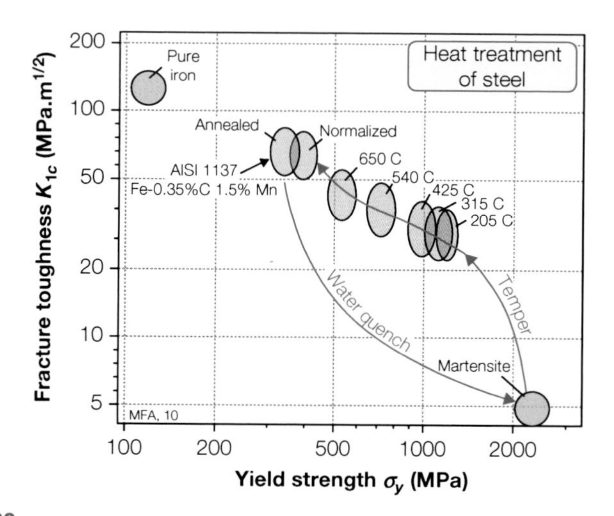

FIGURE 13.20
The changes in fracture toughness and yield strength for a plain carbon steel on heat treatment.

13.5 SYSTEMATIC PROCESS SELECTION

The selection strategy

The strategy for selecting processes parallels that for selecting materials. Figure 13.21 lists the now-familiar steps: *translation, screening, ranking,* and *documentation.*

Translation As we saw in earlier chapters, the *function* of a component dictates the initial choice of material and shape. This choice exerts constraints on the choice of processes. It is helpful to think of two types of constraint: *technical*—can the process do the job at all? And *quality*—can it do so sufficiently well? One technical constraint is always there: the compatibility of material and process. Quality constraints include achieving the desired precision, surface finish, and property profile while avoiding defects. The usual *objective* in processing is to minimize cost. The *free variables* are largely limited to choosing the process itself and its operating parameters (such as temperatures, flow rates, and so on). Table 13.2 summarizes the outcome of the translation stage.

Screening The screening step applies the constraints, eliminating processes that cannot meet them. Some process attributes are simple numeric ranges—the size or mass of component the process can handle, the precision, or the surface smoothness it can achieve. Others are nonnumeric—lists of materials to which the process can be applied, for example. Requirements such as "made of magnesium and weighing about 3 kg" are easily compared with the process attributes to eliminate those that cannot shape magnesium or cannot handle a component as large as 3 kg.

Ranking Ranking, as before, is based on one or more *objectives*, the most obvious of which is that of minimizing cost. In certain demanding applications it may be replaced by the objective of maximizing quality regardless of cost, though more usually it is a trade-off between the two that is sought.

Documentation Screening and ranking do not cope adequately with the less tractable issues of quality and productivity; they are best explored through a search for documentation—that is, design guidelines, best practice guides, case studies, and failure analyses. The most important technical expertise relates to productivity and quality. All types of processing equipment have optimum operating condition ranges under which they work best and produce products with uncompromised quality. Failure to operate within this window can lead to manufacturing defects, such as excessive porosity, cracking, or residual stress. This in turn leads to scrap and lost productivity, and, if passed on to the user, may cause premature failure. Documentation is an essential part of the selection exercise.

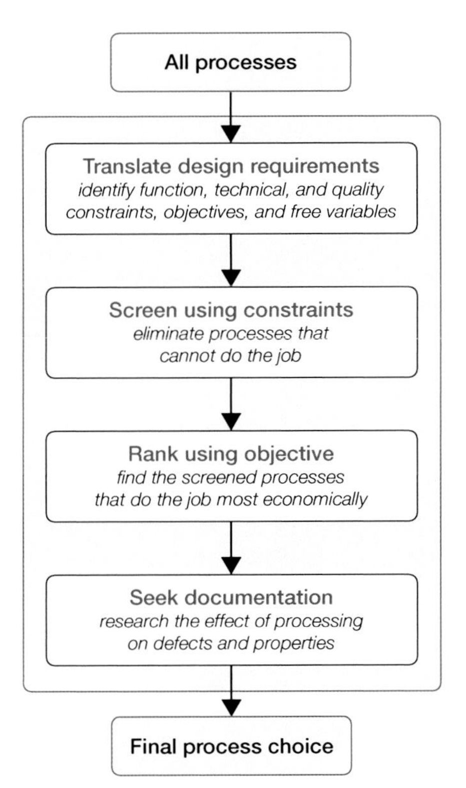

FIGURE 13.21
A flow chart of the procedure for process selection. It parallels that for material selection.

Implementing the strategy

As already explained, each process is characterized by a set of attributes. These are conveniently displayed as simple matrices and bar charts. They provide the selection tools we need for screening. The hard-copy versions, shown here, are necessarily simplified, showing only a limited number of

Table 13.2 Translation of Process Requirements

Function	What must the process do? (Shape? Join? Finish?)
Constraints	What technical limits must be met? (Material and shape compatibility) What quality limits must be met? (Precision, avoidance of defects...)
Objectives	What is to be maximized or minimized? (Cost? Time? Quality?)
Free variables	Choice of process and process-operating conditions

processes and attributes. Computer implementation allows exploration of a much larger number of both.

Material-process compatibility Figure 13.22 shows a process-material compatibility matrix. Shaping processes are at the top, with compatible combinations marked by colored dots that identify the material family. Its use for screening is straightforward: Specify the material and read off the processes or the reverse: specify the process and read off the materials. The diagonal spread of the dots in the matrix reveals that each material class—metals,

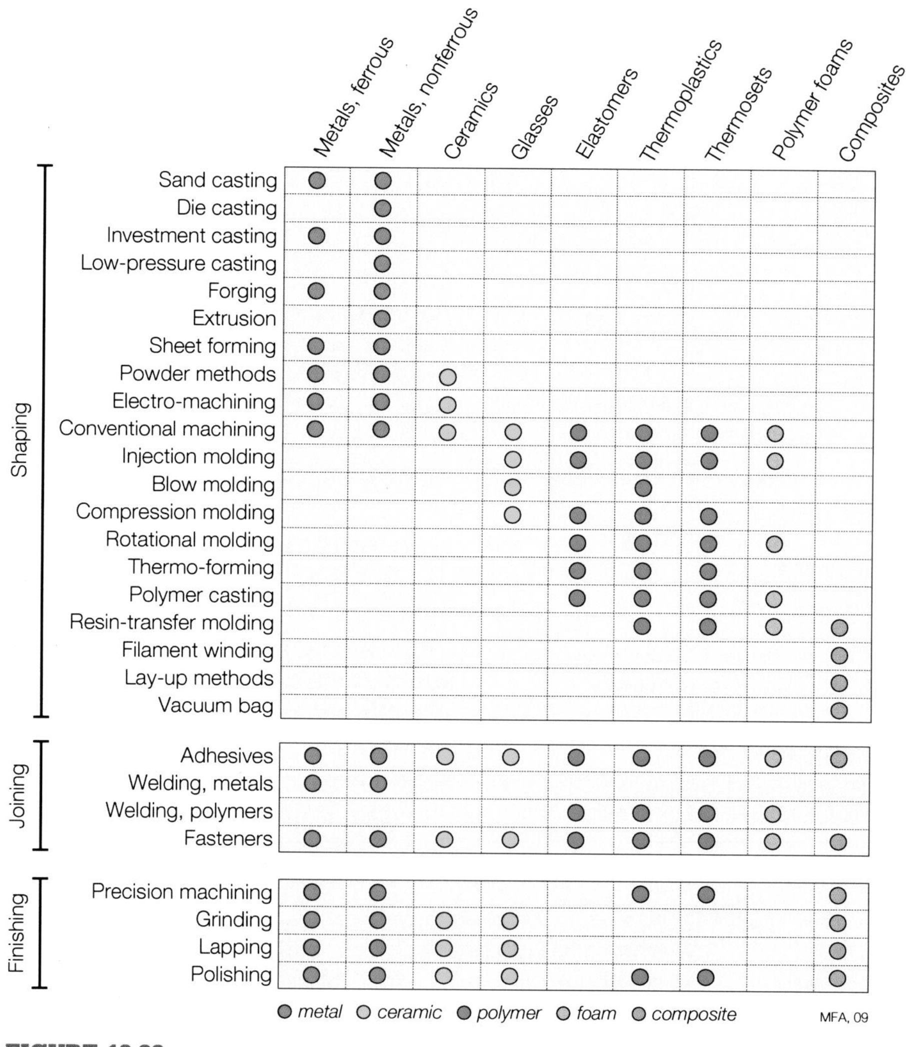

FIGURE 13.22
The process-material matrix. A colored dot indicates that the pair are compatible.

polymers, and such—has its own set of process routes. There are some overlaps —powder methods are compatible with both metals and ceramics, molding with both polymers and glasses. Machining (when used for shaping) is compatible with almost all families. Joining processes using adhesives and fasteners are very versatile and can be used with most materials, whereas welding methods are material-specific. Finishing processes are used primarily for the harder materials, particularly metals; polymers are molded to shape and rarely treated further except for decorative purposes. We will see why later.

Process-shape compatibility Shape is the most difficult attribute to characterize. Many processes involve rotation or translation of a tool or of the material, directing our thinking toward axial symmetry, translational symmetry, uniformity of section, and the like. Turning creates axisymmetric (or circular) shapes; extrusion, drawing, and rolling make prismatic shapes, both circular and noncircular. Sheet-forming processes make flat shapes (stamping) or dished shapes (drawing). Certain processes can make three-dimensional shapes, and among these some can make hollow shapes whereas others cannot. Figure 13.23 illustrates this classification scheme. The prismatic shapes shown on the left, made by rolling, extrusion, or drawing, have a special feature: They can be made in continuous lengths. The other shapes cannot—they are discrete, and the processes that make them are called batch processes. Continuous processes are well suited to long, prismatic products such as railway track or standard stock such as tube, plate, and sheet. Cylindrical rolls produce sheets. Shaped rolls make more complex profiles—rail track is one. Extrusion is a particularly versatile continuous process, since complex prismatic profiles that include internal channels and longitudinal features such as ribs and stiffeners can be manufactured in one step.

The process-shape matrix displays the links between the two. If the process cannot make the shape we want, it may be possible to combine it with a secondary process to give a process chain that adds the additional features: Casting followed by machining is an obvious example. But remember: Every additional process step adds cost.

Shaping processes: Mass and section thickness There are limits to the size of component that a process can make. Figure 13.24 shows the limits. The color coding for material compatibility has been retained, using more than one color when the process can treat more than one material family. Size can be measured by volume or by mass, but since the range of either one covers many orders of magnitude, whereas densities span only a factor of about 50, it doesn't make much difference which we use—big things are heavy, whatever they are made of. Most processes span a mass range of about a factor of 1000 or so. Note that this attribute is most discriminating at the extremes; the vast majority of components are in the 0.1 to 10 kg range, for which virtually any process will work.

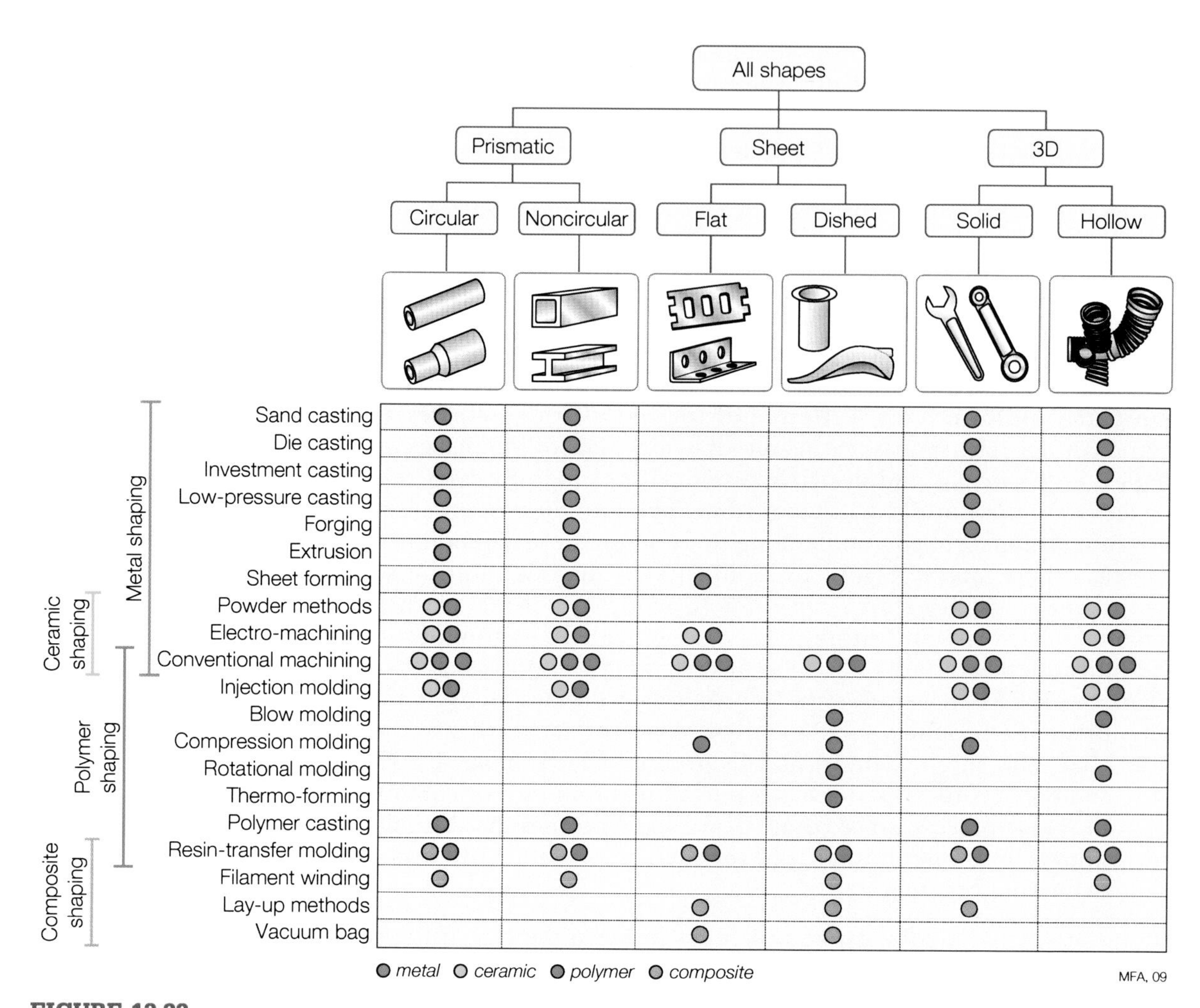

FIGURE 13.23

The process-shape matrix. Information about material compatibility is included at the extreme left.

Each bar spans the size range of which the process is capable without undue technical difficulty. All can be stretched to smaller or larger extremes but at the penalty of extra cost because the equipment is no longer standard. During screening, therefore, it is important to recognize "near misses"—processes that narrowly failed, but that could, if needed, be reconsidered and used.

Figure 13.25 shows a second bar chart: that for the ranges of section thickness of which each shaping process is capable. It is the lower end of the ranges—the minimum section thickness—where the physics of the process imposes limits. The origins of these limits are the subject of the next subsection.

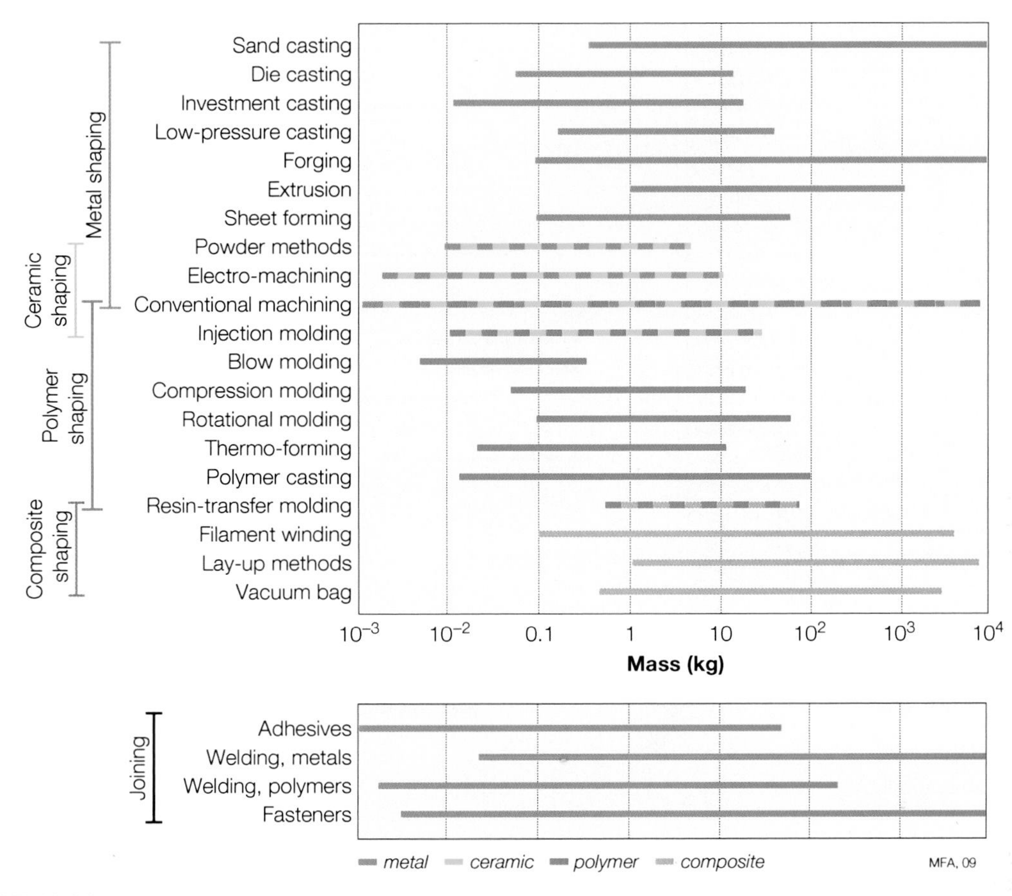

FIGURE 13.24
The process–mass-range chart. The inclusion of joining allows simple process chains to be explored.

Physical limits to size and section thickness Casting and molding both rely on material flow in the liquid or semiliquid state. Lower limits on section thickness are imposed by the physics of flow. Viscosity and surface tension oppose flow through narrow channels, and heat loss from the large surface area of thin sections cools the flowing material, raising the viscosity before the channel is filled (Figure 13.26). Pure metals solidify at a fixed temperature, with a step increase in viscosity, but for alloys solidification happens over a range of temperature, known as the "mushy zone," in which the alloy is part liquid, part solid. The width of this zone can vary from a few degrees centigrade to several hundred—so metal flow in castings depends on alloy composition. In general, higher-pressure die casting and molding methods enable thinner sections to be made, but the equipment costs more and the faster, more turbulent flow can entrap more porosity and cause damage to the molds.

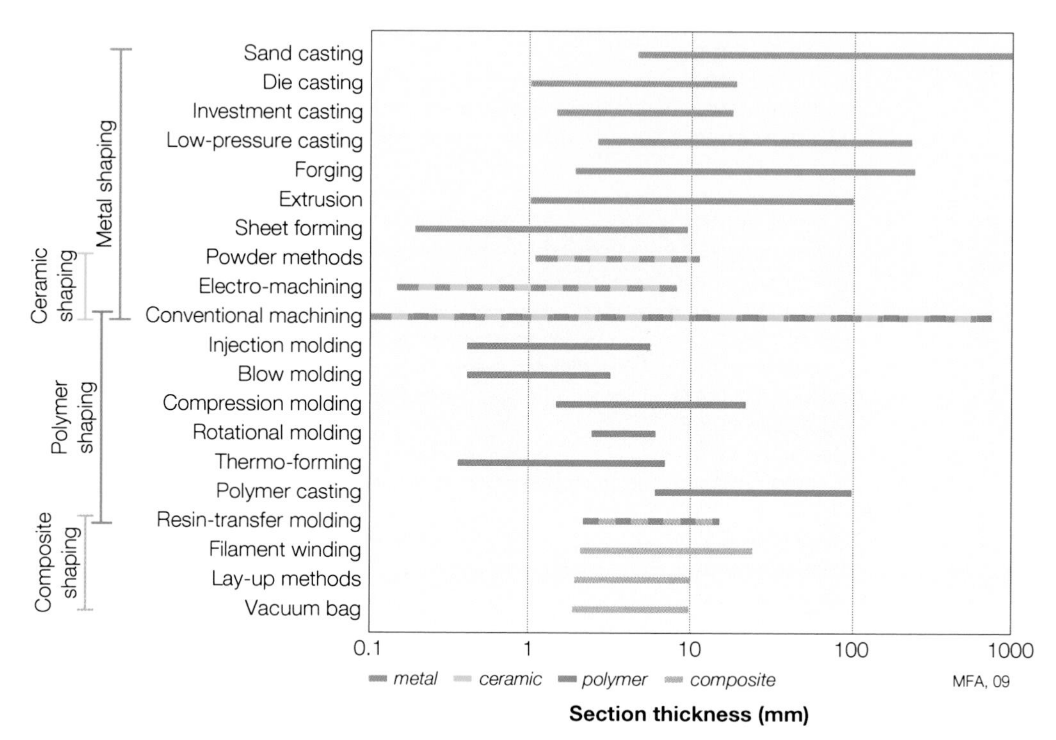

FIGURE 13.25
The process–section thickness chart.

Upper limits to size and section in casting and molding are set by problems of shrinkage. The outer layer of a casting or molding cools and solidifies first, giving it a rigid skin. When the interior subsequently solidifies, the change in volume can distort the product or crack the skin, or cause internal cavitation. Problems of this sort are most severe where there are changes of section, since the constraint introduces tensile stresses that cause *hot tearing*—cracking

Flow in thin channels

A mold for casting an intricate aluminum casting has some channel-like features with a width of only 10 μm. Will an overpressure of 1 atmosphere (0.1 MPa) be sufficient to overcome surface tension, allowing the feature to be filled? The surface tension γ of liquid aluminum is 1.1 J/m^2.

Answer

The pressure required to overcome surface tension and force the metal into a parallel-sided channel of width $2x$ is $p = \gamma/x$. Thus the narrowest channel that can be filled with an overpressure of 1 atmosphere is

$$2x = \frac{2\gamma}{p} = \frac{2.2}{0.1 \times 10^6} 2.2 \times 10^{-5} = 22\,\mu\text{m}$$

An overpressure of 1 atmosphere is not sufficient to fill the channel. An overpressure of 5 atmospheres would do it comfortably.

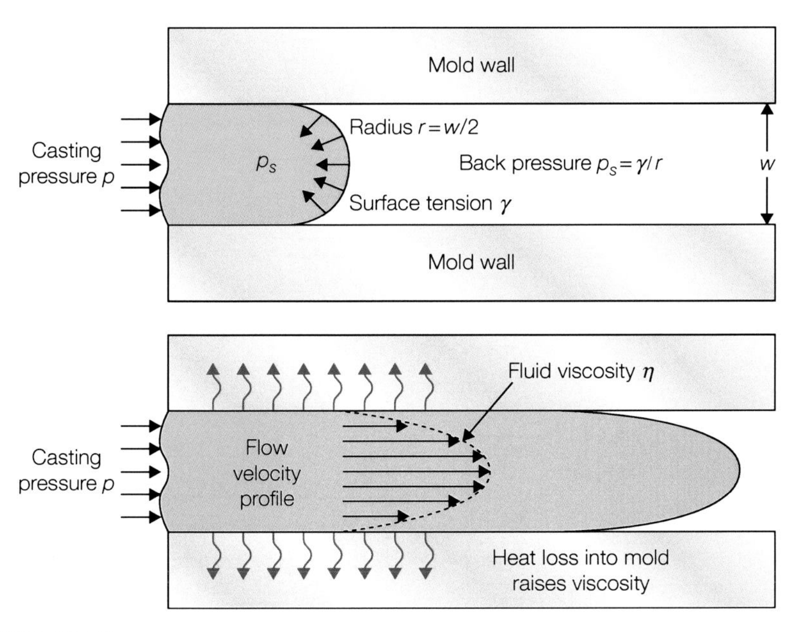

FIGURE 13.26

Flow of liquid metal or polymer into thin sections is opposed by surface tension as at (a) and by viscous forces (b). Loss of heat into the mold increases viscosity and may cause premature solidification.

caused by constrained thermal contraction. Different compositions have different susceptibilities to hot tearing—another example of coupling between material, process, and design detail.

Metal shaping by deformation—hot or cold rolling, forging, or extrusion—also involves flow. The thinness that can be forged, rolled, or extruded is limited by plastic flow in much the same way that the thinness in casting is limited by viscosity: The thinner the section, the greater the required roll pressure or forging force. Figure 13.27 illustrates the problem. Friction changes the pressure distribution on the die and under the rolls. When they are well-lubricated, as in (a), the loading is almost uniaxial and the material flows at its yield stress σ_y. With friction, as in (b), the metal shears at the die interface and the pressure ramps up because the friction resists the lateral spreading, giving a "friction hill." The area under the pressure distribution is the total forming load, so friction increases the load. The greater the aspect ratio of the section (width/thickness), the higher the maximum pressure needed to cause yielding, as in (c). This illustrates the fundamental limit of friction on section thickness—very thin sections simply stick to the tools and will not yield, even with very large pressures.

Friction also limits aspect ratio in powder processing. The externally applied pressure is diminished by die-wall friction (Figure 13.28) with the result

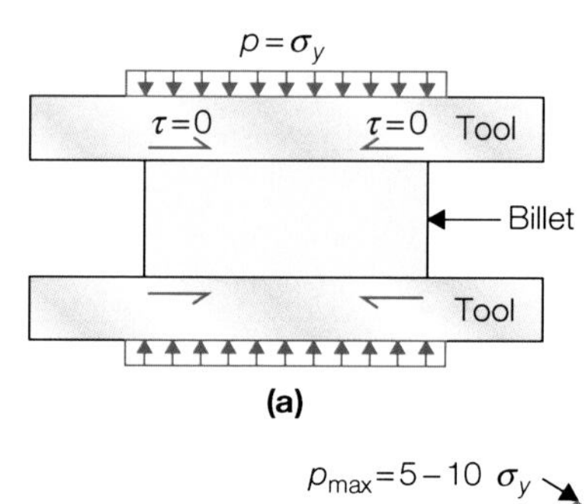

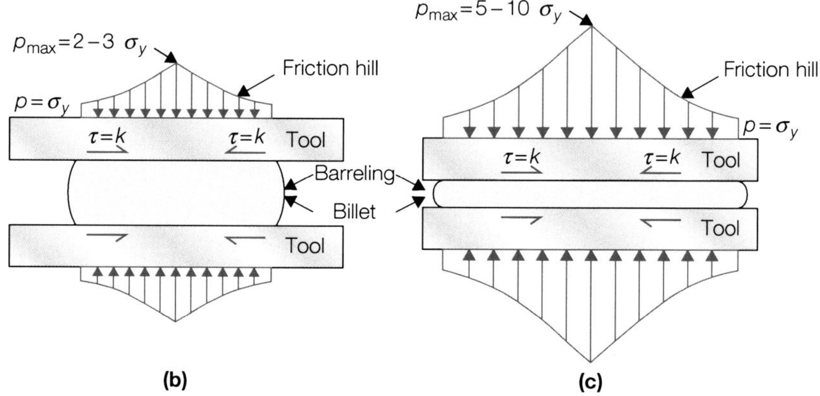

FIGURE 13.27

The influence of friction and aspect ratio on forging: (a) uniaxial compression with very low friction; (b) with sticking friction the contact pressure rises in a friction hill; (c) the greater the aspect ratio, the greater the pressure rise, ultimately limiting the thinness that can be achieved.

Limiting aspect ratio for powder pressing

How large is the pressure drop in a cylindrical powder mass like that in Figure 13.28 caused by die-wall friction if the coefficient of friction on the die wall is $\mu = 0.5$?

Answer

The frictional force opposing sliding in the green band of thickness dx is $2\pi\, r\, \mu\, p\, dx$ where p is the pressure at a distance x below the die face. Dividing this by the cross-section area of the powder pack πr^2 gives the pressure drop

$$dp = -\frac{2}{r}\mu\, p\, dx$$

Integrating from $x = 0$ where $p = p_o$ to $x = x$ where $p = p(x)$ gives

$$p(x) = p_o \exp -\frac{2\mu x}{r}$$

With a coefficient of friction of $\mu = 0.5$, the pressure drops to half its remote value p_o at a depth-to-radius ratio of only

$$\frac{x}{r} = -\frac{1}{2\mu}\ln\left(\frac{1}{2}\right) = 0.69$$

The answer is to lubricate the mold, reducing μ.

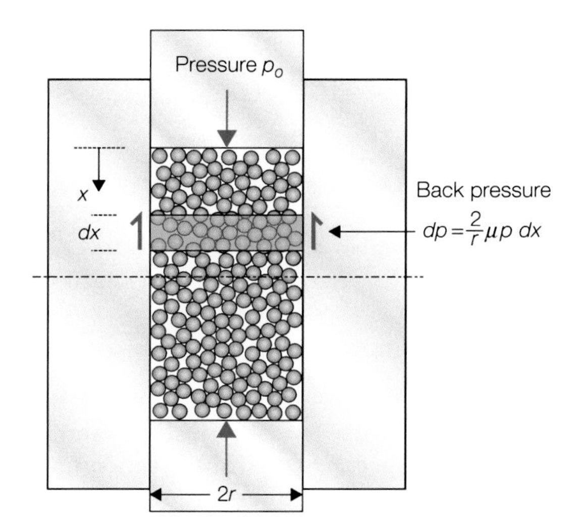

FIGURE 13.28

The height-to-width aspect ratio in powder forging is limited by die-wall friction, which causes the compaction pressure to fall exponentially with distance from the surface.

that, if the aspect ratio is too great, there is insufficient pressure to compact powder in the center of the product.

Tolerance and roughness The precision and surface finish of a component are aspects of its quality. They are measured by the *tolerance* and the *surface roughness R*. When the dimensions of a component are specified the surface quality is specified as well, though not necessarily over the entire surface. Surface quality is critical in contacting surfaces such as the faces of flanges that must mate to form a seal or sliders running in grooves. It is also important for resistance to fatigue crack initiation and for aesthetic reasons. The tolerance T on a dimension y is specified as $y = 100 \pm 0.1$ mm, or as $y = 50^{+0.01}_{-0.001}$ mm, indicating that there is more freedom to oversize than to undersize. Surface roughness is specified as an upper limit, for example, $R < 100\,\mu$m. The typical surface finish required in various products is shown in Table 13.3. The table also indicates typical processes that can achieve these levels of finish.

Surface roughness is a measure of the irregularities of the surface (Figure 13.29). It is defined as the root-mean-square (RMS) amplitude of the surface profile:

$$R^2 = \frac{1}{L}\int_0^L y^2(x)dx \qquad (13.1)$$

One way to measure it is to drag a light, sharp stylus over the surface in the x-direction while recording the vertical profile $y(x)$, like playing a gramophone record. *Optical profilometry*, which is faster and more accurate, uses laser

Table 13.3 Levels of Finish

Finish, μm	Process	Typical Application
$R = 0.01$	Lapping	Mirrors
$R = 0.1$	Precision grind or lap	High-quality bearings
$R = 0.2–0.5$	Precision grinding	Cylinders, pistons, cams, bearings
$R = 0.5–2$	Precision machining	Gears, ordinary machine parts
$R = 2–10$	Machining	Light-loaded bearings, noncritical components
$R = 3–100$	Unfinished castings	Nonbearing surfaces

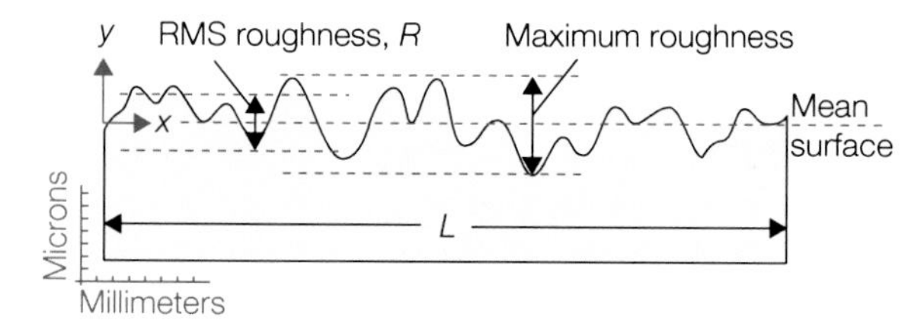

FIGURE 13.29
A section through a surface showing its irregular surface (artistically exaggerated in the vertical direction). The irregularity is measured by the RMS roughness R.

interferometry to map surface irregularity. The tolerance T is obviously greater than $2R$; indeed, since R is the root-mean-square roughness, the peak roughness, and hence the absolute lower limit for tolerance, is more like $5R$. Real processes give tolerances that range from $10R$ to $1000R$.

Figures 13.30 and 13.31 show the characteristic ranges of tolerance and roughness of which processes are capable, retaining the color coding for material families. Data for finishing processes are added below the shaping processes. Sand casting gives rough surfaces; casting into metal dies gives smoother ones. No shaping processes for metals, however, do better than $T = 0.1$ mm and $R = 0.5$ μm. Machining, capable of high dimensional accuracy and surface finish, is commonly used after casting or deformation processing to bring the tolerance or finish up to the desired level, creating a *process chain*. Metals and ceramics can be surface-ground and lapped to a high precision and smoothness: A large telescope reflector has a tolerance approaching 5 μm and a roughness of about 1/100 of this over a dimension of a meter or more. But precision and finish carry a cost: Processing costs increase exponentially as the requirements for both are made more severe. It is an expensive mistake to overspecify precision and finish.

Molded polymers inherit the finish of the molds used to shape them and thus can be very smooth; machining to improve the finish is rarely necessary. Tolerances better than $\pm$ 0.2 mm are seldom possible because internal stresses left by molding cause distortion and because polymers creep in service.

Joining: Material compatibility Processes for joining metals, polymers, ceramics, and glasses differ. A given adhesive will bond to some materials but not to others; welding methods for polymers differ from those for welding metals; and ceramics, which cannot be welded, are joined instead by diffusion bonding or glaze bonding. The process-material matrix (Figure 13.22) included four classes of joining process.

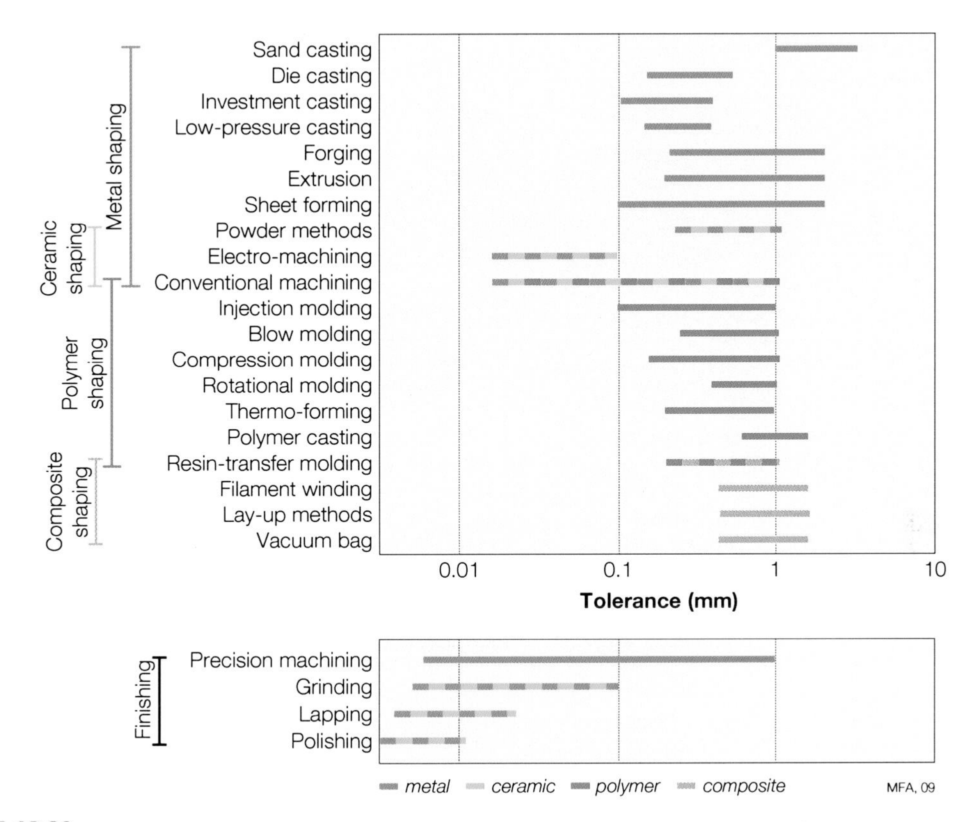

FIGURE 13.30
The process-tolerance chart. The inclusion of finishing processes allows simple process chains to be explored.

When the joint is between dissimilar materials, the process must be compatible with both. Adhesives and fasteners allow joints between different materials; many welding processes do not. If dissimilar metals are joined in such a way that they are in electrical contact, a corrosion couple appears if the joint is wet. This can be avoided by inserting an insulating layer between the surfaces. Thermal-expansion mismatch gives internal stresses in the joint if the temperature changes, with risk of distortion or damage. Identifying good practice in joining dissimilar materials is part of the documentation step.

Joint geometry and mode of loading The geometry of the joint and the way it is loaded (Figure 13.32) influence process choice. Adhesive joints support shear but are poor in peeling—think of stripping off adhesive tape. Adhesives need a large working area—lap joints work well, butt joints do not. Rivets and staples, too, are well adapted for shear loading of lap joints but are less good in tension. Welds and threaded fasteners are more adaptable, but here, too, matching choice of process to geometry and loading is important.

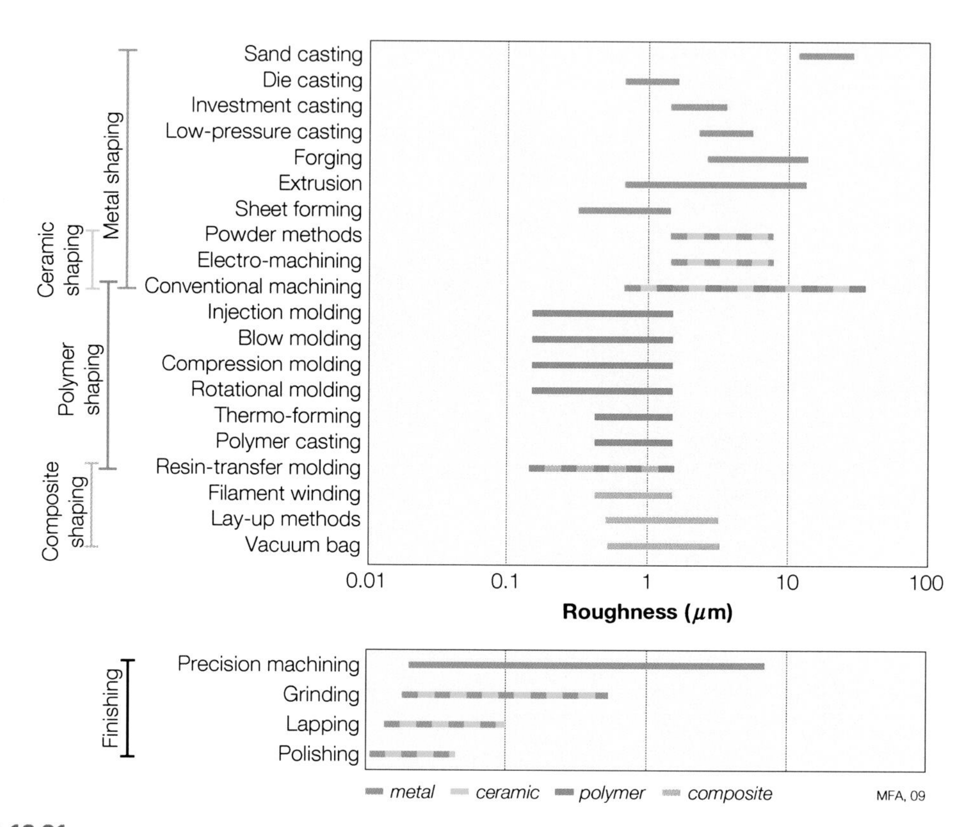

FIGURE 13.31
The process–surface roughness chart. The inclusion of finishing processes allows simple process chains to be explored.

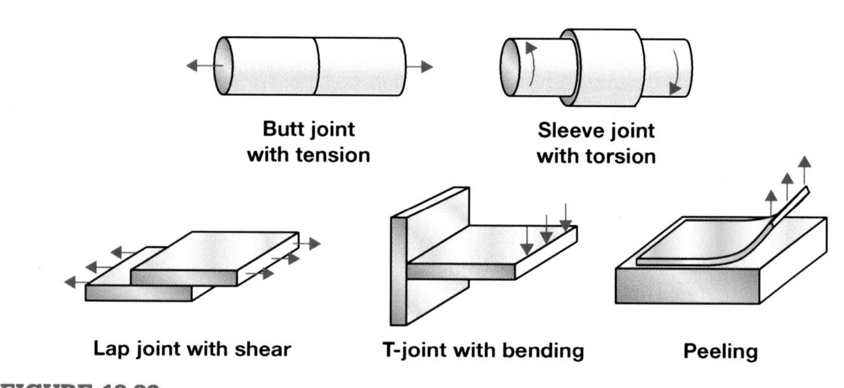

FIGURE 13.32
Joint geometries and modes of loading.

Secondary functions of joints A joint may serve as a seal, and be required to exclude gases or liquids. It may be required to conduct or insulate against the conduction of heat or electricity, or to operate at an elevated operating temperature. It may be permanent or have to be disassembled at the end of product life for recycling and reuse: Threaded fasteners and adhesives, loosened with solvents or heat, permit this.

Surface treatment: Material compatibility Material process compatibility for surface treatments is shown at the bottom of the matrix in Figure 13.22. As noted previously, surface finishing is more important for metals than for polymers.

The purpose of the surface treatment All surface treatment adds cost but the added value can be large. Table 13.4 illustrates the diversity of functions that surface treatments can provide. Some protect, some enhance performance, still others are primarily aesthetic. Protecting a component surface extends product life and increases the interval between maintenance cycles. Coatings on cutting tools enable faster cutting speeds and greater productivity. And surface-hardening processes may enable the substrate alloy to be replaced with a cheaper material—for example, a plain carbon steel with a hard carburized surface or a coating of hard titanium nitride (TiN) instead of a more expensive alloy steel.

Secondary compatibilities Some surface treatments, such as anodizing, leave the dimensions, precision, and roughness of the surface unchanged. Electro- and vapor-deposited coatings change the dimensions a little, but may still leave a perfectly smooth surface. Polymer powder coating builds up a relatively thick, smooth layer; others, such as weld deposition, create a thick layer with a rough surface, requiring refinishing. "Line-of-sight" deposition processes coat only the surface at which they are directed, leaving inaccessible areas uncoated; others, with what is called "throwing power," coat flat, curved, and reentrant surfaces equally well. Many surface treatment processes require heat. These can only be used on materials that can tolerate the rise in temperature. Some paints are applied cold, but many require a bake at up to 150°C. Heat treatments like carburizing or nitriding to give a hard surface layer require prolonged heating at temperatures up to 800°C, which can change the microstructure of the material being coated.

Table 13.4 Functions Provided by Surface Treatments

Corrosion protection, aqueous environments	Thermal insulation
Corrosion protection, gas environments	Electrical insulation
Wear resistance	Magnetic response
Friction control	Decoration
Fatigue resistance	Color
Thermal conduction	Reflectivity

13.6 RANKING: PROCESS COST

Part of the cost of any component is the expense of the material from which it is made. The rest is the cost of manufacture—that is, forming it to a shape and joining and finishing it. Before turning to details, there are four common-sense rules for minimizing cost that the designer should bear in mind, as follows.

Keep things standard If someone already makes the part you want, it will almost certainly be cheaper to buy it than to make it. If nobody does, then it is cheaper to design it to be made from standard stock (sheet, rod, tube) than from nonstandard shapes or from special castings or forgings. Try to use standard materials, and as few of them as possible: This will reduce inventory costs and the range of tooling the manufacturer needs, and it can help with recycling.

Keep things simple If a part has to be machined, it will have to be clamped; the cost increases with the number of times it will have to be rejigged or reoriented, especially if special tools are necessary. If a part is to be welded or brazed, the welder must be able to reach it with his torch and still see what he is doing. If it is to be cast or molded or forged, it should be remembered that high (and expensive) pressures are required to make fluids flow into narrow channels, and that reentrant shapes greatly complicate mold and die design. Think of making the part yourself: Will it be awkward? Could slight redesign make it less awkward?

Make the parts easy to assemble Assembly takes time, and time is money. If the overhead rate is a mere $60 per hour, every minute of assembly time adds another $1 to the cost. Design for assembly (DFA) addresses this problem with a set of common-sense criteria and rules. Briefly, there are three:

1. Minimize part count.
2. Design parts to be self-aligning on assembly.
3. Use joining methods that are fast; snap fits and spot welds are faster than threaded fasteners or, usually, adhesives.

Do not specify more performance than is needed Performance must be paid for. High-strength metals are more heavily alloyed with expensive additions; high-performance polymers are chemically more complex; high-performance ceramics require greater quality control in their manufacture. All of these increase material costs. In addition, high-strength materials are hard to fabricate. The forming pressures (whether for a metal or a polymer) are higher; tool wear is greater; ductility is usually less so that deformation processing can be difficult or impossible. This can mean that new processing routes must be used: investment casting or powder forming instead of conventional casting and mechanical working; more expensive molding equipment operating at higher temperatures and pressures; and so on. The better performance of the high-strength material must be paid for, not only in greater material cost but also in the higher cost of processing. Finally, there are the questions of tolerance and roughness. Cost rises exponentially with demands for precision and surface finish. The message is clear. Performance costs money. Do not overspecify it.

To make further progress, we must examine the contributions to process costs, and their origins.

Economic criteria for selection If you have to sharpen a pencil, you can do it with a knife. If, instead, you had to sharpen a thousand pencils, it would pay to buy an electric sharpener. And if you had to sharpen a million, you might wish to equip yourself with an automatic feeding, gripping, and sharpening system. To cope with pencils of different length and diameter, you could go further and devise a microprocessor-controlled system with sensors to measure pencil dimensions, sharpening pressure, and so on —an "intelligent" system that can recognize and adapt to pencil size. The choice of process, then, depends on the number of pencils you wish to sharpen, that is, on the *batch size.* The best choice is the one that costs the least per pencil sharpened.

Figure 13.33 is a schematic of how the cost of sharpening a pencil might vary with batch size. A knife does not cost much but it is slow, so the labor

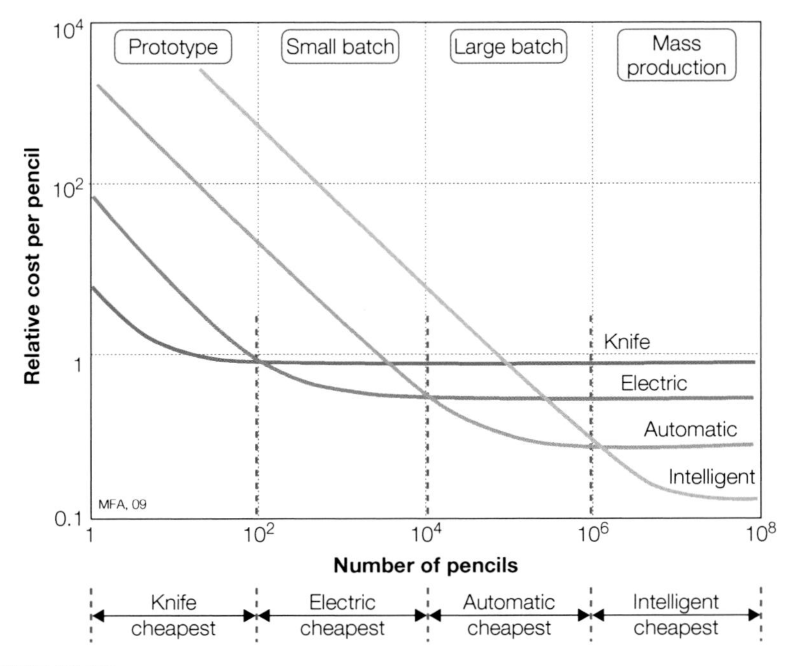

FIGURE 13.33
The cost of sharpening a pencil plotted against batch size for four processes. The curves all have the form of Equation (13.5).

cost is high. The other processes involve progressively greater capital investment but do the job more quickly, reducing labor costs. The balance between capital cost and rate gives the shape of the curves. In this figure the best choice is the lowest curve—a knife for up to 100 pencils; an electric sharpener for 10^2 to 10^4, an automatic system for 10^4 to 10^6, and so on. Each process has an *economic batch size.*

Economic batch size Process cost depends on a large number of independent variables, not all within the control of the modeler. Cost modeling is described in the next section, but—given the disheartening implications of the last sentence—it is comforting to have an alternative, if approximate, way out. The economic batch size provides it. Values for the processes described in this chapter are shown in Figure 13.34. A process with an economic batch size with the range B_1–B_2 is one that is found by experience to be competitive in cost when the output lies in that range, just as the electric sharpener was economic in the range 10^2 to 10^4. The economic batch size is commonly cited for processes. The easy way to introduce economy into the selection is to rank candidate processes by economic batch size and retain those that are economic in the range you want. But do not harbor false illusions: Many variables cannot be rolled into one without loss of discrimination. A cost model gives deeper insight.

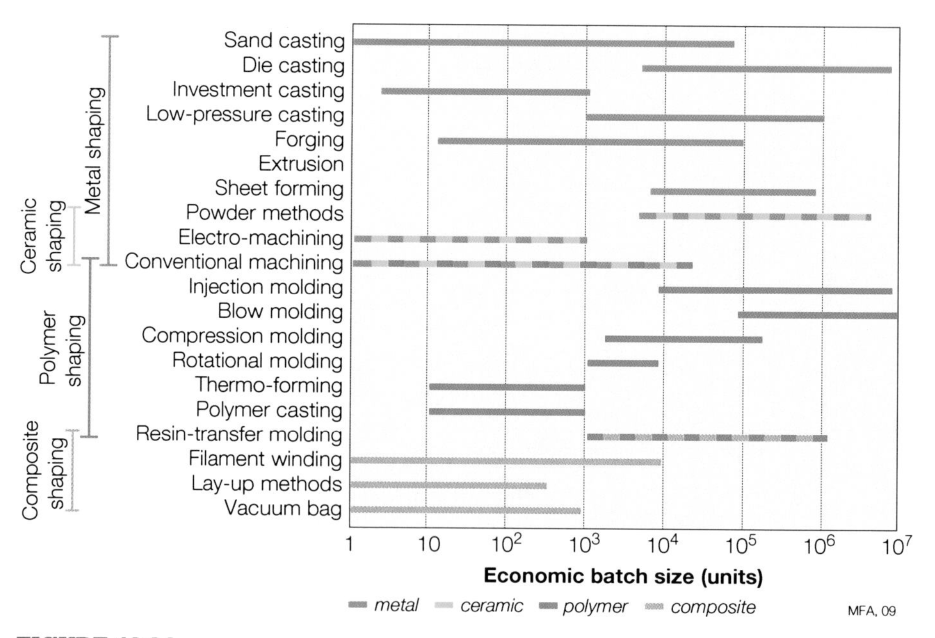

FIGURE 13.34

The economic batch size chart.

Cost modeling

The manufacture of a component consumes resources (Figure 13.35), each of which has an associated cost. The final cost is the sum of expenses of all of the resources it consumes (detailed in Table 13.5). Thus the cost of producing a component of mass m entails the cost Cm (\$/kg) of the materials and feedstocks from which it is made. It involves the cost of dedicated tooling Ct (\$) and that of the capital equipment Cc (\$) in which the tooling will be used. It requires time, chargeable at an overhead rate $\dot{C}_{oh}$ (thus with units of \$/hr), in which we include the cost of labor, administration, and general plant costs. It requires energy, which is sometimes charged against a process step if it is very energy intense but more commonly is treated as part of the overhead and lumped into $\dot{C}_{oh}$, as we shall do here. Finally there is the cost of information, meaning research and development, royalty or license fees; this, too, we view as a cost per unit time and lump it into the overhead.

Think now of the manufacture of a component (the "unit of output") weighing m kg, made of a material costing C_m \$/kg. The first contribution to the unit cost is that of the material mC_m magnified by the factor $1/(1-f)$ where f is the scrap fraction—the fraction of the starting material that ends up as sprues, risers, turnings, rejects, or waste:

$$C_1 = \frac{mC_m}{(1-f)} \qquad (13.2)$$

The cost C_t of a set of tooling—dies, molds, fixtures, and jigs—is what is called a *dedicated cost*: one that must be wholly assigned to the production run of this single component. It is written off against the numerical size n of the production

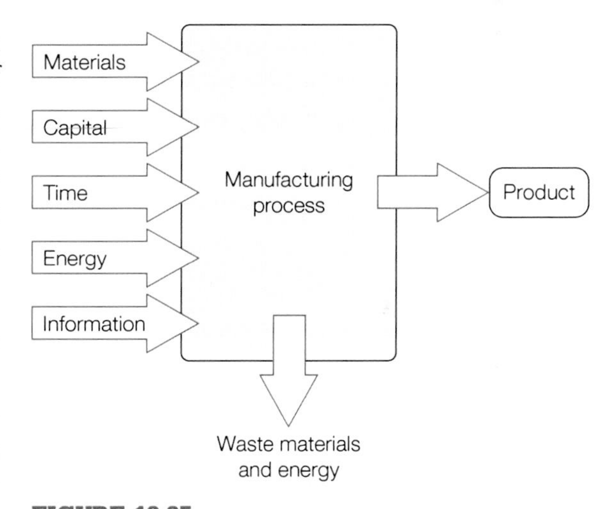

FIGURE 13.35
The inputs to a cost model.

Table 13.5 Symbols, Definitions, and Units

Resource	Symbol	Unit
Materials: including consumables	C_m	\$/kg
Capital: cost of tooling	C_t	\$
cost of equipment	C_c	\$
Time: overhead rate, including labor, administration, rent ...	$\dot{C}_{oh}$	\$/hr
Energy: cost of energy	$\dot{C}_e$	\$/hr
Information: R & D or royalty payments	$\dot{C}_i$	\$/year

run. Tooling wears out. If the run is a long one, replacement will be necessary. Thus tooling cost per unit takes the form

$$C_2 = \frac{C_t}{n}\left\{Int\left(\frac{n}{n_t} + 0.51\right)\right\} \tag{13.3}$$

where n_t is the number of units that a set of tooling can make before it has to be replaced, and *Int* is the integer function. The term in curly brackets simply increments the tooling cost by that of one tool set every time n exceeds n_t.

The capital cost of equipment C_c, by contrast, is rarely dedicated. A given piece of equipment—a powder press, for example—can be used to make many different components by installing different die-sets or tooling. It is usual to convert the capital cost of *nondedicated* equipment, and the cost of borrowing the capital itself, into an overhead by dividing it by a capital write-off time t_{wo} (five years, say) over which it is to be recovered. The quantity C_c/t_{wo} is then a cost per hour—provided the equipment is used continuously. That is rarely the case, so the term is modified by dividing it by a load factor L—the fraction of time for which the equipment is productive. The cost per unit is then this hourly cost divided by the rate $\dot{n}$ at which units are produced:

$$C_3 = \frac{1}{\dot{n}}\left(\frac{C_c}{Lt_{wo}}\right) \tag{13.4}$$

Finally there is the overhead rate $\dot{C}_{oh}$. It becomes a cost per unit when divided by the production rate $\dot{n}$ units per hour (averaged over the production run to allow for downtime):

$$C_4 = \frac{\dot{C}_{oh}}{\dot{n}} \tag{13.5}$$

The total shaping cost per part is the sum of these four terms, taking the form

$$C = \frac{mC_m}{(1-f)} + \frac{C_t}{n}\left\{Int\left(\frac{n}{n_t} + 0.51\right)\right\} + \frac{1}{\dot{n}}\left(\frac{C_c}{Lt_{wo}} + \dot{C}_{oh}\right) \tag{13.6}$$

The equation says: The cost has three essential contributions—a material cost per unit of production that is independent of batch size and rate, a dedicated cost per unit of production that varies as the reciprocal of the production volume $(1/n)$, and a gross overhead per unit of production that varies as the reciprocal of the production rate $(1/\dot{n})$. The equation describes a set of curves relating cost C to batch size n, one for each process. Each has the shape of the pencil-sharpening curves of Figure 13.33.

The use of the model is illustrated more fully in the case studies of Chapter 14.

Technical cost modeling Equation (13.6) is the first step in modeling cost. Greater predictive power is possible with technical cost models that exploit understanding of the way in which the design, the process, and cost interact. The capital cost of equipment depends on size and degree of automation. Tooling cost and production rate depend on complexity. These and many other dependencies can be captured in theoretical or empirical formulae or look-up tables that can be built into the cost model, giving more resolution in ranking competing processes. For more advanced analyses the reader is referred to the literature listed in "Further reading," at the end of this chapter.

13.7 COMPUTER-AIDED PROCESS SELECTION

Screening

If process attributes are stored in a database with an appropriate user interface, selection charts can be created and selection boxes manipulated with much greater freedom. The CES platform, mentioned earlier, is an example of such a system. The database contains records, each describing the attributes of a single process. Example 13.1 shows part of a typical record: that for injection molding. A schematic indicates how the process works; it is supported by a short description. This is followed by a list of attributes: the shapes it can make, the attributes relating to shape and physical characteristics, and those that describe economic parameters; ending with brief documentation in the form of guidelines, technical notes, and typical uses. The numeric attributes are stored as ranges, indicating the range of capability of the process. Each record is linked to records for the materials with which it is compatible, allowing choice of material to be used as a screening criterion, like the material compatibility matrix of Figure 13.22 but with greater resolution. A short list of candidates is extracted in two steps: screening to eliminate processes that cannot meet the design specification, and ranking to order the survivors by economic criteria.

EXAMPLE 13.1

Injection Molding

The Process The most widely used process for shaping thermoplastics is the reciprocating screw injection molding machine, shown in the diagram on the right. Polymer granules are fed into a spiral press where they are heated, mixed, and softened to a dough-like consistency that is forced through one or more channels ("sprues") into the die. The polymer solidifies under pressure and the component is then ejected.

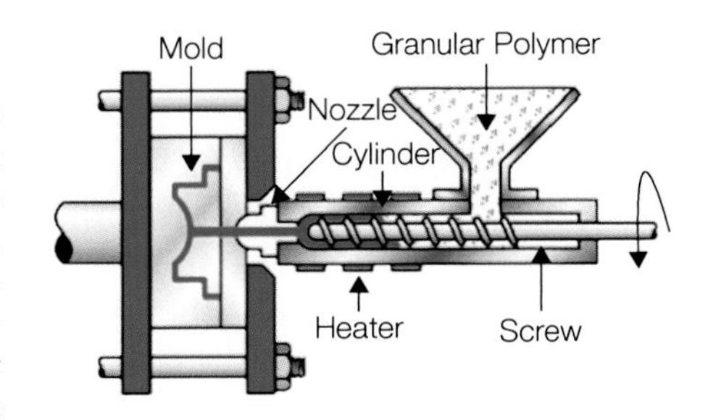

Thermoplastics, thermosets, and elastomers can also be injection molded. Co-injection allows molding of components with different materials, colors, and features. Injection foam molding allows economical production of large molded components by using inert gas or chemical blowing agents to make components that have a solid skin and a cellular inner structure.

Shapes	
Circular prismatic	✓
Noncircular prismatic	✓
Solid 3D	✓
Hollow 3D	✓
Physical Attributes	
Mass range	0.01–25 kg
Range of section thickness	0.4–6.3 mm
Tolerance	0.2–1 mm
Roughness	0.2–1.6 μm
Surface roughness (A = v. smooth)	A
Economic Attributes	
Economic batch size (units)	10^4–10^6
Cost Modeling	
Capital cost	3×10^4–7×10^5 USD
Material utilization fraction	0.6–0.9
Production rate (units)	60–1000/hr
Tooling cost	3000–30,000 USD
Tool life (units)	10^4–10^6

Documentation

Design Guidelines Injection molding is the best way to mass-produce small, precise polymer components with complex shapes. The surface finish is good; texture and pattern can be easily altered in the tool; and fine detail reproduces well. Decorative labels can be molded onto the surface of the component. The only finishing operation is the removal of the sprue.

Technical Notes Most thermoplastics can be injection molded, although those with high melting temperatures (e.g., PTFE) are difficult. Thermoplastic-based composites (short-fiber and particulate filled) can be processed providing the filler-loading is not too large. Large changes in section area are not recommended. Small reentrant angles and complex shapes are possible, though some features (e.g., undercuts, screw threads, inserts) may result in increased tooling costs.

Typical Uses Housings, containers, covers, knobs, tool handles, plumbing fittings, lenses, toys, and so on.

To enable this, the cost model described in Section 13.6 is implemented in the CES software. The records contain approximate data for the ranges of capital and tooling costs (C_c and C_t) and for the rate of production ($\dot{n}$). Equation (13.6) contains other parameters not listed in the record because they are not attributes of the process itself but depend on the design, or the material, or the economics (and thus the location) of the plant in which the processing will be done. The user must provide this information, conveniently entered through a dialog box. The output is a plot of cost against batch size, like that shown earlier in Figure 13.33.

More information about computer-aided selection can be found in the sources listed under Granta Design (2010) in "Further reading."

13.8 SUMMARY AND CONCLUSIONS

A wide range of shaping, joining, and finishing processes is available to the design engineer. Each has certain characteristics, which, taken together, suit it to the processing of certain materials in certain shapes, but disqualify it for others. Faced with the choice, the designer in the past, relied on locally available expertise or on common practice. Neither of these leads to innovation, nor are they well matched to current design methods. The structured, systematic approach of this chapter provides a way forward. It ensures that potentially interesting processes are not overlooked, and guides the user quickly to processes capable of achieving the desired requirements.

The method parallels that for selection of material, using process selection matrices and charts to implement the procedure. A component design dictates a certain, known combination of process attributes. These design requirements are plotted onto the charts, identifying a subset of possible processes. The method lends itself to computer implementation, allowing selection from a large portfolio of processes by screening on attributes and ranking by economic criteria.

There is, of course, much more to process selection than this. It is to be seen, rather, as a first systematic step, replacing a total reliance on local experience and past practice. The narrowing of choice helps considerably: It is now much easier to identify the right source for more expert knowledge and to ask it the right questions. But the final choice still depends on local economic and organizational factors that can only be decided on a case-by-case basis.

13.9 FURTHER READING

ASM Handbook Series (1971–2004). Heat treatment, vol. 4; Surface engineering, vol. 5; Welding, brazing and soldering, vol. 6; Powder metal technologies, vol. 7; Forming and forging, vol. 14; Casting, vol. 15; Machining, vol. 16. ASM International.
A comprehensive set of handbooks on processing, occasionally updated, and now available online at http://products.asminternational.org/hbk/index.jsp

Bralla, J.G. (1998). *Design for manufacturability handbook* (2nd ed.). McGraw-Hill, ISBN 0-07-007139-X.
Turgid reading, but a rich mine of information about manufacturing processes.

Bréchet, Y., Bassetti, D., Landru, D., & Salvo, L. (2001). Challenges in materials and process selection. *Prog. Mat. Sci., 46*, 407–428.
An exploration of knowledge-based methods for capturing material and process attributes.

Budinski, K.G., & Budinski, M.K. (2010). *Engineering materials, properties and selection* (9th ed.). Prentice Hall, ISBN 978-0-13-712842-6.
A well-established text on the processing and use of engineering materials, now in its 9th edition.

Campbell, J. (1991). *Casting*. Butterworth-Heinemann, ISBN 0-7506-1696-2.
The fundamental science and technology of casting processes.

Clark, J.P., & Field, F.R. III (1997). Techno-economic issues in materials selection. In: *ASM Metals Handbook, 20*. American Society for Metals.
A paper outlining the principles of technical cost modeling and its use in the automobile industry.

Dieter, G.E. (1991). *Engineering design, a materials and processing approach* (2nd ed.). McGraw-Hill, ISBN 0-07-100829-2.
A well-balanced and respected text focusing on the place of materials and processing in technical design.

Dieter, G.E., & Schmidt, L.C. (2009). *Engineering design* (4th ed.). McGraw-Hill, ISBN 978-0-07-283703-2.
Professor Dieter is a pioneer in presenting design from a materials perspective. The book contains a notable chapter on conceptualization.

Esawi, A., & Ashby, M.F. (1998). Computer-based selection of manufacturing processes: Methods, software and case studies.. *Proc. Inst. Mech. Eng., 212*, 595–610.
A paper describing the development and use of the CES database for process selection.

Grainger, S., & Blunt, J. (1998). *Engineering coatings, design and application*. Abington Publishing, ISBN 1-85573-369-2.
A handbook of surface treatment processes to improve surface durability—generally meaning surface hardness.

Granta Design (2010). *The CES Edu system* and other teaching resources, accessed from *www.grantadesign.com/education/*

Houldcroft, P. (1990). *Which process?*. Abington Publishing, ISBN 1-85573-008-1.
The title of this useful book is misleading—it deals only with a subset of the joining process: the welding of steels. But here it is good, matching the process to the design requirements.

Kalpakjian, S., & Schmidt, S.R. (2008). *Manufacturing processes for engineering materials* (5th ed.). Prentice Hall, ISBN 978-0-13-227271-1.
A comprehensive and widely used text on material processing.

Kalpakjian, S., & Schmidt, S.R. (2010). *Manufacturing engineering and technology* (6th ed.). Prentice Hall, ISBN 978-0-13-608168-5.
A comprehensive and widely used text on material processing.

Lascoe, O.D. (1988). *Handbook of fabrication processes*. ASM International, ISBN 0-87170-302-5.
A reference source for fabrication processes.

Shackelford, J.F. (2009). *Introduction to materials science for engineers* (7th ed.). Prentice Hall, ISBN 978-0-13-601260-3.
A long-established text on materials from an engineering perspective.

Swift, K.G., & Booker, J.D. (1997). *Process selection, from design to manufacture*. Arnold, ISBN 0-340-69249-9.
Details of 48 processes in a standard format, structured to guide process selection.

Wise, R.J. (1999). *Thermal welding of polymers*. Abington Publishing, ISBN 1-85573-495-8.
An introduction to the thermal welding of thermoplastics.

CHAPTER 14

Case Studies: Process Selection

Aluminum die castings. (Image courtesy of Aluminum Recovery Technologies Kendallville, Indiana.)

CONTENTS

Materials Selection in Mechanical Design. DOI: 10.1016/B978-1-85617-663-7.00014-X

14.1 INTRODUCTION AND SYNOPSIS

The last chapter described a systematic procedure for process selection. The inputs are design requirements; the output is a shortlist of processes capable of meeting them. Where processes compete, a ranking is made possible by a cost model. The case studies in this chapter illustrate the method. The first four make use of hard-copy charts; the last two show how computer-based selection works.

The case studies follow a standard pattern. First, we list the *design requirements*: material, shape, size, minimum section, precision, and finish. Then we plot these requirements onto the process matrices, identifying search areas. The processes that lie within the search areas are capable of making the component to its design specification: They are the candidates. If no one process meets all the design requirements, then processes have to be "stacked": casting followed by machining (to meet the tolerance specification on one surface, for instance); or powder methods followed by grinding.

More details for the most promising are then sought, starting with the data sources listed under "Further reading" and in the more comprehensive compilation of Appendix D. The final choice evolves from this subset, taking into account local factors, often specific to a particular company, geographical area, or country.

14.2 CASTING AN ALUMINUM CON-ROD

Connecting rods link oscillatory motion to rotary motion in IC engines, and rotary to oscillatory motion in pumps. Here we explore competing processes for casting a small aluminum alloy connecting rod (Figure 14.1), using the cost model of Chapter 13 to distinguish between them.

The design requirements The con-rod is a solid 3D shape to be made of a nonferrous alloy. The dimensions of the con-rod are such that its mass is about 0.3 kg and its minimum section 8 mm. The precision and tolerance of the casting are not critical since the bores and their faces will have to be machined to give the required precision. A batch size of 100,000 is envisaged. Table 14.1 summarizes the requirements.

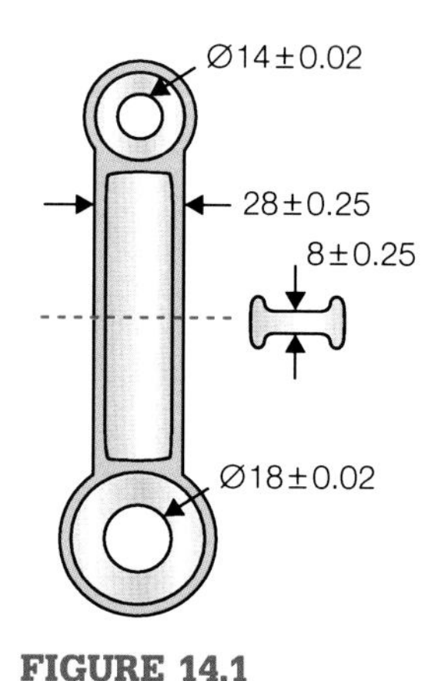

FIGURE 14.1
A connecting rod. The precision of the bores and bore facings is much higher than that of the rest of the body, requiring subsequent machining.

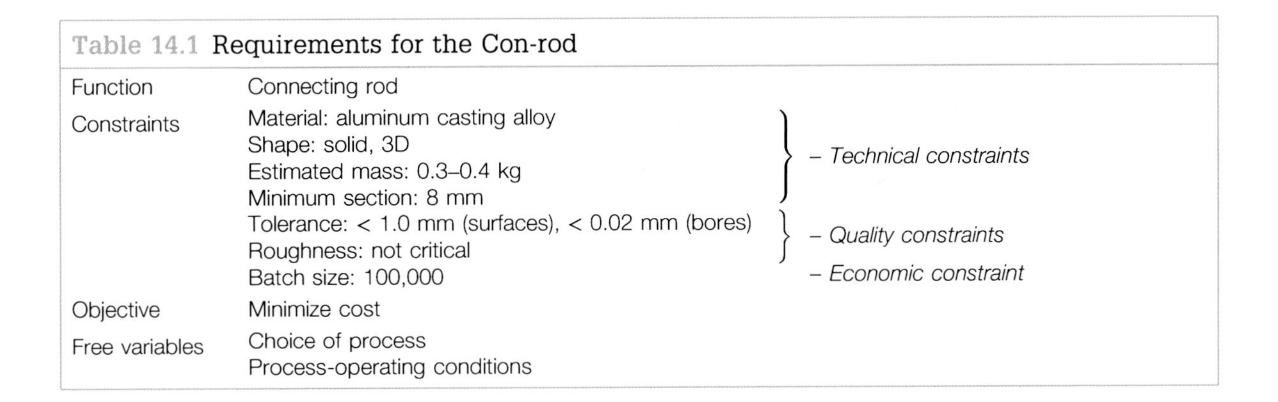

Table 14.1 Requirements for the Con-rod

Function	Connecting rod	
Constraints	Material: aluminum casting alloy Shape: solid, 3D Estimated mass: 0.3–0.4 kg Minimum section: 8 mm	– *Technical constraints*
	Tolerance: < 1.0 mm (surfaces), < 0.02 mm (bores) Roughness: not critical	– *Quality constraints*
	Batch size: 100,000	– *Economic constraint*
Objective	Minimize cost	
Free variables	Choice of process Process-operating conditions	

Table 14.2 Shortlist for Shaping the Con-rod

Process	Comment
Sand casting plus machining Low-pressure casting Die casting	The process–tolerance chart, Figure 14.8, reveals that none of these can meet the requirement on bore tolerance. All will require a subsequent machining operation.

The selection The constraints are plotted on the process matrices and charts in Figures 14.4 through 14.10 in the chapter. The material, the shape, and the required batch size eliminate most of the processes listed on them, leaving three casting processes and machining from solid as options. Machining is rejected on the grounds of material wastage; retained are sand casting, die casting, and low-pressure casting (Table 14.2).

More insight is given by examining their relative cost, using the model of Equation (13.6). The relevant data are assembled in Table 14.3, in which all costs all are normalized by the material cost, the term $mC_m/(1-f)$. The results are plotted in Figure 14.2 as unit cost C versus batch size n curves, following the example in Figure 13.33. At small batch sizes the unit cost is dominated by the "fixed" costs of tooling (the second term on the right of Equation (13.6)). As the batch size n increases, this contribution falls (provided, of course, that the tooling has a life that is greater than n) until it flattens out at a value that is dominated by the "variable" costs of material, labor, and other overheads.

Competing processes usually differ in tooling cost C_t and production rate $\dot{n}$, causing their $C - n$ curves to intersect, as they do here. Sand-casting equipment is inexpensive but the process is slow. The cost of molds for low-pressure casting is greater than for sand casting, and the process is a little faster. Die-casting equipment costs much more but it is also much

Table 14.3 Data for the Cost Equation for the Processes in Table 14.2

Relative Cost*	Sand Casting	Die Casting	Low-pressure Casting	Comment
Material, $mC_m/(1-f)$	1	1	1	*Process-independent parameters*
Basic overhead $\dot{C}_{oh}$ (hr^{-1})	10	10	10	
Capital write-off time *two* (yrs)	5	5	5	
Load factor	0.5	0.5	0.5	
Dedicated tool cost, C_t	210	16,000	2,000	*Process-dependent parameters*
Capital cost, C_c	1800	30,000	8,000	
Batch rate, $\dot{n}$ (hr^{-1})	3	50	10	
Tool life, n_t (number of units)	200,000	1,000,000	500,000	

* *All costs normalized to the material cost.*

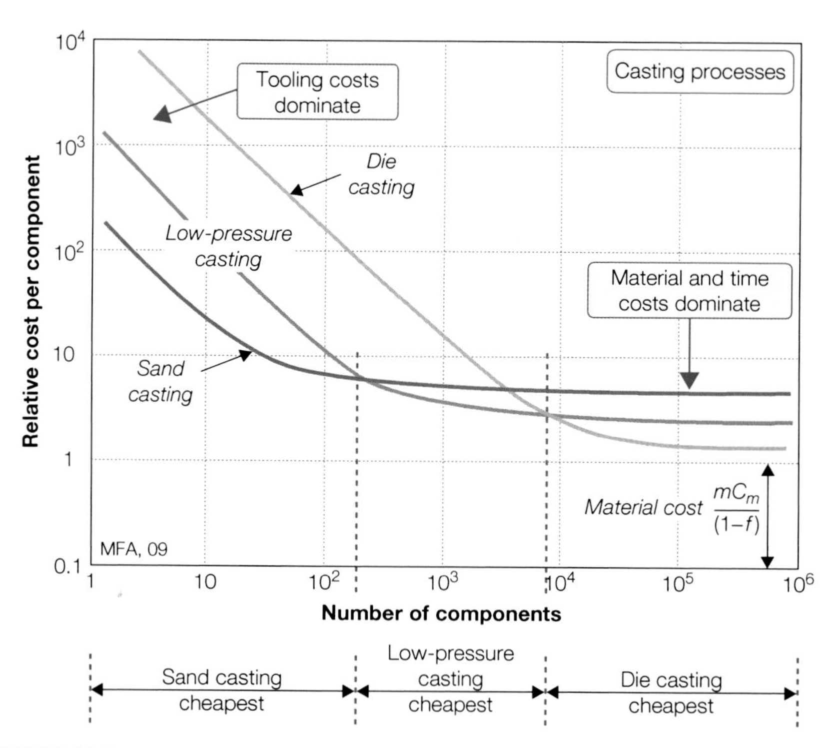

FIGURE 14.2
The relative cost of casting the con-rod as a function of the production run. The costs are normalized by the cost of the material.

faster. The material cost, the labor cost per hour, and the capital write-off time are taken to be the same for all.

The curves for sand, low-pressure, and die casting intersect at a batch size of 200: Below this, sand casting is the more economical. Low-pressure casting becomes marginally less expensive than the others for batches of between 200 and 8000, above which die casting becomes the most economic choice. The best choice for a batch size of 100,000 is die casting. Note that, for small batches, the component cost is dominated by that of the tooling—the material cost hardly matters. But as the batch size grows, the contribution of the second term in the cost equation diminishes; and if the process is fast, the unit cost falls until it is typically about three times that of the material of which the component is made.

Postscript There are other issues of quality besides precision and smoothness that enter into the choice of process. Sand castings tend to trap bubbles and inclusions that act as starting points for fatigue cracks in a cyclically loaded component like a con-rod. Some low-pressure–casting techniques smooth the flow of liquid metal into the die, reducing defect content. Die casting uses higher pressures, and generally gives the highest-quality casting. Considerations such as these can shift the economic switch-points, expanding the economic batch range of the process that offers the highest quality.

14.3 FORMING A FAN

Fans for vacuum cleaners (Figure 14.3) are designed to be cheap, quiet, and efficient, probably in that order. The key to minimizing process costs is to form the fan to its final shape in a single operation, leaving only the central hub to be machined to fit the shaft with which it mates. This means the selection of a single process that can meet the specifications for precision and tolerance, avoiding the need for machining or finishing of the disk or blades.

The design requirements Nylon is the material of choice for the fan. The pumping rate of a fan is determined by its rate of revolution and its size. The designer calculates the need for a fan of radius 60 mm, with 12 profiled blades of average thickness 4 mm. The volume of material in the fan is, roughly, its surface area times its thickness—about 10^{-4} m^3—giving (when multiplied by the density of nylon, 1100 kg/m^3) a weight in the range 0.1 to 0.15 kg. The fan has a fairly complex shape, although its high symmetry simplifies it somewhat. We classify it as 3D solid.

In the designer's view, balance and surface smoothness are what really matter. They (and the geometry) determine the pumping efficiency of the fan and influence the noise it makes. The designer specifies a tolerance of $\pm$ 0.5 mm and a surface roughness of $\leq$ 1 μm. A production run of 10,000 fans is planned.

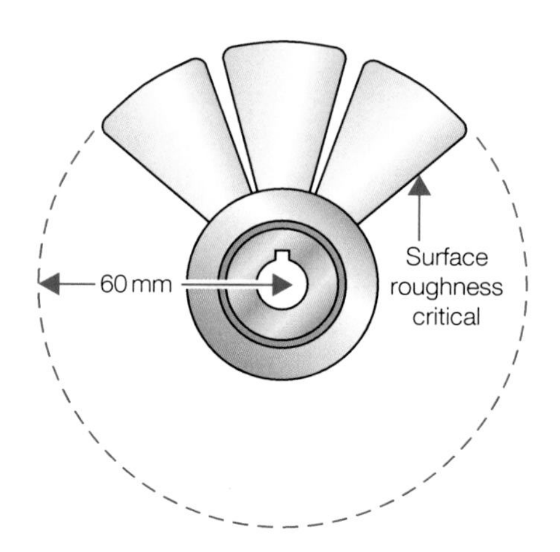

FIGURE 14.3
A fan. It is to be made of nylon, requires low roughness and a certain precision, and is to be produced in large numbers.

Table 14.4 Process Requirements for the Fan

Function	*Fan*	
Constraints	Material: nylon Shape: solid, 3D Estimated mass: 0.1–0.15 kg Minimum section: 4 mm	– *Technical constraints*
	Tolerance: ± 0.5 mm Roughness: < 1 μm	– *Quality constraints*
	Batch size: 100,000	– *Economic constraint*
Objective	Minimize cost	
Free variables	Choice of process Process-operating conditions	

The design requirements are summarized in Table 14.4. What processes can meet them?

The selection We turn first to the material-process and shape-process matrices (Figures 14.4 and 14.5) on which selection boxes have been drawn. The intersection of the two leaves five classes of shaping process—those boxed in the second figure. Screening on mass and section thickness (Figures 14.6 and 14.7) eliminates polymer casting and RTM, leaving the other three.

The constraints on tolerance and roughness are upper limits only (Figures 14.8 and 14.9); all three process classes survive. The planned batch size of 10,000 plotted on the economic batch-size chart (Figure 14.10) eliminates machining from solid. The surviving processes are listed in Table 14.5.

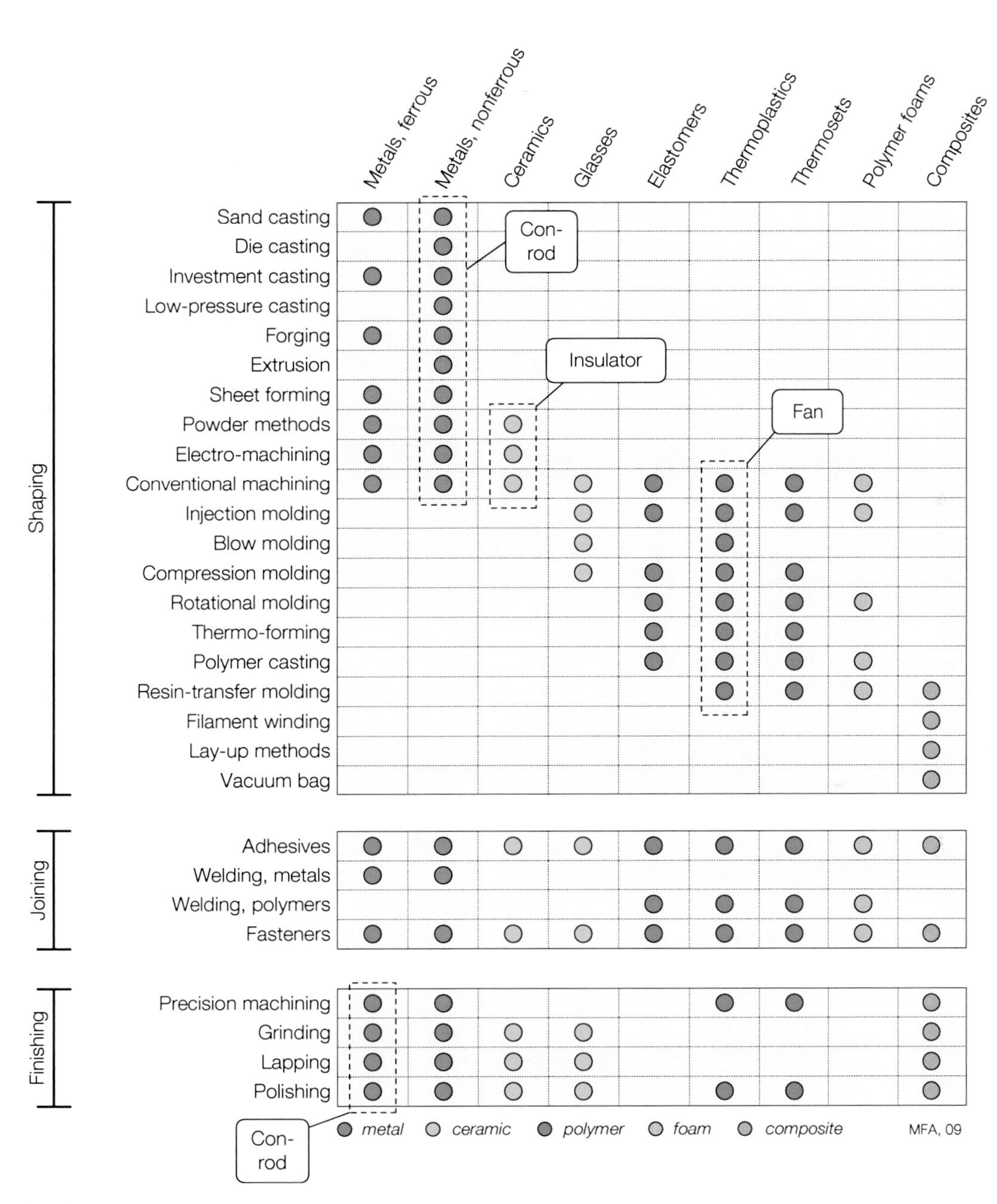

FIGURE 14.4

The process–material compatibility matrix, showing the requirements of the case studies. By including the joining and finishing processes it becomes possible to check that the more restrictive requirements can be met by combining processes.

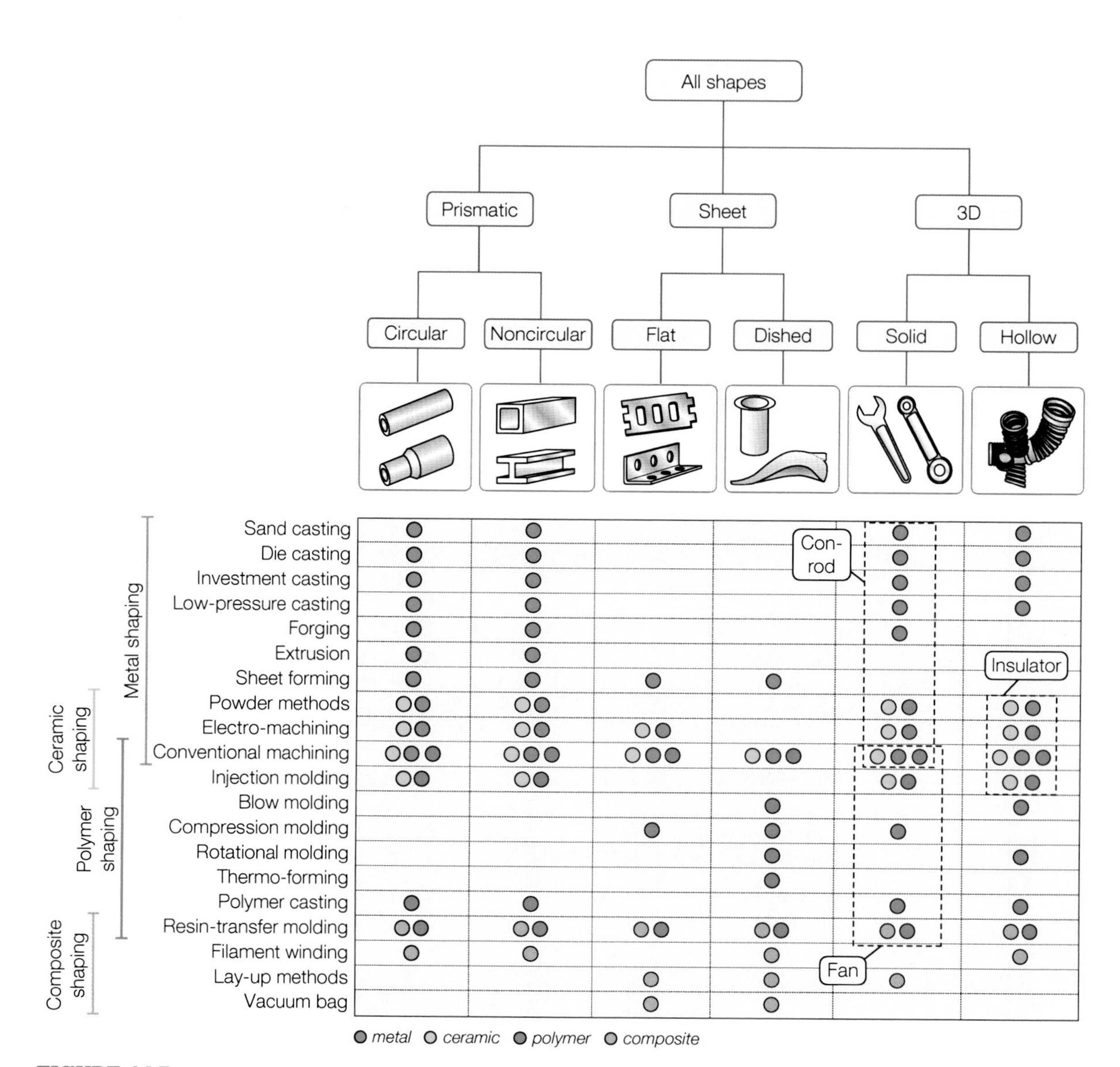

FIGURE 14.5

The process–shape compatibility matrix, showing the requirements of the case studies. A summary of the material compatibility appears at the left. The intersection of this selection stage and the last one narrows the choice.

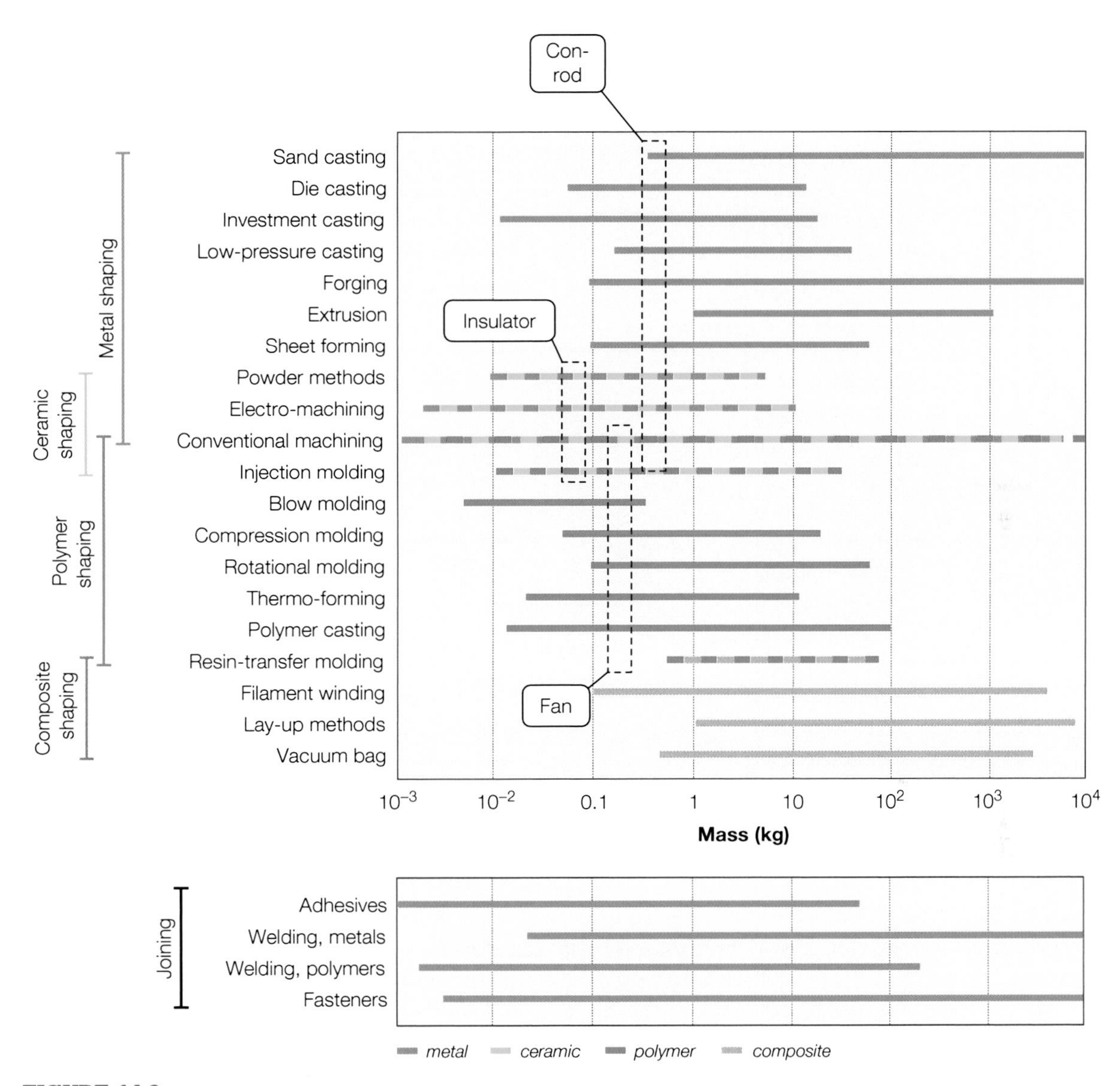

FIGURE 14.6

The process–mass range chart, showing the requirements of the three case studies. The inclusion of joining processes allows the possibility of fabrication of large structures to be explored.

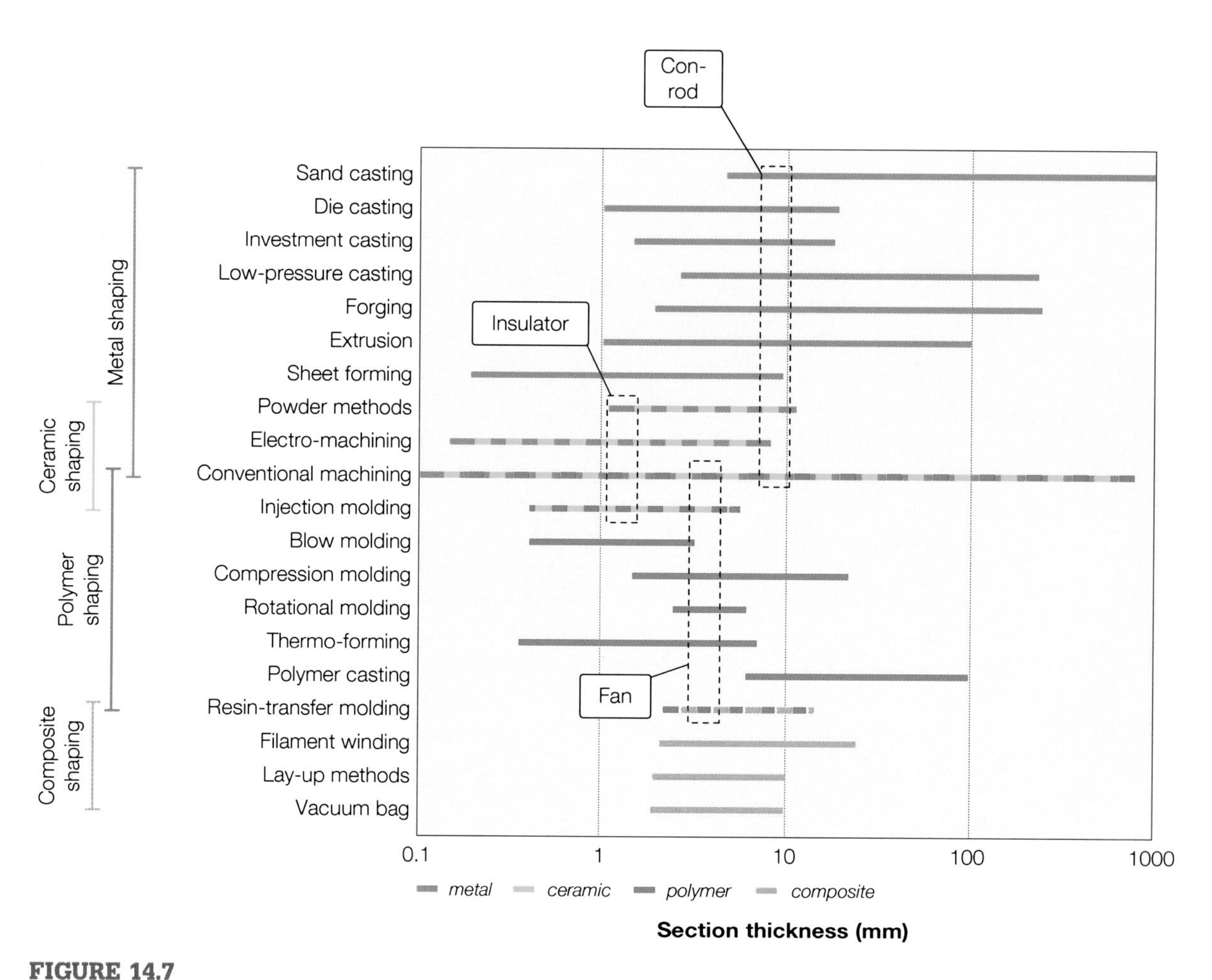

FIGURE 14.7

The process–section thickness chart, showing the requirements of the case studies.

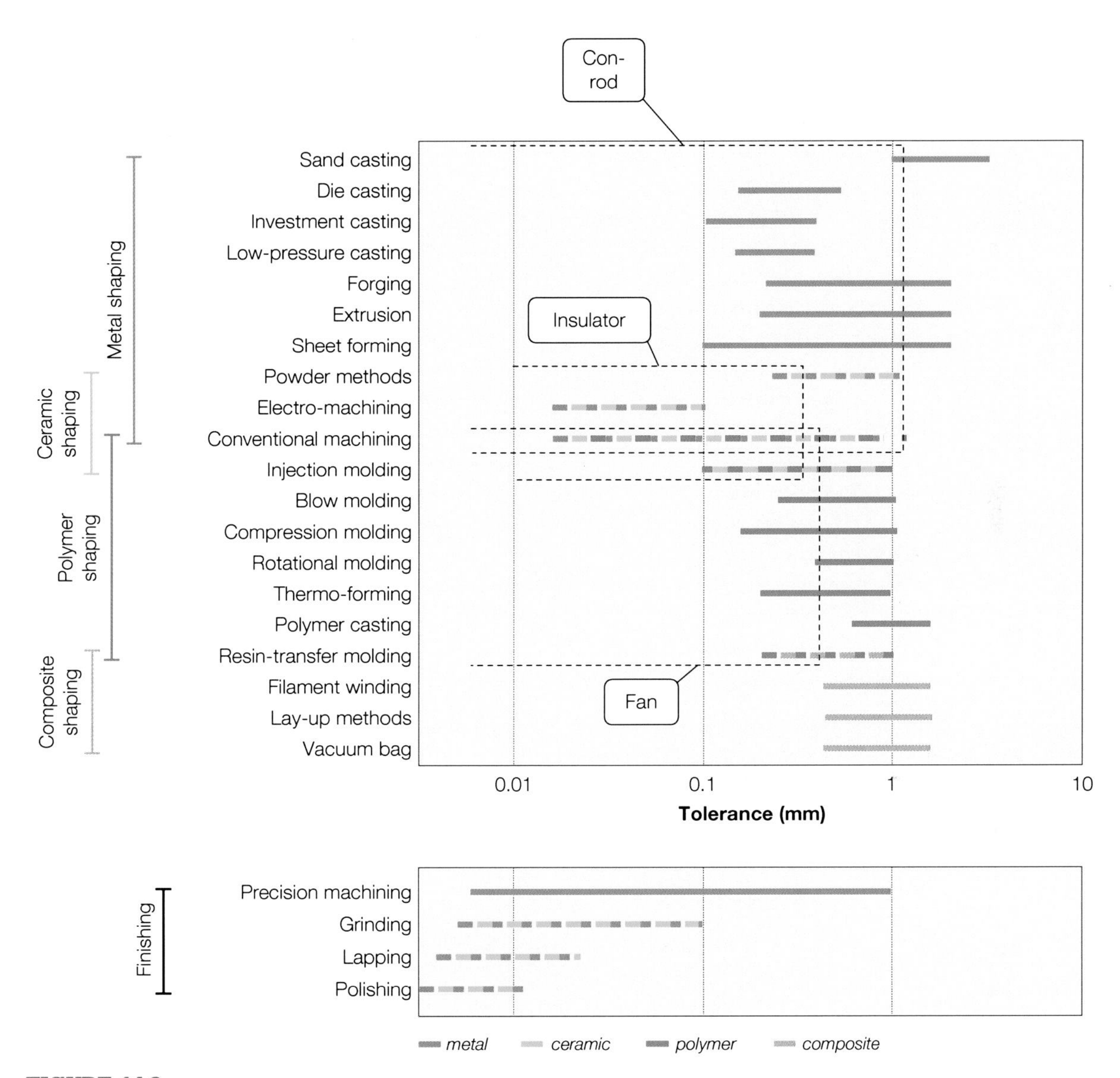

FIGURE 14.8

The process–tolerance chart, showing the requirements of the case studies. The inclusion of joining and finishing processes allows the possibility of fabrication of large structures to be explored. Tolerance and surface roughness are specified as an upper limit only, so the selection boxes (*left*) are open-ended.

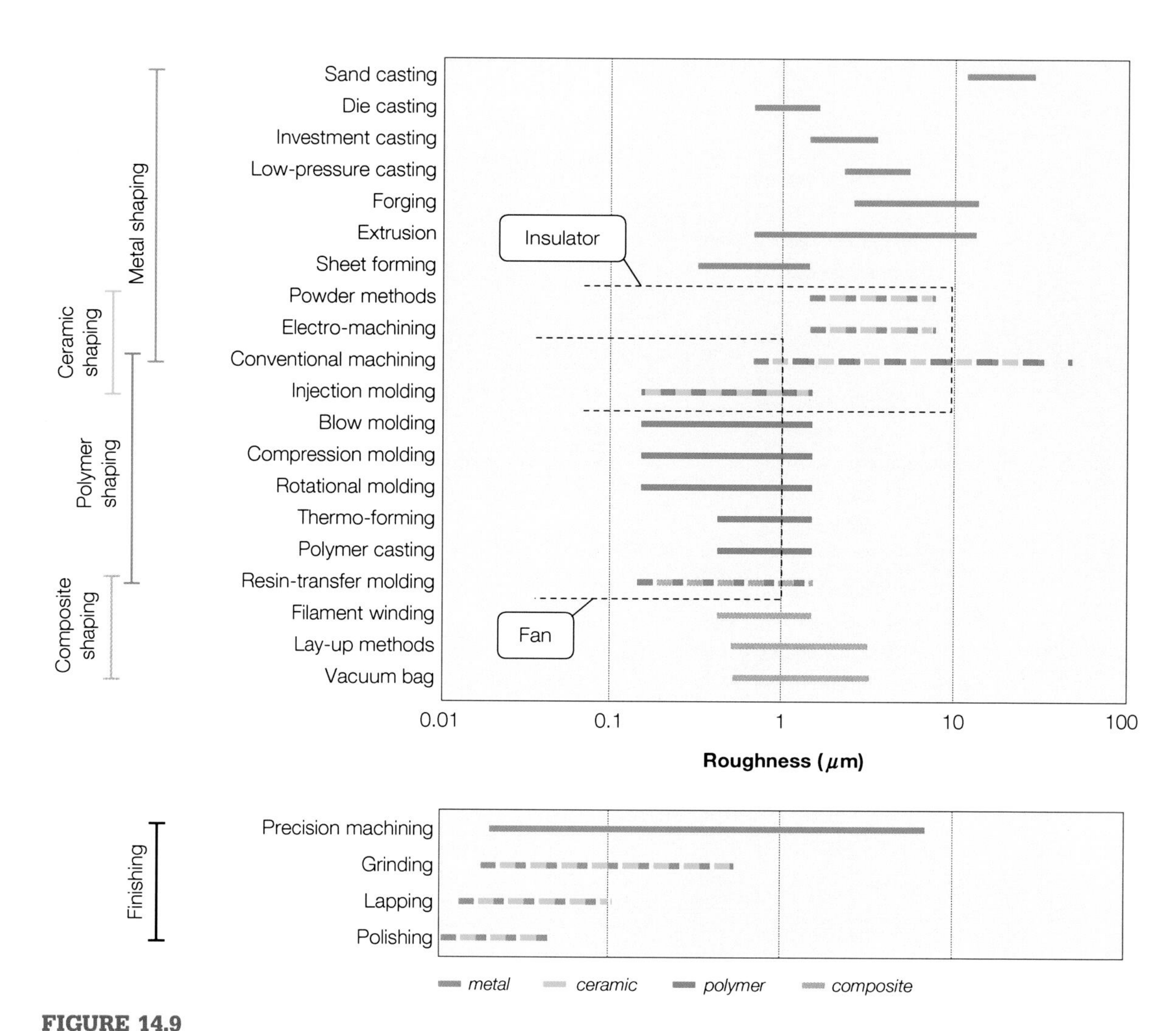

FIGURE 14.9
The process–surface roughness chart. Only one case study—the fan—imposes restrictions on this.

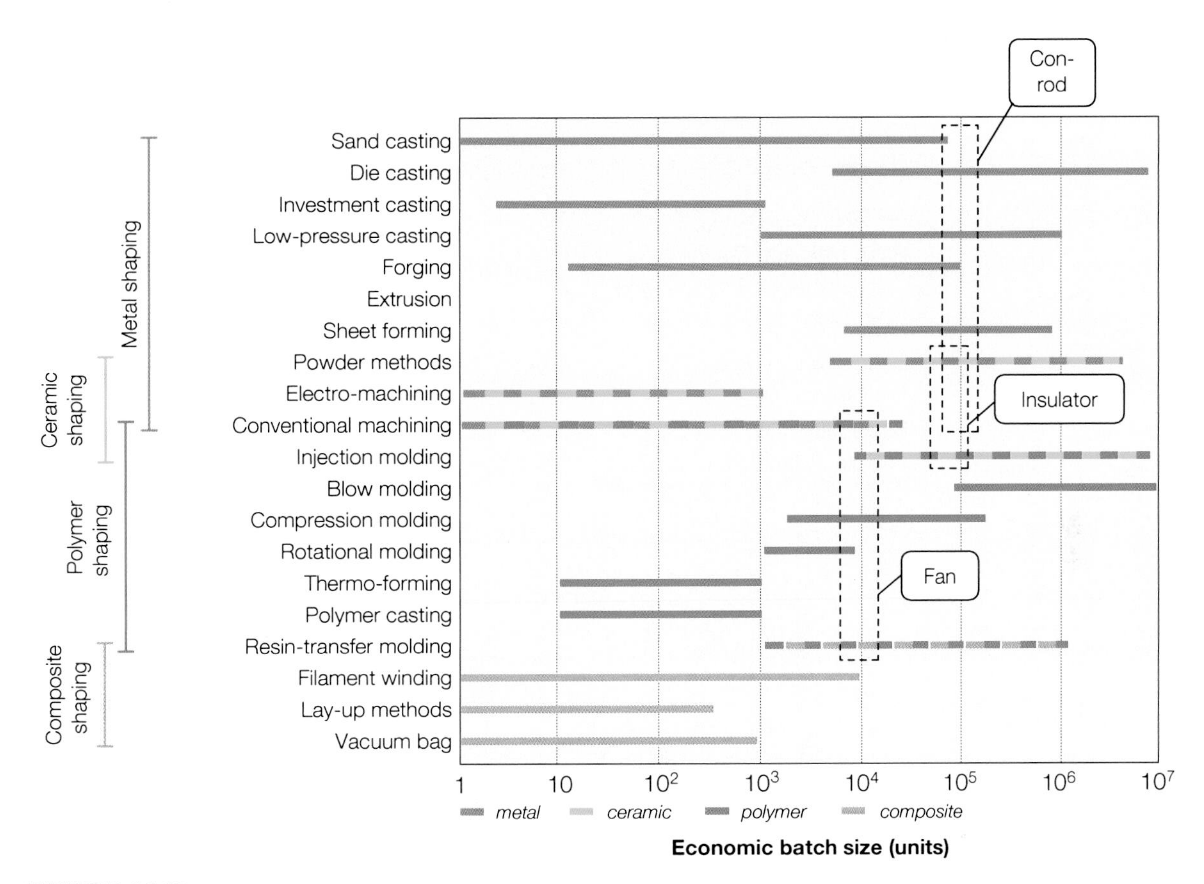

FIGURE 14.10

The process–economic batch-size chart, showing the requirements of the three case studies. The box for the optical table spans the range of possible production volumes listed in the requirements.

Table 14.5 Processes for Forming the Fan

Process	Comment
Injection molding Compression molding	Injection molding meets all the design requirements; compression molding may require further finishing operations

Table 14.6 Data for the Cost Equation for the Processes in Table 14.5

Relative Cost	Compression Molding	Injection Molding	Comment
Material, $mC_m/(1-f)$	1	1	*Process-independent parameters*
Basic overhead $\dot{C}_{oh}$ (hr^{-1})	20	20	
Capital write-off time *two* (yrs)	5	5	
Load factor	0.5	0.5	
Dedicated tool cost, C_t	2000	10,000	*Process-dependent parameters*
Capital cost, C_c	20,000	100,00	
Batch rate, $\dot{n}$ (hr^{-1})	30	150	
Tool life, n_t (number of units)	100,000	200,000	

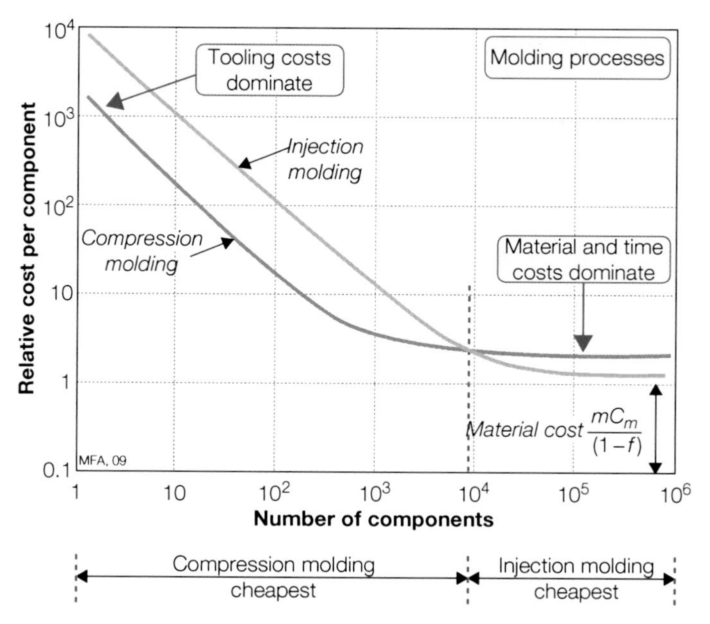

FIGURE 14.11

The relative cost of molding the fan as a function of the production run. The costs are normalized by the cost of the material.

To get further we need the cost model. The data required to implement it[1] are gathered in Table 14.6. Figure 14.11 shows the resulting cost curves. The tooling for compression molding is less expensive than that for injection molding, but it is slower because, by using multiple die cavities, injection

[1] Data from the records in the CES software for these processes.

molding can produce several units in one shot. This makes the two curves intersect at about 10,000 units: Below, compression molding is the less expensive; above, it is injection molding.

Postscript There are (as always) other considerations—the questions of capital investment, local skills, overhead rate, and so forth. The charts cannot answer these. But the procedure has been helpful in narrowing the choice, suggesting alternatives, and providing a background against which a final selection can be made.

Related case studies

6.6 "Materials for flywheels"
14.2 "Casting an aluminum con-rod"

14.4 SPARK PLUG INSULATORS

The difficulties of using hard-copy charts for process selection will, by now, be obvious: The charts have limited resolution and are clumsy to use. They give a helpful overview but they are not the way to get a definitive selection. Computer-based methods overcome both problems.

The CES system, which builds on the methods in Chapter 5, was described earlier. It allows limits to be set on material, shape, mass, section, tolerance, and surface roughness, delivering the subset of processes that meet all the limits. The economics are then examined by plotting the desired batch size on a bar chart of economic batch size, or by implementing the cost model that is built into the software. If the requirements are very demanding, no single process can meet them all. Then the procedure is to relax the most demanding of them (often tolerance and surface roughness), seeking processes that can meet the others. A second process is then sought to add the desired refinement.

The next two case studies show how the method works.

The design requirements The anatomy of a spark plug is sketched in Figure 14.12. It is an assembly of components, one of which is the insulator. This is to be made of a ceramic, *alumina*, with the shape shown in the figure: an axisymmetric, hollow, stepped shape. It weighs about 0.05 kg and has a minimum section of 1.2 mm.

Precision is important, since the insulator is part of an assembly; the design specifies a precision of $\pm$ 0.3 mm and a surface finish of better than 10 μm. The insulators are to be made in large numbers: The projected batch size is 100,000. Cost should be as low as possible. Table 14.7 summarizes the requirements.

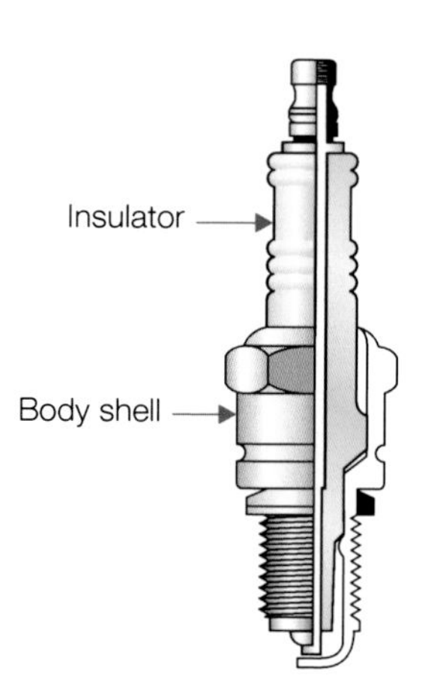

FIGURE 14.12
A spark plug. We seek a process to make the insulator.

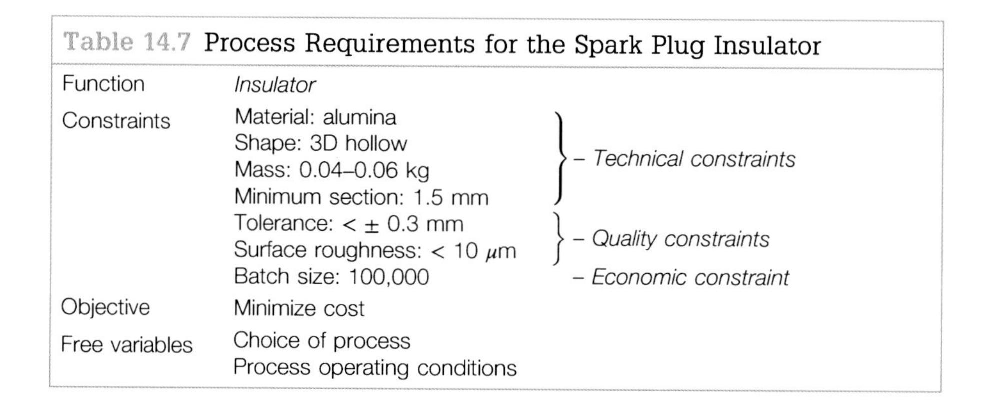

Table 14.7 Process Requirements for the Spark Plug Insulator

Function	*Insulator*	
Constraints	Material: alumina Shape: 3D hollow Mass: 0.04–0.06 kg Minimum section: 1.5 mm	– *Technical constraints*
	Tolerance: < ± 0.3 mm Surface roughness: < 10 μm	– *Quality constraints*
	Batch size: 100,000	– *Economic constraint*
Objective	Minimize cost	
Free variables	Choice of process Process operating conditions	

The selection The constraints are plotted on the compatibility matrices and bar charts in Figures 14.4 through 14.10. Only one process family survives: powder methods. The required batch size of 100,000 lies within its economic range (refer to Figure 14.6).

The family of powder methods is a large one, including die pressing and sintering, hot pressing, powder extrusion, powder injection molding, hot isostatic pressing, and spray deposition. These (and others) could be added to the charts, but to do so makes them cumbersome. It is better to manage the information in a computer-based environment. The CES system does this.

Applying the constraints listed in Table 14.7 eliminates all but two of the processes in its database: die pressing and sintering and powder injection molding (PIM: the shaping of a powder by mixing it with a polymer binder, molding, then burning off the binder during the subsequent sintering). The same system implements the cost model that was described in Chapter 13.

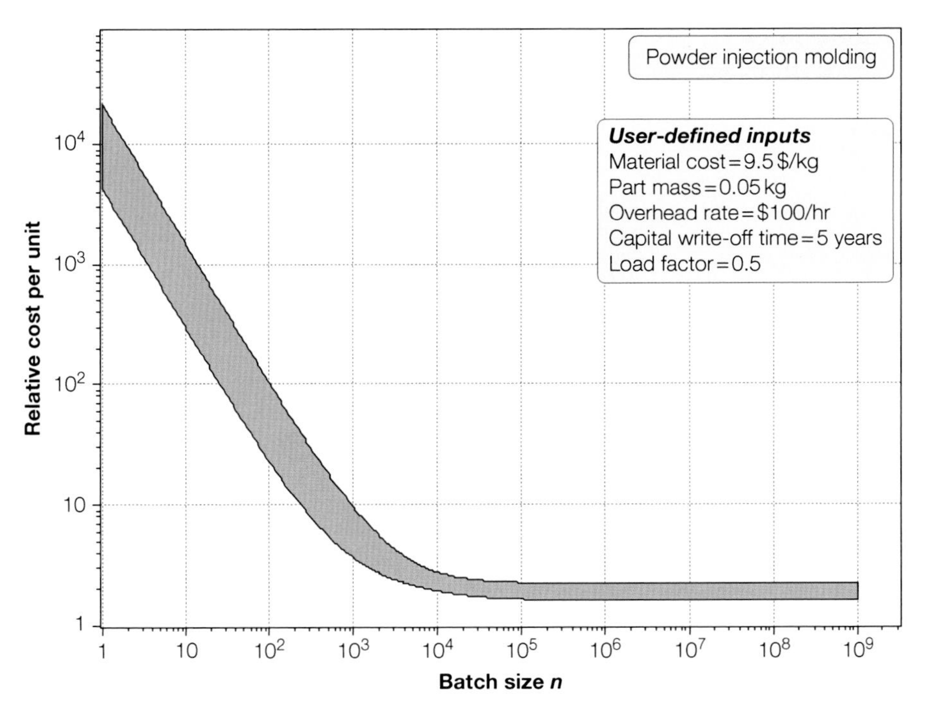

FIGURE 14.13
The unit cost as a function batch size for powder-injection molding of a small insulator. Cost is normalized by the cost of the material.

Its output, showing the cost, in units of material cost for PIM, is shown in Figure 14.13. The user-defined inputs are listed at the top right. The unit cost, high at low batch sizes, falls to about twice that of the material at batch sizes above about 10,000.

Postscript Insulators are made commercially by PIM. Economics, for a mass-produced product like this, are critical—pennies count. Technical cost modeling, described before in Chapter 13, can guide the choice of the best equipment and optimum operating conditions to minimize cost.

Related case studies

6.18 "Heat sinks for hot microchips"
14.5 "A manifold jacket"

14.5 A MANIFOLD JACKET

The design requirements The manifold jacket shown in Figure 14.14 is part of the propulsion system of a space vehicle. It is to be made of nickel. It is large, weighing about 7 kg, and complex, having an unsymmetrical 3D-hollow

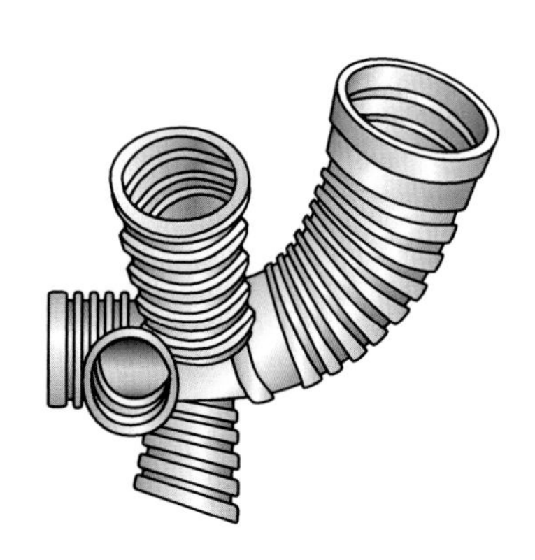

FIGURE 14.14
A manifold jacket. (Redrawn from Bralla, 1986.)

Table 14.8 Process Requirements for the Manifold Jacket

Function	Manifold jacket	
Constraints	Material: nickel Shape: 3D hollow Mass: 7 kg Minimum section: 2–5 mm	– *Technical constraints*
	Tolerance: < ± 0.1 mm Surface roughness: < 20 μm	– *Quality constraints*
	Batch size: 20	– *Economic constraint*
Objective	Minimize cost	
Free variables	Choice of process Process-operating conditions	

shape. The minimum section thickness is between 2 and 5 mm. The requirements for precision and surface finish are severe (tolerance < ± 0.1 mm, roughness < 20 μm). Because of its limited application, only 20 units are to be made. Table 14.8 lists the requirements.

The charts cannot give much guidance in selecting processes to make such a complex component. We turn instead to computer-aided selection.

The selection The constraints are applied by entering them in a limit-selection dialog box like that in Figure 5.12 (*left*). The material constraint (nickel) and the shape constraint (3D hollow) eliminate a large number of processes. Figure 14.15 shows 15 that survive. The subsequent constraints on mass and section thickness eliminate more, leaving only the six that remain in color in the figure. Applying the severe constraints on tolerance and surface roughness eliminates all but one: electroforming. The required batch size lies within its range.

Postscript A search for further information in the sources listed in Appendix D reveals that electroforming of nickel is an established practice and that components as large as 20 kg are routinely made by this process. It looks like a good choice.

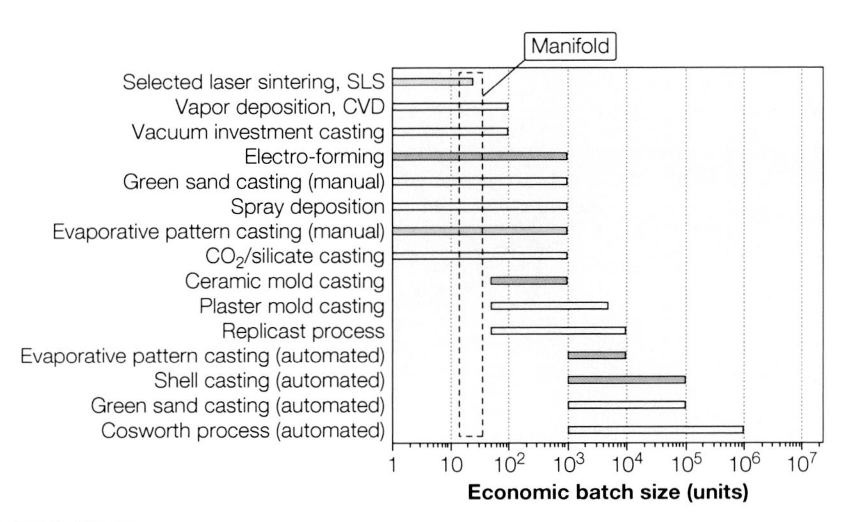

FIGURE 14.15

Output of the computer-based selection of processes to make the manifold. The grayed-out processes have failed one or more constraints.

Related case study

14.4 "Spark plug insulators"

14.6 JOINING A STEEL RADIATOR

Figure 14.16 shows a section through a domestic radiator made from corrugated pressed sheet steel. The task is to choose a joining process for the seams between the sheets.

The design requirements The process must be compatible with the material, here low carbon steel sheet of thickness 1.5 mm. The lap joints carry only low loads in service but handling during installation may impose tension and shear. They must conduct heat, be watertight, and able to tolerate temperatures up to 100°C. There is no need for the joints to be disassembled for recycling at the end of life since the whole thing is steel. Table 14.9 summarizes the translation.

The selection The CES system has records and attributes for 52 joining processes. Applying the material constraint leaves 32 of them—those eliminated are polymer specific or composite specific. Further screening on joint geometry, mode of loading, and section thickness reduces the list to 20. The requirement to conduct heat reduces this further to 12. Water resistance and operating temperature do not change the shortlist further. The processes passing the screening stage are listed in Table 14.10.

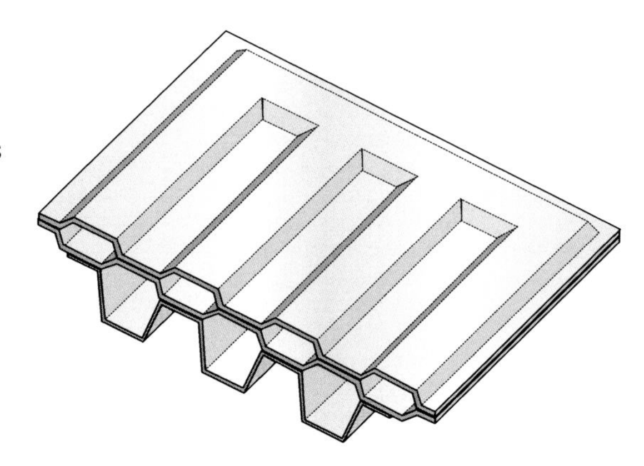

FIGURE 14.16
A section through a domestic radiator. The three pressed steel sections are joined by lap joints.

Table 14.9 Translation for Joining a Steel Radiator

Function	Domestic radiator	
Objective	Minimize cost	
Constraints	Material compatibility: low carbon steel Joint geometry: lap joint Mode of loading: tension, shear (moderate) Sheet thickness: 1–2 mm Joint must conduct heat Joint must be watertight Service temperature > 100°C Disassembly not required	} – *Technical constraints*
Free variables	Choice of shaping process Process operating conditions	

Table 14.10 Shortlist of Processes for Joining a Steel Radiator

Process	Comment
Soldering Brazing	Risk of corrosion arising from dissimilar metals in electrical contact
Laser beam welding Electron beam welding Explosive welding	Expensive, specialized processes
Metal inert gas arc welding (MIG) Tungsten inert gas arc welding (TIG) Manual metal arc welding (MMA) Oxyacetylene welding Projection welding Seam welding	Well-established, conventional welding processes

Postscript At this point we seek documentation for the processes. This reveals that explosive welding requires special facilities and permits (hardly a surprise). Electron and laser welding require expensive equipment, so use of a shared facility might be necessary to make them economic. Resistance spot welding is screened out because it failed the requirement to be watertight, although that requirement only applies to the edge seams.

Quality constraints are not severe, but distortion and residual stress will make fit-up of adjacent joints difficult. Low carbon steels are readily weldable, so it is unlikely that welding will cause cracking or embrittlement—but this can be checked by seeking documentation.

14.7 SURFACE-HARDENING A BALL-BEARING RACE

The balls of ball races run along grooved tracks (Figure 14.17). The life of a ball race is limited by wear and fatigue. Both are suppressed by using hard materials. Hard materials, however, are not tough, incurring the risk that shock loading or mishandling might cause the race to fracture.

The design requirements The solution is to use an alloy steel, which has excellent bulk properties, and to apply a separate surface treatment to increase hardness where it matters. We therefore seek processes to surface-harden alloy steels for wear and fatigue resistance. The precision of both balls and race is critical, so the process must not compromise the dimensions or the surface smoothness. Table 14.11 summarizes the translation.

The selection The CES system contains records for 46 surface treatment processes. Many are compatible with alloy steels. More discriminating is the purpose of the treatment—to impart fatigue and wear resistance—reducing the list to 8. Imposing the requirement for a very smooth surface knocks out processes that coat or deform the surface because these compromise the finish. Adding the further constraint that the curved surface coverage must be *good* or *very good* leaves just five, listed in Table 14.12. The records for these describe and illustrate the process, and summarize their typical uses.

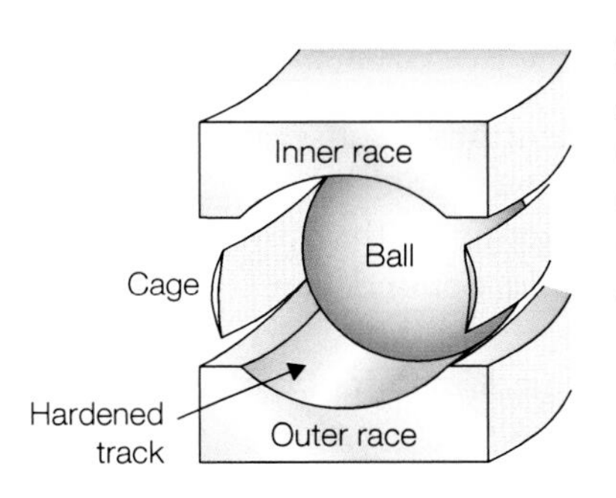

FIGURE 14.17
A section through a ball race. The surfaces of the race are to be hardened to resist wear and fatigue crack nucleation.

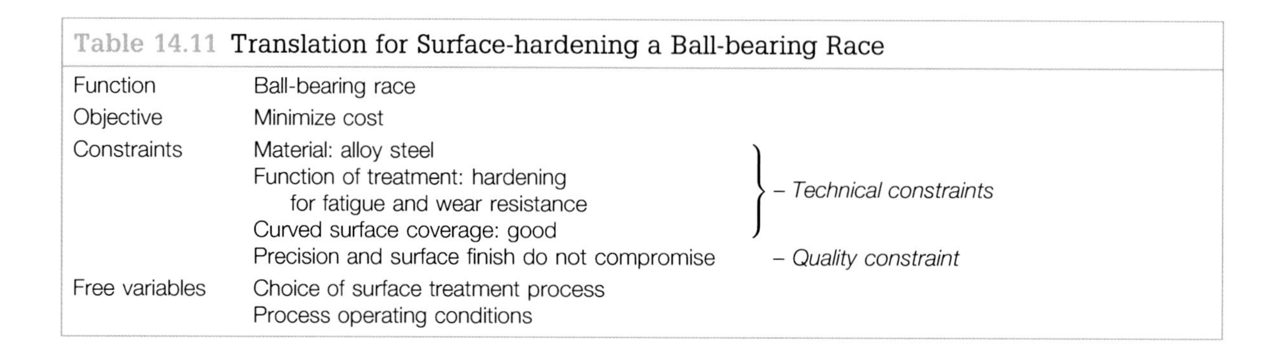

Table 14.11 Translation for Surface-hardening a Ball-bearing Race

Function	Ball-bearing race	
Objective	Minimize cost	
Constraints	Material: alloy steel Function of treatment: hardening for fatigue and wear resistance Curved surface coverage: good	– *Technical constraints*
	Precision and surface finish do not compromise	– *Quality constraint*
Free variables	Choice of surface treatment process Process operating conditions	

Table 14.12 Shortlist of Processes for Hardening an Alloy Steel Ball Race

Process	Comment
Carburizing Carbo-nitriding Nitriding Aluminizing Boriding Chromizing	All are high-temperature, diffusion-controlled processes, and so are slow. Documentation in the records gives details.

Postscript To get further we seek documentation for these processes. The hardness of the surface and the depth of the hardened layer depend on process variables: the time and temperature of the treatment and the composition of the steel in ways that are tabulated. And economics, of course, enters the equation. Ball races are made in enormous numbers and while their sizes vary, their geometry does not. This is where dedicated equipment, even if very expensive, is viable.

14.8 SUMMARY AND CONCLUSIONS

Process selection, at first sight, looks like a black art: The initiated know; the rest of the world cannot even guess how they do it. But this—as the chapter demonstrates—is not really so. The systematic approach, developed in Chapter 13, and illustrated here, identifies a subset of viable processes using design information only: size, shape, complexity, precision, roughness, and material—itself chosen by the systematic methods in Chapter 5. It does not identify the single, best choice; that depends on too many case-specific considerations. But, by identifying candidates, it directs the user to data sources (starting with those listed in Appendix D) that provide the details needed to make a final selection.

CHAPTER 15

Materials and the Environment

CONTENTS

Materials Selection in Mechanical Design. DOI: 10.1016/B978-1-85617-663-7.00015-1

15.1 INTRODUCTION AND SYNOPSIS

All human activity has some impact on the environment. The environment has some capacity to cope with this so that a certain level of impact can be absorbed without lasting damage. But it is clear that current human activities exceed this threshold with increasing frequency, diminishing the quality of the world in which we now live and threatening the well-being of future generations. The manufacture and use of products, with their associated consumption of materials and energy, are among the culprits. The position is dramatized by the following statement: At a global growth rate of 3% per year we will mine, process, and dispose of more "stuff" in the next 25 years than in the entire industrial history of mankind.

Design for the environment is generally interpreted as the effort to adjust our present design methods to correct known, measurable, environmental degradation; the time scale of this thinking is 10 years or so, an average product's expected life. *Design for sustainability* is the longer view: that of adaptation to a lifestyle that meets present needs without compromising the needs of future generations. The time scale here is less clear—it is measured in decades or centuries—and the adaptation required is much greater. This chapter focuses on the role of materials and processes in achieving design for the environment. Sustainability requires social and political changes that are beyond the scope of this book.

The ideas in this chapter are developed further in the text that appears as the first entry in "Further reading."

15.2 THE MATERIAL LIFE-CYCLE

The nature of the problem is brought into focus by examining the material life-cycle shown in the sketch in Figure 15.1. Ore and feedstock, most of them nonrenewable, are processed to give materials; these are manufactured into products that are used and, at the end of their lives, disposed of, a fraction perhaps entering a recycling loop, the rest committed to incineration or landfill. Energy and materials are consumed at each point in this cycle (we shall call them "phases"), with an associated penalty of CO_2 and other emissions—heat and gaseous, liquid, and solid waste.

The problem, crudely put, is that the sum of these unwanted by-products now often exceeds the capacity of the environment to absorb them. Some of the injury is local in scale and its origins can be traced and remedial action taken. For some the scale is national, for others it is global, and here remedial action has wider social and organizational prerequisites. Much present environmental legislation aims at modest reductions in the damaging

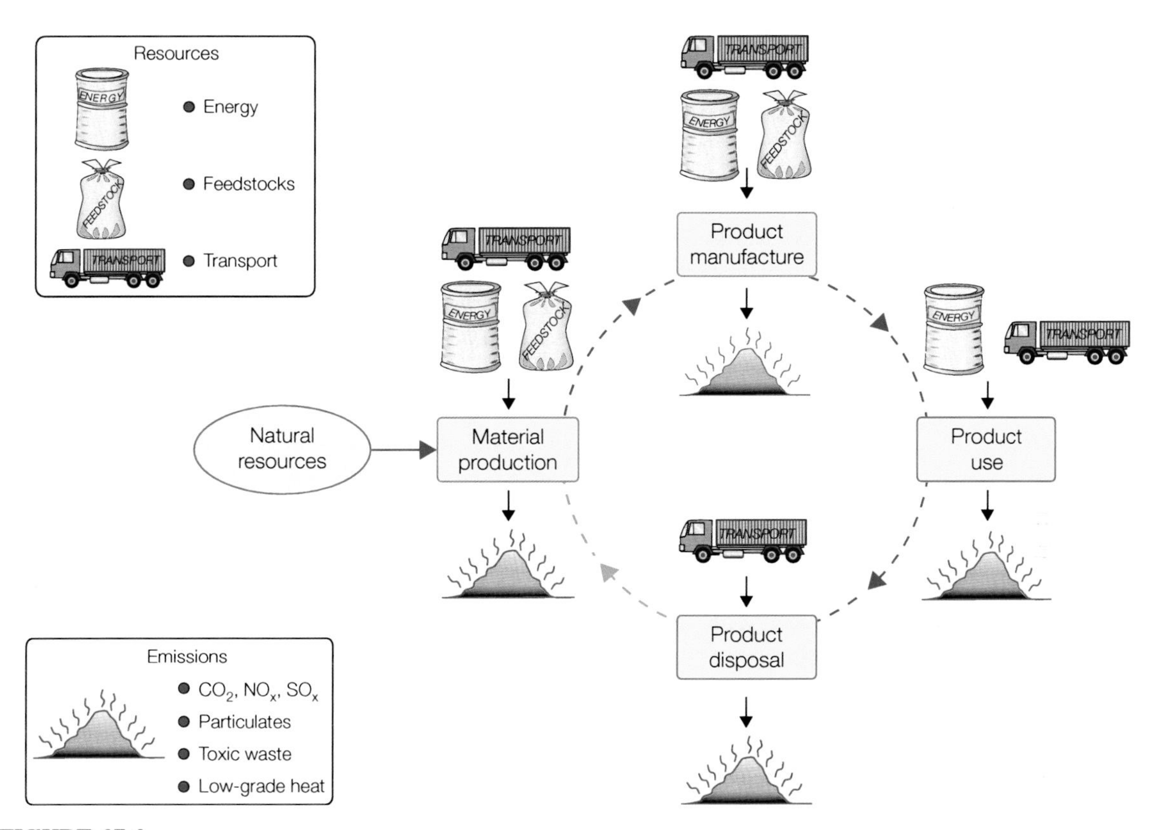

FIGURE 15.1

The material life-cycle. Ore and feedstock are mined and processed to yield a material. These are manufactured into a product that is used and, at the end of its life, discarded or recycled. Energy and materials are consumed in each phase, generating waste heat and solid, liquid, and gaseous emissions.

activity; a regulation requiring a 20% reduction in—say—the average gasoline consumption of passenger cars is seen by car makers as a major challenge.

Sustainability requires solutions of a completely different kind. Even conservative estimates of the adjustment needed to restore long-term equilibrium with the environment envisage a reduction in the flows of Figure 15.1 by a factor of four or more. Population growth and the growth of the expectations of this population more than cancel any modest savings that the developed nations may achieve. It is here that the challenge is greatest, requiring difficult adaptation; it is one for which no generally agreed-on solutions yet exist. But it remains the long-term driver of eco-design, to be retained as background to any creative thinking.

15.3 MATERIAL AND ENERGY-CONSUMING SYSTEMS

It would seem that the obvious ways to conserve materials are to make products smaller, make them last longer, and recycle them when they finally reach the end of their lives. But the seemingly obvious can sometimes be deceptive. Materials and energy form part of a complex and highly interactive system, as illustrated in Figure 15.2. Here primary catalysts of consumption such as *new technology, planned obsolescence, increasing wealth, education,* and *population growth* influence aspects of product use and through these the consumption of materials and energy and the by-products that these produce. The connecting lines indicate influences; a green line suggests positive, broadly desirable influence; a red line suggests negative, undesirable influence; and red-green suggests that the driver has the capacity for both positive and negative influence.

The diagram brings out the complexity. Follow, for instance the lines of influence of *new technology* and its consequences. It offers more material and energy-efficient products, but by also offering new functionality it creates obsolescence and the desire to replace a product that has useful life left in it. Electronic products are prime examples of this: Most are discarded while

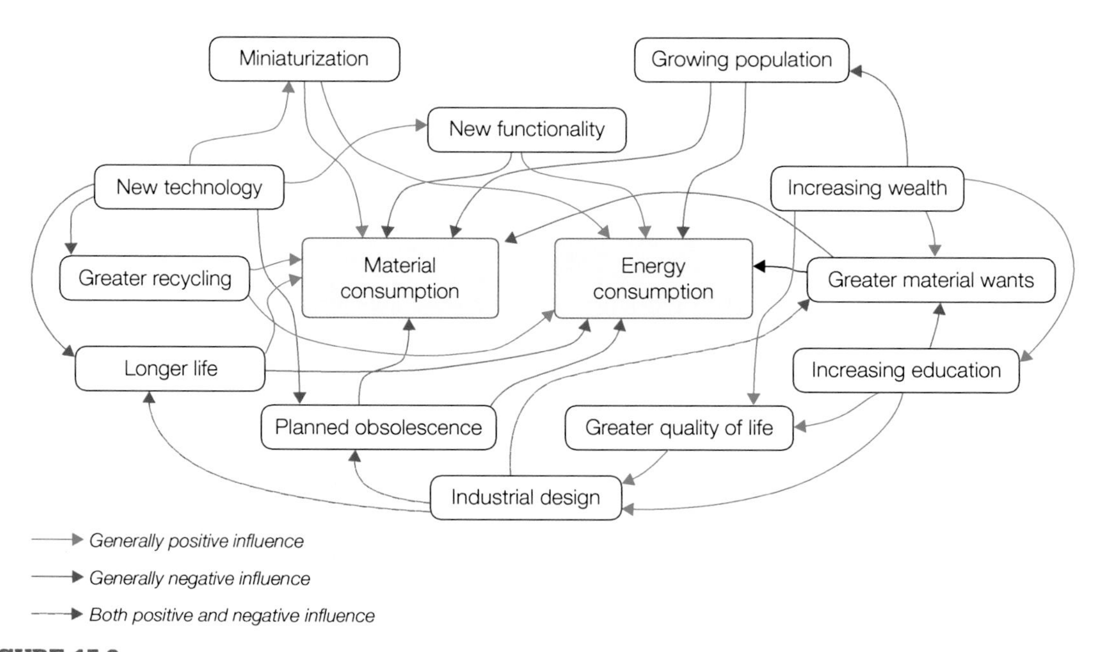

FIGURE 15.2
The influences on consumption of materials and energy. It is essential to see eco-design as a systems problem, not solved by simply choosing "good" and avoiding "bad" materials but rather by matching the material to the system requirements.

still functional. And observe, even at this simple level, the consequences of *longer life*—a seemingly obvious measure. It can certainly help to conserve materials (a positive influence) but, in an era in which new technology delivers more energy-efficient products (particularly cars, electronics, and household appliances today), extending the life of old products can have a negative influence on energy consumption.

As a final example, consider the bivalent influence of industrial design—the subject of Chapter 16. The lasting designs of the past are evidence of industry's ability to create products that are treasured and conserved. But today industrial design is frequently used as a potent tool to stimulate consumption by deliberate obsolescence, creating the perception that "new" is desirable and that even slightly "old" is unappealing.

The use patterns of products Table 15.1 suggests a matrix of product use patterns. Those in the first row require energy to perform their primary function. Those in the second could function without energy but, for reasons of comfort, convenience, or safety, consume energy to provide a secondary function. Those in the third row provide their primary function without any need for energy other than human effort. The load factor across the top is an approximate indicator of the intensity of use—one that will, of course, vary widely.

The choice of materials and processes influences all the phases of Figure 15.1: *production*, through the drainage of resources and the undesired by-products of refinement; *manufacture*, through the level of efficiency and cleanness of the shaping, joining, and finishing processes; *use*, through the ability to conserve energy through lightweight design, higher thermal efficiency, and lower energy consumption; and finally *disposal* through a greater ability to allow reuse, disassembly, and recycling.

Table 15.1 Use Matrix of Product Classes

	High Load Factor	Modest Load *P* Factor	Low Load Factor
Primary power-consuming	Family car Train set Aircraft	Television Freezer	Coffee maker Vacuum cleaner Washing machine
Secondary power-consuming	Housing (heat, light)	Parking lot (light)	Household dishes Clothing (washing)
Nonpower-consuming	Bridges Roads	Furniture Bicycle	Canoe Tent

Energy intense ↕ *Material intense*

High impact ⟵⟶ *Low impact*

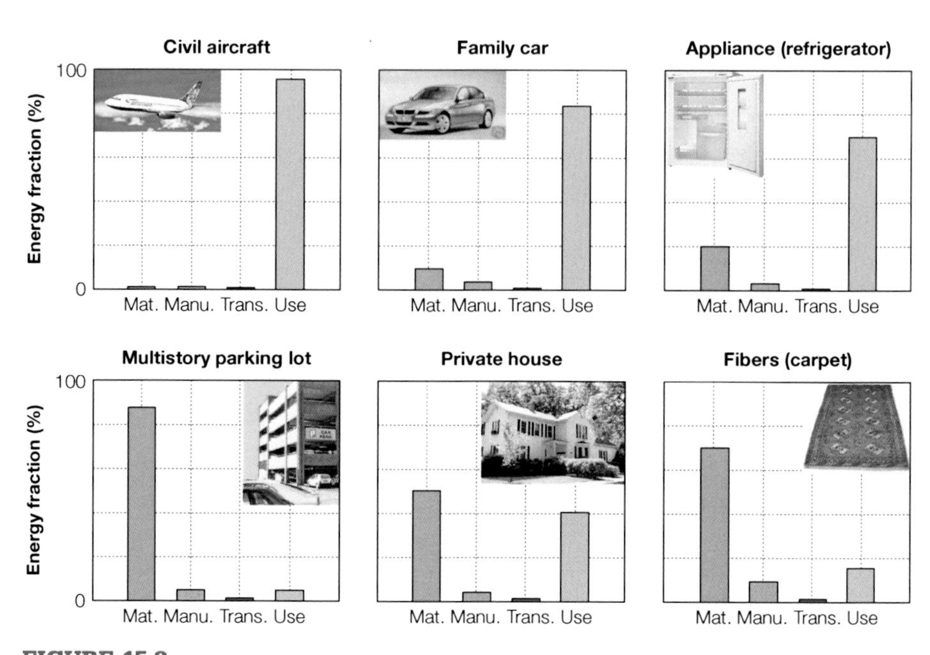

FIGURE 15.3

Approximate values for the energy consumed at each phase of Figure 15.1 for a range of products.

It is generally true that one of the four phases of Figure 15.1 dominates the picture. Simplifying for a moment, let us take *energy consumption* as a measure of both the inputs and undesired by-products of each phase and use it for a character appraisal of use sectors. Figure 15.3 presents the evidence, using this measure. It has two significant features, with important implications. The first feature is that one phase almost always dominates, accounting for 80% or more of the energy, sometimes much more. If large changes are to be achieved, it is this phase that must be the target; a reduction by a factor of 2, even of 10, in any other makes little significant difference to the total. The second feature: When differences are as great as those of Figure 15.3, precision is not the issue—an error of a factor of 2 changes very little. It is the nature of people who *measure* things to wish to do so with precision, and precise data must be the ultimate goal. But it is possible to move forward without precision: Precise judgments can be drawn from imprecise data. This is an important consideration: Much information about eco-attributes is imprecise.

15.4 THE ECO-ATTRIBUTES OF MATERIALS

Material production: Energy and emissions

Most of the energy consumed in the four phases of Figure 15.1 is derived from fossil fuels. Some is consumed in that state—as gas, oil, coal, or coke. Much is first converted to electricity at a European average conversion efficiency of about 38%. Not all electricity is generated from fossil fuels—there are contributions from hydroelectric, nuclear, and wind/wave sources. But with the exception of Norway (70% hydro) and France (80% nuclear), the predominant energy sources are fossil fuels; and since the national grids of European countries are linked, with power flowing from one to another as needed, it is adequate here to think of a European average fossil-fuel energy per kilowatt of delivered electrical power.

The fossil-fuel energy consumed in making one kilogram of material is called its *embodied energy*. Some of the energy is stored in the created material and can be reused, in one sense or another, at the end of life. Polymers made from oil (as most are) contain energy in another sense—that of the oil that enters the production as a primary feedstock. Natural materials such as wood similarly contain "intrinsic" or "contained" energy, this time derived from solar radiation absorbed during growth. Views differ on whether intrinsic energy should be included in embodied energy or not. There is a sense in which not only polymers and woods but also metals carry intrinsic energy that could—by chemical reaction or by burning the metal in the form of finely divided powder—be recovered, so omitting it when reporting production energy for polymers but including it for metals seems inconsistent. For this reason we will include intrinsic energy from nonrenewable resources in reporting embodied energies, which generally lie in the range of 25 to 250 MJ/kg, though a few are much higher. The existence of intrinsic energy has another consequence: The energy to recycle a material is sometimes much less than that required for its first production, because the intrinsic energy is retained. Typical values lie in the range 10 to 100 MJ/kg.

The production of 1 kilogram of material is associated with *undesired gas emissions*, among which CO_2, NO_x, SO_x, and CH_4 cause general concern (global warming, acidification, ozone-layer depletion). The quantities can be large—each kilogram of virgin aluminum produced with energy from fossil fuels creates some 9 kilograms of CO_2, 40 grams of NO_x, and 90 grams of SO_x (Table 15.2). Material production is generally associated with other undesirable outputs, particularly toxic wastes and particulates, but these can, in principle, be dealt with at the source.

Approximate data for embodied energy and CO_2 burden of materials are listed in Appendix A.

Estimates for material processing energies (at an overall conversion efficiency of 15%)

Many processes depend on casting, evaporation, or deformation. It is helpful to have a feel for the approximate magnitudes of energies required by these. The use of primary energy (thought of as the equivalent quantity of oil) for

Table 15.2 The Eco-attributes of One Grade of Wrought 1,000-grade Aluminum Alloy

Material Production: Energy and Emissions	
Embodied energy	195–210 MJ/kg
Carbon dioxide	9–10 kg/kg
Nitrogen oxides	72–79 g/kg
Sulphur oxides	120–140 g/kg
Material Processing Energy	
Casting energy	2.5–2.8 MJ/kg
Minimum energy to vaporization	17–19 MJ/kg
Minimum energy to 90% deformation	2.55–2.8 MJ/kg
End of Life	
Downcycle	Yes
Recycle	Yes
Biodegrade	No
Incinerate	No
Landfill	Acceptable
Recycling energy	17–20 MJ/kg
Recycle fraction	52–58%

processing materials involves several energy conversion steps, each with a conversion efficiency that is less than 1. Many processes use electric energy, generated from primary energy with a conversion efficiency on the order of 38%. The use of this for electroheating or for electroprocessing (like electroforming) involves further losses. Scrap, discarded material, and other wastes carry with them an energy investment that does not appear in the "good" casting, moldings, or forgings that are the output of the processes.

Other process routes involve other energy conversion sequences, all with their losses. It is possible, as we do below, to estimate the energy to melt, mold, vaporize, or deform a material, but to express this in terms of primary energy it must be divided by the product of the conversion efficiencies of each conversion step. For the sake of example we set this overall efficiency at a (realistic) 15%.

Melting To melt a material, it must first be heated to its melting point, requiring a minimum input of the heat $C_p\,(T_m - T_0)$, and then caused to melt, requiring the latent heat of melting L_m:

$$H_{\min} = C_p\,(T_m - T_0) + L_m \tag{15.1}$$

where $H_{\min}$ is the minimum energy per kilogram for melting, C_p is the specific heat, T_m is the melting point, and T_0 is the ambient temperature. A close correlation exists between L_m and $C_p\,T_m$:

$$L_m \approx 0.4\, C_p T_m \tag{15.2}$$

and for metals and alloys $T_m >> T_0$, giving

$$H_{\min} \approx 1.4\, C_p T_m \tag{15.3}$$

Assuming efficiency of 15%, the estimated energy to melt 1 kilogram H_m^* is

$$H_m^* \approx 8.4\, C_p T_m \tag{15.4}$$

the asterisk recalling that it is an estimate. For metals and alloys, the quantity H_m^* lies in the range 1 to 8 MJ/kg.

Vaporization As a rule of thumb, the latent heat of vaporization L_v is larger than that for melting, L_m, by a factor of 24 ± 5, and the boiling point T_b is larger than the melting point T_m by a factor 2.1 ± 0.5. Using the same assumptions as before, we find an estimate for the energy to evaporate 1 kg of material (as in PVD processing) to be

$$H_v^* \approx 76\, C_p T_m \tag{15.5}$$

again assuming an efficiency of 15%. For metals and alloys, the quantity H_v^* lies in the range 6 to 60 MJ/kg.

Deformation Deformation processes like rolling or forging generally involve large strains. Assuming an average flow strength of $(\sigma_y + \sigma_{uts})/2$, a strain of order 1, and an efficiency factor of 15%, we find, the work of deformation per kg, to be

$$W_D^* \approx 3(\sigma_y + \sigma_{uts}) \tag{15.6}$$

where σ_y is the yield strength and σ_{uts} is the tensile strength. For metals and alloys, the quantity W_D^* lies in the range 0.05 to 2 MJ/kg.

We conclude that casting or deformation requires processing energies that are small compared to the production energy of the material being processed, but the larger energies required for vapor-phase processing may become comparable with those for material production.

End of life

Figure 15.4 introduces the options: landfill, combustion for heat recovery, recycling, reengineering, and reuse.

Landfill Much of what we now reject is committed to landfill. Already there is a problem—the land available to "fill" in this way is already, in some European countries, almost full. Administrations react by charging a landfill tax—currently somewhere near €50 per tonne and rising, seeking to divert waste into the other channels of Figure 15.4.

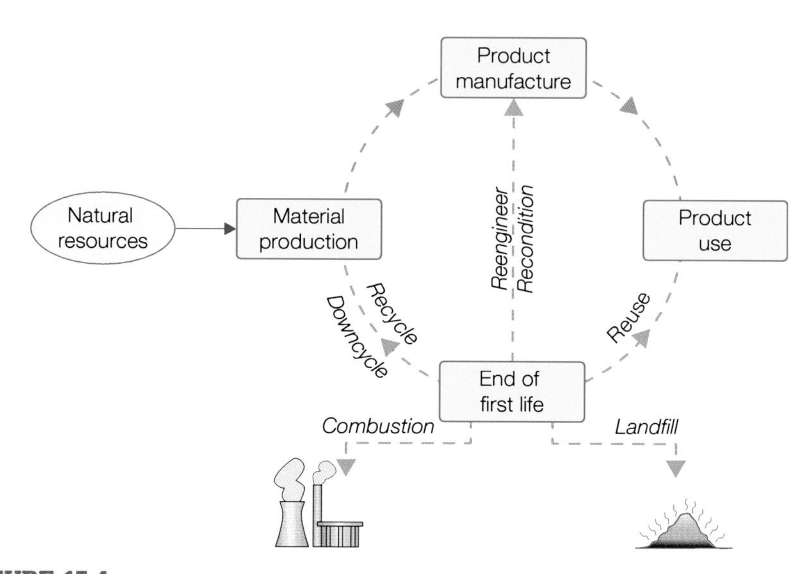

FIGURE 15.4
End of life options: landfill, combustion, recycling, refurbishment or upgrading, and reuse.

Combustion for heat recovery Materials, we know, contain energy. Rather than throw them away it would seem better to retrieve and reuse some of their energy by controlled combustion, capturing the heat. But this is not as easy as it sounds. First there is the need for a primary sorting to separate combustibles from noncombustible material. Then the combustion must be carried out under conditions that do not generate toxic fumes or residues, requiring high temperatures, sophisticated control, and expensive equipment. Energy recovery is imperfect partly because it is incomplete and partly because the incoming waste carries a moisture content that has to be boiled off.

Recycling Recycling requires energy, and this energy carries its burden of gases. But the *recycle energy* is generally small compared to the initial production energy, making recycling—when it is possible at all—an energy-efficient proposition. It may not, however, be one that is cost-efficient; that depends on the degree to which the material has become dispersed. In-house scrap, generated at the point of production or manufacture, is localized and is already recycled efficiently (near 100% recovery). Widely distributed "scrap"—material contained in discarded products—is much more expensive to collect, separate, and clean. Many materials cannot be recycled, although they may still be reused in a lower-grade activity; continuous-fiber composites, for instance, cannot be reseparated economically into fiber and polymer in order to recycle them, though they can be chopped and used as fillers. Most other materials require an input of virgin material to avoid build-up of uncontrollable impurities. Thus the fraction of a material's production that can ultimately reenter the cycle of Figure 15.4 depends both on the material itself and on the product into which it has been incorporated.

Reengineering Reengineering is the refurbishment or upgrading of the product or of its recoverable components. Certain criteria must be met to make it practical. One is that the design of the product is fixed or that the technology on which it is based is evolving so slowly that there remains a market for the restored product. Some examples are housing, office space, and road and rail infrastructure; these are sectors with enormous appetites for materials.

Reuse Reuse is the redistribution of the product to a consumer sector that is willing to accept it in its used state, perhaps to reuse it for its original purpose (e.g., a second-hand car), perhaps to adapt it to another (converting a car into a hot rod or a bus into a mobile home). Charity stores pass on clothing, books, and objects acquired from those for whom they had become waste; the stores sell them to others who perceive them to have value. Reuse is the most benign of the end-of-life scenarios.

15.5 ECO-SELECTION

To select materials to minimize impact on the environment we must first ask—as we did in Section 15.2—which phase of the life-cycle of the product under consideration makes the largest contribution. The answer guides the choice of strategy to ameliorate that contribution (Figure 15.5). The strategies are described in the following sections.

The material production phase If material production is the dominant phase of life, this becomes the first target. Drink containers (Figure 15.6) provide an example: They consume materials and energy during material extraction and production, but apart from transport and refrigeration, which are small, not thereafter. We use the energy consumed in extracting and refining the material (the embodied energy noted in Table 15.2) as the measure; CO_2, NO_x, and SO_x emissions are related to it, although not in a simple way. The energy associated with the production of 1 kilogram of a material is H_p, that per unit volume is $H_p \rho$, where ρ is the density of the material.

The bar charts of Figures 15.7 and 15.8 show these two quantities for ceramics, metals, polymers, and composites (hybrids). On a "per kg" basis (upper chart), glass, the material of the first container, carry the lowest penalty. Steel is slightly higher. Polymers carry a much higher energy penalty than steel. Aluminum and the other light alloys carry the highest penalty of all. But if these same materials are compared on a "per m^3" basis (lower chart) the conclusions change: Glass is still the lowest, but now commodity polymers, such

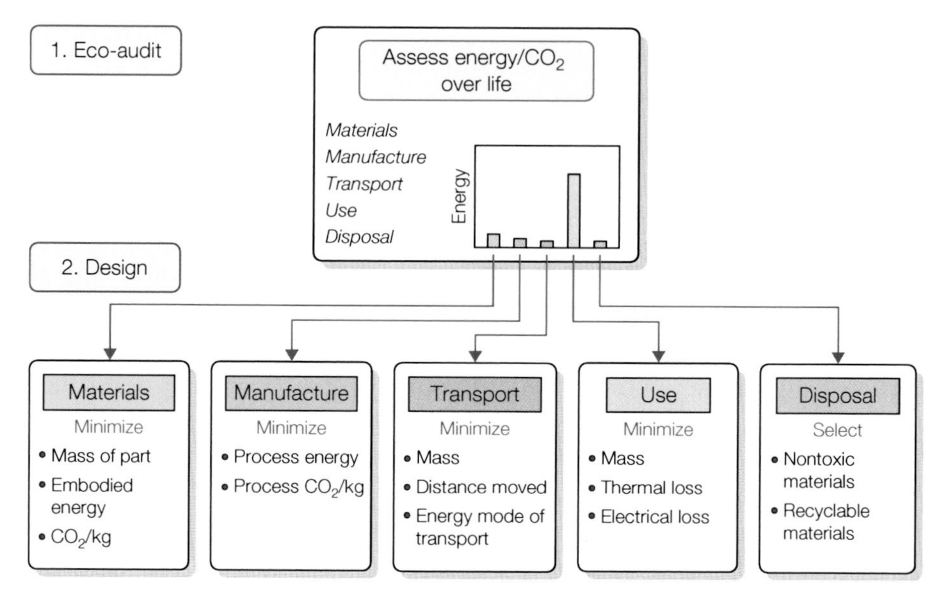

FIGURE 15.5
Rational design for the environment starts with an analysis of the phase of life to be targeted. This decision then guides the method of selection to minimize the impact of the phase on the environment.

FIGURE 15.6
Liquid containers: glass, polyethylene, PET, aluminum, and steel. All can be recycled. Which carries the lowest production energy penalty?

as PE and PP, carry a *lower* burden than steel; the composite GFRP is only a little higher. But is comparison "per kg" or "per m^3" the right way to do it? Rarely. To deal with environmental impact at the production phase properly we must seek to minimize the energy, the CO_2 burden, or the eco-indicator value *per unit of function.*

Performance indices that include energy content are derived in the same way as those for weight or cost (Chapter 5). As an example, consider the

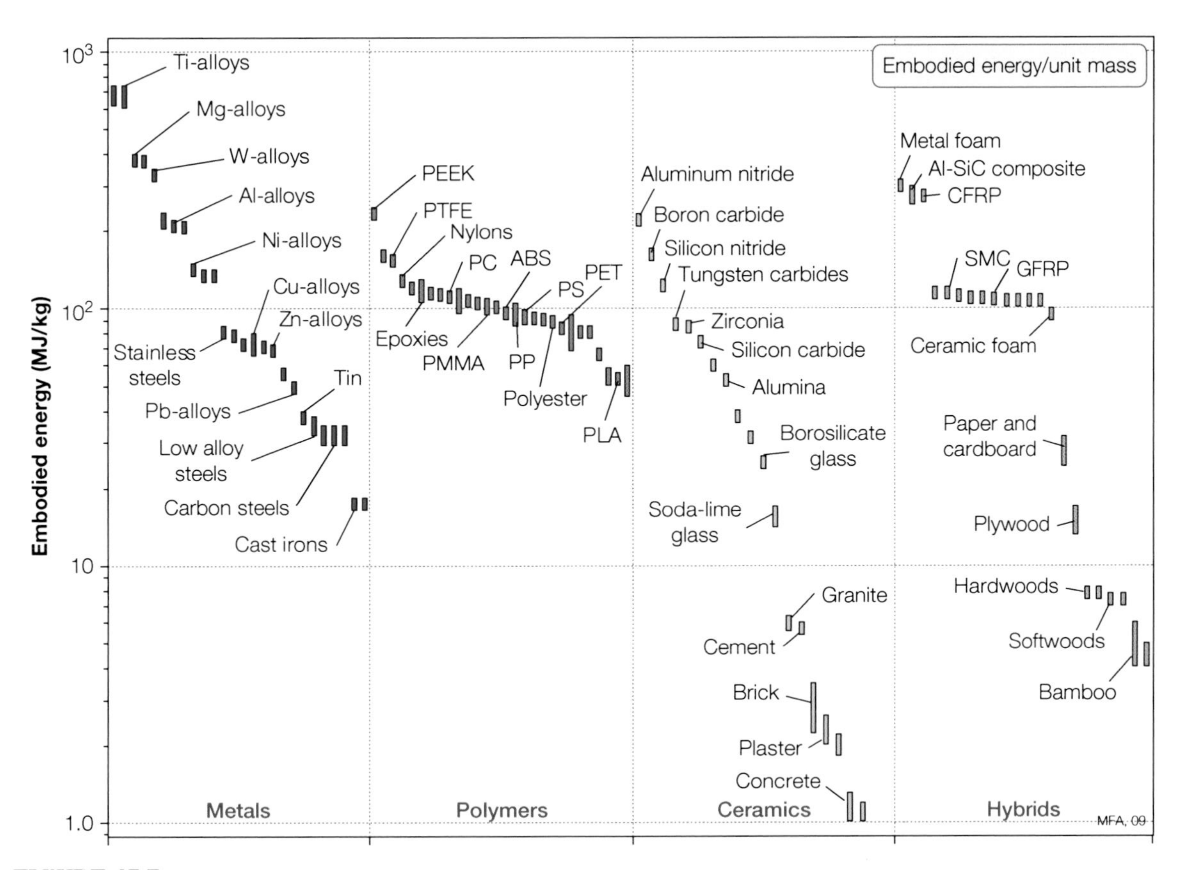

FIGURE 15.7

The embodied energy per unit mass of materials (data in Appendix A).

selection of a material for a beam that must meet a stiffness constraint at minimum energy content. Repeating the derivations of Chapter 5 but with the objective of minimizing embodied energy rather than mass leads to performance equations and material indices that are simply those of Chapter 5 with ρ replaced by $H_p\rho$. Thus the best materials to minimize the embodied energy of a beam of specified stiffness and length are those with large values of the index

$$M_1 = \frac{E^{1/2}}{H_p\rho} \tag{15.7}$$

where E is the modulus of the material of the beam. The stiff tie of minimum energy content is best made of a material of high $E/H_p\rho$; the stiff plate, of a material with high $E^{1/3}/H_p\rho$, and so on.

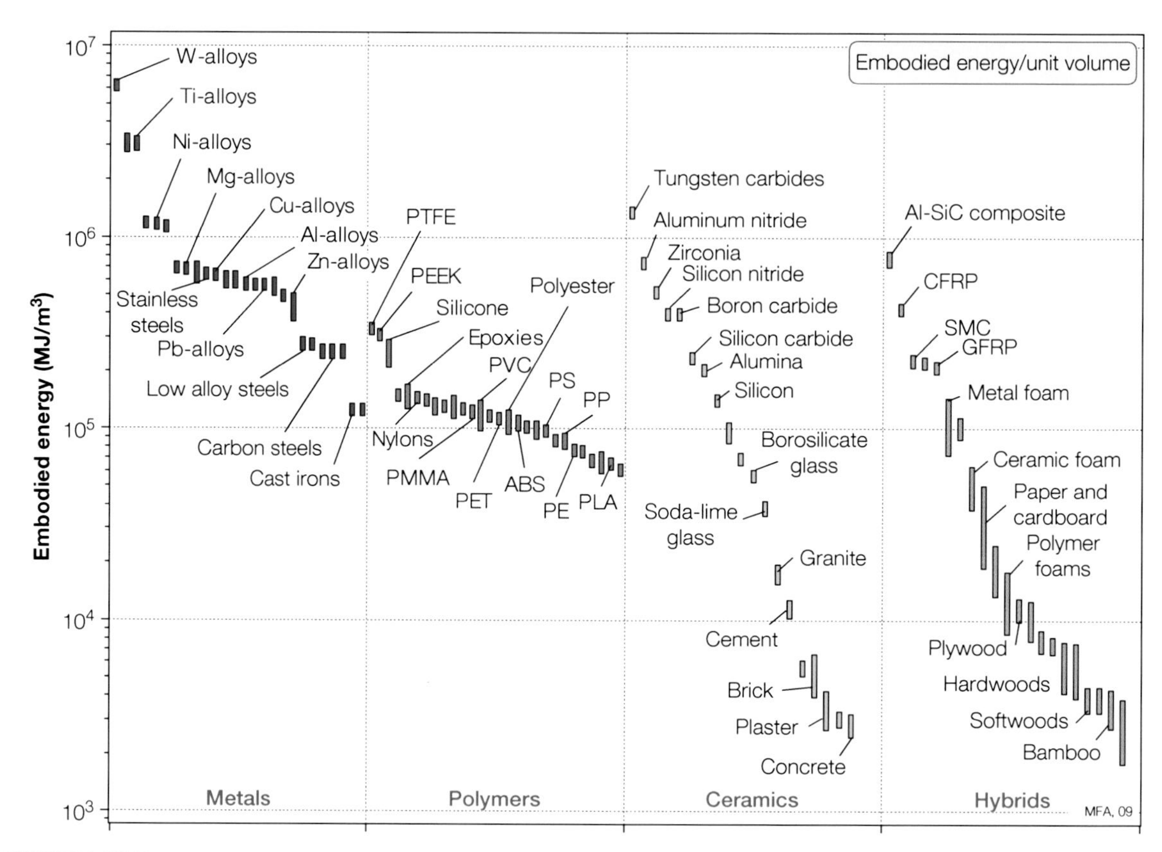

FIGURE 15.8
The embodied energy per unit volume of materials.

Strength works in the same way. The best materials for a beam of specified bending strength and minimum embodied energy are those with large values of

$$M_3 = \frac{\sigma_f^{2/3}}{H_p \rho} \qquad (15.8)$$

where σ_f is the failure strength of the beam material. Other indices follow in a similar way.

Figures 15.9 and 15.10 are a pair of charts for selection to minimize embodied energy H_p per unit of function (similar charts for CO_2 burden can be made using the CES Edu software). The first chart shows modulus E plotted against $H_p\rho$; the guide lines give the slopes for three of the commonest performance indices. The second shows strength σ_f (defined in

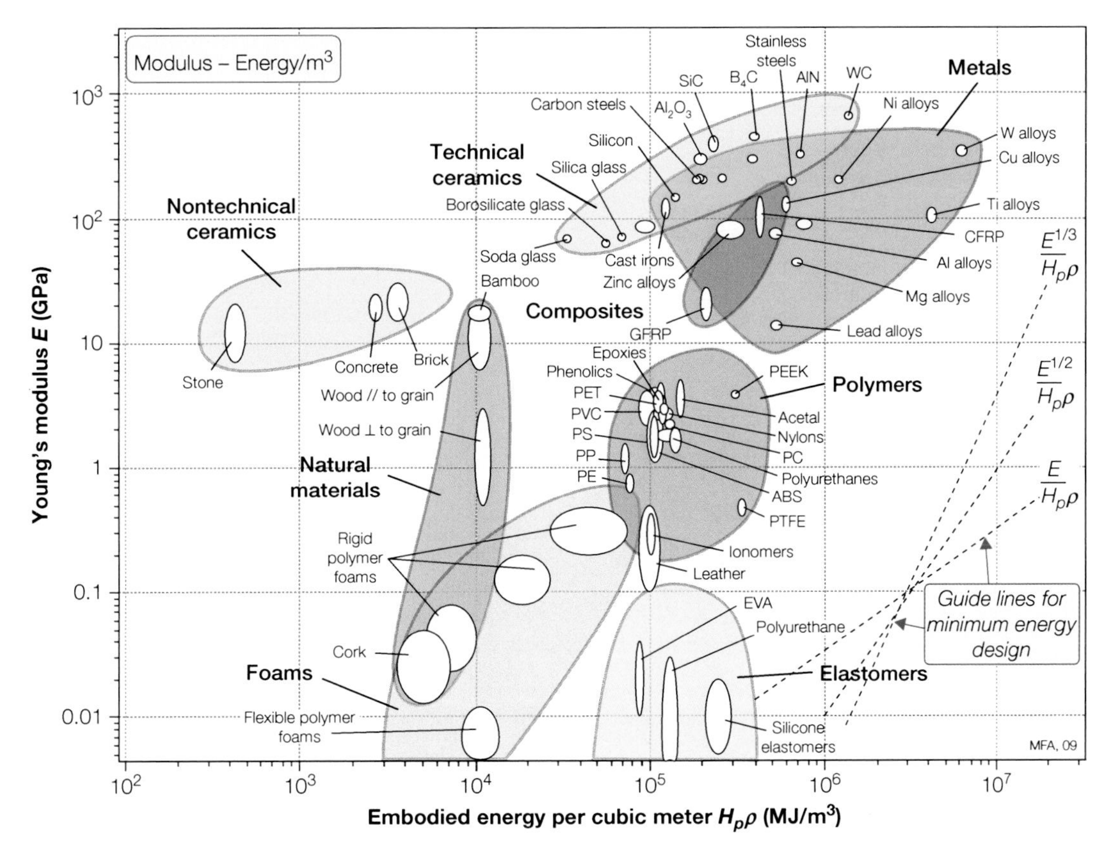

FIGURE 15.9

A selection chart for stiffness with minimum production energy. It is used in the same way as Figure 4.3.

Chapter 4) plotted against $H_p\rho$; again, guide lines give the slopes. The two charts give survey data for minimum energy design. They are used in exactly the same way as the $E - \rho$ and $\sigma_f - \rho$ charts for minimum mass design.

Most polymers are derived from oil. This leads to statements that they are energy-intense, with implications for their future. The two charts in Figures 5.9 and 5.10 show that, per unit of function in bending (the commonest mode of loading), most polymers carry a lower energy penalty than primary aluminum, magnesium, or titanium, and several are competitive with steel. Most of the energy consumed in the production of metals such as steel, aluminum, a magnesium is used to reduce the ore to the elemental metal, so that these materials, when recycled, require much less energy. Efficient collection and recycling makes important contributions to energy saving.

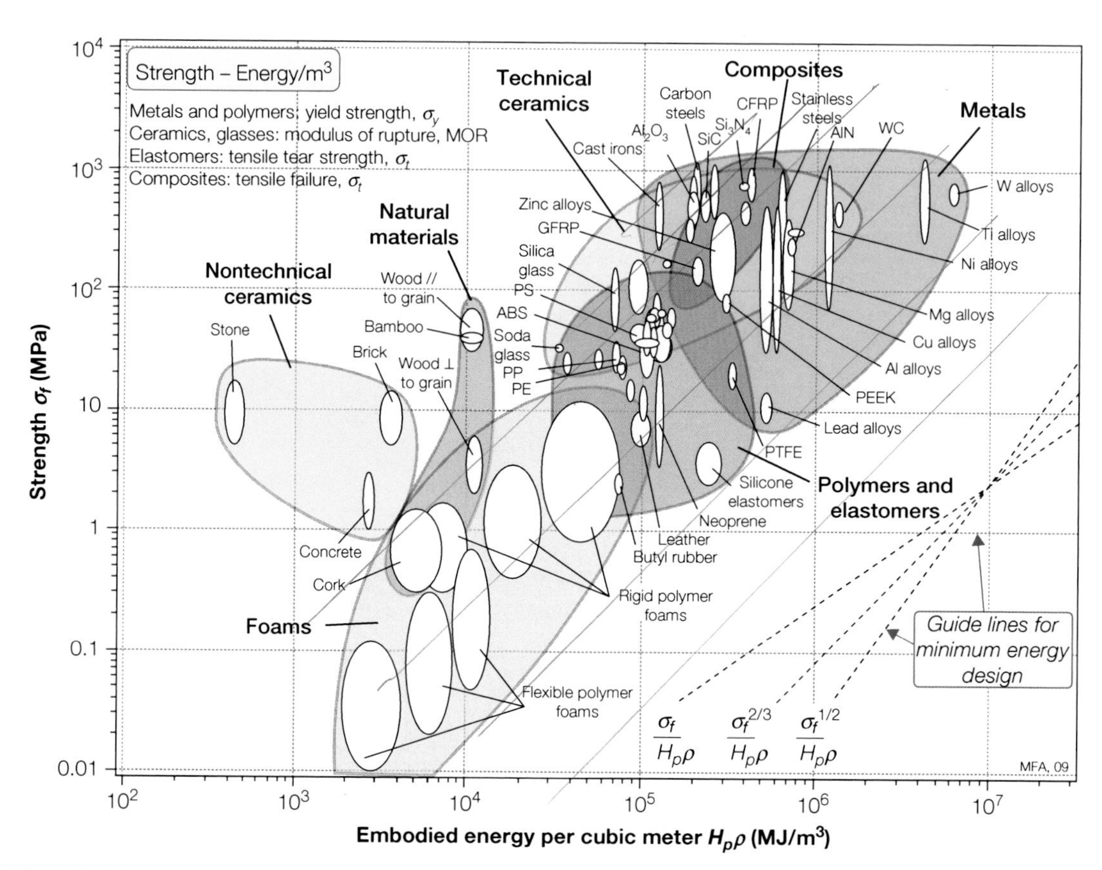

FIGURE 15.10

A selection chart for strength with minimum production energy. It is used in the same way as Figure 4.4.

The product manufacture phase Shaping materials, as we discussed earlier, requires energy. Certainly it is important to save energy in production. But higher priority often attaches to the *local* impact of emissions and toxic waste during manufacture, and this depends crucially on local circumstances. Paper making (to take an example) uses very large quantities of water. Historically the wastewater was heavily polluted with alkalis and particulates, devastating the river systems into which it was dumped. Today, the best paper mills discharge water that is as clean and pure as it was when it entered. Production sites of the former communist-bloc countries are terminally polluted; those producing the same materials elsewhere, using best-practice methods, have no such problems. Clean manufacture is the issue here.

The use phase The eco-impact of the use phase of energy-consuming products has nothing to do with the energy content of the materials themselves—indeed, minimizing this may frequently have the opposite effect on use energy. Use energy depends on mechanical, thermal, and electrical efficiencies; it is minimized by maximizing these. Fuel efficiency in transport systems (measured, say, by MJ/km) correlates closely with the mass of the vehicle itself; the objective then becomes minimizing mass. Energy efficiency in refrigeration or heating systems is achieved by minimizing the heat flux into or out of the system; the objective is then minimizing thermal conductivity or thermal inertia. Energy efficiency in electrical generation, transmission, and conversion is maximized by minimizing the ohmic losses in the conductor; here the objective is minimizing electrical resistance while meeting necessary constraints on strength, cost, and so on. Selection to meet these objectives is exactly what the previous chapters of this book were about.

The product disposal phase The environmental consequences of the final phase of product life has many aspects. Increasingly, legislation dictates disposal procedures, take-back, and recycle requirements, and—through landfill taxes and subsidized recycling—deploys market forces to determine the end-of-life choice.

15.6 CASE STUDIES: DRINK CONTAINERS AND CRASH BARRIERS

The energy content of containers

The containers shown earlier in Figure 15.6 are examples of products for which the first and second phases of life—material production and product manufacture—consume the most energy and generate the most emissions. Thus material selection to minimize energy and consequent gas and particle emissions focuses on these. Table 15.3 summarizes the requirements.

The masses of five competing container types, the material of which they are made, and the specific energy content of each are listed in Tables 15.4 and 15.5. Their production involves molding or deformation; approximate energies for each are listed. All five materials can be recycled. Which container type carries the lowest overall energy penalty per unit of fluid contained?

Table 15.3 Design Requirements for the Containers

Function	Container for cold drink
Constraint	Must be recyclable
Objective	Minimize embodied energy per unit capacity
Free variable	Choice of material

Table 15.4 Details of the Containers

Container Type	Material	Mass, g	Mass/ Liter, g	Energy/ Liter, MJ/Liter
PET 400-ml bottle	PET	25	62	5.4
PE 1-liter milk bottle	High-density PE	38	38	3.2
Glass 750-ml bottle	Soda glass	325	433	8.2
Al 440-ml can	5000 series Al alloy	20	45	9.0
Steel 440-ml can	Plain carbon steel	45	102	2.4

Table 15.5 Data for the Materials of the Containers (from Appendix A)

Material	Embodied Energy, MJ/kg	Forming Method	Forming Energy, MJ/kg
PET	84	Molding	3.1
PE	81	Molding	3.1
Soda glass	15.5	Molding	4.9
5000 series Al alloy	210	Deep drawing	0.13
Plain carbon steel	32	Deep drawing	0.15

The method and results A comparison of the energies in Table 15.5 shows that the energy to shape the container is always less than that to produce the material in the first place. Only in the case of glass is the forming energy significant. The dominant phase is material production. Summing the two energies for each material and multiplying by the container mass per liter of capacity gives the ranking shown in the second to last column of Table 15.4. Steel can carry the lowest energy penalty, glass and aluminum the highest.

Crash barriers

Barriers to protect the driver and passengers in road vehicles are of two types: those that are static—the central divider of a freeway, for instance—and those that move—the bumper of the vehicle itself (Figure 15.11). The static types line tens of thousands of miles of road. Once in place they consume no energy, create no CO_2, and last a long time. The dominant phases of their life in the sense of the life-cycle shown in Figure 15.1 are material production and manufacture. The bumper, by contrast, is part of the vehicle; it adds to its weight and thus to its fuel consumption.

The dominant phase here is use. This means that, if eco-design is the objective, the criteria for selecting materials for the two sorts of barrier will differ. We take as a criterion for the first that of maximizing the energy that the

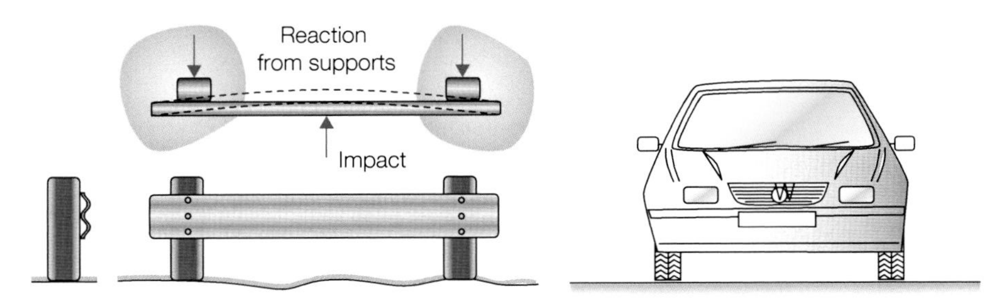

FIGURE 15.11
Two crash barriers: one static, the other—the bumper—attached to something that moves. Different eco-criteria are needed to select materials for each.

Table 15.6 Design Requirements for the Crash Barriers

Function	Energy-absorbing crash barriers
Constraint	Must be recyclable
Objectives	Maximize energy absorbed per unit production energy, or Maximize energy absorbed per unit mass
Free variable	Choice of material

barrier can absorb per unit of production energy; for the second we take absorbed energy per unit mass. Table 15.6 summarizes.

In an impact, the barrier is loaded in bending (Figure 15.11). Its function is to transfer load from the point of impact to the support structure, where reaction from the foundation or from crush elements in the vehicle support or absorb it. To do this the material of the barrier must have high strength, σ_f, be adequately tough and able to be recycled. That of the static barrier must meet these constraints with *minimum embodied energy* as the objective, since this will reduce the overall life energy most effectively. We know from Section 15.5 that this means materials with large values of the index

$$M_3 = \frac{\sigma_f^{2/3}}{H_p \rho} \tag{15.9}$$

where σ_f is the yield strength, ρ the density, and H_p the embodied energy per kg of material. For the car bumper it is mass, not embodied energy, that is the problem. If we change the objective to *minimum mass*, we require materials with high values of the index

$$M_4 = \frac{\sigma_f^{2/3}}{\rho} \tag{15.10}$$

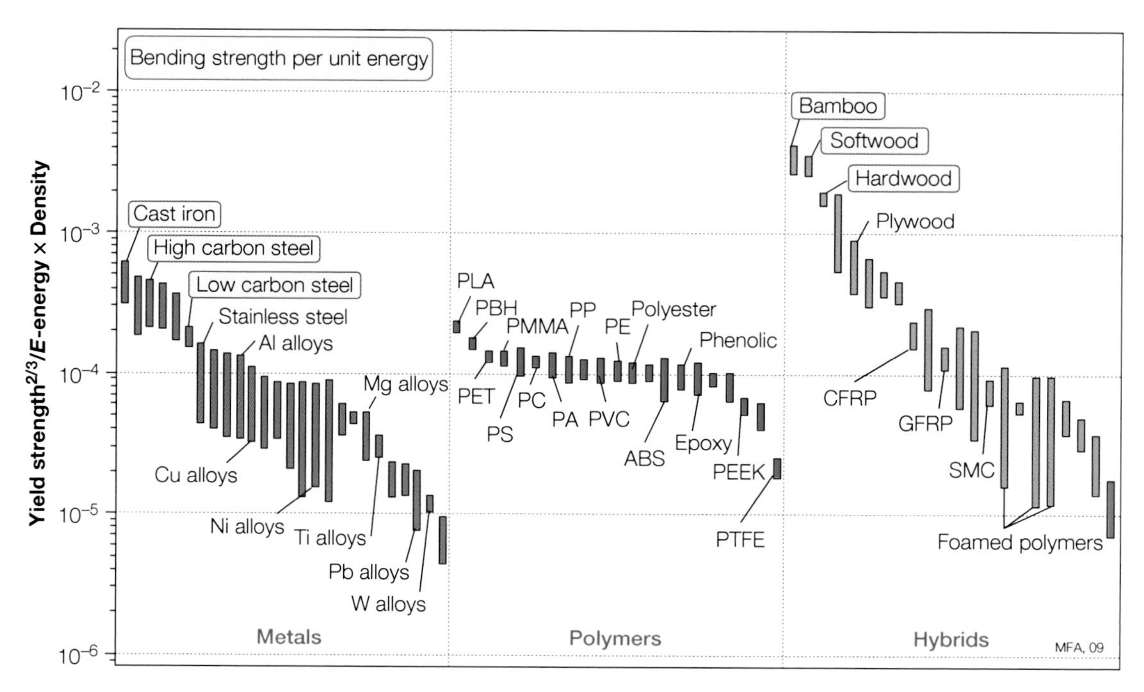

FIGURE 15.12

Material choice for the static barrier; the units are $(MPa)^{2/3}/(MJ/m^3)$. Cast irons, carbon steels, low alloy steels, and wood are the best choices.

These indices can be plotted on the charts shown in Figure 15.10 (see also Figure 4.4), enabling a selection. We leave that as one of the exercises in order to show here an alternative: simply plotting the index itself as a bar chart. Figures 15.12 and 15.13 show the result for metals, polymers, and polymer-matrix composites. The first guides the selection for static barriers. It shows that embodied energy (for a given load-bearing capacity) is minimized by making the barrier out of carbon steel or cast iron or wood; nothing else comes close. The second figure guides selection for the mobile barrier. Here CFRP (continuous fiber carbon epoxy, for instance) excels in its strength per unit weight, but it is not recyclable. Heavier but recyclable are alloys of magnesium, titanium, and aluminum. Polymers, which rank poorly in the first figure, now become candidates—even without reinforcement, they can be as good as steel.

Postscript Metal crash barriers have a profile like that shown in Figure 15.11 (*left*). The curvature increases the second moment of area of the cross-section and through this the bending stiffness and strength. This is an example of combining material choice and section shape (Section 9.5) to optimize a design.

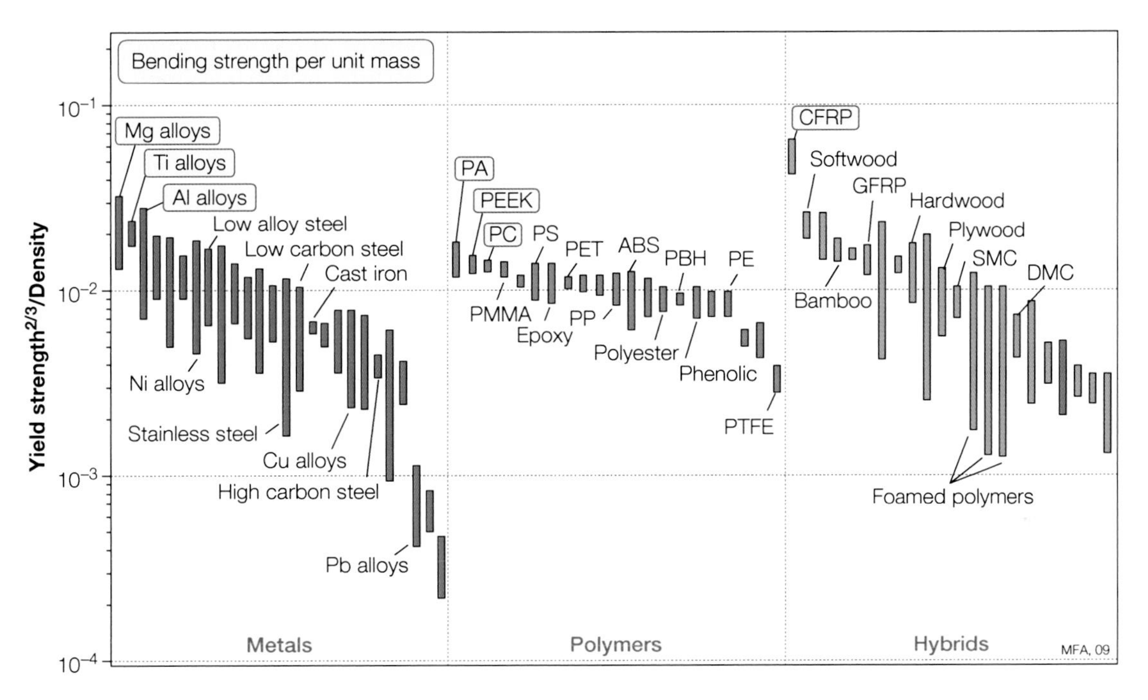

FIGURE 15.13
Material choice for the mobile barrier; the units are $(MPa)^{2/3}/(kg/m^3)$. CFRP and light alloys offer the best performance; nylon and polycarbonate perform as well as steel.

15.7 SUMMARY AND CONCLUSIONS

Rational selection of materials to meet environmental objectives starts by identifying the phase of product life that causes the greatest concern: production, manufacture, use, or disposal. Dealing with all of these requires data not only for the obvious eco-attributes (e.g., energy, CO_2 and other emissions, toxicity, and the ability to be recycled) but also for mechanical, thermal, electrical, and chemical properties. Thus if material production is the phase of concern, selection is based on minimizing production energy or the associated emissions (e.g., CO_2 production). But if it is the use phase that is of concern, selection is based instead on light weight and excellence as a thermal insulator or as an electrical conductor (while meeting other constraints on stiffness, strength, cost, etc.).

This chapter developed methods to deal with these issues. The methods are most effective when implemented in software. The CES Edu system, described in Chapters 5, 6, 13, and 14 contains eco-data for materials and includes an eco-audit tool to analyze product life in the manner of Figure 15.3.

If you found this chapter interesting and would like to read more, you will find the ideas it contains developed more fully in the first book listed in the following section.

15.8 FURTHER READING

Ashby, M.F. (2009). *Materials and the environment.* Butterworth-Heinemann, ISBN 978-1-85617-608-8.

A teaching text that provides the resources—background, methods, data—to enable environmental issues relating to materials to be explored in depth.

CES Edu (2010). *The Cambridge Engineering Selector.* Granta Design.

The material selection platform now has an optional eco-design module. www.grantadesign.com.

Fuad-Luke, A. (2002). *The eco-design handbook.* Thames and Hudson, ISBN 0-500-28343-5.

A remarkable sourcebook of examples, ideas, and materials of eco-design.

Goedkoop, M.J., Demmers, M., & Collignon, M.X. (1995). *Eco-indicator '95 manual.* PRé Consultants and The Netherlands Agency for Energy and the Environment, ISBN 90-72130-80-4.

Goedkoop, M., Effting, S., & Collignon, M. (2000). *The Eco-indicator 99: A damage oriented method for life cycle impact assessment.* Manual for Designers.

PRé Consultants market a leading life-cycle analysis tool and are proponents of the eco-indicator method. http://www.pre.nl.

ISO 14001 (1996) and ISO 14040 (1997, 1998, 1999). Environmental management system-specification with guidance for use. International Organization for Standardization (ISO), Geneva.

Kyoto Protocol (1997). Environmental management—life cycle assessment (and subsections). Framework Convention on Climate Change. Document FCCC/CP1997/7/ADD.1. United Nations, Geneva.

An agreement among the developed nations to limit their greenhouse gas emissions relative to the levels emitted in 1990. The United States agreed to reduce emissions from 1990 levels by 7% during the period 2008–2012; European nations adopted a more stringent agreement.

Lovins, L.H., von Weizsäcker, E., & Lovins, A.B. (1998). *Factor four: Doubling wealth, halving resource use.* Earthscan, ISBN 1-853834-068.

An influential book arguing that resource productivity can and should grow fourfold—that is, the wealth extracted from one unit of natural resources can quadruple, allowing the world to live twice as well yet use half as much.

MacKay, D.J.C. (2008). *Sustainable energy—without the hot air.* Cambridge University: Department of Physics, *www.withouthotair.com/*.

MacKay brings a welcome dose of common sense into the discussion of energy sources and use. Fresh air replacing hot air.

Mackenzie, D. (1997). *Green design: Design for the environment* (2nd ed.). Lawrence King Publishing, ISBN 1-85669-096-2.

A lavishly produced compilation of case studies of eco-design in architecture, packaging, and product design.

Meadows, D.H., Meadows, D.L., Randers, J., & Behrens, W.W. (1972). *The limits to growth—1st report of the club of Rome.* Universe Books, ISBN 0-87663-165-0.

A pivotal publication, alerting the world to the possibility of resource depletion, undermined by the questionable quality of the data used for the analysis; despite that, it was the catalyst for subsequent studies and for views that are now more widely accepted.

Schmidt-Bleek, F. (1997). *How much environment does the human being need—factor 10—the measure for an ecological economy*. Deutscher Taschenbuchverlag.

The author argues that true sustainability requires a reduction in energy and resource consumption by the developed nations by a factor of 10.

Wenzel, H., Hauschild, M., & Alting, L. (1997). *Environmental assessment of products, Vol. 1*. Chapman and Hall, ISBN 0-412-80800-5.

Professor Alting leads Danish research in eco-design.

CHAPTER 16

Materials and Industrial Design

Two cars that are about the same in power and price but very different in style, comfort, and sophistication. How is it that they can compete in the same market? It has to do with aesthetics, association, and perception.

Materials Selection in Mechanical Design. DOI: 10.1016/B978-1-85617-663-7.00016-3

CONTENTS

16.1 INTRODUCTION AND SYNOPSIS

Good design works. Excellent design also gives pleasure.

Pleasure derives from form, color, texture, feel, and the associations that these invoke. Pleasing design says something about itself; generally speaking, honest statements are more satisfying than deception, though eccentric or humorous designs can be appealing too.

Materials play a central role in this. A major reason for introducing new materials is the greater freedom of design that they allow. Metals, during the past century, allowed the building of structures that could not have been built before: cast iron, the Crystal Palace; wrought iron, the Eiffel Tower; drawn steel cable, the Golden Gate Bridge—all undeniably beautiful. Polymers lend themselves to bright colors, satisfying textures, and great freedom of form; they have opened new styles of design, of which some of the best examples are found in household appliances: Food mixers, hair dryers, mobile phones, MP3 players, and vacuum cleaners make extensive and imaginative use of materials to allow styling, weight, feel, and form that give pleasure.

Those who concern themselves with this aesthetic dimension of engineering are known, rather confusingly, as "industrial designers." This chapter introduces some of the ideas of industrial design, emphasizing the role of materials. It ends with case studies. But first a word of caution.

Previous chapters dealt with systematic ways of choosing material and processes. "Systematic" means that if *you* do it and *I* do it, following the same procedure, we will get the same result, and that result, next year, will be the same as it is today. Industrial design is not, in this sense, systematic. Succes, here involves sensitivity to fashion, custom, and educational background, and is influenced (even manipulated) by advertising and association. The views in this chapter are partly those of writers who seem to me to say sensible things, and partly my own. You may not agree with them, but if they make you think about designing to give pleasure, the chapter has done what it should.

The ideas in this chapter are developed more fully in the first text listed in "Further reading."

16.2 THE REQUIREMENTS PYRAMID

The pen with which I am writing this chapter cost $5 (Figure 16.1, *upper image*). If you go to the right shop, you can find a pen that costs well over $1000 (*lower image*). Does it write 200 times better than mine? Unlikely; mine writes perfectly well. Yet there is a market for expensive pens. Why?

A product has a *cost*—the outlay in manufacture and marketing it. It has a *price*—the sum at which it is offered to the consumer. And it has a *value*—a measure of what the consumer thinks it is worth. The expensive pens command the price they do because the consumer perceives their value to justify it. What determines value? Three things.

Functionality, provided by sound technical design, clearly plays a role. The requirements pyramid in Figure 16.2 has this as its base: The product must work properly and be safe and economical. Functionality alone is not enough: The product must be easy to understand and operate, and these are questions of *usability*, the second tier in the figure. The third, completing the pyramid, is the requirement that the product give *satisfaction*: that it enhances the life of those who own or use it.

FIGURE 16.1
Pens, inexpensive and expensive. The chosen material—acrylic in the upper two; gold, silver, and enamel in the lower two—creates the aesthetics and the associations of the pens. (*Bottom* image courtesy of David Nishimura, *Vintagepens.com*.)

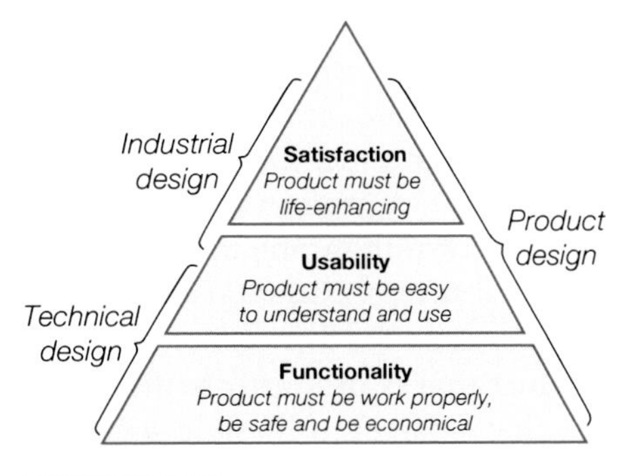

FIGURE 16.2
The requirements pyramid. The lower part of the pyramid tends to be called "technical design," the upper part, "industrial design," suggesting that they are separate activities. It is better to think of all three tiers as part of a single process that we shall call "product design."

The value of a product is a measure of the degree to which it meets (or exceeds) the expectation of the consumer for all three of these—functionality, usability, and satisfaction. Think of this as the product. It is very like human character. An admirable character is one that functions well, interacts effectively, and is rewarding company. An unappealing character is one that does none of these. An odious character is one that behaves in such an unattractive way that you cannot bear to be near them.

Products are the same. All the pens in Figure 16.1 function well and are easy to use. The huge difference in price implies that the lower two provide a degree of satisfaction not offered by the upper two. The most obvious difference between them is the materials of which they are made—the upper pair of molded acrylic, the lower pair of gold, silver, and enamel. Acrylic is the material of toothbrush handles, something you throw away after use. Gold and silver are the materials of precious jewelry; they have associations of craftsmanship, of heirlooms passed from one generation to the next. Well, that's part of the difference, but there is more. To find it we need to address a question—what creates product character?

16.3 PRODUCT CHARACTER

Figure 16.3 shows a way of dissecting product character. It is a map of the ideas we are going to explore; like all maps there is a lot of detail, but we need it to find our way. In the center is information about the *product* itself: the basic design requirements, its function, its features. The way these are thought through and developed is conditioned by the *context*, shown in the circle above the product. The context is set by the answers to the questions in the top box: *Who? Where? When? Why?* Consider the first of these: *Who?* A designer seeking to create a product attractive to women will make choices that differ from those for a product intended for children, or for elderly people, or for sportsmen. *Where?* A product for use in the home requires a different choice of material and form than one to be used—say—in a school or hospital. *When?* One intended for occasional use is designed in a different way from one used all the time; one for formal occasions differs from one for informal use. *Why?* A product that is primarily utilitarian involves different design decisions from one that is largely a lifestyle statement. The context influences and conditions all the decisions that the designer makes in finding a solution. It sets the *mood*.[1]

On the left of the *product* circle of Figure 16.3 lie packages of information about the *materials* and the *processes* used to shape, join, and finish it. Each

[1] Many designers working on a project assemble a *mood board* with images of the sort of people for whom the product is intended, the surroundings in which they suppose it will be used, and other products that the intended user group might own, seeking to capture the flavor of their lifestyle.

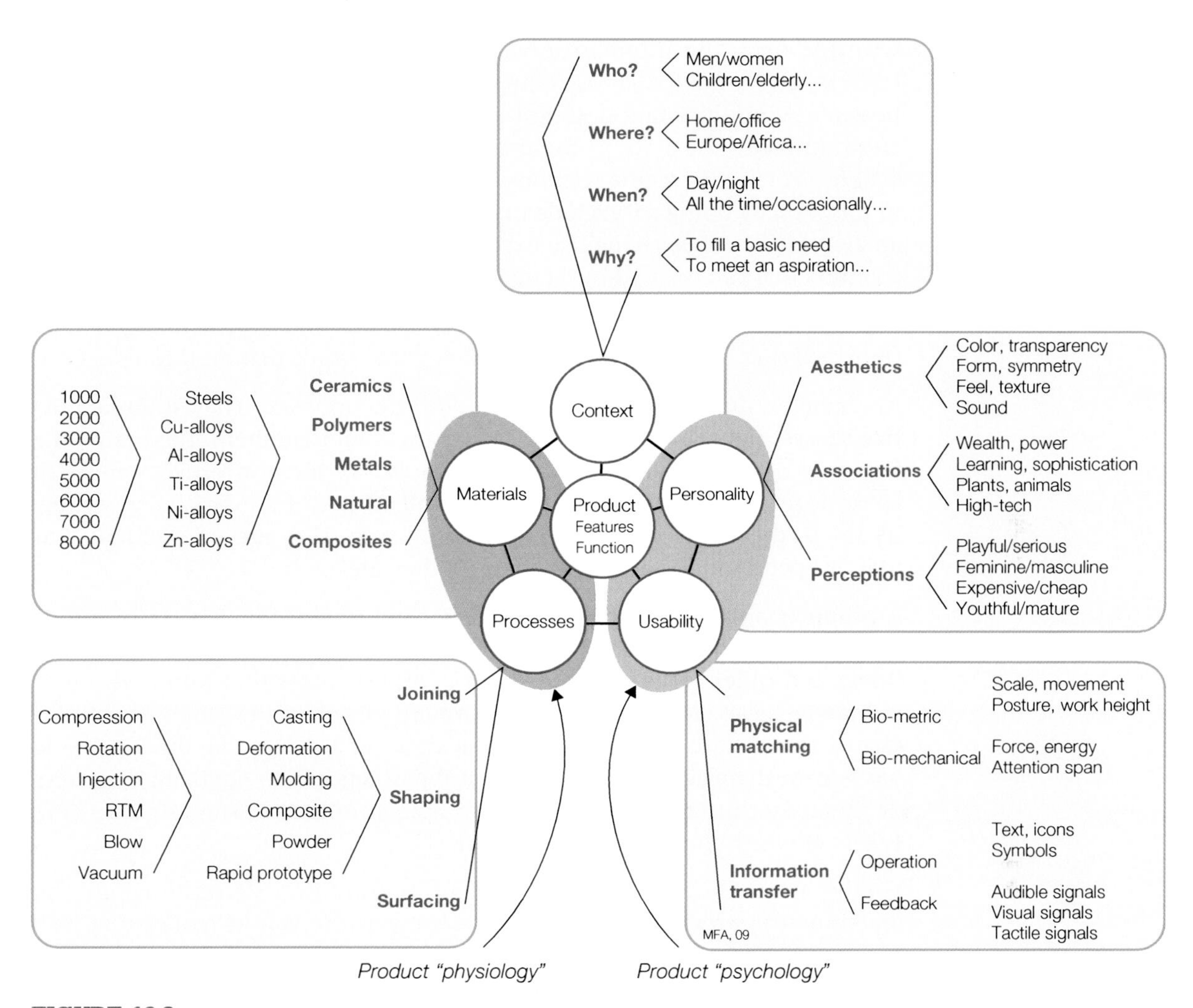

FIGURE 16.3
The breakdown of product character. *Context* defines the intentions or "mood"; *materials* and *processes* create the flesh and bones; the user interface determines *usability*; and the product's aesthetics, associations, and perceptions create its *personality*.

illustrates the library, so to speak, from which the choices can be made. They are the families, classes, and members that we first met in Chapter 3. Choosing these to provide functionality—the bottom of the pyramid—has been the subject of this book so far. Material and process give the product its tangible form, its flesh and bones so to speak; they create the *product physiology*.

On the right in Figure 16.3 are two more packages of information. The lower one—*usability*—characterizes the ways in which the product communicates with the user: the interaction with their sensory, cognitive, and motor functions. Product success requires a mode of operation that, as far as possible, is intuitive and does not require taxing effort, and an interface that communicates the state of the product and its response to user action by visible, acoustic, or tactile response. It is remarkable how many products fail in this and, by doing so, exclude many of their potential users. There is today an awareness of this, giving rise to research on *inclusive design*: design to make products that can be used by a larger spectrum of the population.

One package remains: the one labeled *personality*. Product personality derives from *aesthetics, associations*, and *perceptions*, three words that need explanation.

Anaesthetics dull the senses. Aesthetics do the opposite: They stimulate the five senses: sight, hearing, touch, taste, smell, and via them, the brain. The first row of the personality box elaborates: We are concerned here with color, form, texture, feel, smell, and sound. Think of a new car: its styling, its smell, the sound its doors make on closing. They are no accident. Car makers spend millions to make them as they are.

A products also has associations—the second row of the box. Associations are the things the product reminds you of, the things it suggests. The Land Rover and other SUVs have forms and (often) colors that mimic those of military vehicles. The streamlining of American cars of the 1960s and 1970s carried associations with aerospace. It may be an accident that the VW Beetle has a form that suggests the insect, but the others are no accident; they were deliberately chosen by the designer to appeal to the consumer group (the *Who?*) at which the product was aimed.

Finally, the most abstract quality of all, perceptions. Perceptions are the reactions the product induces in an observer, the way it makes you *feel*.

Table 16.1 Some Perceived Attributes of Products

Perceptions (with Opposites)	
Aggressive—Passive	Extravagant—Restrained
Cheap—Expensive	Feminine—Masculine
Classic—Trendy	Formal—Informal
Clinical—Friendly	Handmade—Mass-produced
Clever—Silly	Honest—Deceptive
Common—Exclusive	Humorous—Serious
Decorated—Plain	Informal—Formal
Delicate—Rugged	Irritating—Lovable
Disposable—Lasting	Lasting—Disposable
Dull—Sexy	Mature—Youthful
Elegant—Clumsy	Retro—Futuristic

Here there is room for disagreement; the perceptions of a product change with time and depend on the culture and background of the observer. Yet in the final analysis it is the perception that causes the consumer, when choosing between a multitude of similar models, to prefer one above the others; it creates the "must have" feeling (see chapter cover image). Table 16.1 lists some perceptions with their opposites in order to sharpen the meaning. They derive from product reviews and magazines specializing in product design; they are a part of a vocabulary, one that is used to communicate views about product character.

16.4 USING MATERIALS AND PROCESSES TO CREATE PRODUCT PERSONALITY

Do materials, of themselves, have a personality? There is a school of thinking that holds as a central tenet that materials must be used "honestly." This means that deception and disguise are unacceptable—each material must be used in ways that expose its intrinsic qualities and natural appearance. It has its roots in the tradition of craftsmanship—potters' use of clays and glazes, carpenters' use of woods, the skills of silversmiths and glass makers in crafting beautiful objects that exploit the unique qualities of the materials with which they work and the integrity of their craft.

This is a view to be respected. But it is not the only one. Design integrity is a quality that consumers value, but they also value other qualities: humor, sympathy, surprise, provocation, even shock. You don't have to look far to find a product that has one of these, and often that quality is achieved by using materials in ways that deceive. Polymers are frequently used in this way—their adaptability invites it. And, of course, it is partly a question of definition—if you say that a characterizing attribute of polymers is their ability to mimic other materials, then using them in this way *is* honest.

Materials and the senses: Aesthetic attributes

Aesthetic attributes are those that relate to the senses: touch, sight, hearing, taste, and smell (Table 16.2). Almost everyone would agree that metals feel "cold"; that cork feels "warm"; that a wine glass, when struck, "rings"; that a pewter mug sounds "dull," even "dead." A polystyrene water glass can look indistinguishable from one made of glass, but pick it up and it feels lighter, less cold, less rigid; tap it and it does not sound the same. It leaves the impression that is so different from glass that, in an expensive restaurant, it would be completely unacceptable. Materials, then, have certain characterizing aesthetic attributes. Let us see if we can pin these down.

Table 16.2 Some Aesthetic Attributes of Materials

Sense	Attribute	Sense	Attribute
Touch	Warm Cold Soft Hard Flexible Stiff	Hearing	Muffled Dull Sharp Resonant Ringing High-pitched Low-pitched
Sight	Optically clear Transparent Translucent Opaque Reflective Glossy Matte Textured	Taste Smell	Bitter Sweet

Touch: Soft–hard/warm–cold Steel is "hard"; so is glass; diamond is harder than both of them. Hard materials do not scratch easily; indeed they can be used to scratch other materials. They generally accept a high polish, resist wear, and are durable. The impression that a material is hard is directly related to its Vickers hardness H. Here is an example of a sensory attribute that relates directly to a technical one.

"Soft" sounds like the opposite of "hard" but it has more to do with modulus E than with hardness H. A soft material deflects when handled, it gives a little, it is squashy; but when it is released it returns to its original shape. Elastomers (rubbers) feel soft; so do polymer foams. Both have moduli that are 100 to 10,000 lower than ordinary "hard" solids; it is this that makes them feel soft. Soft to hard is used as one axis in Figure 16.4. It uses the quantity $\sqrt{EH}$ as a measure.

A material feels "cold" to the touch if it conducts heat away from the finger quickly; it is "warm" if it does not. This has something to do with its thermal conductivity λ but there is more to it than that—it depends also on its specific heat C_p. A measure of the perceived coldness or warmth of a material is the quantity $\sqrt{\lambda C_p \rho}$. It is shown as the other axis in Figure 16.4, which nicely displays the tactile properties of materials. Polymer foams and low-density woods are warm and soft; so are balsa and cork. Ceramics and metals are cold and hard; so is glass. Polymers and composites lie in between.

Sight: Transparency, color, reflectivity Metals are opaque. Most ceramics, because they are polycrystalline and the crystals scatter light, are either opaque or translucent. Glasses, and single crystals of some ceramics, are transparent. Polymers have the greatest diversity of optical transparency,

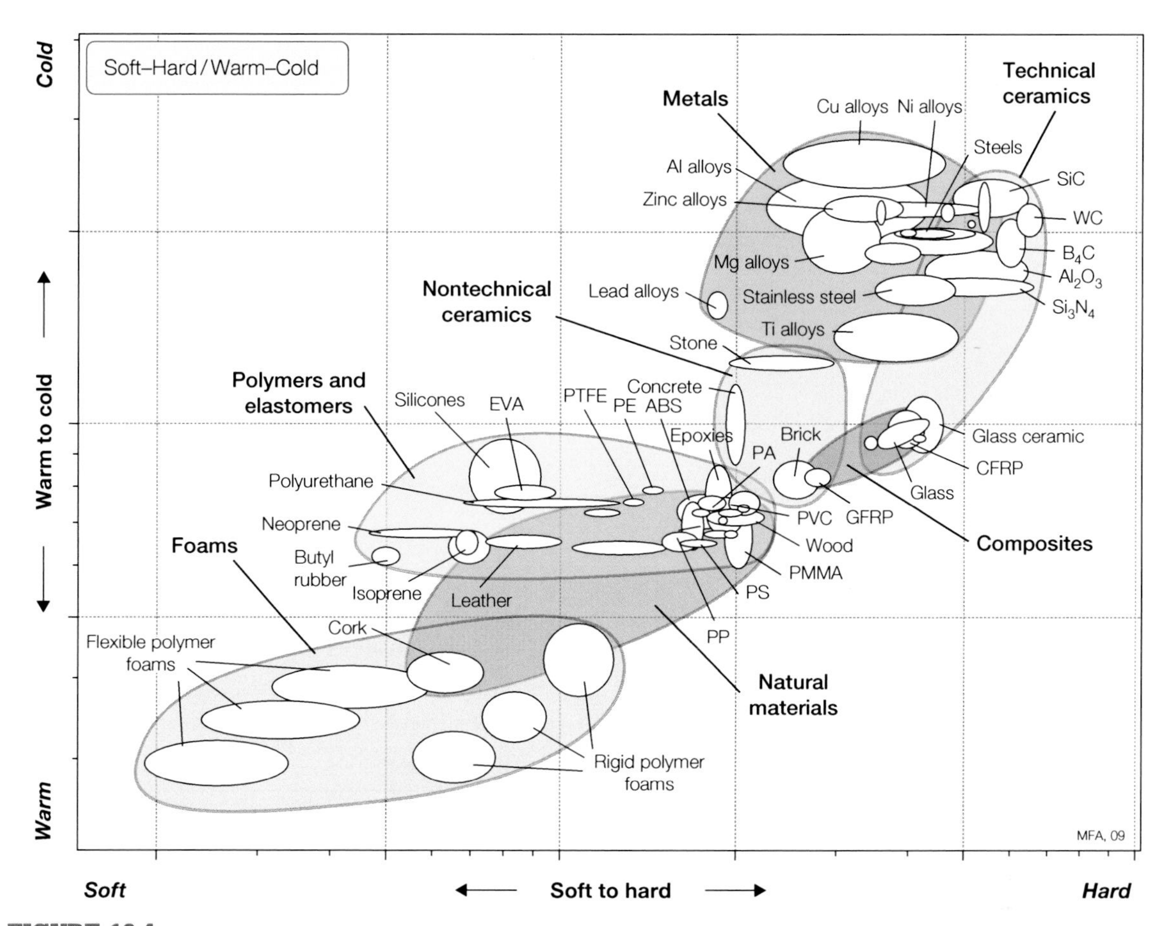

FIGURE 16.4

Tactile qualities of materials. Foams and many natural materials are soft and warm; metals, ceramics, and glasses are hard and cold. Polymers lie in between.

ranging from optical quality to completely opaque. Transparency is commonly described by a four-level ranking that uses easily understood everyday words: "opaque," "translucent," "transparent," and "water-clear." Figure 16.5 ranks the transparency of common materials. So that the data are spread in a useful way—they are plotted against cost. The cheapest materials offering optical-quality transparency ("water clarity") are glass, PS, PET, and PMMA. Epoxies can be transparent but not with water clarity. Nylons are, at best, translucent. All metals, most ceramics, and all carbon-filled or reinforced polymers are opaque.

Color can be quantified by analyzing spectra but this—from a design standpoint—doesn't help much. A more effective method is one of color

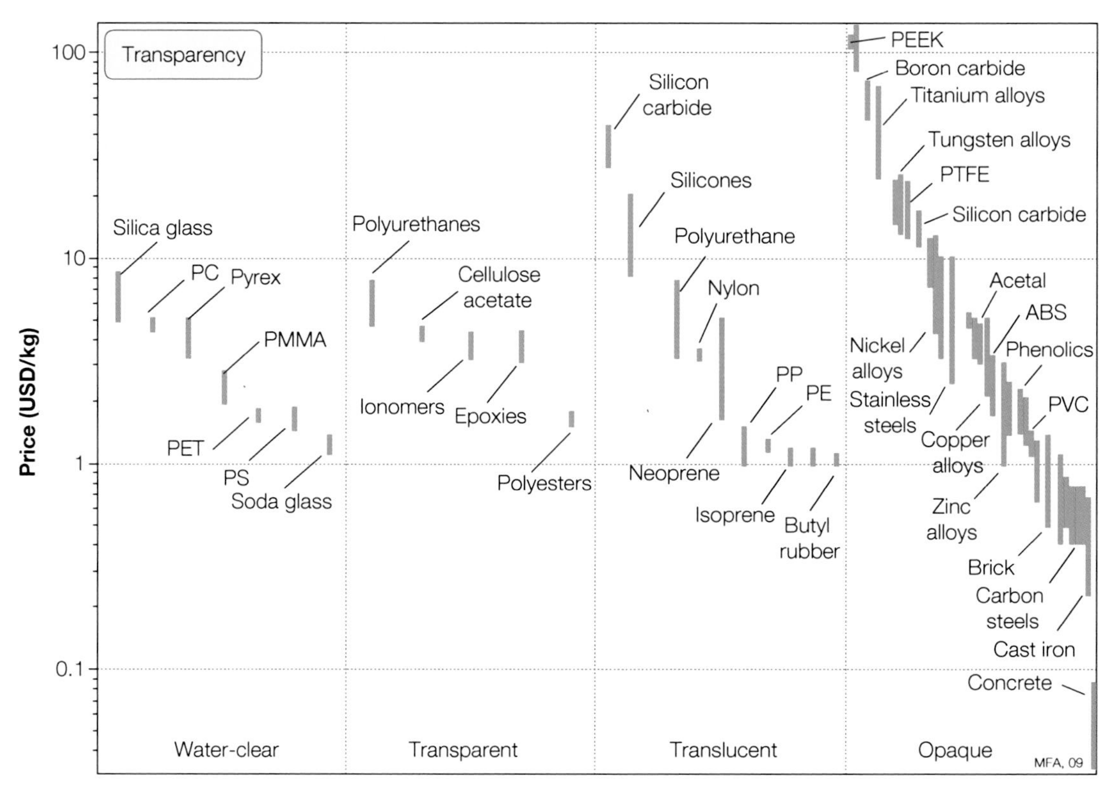

FIGURE 16.5
Here transparency is ranked on a four-point scale, from water-clear to opaque. Water-clear materials are used for windows, display cases, and lenses. Transparent and translucent materials transmit light but diffuse it in doing so. Opaque materials absorb light.

matching, using color charts such as those provided by Pantone[2]; once a match is found it can be described by the code that each color carries. Finally there is reflectivity, an attribute that depends partly on the material and partly on the state of its surface. Like transparency, it is commonly described by a ranking: dead matte, eggshell, semi-gloss, gloss, mirror.

Hearing: Pitch and brightness The frequency of sound (pitch) emitted when an object is struck relates to its material properties. A measure of this pitch $\sqrt{E/\rho}$ is used as one axis in Figure 16.6. Frequency is not the only aspect of acoustic response—the other has to do with the damping or loss coefficient η. A highly damped material sounds dull and muffled; one with low damping rings. Acoustic brightness—the inverse of damping—is used as the other axis in Figure 16.6. It groups materials that have similar acoustic behavior.

[2] Pantone (*www.pantone.com*) provides detailed advice on color selection, including color-matching charts and good descriptions of the associations and perceptions of color.

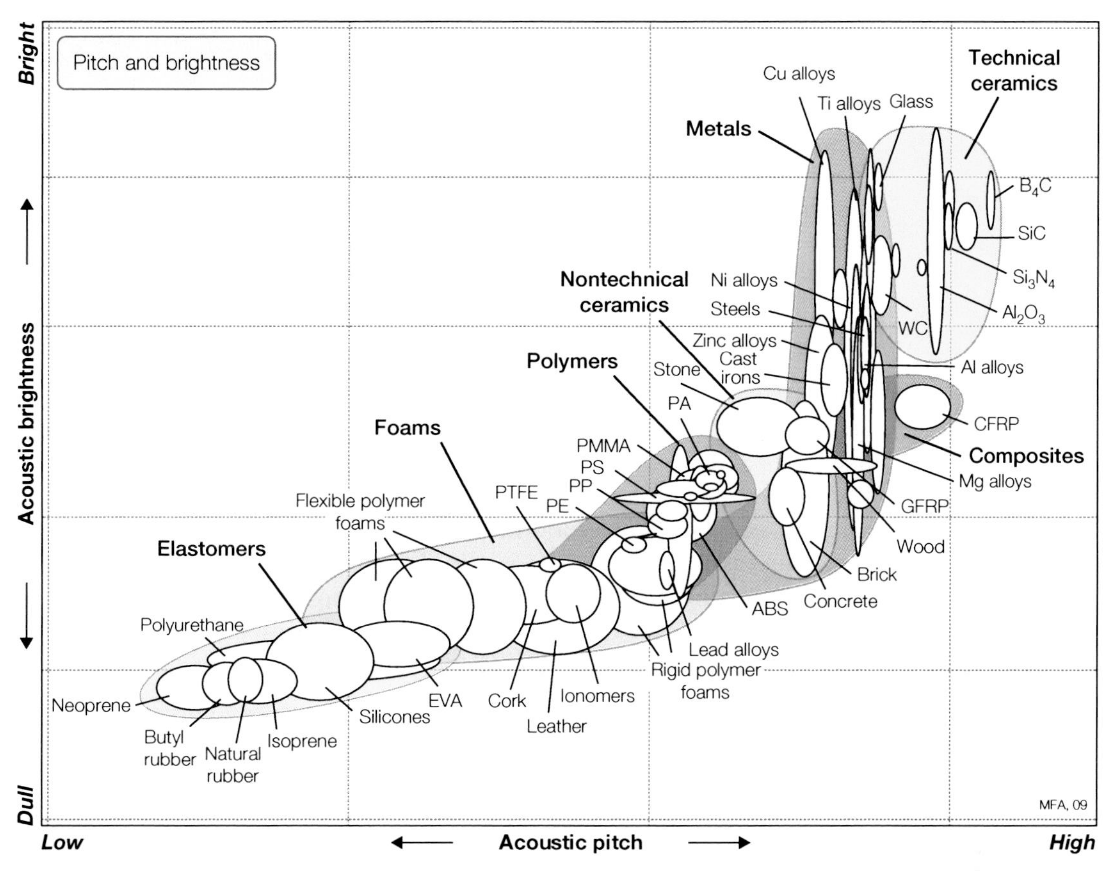

FIGURE 16.6

Acoustic properties of materials. The "ring" of a wine glass occurs because glass is an acoustically bright material with a high natural pitch; the dull "ping" of a plastic glass occurs because polymers are much less bright and—in the same shape—vibrate at a lower frequency. Materials at the top right make good bells; those at the bottom left are good for damping sound.

Bronze, glass, and steel ring when struck, and the sound they emit has—on a relative scale—a high pitch; they are used to make bells. Alumina, on this ranking, has bell-like qualities. Rubber, foams, and many polymers sound dull and, relative to metals, vibrate at low frequencies; they are used for sound damping. Lead, too, is dull and low-pitched; it is used to clad buildings for sound insulation.

The three figures show that each material class has a certain recognizable aesthetic character. Ceramics are hard, cold, high-pitched, and acoustically bright. Metals, too, are relatively hard and cold but although some (e.g., bronze) ring when struck, others (e.g., lead) are dull. Polymers and foams are most nearly like natural materials—warm, soft, low-pitched, and muffled, although some have outstanding optical clarity and almost all can be colored. But their low hardness means that they scratch easily, losing their gloss.

These qualities of a material contribute to a product's personality. The product acquires some of the attributes of the material from which it is made, an effect that designers recognize and use when seeking to create a personality. A stainless steel fascia, whether it is in a car or on a hi-fi system, has a different personality from one of CFRP, of polished wood, or of leather, and that in part is because the product has acquired some of the aesthetic qualities of the material.

Materials and the mind: Associations and perceptions

So a material certainly has aesthetic qualities—but can it be said to have a personality? At first sight, no—it only acquires one when used in a product. Like an actor, it can assume many different personalities, depending on the role it is asked to play. Wood in fine furniture suggests craftsmanship, but in a packing case, cheap utility. Glass in the lens of a camera has associations of precision engineering, but in a beer bottle, disposable packaging. Even gold, so often associated with wealth and power, has different associations when used in microcircuits: technical functionality.

But wait. The object shown in Figure 16.7 has its own somber association. It appears to be made of polished hardwood—the traditional material for such things. If you had to choose one, you would probably not think polished hardwood inappropriate. But suppose I told you it was made of *polystyrene foam*—would you feel the same? Suddenly it becomes a bin, a wastebasket, inappropriate for its dignified purpose. Materials, it seems, *do* have personality.

Expression through material Think of wood. It is a natural material with a grain that has a surface texture, pattern, color, and feel that other materials do not have. It is tactile—it is perceived as warmer than many other materials and seemingly softer. It is associated with characteristic sounds and smells. It has a tradition; it carries associations of craftsmanship. No two pieces are exactly alike; the woodworker selects the piece on which he will work for its grain and texture. Wood enhances value: The interior of very inexpensive cars is plastic, that of more costly ones is burr-walnut and calf leather. And it ages well, acquiring additional character with time—objects made of wood are more highly valued when they

FIGURE 16.7
A coffin. Wood is perceived to be appropriate for its somber, ceremonial function; plastic is perceived to be inappropriate.

are old than when they are new. There is more to this than just aesthetics; there are the makings of a personality, to be brought out by the designer, certainly, but there nonetheless.

Now consider metal. Metals are cold, clean, precise. They ring when struck. They reflect light—particularly when polished. They are accepted and trusted: Machined metal looks strong—its very nature suggests it has been *engineered*. Metals are associated with robustness, reliability, and permanence. Their strength allows slender structures—the cathedral-like space of railway stations or the span of bridges. They can be worked into flowing forms, such as intricate lace or cast, into solid shapes with elaborate, complex detail. The history of humans and of metals is intertwined—the titles "Bronze age" and "Iron age" tell you how important these metals were—and their qualities are so sharply defined that that they have become ways of describing human qualities—an iron will, a silvery voice, a golden touch, a leaden look. And, like wood, metals can age well, acquiring a patina that makes them more attractive than when newly polished—the bronze of sculptures, the pewter of mugs, the lead of roofs.

Ceramics and glass? They have an exceptionally long tradition—think of Greek pottery and Roman glass. They accept almost any color. Their total resistance to scratching, abrasion, discoloration, and corrosion gives them a certain immortality, threatened only by their brittleness. They are—or were—the materials of great craft-based industries: the glass of Venice, the porcelain of Meissen, the pottery of Wedgwood—valued at certain times more highly than silver. But at the same time ceramics and glass can be robust and functional—think of beer bottles. The transparency of glass gives it an ephemeral quality—sometimes you see it, sometimes you don't. It interacts with light, transmitting it, refracting it, reflecting it. And ceramics today have additional associations—those of advanced technology: kitchen stove-tops, high-pressure/high-temperature valves, space shuttle tiles; all are materials for extreme conditions.

And finally polymers. "A cheap, plastic imitation," it used to be said—and that is a hard reputation to live down. It derives from the early use of plastics to simulate the color and gloss of Japanese handmade pottery, much valued in Europe. Commodity polymers *are* cheap. They are easily colored and molded (that is why they are called "plastic"), making imitation easy. Unlike ceramics, their gloss is easily scratched, and their colors fade—they do not age gracefully. You can see where the reputation came from. But is it justified? No other class of material can take on as many characters as polymers: Colored, they look like ceramics; printed, they can look like wood or textile; metallized, they look exactly like metal. They can be as transparent as glass or as opaque as lead, as flexible as rubber or as stiff—when reinforced—as aluminum.

Plastics emulate precious stones in jewelry, glass in drinking glasses and glazing, wood in counter tops, velvet and fur in clothing, even grass. But despite

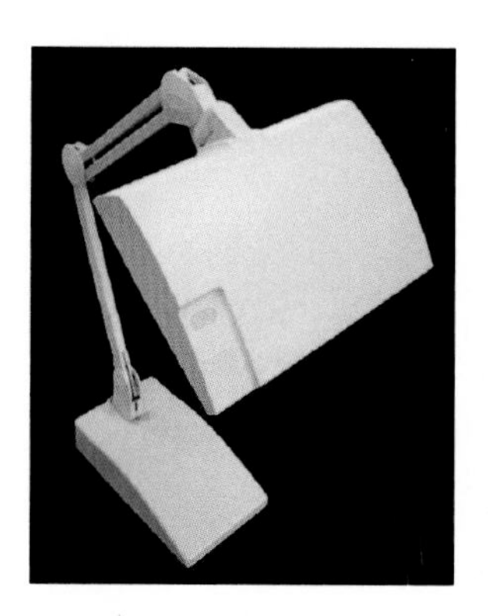

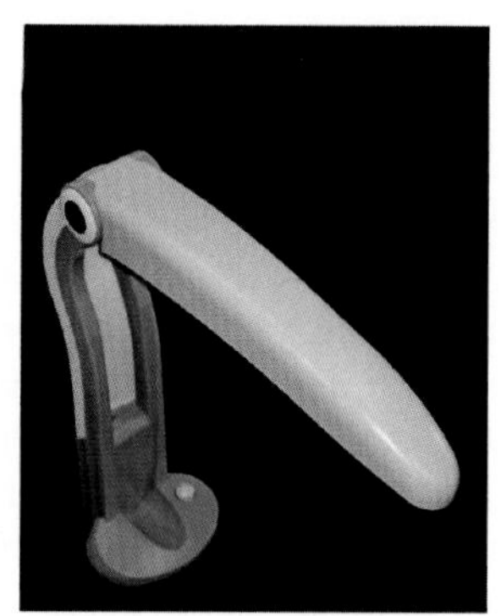

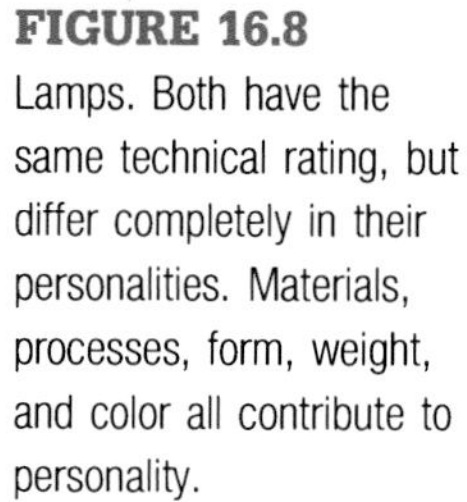

FIGURE 16.8
Lamps. Both have the same technical rating, but differ completely in their personalities. Materials, processes, form, weight, and color all contribute to personality.

this chameleon-like behavior they *do* have a certain personality: They feel warm—much warmer than metal or glass. They are adaptable—that is part of their special character—and they lend themselves particularly to brightly colored, light hearted, even humorous, design. But their very cheapness creates problems as well as benefits: Our streets, countryside, and rivers are littered with discarded plastic bags and packaging that decay only very slowly.

The ways in which material, processes, usability, and personality combine to create a product character tuned to context or "mood" are best illustrated by examples. Figure 16.8 illustrates the first. The lamp shown on the left is designed for the office. It is angular, functional, creamy grey, and heavy. Its form and color echo those of computer consoles and keyboards, creating associations of contemporary office technology. Its form and weight transmit the ideas of stability, robustness, efficiency, and fitness for a task—but for tasks in the workplace, not in the bedroom. Materials and processes have been chosen to reinforce these associations and perceptions. The enameled frame is pressed and folded sheet steel, the base weight is cast iron, and the reflector is stainless steel set in a high-impact ABS enclosure.

The lamp on the right in Figure 16.8 has the same technical rating as the left one, as well as the same functionality and usability. But there the resemblance ends. This product is not designed for the busy executive but for children (and adults that still enjoy being children), to be used in the playroom or bedroom. It has a contoured form and contrasting translucent colors, and it is light. It is made of colored acrylic in translucent and opaque grades so that the outside glows like a neon sign when it is lit. Its form is partly derived from nature, partly from cartoons and comic strips, giving it a light hearted character. I perceive it as playful, funny, cheerful, and clever—but also as eccentric and easily damaged. You may perceive it in other ways—perception is a personal thing; it depends where you are coming from. Skilled designers manipulate perception to appeal to the user group they wish to attract.

Figure 16.9 shows a second example. Here are two contrasting ways of presenting home entertainment systems. On the left: a music center aimed at successful professionals with disposable income, who are comfortable with (or addicted to) advanced technology, and for whom only the best is good enough. The linear form, the use of primitives (rectangles, circles, cylinders, cones), and the matt silver and black proclaim that this product has not just been *made*, it has been *Designed* (big *D*). The formal geometry and finish suggest precision instruments, telescopes, electron microscopes; the shapes resemble those of organ pipes (hence associations of music, of culture). The perception is that of quality cutting-edge technology, a symbol of discriminating taste. The form has much to do with these associations and perceptions, but so too do the materials: brushed aluminum, stainless steel, and black enamel—these are not materials you choose for a cuddly toy.

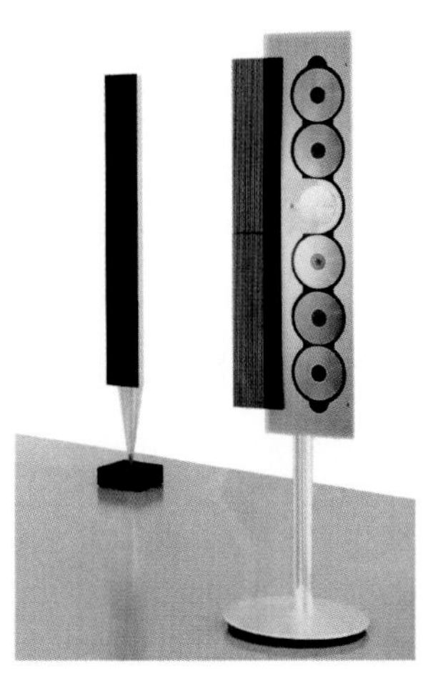

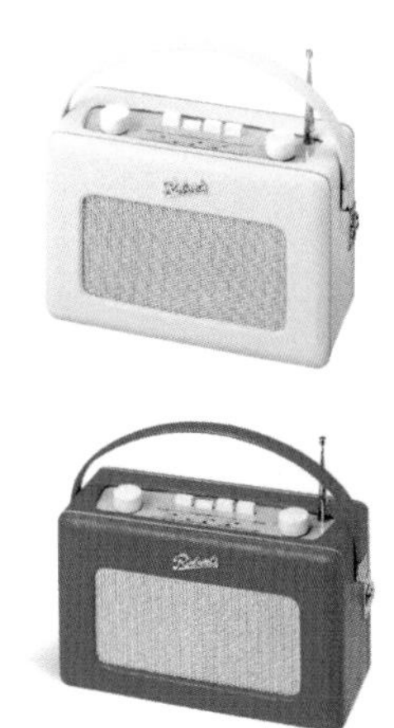

FIGURE 16.9
Consumer electronics. The system on the left is aimed at a different consumer group than the radios on the right. The personalities of each (meaning the combination of aesthetics, associations, and perceptions) have been constructed to appeal to the target group. Materials play a central role here in creating personality. (Image on left courtesy of Bang & Olufsen.)

On the right, electronics are presented in another way. This is a company that has retained market share, even increased it, by *not* changing, at least as far as appearance is concerned. (Fifty years ago I had a radio that looked exactly like this.) The context? Clearly, the home, perhaps aimed at consumers who are uncomfortable with modern technology (though the electronics in these radios is modern enough) or who simply feel

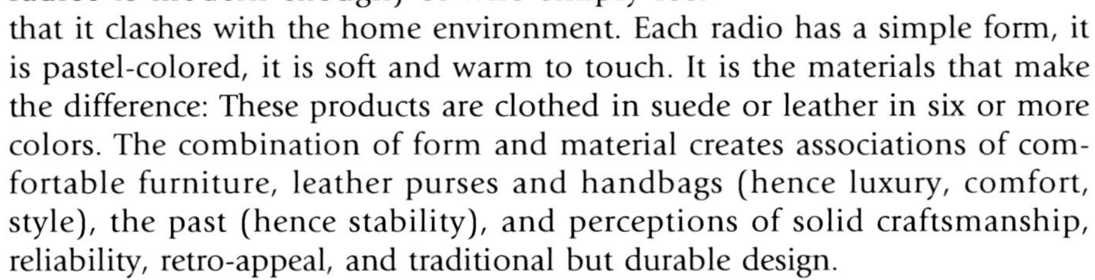

that it clashes with the home environment. Each radio has a simple form, it is pastel-colored, it is soft and warm to touch. It is the materials that make the difference: These products are clothed in suede or leather in six or more colors. The combination of form and material creates associations of comfortable furniture, leather purses and handbags (hence luxury, comfort, style), the past (hence stability), and perceptions of solid craftsmanship, reliability, retro-appeal, and traditional but durable design.

So there is a character hidden in a material even before it has been made into a recognizable form. It is a sort of embedded personality, a shy one, not always obvious, easily concealed or disguised, but one that, when appropriately manipulated, imparts its qualities to the design. It is for this reason that certain materials are so closely linked to certain *design styles*. *Style* is shorthand for a manner of design with a shared set of aesthetics, associations, and perceptions. The Early Industrial style (1800–1890)[3] embraced the technologies of the Industrial Revolution, using cast iron and steel, often elaborately decorated to give it a historical façade. The Arts and Crafts Movement from 1860 to 1910 rejected this, choosing instead natural materials and fabrics to create products with the character of traditional handcrafted quality. Art Nouveau (1890–1918), by contrast, exploited the fluid shapes and durability made possible by wrought iron and cast bronze, the warmth and textures of hardwood, and the transparency of glass to create products of flowing, organic character. Art Deco (1918–1935) extended the range of materials to include for the first time plastics (Bakelite and Catalin), allowing production of both luxury products for the rich and mass-produced products for a wider market. The simplicity and explicit character of Bauhaus

[3] The dates are, of course, approximate. Design styles do not switch on and off on specific dates; they emerge as a development of, or a reaction to, earlier styles with which they often coexist, and they merge into the styles that follow.

designs (1919–1933) are most clearly expressed by the use of chromed steel tubing, glass, and molded plywood. Plastics first reached maturity in product design in the cheeky iconoclastic character of the Pop Art style (1940–1960). Since then the range of materials has continued to increase, and their role in helping to mold product character remains.

16.5 SUMMARY AND CONCLUSIONS

What do we learn? The element of satisfaction is central to contemporary product design. It is achieved through an integration of good technical design to provide functionality, proper consideration of the needs of the user in the design of the interface, and imaginative industrial design to create a product that will appeal to the consumers at whom it is aimed.

Materials play a central role in this. Functionality is dependent on the choice of proper material and process to meet the technical requirements of the design safely and economically. Usability depends on the visual and tactile properties of materials to convey information and respond to user actions. Above all, the aesthetics, associations, and perceptions of the product are strongly influenced by the choice of the material and its processing, imbuing the product with a personality that, to a greater or lesser extent, reflects that of the material itself.

Consumers look for more than functionality in the products they purchase. In the sophisticated markets of developed nations, the "consumer durable" is a thing of the past. The challenge for the designer no longer lies in meeting functional requirements alone but in doing so in a way that also satisfies aesthetic and emotional needs. The product must carry the image and convey the meaning that the consumer seeks: timeless elegance, perhaps, or racy newness. One Japanese manufacturer goes so far as to say: "*Desire* replaces *need* as the engine of design."

Not everyone, perhaps, would wish to accept that idea. So we end with simpler words—the same ones with which we started. Good design works. Excellent design also gives pleasure. The imaginative use of materials provides it.

If you found this chapter interesting and would like to read more, you will find the ideas it contains developed more fully in the first book listed in the next section.

16.6 FURTHER READING

Ashby, M.F., & Johnson, K. (2010). *Materials and design—the art and science of materials selection in product design* (2nd ed.). Butterworth-Heinemann, ISBN 0-7506-5554-2.
A text that develops further the ideas outlined in this chapter.

Clark, P., & Freeman, J. (2000). *Design, a crash course.* The Ivy Press Ltd, Watson-Guptil Publications, BPI Communications, ISBN 0-8230-0983-1.
An entertainingly written scoot through the history of product design from 5000 BC to the present day.

Dormer, P. (1993). *Design since 1945.* Thames and Hudson, ISBN 0-500-20269-9.
A well-illustrated and inexpensive paperback documenting the influence of industrial design in furniture, appliances, and textiles—a history of contemporary design that complements the wider-ranging history of Haufe (1998), q.v.

Forty, A. (1986). *Objects of desire—design in society since 1750.* Thames and Hudson, ISBN 0-500-27412-6.
A refreshing survey of the design history of printed fabrics, domestic products, office equipment, and transport systems. The book is mercifully free of eulogies about designers and focuses on what industrial design does rather than who did it. The black-and-white illustrations are disappointing, mostly drawn from the late nineteenth or early twentieth centuries, with few examples of contemporary design.

Haufe, T. (1998). *Design, a concise history.* Laurence King Publishing (originally in German), ISBN 1-85669-134-9.
An inexpensive paperback. Probably the best introduction to industrial design for students (and anyone else). Concise, comprehensive, clear, and with intelligible layout and good, if small, color illustrations.

Jordan, P.S. (2000). *Designing pleasurable products.* Taylor and Francis, ISBN 0-748-40844-4.
Jordan, Manager of Aesthetic Research at Philips Design, argues that products today must function properly, must be usable, and must also give pleasure. Much of the book is a description of market research methods for eliciting user reactions to products.

Julier, G. (1993). *Encyclopedia of 20th-century design and designers.* Thames and Hudson, ISBN 0-500-20261-3.
A brief summary of design history with good pictures and discussions of the evolution of product designs.

Manzini, E. (1989). *The material of invention.* The Design Council, ISBN 0-85072-247-0.
Intriguing descriptions of the role of material in design and in inventions. The translation from Italian provides interesting—and often inspiring—commentary and vocabulary that is rarely used in traditional writings about materials.

McDermott, C. (1999). *The product book.* D & AD in association with Rotovison.
Fifty essays by respected designers who describe their definition of design, the role of their respective companies, and their approach to product design.

Norman, D.A. (1988). *The design of everyday things.* Doubleday, ISBN 0-385-26774-6.
A book that provides insight into the design of products with particular emphasis on ergonomics and ease of use.

CHAPTER 17

Forces for Change

CONTENTS

Materials Selection in Mechanical Design. DOI: 10.1016/B978-1-85617-663-7.00017-5

17.1 INTRODUCTION AND SYNOPSIS

If there were no forces for change, everything would stay the same. The clear message of Figure 1.1, at the start of this book, is that exactly the opposite is true: Things are changing faster now than ever before. The evolving circumstances of the world in which we live change the boundary conditions for design, and with them those for selecting materials and processes.

These changes are driven by a number of forces. First, there is the *market pull*: the demand from industry for materials that are lighter, stiffer, stronger, tougher, cheaper, and more tolerant of extremes of temperature and environment, and that offer greater functionality. Then there is the *science push*: the curiosity-driven research of materials experts in the laboratories of universities, industries, and government agencies. There is the driving force of what might be called *mega-projects*: historically, the Manhattan Project, the development of nuclear power, the space race, and various defense programs; today, one might think of alternative energy technologies, the problems of maintaining an aging infrastructure of drainage, roads, bridges, and aircraft, and the threat of terrorism. There is the trend to miniaturization while at the same time increasing the functionality of products. There is legislation regulating product safety, and there is the increased emphasis on liability established by recent legal precedent.

This chapter examines forces for change and the directions in which they push materials and their deployment. Figure 17.1 sets the scene.

17.2 MARKET PULL AND SCIENCE PUSH

Market forces and competition

The end users of materials are the manufacturing industries. They decide which materials they will purchase, and adapt their designs to make the best use of them. Their decisions are based on the nature of their products. Materials for large civil structures (which might weigh 100,000 tonnes or more) must be cheap; economy is the overriding consideration. By contrast, the cost of materials for high-tech products (sports equipment, military hardware, space projects, biomedical applications) plays a less important role: For an artificial heart valve, for instance, material cost is almost irrelevant. Performance, not economy, dictates the choice.

The market price of a product has several contributions. One is the cost of the materials of which the product is made, but there is also the cost of the research and development that went into its design, the cost of manufacture and marketing, and the perceived value associated with fashion, scarcity, lack of competition, and such like—as described in Chapter 13. When the

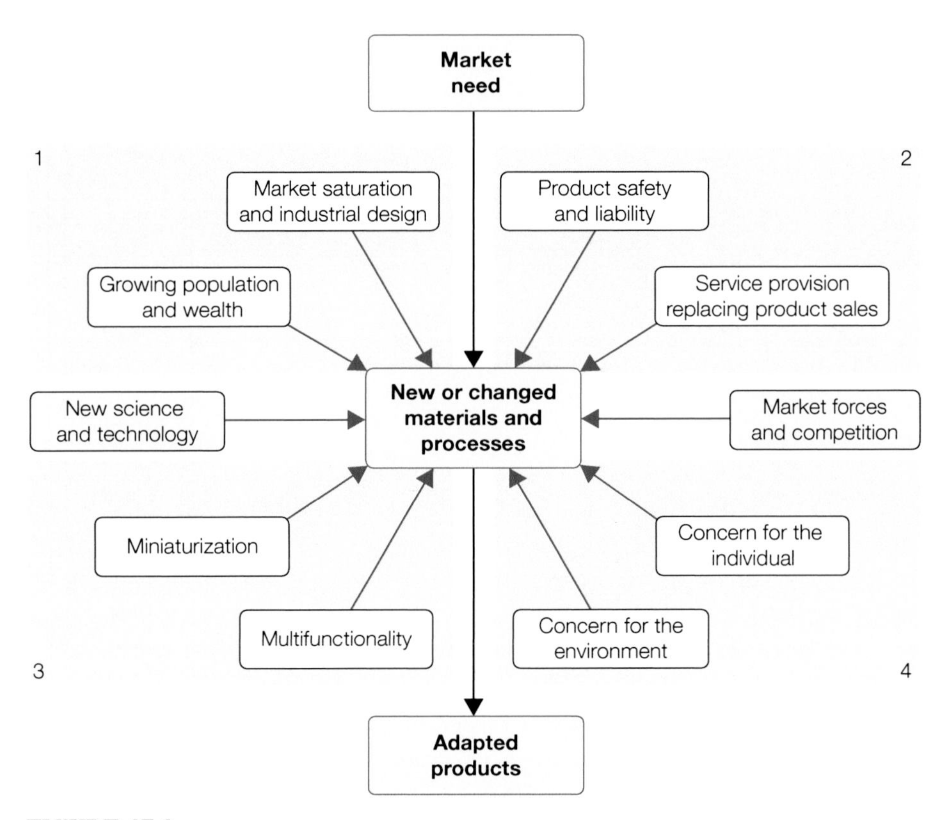

FIGURE 17.1

Forces for change. Each of the influences exerts pressure to change the choice of material and process, and stimulates efforts to develop new ones. Market pull and science push are shown on the extreme *right* and *left* of the figure.

material costs are a large part of the market value (50%, say)—that is, when the value added to the material is small—the manufacturer seeks to economize on material usage to increase profit or market share. When, by contrast, material costs are a tiny fraction of the market value (1%, say), the manufacturer seeks the materials that will most improve the performance of the product, with little concern for their cost.

With this background, examine Figures 17.2 and 17.3. The vertical axis in both is the price per unit weight ($/kg) of materials and products: It gives a common measure by which materials and products can be compared. The measure is a crude one but its great merit is that it is unambiguous, easily determined, and bears some relationship to value added. A product with a price/kg that is only two or three times that of the materials of which it is made is material-intensive and is sensitive to material costs; one with a price/kg that is 100 times that of its materials is insensitive to material

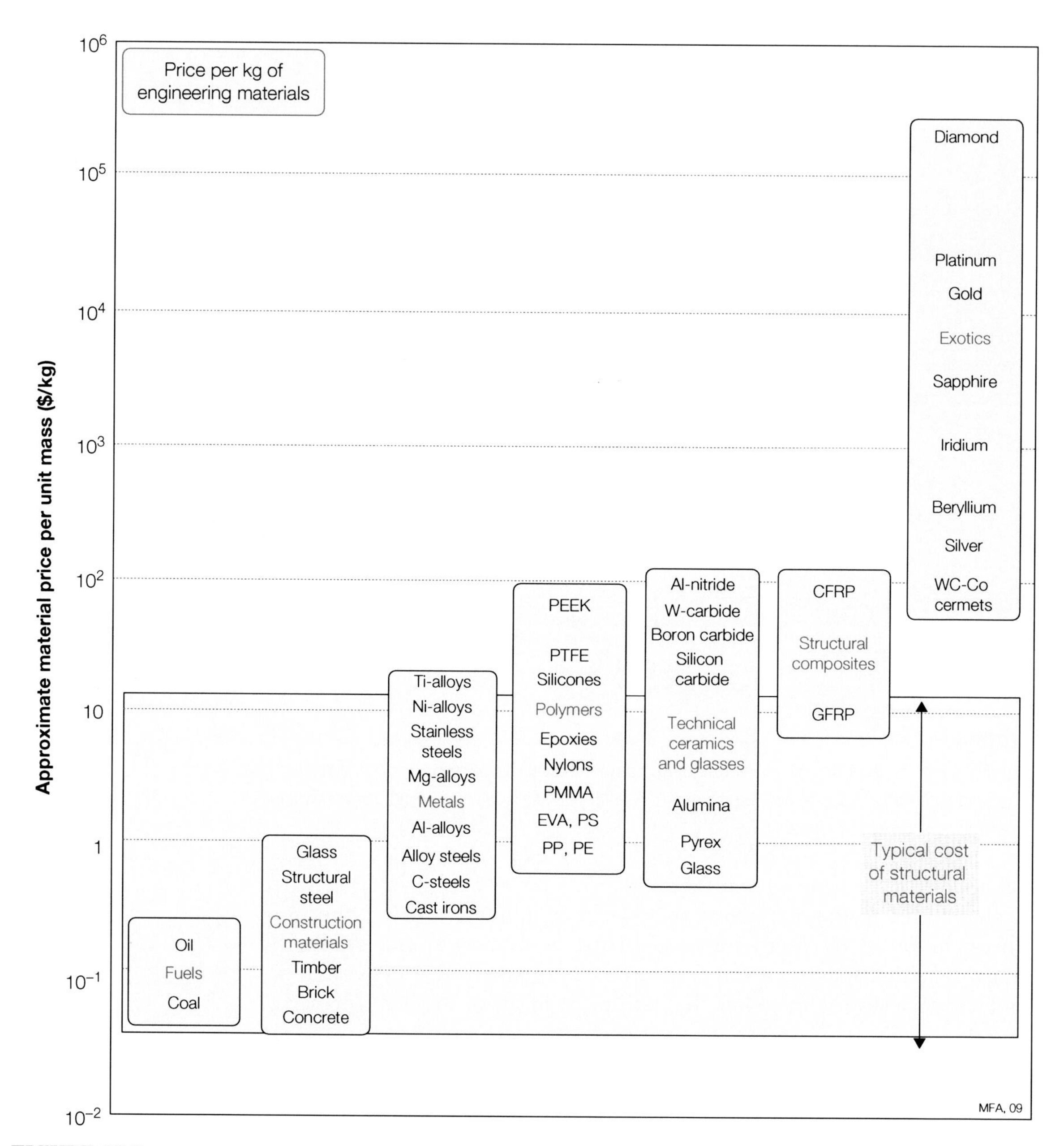

FIGURE 17.2
The price per unit weight for materials. The shaded band spans the range where the most widely used commodity materials of manufacture and construction lie.

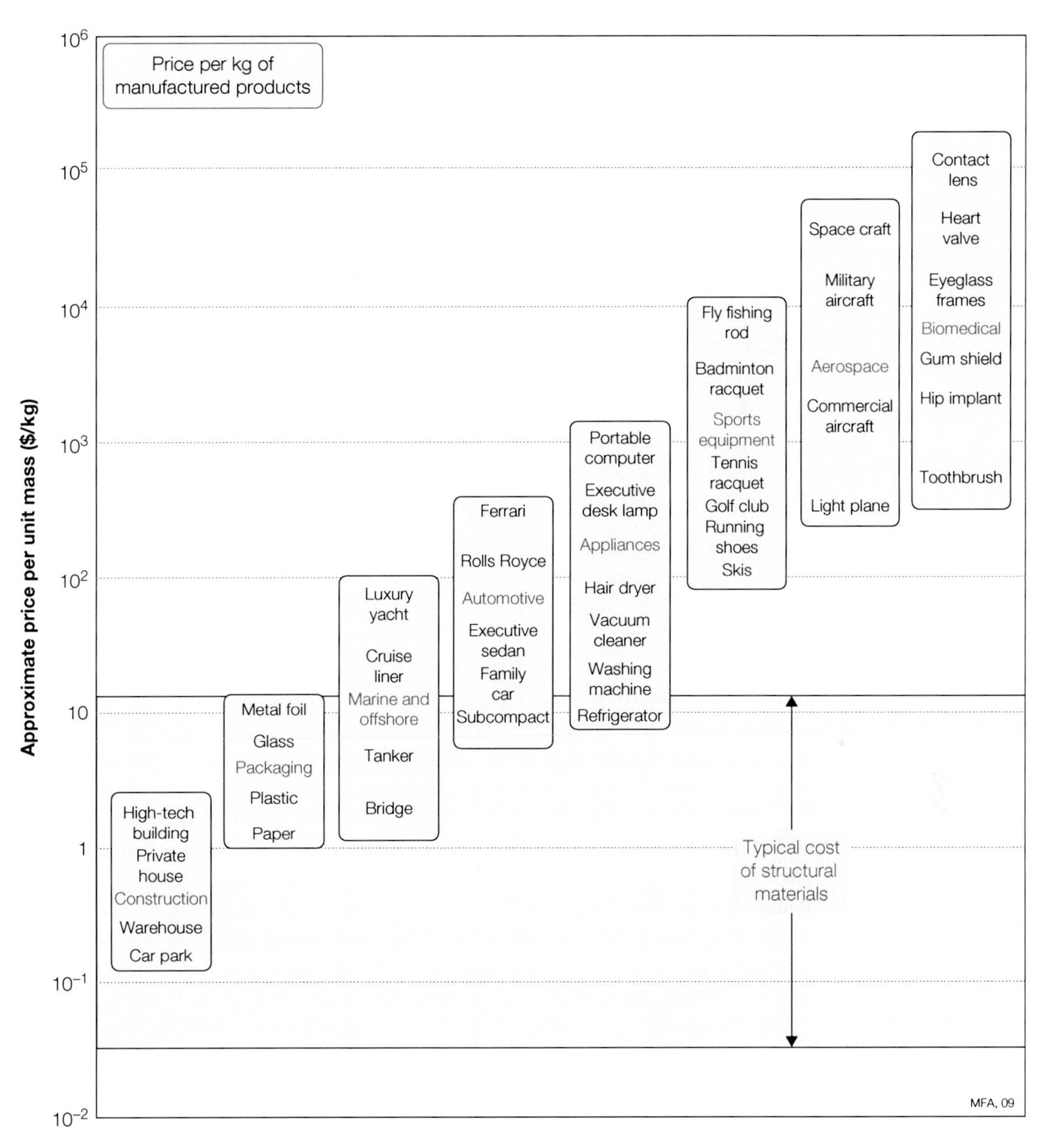

FIGURE 17.3

The price per unit weight for products. The shaded band spans the range in which lies most of the material of which they are made. Products in the shaded band are material-intensive; those above it are not.

costs, and is probably performance-driven rather than cost-driven. On this scale the price per kg of a contact lens differs from that of a glass bottle by a factor of 10^5, even though both are made of almost the same glass; the cost per kg of a heart valve differs from that of a plastic bottle by a similar factor, even though both are made of polyethylene. There is obviously something to be learned here.

Look first at the price per unit weight of materials (refer to Figure 17.2). The bulk "commodity" materials of construction and manufacture lie in the shaded band; they all cost between $0.05 and $20/kg. Construction materials like concrete, brick, timber, and structural steel lie at the lower end; high-tech materials like titanium alloys lie at the upper. Polymers span a similar range: polyethylene at the bottom, polytetrafluorethylene (PTFE) near the top. Composites lie higher, with GFRP at the bottom and CFRP at the top of the range. Engineering ceramics, at present, lie higher still, though this is changing as production increases. Only the low-volume "exotic" materials lie much above the shaded band.

The price per kg of products (refer to Figure 17.3) shows a different distribution. Eight market sectors are shown, covering much of the manufacturing industry. The shaded band on this figure spans the cost of commodity materials, exactly as on the previous figure. Sectors and their products within the shaded band have the characteristic that material cost is a major fraction of product price: about 50% in civil construction, large marine structures, and some consumer packaging, falling to perhaps 20% as the top of the band is approached (family car—around 25%). The value added in converting material to product in these sectors is relatively low, but the market volume is large. These constraints condition the choice of materials: They must meet modest performance requirements at the lowest possible cost. Associated market sectors generate a driving force for improved processing of conventional materials in order to reduce cost without a loss of performance, or to increase reliability with no increase in cost. For these sectors, incremental improvements in well-tried materials are far more important than revolutionary research findings. Slight improvements in steels, in precision manufacturing methods, or in lubrication technology are quickly assimilated and used.

The products in the upper half of the diagram are technically more sophisticated. The materials of which they are made account for less than 10%—sometimes less than 1%—of the price of the product. The value added to the material during manufacture is high. Product competitiveness is closely linked to material performance. Designers in these sectors have greater freedom in their choice of material and there is a readier acceptance of new materials with attractive property profiles. The market pull here is for performance, with cost as a secondary consideration. These smaller-volume, higher-value-added sectors drive the development of new or improved materials with enhanced performance: materials that are lighter, or stiffer, or stronger, or tougher, or expand less, or conduct better—or all of these at once.

The sectors have been ordered to form an ascending sequence, prompting the question: What does the horizontal axis measure? Many factors are involved here, one of which can be identified as "information content." The technical knowledge involved in the manufacture of a contact lens or a heart valve is clearly greater than that needed to make a water glass or a plastic bottle. The sectors on the left make few demands on the materials they employ; those on the right push materials to their limits, and at the same time demand the highest reliability. These features make them information-intensive. But there are other factors: market size, competition (or lack of it), perceived value, advertising, fashion, and taste. For this reason, the diagram should not be over-interpreted: It is a help in structuring information, but it is not a quantitative tool.

The manufacturing industry, even in times of recession, has substantial resources; it is in the interests of governments to support their needs. Market pull is ultimately the strongest force for change.

New science: Curiosity-driven research

Curiosity may kill cats, but it is the oxygen of innovative engineering. Technically advanced countries sustain the flow of new ideas by supporting research in three kinds of organization: universities, government laboratories, and industrial research laboratories. Some of the scientists and engineers working in these institutions are given the freedom to pursue ideas that may have no immediate economic objective but may evolve into the materials and manufacturing methods of future decades. Numerous now-commercial materials started in this way. Aluminum, in the time of Napoleon III, was a scientific wonder—he commissioned a set of aluminum spoons for which he paid more than for those of solid silver. Titanium, more recently, has had a similar history. Amorphous (= noncrystalline) metals, now important in transformer technology and in recording heads of tape decks, were, for years, of only academic interest. Semiconductors and superconductors did not appear in response to market forces; it took long-term curiosity-driven research, rewarded in both cases by Nobel prizes, to reveal the principles on which they are based. Polyethylene was discovered by chemists studying the effect of pressure on chemical reactions, not by the sales or marketing departments of multinational corporations. History is dotted with examples of materials and processes that have developed from the inquisitiveness of individuals.

New materials, products of fundamental research, continue to emerge. Some are already entering commercial use, for others the potential is not yet clear. Some, at least, will provide opportunities for innovation; the best may create new markets.

Monolithic ceramics, now produced in commercial quantities, offer high hardness, chemical stability, wear resistance, and resistance to extreme temperatures. Their use as substrates for microcircuits is established, their use in wear-resistant applications is growing, and their use in heat engines is being explored. The emphasis in the development of *composite materials* is shifting toward those that can support loads at higher temperatures. *Metal-matrix composites* (example: aluminum containing particles or fibers of silicon-carbide) and *intermetallic-matrix composites* (titanium-aluminide or molybdenum-disilicide containing silicon-carbide, for instance) can do this. So, potentially, can *ceramic-matrix composites* (alumina with silicon carbide fibers), though the extreme brittleness of these materials requires new design techniques. *Metallic foams*, up to 90% less dense than the parent metal, promise light, stiff sandwich structures competing with composites. *Aerogels*, ultra-low–density foams, provide exceptionally low thermal conductivities.

New *bio-materials*, designed to be implanted in the human body, have structures to which growing tissue will bond without rejection. New *polymers* that can be used at temperatures up to 350°C allow plastics to replace metals in even more applications—the inlet manifold of the automobile engine is an example. New *elastomers* are flexible but strong and tough; they allow better seals, elastic hinges, and resilient coatings. Techniques for producing *functionally graded materials* can give tailored gradients of composition and structure through a component to make it corrosion resistant on the outer surface, tough in the middle, and hard on the inner surface. *Nanostructured materials* promise unique mechanical, electrical, magnetic, and optical properties. *"Intelligent" materials* that can sense and report their condition (via embedded sensors) allow safety margins to be reduced. *Self-healing materials* have the capacity to repair service damage without human intervention.

Developments in *rapid prototyping* now permit single complex parts to be made quickly without dies or molds from a wide range of materials. *Micron-scale fabrication methods* create miniature electro-mechanical systems (MEMS). New techniques of *surface engineering* allow the alloying, coating, or heat-treating of a thin surface layer of a component, modifying its properties to enhance its performance. They include laser hardening, coatings of well-adhering polymers and ceramics, ion implantation, and even the deposition of ultra-hard carbon films with a structure and properties like those of diamond. New *adhesives* displace rivets and spot-welds; the glue-bonded automobile is a real possibility. And new techniques of *mathematical modeling* and *process control* allow much tighter control of composition and structure in manufacture, reducing cost and increasing reliability and safety.

All these and many more are now realities. They have the potential to enable new designs and to stimulate redesign of products that already have a market, increasing their market share. The designer must stay alert.

17.3 GROWING POPULATION AND WEALTH AND MARKET SATURATION

The world's population continues to grow (see sector 1 in Figure 17.1). Much of this population is also growing in wealth. This wealthier population consumes more products, and as wealth increases they want still more. But thus far the growth of the world's product-producing capacity has grown faster still, with the result that, in developed and developing countries, product markets are saturated. If you want a product—a mobile phone, a refrigerator, a car—you don't have to stand in line with the hope of getting the only one in the store or put your name on a three-year waiting list (as was the case within living memory throughout Europe). Instead, an eager salesperson guides you through the task of selecting, from an array of near-identical products with near-identical prices, the one you think you want.

This has certain consequences. One is the massive and continuing growth in consumption of energy and material resources. This, as a force for change in material selection, was the subject of Chapter 15. Another arises because in a saturated market the designer must seek new ways to attract the consumer. The more traditional reliance on engineering qualities to sell a product are replaced (or augmented) by visual qualities, associations, and carefully managed perceptions created by industrial design. This, too, influences material and process choice; it was the subject of Chapter 16.

17.4 PRODUCT LIABILITY AND SERVICE PROVISION

Legislation now requires that, if a product has a fault, it must be recalled and the fault rectified (see sector 2 in Figure 17.1). Product recalls are exceedingly expensive and they damage the company's image. The higher up the chart shown in Figure 17.3 we go, the more catastrophic a fault becomes. When you buy a packet of six ballpoint pens, it is no great tragedy if one doesn't write properly. The probability ratio 1:6, in a Gaussian distribution, is one standard deviation or "1-sigma." But would you fly in a plane designed on a "1-sigma" basis? Safety-critical systems today are designed on a "6-sigma" reliability rating, meaning that the probability that the component or assembly will fail to meet specification is less than three parts in a million.

This impacts the way materials are produced and processed. Reliability and reproducibility require sophisticated process control and that every aspect of their production is monitored and verified. Clean steels (steels with greatly reduced inclusion account), aluminum alloys with tight composition control, real-time feedback control of processing, new nondestructive testing for quality and integrity, and random sample testing all help ensure that quality is maintained.

This pressure on manufacturers to assume full-life responsibility for their products causes them to consider taking maintenance and replacement out of the hands of the consumer and doing it themselves. In this way they can monitor product use and replace the product, taking back the original, not when it is worn out but when it is optimal to recondition and redeploy it. The producers no longer sell a product; they sell a service. In common with many others, my university does not own copying machines, though we have many. Instead it has a contract with a provider to guarantee a certain copying service—the provision of "copied pages per week," you might call it.

This changes the economic boundary conditions for the designer of the copier. The objective is no longer that of building copiers with the most features at the lowest price, but rather building ones that will, over the long term, provide copies at the lowest cost per page. The result: a priority to design for ease of replacement, reconditioning, and standardization of materials and components between models. The best copier is no longer the one that lasts the longest; it's the one that is the cheapest for the maker to recondition, upgrade, or replace. This is not an isolated example. Aircraft engine makers now adopt a strategy of providing "power by the hour." It is the same strategy that makers of copiers use: They own the engines, monitor their use, and replace them on the plane at the point at which the tradeoff between cost, reliability, and power-providing capacity is optimal.

Legislation exerts other forces for change. Manufacturers must now register the large number of restricted substances that are contained in the products they make or used in their manufacture. If the level of use exceeds a threshold, the material must be replaced. The list includes the metals cadmium, lead, and mercury, along with a large number of their compounds and chemicals, many of them used in material processing. So the restrictions exert forces for change both in material choice and in the way that the material is shaped, joined, and finished.

17.5 MINIATURIZATION AND MULTIFUNCTIONALITY

Today we make increasing use of precision mechanical engineering at the micron scale (see sector 3 in Figure 17.1). CD and DVD players require the ability to position the read-head with micron precision. Hard disk drives are yet more impressive: Think of storing and retrieving a gigabyte of information on a few square millimeters. MEMS (miniature electro-mechanical systems), now universal as the accelerometers that trigger automobile airbags, rely on micron-thick cantilever beams that bend under inertial forces in a (very) sudden stop. MEMS technology promises much more. There are now video projectors that work not by projecting light through an LCD screen but by reflecting it from arrays of electrostatically actuated mirrors, each a few microns across, allowing much higher light intensities. There are even studies of mechanical microprocessors using bi-stable mechanical switching that could compete with semiconductor switching for information processing.

The smaller you make the components of a device, the greater the functionality you can pack into it. Think (as we already did in Section 8.6) of mobile phones, PDAs, MP3 players, and above all portable computers so small that they fit in a jacket pocket. All these involve components with mechanical functions: protection, positioning, actuation, sensing, casings, hard disks, keyboards, thermal and shock protection. Microengineering no longer means Swiss watches. It means almost all the devices with which we interact continuously throughout a working day.

Miniaturization imposes new demands on materials. As devices become smaller, it is mechanical failures that become design-limiting. In the past the allowable size and weight of casings, connectors, keypads, motors, and actuators gave plenty of margin of safety on bending stiffness, strength, wear rates, and corrosion rates. But none of these scale linearly with size. If we measure scale by a characterizing length L, bending stiffness scales as L^3 or L^4, strength as L^2 or L^3, and wear and corrosion, if measured by fractional loss of section per unit time, scales as $1/L$. Thus the smaller the device, the greater the demands placed on the materials of which it is made.

That is the bad news. The better news is that, since the device is smaller, the amount of material needed to make it is less. Add to that the fact that consumers want small size with powerful functionality and will pay more for it. The result: Expensive, high-performance materials that would not merit consideration for bulky products become viable for this new, miniaturized generation. Think of the armature of an airbag actuator: it weighs about 1 mg, and 1 mg of almost any material—even gold—costs, in it raw state, very little.

So, it seems, material cost no longer limits choice. It is processing that imposes restrictions. Making micron-scale products with precision presents new processing challenges. The watchmakers of Switzerland perfected tools to meet their needs, but the digital watch decimated the mechanical watch business—their methods were not transferable to the new form of miniaturization, which requires mass production at low cost. The makers of microprocessors, on the other hand, had already coped with this sort of scaling-down in nonmechanical systems—their microfabrication methods, rather than those of the watchmaker, have been adapted to make micromachines. But adopting these methods means that the menu of materials is drastically curtailed. In their present form these processes can shape silicon, silicon oxides, nitride and carbide, thin films of copper and gold, but little else. The staple materials of large-scale engineering—carbon and alloy steels, aluminum alloys, polyolefins, glass—do not appear. The current challenge is to broaden the range of microfabrication processes to allow a wider choice of materials to be manipulated.

17.6 CONCERN FOR THE ENVIRONMENT AND FOR THE INDIVIDUAL

We live in a carbon-burning economy. Energy from fossil fuels has given the developed countries high standards of living and made them wealthy. Other, much more populous countries aspire to the same standards of life and wealth and are well-advanced in acquiring them. Yet the time is approaching when burning hydrocarbons will no longer meet energy needs, and even if it could the burden it places on the natural environment will force limits to be set (see sector 4 in Figure 17.1). Short-term measures to address this and the implications these have for material selection were the subject of Chapter 15. The changes necessary to allow longer-term sustainable development will have to be much more far-reaching, changing the ways in which we manufacture, transport ourselves and our goods, and live, and having a major impact on the way materials, central to all of these, are used.

It is one thing to point out that this force for change is a potent one, but quite another to predict what its consequences will be. It is possible to point to fuel-cell technology, energy from natural sources, and safe nuclear power (all with material challenges) as ways forward, but—in the sense we are discussing here—even these are short-term solutions. Remember the figure cited in the introduction to Chapter 15: At a global growth rate of 3% per year, we will mine, process, and dispose of more "stuff" in the next 25 years than in the entire industrial history of humans. The question that has no present answer is this: How can a world population growing at about 3% per year, living in countries with economic growth rates of between 2% and 20% per year, continue to meet its aspirations with zero—or even negative—growth in consumption of energy and materials?

Wealth has another dimension: It enables resources to be allocated to health care, for which there is a growing demand in a population with an increasing lifespan. A surprisingly large fraction of the organs in a human body perform predominantly mechanical functions: teeth for cutting and grinding, bone to support structural loads, joints to allow articulation, the heart to pump blood, and arteries to carry it under pressure to the extremities of the body, muscle to actuate, skin to provide flexible protection. Aging or accidental damage frequently causes one or more of these to malfunction. Because there are mechanical, it is possible, in principle, to replace them. One way to do so is to use real body parts, but the shear number of patients requiring such treatment and the ethical and other difficulties of using human replacements drives efforts to develop artificial substitutes. Man-made teeth and bone implants, hip and knee joint replacements, artificial artery and skin, even hearts, exist already and are widely used. But at present they are exceedingly expensive, limiting their availability, and for the most part they are only crude substitutes for the real thing. There is major incentive for change here, stimulating research into affordable materials for artificial organs of all types.

Studies show that in a world that is aging the usability of many products and services is unnecessarily challenging to people, whether they are young or old, able-bodied or less able. Until recently the goal of many designers was to appeal to the youth market (meaning the 15–35 age group) as the mainstay of their business, thereby developing products that often could not be used by others. Astute manufacturers have perceived the problem and have sought to redesign their products and services to make them accessible to a wider clientele. The U.S. company OXO has seen sales of its Good Grips range of kitchen and garden tools grow by 50% year after year, and the U.K. supermarket Tesco has added thousands of Internet customers by getting rid of clever but confusing graphics on its website and instead making its services quick to download and the website simple and intuitive to navigate. In this way they include a wider spectrum of people in online shopping. There is a powerful business case for *inclusive design*—design to ensure that products and services address the needs of the widest possible audience. The social case is equally compelling—at present many people are denied use of electronic and other products because they are unable to understand how to use them or lack the motor skills to do so.

All products exclude some users, sometimes deliberately—the childproof medication bottle for instance—but more usually the exclusion is unintended and unnecessary. Many services rely on the use of products for their delivery, so unusable products deny people access to them too. Video recorders are the standing joke of poor usability, regularly derided in reviews. A much larger number of contemporary products exclude those with disability—limited sight, hearing, physical mobility or strength, or mental acuity. Countering design exclusion is a growing priority of government and is increasingly seen as important by product makers. This thinking influences material choice, changing the constraints and objectives for their selection. The use of materials that, through color or feel, communicate the function of a control, give good grip, insulate, or protect, become priorities.

17.7 SUMMARY AND CONCLUSIONS

Powerful forces drive the development of new and improved materials, encourage substitution, and modify the way in which materials are produced and used. Market forces and military prerogatives, historically the most influential, remain the strongest. The ingenuity of research scientists, too, drives change by revealing a remarkable spectrum of new materials with exciting possibilities, though the time it takes to develop and commercialize them is long: typically 15 years from laboratory to marketplace.

Today additional new drivers influence material development and use. Increasing wealth creates markets for ever more sophisticated products. The trend to products that are smaller in size, lighter in weight, and have greater functionality makes increasing demands on the mechanical properties of the materials used to make them. Greater insistence on product reliability and safety, with the manufacturer held liable for failures or malfunctions, requires materials with consistently reproducible properties and processes that are tightly controlled. Concern for the impact of industrial growth on the natural environment introduces the new objective of selecting materials in such a way as to minimize that impact. And the perception that many products exclude users by their complexity and difficulty of use drives a reassessment of the way in which they are designed and the choice of materials to make them.

The result is that products that were seen as optimal yesterday are no longer optimal today. There is always scope for reassessing designs and the choice of materials to implement them.

17.8 FURTHER READING

Defence and aerospace materials and structures. National Advisory Committee (NAC) Annual Report 2000. *www.iom3.org/foresight/nac/html.*

Ashby, M.F., Ferreira, P.J., & Schodek, D.L. (2009). *Nanomaterials, nanotechnologies and design, an introduction for engineers and architects.* Butterworth-Heinemann, ISBN 978-0-7506-8149-0. *Nanomaterials from an architectural perspective.*

Keates, S., & Clarkson, J. (2004). *Countering design exclusion—An introduction to inclusive design*. Springer-Verlag, ISBN 1-85233-769-9.

An in-depth study of design exclusion and ways to overcome it.

Starke, E.A., & Williams, J.C. (1999). Structural materials: Challenges and opportunities. *The Bridge, 29*(4).

The Bridge *is the journal of the U.S. National Academy of Engineering. This report and the one by Williams can be accessed from their website.*

van Griethuysen, A.J. (Ed.). (1987). *New applications of materials*. Scientific and Technical Publications Ltd, ISBN 0-951362-0-5.

A Dutch study of material evolution.

Williams, J.C. (1998). Engineering competitive materials. *The Bridge, 28*(4).

The Bridge *is the journal of the U.S. National Academy of Engineering. Dr. Williams has long experience in the aerospace industry and in academia.*

Appendix A: Data for Engineering Materials

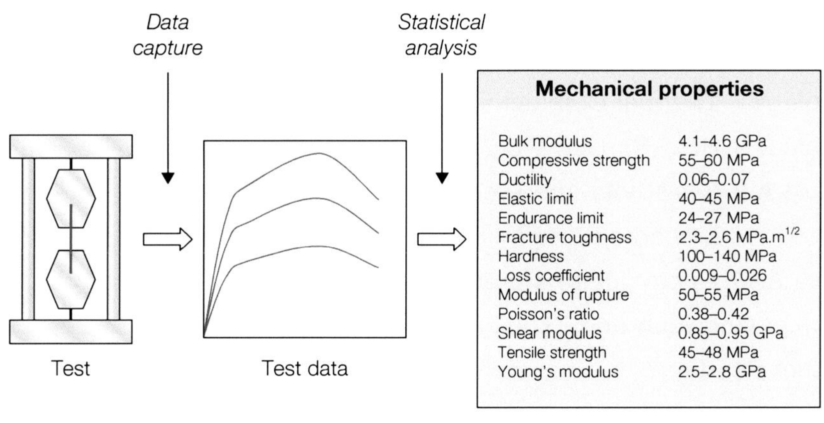

Materials Selection in Mechanical Design. DOI: 10.1016/B978-1-85617-663-7.00018-7

CONTENTS

This appendix lists the names and typical applications common to engineering materials, together with data for their properties.

Table A.1 Names and Applications: Metals and Alloys

Metals	Applications
Ferrous	
Cast irons	Automotive parts, engine blocks, machine tool structural parts, lathe beds
High carbon steels	Cutting tools, springs, bearings, cranks, shafts, railway track
Medium carbon steels	General mechanical engineering (tools, bearings, gears, shafts, bearings)
Low carbon steels	Steel structures ("mild steel")—bridges, oil rigs, ships; reinforcement for concrete; automotive parts, car body panels; galvanized sheet; packaging (cans, drums)
Low alloy steels	Springs, tools, ball bearings, automotive parts (gears connecting rods, etc.)
Stainless steels	Transport, chemical and food processing plant, nuclear plant, domestic ware (cutlery, washing machines, stoves), surgical implements, pipes, pressure vessels, liquid gas containers
Nonferrous	
Aluminum alloys	
Casting alloys	Automotive parts (cylinder blocks), domestic appliances (irons)
Non-heat-treatable alloys	Electrical conductors, heat exchangers, foil, tubes, saucepans, beverage cans, lightweight ships, architectural panels
Heat-treatable alloys	Aerospace engineering, automotive bodies and panels, lightweight structures, ships
Copper alloys	Electrical conductors and wire, electronic circuit boards, heat exchangers, boilers, cookware, coinage, sculptures
Lead alloys	Roof and wall cladding, solder, X-ray shielding, battery electrodes
Magnesium alloys	Automotive castings, wheels, general lightweight castings for transport, nuclear fuel containers; principal alloying addition to aluminum alloys
Nickel alloys	Gas turbines and jet engines, thermocouples, coinage; alloying addition to austenitic stainless steels
Titanium alloys	Aircraft turbine blades; general structural aerospace applications; biomedical implants
Zinc alloys	Die castings (automotive, domestic appliances, toys, handles); coating on galvanized steel

Table A.1 Names and Applications: Polymers and Foams

Polymers	Abbreviation	Applications
Elastomer		
Butyl rubber		Tires, seals, anti-vibration mountings, electrical insulation, tubing
Ethylene-vinyl-acetate	EVA	Bags, films, packaging, gloves, insulation, running shoes
Isoprene	IR	Tires, inner tubes, insulation, tubing, shoes
Natural rubber	NR	Gloves, tires, electrical insulation, tubing
Polychloroprene (neoprene)	CR	Wetsuits, O-rings and seals, footwear
Polyurethane elastomers	el-PU	Packaging, hoses, adhesives, fabric coating
Silicone elastomers		Electrical insulation, electronic encapsulation, medical implants
Thermoplastic		
Acrylonitrile butadiene styrene	ABS	Communication appliances, automotive interiors, luggage, toys, boats
Cellulose polymers	CA	Tool and cutlery handles, decorative trim, pens
Ionomer	I	Packaging, golf balls, blister packs, bottles
Polyamides (nylons)	PA	Gears, bearings, plumbing, packaging, bottles, fabrics, textiles, ropes
Polycarbonate	PC	Safety goggles, shields, helmets, light fittings, medical components
Polyetheretherketone	PEEK	Electrical connectors, racing car parts, fiber composites
Polyethylene	PE	Packaging, bags, squeeze tubes, toys, artificial joints
Polyethylene terephthalate	PET	Blow-molded bottles, film, audio/video tape, sails
Polymethyl methacrylate (acrylic)	PMMA	Aircraft windows, lenses, reflectors, lights, compact discs
Polyoxymethylene (acetal)	POM	Zips, domestic and appliance parts, handles
Polypropylene	PP	Ropes, garden furniture, pipes, kettles, electrical insulation, Astroturf
Polystyrene	PS	Toys, packaging, cutlery, audio cassette/CD cases
Polyurethane thermoplastics	tp-PU	Cushioning, seating, shoe soles, hoses, car bumpers, insulation
Polyvinylchloride	PVC	Pipes, gutters, window frames, packaging
Polytetrafluoroethylene (Teflon)	PTFE	Nonstick coatings, bearings, skis, electrical insulation, tape
Thermoset		
Epoxies	EP	Adhesives, fiber composites, electronic encapsulation
Phenolics	PHEN	Electrical plugs, sockets, cookware, handles, adhesives
Polyester	PEST	Furniture, boats, sporting goods
Polymer foam		
Flexible polymer foam		Packaging, buoyancy, cushioning, sponges, sleeping mats
Rigid polymer foam		Thermal insulation, sandwich panels, packaging, buoyancy

Table A.1 Names and Applications: Composites, Ceramics, Glasses, and Natural Materials

Composites	Applications
Metal polymer	
Aluminum/silicon carbide	Automotive parts, sporting goods
CFRP	Lightweight structural parts (aerospace, bike frames, sporting goods, boat hulls and oars, springs)
GFRP	Boat hulls, automotive parts, chemical plant
Ceramic	
Glass	
Borosilicate glass	Ovenware, laboratory ware, headlights
Glass ceramic	Cookware, lasers, telescope mirrors
Silica glass	High-performance windows, crucibles, high-temperature applications
Soda-lime glass	Windows, bottles, tubing, light bulbs, pottery glazes
Porous	
Brick	Buildings
Concrete	General civil engineering construction
Stone	Buildings, architecture, sculpture
Technical	
Alumina	Cutting tools, spark plugs, microcircuit substrates, valves
Aluminum nitride	Microcircuit substrates and heat sinks
Boron carbide	Lightweight armor, nozzles, dies, precision tool parts
Silicon	Microcircuits, semiconductors, precision instruments, IR windows, MEMS
Silicon carbide	High-temperature equipment, abrasive polishing grits, bearings, armor
Silicon nitride	Bearings, cutting tools, dies, engine parts
Tungsten carbide	Cutting tools, drills, abrasives
Natural	
Bamboo	Building, scaffolding, paper, ropes, baskets, furniture
Cork	Corks and bungs, seals, floats, packaging, flooring
Leather	Shoes, clothing, bags, drive belts
Wood	Construction, flooring, doors, furniture, packaging, sporting goods

Table A.2 Melting Temperature, T_m, and Glass Temperature, T_g

	T_m or T_g (°C)
Metal	
Ferrous	
Cast irons	1130–1250
High carbon steels	1289–1478
Medium carbon steels	1380–1514
Low carbon steels	1480–1526
Low alloy steels	1382–1529
Stainless steels	1375–1450
Nonferrous	
Aluminum alloys	475–677
Copper alloys	982–1082
Lead alloys	322–328
Magnesium alloys	447–649
Nickel alloys	1435–1466
Titanium alloys	1477–1682
Zinc alloys	375–492
Ceramic	
Glass	
Borosilicate glass (*)	450–602
Glass ceramic (*)	563–1647
Silica glass (*)	957–1557
Soda-lime glass (*)	442–592
Porous	
Brick	927–1227
Concrete, typical	927–1227
Stone	1227–1427
Technical	
Alumina	2004–2096
Aluminum nitride	2397–2507
Boron carbide	2372–2507
Silicon	1407–1412
Silicon carbide	2152–2500
Silicon nitride	2388–2496
Tungsten carbide	2827–2920
Composite	
Metal	
Aluminum/silicon carbide	525–627
Polymer	
CFRP	n/a
GFRP	n/a
Natural	
Bamboo (*)	77–102
Cork (*)	77–102

Table A.2 *continued*

	T_m or T_g (°C)
Natural continued	
Leather (*)	107–127
Wood, typical (longitudinal) (*)	77–102
Wood, typical (transverse) (*)	77–102
Polymer	
Elastomer	
Butyl rubber (*)	−73–−63
EVA (*)	−73–−23
Isoprene (IR) (*)	−83–−78
Natural rubber (NR) (*)	−78–−63
Neoprene (CR) (*)	−48–−43
Polyurethane elastomers (elPU) (*)	−73–−23
Silicone elastomers (*)	−123–−73
Thermoplastic	
ABS (*)	88–128
Cellulose polymers (CA) (*)	−9–107
Ionomer (I) (*)	27–77
Nylons (PA) (*)	44–56
Polycarbonate (PC) (*)	142–205
PEEK (*)	143–199
Polyethylene (PE) (*)	−25–−15
PET (*)	68–80
Acrylic (PMMA) (*)	85–165
Acetal (POM) (*)	−18–−8
Polypropylene (PP) (*)	−25–−15
Polystyrene (PS) (*)	74–110
Polyurethane thermoplastics (tpPU) (*)	120–160
PVC	75–105
Teflon (PTFE)	107–123
Thermoset	
Epoxies	n/a
Phenolics	n/a
Polyester	n/a
Polymer foam	
Flexible polymer foam (VLD) (*)	112–177
Flexible polymer foam (LD) (*)	112–177
Flexible polymer foam (MD) (*)	112–177
Rigid polymer foam (LD) (*)	67–171
Rigid polymer foam (MD) (*)	67–157
Rigid polymer foam (HD) (*)	67–171

Note: *The table lists the melting point for crystalline solids and the glass temperature for polymeric and inorganic glasses.*

(*) glass transition temperature.

n/a: not applicable (materials decompose rather than melt).

Table A.3 Density, ρ

	ρ (Mg/m^3)
Metal	
Ferrous	
Cast Irons	7.05–7.25
High carbon steels	7.8–7.9
Medium carbon steels	7.8–7.9
Low carbon steels	7.8–7.9
Low alloy steels	7.8–7.9
Stainless steels	7.6–8.1
Nonferrous	
Aluminum alloys	2.5–2.9
Copper alloys	8.93–8.94
Lead alloys	10–11.4
Magnesium alloys	1.74–1.95
Nickel alloys	8.83–8.95
Titanium alloys	4.4–4.8
Zinc alloys	4.95–7
Ceramic	
Glass	
Borosilicate glass	2.2–2.3
Glass ceramic	2.2–2.8
Silica glass	2.17–2.22
Soda-lime glass	2.44–2.49
Porous	
Brick	1.9–2.1
Concrete, typical	2.2–2.6
Stone	2.5–3
Technical	
Alumina	3.5–3.98
Aluminum nitride	3.26–3.33
Boron carbide	2.35–2.55
Silicon	2.3–2.35
Silicon carbide	3–3.21
Silicon nitride	3–3.29
Tungsten carbide	15.3–15.9
Composite	
Metal	
Aluminum/silicon carbide	2.66–2.9
Polymer	
CFRP	1.5–1.6
GFRP	1.75–1.97

Table A.3 *continued*

	ρ (Mg/m^3)
Natural	
Bamboo	0.6–0.8
Cork	0.12–0.24
Leather	0.81–1.05
Wood, typical (longitudinal)	0.6–0.8
Wood, typical (transverse)	0.6–0.8
Polymer	
Elastomer	
Butyl rubber	0.9–0.92
EVA	0.945–0.955
Isoprene (IR)	0.93–0.94
Natural rubber (NR)	0.92–0.93
Neoprene (CR)	1.23–1.25
Polyurethane elastomers (elPU)	1.02–1.25
Silicone elastomers	1.3–1.8
Thermoplastic	
ABS	1.01–1.21
Cellulose polymers (CA)	0.98–1.3
Ionomer (I)	0.93–0.96
Nylons (PA)	1.12–1.14
Polycarbonate (PC)	1.14–1.21
PEEK	1.3–1.32
Polyethylene (PE)	0.939–0.96
PET	1.29–1.4
Acrylic (PMMA)	1.16–1.22
Acetal (POM)	1.39–1.43
Polypropylene (PP)	0.89–0.91
Polystyrene (PS)	1.04–1.05
Polyurethane thermoplastics (tpPU)	1.12–1.24
PVC	1.3–1.58
Teflon (PTFE)	2.14–2.2
Thermoset	
Epoxies	1.11–1.4
Phenolics	1.24–1.32
Polyester	1.04–1.4
Polymer foam	
Flexible polymer foam (VLD)	0.016–0.035
Flexible polymer foam (LD)	0.038–0.07
Flexible polymer foam (MD)	0.07–0.115
Rigid polymer foam (LD)	0.036–0.07
Rigid polymer foam (MD)	0.078–0.165
Rigid polymer foam (HD)	0.17–0.47

Table A.4 Young's Modulus, E

	E (GPa)
Metal	
Ferrous	
Cast irons	165–180
High carbon steels	200–215
Medium carbon steels	200–216
Low carbon steels	200–215
Low alloy steels	201–217
Stainless steels	189–210
Nonferrous	
Aluminum alloys	68–82
Copper alloys	112–148
Lead alloys	12.5–15
Magnesium alloys	42–47
Nickel alloys	190–220
Titanium alloys	90–120
Zinc alloys	68–95
Ceramic	
Glass	
Borosilicate glass	61–64
Glass ceramic	64–110
Silica glass	68–74
Soda-lime glass	68–72
Porous	
Brick	15–25
Concrete, typical	25–38
Stone	20–60
Technical	
Alumina	215–413
Aluminum nitride	302–348
Boron carbide	400–472
Silicon	140–155
Silicon carbide	300–460
Silicon nitride	280–310
Tungsten carbide	600–720
Composite	
Metal	
Aluminum/silicon carbide	81–100
Polymer	
CFRP	69–150
GFRP	15–28

Table A.4 *continued*

	E (GPa)
Natural	
Bamboo	15–20
Cork	0.013–0.05
Leather	0.1–0.5
Wood, typical (longitudinal)	6–20
Wood, typical (transverse)	0.5–3
Polymer	
Elastomer	
Butyl rubber	0.001–0.002
EVA	0.01–0.04
Isoprene (IR)	0.0014–0.004
Natural rubber (NR)	0.0015–0.0025
Neoprene (CR)	0.0007–0.002
Polyurethane elastomers (elPU)	0.002–0.003
Silicone elastomers	0.005–0.02
Thermoplastic	
ABS	1.1–2.9
Cellulose polymers (CA)	1.6–2
Ionomer (I)	0.2–0.424
Nylons (PA)	2.62–3.2
Polycarbonate (PC)	2–2.44
PEEK	3.5–4.2
Polyethylene (PE)	0.621–0.896
PET	2.76–4.14
Acrylic (PMMA)	2.24–3.8
Acetal (POM)	2.5–5
Polypropylene (PP)	0.896–1.55
Polystyrene (PS)	2.28–3.34
Polyurethane thermoplastics (tpPU)	1.31–2.07
PVC	2.14–4.14
Teflon (PTFE)	0.4–0.552
Thermoset	
Epoxies	2.35–3.075
Phenolics	2.76–4.83
Polyester	2.07–4.41
Polymer foam	
Flexible polymer foam (VLD)	0.0003–0.001
Flexible polymer foam (LD)	0.001–0.003
Flexible polymer foam (MD)	0.004–0.012
Rigid polymer foam (LD)	0.023–0.08
Rigid polymer foam (MD)	0.08–0.2
Rigid polymer foam (HD)	0.2–0.48

Table A.5 Yield Strength, σ_y, and Tensile Strength, σ_{ts}

	σ_y (MPa)	σ_{ts} (MPa)
Metal		
Ferrous		
Cast irons	215–790	350–1000
High carbon steels	400–1155	550–1640
Medium carbon steels	305–900	410–1200
Low carbon steels	250–395	345–580
Low alloy steels	400–1100	460–1200
Stainless steels	170–1000	480–2240
Nonferrous		
Aluminum alloys	30–500	58–550
Copper alloys	30–500	100–550
Lead alloys	8–14	12–20
Magnesium alloys	70–400	185–475
Nickel alloys	70–1100	345–1200
Titanium alloys	250–1245	300–1625
Zinc alloys	80–450	135–520
Ceramic		
Glass		
Borosilicate glass (*)	264–384	22–32
Glass ceramic (*)	750–2129	62–177
Silica glass (*)	1100–1600	45–155
Soda-lime glass (*)	360–420	31–35
Porous		
Brick (*)	50–140	7–14
Concrete, typical (*)	32–60	2–6
Stone (*)	34–248	5–17
Technical		
Alumina (*)	690–5500	350–665
Aluminum nitride (*)	1970–2700	197–270
Boron carbide (*)	2583–5687	350–560
Silicon (*)	3200–3460	160–180
Silicon carbide (*)	1000–5250	370–680
Silicon nitride (*)	524–5500	690–800
Tungsten carbide (*)	3347–6833	370–550
Composite		
Metal		
Aluminum/silicon carbide	280–324	290–365
Polymer		
CFRP	550–1050	550–1050
GFRP	110–192	138–241

Table A.5 *continued*

	σ_y (MPa)	σ_{ts} (MPa)
Natural		
Bamboo	35–44	36–45
Cork	0.3–1.5	0.5–2.5
Leather	5–10	20–26
Wood, typical (longitudinal)	30–70	60–100
Wood, typical (transverse)	2–6	4–9
Polymer		
Elastomer		
Butyl rubber	2–3	5–10
EVA	12–18	16–20
Isoprene (IR)	20–25	20–25
Natural rubber (NR)	20–30	22–32
Neoprene (CR)	3.4–24	3.4–24
Polyurethane elastomers (elPU)	25–51	25–51
Silicone elastomers	2.4–5.5	2.4–5.5
Thermoplastic		
ABS	18.5–51	27.6–55.2
Cellulose polymers (CA)	25–45	25–50
Ionomer (I)	8.3–15.9	17.2–37.2
Nylons (PA)	50–94.8	90–165
Polycarbonate (PC)	59–70	60–72.4
PEEK	65–95	70–103
Polyethylene (PE)	17.9–29	20.7–44.8
PET	56.5–62.3	48.3–72.4
Acrylic (PMMA)	53.8–72.4	48.3–79.6
Acetal (POM)	48.6–72.4	60–89.6
Polypropylene (PP)	20.7–37.2	27.6–41.4
Polystyrene (PS)	28.7–56.2	35.9–56.5
Polyurethane thermoplastics (tpPU)	40–53.8	31–62
PVC	35.4–52.1	40.7–65.1
Teflon (PTFE)	15–25	20–30
Thermoset		
Epoxies	36–71.7	45–89.6
Phenolics	27.6–49.7	34.5–62.1
Polyester	33–40	41.4–89.6
Polymer foam		
Flexible polymer foam (VLD)	0.01–0.12	0.24–0.85
Flexible polymer foam (LD)	0.02–0.3	0.24–2.35
Flexible polymer foam (MD)	0.05–0.7	0.43–2.95
Rigid polymer foam (LD)	0.3–1.7	0.45–2.25
Rigid polymer foam (MD)	0.4–3.5	0.65–5.1
Rigid polymer foam (HD)	0.8–12	1.2–12.4

(*) *NB: For ceramics, yield strength is replaced by* compressive strength, *which is more relevant in ceramic design. Note that ceramics are of the order of 10 times stronger in compression than in tension.*

Table A.6 Fracture Toughness (plane strain), K_{IC}

	K_{IC} (MPa$\sqrt{m}$)
Metal	
Ferrous	
Cast irons	22–54
High carbon steels	27–92
Medium carbon steels	12–92
Low carbon steels	41–82
Low alloy steels	14–200
Stainless steels	62–280
Nonferrous	
Aluminum alloys	22–35
Copper alloys	30–90
Lead alloys	5–15
Magnesium alloys	12–18
Nickel alloys	80–110
Titanium alloys	14–120
Zinc alloys	10–100
Ceramic	
Glass	
Borosilicate glass	0.5–0.7
Glass ceramic	1.4–1.7
Silica glass	0.6–0.8
Soda-lime glass	0.55–0.7
Porous	
Brick	1–2
Concrete, typical	0.35–0.45
Stone	0.7–1.5
Technical	
Alumina	3.3–4.8
Aluminum nitride	2.5–3.4
Boron carbide	2.5–3.5
Silicon	0.83–0.94
Silicon carbide	2.5–5
Silicon nitride	4–6
Tungsten carbide	2–3.8
Composite	
Metal	
Aluminum/silicon carbide	15–24
Polymer	
CFRP	6.1–88
GFRP	7–23

Table A.6 *continued*

	K_{IC} (MPa$\sqrt{m}$)
Natural	
Bamboo	5–7
Cork	0.05–0.1
Leather	3–5
Wood, typical (longitudinal)	5–9
Wood, typical (transverse)	0.5–0.8
Polymer	
Elastomer	
Butyl rubber	0.07–0.1
EVA	0.5–0.7
Isoprene (IR)	0.07–0.1
Natural rubber (NR)	0.15–0.25
Neoprene (CR)	0.1–0.3
Polyurethane elastomers (elPU)	0.2–0.4
Silicone elastomers	0.03–0.5
Thermoplastic	
ABS	1.19–4.30
Cellulose polymers (CA)	1–2.5
Ionomer (I)	1.14–3.43
Nylons (PA)	2.22–5.62
Polycarbonate (PC)	2.1–4.60
PEEK	2.73–4.30
Polyethylene (PE)	1.44–1.72
PET	4.5–5.5
Acrylic (PMMA)	0.7–1.6
Acetal (POM)	1.71–4.2
Polypropylene (PP)	3–4.5
Polystyrene (PS)	0.7–1.1
Polyurethane thermoplastics (tpPU)	1.84–4.97
PVC	1.46–5.12
Teflon (PTFE)	1.32–1.8
Thermoset	
Epoxies	0.4–2.22
Phenolics	0.79–1.21
Polyester	1.09–1.70
Polymer foam	
Flexible polymer foam (VLD)	0.005–0.02
Flexible polymer foam (LD)	0.015–0.05
Flexible polymer foam (MD)	0.03–0.09
Rigid polymer foam (LD)	0.002–0.02
Rigid polymer foam (MD)	0.007–0.049
Rigid polymer foam (HD)	0.024–0.091

Note: K_{IC} is only valid for conditions under which linear elastic fracture mechanics apply (see Chapter 8). The plane-strain toughness, G_{IC}, may be estimated from $K_{IC}^2 = EG_{IC}/(1-\nu^2) \approx EG_{IC}$ (as $\nu^2 \approx 0.1$).

Table A.7 Thermal Conductivity, λ, and Thermal Expansion, α

	λ (W/m.K)	α (10^{-6}/C)
Metal		
Ferrous		
Cast irons	29–44	10–12.5
High carbon steels	47–53	11–13.5
Medium carbon steels	45–55	10–14
Low carbon steels	49–54	11.5–13
Low alloy steels	34–55	10.5–13.5
Stainless steels	11–19	13–20
Nonferrous		
Aluminum alloys	76–235	21–24
Copper alloys	160–390	16.9–18
Lead alloys	22–36	18–32
Magnesium alloys	50–156	24.6–28
Nickel alloys	67–91	12–13.5
Titanium alloys	5–12	7.9–11
Tungsten alloys	100–142	4–5.6
Zinc alloys	100–135	23–28
Ceramic		
Glass		
Borosilicate glass	1–1.3	3.2–4.0
Glass ceramic	1.3–2.5	1–5
Silica glass	1.4–1.5	0.55–0.75
Soda-lime glass	0.7–1.3	9.1–9.5
Porous		
Brick	0.46–0.73	5–8
Concrete, typical	0.8–2.4	6–13
Stone	5.4–6.0	3.7–6.3
Technical		
Alumina	30–38.5	7–10.9
Aluminum nitride	80–200	4.9–6.2
Boron carbide	40–90	3.2–3.4
Silicon	140–150	2.2–2.7
Silicon carbide	115–200	4.0–5.1
Silicon nitride	22–30	3.2–3.6
Tungsten carbide	55–88	5.2–7.1
Composite		
Metal		
Aluminum/silicon carbide	180–160	15–23
Polymer		
CFRP	1.28–2.6	1–4
GFRP	0.4–0.55	8.6–33

Table A.7 *continued*

	λ (W/m.K)	α (10^{-6}/C)
Natural		
Bamboo	0.1–0.18	2.6–10
Cork	0.035–0.048	130–230
Leather	0.15–0.17	40–50
Wood, typical (longitudinal)	0.31–0.38	2–11
Wood, typical (transverse)	0.15–0.19	32–42
Polymer		
Elastomer		
Butyl rubber	0.08–0.1	120–300
EVA	0.3–0.4	160–190
Isoprene (IR)	0.08–0.14	150–450
Natural rubber (NR)	0.1–0.14	150–450
Neoprene (CR)	0.08–0.14	575–610
Polyurethane elastomers	0.28–0.3	150–165
Silicone elastomers	0.3–1.0	250–300
Thermoplastic		
ABS	0.19–0.34	84.6–234
Cellulose polymers (ca)	0.13–0.3	150–300
Ionomer (I)	0.24–0.28	180–306
Nylons (PA)	0.23–0.25	144–150
Polycarbonate (PC)	0.19–0.22	120–137
PEEK	0.24–0.26	72–194
Polyethylene (PE)	0.40–0.44	126–198
PET	0.14–0.15	114–120
Acrylic (PMMA)	0.08–0.25	72–162
Acetal (POM)	0.22–0.35	76–201
Polypropylene (PP)	0.11–0.17	122–180
Polystyrene (PS)	0.12–0.13	90–153
Polyurethane thermoplastics	0.23–0.24	90–144
PVC	0.15–0.29	100–150
Teflon (PTFE)	0.24–0.26	126–216
Thermoset		
Epoxies	0.18–0.5	58–117
Phenolics	0.14–0.15	120–125
Polyester	0.28–0.3	99–180
Polymer foam		
Flexible polymer foam (VLD)	0.036–0.048	120–220
Flexible polymer foam (LD)	0.04–0.06	115–220
Flexible polymer foam (MD)	0.04–0.08	115–220
Rigid polymer foam (LD)	0.023–0.04	20–80
Rigid polymer foam (MD)	0.027–0.038	20–75
Rigid polymer foam (HD)	0.034–0.06	22–70

Table A.8 Heat Capacity, C_p

	C_p (J/kg.K)
Metal	
Ferrous	
Cast irons	439–495
High carbon steels	440–510
Medium carbon steels	440–510
Low carbon steels	460–505
Low alloy steels	410–530
Stainless steels	450–530
Nonferrous	
Aluminum alloys	857–990
Copper alloys	372–388
Lead alloys	122–145
Magnesium alloys	955–1060
Nickel alloys	452–460
Titanium alloys	520–600
Zinc alloys	405–535
Ceramic	
Glass	
Borosilicate glass	760–800
Glass ceramic	600–900
Silica glass	680–730
Soda-lime glass	850–950
Porous	
Brick	750–850
Concrete, typical	835–1050
Stone	840–920
Technical	
Alumina	790–820
Aluminum nitride	780–820
Boron carbide	840–1290
Silicon	668–715
Silicon carbide	663–800
Silicon nitride	670–800
Tungsten carbide	184–292
Composite	
Metal	
Aluminum/silicon carbide	800–900
Polymer	
CFRP	900–1040
GFRP	1000–1200

Table A.8 *continued*

	C_p (J/kg.K)
Natural	
Bamboo	1660–1710
Cork	1900–2100
Leather	1530–1730
Wood, typical	1660–1710
Aluminum nitride	780–820
Polymer	
Elastomer	
Butyl rubber	1800–2500
EVA	2000–2200
Isoprene (IR)	1800–2500
Natural rubber (NR)	1800–2500
Neoprene (CR)	2000–2200
Polyurethane elastomers (elPU)	1650–1700
Silicone elastomers	1050–1300
Thermoplastic	
ABS	1390–1920
Cellulose polymers (CA)	1390–1670
Ionomer (I)	1810–1890
Nylons (PA)	1600–1660
Polycarbonate (PC)	1530–1630
PEEK	1440–1500
Polyethylene (PE)	1810–1880
PET	1420–1470
Acrylic (PMMA)	1490–1610
Acetal (POM)	1360–1430
Polypropylene (PP)	1870–1960
Polystyrene (PS)	1690–1760
Polyurethane thermoplastics (tpPU)	1550–1620
PVC	1360–1440
Teflon (PTFE)	1010–1050
Thermoset	
Epoxies	1490–2000
Phenolics	1470–1530
Polyester	1510–1570
Polymer foam	
Flexible polymer foam (VLD)	1750–2260
Flexible polymer foam (LD)	1750–2260
Flexible polymer foam (MD)	1750–2260
Rigid polymer foam (LD)	1120–1910
Rigid polymer foam (MD)	1120–1910
Rigid polymer foam (HD)	1120–1910

Table A.9 Resistivity and Dielectric Constant

	Resistivity (μohm.cm)	Dielectric Constant
Metal		
Ferrous		
Cast irons	49–56	–
High carbon steels	17–20	–
Medium carbon steels	15–22	–
Low carbon steels	15–20	–
Low alloy steels	15–35	–
Stainless steels	64–107	–
Nonferrous		
Aluminum alloys	2.5–5.0	–
Copper alloys	1.7–24	–
Lead alloys	15–22	–
Magnesium alloys	4.2–15	–
Nickel alloys	6–114	–
Titanium alloys	100–170	–
Tungsten alloys	10.2–14	–
Zinc alloys	5.4–7.2	–
Ceramic		
Glass		
Borosilicate glass	3×10^{21}–3×10^{22}	4.6–6.0
Glass ceramic	2×10^{19}–1×10^{21}	5.3–6.2
Silica glass	1×10^{23}–1×10^{27}	3.7–3.9
Soda-lime glass	8×10^{17}–8×10^{18}	7.0–7.6
Porous		
Brick	1×10^{14}–3×10^{16}	7.0–10
Concrete, typical	1.8×10^{12}–1.8×10^{13}	8.0–12
Stone	1×10^{8}–1×10^{14}	6.0–18
Technical		
Alumina	1×10^{20}–1×10^{22}	6.5–6.8
Aluminum nitride	1×10^{19}–1×10^{21}	8.3–9.3
Boron carbide	1×10^{5}–1×10^{7}	4.8–8.0
Silicon	1×10^{6}–1×10^{12}	11–12
Silicon carbide	1×10^{9}–1×10^{12}	6.3–9.0
Silicon nitride	1×10^{20}–1×10^{21}	7.9–8.1
Tungsten carbide	20–100	–
Composite		
Metal		
Aluminum/silicon carbide	5–12	–
Polymer		
CFRP	1.7×10^{5}–1×10^{6}	–
GFRP	1×10^{16}–2×10^{22}	4.2–5.2

Table A.9 *continued*

	Resistivity (μohm.cm)	Dielectric Constant
Natural		
Bamboo	6×10^{13}–7×10^{14}	5–7
Cork	1×10^{9}–1×10^{11}	6–8
Leather	1×10^{8}–1×10^{10}	5–10
Wood, typical (longitudinal)	6×10^{13}–2×10^{14}	5–6
Wood, typical (transverse)	2×10^{14}–7×10^{14}	5–6
Polymer		
Elastomer		
Butyl rubber	1×10^{15}–1×10^{16}	2.8–3.2
EVA	3.2×10^{21}–1×10^{22}	2.9–3.0
Isoprene (IR)	1×10^{15}–1×10^{16}	2.5–3.0
Natural rubber (NR)	1×10^{15}–1×10^{16}	3.0–4.5
Neoprene (CR)	1×10^{19}–1×10^{23}	6.7–8.0
Polyurethane elastomers	1×10^{18}–1×10^{22}	5.0–9.0
Silicone elastomers	3×10^{19}–1×10^{22}	2.9–4.0
Thermoplastic		
ABS	3×10^{21}–3×10^{22}	2.8–3.2
Cellulose polymers (ca)	1×10^{17}–4×10^{20}	3.0–5.0
Ionomer (I)	3×10^{21}–3×10^{22}	2.2–2.4
Nylons (PA)	1.5×10^{19}–1.1×10^{20}	3.7–3.9
Polycarbonate (PC)	1×10^{20}–1×10^{21}	3.1–3.3
PEEK	3×10^{21}–3×10^{22}	3.1–3.3
Polyethylene (PE)	3×10^{22}–3×10^{24}	2.2–2.4
PET	3×10^{20}–3×10^{21}	3.5–3.7
Acrylic (PMMA)	3×10^{23}–3×10^{24}	3.2–3.4
Acetal (POM)	3×10^{20}–3×10^{21}	3.6–4.0
Polypropylene (PP)	3×10^{22}–3×10^{23}	2.1–2.3
Polystyrene (PS)	1×10^{25}–1×10^{27}	3.0–3.3
Polyurethane thermoplastics	3×10^{18}–3×10^{19}	6.6–7.1
PVC	1×10^{20}–1×10^{22}	3.1–4.4
Teflon (PTFE)	3×10^{23}–3×10^{24}	2.1–2.2
Thermoset		
Epoxies	1×10^{20}–6×10^{21}	3.4–5.7
Phenolics	3×10^{18}–3×10^{19}	4.0–6.0
Polyester	3×10^{18}–3×10^{19}	2.8–3.3
Polymer foam		
Flexible polymer foam (VLD)	1×10^{20}–1×10^{23}	1.1–1.15
Flexible polymer foam (LD)	1×10^{20}–1×10^{23}	1.15–1.2
Flexible polymer foam (MD)	1×10^{20}–1×10^{23}	1.2–1.3
Rigid polymer foam (LD)	1×10^{17}–1×10^{21}	1.04–1.1
Rigid polymer foam (MD)	3×10^{16}–3×10^{20}	1.1–1.19
Rigid polymer foam (HD)	1×10^{16}–1×10^{20}	1.2–1.45

Table A.10 Embodied Energy and CO_2 Footprint

	Energy (MJ/kg)	CO_2 (kg/kg)
Polymer		
Elastomer		
Butyl rubber	95–120	3.6–4.2
EVA	87–96	2.9–3.2
Isoprene (IR)	77–85	2.2–2.4
Natural rubber (NR)	62–70	1.5–1.6
Neoprene (CR)	96–106	3.5–3.9
Polyurethane elastomers	109–120	4.5–4.9
Silicone elastomers	152–168	8.2–9.0
Thermoplastic		
ABS	91–102	3.3–3.6
Cellulose polymers (ca)	108–119	4.4–4.9
Ionomer (I)	102–112	4.0–4.4
Nylons (PA)	121–135	5.5–5.6
Polycarbonate (PC)	105–116	5.4–5.9
PEEK	223–246	13.0–14.0
Polyethylene (PE)	77–85	2.0–2.2
PET	80–88	2.2–2.5
Acrylic (PMMA)	94–110	3.4–3.8
Acetal (POM)	100–110	3.8–4.2
Polypropylene (PP)	85–105	2.6–2.8
Polystyrene (PS)	86–99	2.7–3.0
Polyurethane thermoplastics	113–125	4.6–5.3
PVC	68–95	2.2–2.6
Teflon (PTFE)	145–160	7.1–7.8
Thermoset		
Epoxies	105–130	4.2–4.6
Phenolics	86–95	2.8–3.1
Polyester	84–93	2.7–3.0
Polymer foam		
Flexible polymer foam (VLD)	103–115	4.0–4.8
Flexible polymer foam (LD)	103–115	4.0–4.8
Flexible polymer foam (MD)	103–115	4.0–4.8
Rigid polymer foam (LD)	105–110	3.5–4.0
Rigid polymer foam (MD)	105–110	3.5–4.0
Rigid polymer foam (HD)	105–110	3.5–4.0
Metal		
Ferrous		
Cast irons	16–18	0.97–1.1
High carbon steels	29–35	2.2–2.8
Medium carbon steels	29–35	2.2–2.8
Low carbon steels	29–35	2.2–2.8
Low alloy steels	32–38	2.2–2.8
Stainless steels	77–85	4.9–5.4

Table A.10 *continued*

	Energy (MJ/kg)	CO_2 (kg/kg)
Nonferrous		
Aluminum alloys	200–220	11.2–12.8
Copper alloys	68–74	4.9–5.6
Lead alloys	53–58	3.3–3.7
Magnesium alloys	360–400	22.4–24.8
Nickel alloys	127–140	7.9–9.2
Titanium alloys	600–740	7.9–11
Tungsten alloys	313–346	19.7–21.8
Zinc alloys	70–75	3.7–4.0
Ceramic		
Glass		
Borosilicate glass	24–26	1.3–1.4
Glass ceramic	36–40	2.0–2.2
Silica glass	30–33	1.6–1.8
Soda-lime glass	14–17	0.7–1.0
Porous		
Brick	2.2–3.5	0.2–0.23
Concrete, typical	1–1.3	0.13–0.15
Stone	5.5–6.4	0.3–0.34
Technical		
Alumina	50–55	2.7–3.0
Aluminum nitride	209–231	11.3–12.5
Boron carbide	153–169	8.3–9.1
Silicon (commercial grade)	56–62	3.8–4.2
Silicon carbide	70–78	6.3–6.9
Silicon nitride	116–128	3.1–3.4
Tungsten carbide	82–91	4.4–4.9
Composite		
Metal		
Aluminum/silicon carbide	250–300	7.5–8.3
Polymer		
CFRP	259–286	16.1–18.3
GFRP	107–118	7.5–8.3
Natural		
Bamboo	4–6	0.3–0.33
Cork	4–5	0.19–0.22
Leather	102–113	4.1–4.5
Wood, typical (longitudinal)	7–8	0.4–0.46
Wood, typical (transverse)	7–8	0.45–0.49

Table A.11 Approximate Material Prices, C_m

	C_m ($/kg)
Metal	
Ferrous	
Cast irons	0.57–0.69
High carbon steels	0.72–0.8
Medium carbon steels	0.67–0.74
Low carbon steels	0.63–0.7
Low alloy steels	0.81–0.89
Stainless steels	6.5–7.2
Nonferrous	
Aluminum alloys	1.5–1.7
Copper alloys	3.2–3.5
Lead alloys	3.2–3.5
Magnesium alloys	5.2–5.7
Nickel alloys	33–37
Titanium alloys	67–74
Zinc alloys	1.2–1.3
Ceramic	
Glass	
Borosilicate glass	4.1–6.2
Glass ceramic	2.1–12
Silica glass	6.2–10
Soda-lime glass	1.4–1.7
Porous	
Brick	0.62–1.7
Concrete, typical	0.041–0.062
Stone	0.4–0.6
Technical	
Alumina	18–27
Aluminum nitride	100–179
Boron carbide	60–89
Silicon	9.1–15
Silicon carbide	15–21
Silicon nitride	35–54
Tungsten carbide	19–29
Composite	
Metal	
Aluminum/silicon carbide	6.2–8.3
Polymer	
CFRP	40–44
GFRP	19–21

Table A.11 *continued*

	C_m ($/kg)
Polymer *continued*	
SMC	5.8–6.4
DMC	4.7–5.2
Natural	
Bamboo	1.4–2.1
Cork	2.8–14
Leather	17–21
Wood	0.8–1
Polymer	
Elastomer	
Butyl rubber	3.7–4.1
EVA	1.5–1.7
Isoprene (IR)	2.9–3.2
Natural rubber (NR)	1.7–1.9
Neoprene (CR)	5.1–5.6
Polyurethane elastomers (elPU)	4.9–5.3
Silicone elastomers	13–14
Thermoplastic	
ABS	2.1–2.5
Cellulose polymers (CA)	3.9–4.3
Ionomer (I)	2.8–3.7
Nylons (PA)	3.3–3.6
Polycarbonate (PC)	3.7–4.0
PEEK	100–120
Polyethylene (PE)	1.3–1.4
PET	1.5–1.7
Acrylic (PMMA)	2.6–2.8
Acetal (POM)	3.3–3.8
Polypropylene (PP)	1.2–1.3
Polystyrene (PS)	1.4–1.6
Polyurethane thermoplastics (tpPU)	4.1–5.6
PVC	0.93–1.0
Teflon (PTFE)	11–21
Thermoset	
Epoxies	2.6–2.8
Phenolics	1.7–1.9
Polyester	4.1–4.5
Polymer foam	
Flexible polymer foam (VLD)	2.9–3.1
Flexible polymer foam (LD)	2.9–3.1
Flexible polymer foam (MD)	2.9–3.1
Rigid polymer foam (LD)	12–25
Rigid polymer foam (MD)	12–25
Rigid polymer foam (HD)	12–25

WAYS OF CHECKING AND ESTIMATING DATA

The value of a database of material properties depends on its precision and its completeness—in short, on its quality. One way of maintaining or enhancing its quality is to subject its contents to validating procedures. The property range checks and dimensionless correlations, described below, provide powerful tools for doing this. The same procedures fill a second function: providing estimates for missing data, essential when no direct measurements are available.

Property ranges

Each property of a given class of materials has a characteristic *range*. A convenient way of presenting the information is as a table in which a low (L) and a high (H) value are stored, identified by the material family and class. An example listing Young's modulus, E, is shown in Table A.12, in which E_L is the lower limit and E_H the upper limit.

All properties have characteristic ranges like these. The range becomes narrower if the classes are made more restrictive. For purposes of checking and estimation, described in a moment, it is helpful to break down the family of *metals* into classes of cast irons, steels, aluminum alloys, magnesium alloys, titanium alloys, copper alloys, and so on. Similar subdivisions for polymers (thermoplastics, thermosets, elastomers) and for ceramics and

Table A.12 Ranges of Young's Modulus, E, for Broad Material Classes

Material class	E_L (GPa)	E_H (GPa)
All solids	0.00001	1,000
Classes of solid		
Metals: ferrous	70	220
Metals: nonferrous	4.6	570
Technical ceramics*	91	1,000
Glasses	47	83
Polymers: thermoplastic	0.1	4.1
Polymers: thermosets	2.5	10
Polymers: elastomers	0.0005	0.1
Polymeric foams	0.00001	2
Composites: metal-matrix	81	180
Composites: polymer-matrix	2.5	240
Woods: parallel to grain	1.8	34
Woods: perpendicular to grain	0.1	18

* *Technical ceramics are dense, monolithic ceramics such as SiC, Al_2O_3, ZrO_2, and so on.*

glasses (engineering ceramics, whiteware, silicate glasses, minerals) increase resolution here also.

Correlations between material properties

Materials that are stiff have high melting points. Solids with low densities have high specific heats. Metals with high thermal conductivities have high electrical conductivities. These rules of thumb describe correlations between two or more material properties that can be expressed more quantitatively as limits for the values of *dimensionless property groups*. They take the form

$$C_L < P_1 P_2{}^n < C_H \tag{A.1}$$

or

$$C_L < P_1 P_2{}^n P_3{}^m < C_H \tag{A.2}$$

(or larger groupings), where P_1, P_2, P_3 are material properties, n and m are powers (usually −1, −1/2, +1/2, or +1), and C_L and C_H are dimensionless constants—the lower and upper limits between which the values of the property group lies. The correlations exert tight constraints on the data, giving the "patterns" of property envelopes that appear on the material selection charts. An example is the relationship between expansion coefficient, α (units: K^{-1}), and the melting point, T_m (units: K) or, for amorphous materials, the glass temperature, T_g:

$$C_L \leq \alpha T_m \leq C_H \tag{A.3a}$$

$$C_L \leq \alpha T_g \leq C_H \tag{A.3b}$$

a correlation with the form of Equation (A.1). Values for the dimensionless limits C_L and C_H for this group are listed in Table A.13 for a number of material classes. The values span a factor of 2 to 10 rather than the factor of 10 to 100 of the property ranges. There are many such correlations. They form the basis of a hierarchical scheme for data checking and estimating (one used in preparing the charts in this book), described next.

Data checking

Data checks proceed in three steps. Each datum is first associated with a material class or, at a higher level of checking, with a subclass. This identifies the values of the property range and correlation limits against which it will be checked. The datum is then compared with the range limits L and H for that class and property. If it lies within the range limits, it is accepted; if it does not, it is flagged for checking.

Table A.13 Limits for the Groups αT_m and αT_g for Broad Material Classes

Correlation* $C_L < \alpha T_m < C_H$	C_L ($\times 10^{-3}$)	C_H ($\times 10^{-3}$)
All solids	0.1	56
Classes of solid		
Metals: ferrous	13	27
Metals: nonferrous	2	21
Fine ceramics*	6	24
Glasses	0.3	3
Polymers: thermoplastic	18	35
Polymers: thermosets	11	41
Polymers: elastomers	35	56
Polymeric foams	16	37
Composites: metal-matrix	10	20
Composites: polymer-matrix	0.1	10
Woods: parallel to grain	2	4
Woods: perpendicular to grain	6	17

* *For amorphous solids, the melting point, T_m, is replaced by the glass temperature, T_g.*

Why bother with such low-level stuff? Because it provides a sanity check. The commonest error in handbooks and other compilations of material or process properties is that of a value expressed in the wrong units or, for less obvious reasons, in error by one or more orders of magnitude (slipped decimal point, for instance). Range checks catch errors of this sort. If a demonstration of this is needed, it can be found by applying them to the contents of almost any standard reference data-book; none among those we have tried has passed without errors.

In the third step, each of the dimensionless groups of properties like the group in Table A.13 is formed in turn, and compared with the range bracketed by the limits C_L and C_H. If the value lies within its correlation limits, it is accepted; if not, it is checked. Correlation checks are more discerning than range checks and catch subtler errors, allowing the quality of the data to be further enhanced.

Data estimation

The relationships have another, equally useful, function. There remain gaps in our knowledge of material properties. The fracture toughness of many materials has not yet been measured, nor has the electric breakdown potential; even moduli are not always known. The absence of a datum for a material would falsely eliminate it from a screening exercise that used that property, even though the material might be a viable candidate. This difficulty is avoided by using the correlation and range limits to estimate a value for the missing datum, adding a flag to alert the user that it is an estimate.

In estimating property values, the procedure used for checking is reversed: The dimensionless groups are used first because they are the more accurate. They can be surprisingly good. As an example, consider estimating the expansion coefficient, α, of polycarbonate from its glass temperature, T_g. Inverting Equation (A.3b) gives the estimation rule:

$$\frac{C_L}{T_g} \leq \alpha \leq \frac{C_H}{T_g} \tag{A.4}$$

Inserting values of C_L and C_H from Table A.13, and the value T_g = 420 K for a particular sample of polycarbonate gives the mean estimate

$$\overline{\alpha} = 63 \times 10^{-3}\,\mathrm{K}^{-1} \tag{A.5}$$

The reported value for polycarbonate is

$$\alpha = 54\text{–}62 \times 10^{-3}\,\mathrm{K}^{-1}$$

The estimate is within 9% of the mean of the measured values, perfectly adequate for screening purposes. That it is an estimate must not be forgotten, however: If thermal expansion is crucial to the design, better data or direct measurements are essential.

Only when the potential of the correlations is exhausted are the property ranges invoked. They provide a crude first estimate of the value of the missing property, far less accurate than the correlations but still useful in providing guide values for screening.

FURTHER READING

Ashby, M.F. (1998). Checks and estimates for material properties. Cambridge University Engineering Department. *Proc. Roy. Soc. A., 454,* 1301–1321.

This and the next reference detail the data-checking and data-estimating methods, listing the correlations.

Bassetti, D., Brechet, Y., & Ashby, M.F. (1998). Estimates for material properties: The method of multiple correlations. *Proc. Roy. Soc. A., 454,* 1323–1336.

This and the previous reference detail the data-checking and data estimating methods, with examples.

Appendix B: Useful Solutions for Standard Problems

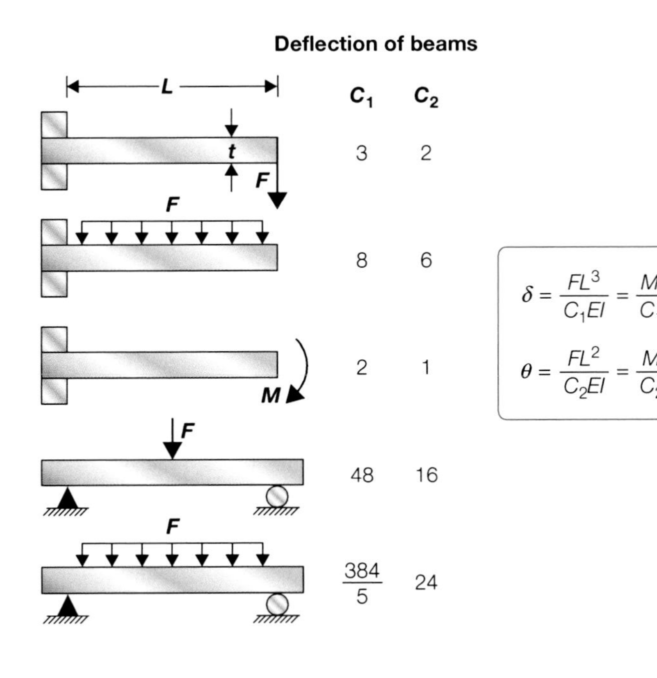

Materials Selection in Mechanical Design. DOI: 10.1016/B978-1-85617-663-7.00019-9

CONTENTS

INTRODUCTION AND SYNOPSIS

Modeling is a key part of design. In the early stage, approximate modeling establishes whether the concept will work at all, and identifies the combination of material properties that maximize performance. At the embodiment stage, more accurate modeling brackets values for the forces, the displacements, the velocities, the heat fluxes, and the dimensions of the components. And in the final stage, modeling gives precise values for stresses, strains and failure probability in key components: power, speed, efficiency, and so on.

Many components with simple geometries and loads have been modeled already. Many more complex components can be modeled approximately by idealizing them as one of these. There is no need to reinvent the beam or the column or the pressure vessel; their behavior under all common types of loading has already been analyzed. The important thing is to know that the results exist and where to find them.

This appendix summarizes the results of modeling a number of standard problems. The usefulness of these results cannot be overstated. Many problems of conceptual design can be treated, with adequate precision, by patching together the solutions given here; even the detailed analysis of noncritical components can often be tackled in the same way. Even when this approximate approach is not sufficiently accurate, the insight it gives is valuable.

This appendix contains 15 double-page sections that list, with a short commentary, results for constitutive equations; for the loading of beams, columns, and torsion bars; for contact stresses, cracks, and other stress concentrations; for pressure vessels, vibrating beams, and plates; and for the flow of heat and matter. They are drawn from numerous sources, listed in Section B.16.

B.1 CONSTITUTIVE EQUATIONS FOR MECHANICAL RESPONSE

The behavior of a component when it is loaded depends on the *mechanism* by which it deforms. A beam loaded in bending may deflect elastically; it may yield plastically; it may deform by creep; and it may fracture in a brittle or ductile way. The equation that describes the material response is known as a *constitutive equation*. Each mechanism is characterized by a different constitutive equation, which contains one or more than one *material property*: Young's modulus, E, and Poisson's ratio, ν, are the material properties that enter the constitutive equation for linear-elastic deformation; the yield strength, σ_y, is the material property that enters the constitutive equation for plastic flow; creep constants, $\dot{\varepsilon}_o$, σ_o, and n enter the equation for creep; fracture toughness, K_{1c}, enters that for brittle fracture.

The common constitutive equations for mechanical deformation are listed on the facing page. In each case the equation for uniaxial loading by a tensile stress, σ is given first; below it is the equation for multiaxial loading by principal stresses σ_1, σ_2, and σ_3, always chosen so that σ_1 is the most tensile and σ_3 the most compressive (or least tensile). They are the basic equations that determine mechanical response.

Elastic

Uniaxial	$\epsilon = \frac{\sigma}{E}$	
General	$\varepsilon_1 = \frac{\sigma_1}{E} - \nu(\sigma_2 + \sigma_3)$	etc.

Plastic

Uniaxial	$\sigma \geq \sigma_y$		
	$\sigma_1 - \sigma_3 \geq \sigma_y$	$(\sigma_1 > \sigma_2 > \sigma_3)$	or
General	$[\frac{1}{2}\{(\sigma_1 - \sigma_2)^2 + (\sigma_2 - \sigma_3)^2 + (\sigma_3 - \sigma_1)^2\}]^{\frac{1}{2}} > \sigma_y$		

Creep

Uniaxial	$\dot{\varepsilon} = \dot{\varepsilon}_o \left(\frac{\sigma}{\sigma_o}\right)^n$	
General	$\dot{\varepsilon}_1 = \dot{\varepsilon}_o \frac{\sigma_o^{n-1}}{\sigma_o^n} \{\sigma_1 - \frac{1}{2}(\sigma_2 + \sigma_3)\}$	etc.

Fracture

Uniaxial	$\sigma \geq C K_{1c}/(\pi a)^{1/2}$	
General	$\sigma_1 \geq C K_{1c}/(\pi a)^{1/2}$	$(\sigma_1 > \sigma_2 > \sigma_3)$

E = Young's modulus

n = Poisson's ratio

σ_y = Plastic yield strength

n, ε_o, σ_o = Creep constants

K_{1c}, a = Fracture toughness, crack length

C ≈ 1 Tension

C ≈ 15 Compression

FIGURE B.1

B.2 MOMENTS OF SECTIONS

A beam of uniform section, loaded in simple tension by a force, F, carries a stress

$$\sigma = F/A$$

where A is the area of the section. Its response is calculated from the appropriate constitutive equation. Here the important characteristic of the section is its area, A. For other modes of loading, higher moments of the area are involved. Those for various common sections are given on the facing page. They are defined as follows.

The second moment, I, measures the resistance of the section to bending about a horizontal axis (shown as a broken line). It is

$$I = \int_{\text{section}} y^2\, b(y)\, dy$$

where y is measured vertically and $b(y)$ is the width of the section at y. The moment, K, measures the resistance of the section to twisting. It is equal to the polar moment J for circular sections, where

$$J = \int_{\text{section}} 2\pi r^3 dr$$

where r is measured radially from the center of the circular section. For noncircular sections, K is less than J.

The section modulus $Z = I/y_m$ (where y_m is the normal distance from the neutral axis of bending to the outer surface of the beam) measures the surface stress generated by a given bending moment, M:

$$\sigma = \frac{M y_m}{I} = \frac{M}{Z}$$

Finally, the moment Z_p, defined by

$$Z_p = \int_{\text{section}} y\, b(y)\, dy$$

measures the resistance of the beam to fully plastic bending. The fully plastic moment for a beam in bending is

$$M_p = Z_p \sigma_y$$

Section Shape	Area A m^2	Moment I m^4	Moment K m^4	Moment Z m^3	Moment Z_p m^3
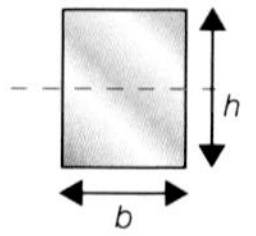 h, b	bh	$\frac{bh^3}{12}$	$\frac{bh^3}{3}(1-0.58\frac{b}{h})$ $(h>b)$	$\frac{bh^2}{6}$	$\frac{bh^2}{4}$
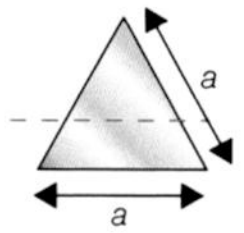 a, a	$\frac{\sqrt{3}}{4}a^2$	$\frac{a^4}{32\sqrt{3}}$	$\frac{\sqrt{3}a^4}{80}$	$\frac{a^3}{32}$	$\frac{3a^3}{64}$
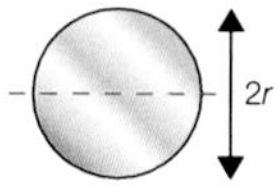 $2r$	πr^2	$\frac{\pi}{4}r^4$	$\frac{\pi}{2}r^4$	$\frac{\pi}{4}r^3$	$\frac{\pi}{3}r^3$
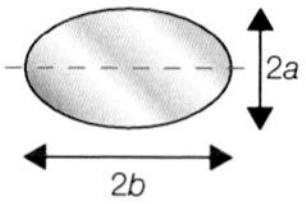 $2a$, $2b$	πab	$\frac{\pi}{4}a^3b$	$\frac{\pi a^3b^3}{(a^2+b^2)}$	$\frac{\pi}{4}a^2b$	$\frac{\pi}{3}a^2b$
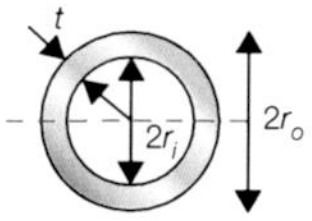 t, $2r_i$, $2r_o$	$\pi(r_o^2-r_i^2)$ $\approx 2\pi rt$	$\frac{\pi}{4}(r_o^4-r_i^4)$ $\approx \pi r^3t$	$\frac{\pi}{2}(r_o^4-r_i^4)$ $\approx 2\pi r^3t$	$\frac{\pi}{4r_o}(r_o^4-r_i^4)$ $\approx \pi r^2t$	$\frac{\pi}{3}(r_o^3-r_i^3)$ $\approx \pi r^2t$
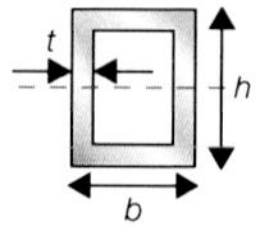 t, h, b	$2t(h+b)$ $(h, b>>t)$	$\frac{1}{6}h^3t(1+3\frac{b}{h})$	$\frac{2tb^2h^2}{(h+b)}(1-\frac{t}{h})^4$	$\frac{1}{3}h^2t(1+3\frac{b}{h})$	$bht(1+\frac{h}{2b})$
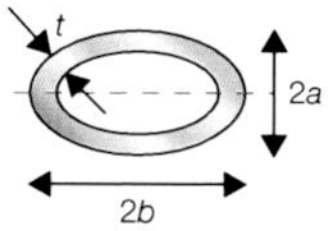 t, $2a$, $2b$	$\pi(a+b)t$ $(a, b>>t)$	$\frac{\pi}{4}a^3t(1+\frac{3b}{a})$	$\frac{4\pi(ab)^{5/2}t}{(a^2+b^2)}$	$\frac{\pi}{4}a^2t(1+\frac{3b}{a})$	$\pi abt(2+\frac{a}{b})$
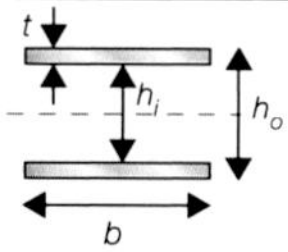 t, h_i, h_o, b	$b(h_o-h_i)$ $\approx 2bt$ $(h, b>>t)$	$\frac{b}{12}(h_o^3-h_i^3)$ $\approx \frac{1}{2}bth_o^2$	–	$\frac{b}{6h_o}(h_o^3-h_i^3)$ $\approx bth_o$	$\frac{b}{4}(h_o^2-h_i^2)$ $\approx bth_o$
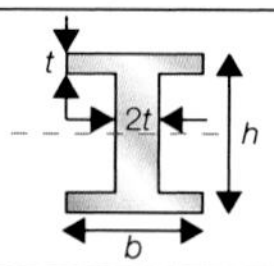 t, $2t$, h, b	$2t(h+b)$ $(h, b>>t)$	$\frac{1}{6}h^3t(1+3\frac{b}{h})$	$\frac{2}{3}bt^3(1+4\frac{h}{b})$	$\frac{1}{3}h^2t(1+3\frac{b}{h})$	$bht(1+\frac{h}{2b})$
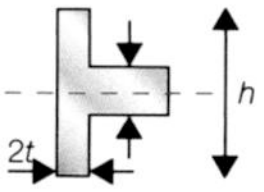 $2t$, h	$2t(h+b)$ $(h, b>>t)$	$\frac{t}{6}(h^3+4bt^2)$	$\frac{t^3}{3}(8b+h)$	$\frac{t}{3h}(h^3+4bt^2)$	$\frac{th^2}{2}\{1+\frac{2t(b-2t)}{h^2}\}$

FIGURE B.2

B.3 ELASTIC BENDING OF BEAMS

When a beam is loaded by a force, F, or moments, M, the initially straight axis is deformed into a curve. If the beam is uniform in section and properties, long in relation to its depth, and nowhere stressed beyond the elastic limit, the deflection, δ, and the angle of rotation, θ, can be calculated using elastic beam theory (see Section B.16). The basic differential equation describing the curvature of the beam at a point x along its length is

$$EI\frac{d^2y}{dx^2} = M$$

where y is the lateral deflection, and M is the bending moment at the point x on the beam. E is Young's modulus, and I is the second moment of area (Section B.2). When M is constant this becomes

$$\frac{M}{I} = E\left(\frac{1}{R} - \frac{1}{R_o}\right)$$

where R_o is the radius of curvature before applying the moment and R is the radius after it is applied. Deflections δ and rotations θ are found by integrating these equations along the beam.

The stiffness of the beam is defined by

$$S = \frac{F}{\delta} = \frac{C_1 EI}{L^3}$$

It depends on Young's modulus, E, for the material of the beam, on its length, L, and on the second moment of its section, I. The end slope of the beam, θ, is given by

$$\theta = \frac{FL^2}{C_2 EI}$$

Equations for the deflection, δ, and end slope, θ, of beams for various common modes of loading are shown on the facing page together with values of C_1 and C_2.

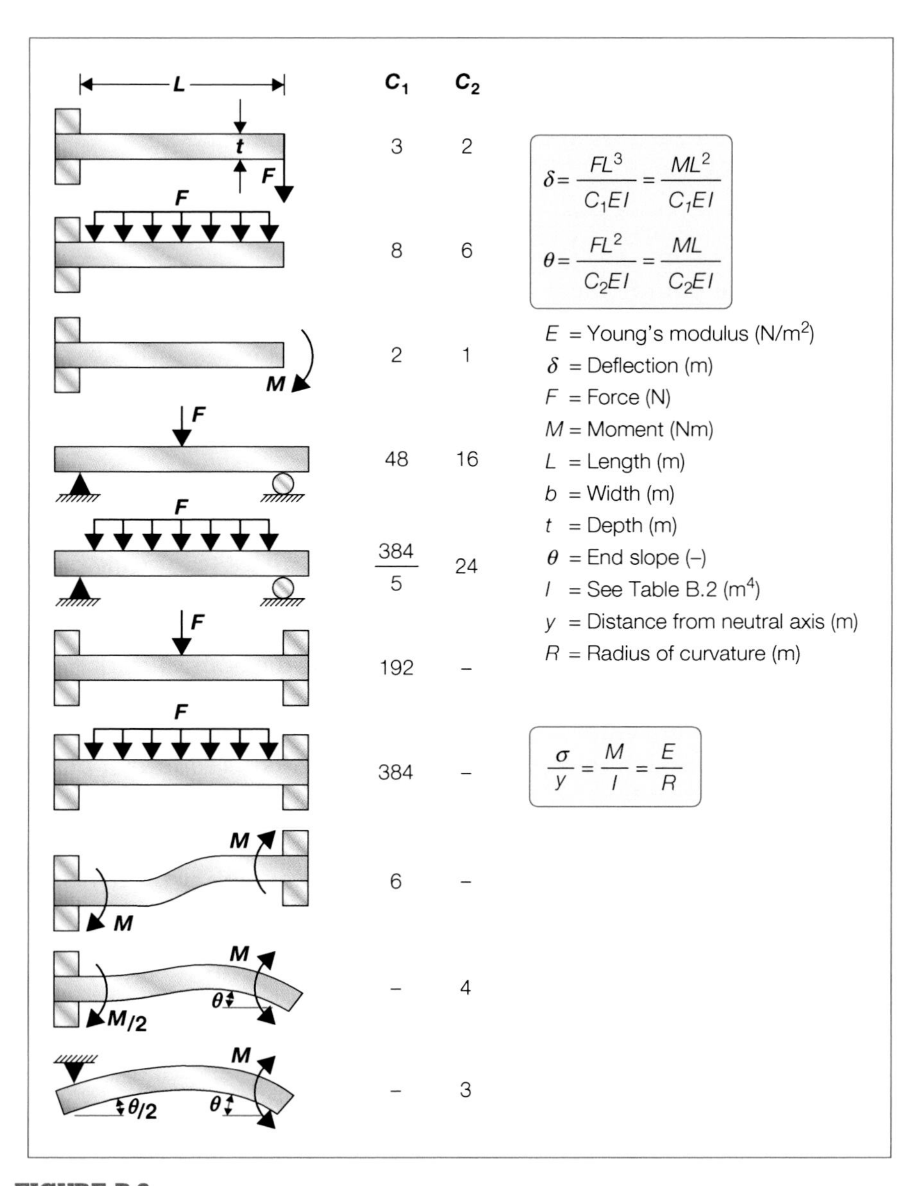
L
t
F
M
M/2
θ
θ/2
C_1
C_2
3 2
8 6
2 1
48 16
384/5 24
192 –
384 –
6 –
– 4
– 3
$\delta = \frac{FL^3}{C_1EI} = \frac{ML^2}{C_1EI}$
$\theta = \frac{FL^2}{C_2EI} = \frac{ML}{C_2EI}$
E = Young's modulus (N/m²)
δ = Deflection (m)
F = Force (N)
M = Moment (Nm)
L = Length (m)
b = Width (m)
t = Depth (m)
θ = End slope (–)
I = See Table B.2 (m⁴)
y = Distance from neutral axis (m)
R = Radius of curvature (m)
$\frac{\sigma}{y} = \frac{M}{I} = \frac{E}{R}$

FIGURE B.3

B.4 FAILURE OF BEAMS AND PANELS

The longitudinal (or "fiber") stress, σ, at a point, y, from the neutral axis of a uniform beam loaded elastically in bending by a moment, M, is

$$\frac{\sigma}{y} = \frac{M}{I} = E\left(\frac{1}{R} - \frac{1}{R_o}\right)$$

where I is the second moment of area (Section B.2), E is Young's modulus, Ro is the radius of curvature before applying the moment, and R is the radius after it is applied. The tensile stress in the outer fiber of such a beam is

$$\sigma = \frac{My_m}{I} = \frac{M}{Z}$$

where y_m is the perpendicular distance from the neutral axis to the outer surface of the beam. If this stress reaches the yield strength, σ_y, of the material of the beam, small zones of plasticity appear at the surface (*top*, facing page). The beam is no longer elastic and, in this sense, has failed. If, instead, the maximum fiber stress reaches the brittle fracture strength, σ_f (the "modulus of rupture," often shortened to MOR) of the material of the beam, a crack nucleates at the surface and propagates inward (*middle*); in this case, the beam has certainly failed. A third criterion for failure is often important: that the plastic zones penetrate through the section of the beam, linking to form a plastic hinge (*bottom*).

The failure moments and failure loads, for each of these three types of failure, and for each of several geometries of loading, are given in the diagram. The formulae labeled "onset" refer to the first two failure modes; those labeled "full plasticity" refer to the third. Two new functions of section shape are involved. Onset of failure involves the quantity Z; full plasticity involves the quantity H. Both are listed in the table in Section B.2, and defined in the text that accompanies it.

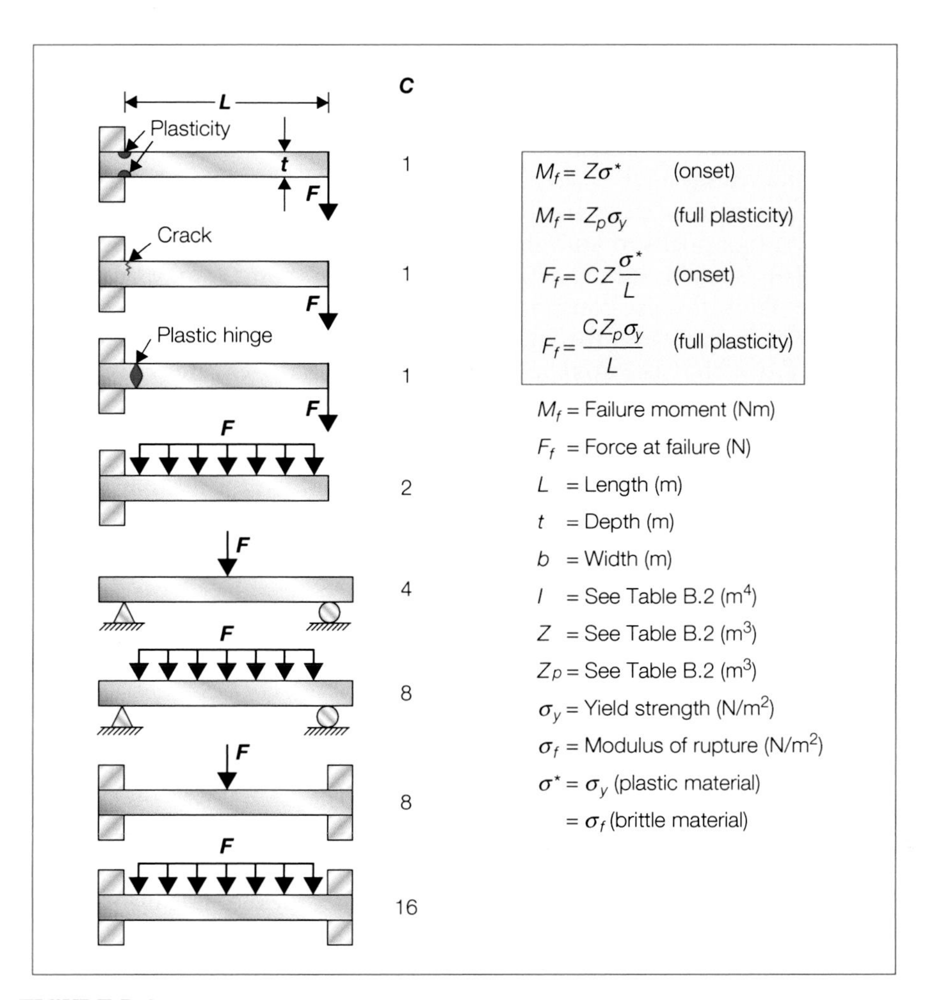
C
L
Plasticity
t
F
1
Crack
F
1
Plastic hinge
F
1
F
2
F
4
F
8
F
8
F
16
$M_f = Z\sigma^*$ (onset)
$M_f = Z_p\sigma_y$ (full plasticity)
$F_f = CZ\frac{\sigma^*}{L}$ (onset)
$F_f = \frac{CZ_p\sigma_y}{L}$ (full plasticity)
M_f = Failure moment (Nm)
F_f = Force at failure (N)
L = Length (m)
t = Depth (m)
b = Width (m)
I = See Table B.2 (m^4)
Z = See Table B.2 (m^3)
Z_p = See Table B.2 (m^3)
σ_y = Yield strength (N/m^2)
σ_f = Modulus of rupture (N/m^2)
$\sigma^* = \sigma_y$ (plastic material)
$= \sigma_f$ (brittle material)

FIGURE B.4

B.5 BUCKLING OF COLUMNS, PLATES, AND SHELLS

If sufficiently slender, an elastic column loaded in compression fails by elastic buckling at a critical load, F_{crit}. This load is determined by the end constraints, of which four extreme cases are illustrated on the facing page: An end may be constrained in a position and direction; it may be free to rotate but not to translate (or "sway"); it may sway without rotation; and it may both sway and rotate. Pairs of these constraints applied to the ends of the column lead to the five cases shown on the facing page. Each is characterized by a value of the constant n that is equal to the number of half-wavelengths of the buckled shape.

The addition of the bending moment, M, reduces the buckling load by the amount shown in the second box. A negative value of F_{crit} means that a tensile force is necessary to prevent buckling.

An elastic foundation is one that exerts a lateral restoring pressure, p, proportional to the deflection:

$$p = ky$$

where k is the foundation stiffness per unit depth and y is the local lateral deflection. Its effect is to increase F_{crit}, by the amount shown in the third box.

A thin-walled elastic tube will buckle inward under an external pressure, p', given in the last box. Here I refers to the second moment of area of a section of the tube wall cut parallel to the tube axis.

Thin or slender shapes may buckle locally before they yield or fracture. It is this that sets a practical limit to the thinness of tube walls and webs (Chapter 9).

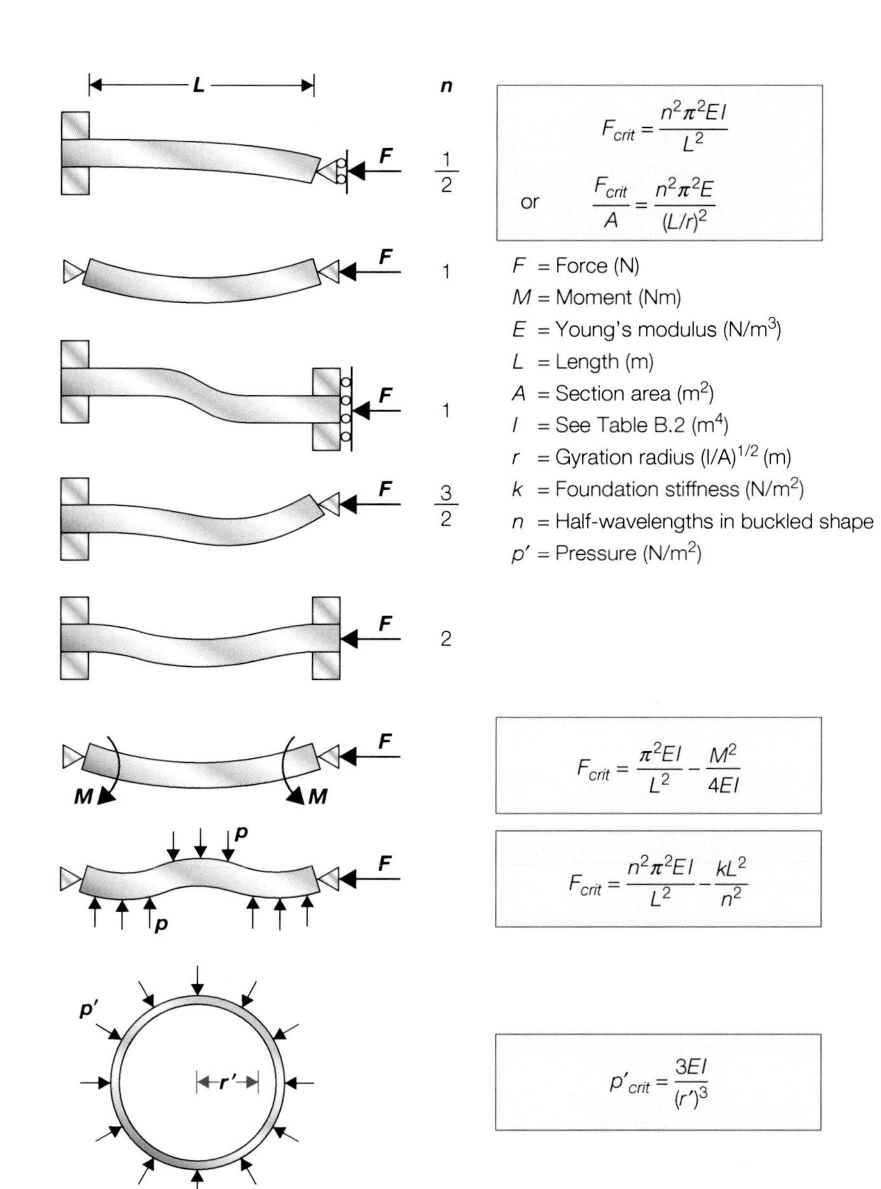
L
n
F
1/2
1
1
3/2
2
M
M
p
p
p'
r'
$F_{crit} = \frac{n^2\pi^2 EI}{L^2}$
or $\frac{F_{crit}}{A} = \frac{n^2\pi^2 E}{(L/r)^2}$
F = Force (N)
M = Moment (Nm)
E = Young's modulus (N/m^3)
L = Length (m)
A = Section area (m^2)
I = See Table B.2 (m^4)
r = Gyration radius (I/A)$^{1/2}$ (m)
k = Foundation stiffness (N/m^2)
n = Half-wavelengths in buckled shape
p' = Pressure (N/m^2)
$F_{crit} = \frac{\pi^2 EI}{L^2} - \frac{M^2}{4EI}$
$F_{crit} = \frac{n^2\pi^2 EI}{L^2} - \frac{kL^2}{n^2}$
$p'_{crit} = \frac{3EI}{(r')^3}$

FIGURE B.5

B.6 TORSION OF SHAFTS

A torque, T, applied to the ends of an isotropic bar of uniform section and acting in the plane normal to the axis of the bar produces an angle of twist, θ. The twist is related to the torque by the first equation on the facing page, in which G is the shear modulus. For round bars and tubes of circular section, the factor K is equal to J, the polar moment of inertia of the section, defined in Section B.2. For any other section shape, K is less than J. Values of K are given in Section B.2.

If the bar ceases to deform elastically, it is said to have failed. This will happen if the maximum surface stress exceeds either the yield strength, σ_y, of the material or the stress at which it fractures, σ_{fr}. For circular sections, the shear stress at any point a distance, r, from the axis of rotation is

$$\tau = \frac{Tr}{K} = \frac{G\theta r}{K}$$

The maximum shear stress, $\tau_{\max}$, and the maximum tensile stress, $\sigma_{\max}$, are at the surface and have the values

$$\tau_{\max} = \sigma_{\max} = \frac{Td_o}{2K} = \frac{G\theta d_o}{2L}$$

If $\tau_{\max}$ exceeds $\sigma_y/2$ (using a Tresca yield criterion), or if $\sigma_{\max}$ exceeds σ_{fr}, the bar fails, as shown in the figure. The maximum surface stress for the solid ellipsoidal, square, rectangular, and triangular sections is at the points on the surface closest to the centroid of the section (the midpoints of the longer sides). It can be estimated approximately by inscribing the largest circle that can be contained within the section and calculating the surface stress for a circular bar of that diameter. More complex section shapes require special consideration and, if thin, may additionally fail by buckling.

Helical springs are a special case of torsional deformation. The extension of a helical spring of n turns of radius R, under a force, F, and the failure force, F_{crit}, are given on the facing page.

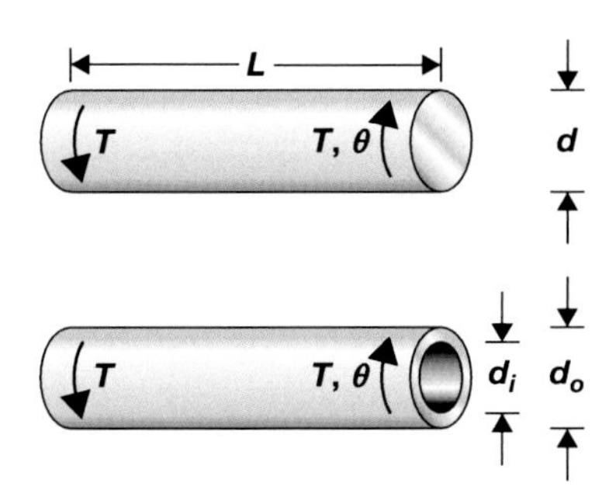

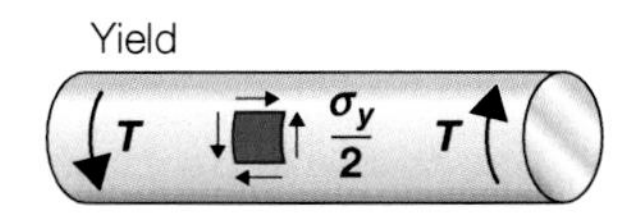

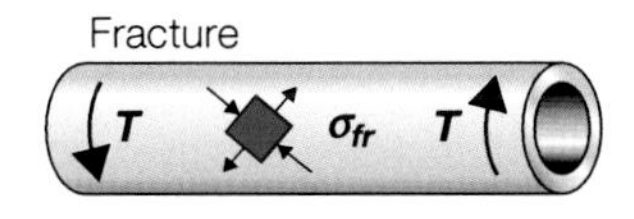

Elastic deflection

$$\theta = \frac{LT}{KG}$$

Failure

$$T_f = \frac{K\sigma_y}{d_o} \text{ (onset of yield)}$$

$$T_f = \frac{2K\sigma_f}{d_o} \text{ (brittle fracture)}$$

T = Torque (Nm)
θ = Angle of twist
G = Shear modulus (N/m^2)
L = Length (m)
d = Diameter (m)
K = See Table B.2 (m^4)
σ_y = Yield strength (N/m^2)
σ_f = Modulus of rupture (N/m^2)

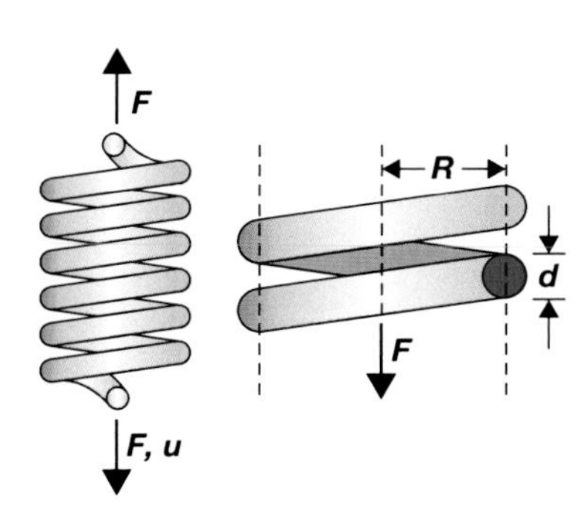

Spring deflection and failure

$$u = \frac{64FR^3n}{Gd^4}$$

$$F_f = \frac{\pi}{32}\frac{d^3\sigma_y}{R}$$

F = Force (N)
u = Deflection (m)
R = Coil radius (m)
n = Number of turns

FIGURE B.6

B.7 STATIC AND SPINNING DISKS

A thin disk deflects when a pressure difference, Δp, is applied across its two surfaces. The deflection causes stresses to appear in the disk. The first box on the facing page gives deflection and maximum stress (important in predicting failure) when the edges of the disk are simply supported. The second gives the same quantities when the edges are clamped. The results for a thin horizontal disk deflecting under its own weight are found by replacing Δp by the mass-per-unit area, $\rho\ g\ t$, of the disk (here ρ is the density of the material of the disk and g is the acceleration due to gravity). Thick disks are more complicated; for those, see "Further reading" (Section B.16).

Spinning disks, rings, and cylinders store kinetic energy. Centrifugal forces generate stresses in the disk. The two boxes list the kinetic energy and the maximum stress, σ_{max}, in disks and rings rotating at an angular velocity ω (radians/sec). The maximum rotation rate and energy are limited by the burst strength of the disk. They are found by equating the maximum stress in the disk to the strength of the material.

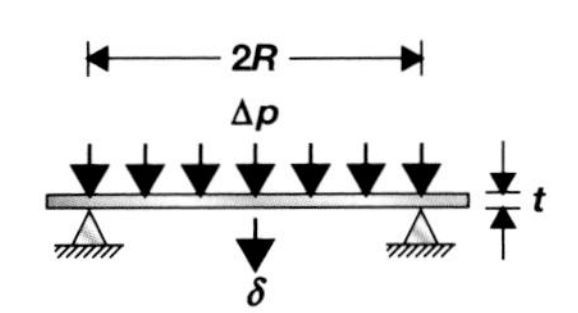

Simply supported

$$\delta = \frac{3}{4}(1-\nu^2)\frac{\Delta p R^4}{E t^3}$$

$$\sigma_{max} = \frac{3}{8}(3+\nu)\frac{\Delta p R^2}{t^2}$$

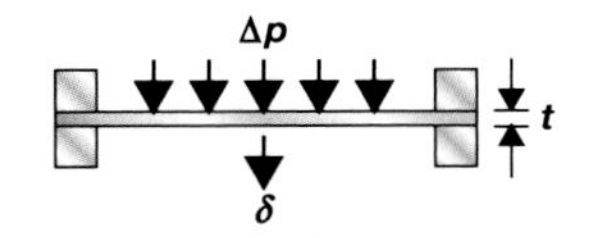

Clamped

$$\delta = \frac{3}{16}(1-\nu^2)\frac{\Delta p R^4}{E t^3}$$

$$\sigma_{max} = \frac{3}{8}(1+\nu)\frac{\Delta p R^2}{t^2}$$

δ = Deflection (m)
E = Young's modulus (N/m^2)
Δp = Pressure difference (N/m^2)
ν = Poisson's ratio

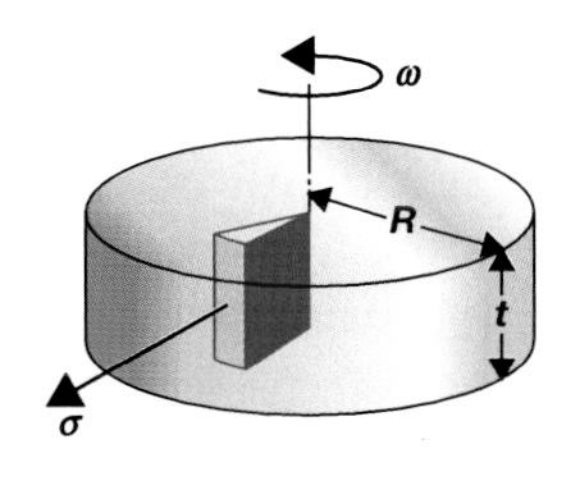

Disk

$$U = \frac{\pi}{4}\rho t \omega^2 R^4$$

$$\sigma_{max} = \frac{1}{8}(3+\nu)\rho\omega^2 R^2$$

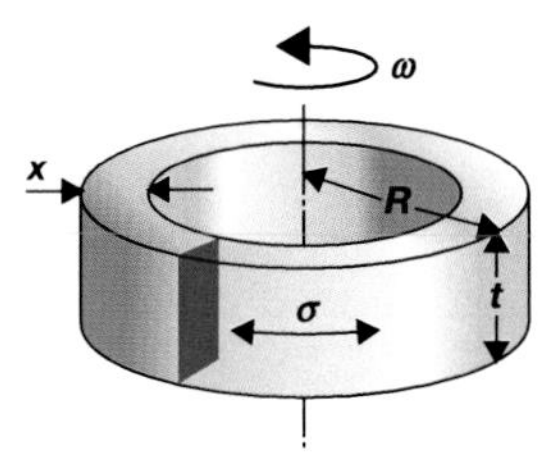

Ring

$$U = \pi \rho t \omega^2 R^3 x$$

$$\sigma_{max} = \rho\omega^2 R^2$$

U = Energy (J)
ω = Angular velocity (rad/s)
ρ = Density (kg/m^3)

FIGURE B.7

B.8 CONTACT STRESSES

When surfaces are placed in contact, they touch at one or a few discrete points. If the surfaces are loaded, the contacts flatten elastically and the contact areas grow until failure of some sort occurs: failure by crushing (caused by the compressive stress, σ_c), tensile fracture (caused by the tensile stress, σ_t), or yielding (caused by the shear stress, σ_s). The boxes on the facing page summarize the important results for the radius, a, of the contact zone, the center-to-center displacement, u, and the peak values of σ_c, σ_t, and σ_s.

The first box shows results for a sphere on a flat, when both have the same moduli and Poisson's ratio has the value 1/3. Results for the more general problem (the "Hertzian indentation" problem) are shown in the second box: Two elastic spheres (radii R_1 and R_2, moduli and Poisson's ratios E_1, ν_1 and E_2, ν_2) are pressed together by a force, F.

If the shear stress, σ_s, exceeds the shear yield strength, $\sigma_y/2$, a plastic zone appears beneath the center of the contact at a depth of about $a/2$ and spreads to form the fully plastic field shown in the two lower figures. When this state is reached, the contact pressure is approximately three times the yield stress, as shown in the bottom box.

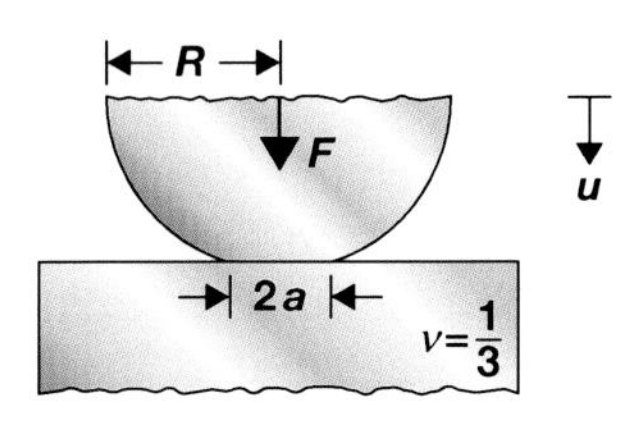

$$a = 0.7\left(\frac{FR}{E}\right)^{\frac{1}{3}}$$

$$u = 1.0\left(\frac{F^2}{E^2R}\right)^{\frac{1}{3}}$$

$\nu = 0.33$

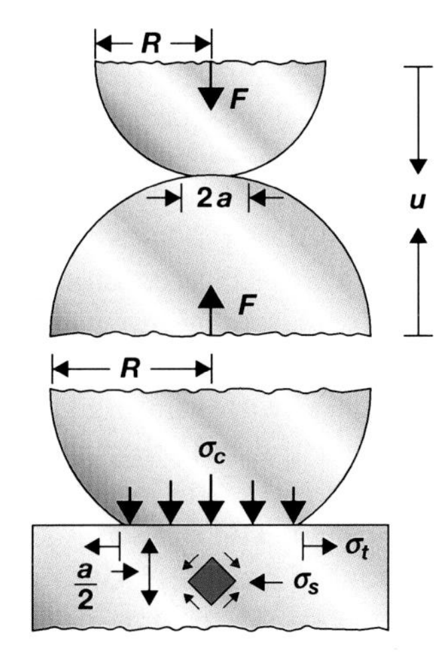

$$a = \left(\frac{3}{4}\frac{F}{E^*}\frac{R_1R_2}{(R_1+R_2)}\right)^{\frac{1}{3}}$$

$$u = \left(\frac{9}{16}\frac{F^2}{(E^*)^2}\frac{(R_1+R_2)}{R_1R_2}\right)^{\frac{1}{3}}$$

$$(\sigma_c)_{max} = \frac{3F}{2\pi a^2}$$

$$(\sigma_s)_{max} = \frac{F}{2\pi a^2}$$

$$(\sigma_t)_{max} = \frac{F}{6\pi a^2}$$

R_1, R = Radii of spheres (m)

E_1, E_2 = Moduli of spheres (N/m^2)

ν_1, ν_2 = Poisson's ratio

F = Load (N)

a = Radius of contact (m)

u = Displacement (m)

σ = Stresses (N/m^2)

σ_y = Yield stress (N/m^2)

$$E^* = \left(\frac{1-\nu_1^2}{E_1} + \frac{1-\nu_2^2}{E_2}\right)^{-1}$$

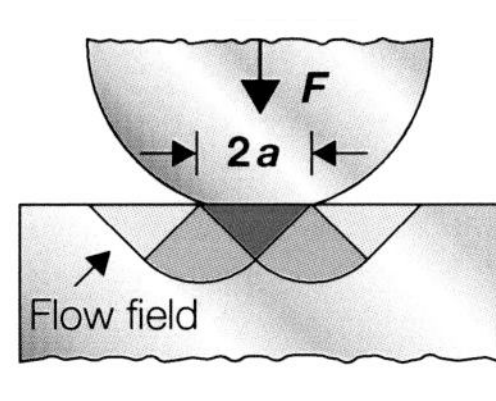

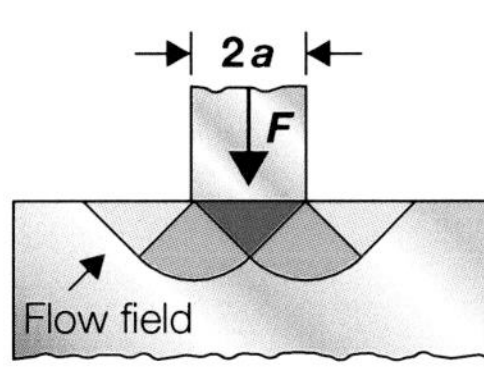

$$\frac{F}{\pi a^2} = 3\sigma_y$$

FIGURE B.8

B.9 ESTIMATES FOR STRESS CONCENTRATIONS

Stresses and strains are concentrated at holes, slots, or changes of section in elastic bodies. Plastic flow, fracture, and fatigue cracking start at these places. The local stresses at the stress concentrations can be computed numerically, but this is often unnecessary. Instead, they can be estimated using the equation shown on the facing page.

The stress concentration caused by a change in section dies away at distances on the order of the characteristic dimension of the section change (defined more fully below), an example of St Venant's principle at work. This means that the maximum local stresses in a structure can be found by determining the nominal stress distribution, neglecting local discontinuities (such as holes or grooves), and then multiplying the nominal stress by a stress concentration factor. Elastic stress concentration factors are given approximately by the equation. In it, σ_{nom} is defined as the load divided by the minimum cross-section of the part, ρ is the minimum radius of curvature of the stress-concentrating groove or hole, and c is the characteristic dimension: either the half-thickness of the remaining ligament, the half-length of a contained crack, the length of an edge-crack, or the height of a shoulder, whichever is *least*. The drawings show examples of each such situation. The factor α is roughly 2 for tension, but is nearer 1/2 for torsion and bending. Though inexact, the equation is an adequate working approximation for many design problems.

The maximum stress is limited by plastic flow or fracture. When plastic flow starts, the strain concentration grows rapidly while the stress concentration remains constant. The strain concentration becomes the more important quantity, and may not die out rapidly with distance (St Venant's principle no longer applies).

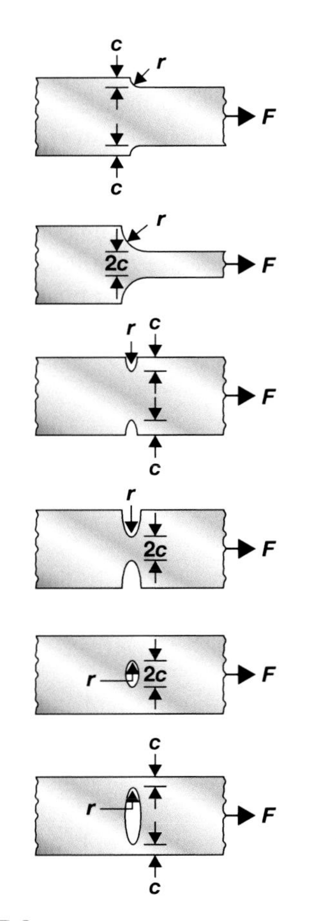

$$\frac{\sigma_{max}}{\sigma_{nom}} = 1 + \alpha \left(\frac{c}{r}\right)^{\frac{1}{2}}$$

F = Force (N)

A_{min} = Minimum section (m^2)

$\sigma_{nom} = F/A_{min}$ (N/m^2)

r = Radius of curvature (m)

c = Characteristic length (m)

α ≈ 0.5 (torsion)

α ≈ 2.0 (tension)

FIGURE B.9

B.10 SHARP CRACKS

Sharp cracks (that is, stress concentrations with a tip radius of curvature of atomic dimensions) concentrate stress in an elastic body more acutely than rounded stress concentrations do. To a first approximation, the local stress falls off as $1/r^{1/2}$ with radial distance r from the crack tip. A tensile stress, σ, applied normal to the plane of a crack of length $2a$ contained in an infinite plate (as in the top figure on the facing page) gives rise to a local stress field, σ_ℓ, that is tensile in the plane containing the crack and given by

$$\sigma_\ell = \frac{C\sigma\sqrt{\pi a}}{\sqrt{2\pi r}}$$

where r is measured from the crack tip in the plane $\theta = 0$ and C is a constant. The mode 1 *stress intensity factor*, K_I, is defined as

$$K_1 = C\sigma\sqrt{\pi a}$$

Values of the constant C for various modes of loading are given in the figure. The stress σ for point loads and moments is given by the equations at the bottom. The crack propagates when $K_1 > K_{1c}$, the *fracture toughness*.

When the crack length is very small compared with all specimen dimensions and compared with the distance over which the applied stress varies, C is equal to 1 for a contained crack and 1.1 for an edge crack. As the crack extends in a uniformly loaded component, it interacts with the free surfaces, giving the correction factors shown opposite. If, in addition, the stress field is nonuniform (as it is in an elastically bent beam), C differs from 1; two examples are given in the figure. The factors, C, given here, are approximate only, good when the crack is short but not when the crack tips are very close to the boundaries of the sample. They are adequate for most design calculations. More accurate approximations and other less common loading geometries can be found in the references listed in "Further reading" (Section B.16).

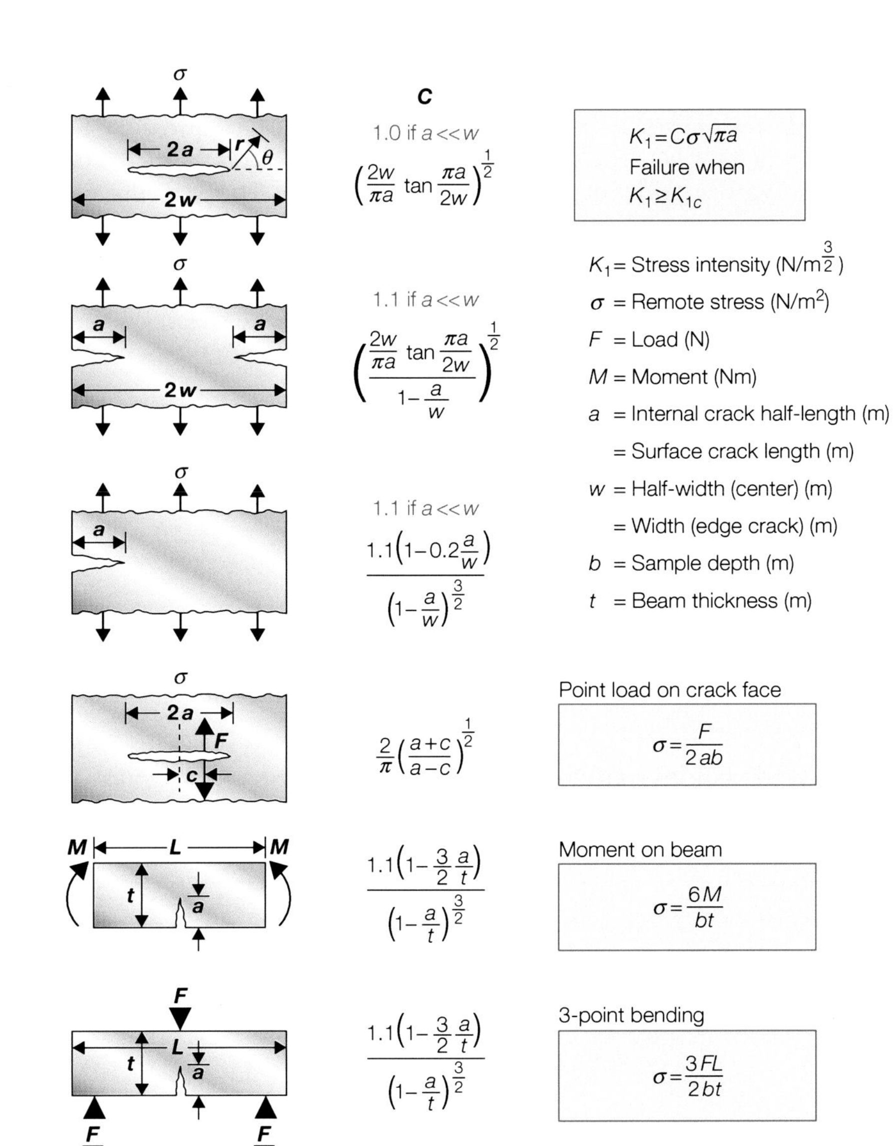
σ
2a
r
θ
2w
C
1.0 if a<<w
$\left(\frac{2w}{\pi a}\tan\frac{\pi a}{2w}\right)^{\frac{1}{2}}$
$K_1 = C\sigma\sqrt{\pi a}$
Failure when
$K_1 \geq K_{1c}$
σ
a
a
2w
1.1 if a<<w
$\left(\frac{\frac{2w}{\pi a}\tan\frac{\pi a}{2w}}{1-\frac{a}{w}}\right)^{\frac{1}{2}}$
K_1 = Stress intensity (N/m$^{\frac{3}{2}}$)
σ = Remote stress (N/m^2)
F = Load (N)
M = Moment (Nm)
a = Internal crack half-length (m)
= Surface crack length (m)
w = Half-width (center) (m)
= Width (edge crack) (m)
b = Sample depth (m)
t = Beam thickness (m)
σ
a
1.1 if a<<w
$\frac{1.1\left(1-0.2\frac{a}{w}\right)}{\left(1-\frac{a}{w}\right)^{\frac{3}{2}}}$
σ
2a
F
c
$\frac{2}{\pi}\left(\frac{a+c}{a-c}\right)^{\frac{1}{2}}$
Point load on crack face
$\sigma = \frac{F}{2ab}$
M
L
M
t
a
$\frac{1.1\left(1-\frac{3}{2}\frac{a}{t}\right)}{\left(1-\frac{a}{t}\right)^{\frac{3}{2}}}$
Moment on beam
$\sigma = \frac{6M}{bt}$
F
L
t
a
F/2
F/2
$\frac{1.1\left(1-\frac{3}{2}\frac{a}{t}\right)}{\left(1-\frac{a}{t}\right)^{\frac{3}{2}}}$
3-point bending
$\sigma = \frac{3FL}{2bt}$

FIGURE B.10

B.11 PRESSURE VESSELS

Thin-walled pressure vessels are treated as membranes. The approximation is reasonable when $t < b/4$. The stresses in the wall are given on the facing page; they do not vary significantly with radial distance, r. Those in the plane tangent to the skin, σ_θ and σ_z for the cylinder and σ_θ and σ_φ for the sphere, are just equal to the internal pressure amplified by the ratio b/t or $b/2t$, depending on geometry. The radial stress, σ_r, is equal to the mean of the internal and external stress, $p/2$ in this case. The equations describe the stresses when an external pressure, p_e, is superimposed if p is replaced by $(p - p_e)$.

In thick-walled vessels, the stresses vary with radial distance, r, from the inner to the outer surfaces, and are greatest at the inner surface. The equations can be adapted for the case of both internal and external pressures by noting that when the internal and external pressures are equal, the state of stress in the wall is

$$\sigma_\theta = \sigma_r = -p \text{ (cylinder)}$$

or

$$\sigma_\theta = \sigma_\phi = \sigma_r = -p \text{ (sphere)}$$

allowing the term involving the external pressure to be evaluated. It is not valid to just replace p by $(p - p_e)$.

Pressure vessels fail by yielding when the von Mises equivalent stress first exceeds the yield strength, σ_y. They fail by fracture if the largest tensile stress exceeds the fracture stress, σ_{fr}, where

$$\sigma_{fr} = \frac{CK_{1c}}{\sqrt{\pi a}}$$

and K_{1c} is the fracture toughness, a is the half-crack length, and C is a constant given in Section B.10.

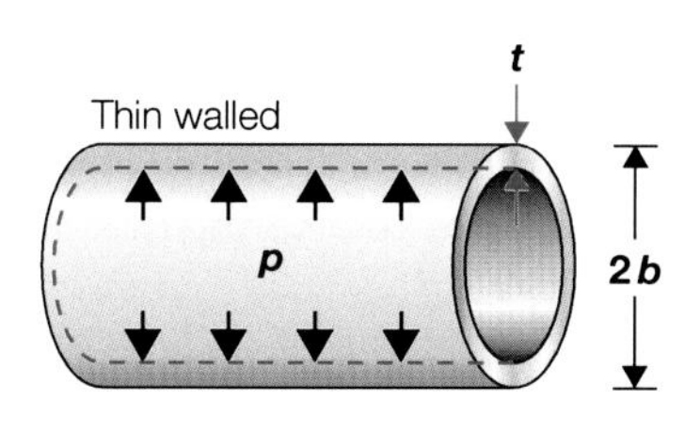

Cylinder

$$\sigma_\theta = \frac{pb}{t}$$

$$\sigma_r = -\frac{p}{2}$$

$$\sigma_z = \frac{pb}{2t} \quad \text{(Closed ends)}$$

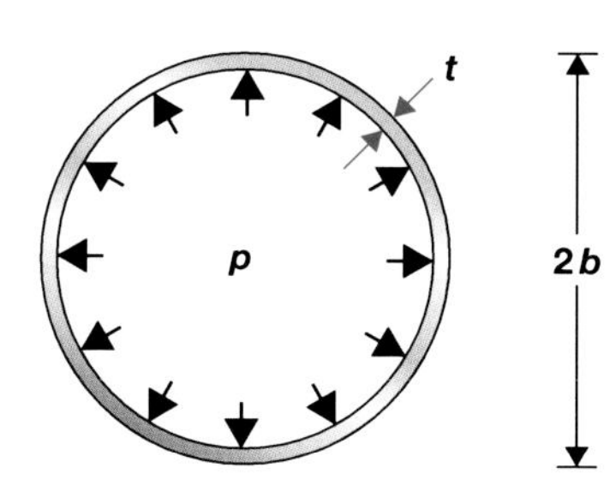

Sphere

$$\sigma_\theta = \sigma_\phi = \frac{pb}{2t}$$

$$\sigma_r = -\frac{p}{2}$$

p = Pressure (N/m^2)

t = Wall thickness (m)

a = Inner radius (m)

b = Outer radius (m)

r = Radial coordinate (m)

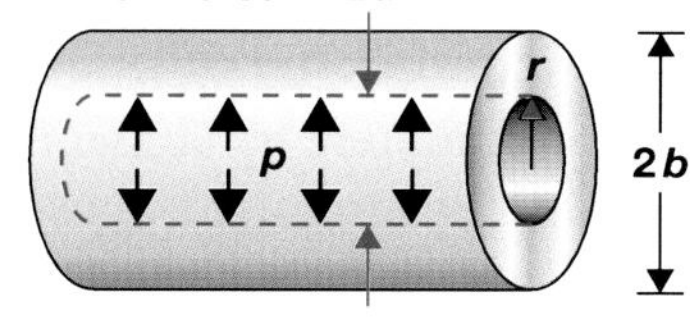

Cylinder

$$\sigma_\theta = \frac{pa^2}{r^2}\left(\frac{b^2 - r^2}{b^2 - a^2}\right)$$

$$\sigma_r = -\frac{pa^2}{r^2}\left(\frac{b^2 + r^2}{b^2 - a^2}\right)$$

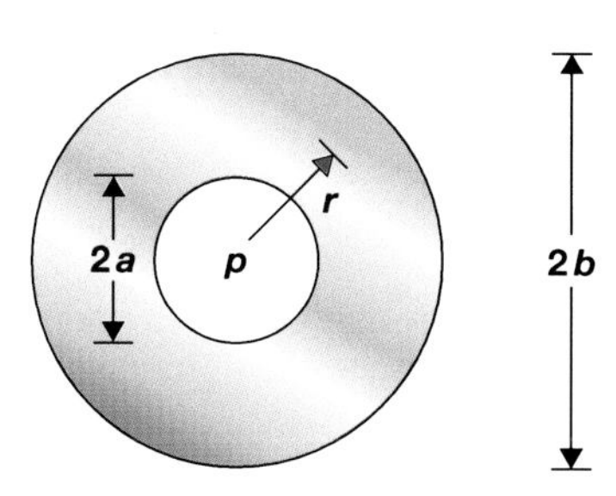

Sphere

$$\sigma_\theta = \sigma_\phi = \frac{pa^3}{2r^3}\left(\frac{b^3 + 2r^3}{b^3 - a^3}\right)$$

$$\sigma_r = -\frac{pa^3}{r^3}\left(\frac{b^3 - r^3}{b^3 - a^3}\right)$$

FIGURE B.11

B.12 VIBRATING BEAMS, TUBES, AND DISKS

Any undamped system vibrating at one of its natural frequencies can be reduced to the simple problem of a mass, m, attached to a spring of stiffness K. The lowest natural frequency of such a system is

$$f = \frac{1}{2\pi}\sqrt{\frac{K}{m}}$$

Specific cases require specific values for m and K. They can often be estimated with sufficient accuracy to be useful in approximate modeling. Higher natural frequencies are simple multiples of the lowest.

The first box on the facing page gives the lowest natural frequencies of the flexural modes of uniform beams with various end constraints. As an example, the first can be estimated by assuming that the effective mass of the beam is one quarter of its real mass, so that

$$m = \frac{m_o L}{4}$$

where m_o is the mass per unit length of the beam, and K is the bending stiffness (given by F/δ from Section B.3); the estimate differs from the exact value by 2%. Vibrations of a tube have a similar form, using I and m_o for the tube. Circumferential vibrations can be found approximately by "unwrapping" the tube and treating it as a vibrating plate, simply supported at two of its four edges.

The second box gives the lowest natural frequencies for flat circular disks with simply supported and clamped edges. Disks with doubly curved faces are stiffer and have higher natural frequencies.

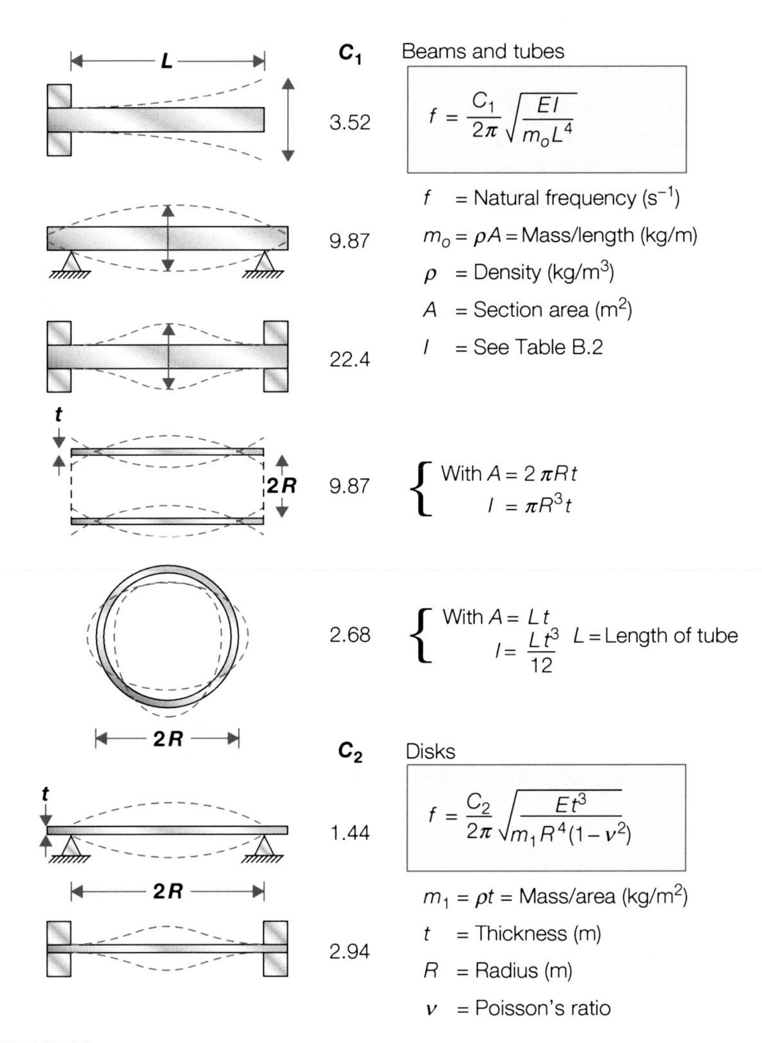

L
C_1
Beams and tubes
3.52
$f = \frac{C_1}{2\pi}\sqrt{\frac{EI}{m_o L^4}}$
f = Natural frequency (s^{-1})
9.87
$m_o = \rho A$ = Mass/length (kg/m)
ρ = Density (kg/m^3)
A = Section area (m^2)
22.4
I = See Table B.2
t
2R
9.87
With $A = 2\pi Rt$
$I = \pi R^3 t$
2.68
With $A = Lt$
$I = \frac{Lt^3}{12}$
L = Length of tube
2R
C_2
Disks
t
1.44
$f = \frac{C_2}{2\pi}\sqrt{\frac{Et^3}{m_1 R^4(1-\nu^2)}}$
2R
$m_1 = \rho t$ = Mass/area (kg/m^2)
2.94
t = Thickness (m)
R = Radius (m)
ν = Poisson's ratio

FIGURE B.12

B.13 CREEP AND CREEP FRACTURE

At temperatures above 1/3 T_m (where T_m is the absolute melting point), materials creep when loaded. It is convenient to characterize the creep of a material by its behavior under a tensile stress, σ, at a temperature T_m. Under these conditions the steady-state tensile strain rate, $\dot{\varepsilon}_{ss}$, is often found to vary as a power of the stress and exponentially with temperature:

$$\dot{\varepsilon}_{ss} = A\left(\frac{\sigma}{\sigma_o}\right)^n exp - \frac{Q}{RT}$$

where Q is an activation energy, A is a kinetic constant, and R is the gas constant. At constant temperature this becomes

$$\dot{\varepsilon}_{ss} = \dot{\varepsilon}_o (\frac{\sigma}{\sigma_o})n$$

where $\dot{\varepsilon}_o$ (s^{-1}), σ_o(N/m^2), and n are creep constants.

The behavior of creeping components is summarized on the facing page, which give the deflection rate of a beam, the displacement rate of an indenter and the change in relative density of cylindrical and spherical pressure vessels in terms of the tensile creep constants.

Prolonged creep causes the accumulation of creep damage that ultimately leads, after a time t_f, to fracture. To a useful approximation

$$t_f \dot{\varepsilon}_{ss} = C$$

where C is a constant characteristic of the material. Creep-ductile materials have values of C between 0.1 and 0.5; creep-brittle materials have values of C as low as 0.01.

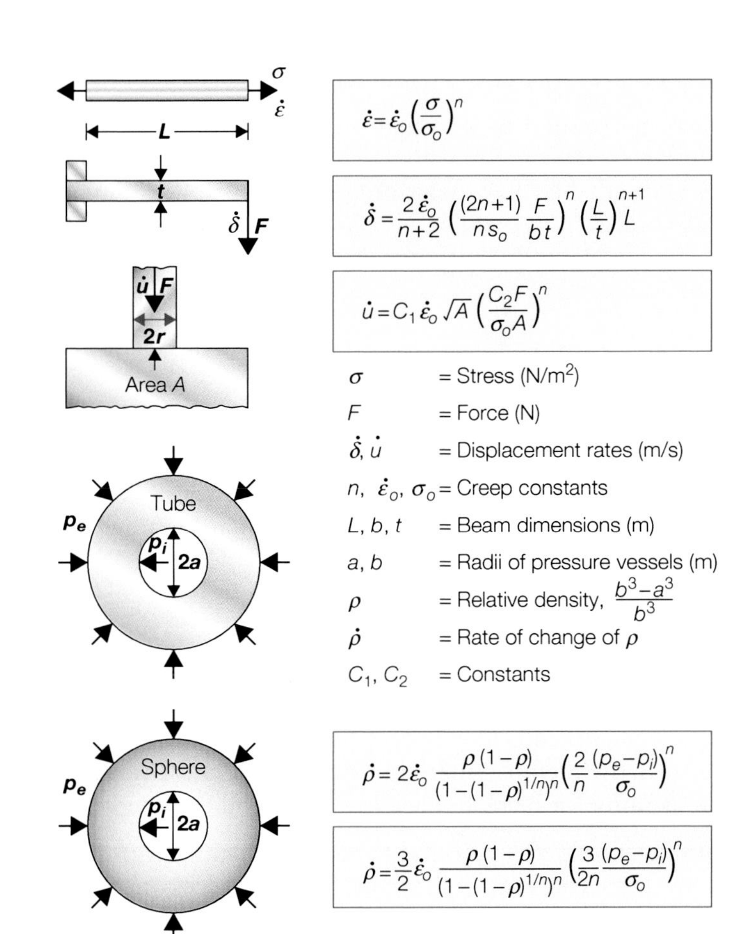

$$\dot{\varepsilon} = \dot{\varepsilon}_o \left(\frac{\sigma}{\sigma_o}\right)^n$$

$$\dot{\delta} = \frac{2\,\dot{\varepsilon}_o}{n+2}\left(\frac{(2n+1)}{n s_o}\,\frac{F}{bt}\right)^n \left(\frac{L}{t}\right)^{n+1} L$$

$$\dot{u} = C_1\,\dot{\varepsilon}_o\,\sqrt{A}\left(\frac{C_2 F}{\sigma_o A}\right)^n$$

σ = Stress (N/m^2)

F = Force (N)

$\dot{\delta}, \dot{u}$ = Displacement rates (m/s)

$n,\ \dot{\varepsilon}_o,\ \sigma_o$ = Creep constants

L, b, t = Beam dimensions (m)

a, b = Radii of pressure vessels (m)

ρ = Relative density, $\frac{b^3 - a^3}{b^3}$

$\dot{\rho}$ = Rate of change of ρ

C_1, C_2 = Constants

$$\dot{\rho} = 2\dot{\varepsilon}_o\,\frac{\rho\,(1-\rho)}{(1-(1-\rho)^{1/n})^n}\left(\frac{2}{n}\,\frac{(p_e - p_i)}{\sigma_o}\right)^n$$

$$\dot{\rho} = \frac{3}{2}\dot{\varepsilon}_o\,\frac{\rho\,(1-\rho)}{(1-(1-\rho)^{1/n})^n}\left(\frac{3}{2n}\,\frac{(p_e - p_i)}{\sigma_o}\right)^n$$

FIGURE B.13

B.14 FLOW OF HEAT AND MATTER

Heat flow can be limited by conduction, by convection, or by radiation. The constitutive equations for each are listed on the facing page. The first equation is Fourier's first law, describing steady-state heat flow; it contains the thermal conductivity, λ. The second is Fourier's second law, which treats transient heat-flow problems; it contains the thermal diffusivity, a, defined by

$$a = \frac{\lambda}{\rho C}$$

where ρ is the density and C is the specific heat at constant pressure. Solutions to these two differential equations are given in Section B.15.

The third equation describes convective heat transfer. It, rather than conduction, limits heat flow when the Biot number is

$$B_i = \frac{hs}{\lambda} < 1$$

where h is the heat-transfer coefficient and s is a characteristic dimension of the sample. When, instead, $B_i > 1$, heat flow is limited by conduction. The final equation is the Stefan-Boltzmann law for radiative heat transfer. The emissivity, ε, is unity for black bodies; less for all other surfaces.

Diffusion of matter follows a pair of differential equations with the same form as Fourier's two laws, and with similar solutions. They are commonly written

$$J = -D\nabla C = -D\frac{dC}{dx} \text{ (steady state)}$$

and

$$\frac{\partial C}{\partial t} = D\nabla^2 C = D\frac{\partial^2 C}{\partial x^2} \text{ (time-dependent flow)}$$

where J is the flux, C is the concentration, x is the distance, and t is time. Solutions are given in Section B.15.

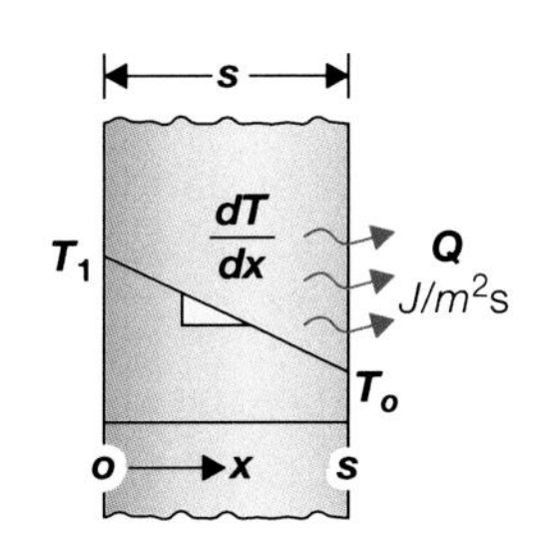

$$Q = -\lambda \nabla T = -\lambda \frac{dT}{dx}$$

Q = Heat flux (J/m²s)

T = Temperature (K)

x = Distance (m)

λ = Thermal conductivity (W/mK)

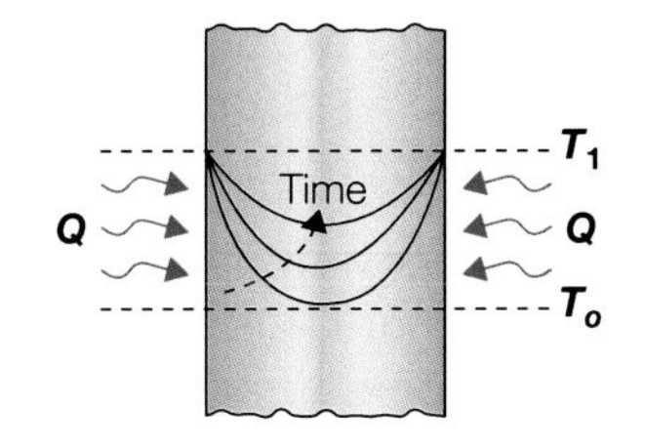

$$\frac{\delta T}{\delta t} = a \nabla^2 T = a \frac{\delta^2 T}{\delta x^2}$$

t = Time (s)

ρ = Density (kg/m³)

C_p = Specific heat (J/kg.K)

a = Thermal diffusivity, $\frac{\lambda}{\rho Cp}$ (m²/s)

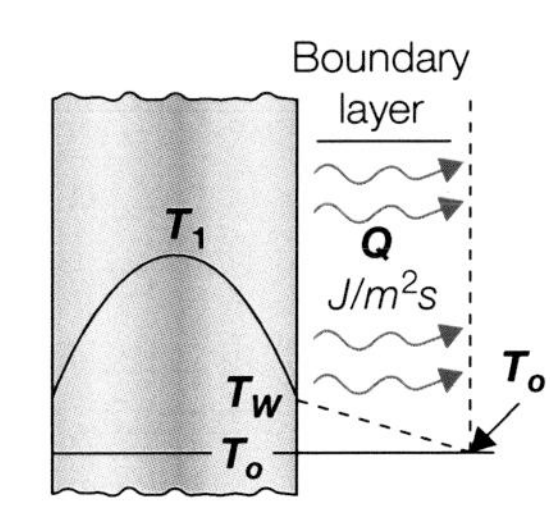

$$Q = h(T_w - T_o)$$

T_w = Surface temperature (K)

T = Fluid temperature (K)

h = Heat transfer coefficient (W/m²K)

= 5–50 W/m².K in air

= 1,000–5,000 W/m².K in water

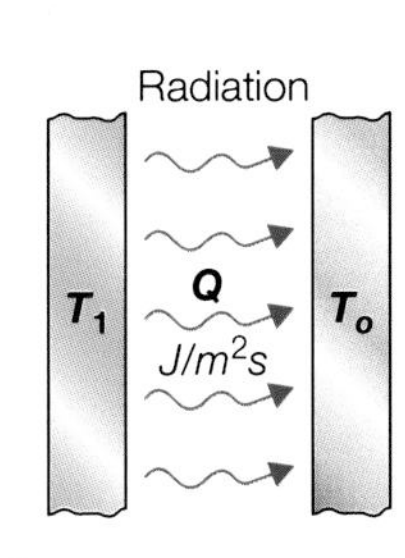

$$Q = \varepsilon \sigma (T_1^4 - T_o^4)$$

ε = Emissivity (1 for black body)

σ = Stefan's constant

= 5.67×10^{-8} W/m².K⁴

FIGURE B.14

B.15 SOLUTIONS FOR DIFFUSION EQUATIONS

Solutions exist for the diffusion equations for a number of standard geometries. They are worth knowing because many real problems can be approximated by one of these.

At steady state the temperature or concentration profile does not change with time. This is expressed by the boxed equations within the first box at the top of the facing page. Solutions for these are given below for uniaxial flow, radial flow in a cylinder, and radial flow in a sphere. The solutions are fitted to individual cases by matching the constants A and B to the boundary conditions. Solutions for matter flow are found by replacing temperature, T, by concentration, C, and replacing conductivity, λ, by diffusion coefficient, D.

The box within the second large box summarizes the governing equations for time-dependent flow, assuming that the diffusivity (a or D) is not a function of position. Solutions for the temperature or concentration profiles, $T(x,t)$ or $C(x,t)$, are given below. The first equation gives the "thin-film" solution: A thin slab at temperature T_1 or concentration C_1 is sandwiched between two semi-infinite blocks at T_o or C_o, at $t = 0$, and flow allowed. The second result is for two semi-infinite blocks, initially at T_1 and T_o (or C_1 or C_o), brought together at $t = 0$. The last is for a T or C profile that is sinusoidal, of wavelength $\lambda/2\pi$ and amplitude A at $t = 0$.

Note that all transient problems end up with a characteristic time constant, t^*, with

$$t^* = \frac{x^2}{\beta a} \quad or \quad \frac{x^2}{\beta D}$$

where x is a dimension of the specimen; or they end up with a characteristic length, x^*, with

$$x^* = \sqrt{\beta a t} \quad or \quad \sqrt{\beta D t}$$

where t is the time scale of observation, with $1 < \beta < 4$, depending on geometry.

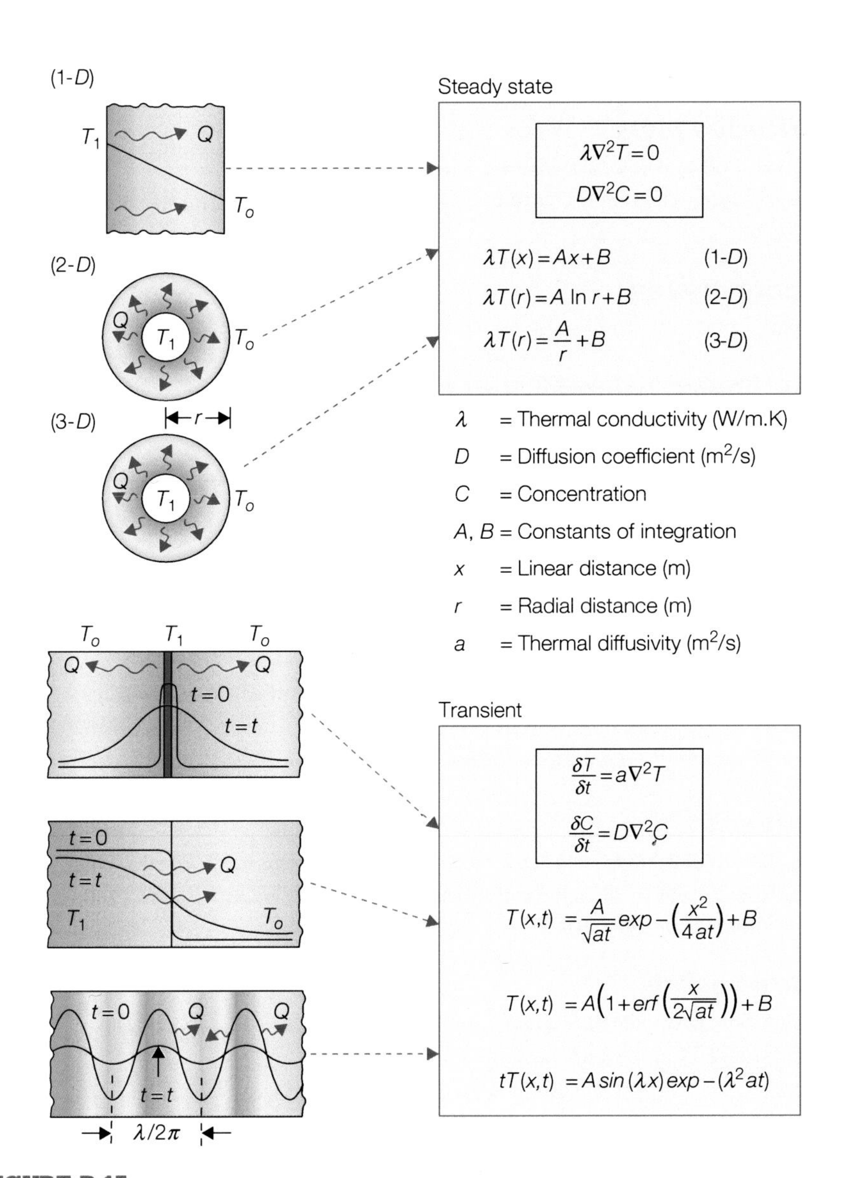
(1-D)
T_1
Q
T_o
(2-D)
Q
T_1
T_o
(3-D)
r
Q
T_1
T_o
Steady state
$\lambda \nabla^2 T = 0$
$D \nabla^2 C = 0$
$\lambda T(x) = Ax + B$ (1-D)
$\lambda T(r) = A \ln r + B$ (2-D)
$\lambda T(r) = \frac{A}{r} + B$ (3-D)
λ = Thermal conductivity (W/m.K)
D = Diffusion coefficient (m^2/s)
C = Concentration
A, B = Constants of integration
x = Linear distance (m)
r = Radial distance (m)
a = Thermal diffusivity (m^2/s)
T_o
T_1
T_o
Q
Q
$t=0$
$t=t$
$t=0$
$t=t$
T_1
Q
T_o
$t=0$
Q
Q
$t=t$
$\lambda/2\pi$
Transient
$\frac{\delta T}{\delta t} = a \nabla^2 T$
$\frac{\delta C}{\delta t} = D \nabla^2 C$
$T(x,t) = \frac{A}{\sqrt{at}} exp - \left(\frac{x^2}{4at}\right) + B$
$T(x,t) = A\left(1 + erf\left(\frac{x}{2\sqrt{at}}\right)\right) + B$
$tT(x,t) = A \sin(\lambda x) exp - (\lambda^2 at)$

FIGURE B.15

B.16 FURTHER READING

Constitutive laws

Cottrell, A.H. (1964). *Mechanical properties of matter.* Wiley.

Gere, J.M., & Timoshenko, S.P. (1985). *Mechanics of materials* (2nd SI ed.). Wadsworth International.

Moments of area

Young, W.C. (1989). *Roark's formulas for stress and strain* (6th ed.). McGraw-Hill.

Beams, shafts, columns, and shells

Calladine, C.R. (1983). *Theory of shell structures.* Cambridge University Press.

Gere, J.M., & Timoshenko, S.P. (1985). *Mechanics of materials* (2nd ed.). Wadsworth International.

Timoshenko, S.P., & Goodier, J.N. (1970). *Theory of elasticity* (3rd ed.). McGraw-Hill.

Timoshenko, S.P., & Gere, J.M. (1961). *Theory of elastic stability* (2nd ed.). McGraw-Hill.

Young, W.C. (1989). *Roark's formulas for stress and strain* (6th ed.). McGraw-Hill.

Contact stresses and stress concentration

Timoshenko, S.P., & Goodier, J.N. (1970). *Theory of elasticity* (3rd ed.). McGraw-Hill.

Hill, R. (1950). *Plasticity.* Oxford University Press.

Johnson, K.L. (1985). *Contact mechanics.* Oxford University Press.

Sharp cracks

Hertzberg, R.W. (1989). *Deformation and fracture of engineering materials* (3rd ed.). Wiley.

Tada, H., Paris, P.C., & Irwin, G.R. (1985). *The stress analysis of cracks handbook* (2nd ed.). Paris Productions and Del Research Group.

Pressure vessels

Timoshenko, S.P., & Goodier, J.N. (1970). *Theory of elasticity* (3rd ed.). McGraw-Hill.

Hill, R. (1950). *Plasticity.* Oxford University Press.

Young, W.C. (1989). *Roark's formulas for stress and strain* (6th ed.). McGraw-Hill.

Vibration

Young, W.C. (1989). *Roark's formulas for stress and strain* (6th ed.). McGraw-Hill.

Creep

Finnie, I., & Heller, W.R. (1976). *Creep of engineering materials.* McGraw-Hill.

Heat and matter flow

Hollman, J.P. (1981). *Heat transfer* (5th ed.). McGraw-Hill.

Carslaw, H.S., & Jaeger, J.C. (1959). *Conduction of heat in solids* (2nd ed.). Oxford University Press.

Shewmon, P.G. (1989). *Diffusion in solids* (2nd ed.). TMS.

Appendix C: Material Indices

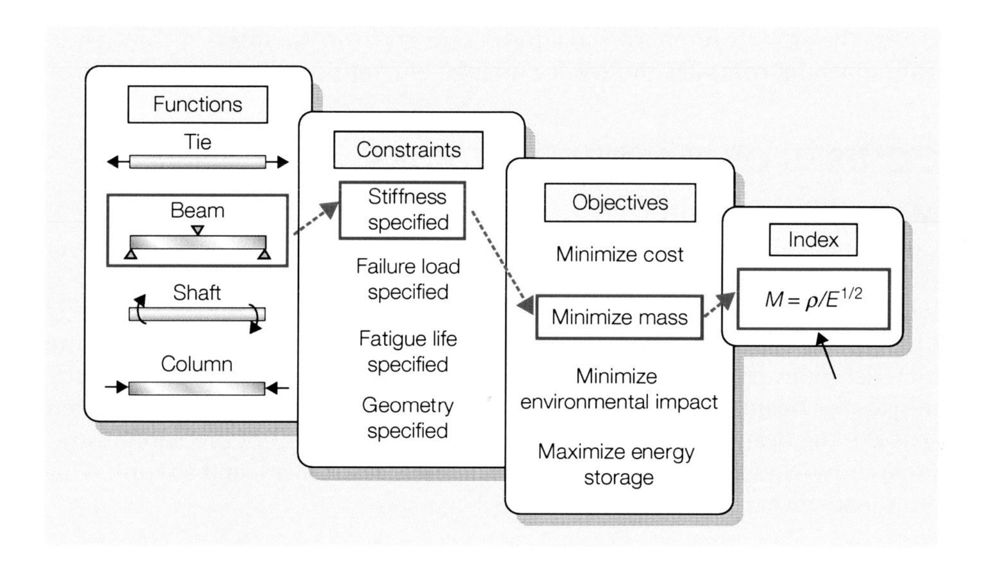

CONTENTS

Materials Selection in Mechanical Design. DOI: 10.1016/B978-1-85617-663-7.00020-5

C.1 INTRODUCTION AND SYNOPSIS

The performance, P, of a component is characterized by a performance equation. The performance equation contains groups of material properties. These groups are the material indices. Sometimes the "group" is a single property; thus if the performance of a beam is measured by its stiffness, the performance equation contains only one property, the elastic modulus, E. It is the material index for this problem. More commonly the performance equation contains a group of two or more properties. Familiar examples are the specific stiffness, E/ρ, and the specific strength, σ_y/ρ (where σ_y is the yield strength or elastic limit, and ρ is the density), but there are many others. They help to determine the optimal selection of materials. Details of the method, with numerous examples, are given in Chapters 5 through 8. This appendix compiles indices for a range of common applications.

C.2 USES OF MATERIAL INDICES

Material selection

Components have functions: to carry loads safely, to transmit heat, to store energy, to insulate, and so forth. Each function has an associated material index. Materials with high values of the appropriate index maximize that aspect of the performance of the component. For reasons given in Chapter 5, the material index is generally independent of the details of the design. Thus the indices for beams in the tables that follow are independent of the detailed shape of the beam; that for minimizing thermal distortion of precision instruments is independent of the configuration of the instrument, and so forth. This gives them great generality.

Material deployment or substitution

A new material may have potential application in functions for which its indices have unusually high or low values. Fruitful applications for a new material can be identified by evaluating its indices and comparing them with those of existing, established materials. Similar reasoning points the way to identifying viable substitutes for an incumbent material in an established application.

How to read the tables

The indices listed in Tables C.1 through C.7 are, for the most part, based on the objective of minimizing mass. To minimize material cost, use the index for minimum mass, replacing the density, ρ, with the cost per unit volume, $C_m\rho$, where C_m is the cost per kg. To minimize material-embodied energy or CO_2 burden, replace ρ with $H_pa \cdot \rho$ or by $CO_2 \cdot \rho$, where H_p is the production energy per kg and CO_2 is the CO_2 burden per kg (Table A.10).

The symbols

The symbols used in Tables C.1 through C.7 are defined in the following table.

Class	Property	Symbol and Units
General	Density	ρ (kg/m^3)
	Price	C_m (\$/kg)
Mechanical	Elastic moduli (Young's, shear, bulk)	E, G, K (GPa)
	Poisson's ratio	ν (–)
	Failure strength (yield, fracture)	σ_f (MPa)
	Fatigue strength	σ_e (MPa)
	Hardness	H (Vickers)
	Fracture toughness	K_{1c} (MPa.m$^{1/2}$)
	Loss coefficient (damping capacity)	η (–)
Thermal	Thermal conductivity	λ (W/m.K)
	Thermal diffusivity	a (m^2/s)
	Specific heat	C_p (J/kg.K)
	Thermal expansion coefficient	α (°K^{-1})
	Difference in thermal conductivity	$\Delta\alpha$ (°K^{-1})
Electrical	Electrical resistivity	ρ_e ($\mu\Omega$.cm)
Eco-properties	Energy/kg to produce material per kg CO_2/kg burden of material production per kg	E_p (MJ/kg) CO_2 (kg/kg)

Table C.1 Stiffness-limited Design at Minimum Mass

Function and Constraints	Maximize
TIE (tensile strut)	
Stiffness, length specified; section area free	E/ρ
SHAFT (loaded in torsion)	
Stiffness, length, shape specified; section area free	$G^{1/2}/\rho$
Stiffness, length, outer radius specified; wall thickness free	G/ρ
Stiffness, length, wall thickness specified; outer radius free	$G^{1/3}/\rho$
BEAM (loaded in bending)	
Stiffness, length, shape specified; section area free	$E^{1/2}/\rho$
Stiffness, length, height specified; width free	E/ρ
Stiffness, length, width specified; height free	$E^{1/3}/\rho$
COLUMN (compression strut, failure by elastic buckling)	
Buckling load, length, shape specified; section area free	$E^{1/2}/\rho$
PANEL (flat plate, loaded in bending)	
Stiffness, length, width specified; thickness free	$E^{1/3}/\rho$
PLATE (flat plate, compressed in-plane, buckling failure)	
Collapse load, length, width specified; thickness free	$E^{1/3}/\rho$
CYLINDER WITH INTERNAL PRESSURE	
Elastic distortion, pressure, radius specified; wall thickness free	E/ρ
SPHERICAL SHELL WITH INTERNAL PRESSURE	
Elastic distortion, pressure, radius specified; wall thickness free	$E/(1-\nu)\rho$

Table C.2 Strength-limited Design at Minimum Mass

Function and Constraints	Maximize
TIE (tensile strut)	
Stiffness, length specified; section area free	σ_f/ρ
SHAFT (loaded in torsion)	
Load, length, shape specified; section area free	$\sigma_f^{2/3}/\rho$
Load, length, outer radius specified; wall thickness free	σ_f/ρ
Load, length, wall thickness specified; outer radius free	$\sigma_f^{1/2}/\rho$
BEAM (loaded in bending)	
Load, length, shape specified; section area free	$\sigma_f^{2/3}/\rho$
Load length, height specified; width free	σ_f/ρ
Load, length, width specified; height free	$\sigma_f^{1/2}/\rho$
COLUMN (compression strut)	
Load, length, shape specified; section area free	σ_f/ρ
PANEL (flat plate, loaded in bending)	
Stiffness, length, width specified; thickness free	$\sigma_f^{1/2}/\rho$
PLATE (flat plate, compressed in-plane, buckling failure)	
Collapse load, length, width specified; thickness free	$\sigma_f^{1/2}/\rho$
CYLINDER WITH INTERNAL PRESSURE	
Elastic distortion, pressure, radius specified; wall thickness free	σ_f/ρ
SPHERICAL SHELL WITH INTERNAL PRESSURE	
Elastic distortion, pressure, radius specified; wall thickness free	σ_f/ρ
FLYWHEELS, ROTATING DISKS	
Maximum energy storage per unit volume; given velocity	ρ
Maximum energy storage per unit mass; no failure	σ_f/ρ

Table C.3 Strength-limited Design: Springs and Hinges

Function and Constraints	Maximize
SPRINGS	
Maximum stored elastic energy per unit volume; no failure	σ_f^2/E
Maximum stored elastic energy per unit mass; no failure	$\sigma_f^2/E\rho$
ELASTIC HINGES	
Radius of bend to be minimized (max flexibility without failure)	σ_f/E
KNIFE EDGES, PIVOTS	
Minimum contact area, maximum bearing load	σ_f^3/E^2 and H
COMPRESSION SEALS AND GASKETS	
Maximum conformability; limit on contact pressure	$\sigma_f^{3/2}/E$ and $1/E$
DIAPHRAGMS	
Maximum deflection under specified pressure or force	$\sigma_f^{3/2}/E$
ROTATING DRUMS AND CENTRIFUGES	
Maximum angular velocity; radius fixed; wall thickness free	σ_f/ρ

Table C.4 Vibration-limited Design

Function and Constraints	Maximize
TIES, COLUMNS	
Maximum longitudinal vibration frequencies	E/ρ
BEAMS, all dimensions prescribed	
Maximum flexural vibration frequencies	E/ρ
BEAMS, length and stiffness prescribed	
Maximum flexural vibration frequencies	$E^{1/2}/\rho$
PANELS, all dimensions prescribed	
Maximum flexural vibration frequencies	E/ρ
PANELS, length, width, stiffness prescribed	
Maximum flexural vibration frequencies	$E^{1/3}/\rho$
TIES, COLUMNS, BEAMS, PANELS (stiffness prescribed)	
Minimum longitudinal excitation from external drivers, ties	$\eta E/\rho$
Minimum flexural excitation from external drivers, beams	$\eta E^{1/2}/\rho$
Minimum flexural excitation from external drivers, panels	$\eta E^{1/3}/\rho$

Table C.5 Damage-tolerant Design

Function and Constraints	Maximize
TIES (tensile member)	
Maximum flaw tolerance and strength, load-controlled design	K_{1c} and σ_f
Maximum flaw tolerance and strength, displacement control	K_{1c}/E and σ_f
Maximum flaw tolerance and strength, energy control	K_{1c}^2/E and σ_f
SHAFTS (loaded in torsion)	
Maximum flaw tolerance and strength, load-controlled design	K_{1c} and σ_f
Maximum flaw tolerance and strength, displacement control	K_{1c}/E and σ_f
Maximum flaw tolerance and strength, energy control	K_{1c}^2/E and σ_f
BEAMS (loaded in bending)	
Maximum flaw tolerance and strength, load-controlled design	K_{1c} and σ_f
Maximum flaw tolerance and strength, displacement control	K_{1c}/E and σ_f
Maximum flaw tolerance and strength, energy control	K_{1c}^2/E and σ_f
PRESSURE VESSEL	
Yield-before-break	K_{1c}/σ_f
Leak-before-break	K_{1c}^2/σ_f

Table C.6 Electro-mechanical Design

Function and Constraints	Maximize
BUS BARS	
Minimum life cost; high current conductor	$1/\rho_e \rho C_m$
ELECTRO-MAGNET WINDINGS	
Maximum short-pulse field; no mechanical failure	σ_f
Maximum field and pulse length; limit on temperature rise	$C_p \rho / \rho_e$
WINDINGS, HIGH-SPEED ELECTRIC MOTORS	
Maximum rotational speed; no fatigue failure	σ_e / ρ_e
Minimum ohmic losses; no fatigue failure	$1/\rho_e$
RELAY ARMS	
Minimum response time; no fatigue failure	$\sigma_e / E \rho_e$
Minimum ohmic losses; no fatigue failure	$\sigma_e^2 / E \rho_e$

Table C.7 Thermal and Thermo-mechanical Design

Function and Constraints	Maximize
THERMAL INSULATION MATERIALS	
Minimum heat flux at steady state; thickness specified	$1/\lambda$
Minimum temp rise in specified time; thickness specified	$1/a = \rho C_p / \lambda$
Minimize total energy consumed in thermal cycle (kilns, etc.)	$\sqrt{a}/\lambda = \sqrt{1/\lambda \rho C_p}$
THERMAL STORAGE MATERIALS	
Maximum energy stored/unit material cost (storage heaters)	C_p / C_m
Maximize energy stored for given temperature rise and time	$\lambda / \sqrt{a} = \sqrt{\lambda \rho C_p}$
PRECISION DEVICES	
Minimize thermal distortion for given heat flux	λ / α
THERMAL SHOCK RESISTANCE	
Maximum change in surface temperature; no failure	$\sigma_f / E \alpha$
HEAT SINKS	
Maximum heat flux per unit volume; expansion limited	$\lambda / \Delta \alpha$
Maximum heat flux per unit mass; expansion limited	$\lambda / \rho \Delta \alpha$
HEAT EXCHANGERS (pressure-limited)	
Maximum heat flux per unit area; no failure under Δp	$\lambda \sigma_f$
Maximum heat flux per unit mass; no failure under Δp	$\lambda \sigma_f / \rho$

Appendix D: Data Sources for Documentation

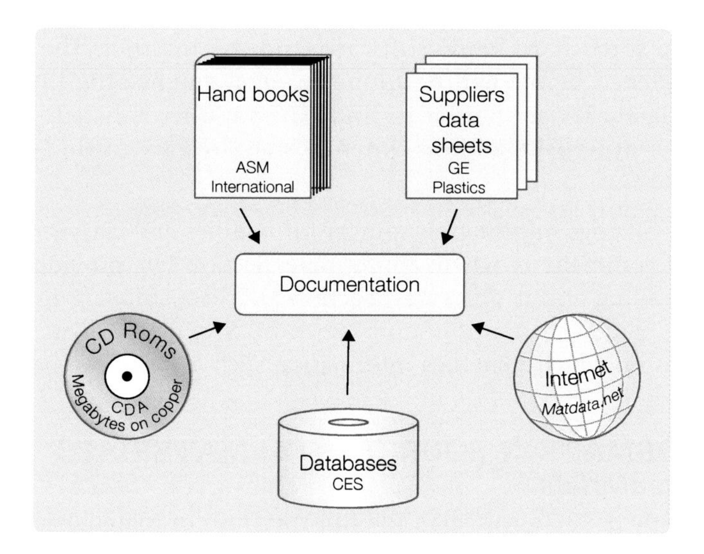

Materials Selection in Mechanical Design. DOI: 10.1016/B978-1-85617-663-7.00021-7

CONTENTS

D.1 INTRODUCTION

This appendix tells you where to look to find information, both structured and unstructured, for material attributes. The sources, broadly speaking, are of three sorts: hard copy, software, and the Internet. The hard-copy documents listed below will be found in most engineering libraries. The computer databases are harder to find: The supplier is listed, along with the most recent website address. Internet sites are easy to find but can be frustrating to use.

Section D.2 catalogs information sources for families and classes of material, with a brief commentary where appropriate. Section D.3 provides a starting point for reading about processes. Section D.4 lists software for materials and process data, information, and selection. Section D.5 lists additional Internet sites on which materials information such as price can be found.

D.2 INFORMATION SOURCES FOR MATERIALS

D2.1 All materials

Few hard-copy data sources span the full spectrum of materials and properties. Ten that, in different ways, attempt to do so are listed here.

Materials Selector (1997). *Materials Engineering, Special Issue*. Penton Publishing.
Tabular data for a broad range of metals, ceramics, polymers, and composites.

Waterman, N.A., & Ashby, M.F. (Eds.) (1996). *The Chapman and Hall materials selector*. Chapman and Hall.
A three-volume compilation of data for all materials, with selection and design guide. Basic reference work.

ASM handbook series (1980–2010). ASM International.
More than 20 volumes of information about materials, emphasizing metals. Now available online.

Bauccio, M.L. (Ed.) (1994). *ASM engineered materials reference book* (2nd ed.). ASM International.
Compact compilation of numeric data for metals, polymers, ceramics, and composites.

Materials selector and design guide (1974). Design Engineering, Morgan-Grampian Ltd.
Resembles the Materials Engineering Materials Selector, *but less detailed and now rather dated.*

Handbook of industrial materials (2nd ed.) (1992). Elsevier.
A compilation of data remarkable for its breadth: metals, ceramics, polymers, composites, fibers, sandwich structures, leather ...

Brady, G.S., & Clauser, H.R. (Eds.) (1986). *Materials handbook* (12th ed.). McGraw-Hill.
A broad survey, covering metals, ceramics, polymers, composites, fibers, sandwich structures, and more.

Cardarelli, F. (1999). *Materials handbook*. Springer-Verlag.
A remarkable compilation of material data spanning mechanical, thermal, and electrical properties of all material classes.

Goldsmith, A., Waterman, T.E., & Hirschhorn, J.J. (1961). *Handbook of thermophysical properties of solid materials*. MacMillan.
Thermophysical and thermochemical data for elements and compounds.

Bittence, J.C. (Ed.) (1994). *Guide to engineering materials producers*. ASM International.
A comprehensive catalog of addresses for material suppliers.

Internet sources of information on all classes of materials

ASM Handbooks online, *www.asminternational.org/hbk/index.jsp*

ASM Alloy center, *www.asminternational.org/alloycenter/index.jsp*

ASM Materials Information, *www.asminternational.org/matinfo/index.jsp*

A to Z of Materials, *www.azom.com*

Design InSite, *www.designinsite.dk*

Goodfellow, *www.goodfellow.com*

K&K Associate's thermal connection, *www.tak2000.com*

Corrosion Source (databases), *www.corrosionsource.com*

Material Data Network, *www.matdata.net*

Materials (Research): Alfa Aesar, *www.alfa.com*

MatWeb, *www.matweb.com*

MSC datamart, *www.mscsoftware.com*

NASA Long Duration Exposure Facility, SETAS, *http://setas-www.larc.nasa.gov/LDEF/*

NPL MIDAS, *midas.npl.co.uk/midas/index.jsp*

D2.2 All metals

Metals and alloys conform to national and (sometimes) international standards. One consequence is the high quality of data. Hard-copy sources for metals data are generally comprehensive, well-structured, and easy to use.

ASM metals handbook (9th and 10th eds.) (1986, 1990). ASM International.
Basic reference work, continuously upgraded and expanded. The 10th edition contains Vol. 1: Irons and steels; Vol. 2: Non-ferrous alloys; Vol. 3: Heat treatment; Vol. 4: Friction, lubrication, and wear; Vol. 5: Surface finishing and coating; Vol. 6: Welding and brazing; Vol. 7: Microstructural analysis. *More volumes have been released.*

Bauccio, M.L. (Ed.) (1993). *ASM metals reference book* (3rd ed.). ASM International.
Consolidates data for metals from a number of ASM publications. Basic reference work.

Smithells, C.J. (1992). *Metals reference book* (7th ed.). E.A. Brandes & G.B. Brook (Eds.). Butterworths.
A comprehensive compilation of data for metals and alloys. Basic reference work.

Robb, C. (1990). *Metals databook*. The Institute of Metals.
A concise collection of data on metallic materials covered by U.K. specifications only.

ASM guide to materials engineering data and information (1986). ASM International.
A directory of suppliers, trade organizations, and publications on metals.

Bringas, J.E. (Ed.) (1992). *The metals black book, vol. 1, steels*. Casti Publishing.
A compact book of data for steels.

Bringas, J.E. (Ed.) (1993). *The metals red book, vol. 2, nonferrous metals*. Casti Publishing.

Internet sources for two or more metals classes

ASM International Handbooks, *www.asminternational.org/hbk/index.jsp*

ASM International, Alloy Center, *www.asminternational.org/alloycenter/index.jsp*

Carpenter Technology home page, *www.cartech.com*

CASTI Publishing site catalog, *www.casti-publishing.com/intsite.htm*

CMW Inc. home page, *www.cmwinc.com*

Eurometaux, European Association of Metals, *www.eurometaux.org*

Materials (high performance): MatTech, *www.mat-tech.com*

Metsteel.com, *www.metsteel.com*

Rare earths: Pacific Industrial Development Corp., *www.pidc.com*

Rare metals: Stanford Materials Inc., *www.stanfordmaterials.com*

Refractory metals: Teledyne Wah Chang, *www.twca.com*

D2.3 Nonferrous metals and alloys

In addition to the references listed in Section D2.2, the following sources give data for specific metals and alloys

Pure metals

Most of the sources listed in the previous section contain some information on pure metals. However, the publiations listed below are particularly useful in this respect.

Emsley, J. (1989). *The elements*. Oxford University Press.
A book aimed more at chemists and physicists than engineers, with good coverage of chemical, thermal, and electrical properties but not mechanical properties.

Brandes, E.A., & Brook, G.B. (Eds.) (1992). *Smithells metals reference book* (7th edition), Butterworth-Heinemann.
Data for the mechanical, thermal, and electrical properties of pure metals.

Goodfellow catalogue (1995–96). Goodfellow Cambridge Limited.
Useful though patchy data for mechanical, thermal, and electrical properties of pure metals in a tabular format. Free.

Alfa Aesar catalog (1995–96). Johnson Matthey Catalog Co. Inc.
Coverage similar to that of the Goodfellow catalogue. Free.

Samsonov, G.V. (Ed.) (1968). *Handbook of the physiochemical properties of the elements*. Oldbourne.
An extensive compilation of data from Western and Eastern sources. Contains a number of inaccuracies, but also contains a large quantity of data on the rarer elements, hard to find elsewhere.

Gschneidner, K.A. (1964). Physical properties and interrelationships of metallic and semimetallic elements. *Solid State Physics, 16*, 275–426.
Probably the best source of its time, this work is very well referenced, and full explanations are given of estimated or approximate data.

Internet sources

Winter, M. WebElements. University of Sheffield, *www.webelements.com*
A comprehensive source of information on all the elements in the Periodic Table. If it has a weakness, it is in the definitions and values of some mechanical properties.

Martindale's: physics web pages, *www.martindalecenter.com/Calculators3A.html*

Aluminum alloys

Aluminum standards and data. (1990). The Aluminum Association.

The properties of aluminum and its alloys. (1981). The Aluminum Federation.

Technical data sheets. (1993). ALCAN International Ltd.

Technical data sheets. (1993). ALCOA.

Technical data sheets. (1994). Aluminum Pechiney, France.

Internet sources

Aluminum Federation, *www.alfed.org.uk*

Aluminum World, *www.sovereign-publications.com/aluminum.htm*

International Aluminum Institute, *www.world-aluminum.org*

Babbitt Metal

The term "Babbitt Metal" denotes a series of lead-tin-antimony–bearing alloys, the first of which was patented in the United States by Isaac Babbitt in 1839. Subsequent alloys are all variations on his original composition.

ASTM Standard B23-83: White metal bearing alloys (known commercially as "Babbitt Metal"), *ASTM Annual Book of Standards*, Vol. 02.04.

Beryllium

Designing with beryllium (1996). Brush Wellman.

Beryllium optical materials (1996). Brush Wellman.

Internet sources

Brush Wellman, *www.brushwellman.com*

Cadmium

Cadmium production, properties and uses (1991). International Cadmium Association.

Chromium

Castings, chromium-nickel alloy, ASTM Standard A560-89: *ASTM Annual Book of Standards, 01*(02).

Cobalt alloys

Betteridge, W. (1982). *Cobalt and its alloys.* Ellis Horwood.
A good general introduction to the subject.

Columbium alloys

See Niobium alloys.

Copper alloys

ASM Metals Handbook (10th ed.) (1990). ASM International

West, E.G. (1979). *The selection and use of copper-rich alloys.* Oxford University Press for the Design Council, the British Standards Institution, and the Council of Engineering Institutions.

Copper Development Association data sheets, 26 (1988), 27 (1981), 31 (1982), 40 (1979), and 82 (1982). Copper Development Association, and The Copper Development Association.

Megabytes on coppers (1994). The Copper Development Association and Granta Design Limited.

Smithells metals reference book, (7th ed.). (1992). E.A. Brandes & G.B. Brook (Eds.). Butterworth-Heinemann.

Internet sources

Copper Development Association, *www.cda.org.uk*

Copper page, *www.copper.org*

Gold and dental alloys

Gold: Art, science and technology (1992). *Interdisciplinary Science Reviews, 17*(3). The Institute of Materials, ISSN 0308-0188.

Focus on gold (1992). *Interdisciplinary Science Reviews, 17*(4). The Institute of Materials, ISSN 0308-0188.

ISO Standard 1562: Dental casting gold alloys (1993). International Standards Organization.

ISO Standard 8891: Dental casting alloys with Noble metal content of 25% up to but not including 75% (1993). International Standards Organization.

Internet sources

Goodfellow Metals, *www.goodfellow.com*
Supplier of pure and precious metals, mainly for laboratory use. Website contains price and property information for its entire stock.

Jeneric Pentron, Casting alloys, *www.jeneric.com/casting*
An informative commercial site. An extensive source of information, both for natural biological materials and for metals used in dental treatments.

Rand Refinery Limited, Chamber of Mines website, *www.bullion.org.za/associates/rr.htm*
Contains useful information on how gold is processed to varying degrees of purity.

Indium

The Indium Info Center, Indium Corp. of America. *www.indium.com*

Lead

ASTM Standard B29-79: Pig lead. *ASTM Annual Book of Standards, 02*(04).

ASTM Standard B102-76: Lead- and tin-alloy die castings. *ASTM Annual Book of Standards, 02*(04).

ASTM Standard B749-85: Lead and lead alloy strip, sheet, and plate products, *ASTM Annual Book of Standards, 02*(04).

Lead for corrosion resistant applications. Lead Industries Association.

ASM metals handbook (9th ed.). (1986). Vol. 2, 500–510.
See also Babbitt Metal (listed previously).

Internet sources

Lead Development Association International, *www.ldaint.org*

India Lead Zinc Development Association, *www.ilzda.com*

Magnesium alloys

Technical data sheets (1994). Magnesium Elektron Ltd.

Technical literature (1994). Magnesium Corp. of America.

Molybdenum

ASTM Standard B386-85: Molybdenum and molybdenum alloy plate, sheet, strip and foil. *ASTM Annual Book of Standards, 02*(04).

ASTM Standard B387-85: Molybdenum and molybdenum alloy bar, rod and wire. *ASTM Annual Book of Standards, 02*(04).

Nickel

A major data source for nickel and its alloys is the Nickel Development Institute (NIDI), a global organization with offices on every continent except Africa. NIDI freely gives away large quantities of technical reports and data compilations, not only for nickel and high-nickel alloys but also for other nickel-bearing alloys, such as stainless steel.

ASTM Standard A297-84: Steel castings, iron-chromium and iron-chromium-nickel, heat resistant, for general application. *ASTM Annual Book of Standards, 01*(02).

ASTM Standard A344-83, Drawn or rolled nickel-chromium and nickel-chromium-iron alloys for electrical heating elements. *ASTM Annual Book of Standards, 02*(04).

ASTM Standard A494-90: Castings, nickel and nickel alloy. *ASTM Annual Book of Standards, 02*(04).

ASTM Standard A753-85: Nickel-iron soft magnetic alloys. *ASTM Annual Book of Standards, 03*(04).

Betteridge, W. (1984). *Nickel and its alloys.* Ellis Horwood.
A good introduction to the subject.

INCO (1995). *High-temperature, high-strength nickel base alloys.* Nickel Development Institute.
Tabular data for over 80 alloys.

Elliott, P. (1990). *Practical guide to high-temperature alloys.* Nickel Development Institute.

INCO (1978). *Heat & corrosion resistant castings.* Nickel Development Institute.

INCO (1969). *Engineering properties of some nickel copper casting alloys.* Nickel Development Institute.

INCO (1968). *Engineering properties of IN-100 alloy.* Nickel Development Institute. INCO (1969). *Engineering properties of nickel-chromium alloy 610 and related casting alloys.* Nickel Development Institute.

INCO (1968). *Alloy 713C: technical data.* Nickel Development Institute.

INCO (1981). *Alloy IN-738: technical data.* Nickel Development Institute.

INCO (1976). *36% nickel-iron alloy for low temperature service.* Nickel Development Institute.

ASTM Standard A658 (discontinued 1989): Pressure vessel plates, alloy steel, 36 percent nickel. *ASTM Annual Book of Standards,* pre-1989 editions.

ASM Metals Handbook (9th ed.) (1986). Vol. 3, 125–178.

Internet sources

Carpenter Technology Corp, *www.cartech.com*

Nickel Development Institute, *www.nidi.org*

Special Metals Corp., *www.specialmetals.com*

Steel & nickel based alloys, *www.superalloys.co.uk*

Niobium (Columbium) alloys

ASTM Standard B391-89: Niobium and niobium alloy ingots. *ASTM Annual Book of Standards, 02*(04).

ASTM Standard B392-89: Niobium and niobium alloy bar, rod and wire. *ASTM Annual Book of Standards, 02*(04).

ASTM Standard B393-89: Niobium and niobium alloy strip, sheet and plate. *ASTM Annual Book of Standards, 02*(04).

ASTM Standard B652-85: Niobium-hafnium alloy ingots. *ASTM Annual Book of Standards, 02*(04).

ASTM Standard B654-79: Niobium-hafnium alloy foil, sheet, strip and plate. *ASTM Annual Book of Standards, 02*(04).

ASTM Standard B655-85: Niobium-hafnium alloy bar, rod and wire. *ASTM Annual Book of Standards, 02*(04).

Internet sources

Husted, R., Los Alamos National Laboratory. *www-c8.lanl.gov/infosys*

An overview of Niobium and its uses.

Palladium

ASTM Standard B540-86: Palladium electrical contact alloy. *ASTM Annual Book of Standards, 03*(04).

ASTM Standard B563-89: Palladium-silver-copper electrical contact alloy. *ASTM Annual Book of Standards, 03*(04).

ASTM Standard B589-82: Refined palladium. *ASTM Annual Book of Standards, 02*(04).

ASTM Standard B683-90: Pure palladium electrical contact material. *ASTM Annual Book of Standards, 03*(04).

ASTM Standard B685-90: Palladium-copper electrical contact material. *ASTM Annual Book of Standards, 03*(04).

ASTM Standard B731-84: 60% palladium-40% silver electrical contact material. *ASTM Annual Book of Standards, 03*(04).

Internet sources

Jeneric Pentron, Casting alloys. *www.jeneric.com/casting*
An informative commercial site, limited to dental alloys.

Goodfellow Metals, *www.goodfellow.com*
Supplier of pure and precious metals, mainly for laboratory use. Website contains price and property information for its entire stock.

Platinum alloys

ASTM Standard B684-81: Platinum-iridium electrical contact material. *ASTM Annual Book of Standards, 03*(04).

Elkonium Series 400 datasheets, CMW.

ASM Metals Handbook (9th ed.) (1986). Vol. 2, 688–698.

Silver alloys

ASTM Standard B413-89: Refined silver. *ASTM Annual Book of Standards, 02*(04).

ASTM Standard B617-83: Coin silver electrical contact alloy. *ASTM Annual Book of Standards, 03*(04).

ASTM Standard B628-83: Silver-copper eutectic electrical contact alloy. *ASTM Annual Book of Standards, 03*(04).

ASTM Standard B693-87: Silver-nickel electrical contact materials. *ASTM Annual Book of Standards, 03*(04).

ASTM Standard B742-90: Fine silver electrical contact fabricated material. *ASTM Annual Book of Standards, 03*(04).

ASTM Standard B780-87: 75% silver, 24.5% copper, 0.5% nickel electrical contact alloy. *ASTM Annual Book of Standards, 03*(04).

Elkonium Series 300 datasheets (1996). CMW.

Elkonium Series 400 datasheets (1996). CMW.

Internet sources

The Silver Institute, *www.silverinstitute.org*

Jeneric Pentron, Casting alloys. *www.jeneric.com/casting*
An informative commercial site, limited to dental alloys.

Goodfellow Metals, *www.goodfellow.com*
Supplier of pure and precious metals, mainly for laboratory use. Website contains price and property information for its entire stock.

Tantalum alloys

ASTM Standard B365-86: Tantalum and tantalum alloy rod and wire. *ASTM Annual Book of Standards, 02*(04).

ASTM Standard B521-86: Tantalum and tantalum alloy seamless and welded tubes. *ASTM Annual Book of Standards, 02*(04).

ASTM Standard B560-86: Unalloyed tantalum for surgical implant applications. *ASTM Annual Book of Standards, 15*(01).

ASTM Standard B708-86: Tantalum and tantalum alloy plate, sheet and strip. *ASTM Annual Book of Standards, 02*(04).

Tantalum data sheet (1996). The Rembar Company.

ASM Handbook (9th ed.) (1986). Vol. 3, 323–325, 343–347.

Tin alloys

ASTM Standard B32-89: Solder metal. *ASTM Annual Book of Standards, 02*(04).

ASTM Standard B339-90: Pig tin. *ASTM Annual Book of Standards, 02*(04).

ASTM Standard B560-79: Modern pewter alloys. *ASTM Annual Book of Standards, 02*(04).

Barry, B.T.K. & Thwaites, C.J. (1983). *Tin and its alloys and compounds*. Ellis Horwood.

ASM Metals Handbook (9th ed.) (1986). Vol. 2, 613–625
See also Babbitt Metal (earlier).

Tin Research Association, *www.tintechnology.com*

Titanium alloys

Technical data sheets (1993). Titanium Development Association.

Technical data sheets (1993). The Titanium Information Group.

Technical data sheets (1995). IMI Titanium.

Internet sources

Titanium Information Group, *www.titaniuminfogroup.co.uk*

The International Titanium Association, *www.titanium.org*
This association site has a large list of member companies and comprehensive information on titanium and its alloys.

Tungsten alloys

ASTM Standard B777-87: Tungsten base, high-density metal. *ASTM Annual Book of Standards, 02*(04).

Yih, S.W.H. & Wang, C.T. (1979). *Tungsten.* Plenum Press.

ASM Metals Handbook (9th ed.) (1986). Vol. 7, p. 476.

Tungsten data sheet (1996). The Rembar Company.

Royal Ordnance speciality metals datasheet (1996). British Aerospace Defence Ltd.

CMW datasheets (1996). CMW.

Internet sources

North American Tungsten, *www.northamericantungsten.com*

Uranium

Uranium Information Centre, Australia, *www.uic.com.au*

UXC Jan 96 uranium indicator update, *www.uxc.com/review/uxc_prices/aspx*

Vanadium

Teledyne Wah Chang (1996). *Vanadium brochure.* TWC .

Zinc

ASTM Standard B6-87: Zinc. *ASTM Annual Book of Standards, 02*(04).

ASTM Standard B69-87: Rolled zinc. *ASTM Annual Book of Standards, 02*(04).

ASTM Standard B86-88: Zinc-alloy die castings. *ASTM Annual Book of Standards, 02*(02).

ASTM Standard B418-88: Cast and wrought galvanic zinc anodes. *ASTM Annual Book of Standards, 02*(04).

ASTM Standard B791-88: Zinc-aluminum alloy foundry and die castings. *ASTM Annual Book of Standards, 02*(04).

ASTM Standard B792-88: Zinc alloys in ingot form for slush casting. *ASTM Annual Book of Standards, 02*(04).

ASTM Standard B793-88: Zinc casting alloy ingot for sheet metal forming dies. *ASTM Annual Book of Standards, 02*(04).

Goodwin, F.E., & Ponikvar, A.L. (Eds.) (1989). *Engineering properties of zinc alloys* (3rd ed.). International Lead Zinc Research Organization.
An excellent compilation of data, covering all industrially important zinc alloys.

Chivers, A.R.L. (1981). Zinc diecasting. *Engineering Design Guide*, 41, OUP.
A good introduction to the subject.

ASM Metals Handbook (9th ed.) (1986). Properties of zinc and zinc alloys. Vol.2, 638–645.

Internet sources

International Zinc Association, *www.iza.com*

Zinc Industrias Nacionales S.A.-Peru, *www.zinsa.com*

Zinc: Eastern Alloys, *www.eazall.com*

India Lead Zinc Development Association, *www.ilzda.com*

Zirconium

ASTM Standard B350-80: Zirconium and zirconium alloy ingots for nuclear application. *ASTM Annual Book of Standards, 02*(04).

ASTM Standards B352-85, B551-83, and B752-85: Zirconium and zirconium alloys. *ASTM Annual Book of Standards, 02*(04).

Teledyne Wah Chang (1996). *Zircadyne: properties & applications*. TWC

ASM Metals Handbook (9th ed.) (1986). Vol. 2, 826–831.

D2.4 Ferrous metals

Ferrous metals are probably the most thoroughly researched and documented class of materials. Nearly every developed country has its own system of standards for irons and steels. Recently, continental and worldwide standards have been developed, which have achieved varying levels of acceptance. There is a large and sometimes confusing literature on the subject. This section is intended to provide the user with a guide to some of the better information sources.

Ferrous metals, general data sources

Bringas, J.E. (Ed.) (1995). *The metals black book – ferrous metals* (2nd ed.). CASTI Publishing.
An excellent short reference work.

ASM Metals Handbook (10th ed.) (1990). Vol. 1. ASM International.
Authoritative reference work for North American irons and steels.

ASM Metals Handbook (desk ed.) (1985). ASM International.
A summary of the multi-volume ASM Metals Handbook.

Wegst, C.W. (2010) *Stahlschlüssel* (in English: *Key to steel*). Verlag Stahlschlüssel Wegst GmbH.
Published every three years, in German, French, and English. Excellent coverage of European products and manufacturers.

Woolman, J., & Mottram, R.A. (1966). *The mechanical and physical properties of the British standard in steels*. Pergamon Press.
Still highly regarded, but based around a British Standard classification system that has been officially abandoned.

Brandes, E.A., & Brook, G.R. (Eds.) (1992). *Smithells metals reference book* (7th ed.). Butterworth-Heinemann.
An authoritative reference work, covering all metals.

Waterman, N.A. & Ashby, M.F. (Eds.) (1996). *Materials Selector.* Chapman and Hall.
Covers all materials; irons and steels are in vol. 2.

Sharpe, C. (Ed.) (1993). *Kempe's engineering year book* (98th ed.). Benn.
Updated each year; has good sections on irons and steels.

Irons and steels standards

Increasingly, national and international standards organizations are providing a complete catalog of their publications on the web. Two of the most comprehensive printed sources are listed here.

Iron and steel specifications (9th ed.) (1998). British Iron and Steel Producers Association (BISPA).
Comprehensive tabulations of data from British Standards on irons and steels, as well as some information on European and North American standards. The same information is available on a searchable CD.

ASTM Annual Book of Standards, Vols. 01(01) through 01(07).
The most complete set of American irons and steels standards. Summaries of the standards can be found on the web at www.astm.org

Cross-referencing of similar international standards and grades

It is difficult to match, even approximately, equivalent grades of iron and steel between countries. No coverage of this subject can ever be complete, but the references listed here are helpful.

Gensure, J.G., & Potts, D.L. (1988). *International metallic materials cross reference* (3rd ed.). Genium Publishing.
Comprehensive worldwide coverage of the subject, well indexed.

Bringas, J.E. (Ed.) (1995). *The metals black book–ferrous metals* (2nd ed.). CASTI Publishing.
Easy-to-use tables for international cross-referencing. (*See "General" section for more information.*)

Unified numbering system for metals and alloys (2nd ed.), (1977). Society of Automotive Engineers.
An authoritative reference work, providing a unifying structure for all standards published by U.S. organizations. No coverage of the rest of the world.

Iron and steel specifications (7th ed.) (1989). British Steel.
Lists "Related Specifications" for France, Germany, Japan, Sweden, U.K. and U.S.A.

Cast irons

Scholes, J.P. (1979). *The selection and use of cast irons.* Engineering Design Guides, OUP.

Angus, H.T. (1976). *Cast iron: physical and engineering properties.* Butterworths.

Cast irons, American standards

These can all be found in the *Annual Book of ASTM Standards, 01*(02).

ASTM A220M-88: Pearlitic malleable iron, *www.astm.org*

ASTM A436-84: Austenitic gray iron castings; *www.astm.org*

ASTM A532: Abrasion-resistant cast irons; *www.astm.org*

ASTM A602-70 (reapproved 1987): Automotive malleable iron castings.

Cast irons, international standards

These are available from the ISO Central Secretariat, Geneva, Switzerland.

ISO 185:1988 Grey cast iron – classification.

ISO 2892:1973 Austenitic cast iron.

ISO 5922:1981 Malleable cast iron.

Cast irons, British standards

Compared with steels, there are relatively few standards for cast iron, which makes it feasible to list them all. Standards are available from BSI Customer Services, London.

BS 1452:1990 Flake graphite cast iron.

BS 1591:1975 Specification for corrosion resisting high silicon castings.

BS 2789:1985 Iron castings with spheroidal or nodular graphite.

BS 3468:1986 Austenitic cast iron.

BS 4844:1986 Abrasion resisting white cast iron.

BS 6681:1986 Specification for malleable cast iron.

Carbon and low alloy steels

ASM metals handbook (10th ed.) (1990). Vol. 1. ASM International.
Authoritative reference work for North American irons and steels.

Fox, J.H.E. (1980). An introduction to steel selection: part 1, carbon and low-alloy steels. *Engineering Design Guide*, no. 34. Oxford University Press.

Stainless steels

ASM metals handbook (10th ed.) (1990). Vol. 1. ASM International.
Authoritative reference work for North American irons and steels.

Elliott, D., & Tupholme, S.M. (1981). An introduction to steel selection: part 2, stainless steels. *Engineering Design Guide*, no. 43. Oxford University Press.

Peckner, D., & Bernstein, I.M. (1977). *Handbook of stainless steels*. McGraw-Hill.

Design guidelines for the selection and use of stainless steel (1991). *Designers' Handbook Series*, no. 9014. Nickel Development Institute.
The Nickel Development Institute (NIDI) is a worldwide organization that gives away a large variety of free literature about nickel-based alloys, including stainless steels. NIDI European Technical Information Centre.

General Internet sites for ferrous metals

British Constructional Steel Work Association, *www.steelconstruction.org*

International Iron & Steel Institute, *www.worldsteel.org*

Iron & Steel Trades Confederation, *www.istc-tu.org*

National Association of Steel Stock Holders, *www.nass.org.uk*

Steel Manufacturers Association, *www.steelnet.org*

Steel: Bethlehem Steel, *www.bethsteel.com* (steel)

Steel: Corus Group, *www.corusgroup.com(steel)*

Steel: Automotive Steel Library, *www.autosteel.org(steel)*

Steel: Great Plains Stainless, *www.gpss.com(steel)*

SteelSpec, *www.steelspec.org.uk/index.htm(steel)*

D2.5 Polymers and elastomers

Polymers are not subject to the same strict specifications as metals. Data tend to be producer-specific. Sources are consequently scattered, incomplete, and poorly presented. Saechtling is the best; although no single hard-copy source is completely adequate, all those listed here are worth consulting. See also Section D.4, Databases and Expert Systems as software; some are good on polymers.

Saechtling, H. (Ed.) (1983). *Saechtling: International plastics handbook*. MacMillan.
The most comprehensive of the hard-copy data sources for polymers.

Seymour, R.B. (1987). *Polymers for engineering applications*. ASM International.
Property data for common polymers. A starting point, but insufficient detail for accurate design or process selection.

Murphy, J (Ed.) (1991). *New horizons in plastics, a handbook for design engineers*. WEKA Publishing.

ASM engineered materials handbook, Vol 2. engineering plastics (1989). ASM International.

Harper, C.A. (Ed.) (1975). *Handbook of plastics and elastomers*. McGraw-Hill.

International plastics selector, plastics (9th ed.) (1987). International Plastics Selector.

Domininghaus, H. (Ed.) (1992). *Die kunststoffe and ihre eigenschaften*. VDI Verlag.

van Krevelen, D.W. (1990). *Properties of polymers* (3rd ed.). Elsevier.
Correlation of properties with structure; estimation from molecular architecture.

Bhowmick, A.K., & Stephens, H.L. (1988). *Handbook of elastomers*. Marcel Dekker.

ICI technical service notes (1981). ICI Plastics Division, Engineering Plastics Group.

Technical data sheets (1995). Malaysian Rubber Producers Research Association.
Data sheets for numerous blends of natural rubber.

Ossvald, T.A., Baur, E., Brinkmann, F., Oberbach, K., & Schmachtenberg, E. (2006). *International plastics handbook*, Carl Hanser Verlag.

Software and Internet sources

CAMPUS Plastics database, *www.campusplastics.com*

CES Polymers database, *www.grantadesign.com*

GE Plastics, *www.ge.com/en/company/businesses/ge_plastics.htm*

Harboro Rubber Co., *www.harboro.co.uk*

IDES Resin Source, *www.ides.com*

MERL, *www.merl-ltd.co.uk*

Plastics.com, *www.plastics.com*

Matweb.com, *www.matweb.com*

M-base engineering, *www.m-base.de/main*

D2.6 Ceramics and glasses

Sources of data for ceramics and glasses, other than the suppliers' data sheets, are limited. Texts and handbooks—for example, the ASM's (1991) *Engineered Materials Handbook, Vol. 4*, Morell's (1985) compilations, Neville's (1996) book on concrete, Boyd and Thompson's (1980) *Handbook on Glass*, and Sorace's (1996) treatise on stone are useful starting points. The CES Ceramics Database contains recent data for ceramics and glasses. In the end, however, it is the manufacturer to whom one has to turn: The data sheets for their products are the most reliable sources of information.

Ceramics and ceramic-matrix composites

ASM engineered materials handbook, Vol. 4, Ceramics and glasses (1991). ASM International.

Waterman, N., & Ashby, M.F. (Eds.) (1996). *Materials selector*. Chapman and Hall.

Brook, R.J. (Ed.) (1991). *Concise encyclopedia of advanced ceramic materials*. Pergamon Press.

Cheremisinoff, N.P. (Ed.) (1990). *Handbook of ceramics and composites*, 3 Vols. Marcel Dekker.

Clark, S.P. (Ed.) (1996). *Handbook of physical constants*, memoir 97. Geological Society of America.

Schwartz, M.M. (1992). *Handbook of structural ceramics*. McGraw-Hill.
A great deal of data and information on processing and applications.

Kaye, G.W.C., & Laby, T.H. (1986). *Tables of physical & chemical constants* (15th ed.). Longman.

Kingery, W.D., Bowen, H.K., & Uhlmann, D.R. (1976). *Introduction to ceramics* (2nd ed.). Wiley.

Materials Engineering (1992). *Materials selector*. Penton Press.

Morrell, R. (1985). *Handbook of properties of technical & engineering ceramics, Parts I and II*. National Physical Laboratory.

Musikant, S. (1991). *What every engineer should know about ceramics*. Marcel Dekker.
Good on data.

Richerson, D.W. (1992). *Modern ceramic engineering* (2nd ed.). Marcel Dekker.

Brandes, E.A., & Brook, G.B. (Eds.) (1992). *Smithells metals reference book* (7th ed). Butterworth-Heinemann.

Harper, C.A. (Ed.) (2001). *Handbook of ceramics, glasses and diamonds*. McGraw-Hill.
A comprehensive compilation of data and design guidelines.

Richerson, D.W. (2000). *The magic of ceramics*. American Ceramics Society.
A readable introduction to ceramics, both old and new.

Glasses

ASM engineered materials handbook, Vol. 4, Ceramics and glasses (1991). ASM International.

Boyd D.C., & Thompson D.A. (1980). *Glass*. Reprinted from Kirk-Othmer: *Encyclopedia of chemical technology, Vol. 11* (3rd ed.), pp 807–880. Wiley.

Oliver, D.S. (1975). *Engineering design guide 05: The use of glass in engineering*. Oxford University Press.

Bansal, N.P., & Doremus, R.H. (1966). *Handbook of glass properties*. Academic Press.

Cement and concrete

Cowan, H.J., & Smith, P.R. (1988). *The science and technology of building materials.* Van Nostrand-Reinhold.

Illston, J.M., Dinwoodie, J.M., & Smith, A.A. (1979). *Concrete, timber and metals.* Van Nostrand-Reinhold.

Neville, A.M. (1996). *Properties of concrete* (4th ed.). Longman Scientific and Technical.
An excellent introduction to the subject.

D2.7 Composites: PMCs, MMCs, and CMCs

The fabrication of composites allows so many variants that no hard-copy data source can capture them all; instead, they list properties of matrix and reinforcement and of certain generic lay-ups or types. The *Engineers Guide* and the *Composite Materials Handbook*, listed first, are particularly recommended.

Composites, general

Weeton, J.W., Peters, D.M., & Thomas, K.L. (Eds.) (1987). *Engineers guide to composite materials.* ASM International.
The best starting point: data for all classes of composites.

Schwartz, M.M. (Ed.) (1992). *Composite materials handbook* (2nd ed.). McGraw-Hill.
Lots of data on PMCs, less on MMCs and CMCs, processing, fabrication, applications, and design information.

ASM engineered materials handbook, Vol 1: Composites (1987). ASM International.

Seymour, R.B. (1991). *Reinforced plastics, properties and applications.* ASM International.

Cheremisinoff, N.P. (Ed.) (1990). *Handbook of ceramics and composites, Vols. 1–3.* Marcel Dekker.

Kelly, A. (Ed.) (1989). *Concise encyclopedia of composited materials.* Pergamon Press.

Middleton, D.H. (1990). *Composite materials in aircraft structures.* Longman Scientific and Technical Publications.

Smith, C.S. (1990). *Design of marine structures in composite materials.* Elsevier Applied Science.

Metal matrix composites

See, first, the sources listed under "All Composite Types; for more detail, go to

ASM engineered materials handbook, Vol. 1: Composites (1987). ASM International.

Technical data sheets (1995). Duralcan USA.

Technical data sheets (1995). 3M Company.

D2.8 Foams and cellular solids

Many of the references given in Section D2.5 for polymers and elastomers mention foam. The references given here contain much graphical data and simple formulae that allow properties of foam to be estimated from the foam's density and the properties of the solid of which it is made, but in the end it is necessary to contact suppliers. See also Section D.4, Databases and Expert Systems as Software; some are good on foams. For woods and wood-based composites, see Section D2.10.

Cellular polymers (1981–1996). RAPRA Technology.

Encyclopedia of chemical technology, Vol. 2 (3rd ed.), pp. 82–126 (1980). Wiley.

Encyclopedia of polymer science and engineering, Vol. 3 (2nd ed.), section C (1985). Wiley.

Gibson, L.J., & Ashby, M.F. (1997). *Cellular solids.* Cambridge University Press.
Basic text on foamed polymers, metals, ceramics, and glasses, as well as and natural cellular solids.

Handbook of industrial materials (2nd ed.), pp. 537–556 (1992). Elsevier Advanced Technology.

Hilyard, N.C., & Cunningham, A. (Eds.) (1994). *Low density cellular plastics —physical basis of behaviour.* Chapman and Hall.
Specialized articles on aspects of polymer-foam production, properties, and uses.

Plascams, Version 6 (1995). *Plastics computer-aided materials selector.* RAPRA Technology Limited.

Saechtling, H. (Ed.) (1983). *Saechtling: International plastics handbook.* MacMillan.

Seymour, R.P. (1987). *Polymers for engineering applications.* ASM International.

D2.9 Stone, rocks, and minerals

There is an enormous literature on rocks and minerals. Start with the handbooks listed below; then ask a geologist for guidance.

Atkinson B.K. (1987). *The fracture mechanics of rock.* Academic Press.

Clark, S.P., Jr. (Ed.) (1966). *Handbook of physical constants.* Memoir 97, The Geological Society of America.
Old but trusted compilation of property data for rocks and minerals.

Lama, R.E., & Vutukuri, V.S. (Eds.) (1978). *Handbook on mechanical properties of rocks, Vols. 1–4.* Trans Tech Publications.

Griggs, D., & Handin, J. (Eds.) (1960). *Rock deformation.* Memoir 79, The Geological Society of America.

Sorace, S. (1966). Long-term tensile and bending strength of natural building stones. *Materials and Structures, 29,* 426–435.

D2.10 Timber woods and wood-based composites

Woods, like composites, are anisotropic; useful sources list properties along and perpendicular to the grain. The U.S. Forest Products Laboratory's *Wood Handbook* and Kollmann and Coté's *Principles of Wood Science and Technology* are particularly recommended.

Woods and timbers, general

Bodig, J., & Jayne, B.A. (1982). *Mechanics of wood and wood composites.* Van Nostrand Reinholt.

Dinwoodie, J.M. (1989). *Wood, nature's cellular polymeric fiber composite.* The Institute of Metals.

Dinwoodie, J.M. (1981). *Timber, its nature and behaviour.* Van Nostrand-Reinhold.
Basic text on wood structure and properties. Not much data.

Gibson, L.J., & Ashby, M.F. (1997). *Cellular solids* (2nd ed.). Cambridge University Press.

Jane, F.W. (1970). *The structure of wood* (2nd ed.). A. and C. Black.

Kollmann, F.F.P., & Côté, W.A., Jr. (1968). *Principles of wood science and technology, Vol. 1, Solid wood*. Springer-Verlag.
The bible.

Kollmann, F., Kuenzi, E., & Stamm, A. (1968). *Principles of wood science and technology, Vol. 2, Wood-based materials*. Springer-Verlag.

Schniewind, A.P. (Ed.) (1989). *Concise encyclopedia of wood and wood-based materials*. Pergamon Press.

Timber: Data compilations

BRE (1996). *BRE information papers*. Building Research Establishment (BRE).

Forest Products Laboratory (1989). *Handbook of wood and wood-based materials*. Forest Service, U.S. Department of Agriculture. Hemisphere Publishing Corporation.
A massive compilation of data for North American woods.

Informationsdienst Holz (1996). Merkblattreihe Holzarten.

TRADA (1978/1979). *Timbers of the world, Vols. 1– 9*. Timber Research and Development Association.

TRADA (1991). *Information sheets*. Timber Research and Development Association.

Timber and wood-composite standards

Great Britain

British Standards Institution (BSI), *www.bsigroup.com*

Germany

Deutsches Institut für Normung (DIN), *www.din.de*

United States

American Society for Testing and Materials (ASTM), *www.astm.org*

Software and Internet data sources

CES woods database, *www.grantadesign.com*

PROSPECT, Oxford Forestry Institute, Department of Plant Sciences, Oxford University *www.plants.ox.ac.uk/ofi/prospect/index*
A database of the properties of tropical woods of interest to a wood user; includes information about uses, workability, treatments, and origins.

Woods of the world (1994). Tree Talk.
A CD-ROM of woods, with illustrations of structure, information about uses, origins, habitat, and so on.

WoodWeb, *www.woodweb.com*
Woodwork industry web service. Contains information about woodworking-related companies.

D2.11 Natural fibers and other natural materials

Natural materials, general

Alexander, R.M. (1983). *Animal mechanics* (2nd ed.). Blackwell Scientific.

Brown, C.H. (1975). *Structural materials in animals*. Pitman.

Currey, J.D., Wainwright, S.A., & Biggs, W.D. (1982). *Mechanical design in organisms*. Princeton University Press.

Fung, Y.C. (1993). *Biomechanics: Mechanical properties of living tissues* (2nd ed.). Springer-Verlag.

Gibson, L.J., & Ashby, M.F. (1997). *Cellular solids: structure and properties* (2nd ed.). Cambridge University Press.

McMahon, T.A. (1984). *Muscles, reflexes, and locomotion.* Princeton University Press.

Silver, F.H. (1987). *Biological materials: Structure, mechanical properties, and modeling of soft tissues.* New York University Press.

Vincent, J.F.V., & Currey, J.D. (1980). The mechanical properties of biological materials. *Proceedings of the Symposia of the Society for Experimental Biology* (no.34). Cambridge University Press for the Society for Experimental Biology.

Vogel, S., & Calvert, R.A (illustrator) (1988). *Life's devices: The physical world of animals and plants.* Princeton University Press.

Yamada, H. (1970). *Strength of biological materials.* Williams & Wilkins.

Wood and wood-like materials (wood cell wall, wood, cork, bamboo, palm, rattan)

Amada, S., Munekata, T., Nagase, Y., Ichikawa, Y., Kirigai, A., & Yang, Z.F. (1996). The mechanical structures of bamboos in viewpoint of functionally gradient and composite materials, *J. Compos. Mater.*, *30*(7), 800–819.

Bhat, K., & Mathew, A. (1995). Structural basis of rattan biomechanics. *Biomimetics*, *3*(2), 67–80.

Frühwald, A., Peek, R.-D., & Schulte, M. (1992). Nutzung von kokospalmenholz am beispiel von nordsulawesi, Indonesien, technical report. *Mitteilungen der Bundesforschungsanstalt für Forst- und Holzwirtschaft, 171.* Kommissionsverlag Max Wiederbusch.

Gibson, L.J., Easterling, K.E., & Ashby, M.F. (1981). The structure and mechanics of cork. *P. Roy. Soc. Lond. A*, *377*(1769), 99–117.

Godbole, V.S., & Lakkad, S.C. (1986). Effect of water-absorption on the mechanical properties of bamboo. *J. Mater. Sci. Lett.*, *5*(3), 303–304.

Janssen, J.J.A. (1991). Mechanical properties of bamboo *Forestry Sciences*, Vol. 37, Kluwer Academic Publishers.

Killmann, W. (1983). Some physical-properties of the coconut palm stem. *Wood Sci. Tech.*, *17*(3), 167–185.

Killmann, W. (1993). *Struktur, eigenschaften and nutzung von stämmen wirtschaftlich wichtiger palmen.* PhD thesis, Universität Hamburg.

Kloot, N. (1952). Mechanical and physical properties of coconut palm. *Austr. J. Appl. Sci.*, *3*(4), 293–322.

Lakkad, S.C., & Patel, J.M. (1981). Mechanical properties of bamboo, a natural composite. *Fibre Sci. Tech.*, *14*(4), 319–322.

Mark, R.E. (1967). *Cell wall mechanics of tracheids.* Yale University Press.

Rich, P.M. (1987). Mechanical structure of the stem of arborescent palms. *Bot. Gazette*, *148*(1), 42–50.

Rosa, M.E., & Fortes, M.A. (1988a). Stress-relaxation and creep of cork. *J. Mater. Sci.*, *23*(1), 35–42.

Rosa, M.E., & Fortes, M.A. (1988b). Rate effects on the compression and recovery of dimensions of cork. *J. Mater. Sci.*, *23*(3), 879–885.

Rosa, M.E., & Fortes, M.A. (1991). Deformation and fracture of cork in tension. *J. Mater. Sci.*, *26*(2), 341–348.

Plant tissue (apple and potato parenchyma, fruit skins, seaweed, nuts)

Blahovec, J. (1988). Mechanical properties of some plant materials. *J. Mater. Sci., 23*(10), 3588–3593.

Gibson, L.J., & Ashby, M.F. (1997). *Cellular Solids: Structure and Properties* (2nd ed.). Cambridge University Press.

Greenberg, A.R., Mehling, A., Lee, M., & Bock, J.H. (1989). Tensile behavior of grass. *J. Mater. Sci., 24*(7), 2549–2554.

Jennings, J.S., & Macmillan, N.H. (1986). A tough nut to crack. *J. Mater. Sci., 21*(5), 1517–1524.

Venkataswamy, M.A., Pillai, C.K.S., Prasad, V.S., & Satyanarayana, K.G. (1987). Effect of weathering on the mechanical properties of midribs of coconut leaves. *J. Mater. Sci., 22*(9), 3167–3172.

Vincent, J.F.V. (1989). Relationship between density and stiffness of apple flesh. *J. Sci. Food. Agr., 47*(4), 443–462.

Vincent, J.F.V. (1990). Fracture properties of plants. *Adv. Bot. Res., 17*, 235–287.

Wang, C.H., Zhang, L.C., & Mai, Y.W. (1994). Deformation and fracture of macadamia nuts. 1. Deformation analysis of nut-in-shell. *Int. J. Fracture, 69*(1), 51–65.

Wang, C.H., & Mai, Y.W. (1994). Deformation and fracture of macadamia nuts. 2. Microstructure and fracture mechanics analysis of nutshell. *Int. J. Fracture, 69*(1), 67–85.

Cellulose, cotton, flax, hemp, jute, ramie

Currey, J.D., Wainwright, S.A., & Biggs, W.D. (1982). *Mechanical design in organisms*. Princeton University Press.

Gibson, L.J., & Ashby, M.F. (1997). *Cellular Solids: Structure and Properties* (2nd ed.). Cambridge University Press.

Vincent, J.F.V. (1990). *Structural biomaterials* (revised edition). Princeton University Press.

Wainwright, S.A, Gosline, J.M., & Biggs, W.D. (1992). *Mechanical design in organisms*. Princeton University Press

Wainwright, S.A. (1980). Adaptive materials: a view from the organism. In J.F.V. Vincent, & J.D. Currey (Eds.), The mechanical properties of biological materials. *Proceedings of the Symposia of the Society for Experimental Biology, 34*, 483–453. Cambridge University Press for the Society for Experimental Biology.

Silk

Calvert, P. (1989). Natural polymers—spinning ties that bind. *Nature, 340*, 266.

Denny, M. (1980). Silks—their properties and functions. In J.F.V. Vincent & J.D. Currey (Eds.), The mechanical properties of biological materials. *Proceedings of the Symposia of the Society for Experimental Biology, 34*, 247–272. Cambridge University Press for the Society for Experimental Biology.

Gosline, J.M., Demont, M.E., & Denny, M.W. (1986). The structure and properties of spider silk. *Endeavour, 10*(1), 37–43.

Elastin and resilin

Goseline, J.M. (1980). The elastic properties of rubber-like proteins and highly extensible tissues. In J.F.V. Vincent & J.D. Currey (Eds.), The mechanical properties of biological materials. *Proceedings of the Symposia of the Society for Experimental Biology, 34*, 331–357. Cambridge University Press for the Society for Experimental Biology.

Oxlund, H., Manschot, J., & Viidik, A. (1988). The role of elastin in the mechanical properties of skin. *J. Biomech., 21*(3), 213–218.

Collagen, ligament, tendon

Andersen, K.L., Pedersen, E.H., & Melsen, B. (1991). Material parameters and stress profiles within the periodontal ligament. *Am. J. Orthod. Dentofac. Orthop., 99*(5), 427–440.

Bennett, M.B., Ker, R.F., Dimery, N.J., & Alexander, R.M. (1986). Mechanical properties of various mammalian tendons. *J. Zool., 209*(4), 537–548.

Bosch, U., Decker, B., Kasperczyk, W., Nerlich, A., Oestern, H.J., & Tscherne, H. (1992). The relationship of mechanical properties to morphology in patellar tendon autografts after posterior cruciate ligament replacement in sheep. *J. Biomech., 25*(8), 821–830.

Butler, D.L., Kay, M.D., & Stouffer, D.C. (1986). Comparison of material properties in fascicle-bone units from human patellar tendon and knee ligaments. *J. Biomech., 19*(6), 425–432.

Grieshaber, F.A., & Faust, U. (1992). Mechanische kenngrößen von biologischem weichgewebe. *Biomed. Technik, 37*(12), 278–286.

Kato, Y.P., Christiansen, D.L., Hahn R.A., Shieh, S.J., Goldstein, J.D., & Silver, F.H. (1989). Mechanical properties of collagen fibres: A comparison of reconstituted and rat tendon fibres. *Biomat., 10*, 38–41.

Ker, R.F. (1981). Dynamic tensile properties of the plantaris tendon of sheep (*Ovis aries*). *J. Exp. Biol., 93*(Aug.), 283–302.

Ker, R.F., Alexander, R.M., & Bennett, M.B. (1988). Why are mammalian tendons so thick. *J. Zool., 216*(2), 309–324.

Park, J.B., & Lakes, R.S. (1992). *Biomaterials* (2nd ed.). Plenum Press.

Rogers, G.J., Milthorpe, B.K., Muratore, A., & Schindhelm, K. (1990). Measurement of the mechanical properties of the ovine anterior cruciate ligament bone complex—a basis for prosthetic evaluation. *Biomaterials, 11*(2), 89–96.

Silver, F.H., Kato, Y.P., Ohno, M., & Wasserman, A.J. (1992). Analysis of mammalian connective-tissue—relationship between hierarchical structures and mechanical properties. *J. Long Term Effects Med. Impl., 2*(2–3), 165–198.

Vincent, J.F.V. (1990). *Structural Biomaterials* (revised edition). Princeton University Press.

Wainwright, S. (1980). Adaptive materials: A view from the organism. In J.F.V. Vincent & J.D. Currey (Eds.), The mechanical properties of biological materials. *Proceedings of the Symposia of the Society for Experimental Biology, 34*, 483–453. Cambridge University Press for the Society for Experimental Biology.

Wang, X.T., & Ker, R.F. (1995a). Creep-rupture of wallaby tail tendons. *J. Exp. Biol., 198*(3), 831–845.

Wang, X.T., Ker, R.F., & Alexander, R.M. (1995b). Fatigue-rupture of wallaby tail tendons. *J. Exp. Biol., 198*(3), 847–852.

Muscle

Dobrunz, L.E., Pelletier, D.G., & McMahon, T.A. (1990). Muscle-stiffness measured under conditions simulating natural sound production. *Biophys. J., 58*(2), 557–565.

Yamada, H. (1970). *Strength of biological materials*. Williams & Wilkins.

Skin

Bauer, A.M., Russell, A.P., & Shadwick, R.E. (1989). Mechanical properties and morphological correlates of fragile skin in gekkonid lizards. *J. Exp. Biol., 145*, 79–102.

Manschot, J.F.M., & Brakkee, A.J.M. (1986a). The measurement and modeling of the mechanical properties of human skin invivo. 1. The measurement. *J. Biomech., 19*(7), 511–515.

Manschot, J.F.M., & Brakkee, A.J.M. (1986b). The measurement and modeling of the mechanical properties of human skin invivo. 2. The model. *J. Biomech., 19*(7), 517–521.

Oxlund, H., Manschot J., & Viidik, A. (1988). The role of elastin in the mechanical properties of skin. *J. Biomech., 21*(3), 213–218.

Silver, F.H., Kato, Y.P., Ohno, M., & Wasserman, A.J. (1992). Analysis of mammalian connective tissue—relationship between hierarchical structures and mechanical properties. *J. Long Term Effects Med. Impl., 2*(2–3), 165–198.

Swartz, S.M., Groves, M.S., Kim, H.D., & Walsh, W.R. (1996). Mechanical properties of bat wing membrane skin. *J. Zool., 239*(2), 357–378.

Leather

Attenburrow, G.E., & Wright, D.M. (1994). Studies of the mechanical behavior of partially processed leather. *J. Am. Leather Chem. Assoc., 89*(12), 391–402.

Kronick, P., & Maleef, B. (1992). Nondestructive failure testing of bovine leather by acoustic-emission. *J. Am. Leather Chem. Assoc., 87*(7), 259–265.

Lin, J., & Hayhurst, D.R. (1993a). The development of a bi-axial tension test facility and its use to establish constitutive equations for leather. *Eur. J. Mech. A—Solids, 12*(4), 493–507.

Lin, J., & Hayhurst, D.R. (1993b). Constitutive equations for multiaxial straining of leather under uniaxial stress. *Eur. J. Mech. A—Solids, 12*(4,), 471–492.

Cartilage

Rains, J.K., Bert, J.L., Roberts, C.R., & Pare, P.D. (1992). Mechanical properties of human tracheal cartilage. *J. Appl. Physiol., 72*(1), 219–225.

Silver, F.H., Kato, Y.P., Ohno, M., & Wasserman, A.J. (1992). Analysis of mammalian connective tissue—relationship between hierarchical structures and mechanical properties. *J. Long Term Effects Med. Impl., 2*(2-3), 165–198.

Swanson, S. (1980). The elastic properties of rubber-like protein and highly extensible tissue. In J.F.V. Vincent & J.D. Currey (Eds.), The mechanical properties of biological materials. *Proceedings of the Symposia of the Society for Experimental Biology, 34*, 377–395. Cambridge University Press for the Society for Experimental Biology.

Bone (including antler)

Ashman, R.B., Corin, J.D., & Turner, C.H. (1987). Elastic properties of cancellous bone—measurement by an ultrasonic technique. *J. Biomech., 20*(10), 979–986.

Ashman, R.B., & Rho, J.Y. (1988). Elastic modulus of trabecular bone material. *J. Biomech., 21*(3), 177–181.

Currey, J.D. (1984). *The mechanical adaptations of bones.* Princeton University Press.

Currey, J.D. (1988). The effects of drying and re-wetting on some mechanical properties of cortical bone. *J. Biomech., 21*(5), 439–441.

Currey, J.D. (1990). Physical characteristics affecting the tensile failure properties of compact bone. *J. Biomech., 23*(8), 837–844.

Dickenson, R.P., Hutton, W.C., & Stott, J.R.R. (1981). The mechanical properties of bone in osteoporosis. *J. Bone Joint Surg.—Brit. Vol.*, *63*(2), 233–238.

Gibson, L.J. (1985). The mechanical behavior of cancellous bone. *J. Biomech.*, *18*(5), 317 et seq.

Goldstein, S.A. (1987). The mechanical properties of trabecular bone—dependence on anatomic location and function. *J. Biomech.*, *20*(11–12), 1055–1061.

Swanson, S. (1980). The elastic properties of rubber-like protein and highly extensible tissue. In J.F.V. Vincent & J.D. Currey (Eds.), The mechanical properties of biological materials. *Proceedings of the Symposia of the Society for Experimental Biology*, *34*, 377–395. Cambridge University Press for the Society for Experimental Biology.

Kitchener, A. (1991). The evolution and mechanical design of horns and antlers. In J. Rayner & R. Wootton (Eds.), *Society for experimental biology seminar series: no. 36. Biomechanics in evolution* (pp. 229–253). Cambridge University Press.

Linde, F., & Hvid, I. (1987). Stiffness behavior of trabecular bone specimens. *J. Biomech.*, *20*(1), 83–89.

Lotz, J.C., Gerhart, T.N., & Hayes, W.C. (1991). Mechanical properties of metaphyseal bone in the proximal femur. *J. Biomech.*, *24*(5), 317–329.

Moyle, D.D., & Gavens, A.J. (1986). Fracture properties of bovine tibial bone. *J. Biomech.*, *19* (11), 919–927.

Moyle, D.D., & Walker, M.W. (1986). The effects of a calcium deficient diet on the mechanical-properties and morphology of goose bone. *J. Biomech.*, *19*(8), 613–625.

Sharp, D.J., Tanner, K.E., & Bonfield, W. (1990). Measurement of the density of trabecular bone. *J. Biomech.*, *23*(8), 853–857.

Sun, J.J., & Geng, J. (1987). A study of Haversian systems. *J. Biomech.*, *20*(8), 815.

Watkins, M. (1987). *The development of a tough artificial composite based on antler bone*. PhD thesis, University of Reading.

Dentine, enamel

Brear, K., Currey, J.D., Pond, C.M., & Ramsay, M.A. (1990). The mechanical properties of the dentin and cement of the tusk of the narwhal *Monodon monoceros* compared with those of other mineralized tissues. *Arch. Oral Biol.*, *35*(8), 615–621.

Coral

Chamberlain, J. (1978). Mechanical properties of coral skeleton: compressive strength and its adaptive significance. *Paleobiol.*, *4*, 419–435.

Kim, K., Goldberg, W.M., & Taylor, G.T. (1992). Architectural and mechanical properties of the black coral skeleton (*Coelenterata, Antipatharia*)—a comparison of 2 species. *Biol. Bull.*, *182*(2), 195–209.

Scott, P.J.B., & Risk, M.J. (1988). The effect of *Lithophaga* (*Bivalvia, Mytilidae*) boreholes on the strength of the coral *Porites lobata*. *Coral Reefs*, *7*(3), 145–151.

Vosburgh, F. (1982). *Acropora reticulata*—structure, mechanics and ecology of a reef coral. *P. Roy. Soc. Lond. B*, *214*(1197), 481–499.

Alpha-keratins

Bertram, J.E.A., & Gosline, J.M. (1986). Fracture toughness design in horse hoof keratin. *J. Exp. Biol.*, *125*, 29–47.

Bertram, J.E.A., & Gosline, J.M. (1987). Functional design of horse hoof keratin—the modulation of mechanical properties through hydration effects. *J. Exp. Biol.*, *130*, 121–136.

Fraser, R., & Macrae, T. (1980). Molecular structure and mechanical properties of keratins. In J.F.V. Vincent & J.D. Currey (Eds.), The mechanical properties of biological materials. *Proceedings of the Symposia of the Society for Experimental Biology, 34*, 211–246. Cambridge University Press for the Society for Experimental Biology.

Kitchener, A. (1991). The evolution and mechanical design of horns and antlers. In J. Rayner & R, Wootton (Eds.), *Society for experimental biology seminar series: no.36. Biomechanics in evolution* (pp. 229–253). Cambridge University Press.

Insect cuticle

Alexander, D.E., Blodig, J., & Hsieh, S.Y. (1995). Relationship between function and mechanical properties of the pleopods of isopod crustaceans. *Invertebrate Biol., 114*(2), 169–179.

Gunderson, S., & Whitney, J. (1992). Insect cuticle microstructure and its application to advanced composites. *Biomimetics, 1*(2), 177–197.

Vincent, J. (1980). Insect cuticle: A paradigm for natural composites. In J.F.V. Vincent & J.D. Currey (Eds.), The mechanical properties of biological materials. *Proceedings of the Symposia of the Society for Experimental Biology, 34*, 183–210. Cambridge University Press for the Society for Experimental Biology.

Mollusk shell

Jackson, A.P., Vincent, J.F.V., & Turner, R.M. (1988). The mechanical design of nacre. *P. Roy. Soc. Lond. B, 234*(1277), 415 et seq.

Jackson, A.P., Vincent, J.F.V., & Turner, R.M. (1990). Comparison of nacre with other ceramic composites. *J. Mater. Sci., 25*(7), 3173–3178.

D2.12 Measures of environmental impact

Aggregain (2007). The Waste and Resources Action Program (WRAP). *www.wrap.org.uk*
Data and an Excel-based tool to calculate energy and carbon footprint of recycled road-bed materials.

AMC (2006). Australian Magnesium Corporation, *www.aph.gov.au/house/committee/environ/greenhse/gasrpt/Sub65-dk.pdf*

APME (1997 ,1998, 1999, 2000). Eco–profiles of the European plastics industry. Association of Plastics Manufacturers in Europe. *www.plasticseurope.org*

Ashby, M.F. (2009). *Materials and the environment*. Butterworth-Heinemann.

BCA (2007). *A carbon strategy for the cement industry.* British Cement Association, *www.cementindustry.co.uk*

Boustead Model 5 (2007). Boustead Consulting. *www.boustead-consulting.co.uk*
An established life-cycle assessment tool.

Boustead, I. (1999–2006). APME Association of Plastics Manufacturers in Europe, Report series.

British Metals Recycling Association, *www.britmetrec.org.uk*

Building Research Establishment (2006). BRE Environmental Profiles database. BRE Environment Division.

BUWAL (1996). Bundesamt für umwelt, wald und landwirtschaft, Environmental series No. 250, *Life Cycle Inventories for Packaging, Vols. I and II.*

Chapman, P.F., & Roberts, F. (1993). *Metals resources and energy.* Butterworths.

Chemlink Australasia (1997). *www.chemlink.com.au/mag&oxide.htm*

ELCD (2008). *http://lca.jrc.ec.europa.eu*
A high-quality Life Cycle Inventory (LCI) core data sets of this first version of the Commission's European Reference Life Cycle Data System (ELCD).

Energy Information Association (2008). *www.eia.doe.gov*
Official energy statistics from the U.S. Government.

European Aluminium Association (2000). *www.eaa.net*

Green Design Initiative, *www.ce.cmu.edu/GreenDesign*

GREET (2007). Argonne National Laboratory and the U.S. Department of Transport, *www.transportation.anl.gov*
Software for analyzing vehicle energy use and emissions.

Hammond, G., & Jones, C. (2006). Inventory of carbon and energy (ICE). Department of Mechanical Engineering, University of Bath.

IDEMAT, Environmental Materials Database, *www.idemat.nl*

International Aluminum Institute (2000). Life cycle inventory of the worldwide aluminum industry Part 1—automotive, *www.world-aluminum.org*

Kemna, R., van Elburg, M., Li, W., & van Holsteijn, R. (2005). Methodology study of eco-design of energy-using products. Van Holsteijn en Kemna BV (VHK).

Kennedy, J. (1997). Energy minimisation in road construction and maintenance. Best Practice Report for the U.K. Department of the Environment.

Lafarge Cement, Lafarge Cement UK, *www.lafarge.com*

Lawson, B. (1996). Building materials, energy and the environment: Towards ecologically sustainable development. RAIA, Canberra, *www.climatechange.gov.au*

Lime Technology (2007). *www.limetechnology.co.uk*

MEEUP Methodology Report, final (2005).VHK. *www.pre.nl*
A report by the Dutch consultancy VHK commissioned by the European Union, detailing its implementation of an LCA tool designed to meet the EU Energy-Using Products directive.

Pentron, *www.pentron.com*

Pilz, H., Schweighofer, J., & Kletzer, E. (2005). The contribution of plastic products to resource efficiency. *Gesellschaft fur umfassende analysen (GUA)*, Vienna.

Schlesinger, M.E. (2007). *Aluminum recycling.* CRC Press.

Stiller, H. (1999). *Material intensity of advanced composite materials.* Wuppertal Institut fur Klima, Umvelt, Energie.

Sustainable concrete (2008). *www.sustainableconcrete.org.uk/*
A website representing U.K. concrete producers carrying useful information about carbon footprint.

Szargut, J., Morris, D.R., Steward, F.R. (1988). *Energy analysis of thermal chemical and metallurgical processes.* Hemisphere.

Szokolay, S.V. (1980). *Environmental science handbook: For architects and builders.* Construction Press.

The Nickel Institute North America (2007). *www.nickelinstitute.org.*

Waste Online (2008). *www.wasteonline.org.uk/resources/InformationSheets/Plastics.htm*

D.3 INFORMATION FOR MANUFACTURING PROCESSES

Alexander, J.M., Brewer, R.C., & Rowe, G.W. (1987). *Manufacturing technology, Vol. 2: Engineering processes.* Ellis Norwood.

Bralla, J.G. (1986). *Handbook of product design for manufacturing.* McGraw-Hill.

Budinski, K.G., & Budinski, M.K. (2010). *Engineering materials, properties and selection* (9th ed.). Prentice Hall.
A well-established materials text that deals well with both material properties and processes.

Waterman, N.A., & Ashby, M.F. (Eds.) (1996). *Materials selector.* Chapman and Hall.

Dieter, G.E. (1999). *Engineering design, a materials and processing approach* (3rd ed.). McGraw-Hill.
A well-balanced and respected text focusing on the place of materials and processing in technical design.

Kalpakjian, S. (1984). *Manufacturing processes for engineering materials.* Addison Wesley.

Lascoe, O.D. (1989). *Handbook of fabrication processes.* ASM International.

Schey, J.A. (1977). *Introduction to manufacturing processes.* McGraw-Hill.

Suh, N.P. (1990). *The principles of design.* Oxford University Press.

Software

CES: Cambridge Engineering Selector (2010). *www.grantadesign.com.*
Comprehensive selection system for all classes of materials and manufacturing processes. Variety of optional reference data sources, connects directly to www.matdata.net. Windows and web format. Modest price.

Internet sources

Cast Metals Federation, *www.castmetalsfederation.com*

Castings Technologies International, *www.castingstechnology.com*

Confederation of British Metalforming, *www.britishmetalforming.com*

National Center for Excellence in Metalworking Technology, *www.ncemt.ctc.com*

PERA, *www.pera.com*

The British Manufacturing Plant Constructors' Association, *www.bmpca.org.uk*

TWI (The Welding Institute), *www.twi.co.uk*

D.4 DATABASES AND EXPERT SYSTEMS AS SOFTWARE

The number and quality of computer-based materials information systems is growing rapidly. A selection of these, with comment and source, is given here. The databases are listed in alphabetical order.

Active Library on Corrosion. ASM International.
PC format requiring CD-ROM drive. Graphical, numerical, and textual information on corrosion of metals. Modest price.

ASM Handbooks Online. ASM International, *www.asminternational.org*
Annual license fee.

Alloycenter. ASM International, *www.asminternational.org*
Annual license fee.

ALUSELECT P1.0 (1992). *Engineering property data for wrought aluminum alloys.* European Aluminum Association.
PC format, DOS environment. Mechanical, thermal, electrical, and environmental properties of wrought aluminum alloys. Inexpensive.

CAMPUS: Computer Aided Material Preselection by Uniform Standards (1995), *www.campusplastics.com*
Polymer data from approximately 30 suppliers, measured to set standards.

CETIM-EQUIST II: Centre Technique des Industries Mécaniques (1997). Senlis: CETIM.
PC format, DOS environment. Compositions and designations of steels.

CETIM-Matériaux: Centre Technique des Industries Mécaniques (1997). Senlis: CETIM.
Online system. Compositions and mechanical properties of materials.

CETIM-SICLOP: Centre Technique des Industries Mécaniques (1997). Senlis: CETIM.
Online system. Mechanical properties of steels.

Cambridge Engineering Selector (2010), *www.grantadesign.com*
Comprehensive selection system for all classes of materials and manufacturing processes. Variety of optional reference data sources; connects directly to www.matdata.net. Windows and web format. Modest price.

CUTDATA: Machining Data System. Cincinnati: Metcut Research Associates, Manufacturing Technology Division.
A PC-based system that guides the choice of machining conditions: tool materials, geometries, feed rates, cutting speeds, and so forth. Modest price.

EASel: Engineering Adhesives Selector Program (1986). The Design Centre.
A knowledge-based program to select industrial adhesives for joining surfaces. Modest price.

SF-CD (replacing ELBASE): Metal Finishing/Surface Treatment Technology (1992). Metal Finishing Information Services.
Comprehensive information on published data related to surface treatment technology. Regularly updated. Modest price.

Material Data Network, *www.matdata.net*
The Material Data Network provides integrated access to a variety of quality information source, from ASM International, Granta Design, TWI, NPL, UKSteel, Matweb, IDES, etc.

PAL II: Permabond Adhesives Locator (1996). Eastleigh: Permabond.
A knowledge-based PC system for adhesive selection among Permabond adhesives. An impressive example of an expert system that works. Modest price.

PROSPECT: Version 1.1 (1995). Oxford Forestry Institute, Department of Plant Sciences, Oxford University.
A database of the properties of tropical woods of interest to a wood user; includes information about uses, workability, treatments, origins.

OPS (Optimal Polymer Selector), *www.grantadesign.com*
Integrates Granta Design's generic PolymerUniverse database, with chemical resistance data from RAPRA and grade-specific data for approximately 6000 polymers from CAMPUS, to produce the only comprehensive polymer selection system available.

teCal, Steel Heat-Treatment Calculations. ASM International.
Computes the properties resulting from defined heat treatments of low-alloy steels, using the composition as input. Modest price.

SOFINE PLASTICS (1997). Villeurbanne Cedex: Société CERAP.
Database of polymer properties. Environment and price unknown.

TAPP 2.0, Thermochemical and Physical Properties (1994). ES Microware.
+PC format, CD-ROM, Windows environment. A database of thermochemical and physical properties of solids, liquids, and gases, including phase diagrams neatly packaged with good user manual. Modest price.

UNSearch: Unified Metals and Alloys Composition Search. Materials Park: ASTM.
PC format, DOS environment. A database of information about composition, U.S. designation, and specification of common metals and alloys. Modest price.

Wegst, C.W. (1997). *Stahlschlüssel (Key to Steel)* (17th ed.). Verlag Stahlschlüssel Wegst GmbH.
CD-ROM, PC format. Excellent coverage of European products and manufacturers.

Woods of the World (1994). Tree Talk, Inc.
A CD-ROM of woods, with illustrations of structure, information about uses, origins, habitat, etc. PC format, requiring CD drive; Windows environment.

D.5 ADDITIONAL USEFUL INTERNET SITES

Metals prices and economic reports

American Metal Market, *www.amm.com*

Business Communications Company, *www.bccresearch.com*

Daily Economic Indicators, *www.bullion.org.za*

Iron and Steel Statistics Bureau, *www.issb.co.uk*

Kitco Inc Gold & Precious Metal Prices, *www.kitco.com/market*

London Metal Exchange, *www.lme.co.uk*

Metal Bulletin, *www.metalbulletin.plc.uk*

Metal Powder Report, *www.metal-powder.net*

Metallurgia, *www.metallurgia-italiana.net*

Minerals Information, *http://minerals.usgs.gov/minerals*

Roskill Reports *www.roskill.com*

The Precious Metal and Gem Connection, *www.thebulliondesk.com*

Supplier registers, government organizations, professional societies and trade associations, and standards organizations

Supplier registers

IndustryLink, *www.industrylink.com*

Kellysearch, *www.kellysearch.com*

Thomas Register, *www.thomasregister.com*

Government organizations

Commonwealth Scientific and Industrial Research Organization (Australia), *www.csiro.au*

National Academies Press, *www.nap.edu*

National Institute of Standards and Technology (U.S.A.), *www.nist.gov*

The Fraunhofer Institute, *www.fhg.de*

Professional societies and trade associations

Aluminium Federation, *www.alfed.org.uk*

ASM International, *www.asminternational.org*

ASME International, *www.asme.org*

British Constructional Steel Work Association, *www.steelconstruction.org*

Cast Metals Federation, *www.castmetalsfederation.com*

Community Trade Union, *www.community-tu.org*

Confederation of British Metalforming, *www.britishmetalforming.com*

Copper Development Association, *www.cda.org.uk*

Institute of Cast Metal Engineers, *www.icme.org.uk*

Institute of Spring Technology, *www.ist.org.uk*

International Aluminium Institute, *www.world-aluminum.org*

International Council on Mining and Metals, *www.icmm.com*

International Titanium Association, *www.titanium.org*

International Zinc Association, *www.iza.com*

Lead Development Association International, *www.ldaint.org*

Materials Information Service, *www.iom3.org*

National Association. of Steel Stock Holders, *www.nass.org.uk*

Nickel Institute, *www.nidi.org*

Pentron, *www.pentron.com*

Society of Automotive Engineers (SAE), *www.sae.org*

Steel Manufacturers Association, *www.steelnet.org*

Steel Construction Institute, *www.steel-sci.org*

The American Ceramics Society, *http://ceramics.org*

The British Manufacturing Plant Constructors'Association, *www.bmpca.org.uk*

The Institute of Materials, Minerals and Mining, *www.instmat.co.uk*

The Minerals, Metals & Materials Society, *www.tms.org/TMSHome.html*

UK Steel, *www.eef.org.uk/uksteel*

World Steel Association, *www.worldsteel.org*

Standards organizations

ASTM, *www.astm.org*

BSI, *www.bsi.org.uk*

DIN Deutsches Institut für Normung e.V, *www.din.de*

International Standards Organisation, *www.iso.org*

National Standards Authority of Ireland, *www.nsai.ie*

Miscellaneous

Hazardous materials (1), *www.phmsa.dot.gov/hazmat*

Hazardous materials (2), *http://ull.chemistry.uakron.edu/erd*

Information for Physical Chemists, *www.liv.ac.uk/Chemistry/Links/links.html*

K&K Associates' thermal connection, *www.tak2000.com*

Appendix E: Exercises

CONTENTS

E.1 INTRODUCTION TO EXERCISES

These exercises are designed to develop facility in selecting materials, processes, and shape, and in devising hybrid materials when no monolithic material completely meets the design requirements. Each exercise is accompanied by a worked solution. They are organized into the 12 sections listed in the Contents.

The early exercises are easy. Those that follow lead the reader through the use of *material properties* and *simple solutions to mechanics problems*, drawing on data and results contained in Appendices A and B; the use of *material property charts*; techniques for the translation of design requirements to

Materials Selection in Mechanical Design. DOI: 10.1016/B978-1-85617-663-7.00022-9

identify *constraints* and *objectives*; the derivation of *indices, screening, ranking, and multiobjective optimization*; coupled choice of *material* and *shape*; devising *hybrids*; and the choice of materials to meet *environmental criteria*.

Three important points:

1. Selection problems are open-ended and, generally, underspecified; there is seldom a single, correct answer. The proper answer is sensible translation of the design requirements into material constraints and objectives, applied to give a "short list" of potential candidates with commentary suggesting what supporting information would be needed to narrow the choice further.
2. The positioning of selection lines on charts is a matter of judgment. The goal is to place the lines such that they leave an adequately large "short list" of candidates (aim for four or so), drawn, if possible, from more than one class of material.
3. A request for a selection based on one material index alone (such as $M = E^{1/2}/\rho$) is correctly answered by listing the subset of materials that maximize this index. But a request for a selection of materials for a component—say a wing spar (a light, stiff beam for which the index is $M = E^{1/2}/\rho$)—requires more: Some materials with high $E^{1/2}/\rho$, such as silicon carbide, are unsuitable for obvious reasons. It is a poor answer that ignores common sense and experience and fails to add further constraints to incorporate them. Students should be encouraged to discuss the implications of their selection and to suggest further selection stages.

The best way to use the charts that are a feature of the book is to make clean copies (or download them from *www.grantadesign.com*) on which you can draw, try out alternative selection criteria, write comments, and so forth. Although the book itself is copyrighted, the reader is authorized to make copies of the charts and to reproduce these, with proper reference to their source, as he or she wishes.

All the materials selection problems can be solved using the CES software, which is particularly effective when multiple criteria and unusual indices are involved.

E.2 MATERIAL EVOLUTION IN PRODUCTS (CHAPTER 1)

E2.1 Use Google to research the history and uses of one of the following materials:

- Tin
- Glass
- Cement

- Titanium
- Carbon fiber

Present the result as a short report of about 100 to 200 words (roughly half a page).

E2.2 Research, at the level of the mini-case studies in this chapter, the evolution of material use in

- Writing implements
- Watering cans
- Bicycles
- Small-boat building
- Book binding

E.3 DEVISING CONCEPTS (CHAPTER 2)

These two examples illustrate the way in which concepts are generated. The left part of each diagram describes a physical principle according to which the need might be met; the right part elaborates, suggesting how the principle might be used.

E3.1 Concepts and embodiments for dust removers We met the need for a "device to remove household dust" in Chapter 1, with examples of established solutions. Now it is time for more creative thinking. Devise as many concepts to meet this need as you can. Nothing, at the concept stage, is too far-fetched; decisions about practicality and cost come later, at the detail stage, so think along the lines shown in Figure 2.2 of the main text and list concepts and outline embodiments as block diagrams like those shown in the figure.

E3.2 Cooling power electronics Microchips, particularly those for power electronics, get hot. If they get too hot they cease to function. The need: a scheme for removing heat from power microchips. Devise concepts to meet the need and sketch an embodiment of one of them, laying out your ideas in the way suggested in Exercise E2.1.

E.4 USING MATERIAL PROPERTIES (CHAPTER 3)

These exercises introduce the reader to two useful resources: the data sheets of Appendices A and B.

E4.1 A cantilever beam has a length $L = 50$ mm, a width $b = 5$ mm, and a thickness $t = 1$ mm. It is made of an aluminum alloy. By how much will the end deflect under an end-load of 5 N? Use data from Appendix A.4 for the (mean) value of Young's modulus of aluminum alloys, the equation for the elastic deflection of a cantilever from Appendix B.3, and for the second moment of a beam from Appendix B.2 to find out.

E4.2 A spring, wound from stainless steel wire with a wire diameter $d = 1$ mm, has $n = 20$ turns of radius $R = 10$ mm. How much will it extend when loaded with a mass P of 1 kg? Assume the shear modulus G of stainless steel to be $3/8E$, where E is Young's modulus; retrieve this from Appendix A.4, and use the expression for the extension of springs from Appendix B.6 to find out.

E4.3 A thick-walled tube has an inner radius $r_i = 10$ mm and an outer radius $r_o = 15$ mm. It is made from polycarbonate, PC. What is the maximum torque that the tube can carry without the onset of yield? Retrieve the (mean) yield strength σ_y of PC from Appendix A.5, the expression for the torque at onset of yield from Appendix B.6, and that for the polar moment of a thick-walled tube from Appendix B.2 to find out.

E4.4 A round bar, 20 mm in diameter, has a shallow circumferential notch with a depth $c = 1$ mm and a root radius $r = 10$ microns. The bar is made of a low carbon steel with a yield strength of $\sigma_y = 250$ MPa. It is loaded axially with a nominal stress σ_{nom} (the axial load divided by the unnotched area). At what value of σ_{nom} will yield first commence at the root of the notch? Use the stress concentration estimate of Appendix B.9 to find out.

E4.5 An acrylic (PMMA) window is clamped in a low carbon steel frame at $T = 20°C$. The temperature falls to $T = -20°C$, putting the window under tension because the thermal expansion coefficient of PMMA is larger than that of steel. If the window was stress-free at 20°C, what stress does it carry at −20°C? Use the result that the biaxial stress caused by a biaxial strain difference $\Delta\varepsilon$ is

$$\sigma = \frac{E\Delta\varepsilon}{1-\nu}$$

where E is Young's modulus for PMMA, and Poisson's ratio $\nu = 0.33$. You will find data for expansion coefficients in Table A.7 and for moduli in Table B.5. Use mean values.

E4.6 The PMMA window described in Exercise E4.5 has a contained crack of length $2a = 0.5$ mm. If the maximum tensile stress that is anticipated in the window is $\sigma = 20$ MPa, will the crack propagate? Choose an appropriate equation for crack propagation from Appendix B.10 and data for the fracture toughness K_{1C} of PMMA from Appendix A.6 to calculate the length of crack that is just unstable under this tensile stress.

E4.7 A flywheel with a radius $R = 200$ mm is designed to spin up to 8000 rpm. It is proposed to make it out of cast iron, but the casting shop can guarantee only that it will have no crack-like flaws greater than $2a = 2$ mm in length. Use the expression for the maximum stress in a spinning disk in Appendix B.7, that for the stress intensity at a small enclosed crack from

Appendix B.10, and data for cast iron from Appendices A.3 and A.6 to establish if the flywheel is safe. Take Poisson's ratio ν for cast iron to be 0.33.

E4.8 You wish to assess, approximately, the thermal conductivity λ of polyethylene (PE). To do so you block one end of a PE pipe with a wall thickness of $x = 3$ mm and diameter of 30 mm and fill it with boiling water while clutching the outside with your other hand. You note that the outer surface of the pipe first becomes appreciably hot at time $t \approx 18$ seconds after filling the inside with water. Use this information, plus data for specific heat C_p and density ρ of PE from Appendices A.3 and A.8 to estimate λ for PE. How does your result compare with the listed value in Table A.7?

E4.9 The capacitance C of a condenser with two plates each of area A separated by a dielectric of thickness t is

$$C = \varepsilon_r \varepsilon_o \frac{A}{t}$$

where ε_o is the permittivity of free space and ε_r is the dielectric constant of the material between the plates. Select a dielectric by scanning data in Appendix A.9 (a) to maximize C and (b) to minimize it, for a given A and t.

E4.10 It is proposed to replace the cast iron casing of a power tool with one with precisely the same dimension molded from nylon. Will the material cost of the nylon casing be greater or less than that made of cast iron? Use data from Appendices A.3 and A.11 to find out.

E.5 USING MATERIAL PROPERTY CHARTS (CHAPTER 4)

The 20 exercises in this section involve the simple use of the charts in Chapter 4 to find materials with given property profiles. They are answered by placing selection lines on the appropriate chart and reading off the materials that lie on the appropriate side of the line. It is a good idea to present the results as a table. All can be solved by using the printed charts.

If the CES Edu Materials Selection software is available, these exercises can be solved by its use. This involves first creating the chart, then applying the appropriate box or line selection. The results, at Level 1 or 2, are the same as those read from the hard-copy charts (most of which were made using the Level 1/2 database). The software offers links to processes, allows a wider search by using the Level 3 database, and gives access to supporting information via the "Search Web" function.

E5.1 A component is at present made from brass, a copper alloy. Use the Young's modulus–density (E–ρ) chart shown in Figure 4.3 to suggest three other metals that, in the same shape, would be stiffer. "Stiffer" means a higher value of Young's modulus.

E5.2 Use the Young's modulus–density (E–ρ) chart shown in Figure 4.3 to identify materials with both a modulus $E > 50$ GPa and a density $\rho < 2000$ kg/m^3.

E5.3 Use the Young's modulus–density (E–ρ) chart shown in Figure 4.3 to find (a) metals that are stiffer and less dense than steels and (b) materials (not just metals) that are both stiffer and less dense than steel.

E5.4 Use the E–ρ chart shown in Figure 4.3 to identify metals with both $E > 100$ GPa and $E/\rho > 0.02$ GPa/(kg/m^3).

E5.5 Use the E–ρ chart shown in Figure 4.3 to identify materials with both $E > 100$ GPa and $E^{1/3}/\rho > 0.003$ (GPa)$^{1/3}$/(kg/m^3). Remember that, on taking logs, the index $M = E^{1/3}/\rho$ becomes

$$Log\,(E) = 3Log\,(\rho) + 3Log\,(M)$$

and that this plots as a line of slope 3 on the chart, passing through the point $E = 27$ when $\rho = 1000$ in the units on the chart.

E5.6 Use the E–ρ chart shown in Figure 4.3 to establish whether woods have a higher specific stiffness E/ρ than epoxies.

E5.7 Do titanium alloys have a higher or lower specific strength (strength/density, σ_f/ρ) than the best steels? This is important when you want strength at low weight (landing gear of aircraft, mountain bikes). Use the σ_f/ρ chart of Figure 4.4 to decide.

E5.8 Use the modulus–strength (E–σ_f) chart shown in Figure 4.5 to find materials that have $E > 10$ GPa and $\sigma_f \geq 1000$ Mpa.

E5.9 Are the fracture toughnesses K_{1C} of the common polymers polycarbonate, ABS, and polystyrene larger or smaller than the engineering ceramic alumina, Al_2O_3? Are their toughnesses $G_{1C} = K_{1C}^2/E$ larger or smaller? The K_{1C}–E chart in Figure 4.7 will help.

E5.10 Use the fracture toughness–modulus chart (Figure 4.7) to find materials that have a fracture toughness K_{1C} greater than 100 MPa.m$^{1/2}$ and a toughness $G_{1C} = K_{1C}^2/E$ (shown as contours in Figure 4.7) greater than 10 kJ/m^3.

E5.11 The elastic deflection at fracture (the "resilience") of an elastic-brittle solid is proportional to the failure strain, $\varepsilon_{fr} = \sigma_{fr}/E$, where σ_{fr} is the stress that will cause a crack to propagate:

$$\sigma_{fr} = \frac{K_{1C}}{\sqrt{\pi c}}$$

Here K_{1C} is the fracture toughness and c is the length of the longest crack the materials may contain. Thus

$$\varepsilon_{fr} = \frac{1}{\sqrt{\pi c}} \left(\frac{K_{1C}}{E} \right)$$

Materials that can deflect elastically without fracturing are therefore those with large values of K_{1C}/E. Use the K_{1C}–E chart shown in Figure 4.7 to identify the class of materials with $K_{1C} > 1$ MPa.m$^{1/2}$ and high values of K_{1C}/E.

E5.12 One criterion for design of a safe pressure vessel is that it should leak before it breaks: The leak can be detected and the pressure released. This is achieved by designing the vessel to tolerate a crack of length equal to the thickness t of the pressure vessel wall, without failing by fast fracture. The safe pressure p is then

$$p \leq \frac{4}{\pi} \frac{1}{R} \left(\frac{K_{1C}^2}{\sigma_f} \right)$$

where σ_f is the elastic limit, K_{1C} is the fracture toughness, and R is the vessel radius. The pressure is maximized by choosing the material with the greatest value of

$$M = \frac{K_{1C}^2}{\sigma_y}$$

Use the $K_{1C} - \sigma_f$ chart shown in Figure 4.8 to identify three alloys that have particularly high values of M.

E5.13 A material is required for the blade of a rotary lawn mower. Cost is a consideration. For safety reasons, the designer specified a minimum fracture toughness for the blade: It is $K_{1C} > 30$ MPa m$^{1/2}$. The other mechanical requirement is for high hardness, H, to minimize blade wear. Hardness, in applications like this one, is related to strength:

$$H \approx 3\sigma_y$$

where σ_f is the strength (Chapter 4 gives a fuller definition). Use the K_{1C}–σ_f chart shown in Figure 4.8 to identify three materials that have $K_{1C} > 30$ MPa m$^{1/2}$ and the highest possible strength. To do this, position a "K_{1C}" selection line at 30 MPa m$^{1/2}$ and then adjust a "strength" selection line such that it just admits three candidates. Use the cost chart shown in Figure 4.19 to rank your selection by material cost, hence making a final selection.

E5.14 Bells ring because they have a low loss (or damping) coefficient, η; a high damping gives a dead sound. Use the loss coefficient–modulus (η–E) chart shown in Figure 4.9 to identify material that should make good bells.

E5.15 Use the loss coefficient–modulus (η–E) chart (Figure 4.9) to find metals with the highest possible damping.

E5.16 Use the thermal conductivity–electrical resistivity (λ–ρ_e) chart (Figure 4.10) to find three materials with high thermal conductivity, λ, and high electrical resistivity ρ_e.

E5.17 The window through which the beam emerges from a high-powered laser must obviously be transparent to light. Even then, some of the energy of the beam is absorbed in the window and can cause it to heat and crack. This problem is minimized by choosing a window material with a high thermal conductivity λ (to conduct the heat away) and a low expansion coefficient α (to reduce thermal strains); that is, by seeking a window material with a high value of

$$M = \lambda/\alpha$$

Use the α–λ chart shown in Figure 4.12 to identify the best material for an ultra-high-powered laser window.

E5.18 Use the maximum service temperature (T_{max}) chart (Figure 4.14) to find polymers that can be used above 200°C.

E5.19

a. Use the Young's modulus–relative cost (E–$C_{v,R}$) chart (Figure 4.18) to find the cheapest materials with a modulus, E, greater than 100 GPa.

b. Use the strength–relative cost (σ_f–$C_{v,R}$) chart (Figure 4.19) to find the cheapest materials with a strength, σ_f, above 100 MPa.

E5.20 Use the friction coefficient chart (Figure 4.15) to find two materials with exceptionally low coefficients of friction.

E.6 TRANSLATION: CONSTRAINTS AND OBJECTIVES (CHAPTERS 5 AND 6)

Translation is the task of re-expressing design requirements in terms that enable material and process selection. Tackle the exercises by formulating the answers to the questions in this table. Don't try to model the behavior at this point (that comes in later exercises). Just think out what the component does, and list the constraints that this imposes on material choice, including processing requirements.

Function	What does the component do?
Constraint	Which essential conditions must be met?
Objective	What is to be maximized or minimized?
Free variable	Which parameters of the problem is the designer free to change?

Here it is important to recognize the distinction between constraints and objectives. As the table says, a constraint is an essential condition that must be met, usually expressed as a **limit** on a material or process attribute. An objective is a quantity for which an **extremum** (a maximum or minimum) is sought, frequently cost, mass or volume, but there are others, several of which appear in the exercises that follow. Take the example of a bicycle frame. It must have a certain stiffness and strength. If it is not stiff and strong enough it will not work, but it is never required to have infinite stiffness or strength. Stiffness and strength are therefore constraints that become limits on modulus, elastic limit, and shape. If the bicycle is for sprint racing, it should be as light as possible—if you could make it infinitely light, that would be best of all. Minimizing mass, here, is the objective, perhaps with an upper limit (a constraint) on cost. If instead it is a shopping bike to be sold through supermarkets, it should be as cheap as possible—the less expensive it is, the more will be sold. This time minimizing cost is the objective, possibly with an upper limit (a constraint) on mass. For most bikes, of course, minimizing mass and cost are both objectives, and then trade-off methods are needed. They come later. For now use judgment to choose the single most important objective and make all others into constraints.

Here are two rules-of-thumb, useful in many "translation" exercises. Many applications require sufficient fracture toughness for the component to survive mishandling and accidental impact during service; a totally brittle material (like untoughened glass) is unsuitable. Then a necessary constraint is that of "adequate toughness." This is achieved by requiring that the fracture toughness K_{1C}15 MPa. $m^{1/2}$. Other applications require some ductility, sufficient to allow stress redistribution under loading points, and some ability to bend or shape the material plastically. This is achieved by requiring that the (tensile) ductility ε_f be 2%.

(If the CES software is available, it can be used to impose the constraints and to rank the survivors using the objective.)

E6.1 A material is required for the windings of an electric air furnace capable of temperatures up to 1000°C. Think out what attributes a material must have if it is to be made into windings and function properly in a furnace. List the function and the constraints; set the objective to "minimize cost" and the free variables to "choice of material."

E6.2 A material is required to manufacture office scissors. Paper is an abrasive material, and scissors sometimes encounter hard obstacles like staples. List the function and the constraints; set the objective to "minimize cost" and the free variable to "choice of material."

E6.3 A material is required for a heat exchanger to extract heat from geothermally heated, saline water at 120°C (and thus under pressure). List the function and the constraints; set the objective to "minimize cost" and the free variable to "choice of material."

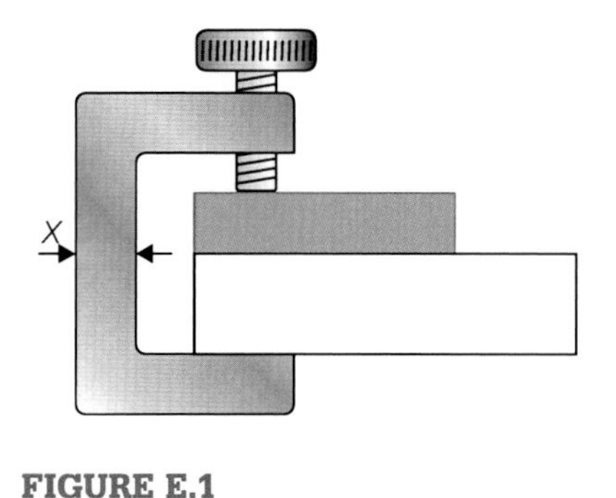

FIGURE E.1

E6.4 C-clamp is required for processing electronic components at temperatures up to 450°C (Figure E.1). It is essential that the clamp have as low a thermal inertia as possible so that it reaches temperature quickly, and it must not charge up when exposed to an electron beam. The time t it takes a component of thickness x to reach thermal equilibrium when the temperature is suddenly changed (which is a transient heat flow problem) is

$$t \approx \frac{x^2}{2a}$$

where the thermal diffusivity $a = \lambda/\rho C_p$ and λ is the thermal conductivity, ρ the density, and C_p the specific heat.

List function, constraints, and objective; set the free variables to "choice of material."

E6.5 A furnace is required to sinter powder-metal parts. It operates continuously at 650°C while the parts are fed through on a moving belt. You are asked to select a material for furnace insulation to minimize heat loss and thus make the furnace as energy efficient as possible. For reasons of space the insulation is limited to a maximum thickness of $x = 0.2$ m. List the function, constraints, objective, and free variable.

E6.6 Ultraprecise bearings that allow a rocking motion make use of knife-edges or pivots (Figure E.2). As the bearing rocks, it rolls, translating sideways by a distance that depends on the radius of contact. The further it rolls, the less precise is its positioning, so the smaller the radius of contact R the better. But the smaller the radius of contact, the greater is the contact pressure (F/A). If this exceeds the hardness H of either face of the bearing, it will be damaged. Elastic deformation is bad too: It flattens the contact, increasing the contact area and the roll.

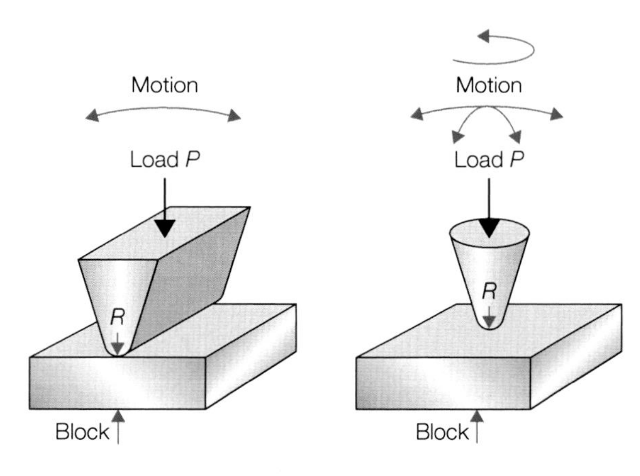

FIGURE E.2

A rocking bearing is required to operate in a micro-chip fabrication unit using fluorine gas at 100°C, followed by e-beam processing requiring that all structural parts of the equipment can be earthed to prevent stray charges. Translate the requirements into material selection criteria, listing function, constraints, objective, and free variable.

E6.7 The standard CD ("jewel" case) cracks easily and, if broken, can scratch the CD. Jewel cases are made of injection-molded polystyrene, chosen because it is transparent, cheap, and easy to mold. A material is sought to make CD cases that do not crack so easily. The case must still be transparent, able to be injection-molded, and able to compete with polystyrene in cost.

E6.8 A storage heater captures heat over a period of time, then releases it, usually to an air stream, when required. Those for domestic heating store solar energy or energy from cheap off-peak electricity and release it slowly during the cold part of the day. Those for research release the heat to a supersonic air stream to test system behavior in supersonic flight. What is a good material for the core of a compact storage material capable of temperatures up to 120°C?

E.7 DERIVING AND USING MATERIAL INDICES (CHAPTERS 5 AND 6)

The exercises in this section give practice in deriving indices.

a. Start each by listing function, constraints, objectives, and free variables; without having those straight, you will get in a mess. Then write down an equation for the objective. Consider whether it contains a free variable other than material choice; if it does, identify the constraint that limits it, substitute, and read off the material index.

b. If the CES Edu software is available, use it to apply the constraints and rank the survivors using the index (start with the Level 2 database). Are the results sensible? If not, what constraint has been overlooked or incorrectly formulated?

E7.1 Aperture grills for cathode ray tubes There are two types of cathode ray tube (CRT). In the older technology, color separation is achieved by using a *shadow mask*: a thin metal plate with a grid of holes that allow only the correct beam to strike a red, green, or blue phosphor. A shadow mask can heat up and distort at high brightness levels ("doming"), causing the beams to miss their target, and giving a blotchy image.

To avoid this, shadow masks are made of Invar, a nickel alloy with a near-zero expansion coefficient between room temperature and 150°C. It is a consequence of shadow-mask technology that the glass screen of the CRT curves inward on all four edges, increasing the probability of reflected glare.

Sony's "Trinitron" technology overcame this problem and allowed greater brightness by replacing the shadow mask with an *aperture grill* of fine vertical wires, each about 200 μm in thickness, that allows the intended beam to

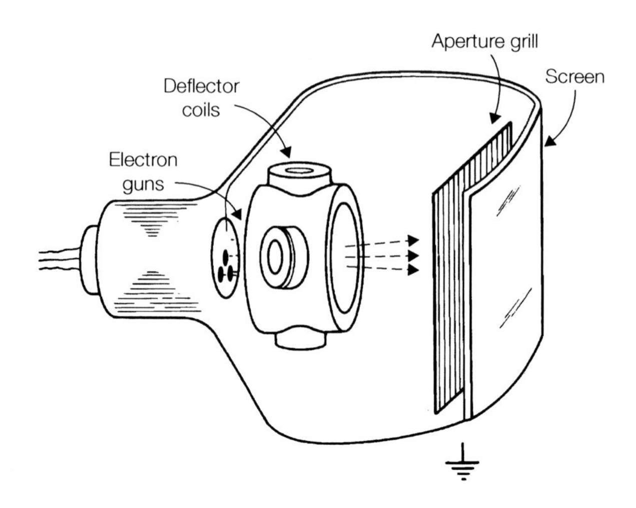

FIGURE E.3

strike either the red, the green, or the blue phosphor to create the image (see Figure E.3). The glass face of the Trinitron tube was curved in one plane only, reducing glare.

The wires of the aperture grill are tightly stretched, so that they remain taut even when hot—it is this tension that allows the greater brightness. What index guides the choice of material to make them? The table summarizes the requirements.

Function	Aperture grill for CRT
Constraints	Wire thickness and spacing specified Must carry pre-tension without failure Electrically conducting to prevent charging Able to be drawn to wire
Objective	Maximize permitted temperature rise without loss of tension
Free variable	Choice of material

E7.2 Material indices for elastic beams with differing constraints (Figure E.4) Start each of the four parts of this problem by listing the function, the objective, and the constraints. You will need the equations for the deflection of a cantilever beam with a square cross-section $t \times t$, given in Appendix B.3. The two that matter are that for the deflection δ of a beam of length L under an end load F:

$$\delta = \frac{FL^3}{3EI}$$

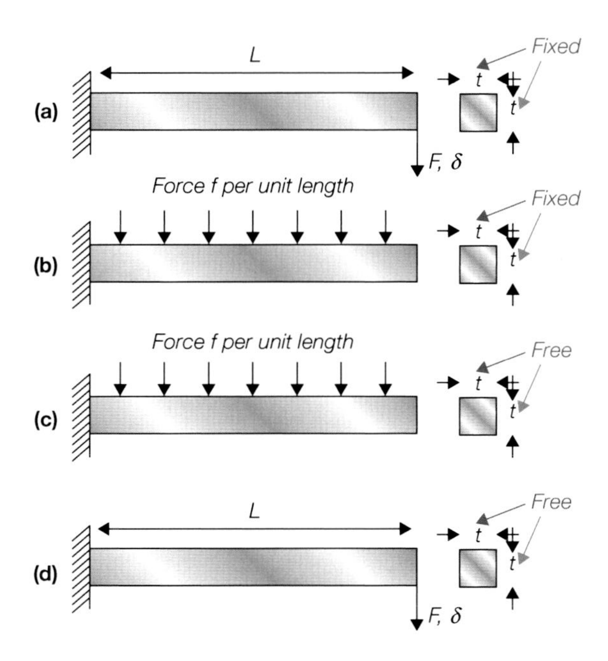

FIGURE E.4

and that for the deflection of a beam under a distributed load f per unit length:

$$\delta = \frac{1}{8}\frac{f\,L^4}{E\,I}$$

where $I = t^4/12$. For a self-loaded beam $f = \rho\, Ag$, where ρ is the density of the material of the beam, A its cross-sectional area, and g the acceleration due to gravity.

a. Show that the best material for a cantilever beam of given length L and *given* (i.e., fixed) square cross-section ($t \times t$) that will deflect least under a given end load F is that with the largest value of the index $M = E$, where E is Young's modulus (neglect self-weight). (See Figure E.4(a).)

b. Show that the best material choice for a cantilever beam of given a length L and with a given section ($t \times t$) that will deflect least under its own self-weight is that with the largest value of $M = E/\rho$, where ρ is the density. (Figure E.4(b).)

c. Show that the material index for the lightest cantilever beam of length L and square section (not given, i.e., the area is a free variable) that will not deflect by more than δ under its own weight is $M = E/\rho^2$. (See Figure E.4(c).)

d. Show that the lightest cantilever beam of length L and square section (area free) that will not deflect by more than δ under an end load F is that made of the material with the largest value of $M = E^{1/2}/\rho$ (neglect self weight). (See Figure E.4(d).)

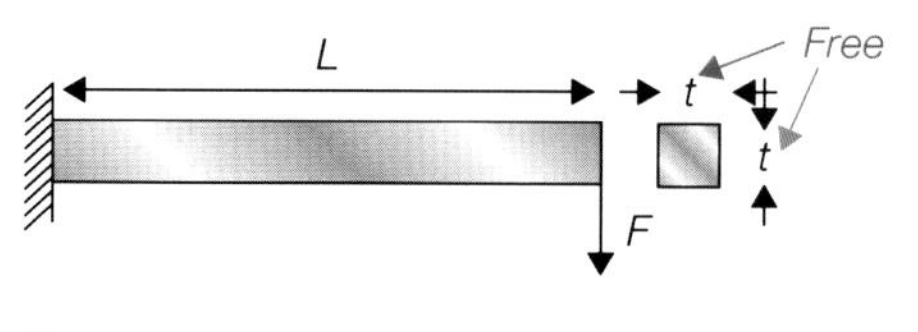

FIGURE E.5

E7.3 Material index for a light, strong beam In stiffness-limited applications, it is elastic deflection that is the active constraint: It limits performance (Figure E.5). In strength-limited applications, deflection is acceptable provided the component does not fail; strength is the active constraint. Derive the material index for selecting materials for a beam of length *L*, specified strength, and minimum weight. For simplicity, assume the beam to have a solid square cross-section $t \times t$. You will need the equation for the failure load of a beam (Appendix B, Section B.4). It is

$$F_f = \frac{I\sigma_f}{y_m L}$$

where y_m is the distance between the neutral axis of the beam and its outer filament and $I = t^4/12 = A^2/12$ is the second moment of the cross-section. The table itemizes the design requirements.

Function	Beam
Constraints	Length *L* is specified Beam must support a bending load *F* without yield or fracture
Objective	Minimize the mass of the beam
Free variables	Cross-section area, *A* Choice of material

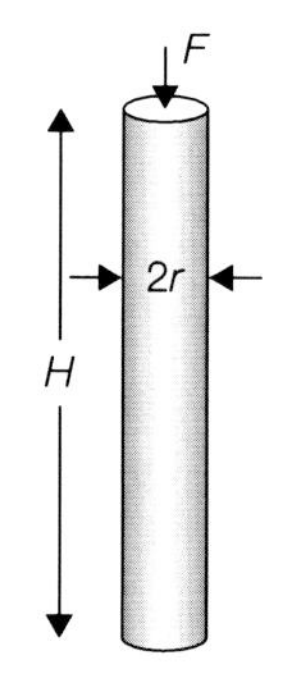

FIGURE E.6

E7.4 Material index for a cheap, stiff column (Figure E.6) In the preceding two exercises the objective has been that of minimizing weight. There are many others. In the selection of a material for a spring, the objective is that of maximizing the elastic energy it can store. In seeking materials for thermal-efficient insulation for a furnace, the best are those with the lowest thermal conductivity and heat capacity. And most common of all is the wish to minimize cost. So here is an example involving cost.

Columns support compressive loads: the legs of a table; the pillars of the Parthenon. Derive the index for selecting materials for the cheapest cylindrical column of specified height, H, that will safely support a load F without buckling elastically. You will need the equation for the load F_{crit} at which a slender column buckles. It is

$$F_{crit} = \frac{n^2 \pi^2 E I}{H^2}$$

where n is a constant that depends on the end constraints and $I = \pi r^4/4 = A^2/4\pi$ is the second moment of area of the column (see Appendix B for both). The following table lists the requirements.

Function	Cylindrical column
Constraints	Length L specified Column must support a compressive load F without buckling
Objective	Minimize the material cost of the column
Free variables	Cross-section area A Choice of material

E7.5 Indices for stiff plates and shells Aircraft and space structures make use of plates and shells (Figure E.7). The index depends on the configuration. Here you are asked to derive the material index for

a. a circular plate of radius a carrying a central load W with a prescribed stiffness $S = W/\delta$ and of minimum mass

b. a hemispherical shell of radius a carrying a central load W with a prescribed stiffness $S = W/\delta$ and of minimum mass, as shown in the figures

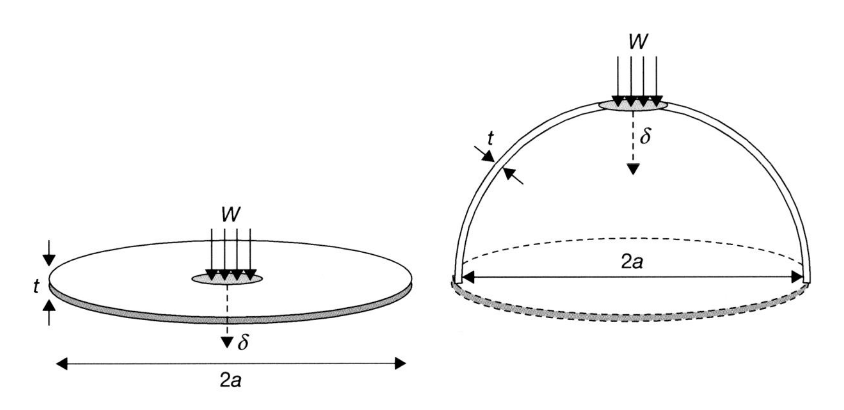

FIGURE E.7

Use the two results listed below for the mid-point deflection δ of a plate or spherical shell under a load W applied over a small central, circular area.

$$\text{Circular plate: } \delta = \frac{3}{4\pi}\frac{Wa^2}{Et^3}(1-\nu^2)\left(\frac{3+\nu}{1+\nu}\right)$$

$$\text{Hemispherical shell: } \delta = A\frac{Wa}{Et^2}(1-\nu^2)$$

in which $A \approx 0.35$ is a constant. Here E is Young's modulus, t is the thickness of the plate or shell, and ν is Poisson's ratio. Poisson's ratio is almost the same for all structural materials and can be treated as a constant. The table summarizes the requirements.

Function	Stiff circular plate or stiff hemispherical shell
Constraints	Stiffness S under central load W specified Radius a of plate or shell specified
Objective	Minimize the mass of the plate or shell
Free variables	Plate or shell thickness t Choice of material

E7.6 The C-clamp in more detail (Figure E.8) Exercise E4.4 introduced the C-clamp for processing of electronic components. The clamp has a square cross-section of width x and given depth b. It is essential that the clamp have low thermal inertia so that it reaches temperature quickly. The time t it takes a component of thickness x to reach thermal equilibrium when the temperature is suddenly changed (a transient heat flow problem) is

$$t \approx \frac{x^2}{2a}$$

where the thermal diffusivity $a = \lambda/\rho C_p$ and λ is the thermal conductivity, ρ the density, and C_p the specific heat.

The time to reach thermal equilibrium is reduced by making the section x thinner, but it must not be so thin that it fails in service. Use this constraint to eliminate x in the equation above, thereby deriving a material index for the clamp. Use the fact that the clamping

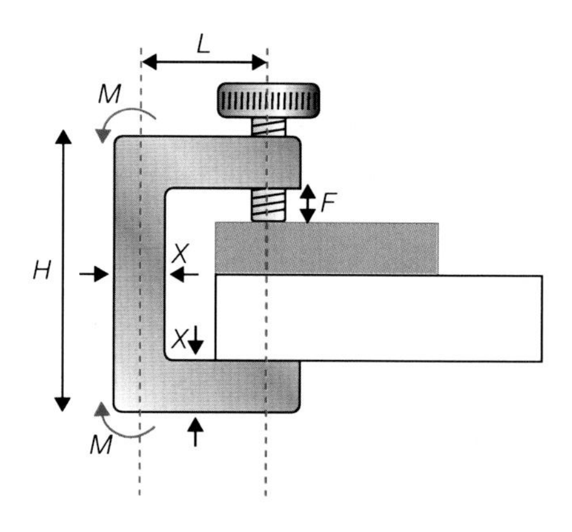

FIGURE E.8

force F creates a moment on the body of the clamp of $M = FL$, and that the peak stress in the body is given by

$$\sigma = \frac{x}{2}\frac{M}{I}$$

where $I = bx^3/12$ is the second moment of area of the body. The table summarizes the requirements.

Function	C-clamp of low thermal inertia
Constraints	Depth b specified Must carry clamping load F without failure
Objective	Minimize time to reach thermal equilibrium
Free variables	Width of clamp body x Choice of material

E7.7 Springs for trucks For the design of vehicle suspension it is desirable to minimize the mass of all components. You have been asked to select a material and dimensions for a light spring to replace the steel leaf spring of an existing truck suspension. The existing leaf-spring is a beam, shown schematically in the Figure E.9. The new spring must have the same length L and stiffness S as the existing one, and must deflect through a maximum safe displacement δ_{max} without failure. The width b and thickness t are free variables.

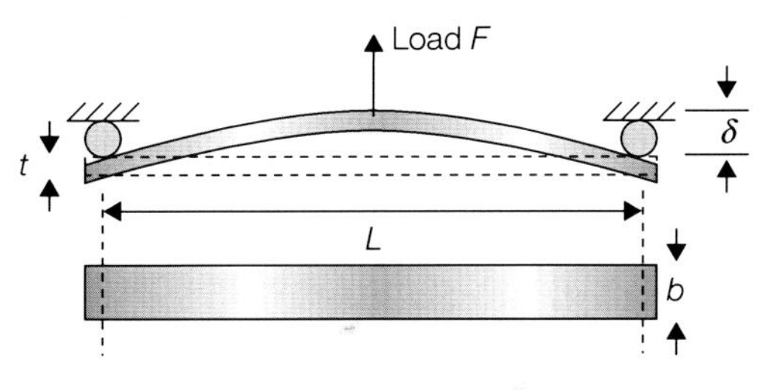

FIGURE E.9

Derive a material index for the selection of a material for this application. Note that this is a problem with two free variables: b and t; and there are two constraints, one on safe deflection δ_{max} and the other on stiffness S. Use the two constraints to fix free variables. The following table catalogs the requirements.

Function	Leaf spring for truck
Constraints	Length L specified Stiffness S specified Maximum displacement δ_{max} specified
Objective	Minimize mass
Free variables	Spring thickness t Spring width b Choice of material

You will need the equation for the mid-point deflection of an elastic beam of length L loaded in three-point bending by a central load F:

$$\delta = \frac{1}{48}\frac{FL^3}{EI}$$

and that for the deflection at which failure occurs:

$$\delta_{\max} = \frac{1}{6}\frac{\sigma_f L^2}{tE}$$

where I is the second moment of area; for a beam of rectangular section, $I = bt^3/12$ and E and σ_f are the modulus and failure stress of the material of the beam. (See Appendix B.)

E7.8 Fin for a rocket A tube-launched rocket has stabilizing fins at its rear (Figure E.10). During launch the fins experience hot gas at $T_g = 1700°$ C for a time $t = 0.3$ seconds. It is important that the fins survive launch without surface melting. Suggest a material index for selecting a material for the fins. The following table summarizes the requirements.

Function	High-heat-transfer rocket fins
Constraints	All dimensions specified Must not suffer surface melting during exposure to gas at 1700°C for 0.3 sec
Objective	Minimize the surface temperature rise during firing Maximize the melting point of the material
Free variable	Choice of material

This is tricky. Heat enters the surface of the fin by transfer from the gas. If the heat transfer coefficient is h, the heat flux per unit area is

$$q = h(T_g - T_s)$$

where T_s is the surface temperature of the fin—the critical quantity we wish to minimize. Heat diffuses into the fin surface by thermal conduction.

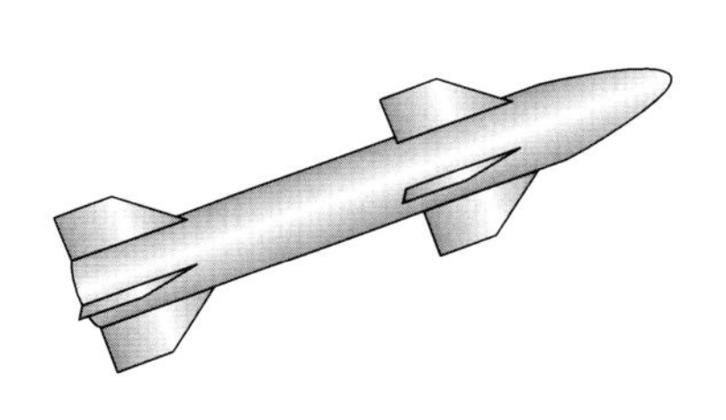

FIGURE E.10

If the heating time is small compared with the characteristic time for heat to diffuse through the fin, a quasi-steady state exists in which the surface temperature adjusts itself such that the heat entering from the gas is equal to that diffusing inward by conduction. This second is equal to

$$q = \lambda\frac{(T_s - T_i)}{x}$$

where λ is the thermal conductivity, T_i is the temperature of the (cold) interior of the fin, and x is a characteristic heat-diffusion length. When the heating time is short (as here) the thermal front, after a time t, has penetrated a distance

$$x \approx (2at)^{1/2}$$

where $a = \lambda/\rho C_p$ is the thermal diffusivity. Substituting this value of x in the previous equation gives

$$q = (\lambda \rho C_p)^{1/2} \frac{(T_s - T_i)}{\sqrt{2t}}$$

where ρ is the density and C_p the specific heat of the material of the fin.

Proceed by equating the two equations for q, solving for the surface temperature T_s to give the objective function. Read off the combination of properties that minimize T_s; it is the index for the problem.

The selection is made by seeking materials with large values of the index and with a high melting point, T_m. If the CES software is available, make a chart with these two as axes and identify materials with high values of the index that also have high melting points.

E.8 MULTIPLE CONSTRAINTS AND OBJECTIVES (CHAPTERS 7 AND 8)

Over constrained problems are normal in materials selection. Often it is just a case of applying each constraint in turn, retaining only those solutions that meet them all. But when constraints are used to eliminate free variables in an objective function (as discussed in Section 9.5 of the text), the "active constraint" method must be used. The first three exercises in this section illustrate problems with multiple constraints.

The remaining two concern multiple objectives and trade-off methods. When a problem has two objectives—minimizing both mass m and cost C of a component, for instance—a conflict arises: The cheapest solution is not the lightest and vice versa. The best combination is sought by constructing a trade-off plot using mass as one axis and cost as the other. The lower envelope of the points on this plot defines the trade-off surface. The solutions that offer the best compromise lie on this surface. To get further we need a penalty function. Define the penalty function

$$Z = C + \alpha\, m$$

where α is an exchange constant describing the penalty associated with unit increase in mass, or, equivalently, the value associated with a unit decrease. The best solutions are found where the line defined by this equation is tangential to the trade-off surface. (Remember that objectives must be expressed in a form such that a minimum is sought; then a low value of Z, not a high one, is desirable).

When a substitute is sought for an existing material it is better to work with ratios. Then the penalty function becomes

$$Z^* = \frac{C}{C_o} + \alpha^* \frac{m}{m_o}$$

in which the subscript o means properties of the existing material and the asterisk on Z^* and α^* is a reminder that both are now dimensionless. The relative exchange constant α^* measures the fractional gain in value for a given fractional gain in performance.

E8.1 Multiple constraints: A light, stiff, strong tie (Figure E.11) A tie of length L loaded in tension is to support a load F, at minimum weight without failing (implying a constraint on strength) or extending elastically by more than δ (implying a constraint on stiffness, F/δ). The table summarizes the requirements.

a. Follow the method of Chapter 7 to establish two performance equations for the mass, one for each constraint, from which two material indices and one coupling equation linking them are derived. Show that the two indices are

$$M_1 = \frac{\rho}{E} \quad \text{and} \quad M_2 = \frac{\rho}{\sigma_y}$$

and that a minimum is sought for both.

b. Use these and the material chart shown in Figure E.12, which has the indices as axes, to identify candidate materials for the tie (1) when $\delta/L = 10^{-3}$ and (2) when $\delta/L = 10^{-2}$.

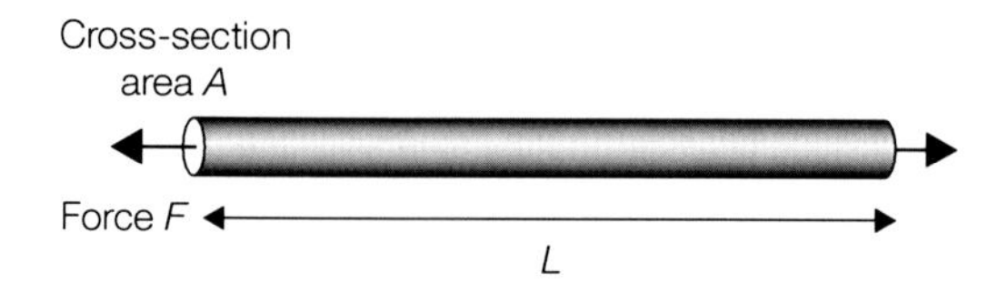

FIGURE E.11

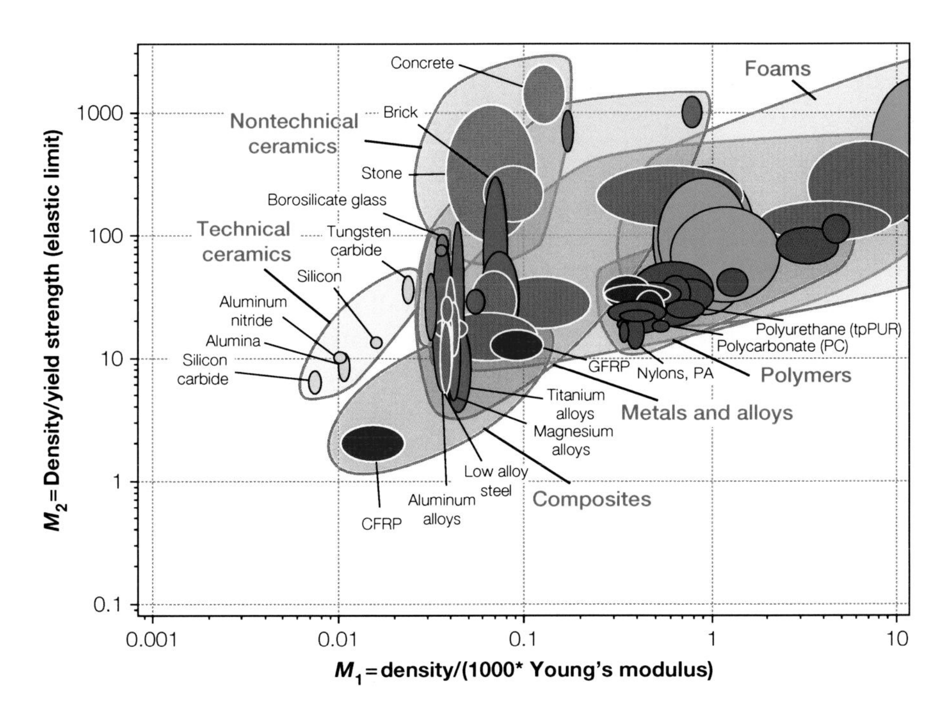

FIGURE E.12

E8.2 Multiple constraints: A light, safe pressure vessel (Figure E.13) When a pressure vessel has to be mobile, its weight becomes important. Aircraft bodies, rocket casings and liquid natural gas containers are examples; they must be light, and at the same time they must be safe; that means that they must not fail by yielding or by fast fracture. What are the best materials for their construction? The following table summarizes the requirements.

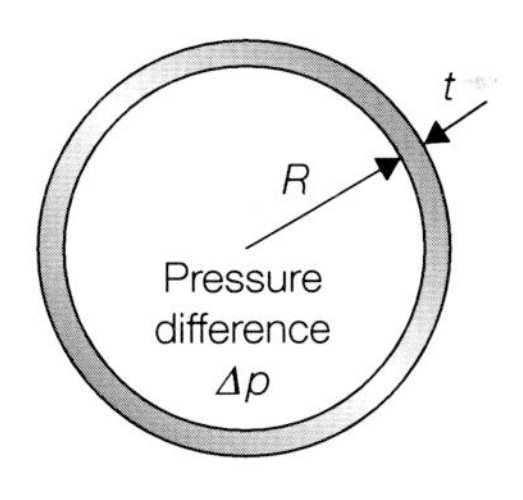

FIGURE E.13

Function	Pressure vessel
Constraints	Must not fail by yielding Must not fail by fast fracture Diameter 2R and pressure difference Δp specified
Objective	Minimize mass m
Free variables	Wall thickness t Choice of material

a. First write a performance equation for the mass m of the pressure vessel. Assume, for simplicity, that it is spherical, of specified radius R, and that the wall thickness t (the free variable) is small compared with R. Then the tensile stress in the wall is

$$\sigma = \frac{\Delta p R}{2t}$$

where Δp, the pressure difference across this wall, is fixed by the design. The first constraint is that the vessel should not yield—that is, that the tensile stress in the wall should not exceed σ_y. The second is that it should not fail by fast fracture; this requires that the wall stress be less than $K_{1C}\sqrt{\pi c}$, where K_{1C} is the fracture toughness of the material of which the pressure vessel is made and c is the length of the longest crack that the wall might contain. Use each of these in turn to eliminate t in the equation for m; use the results to identify two material indices:

$$M_1 = \frac{\rho}{\sigma_y} \quad \text{and} \quad M_2 = \frac{\rho}{K_{1C}}$$

and a coupling relation between them. It contains the crack length, c.

b. Figure E.14 shows the chart you will need with the two material indices as axes. Plot the coupling equation onto this figure for two values of c: one of 5 mm, the other of 5 μm. Identify the lightest candidate materials for the vessel for each case.

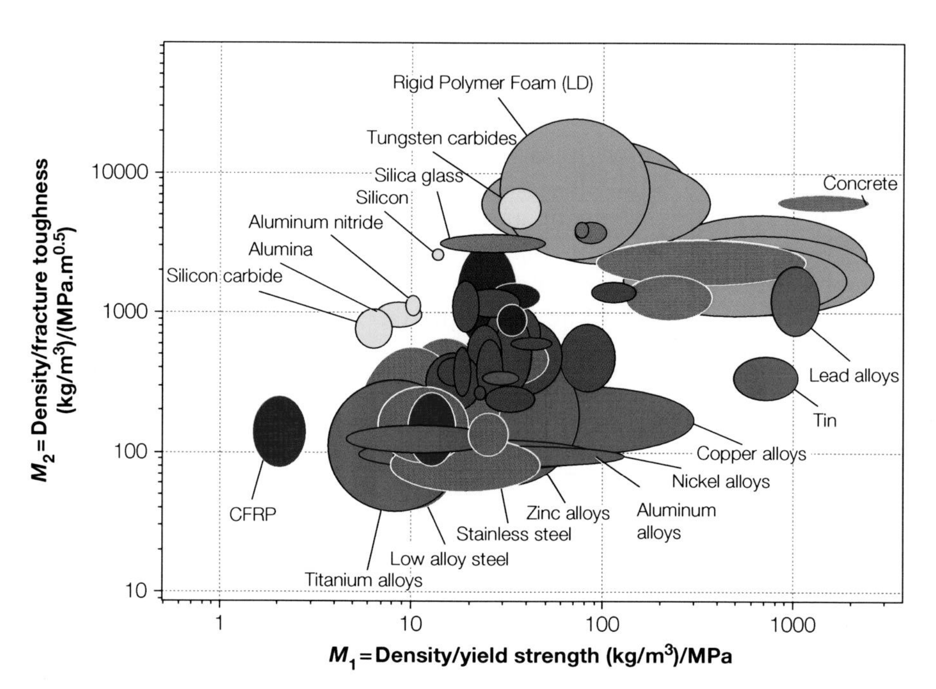

FIGURE E.14

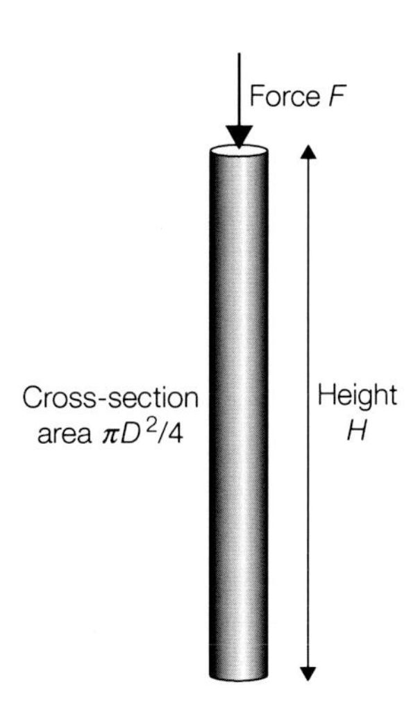

FIGURE E.15

E8.3 A cheap column that must not buckle or crush (Figure E.15) The best choice of material for a light, strong column depends on its aspect ratio: the ratio of its height H to its diameter D. This is because short, fat columns fail by crushing; tall, slender columns buckle instead.

Derive two performance equations for the material cost of a column of a solid circular section and of a specified height H, designed to support a load F (i.e., large compared to its self-load), one using the constraints that the column must not crush, the other that it must not buckle. The following table summarizes the needs.

Function	Column
Constraints	Must not fail by compressive crushing Must not buckle Height H and compressive load F specified
Objective	Minimize material cost C
Free variables	Diameter D Choice of material

a. Proceed as follows

1. Write down an expression for the material cost of the column—its mass times its cost per unit mass C_m.
2. Express the two constraints as equations, and use them to substitute for the free variable to find the cost of the column that will just support the load without failing by either mechanism.
3. Identify the material indices M_1 and M_2 that enter the two equations for the mass, showing that they are

$$M_1 = \left(\frac{C_m \rho}{\sigma_c}\right) \text{ and } M_2 = \left[\frac{C_m \rho}{E^{1/2}}\right]$$

where C_m is the material cost per kg, ρ the material density, σ_c its crushing strength, and E its modulus.

b. Data for six possible candidates for the column are listed in the table that follows. Use these to identify candidate materials when $F = 10^5$ N and $H = 3$ m. Ceramics are admissible here because they have high strength in compression.

Data for Candidate Materials for the Column				
Material	**Density** ρ **(kg/m^3)**	**Cost/kg** C_m **($/kg)**	**Modulus** E **(MPa)**	**Compression Strength** σ_c **(MPa)**
Wood (spruce)	700	0.5	10,000	25
Brick	2100	0.35	22,000	95
Granite	2600	0.6	20,000	150
Poured concrete	2300	0.08	20,000	13
Cast iron	7150	0.25	130,000	200
Structural steel	7850	0.4	210,000	300
Al alloy 6061	2700	1.2	69,000	150

c. Figure E.16 shows a material chart with the two indices as axes. Identify and plot coupling lines for selecting materials for a column with $F = 10^5\ N$ and $H = 3\ m$ (the same conditions as previously noted), and for a second column with $F = 10^3\ N$ and $H = 20\ m$.

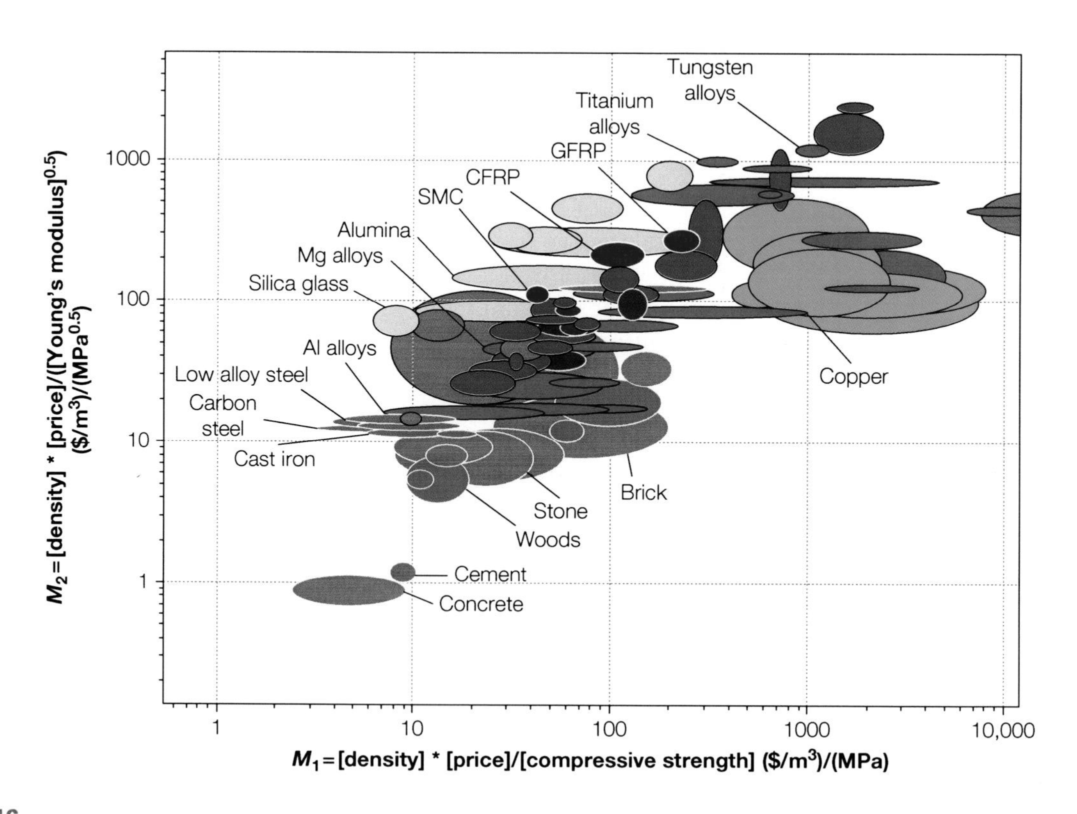

FIGURE E.16

E8.4 A truck's air cylinder Trucks rely on compressed air for braking and other power-actuated systems. The air is stored in one or a cluster of cylindrical pressure tanks like that shown in Figure E.17 (length L, diameter $2R$, hemispherical ends). Most are made of low carbon steel and are heavy. The task: to explore the potential of alternative materials for lighter air tanks, recognizing that there must be a trade-off between mass and cost—if it is too expensive, the truck owner will not want it even if it *is* lighter. The following table summarizes the design requirements.

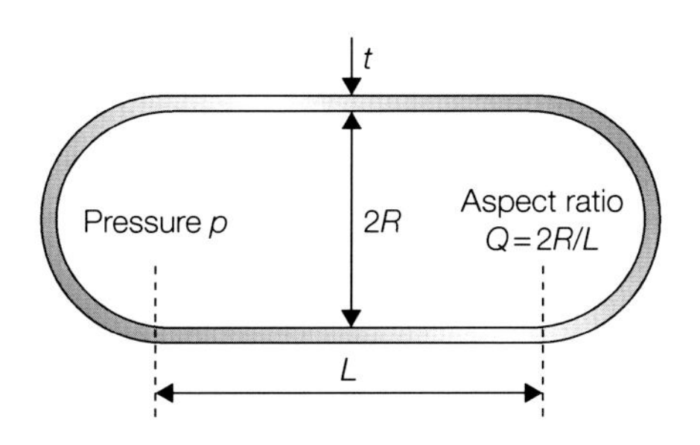

FIGURE E.17

Function	Air cylinder for truck
Constraints	Must not fail by yielding Diameter $2R$ and length L specified, so the ratio $Q = 2R/L$ is fixed
Objectives	Minimize mass m Minimize material cost C
Free variables	Wall thickness t Choice of material

a. Show that the mass and material cost of the tank relative to one made of low carbon steel are given by

$$\frac{m}{m_o} = \left(\frac{\rho}{\sigma_y}\right)\left(\frac{\sigma_{y,o}}{\rho_o}\right) \text{ and } \frac{C}{C_o} = \left(\frac{C_m \rho}{\sigma_y}\right)\left(\frac{\sigma_{y,o}}{C_{m,o}\rho_o}\right)$$

where ρ is the density, σ_y the yield strength, C_m the cost per kg of the material, and the subscript o indicates values for mild steel.

b. Explore the trade-off between relative cost and relative mass, considering the replacement of a mild steel tank with one made, first, of low alloy steel and, second, one made of filament-wound CFRP, using the material properties in the table below. Define a relative penalty function:

$$Z^* = \alpha^* \frac{m}{m_o} + \frac{C}{C_o}$$

where α^* is a relative exchange constant, and evaluate Z^* for $\alpha^* = 1$ and for $\alpha^* = 100$.

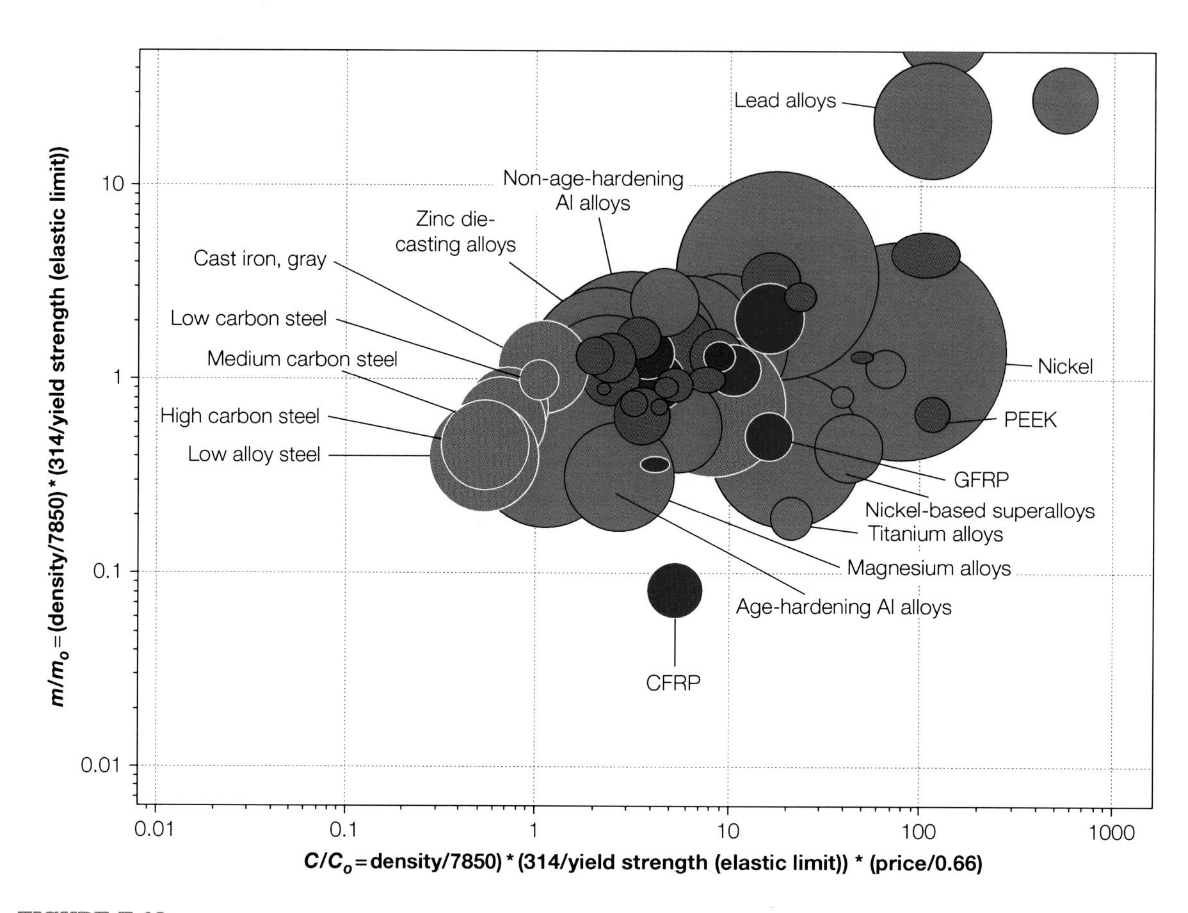

FIGURE E.18

Material	Density ρ (kg/m^3)	Yield Strength σ_c (MPa)	Price per kg C_m (\$/kg)
Mild steel	7850	314	0.66
Low alloy steel	7850	775	0.85
CFRP	1550	760	42.1

c. Figure E.18, and the chart below it, shows the axes of m/m_o and C/C_o. Mild steel (here labeled "Low carbon steel") lies at the coordinates (1,1).

Sketch a trade-off surface and plot contours of Z^* that are approximately tangent to the trade-off surface for $\alpha^* = 1$ and for $\alpha^* = 100$. What selections do these suggest?

E8.5 Insulating walls for freezers Freezers and refrigerated trucks (see Figure E.19)have panel walls that provide thermal insulation, and at the

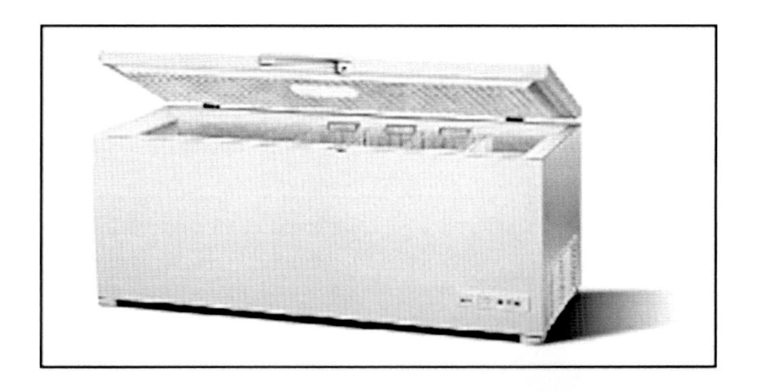

FIGURE E.19

same time are stiff, strong, and light (stiffness to suppress vibration, strength to tolerate rough usage). To achieve this the panels are usually of sandwich construction, with two skins of steel, aluminum, or GFRP (providing the strength) separated by, and bonded to, a low-density insulating core. In choosing the core we seek to minimize thermal conductivity λ and at the same time maximize stiffness, because this allows thinner steel faces and thus a lighter panel while still maintaining the overall panel stiffness. The following table summarizes the design requirements.

Function	Foam for panel-wall insulation
Constraint	Panel-wall thickness specified
Objectives	Minimize foam thermal conductivity λ Maximize foam stiffness, meaning Young's modulus E
Free variable	Choice of material

Figure E.20 shows the thermal conductivity λ of foams plotted against their elastic compliance $1/E$ (the reciprocal of their Young's moduli E, since we

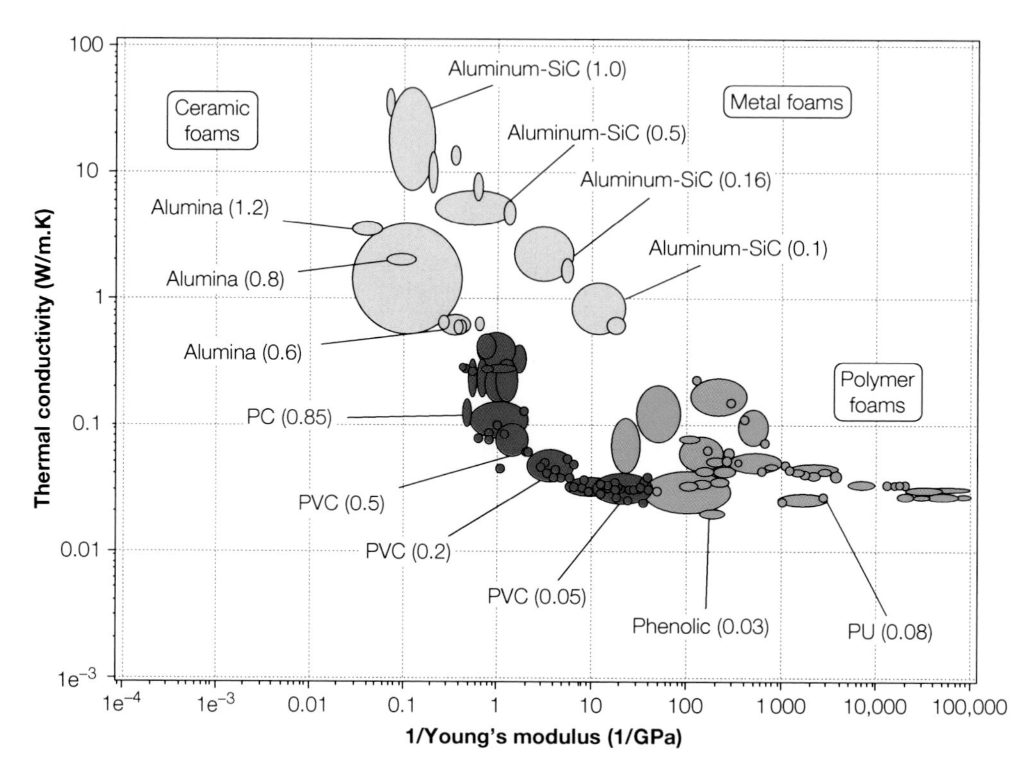

FIGURE E.20

must express the objectives in a form that requires minimization). The numbers in brackets are the densities of the foams in Mg/m^3. The foams with the lowest thermal conductivity are the least stiff; the stiffest have the highest conductivity. Explain the reasoning you would use to select a foam for the truck panel using a penalty function.

E.9 SELECTING MATERIAL AND SHAPE (CHAPTERS 9 AND 10)

The examples in this section relate to the analysis of material and shape from Chapters 9 and 10. They cover the derivation of shape factors and of indices that combine material and shape, and the use of the 4-quadrant chart arrays to explore material and shape combinations. For this last purpose it is useful to have clean copies of the chart arrays shown in Figures 9.9 and 9.12. Like the material property charts, they can be copied from the text without the restrictions of copyright.

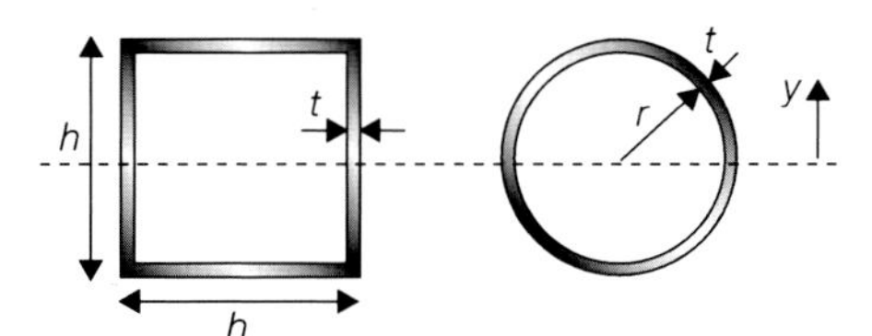

FIGURE E.21

E9.1 Shape factors for tubes (Figure E.21)

a. Evaluate the shape factor φ_B^e for stiffness-limited design in bending of a square box section of outer edge length $h = 100$ mm and wall thickness $t = 3$ mm. Is this shape more efficient than one made of the same material in the form of a tube of diameter $2r = 100$ mm and wall thickness $t = 3.82$ mm (giving it the same mass per unit length m/L)? Treat both as thin-walled shapes.

b. Make the same comparison for the shape factor φ_B^f for strength-limited design.

Use the expressions given in Table 9.3 of the text for the shape factors φ_B^e and φ_B^f.

E9.2 Deriving shape factors for stiffness-limited design

Derive the expression for the shape-efficiency factor for φ_B^e

a. A stiffness-limited design for a beam loaded in bending with each of the three sections shown in Figure E.22 do not assume that the thin-wall approximations is valid

b. A closed circular tube of outer radius $5t$ and wall thickness t

c. a channel section of thickness t, overall flange width $5t$ and overall depth $10t$, bent about its major axis

d. a box section of wall thickness t, and height and width $h_1 = 10t$

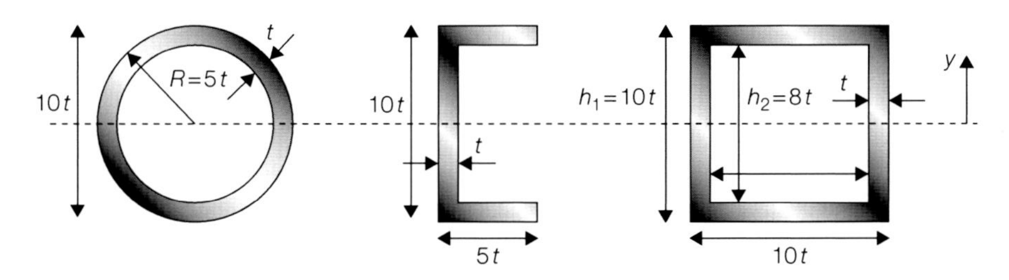

FIGURE E.22

E9.3 Deriving shape factors for strength-limited design

a., b., and c. Determine the shape-efficiency factor φ_B^f for strength limited design in bending, for the same three sections shown in Figure E.22. You will need the results for I for the sections derived in Exercise E9.2.

d. A beam of length L, loaded in bending, must support a specified bending moment M without failing and be as light as possible. Show that to minimize the mass of the beam per unit length m/L, one should select a material and a section shape to maximize the quantity:

$$M = \frac{(\varphi_B^f \sigma_f)^{2/3}}{\rho}$$

where σ_f is the failure stress and ρ the density of the material of the beam; φ_B^f is the shape-efficiency factor for failure in bending.

E9.4 Determining shape factors from stiffness data The elastic shape factor measures the gain in stiffness by shaping, relative to a solid square section of the same area. Shape factors can be determined by experiment. Equation (9.19) in the text gives the mass m of a beam of length L and prescribed bending stiffness S_B with a section of efficiency φ_B^e as

$$m = \left(\frac{12 S_B}{C_1}\right)^{1/2} L^{5/2} \left[\frac{\rho}{(\varphi_B^e E)^{1/2}}\right]$$

where C_1 is a constant that depends on the distribution of load on the beam. Inverting the equation gives

$$\varphi_B^e = \left(\frac{12 L^5 S_B}{C_1 m^2}\right)\left(\frac{\rho^2}{E}\right)$$

Thus if the bending stiffness S_B, mass m, and length L are measured, and the modulus E and density ρ of the material of the beam are known, φ_B^e can be calculated.

a. Calculate the shape factor φ_B^e from the following experimental data, measured on an aluminum alloy beam loaded in 3-point bending (for which $C_1 = 48$; see Appendix B.3) using the data shown in the following table.

Attribute	Value
Beam stiffness S_B	7.2×10^5 N/m
Mass/unit length m/L	1 kg/m
Beam length L	1 m
Beam material	6061 aluminum alloy
Material density ρ	2670 kg/m^3
Material modulus E	69 GPa

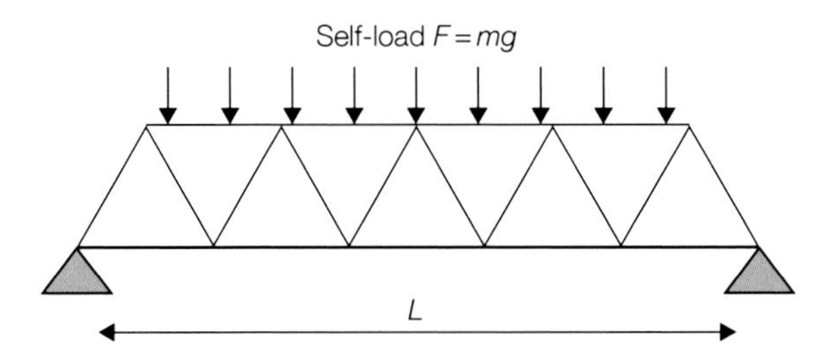

FIGURE E.23

b. A steel truss bridge shown in Figure E.23 has a span L and is simply supported at both ends. It weighs m tonnes. As a rule of thumb, bridges are designed with a stiffness S_B such that the central deflection δ of a span under its self-weight is less than 1/300 of the length L (thus $S_B \geq 300\,mg/L$ where g is the acceleration due to gravity, 9.81 m/s^2). Use the information listed in the table to calculate the minimum shape factor φ_B^e of the three steel truss bridge spans. Take the density ρ of steel to be 7900 kg/m^3 and its modulus E to be 205 GPa. The constant $C_1 = 384/5 = 76.8$ for uniformly distributed load (Appendix B.3).

Bridge and Date of Construction*	Span L (m)	Mass m (tonnes)
Royal Albert Bridge, Saltash, U.K. (1857)	139	1060
Carquinez Strait Bridge, CA (1927)	132	650
Chesapeake Bay Bridge, MD (1952)	146	850

* *Data from the* Bridges Handbook.

E9.5 Deriving indices for bending and torsion

a. A beam, loaded in bending, must support a specified bending moment M^* without failing and be as light as possible. Section shape is a variable, and "failure" here means the first onset of plasticity. Derive the material index. The following table summarizes the requirements.

Function	Lightweight beam
Constraints	Specified failure moment M^* Length L specified
Objective	Minimum mass m
Free variables	Choice of material Section shape and scale

b. A shaft of length L, loaded in torsion, must support a specified torque T^* without failing and be as cheap as possible. Section shape is a variable and "failure" again means the first onset of plasticity. Derive the material index. The table summarizes the requirements.

Function	Cheap shaft
Constraints	Specified failure torque T^* Length L specified
Objective	Minimum material cost C
Free variables	Choice of material Section shape and scale

E9.6 Use of the 4-quadrant chart for stiffness-limited design

a. Use the 4-quadrant chart for stiffness-limited design shown in Figure 9.9 to compare the mass per unit length m/L of a section with $EI = 10^5$ Nm^2 made from

1. structural steel with a shape factor φ_B^e of 20, modulus $E = 210$ GPa, and density $\rho = 7900$ kg/m^3
2. carbon-fiber–reinforced plastic with a shape factor φ_B^e of 10, modulus $E = 70$ GPa, and density $\rho = 1600$ kg/m^3
3. structural timber with a shape factor φ_B^e of 2, modulus $E = 9$ GPa, and density $\rho = 520$ kg/m^3.

The schematic in Figure E.24 illustrates the method.

b. Show, by direct calculation, that the conclusions of part (a) are consistent with the idea that to minimize mass for a given stiffness one should maximize $\sqrt{E^*/\rho^*}$ with $E^* = E/\varphi_B^e$ and $\rho^* = \rho/\varphi_B^e$.

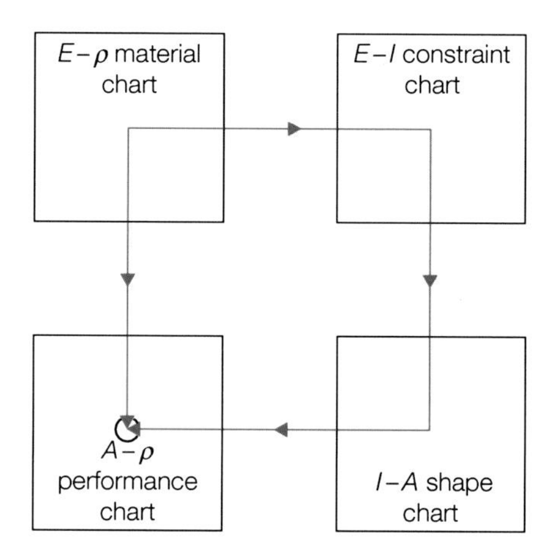

FIGURE E.24

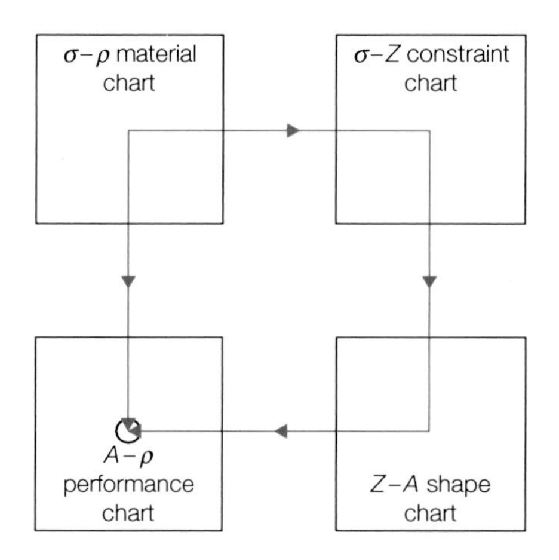

FIGURE E.25

E9.7 Use of the 4-quadrant chart for strength

a. Use the 4-quadrant chart for strength-limited design shown in Figure 9.12 to compare the mass per unit length m/L of a section with $Z_{\sigma f} = 10^4$ Nm (where Z is the section modulus) made from

1. mild steel with a shape factor φ_B^f of 10, strength $\sigma_f = 200$ MPa, and density $\rho = 7900$ kg/m^3
2. 6061-grade aluminum alloy with a shape factor φ_B^f of 3, strength $\sigma_f = 200$ MPa, and density $\rho = 2700$ kg/m^3
3. a titanium alloy with a shape factor φ_B^f of 10, strength $\sigma_f = 480$ MPa, and density $\rho = 4420$ kg/m^3

The schematic in Figure E.25 illustrates the method.

b. Show, by direct calculation, that the conclusions of part (a) are consistent with the idea that to minimize mass for a given stiffness one should maximize

$$\frac{(\varphi_B^f \sigma_f)^{2/3}}{\rho}$$

E9.8 A lightweight display stand Figure E.26 shows a concept for a lightweight display stand. The stalk must support a mass m of 100 kg, to be placed on its upper surface at a height h, without failing by elastic buckling. It is to be made of stock tubing and must be as light as possible. Use the methods in Chapter 9 to derive a material index for the tubular material of the stand that meets these requirements and that includes the shape of the section, described by the shape factor

$$\varphi_B^e = \frac{12I}{A^2}$$

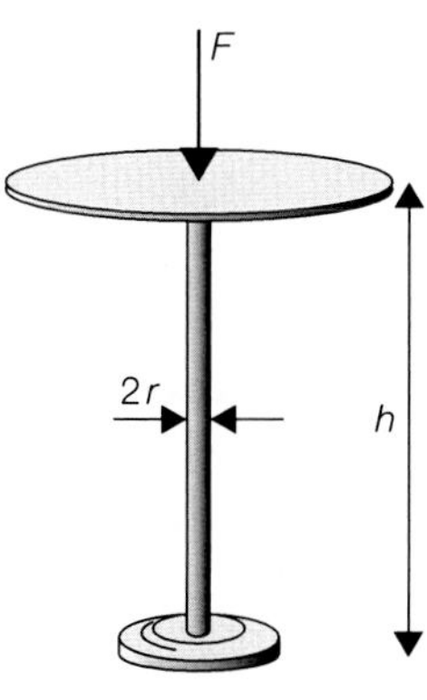

FIGURE E.26

where I is the second moment of area and A is the section area. The following table summarizes the requirements.

Function	Lightweight column
Constraints	Specified buckling load *F* Height *h* specified
Objective	Minimum mass *m*
Free variables	Choice of material Section shape and scale

Cylindrical tubing is available from stock in the following materials and sizes. Use this information and the material index to identify the best stock material for the column of the stand.

Material	Modulus *E* (GPa)	Tube Radius *r*	Wall Thickness/ Tube Radius *t/r*
Aluminum alloys	69	25 mm	0.07 to 0.25
Steel	210	30 mm	0.045 to 0.1
Copper alloys	120	20 mm	0.075 to 0.1
Polycarbonate (PC)	3.0	20 mm	0.15 to 0.3
Various woods	7–12	40 mm	Solid circular sections only

E9.9 Microscopic shape: Tube arrays Calculate the gain in bending efficiency ψ_B^e when a solid is formed into small, thin-walled tubes of radius r and wall thickness t that are then assembled and bonded into a large array, part of which is shown in Figure E.27. Let the solid of which the tubes are made have modulus E_s and density ρ_s. Express the result in terms of r and t.

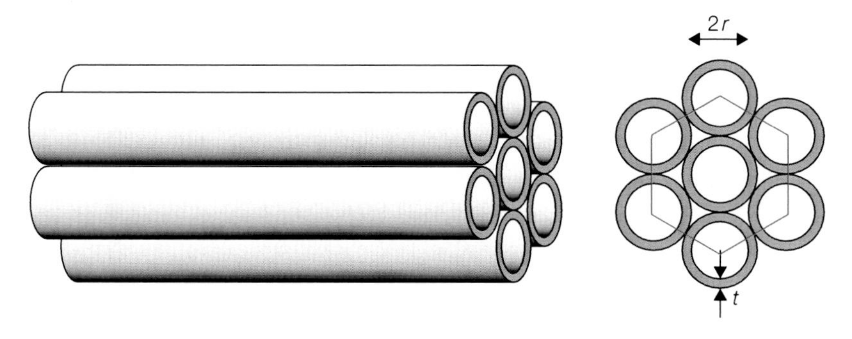

FIGURE E.27

E9.10 The structural efficiency of foamed panels (Figure E.28) Calculate the change in structural efficiency for both bending stiffness and strength when a solid flat panel of unit area and thickness t is foamed to give a foam panel of unit area and thickness h at constant mass. The modulus E and strength σ_f of foams scale with relative density ρ/ρ_s as

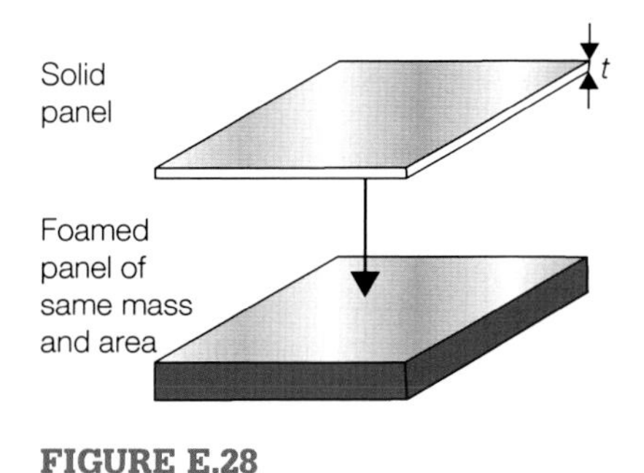

FIGURE E.28

$$E = \left(\frac{\rho}{\rho_s}\right)^2 E_s \text{ and } \sigma_f = \left(\frac{\rho}{\rho_s}\right)^{3/2} \sigma_{f,s}$$

where E, σ_f, and ρ are the modulus, strength, and density of the foam and E_s, $\sigma_{f,s}$, and ρ_s those of the solid panel.

E.10 HYBRID MATERIALS (CHAPTERS 11 AND 12)

The examples in this section relate to the design of hybrid material described in Chapters 11 and 12. The first three involve the use of bounds for evaluating the potential of composite systems. The fourth is an example of hybrid design to fill holes in property space. Then comes one involving a sandwich panel. The next three make use of the charts for natural materials. The last is a challenge: to explore the potential of hybridizing two very different materials.

E10.1 Concepts for light, stiff composites Figure E.29 shows a chart for exploring stiff composites with light alloy or polymer matrices. A construction like that shown in Figure 11.7 of the text allows the potential of any given matrix-reinforcement combination to be assessed. Four matrix materials are shown, highlighted in red. The materials shown in gray are available as fibers (f), whiskers (w), or particles (p). The criteria of excellence (the indices E/ρ, $E^{1/2}/\rho$, and $E^{1/3}/\rho$ for light, stiff structures) are shown; they increase in value toward the top left. Use the chart to compare the performance of a titanium-matrix composite reinforced with (a) zirconium carbide, ZrC, (b) Saffil alumina fibers, and (c) nicalon silicon carbide fibers. Keep it simple: Use Equations (11.1) and (11.2) to calculate the density and upper and lower bounds for the modulus at a volume fraction of $f = 0.5$ and plot these points. Then sketch arcs of circles from the matrix to the reinforcement to pass through them. In making your judgment, assume that $f = 0.5$ is the maximum practical reinforcement level.

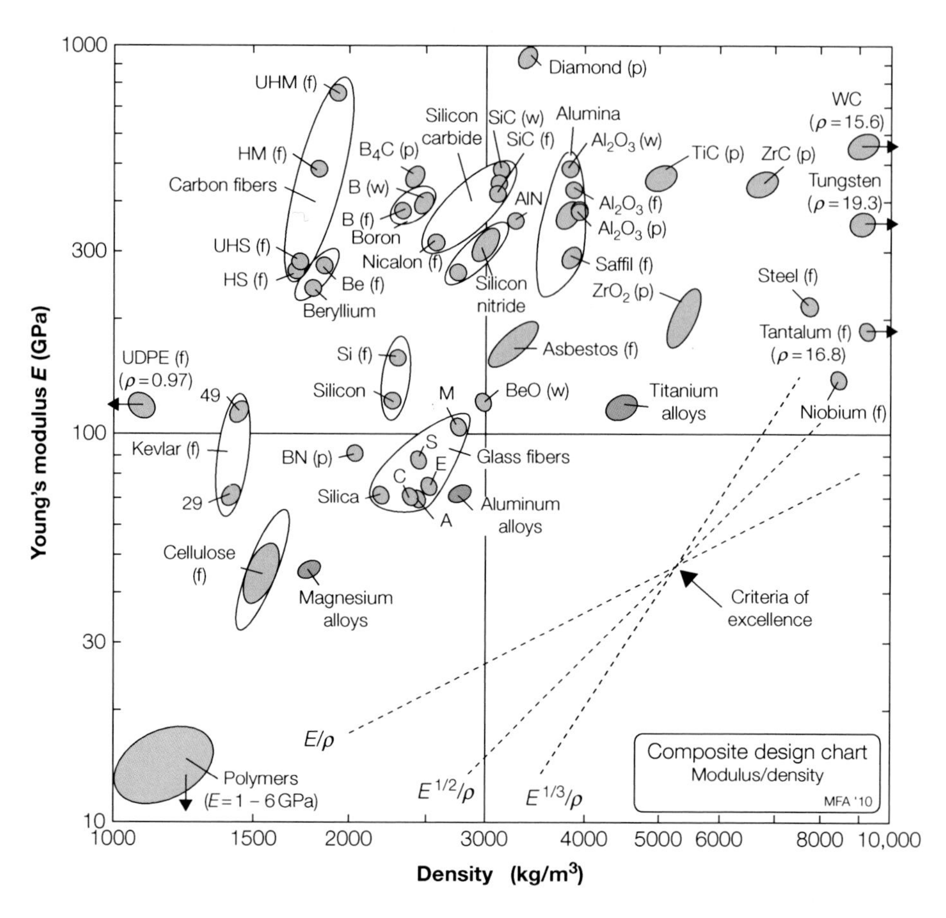

FIGURE E.29

E10.2 Use the chart shown in Figure E.29 to explore the relative potential of magnesium–E-glass-fiber composites and magnesium–beryllium composites for light, stiff structures.

E10.3 Concepts for composites with tailored thermal properties Figure E.30 shows a chart for exploring the design of composites with desired combinations of thermal conductivity and expansion, using light alloy or polymer matrices. A construction like that shown in Figure 11.10 allows the potential of any given matrix-reinforcement combination to be assessed. One criterion of excellence (the index for materials to minimize thermal distortion λ/α) is shown; it increases in value toward the bottom right. Use the chart to compare the performance of a magnesium AZ63 alloy-matrix composite reinforced with (a) alloy steel fibers, (b) silicon carbide fibers, SiC (f), and (c) diamond-structured carbon particles. Keep it simple: Use Equations (11.7) through (11.10) to calculate the upper and lower bounds for α and λ at a volume fraction of $f = 0.5$ and plot

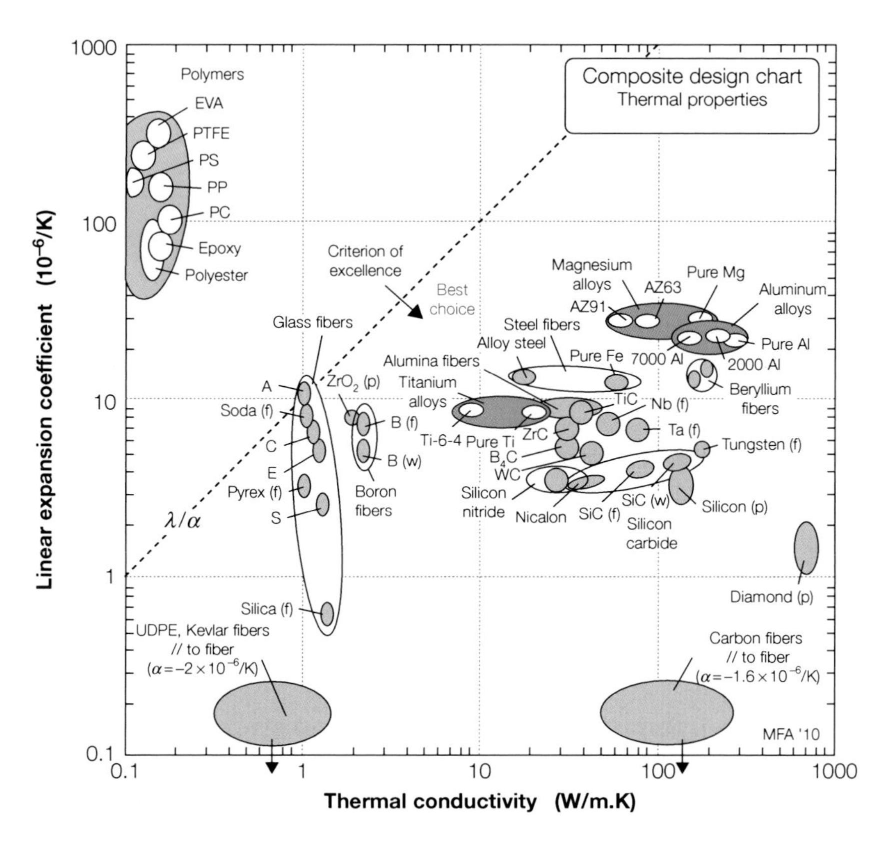

FIGURE E.30

these points. Then sketch curves linking matrix-to-reinforcement to pass through the outermost of the points. In making your judgment, assume that $f = 0.5$ is the maximum practical reinforcement level.

E10.4 Hybrids with exceptional combinations of stiffness and damping The loss-coefficient–modulus (η–E) chart shown in Figure 4.9 is populated only along one diagonal band. (The loss coefficient η measures the fraction of the elastic energy that is dissipated during a load-unload cycle.) Monolithic materials with low E have high η, those with high E have low η. The challenge here is to devise hybrids to fill the holes, with the following applications in mind.

a. Sheet steel (as used in car body panels, for instance) is prone to lightly damped flexural vibration. Devise a hybrid sheet that combines the high stiffness E of steel with high loss coefficient η.

b. High loss coefficient means that energy is dissipated on mechanical cycling. This energy appears as heat, sometimes with undesirable consequences. Devise a hybrid with low modulus E and low loss coefficient η.

E10.5 Sandwich panels An aircraft-quality sandwich panel has the characteristics listed in the following table.

a. Use the data and Equations (11.13) through (11.15) of the text to calculate the equivalent density, flexural modulus, and strength.
b. Plot these on a modulus–density and a modulus–strength chart. Do the flexural properties of the panel lie in a region of property space not filled by monolithic materials?

Data for Glass Fiber/Aluminum Honeycomb Sandwich Panel	
Face material	0.38-mm glass fiber/epoxy
Core material	3.2-mm cell, 97.8 kg/m^3, 5052 Alu honeycomb
Panel weight per unit area m_a	2.65 kg/m^2
Panel length L	510 mm
Panel width b	51 mm
Panel thickness d	10.0 mm
Flexural stiffness EI	67 N.m^2
Failure moment M_f	160 Nm

E10.6 Natural hybrids that are light, stiff, and strong

a. Plot aluminum alloys, steels, CFRP, and GFRP onto a copy of the E–ρ chart for natural materials (refer to Figure 12.13 in the text), where E is Young's modulus and ρ is the density. How do they compare, using the flexural stiffness index $E^{1/2}/\rho$ as the criterion of excellence?
b. Do the same thing for strength using a copy of Figure 12.14, using the flexural strength index $\sigma_f^{2/3}/\rho$ (where σ_f is the failure stress) as the criterion of excellence.

The following table lists the necessary data.

Material	Young's Modulus E (GPa)	Density ρ (kg/m^3)	Strength, σ_f (MPa)
Aluminum alloy	74	2700	335
Steel	210	7850	700
CFRP	100	1550	760
GFRP	21	1850	182

E10.7 Natural hybrids that act as springs The table that follows lists the moduli and strengths of spring materials. Plot these onto a copy of the E–σ_f chart for natural materials shown in Figure 12.15, and compare their energy-storing performance with that of natural materials, using σ_f^2/E as the criterion of choice. Here σ_f is the failure stress, E is Young's modulus, and ρ is the density.

Material	Modulus (GPa)	Strength (MPa)	Density (kg/m^3)
Spring steel	206	1100	7850
Copper-2% beryllium	130	980	8250
CFRP filament wound	68	760	1580

E10.8 Finding a substitute for bone Find an engineering material that most closely resembles the longitudinal strength of compact bone (*cortical bone* (*L*)) in its strength/weight (σ_f/ρ) characteristics by plotting data for this material, read from Figure 12.14, onto a copy of the σ_f–ρ chart for engineering materials (refer to Figure 4.4). Here σ_f is the failure stress and ρ is the density.

E10.9 Creativity: What could you do with *x*? The same 68 materials appear on all of the charts in Chapter 4. These can be used as the starting point for "What if…?" exercises. As a challenge, use any chart or combination of charts to explore what might be possible by hybridizing any pair of the materials listed next, in any configuration you care to choose.

a. Cement
b. Wood
c. Polypropylene
d. Steel
e. Copper

E.11 SELECTING PROCESSES (CHAPTERS 13 AND 14)

The exercises in this section use the process selection charts in Chapters 13 and 14. They are useful in giving a feel for process attributes and the way in which process choice depends on material and shape. Here the CES software offers greater advantages: what is cumbersome and of limited resolution with the charts is easy with the software, which offers much greater resolution.

Each exercise has two parts, labeled (a) and (b). The first involves translation. The second uses the selection charts in Chapter 13 (which you are free to copy) in the way that was illustrated in Chapter 14.

E11.1 Elevator control quadrant The quadrant sketched in Figure E.31 is part of the control system for the wing elevator of a commercial aircraft. It is to be made of a light alloy (aluminum or magnesium) with the shape shown in the figure. It weighs about 5 kg. The minimum section thickness is 5 mm, and

FIGURE E.31

—apart from the bearing surfaces—the requirements on surface finish and precision are not strict: surface finish ≤ 10 μm and precision ≤ 0.5 mm. The bearing surfaces require a surface finish ≤ 1 μm and a precision ≤ 0.05 mm. A production run of 100 to 200 is planned.

- **a.** Itemize the function and constraints, leave the objective blank, and enter "Choice of process" for the free variable.
- **b.** Use copies of the charts from Chapter 13 in succession to identify processes to shape the quadrant.
- **c.** If the CES software is available, apply the constraints and identify in more detail the viable processes.

FIGURE E.32

E11.2 Casing for an electric plug The electric plug is perhaps the commonest of electrical products (Figure E.32). It has a number of components, each performing one or more functions. The most obvious are the casing and the pins, although there are many more (i.e., connectors; a cable clamp; fasteners; and, in some plugs, a fuse). The task is to investigate the processes for shaping the two-part insulating casing, the thinnest part of which is 2 mm thick. Each part weighs about 30 grams and is to be made in a single step from a thermoplastic or thermosetting polymer with a planned batch size of 5×10^4 -2×10^6. The required tolerance of

0.3 mm and surface roughness of 1 μm must be achieved without secondary operations.

a. Itemize the function and constraints, leave the objective blank, and enter "Choice of process" for the free variable.
b. Use the charts from Chapter 13 successively to identify possible processes to make the casing.

E11.3 Ceramic valves for taps Few things are more irritating than a dripping tap. Taps drip because the rubber washer is worn or the brass seat is pitted by corrosion, or both. Ceramics have good wear resistance, and they have excellent corrosion resistance in both pure and saltwater. Many household taps now use ceramic valves.

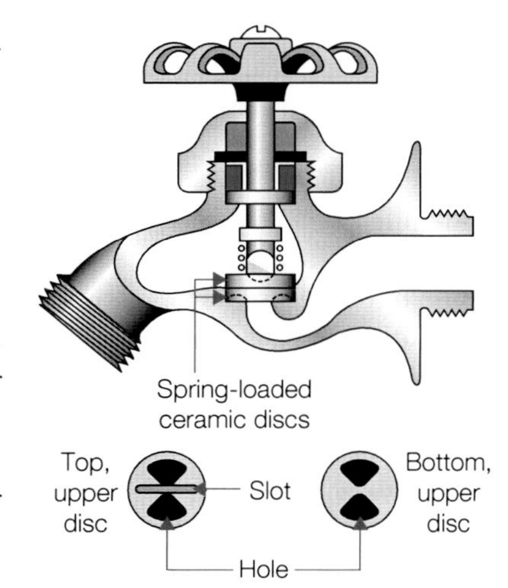

FIGURE E.33

The sketch in Figure E.33 shows how they work. A ceramic valve consists of two disks mounted one above the other, spring-loaded so that their faces are in contact. Each disk has a diameter of 20 mm and a thickness of 3 mm, and weighs about 10 grams. In order to seal well, the mating surfaces of the two disks must be flat and smooth, requiring high levels of precision and surface finish; typically tolerance < 0.02 mm and surface roughness < 0.1 μm. The outer face of each has a slot that registers it and allows the upper disc to be rotated through 90° (a quarter-turn). In the off position the holes in the upper disc are blanked off by the solid part of the lower one; in the on position the holes are aligned. A production run of 10^5–10^6 is envisaged.

a. List the function and constraints, leave the objective blank, and enter "Choice of process" for the free variable.
b. Use the charts from Chapter 13 to identify possible processes to make the casing.

E11.4 Shaping plastic bottles Plastic bottles are used to contain fluids as various as milk and engine oil (Figure E.34). A typical polyethylene bottle weighs about 30 grams and has a wall thickness of about 0.8 mm. The shape is 3D hollow. The batch size is large (1,000,000 bottles). What process could be used to make them?

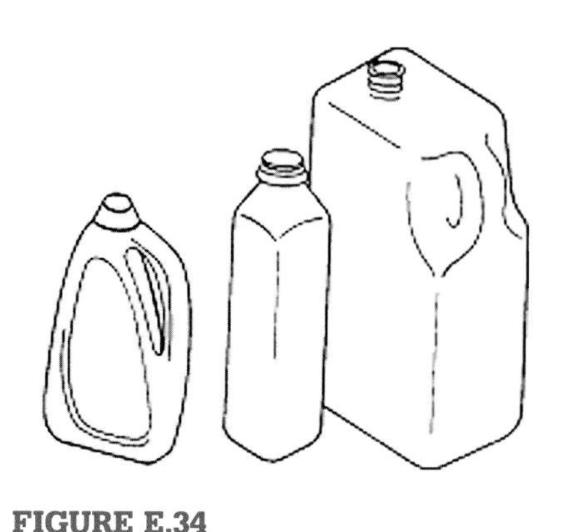

FIGURE E.34

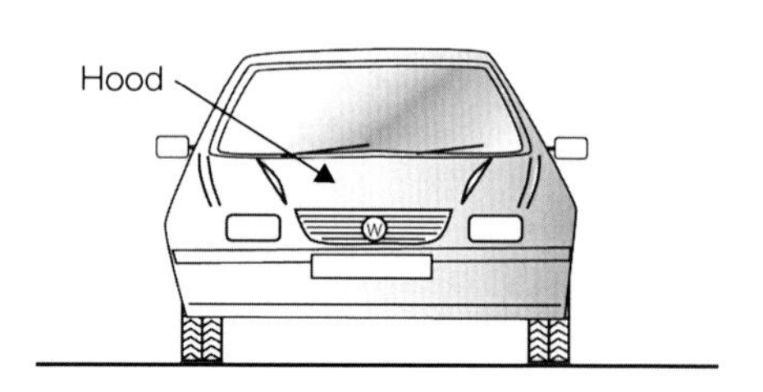

FIGURE E.35

a. List the function and constraints, leave the objective blank, and enter "Choice of process" for the free variable.
b. Use the charts from Chapter 13 to identify possible processes to make the casing.

E11.5 Car hood As weight saving assumes greater importance in automobile design, the replacement of steel parts with polymer-composite substitutes becomes increasingly attractive. Weight can be saved by replacing a steel hood with one made from a thermosetting composite (Figure E.35). The weight of the hood depends on the car model: A typical composite hood weighs is 8 to 10 kg. The shape is a dished sheet and the requirements on tolerance and roughness are 1 mm and 2 μm, respectively. A production run of 100,000 is envisaged.

a. List the function and constraints, leave the objective blank, and enter "Choice of process" for the free variable.
b. Use the charts from Chapter 13 to identify possible processes to make the casing.

E11.6 Complex structural channels Channel sections for window frames, for slide-together sections for versatile assembly, and for ducting for electrical wiring can be complex in shape. Figure E.36 shows an example. The order is for 10,000 such sections, each 1 m in length and weighing 1.2 kg, with a minimum section of 4 mm. A tolerance of 0.2 mm and a surface roughness of less than 1 μm must be achieved without any additional finishing operation.

a. List the function and constraints, leave the objective blank, and enter "Choice of process" for the free variable.
b. Use the charts from Chapter 11 to identify possible processes to make the casing.

FIGURE E.36

E11.7 Selecting joining processes This exercise and the next require the use of the CES software.

a. Use the software from CES to select a joining process to meet the following requirements.

Function	Create a permanent butt joint between steel plates
Constraints	Material class: carbon steel Joint geometry: butt joint Section thickness: 8 mm Permanent Watertight
Objective	–
Free variable	Choice of process

b. Use the software from CES to select a joining process to meet the following requirements.

Function	Create a watertight, demountable lap joint between glass and polymer
Constraints	Material class: glass and polymers Joint geometry: lap joint Section thickness: 4 mm Demountable Watertight
Objective	–
Free variable	Choice of process

E11.8 Selecting surface treatment processes This exercise, like the previous one, requires the use of the CES software.

a. Use CES to select a surface treatment process to meet the following requirements.

Function	Increase the surface hardness and wear resistance of a high carbon steel component
Constraints	Material class: carbon steel Purpose of treatment: increase surface hardness and wear resistance
Objective	–
Free variable	Choice of process

b. Use CES to select a surface treatment process to meet the following requirements.

Function	Apply color and pattern to the curved surface of a polymer molding
Constraints	Material class: thermoplastic Purpose of treatment: aesthetics, color Curved surface coverage: good or very good
Objective	–
Free variable	Choice of process

E.12 MATERIALS AND THE ENVIRONMENT (CHAPTER 15)

E12.1 Use the E–$H_p\rho$ chart shown in Figure 15.9 to find the polymer with a modulus E greater than 1 GPa and the lowest embodied energy per unit volume.

E12.2 A maker of polypropylene (PP) garden furniture is concerned that the competition is stealing part of his market by claiming that the "traditional" material for garden furniture, cast iron, is much less energy and CO_2-intensive than PP. A typical PP chair weighs 1.6 kg; one made of cast iron weighs 8.5 kg. Use the data for these two materials in Appendix A, Table A.10, to find out who is right; are the differences significant if the data for embodied energy are only accurate to ±20%?

If the PP chair lasts 5 years and the cast iron chair lasts 25 years, does the conclusion change?

E12.3 Identical casings for a power tool could be die-cast in aluminum or molded in ABS or polyester GFRP. Use the bar chart of embodied energy per unit volume shown in Figure 15.8 to decide which choice minimizes the material embodied energy, assuming the same volume of material is used for each casing.

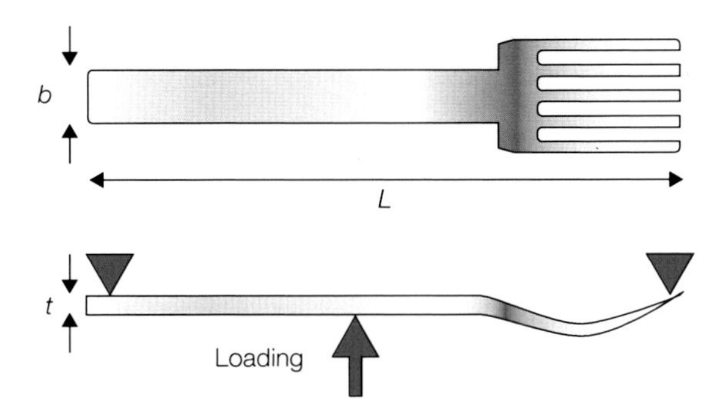

FIGURE E.37

E12.4 Disposable knives and forks Disposable knives and forks are ordered by an environmentally conscious pizza house (Figure E.37). The shape of each (and thus the length, width, and profile) are fixed, but the thickness is free: It is chosen to give enough bending stiffness to cut and impale the pizza without excessive flexure. The pizzeria proprietor wishes to enhance the greenness of her image by minimizing the energy content of the throw-away tableware, which could be molded from polystyrene (PS) or stamped from aluminum sheet.

Establish an appropriate material index for selecting materials for energy-economic forks. Model the eating implement as a beam of fixed length L and width w, but with a thickness t that is free, and loaded in bending, as in the figure. The objective function is the volume of material in the fork times its energy content $H_p\rho$ per unit volume (H_p is the embodied energy per kg, and ρ the density). The limit on flexure imposes a stiffness constraint (Appendix B.3). Use this information to develop the index.

Flexure in cutlery is an inconvenience. Failure—whether by plastic deformation or by fracture—is more serious: It causes loss of function; it might even cause hunger. Repeat the analysis, deriving an index when a required strength is the constraint.

E12.5 Show that the index for selecting materials for a strong panel with the dimensions shown in Figure E.38, loaded in bending, with minimum embodied energy content is that with the largest value of

$$M = \frac{\sigma_y^{1/2}}{H_p\rho}$$

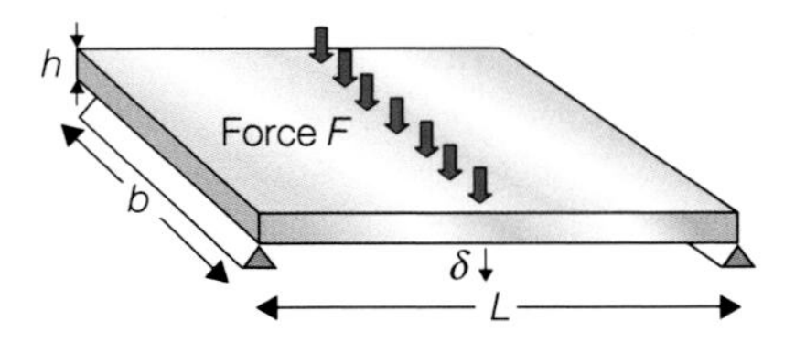

FIGURE E.38

where H_p is the embodied energy of the material, ρ its density, and σ_y its yield strength. To do so, rework the panel derivation in Chapter 5, Equation (5.9), replacing the stiffness constraint with a constraint on failure load F requiring that it exceed a chosen value F^* where

$$F = C_2 \frac{I\sigma_y}{hL} > F^*$$

where C_2 is a constant and I is the second moment of area of the panel $I = \frac{bh^2}{12}$.

E12.6 Use the indices for the crash barriers (Equations (15.9) and (15.10)) with the charts for strength and density (refer to Figure 4.4) and strength and embodied energy (refer to Figure 15.10) to select materials for each of the barriers. Position your selection line to include one metal for each. Reject ceramics and glass on the grounds of brittleness. List what you find for each barrier.

E12.7 The makers of a small electric car wish to make bumpers out of a molded thermoplastic. Which index is the one to guide this selection if the aim is to maximize the range for a given battery storage capacity? Plot it on the strength–density chart in Figure 4.4, and make a selection.

E12.8 Energy-efficient floor joists Floor joists are beams loaded in bending. They can be made of wood, of steel, or of steel-reinforced concrete, with the shape factors listed in the following table. For a given bending stiffness and strength, which of these carries the lowest production-energy burden? The relevant data, drawn from the tables in Appendix A, are listed.

a. Start with stiffness. Locate from Equation (9.20) in the text the material index for stiffness-limited, shaped beams of minimum mass. Adapt this to make the index for stiffness-limited, shaped beams of embodied energy by multiplying density ρ by embodied energy/kg H_p. Use the modified index to rank the three beams.

b. Repeat the procedure, this time for strength, creating the appropriate index for strength-limited shaped beams at minimum energy content by adapting Equation (9.28).

What do you conclude about the relative energy penalty of design with wood and with steel?

Material	Density ρ kg/m^3	Modulus E GPa	Strength σ_f MPa	Energy H_p MJ/kg	ϕ_B^e	ϕ_B^f
Soft wood	700	10	40	7.5	2	1.4
Reinforced concrete	2900	35	10	5	2	1.4
Steel	7900	210	200	30	15	4

Index

T

U

V

W